Low-Dimensional Conductors and Superconductors

NATO ASI Series

Advanced Science Institutes Series

A series presenting the results of activities sponsored by the NATO Science Committee, which aims at the dissemination of advanced scientific and technological knowledge, with a view to strengthening links between scientific communities.

The series is published by an international board of publishers in conjunction with the NATO Scientific Affairs Division

A Life Sciences **B Physics**	Plenum Publishing Corporation New York and London
C Mathematical and Physical Sciences	D. Reidel Publishing Company Dordrecht, Boston, and Lancaster
D Behavioral and Social Sciences **E Engineering and Materials Sciences**	Martinus Nijhoff Publishers The Hague, Boston, Dordrecht, and Lancaster
F Computer and Systems Sciences **G Ecological Sciences** **H Cell Biology**	Springer-Verlag Berlin, Heidelberg, New York. London, Paris, and Tokyo

Recent Volumes in this Series

Volume 150—Particle Physics: *Cargèse 1985*
edited by Maurice Lévy, Jean-Louis Basdevant, Maurice Jacob, David Speiser, Jacques Weyers, and Raymond Gastmans

Volume 151—Giant Resonances in Atoms, Molecules, and Solids
edited by J. P. Connerade, J. M. Esteva, and R. C. Karnatak

Volume 152—Optical Properties of Narrow-Gap Low-Dimensional Structures
edited by C. M. Sotomayor Torres, J. C. Portal, J. C. Mann, and R. A. Stradling

Volume 153—Physics of Strong Fields
edited by W. Greiner

Volume 154—Strongly Coupled Plasma Physics
edited by Forrest J. Rogers and Hugh E. Dewitt

Volume 155—Low-Dimensional Conductors and Superconductors
edited by D. Jérome and L. G. Caron

Volume 156—Gravitation in Astrophysics: *Cargèse 1986*
edited by B. Carter and J. B. Hartle

Series B: Physics

Low-Dimensional Conductors and Superconductors

Edited by

D. Jérome

University of Paris-South
Orsay, France

and

L. G. Caron

University of Sherbrooke
Sherbrooke, Quebec, Canada

Springer Science+Business Media, LLC

Proceedings of a NATO Advanced Study Institute on Low-Dimensional Conductors and Superconductors, held August 24–September 6, 1986, in Magog, Canada

Library of Congress Cataloging in Publication Data

NATO Advanced Study Institute on Low-Dimensional Conductors and Superconductors (1986: Magog, Quebec)
Low-dimensional conductors and superconductors.

(NATO ASI series. Series B, Physics; v. 155)
"Proceedings of a NATO Advanced Study Institute on Low-Dimensional Conductors and Superconductors, held August 24–September 6, 1986, in Magog, Canada"—T.p. verso.
"Published in cooperation with NATO Scientific Affairs Division."
Bibliography: p.
Includes index.
1. One-dimensional conductors—Congresses. 2. Superconductors—Congresses. I. Jérome, D. II. Caron, L. G. III. Title. IV. Series.
QC176.8.E4N36 1986 537.6′23 87-14182

Originally published by Plenum Press, New York in 1987.
Softcover reprint of the hardcover 1st edition 1987

ISBN 978-1-4899-3613-4 ISBN 978-1-4899-3611-0 (eBook)
DOI 10.1007/978-1-4899-3611-0

PREFACE

Research activities in low dimensional conductors have shown a rapid growth since 1972 and have led to the discovery of new and remarkable physical properties unique to both molecular and inorganic conductors exhibiting one-dimensional transport behaviour. This NATO Institute was a continuation of a series of NATO Advanced Study Institutes of Worshops which took place at regular intervals till 1979. This is the first time, however, that charge density wave transport and electronic properties of low dimensional organic conductors are treated on an equal footing.

The program of the Institute was framed by tutorial lectures in the theories and experiments of low dimensional conductors.

The bulk of the course covered two series of low-dimensional materials with their respective properties.

1) The 1-D inorganic conductors exhibiting the phenomena of sliding charge density waves, narrow band noise, memory effects, etc...

2) Low-dimensional crystallized organic conductors giving rise to various possibilities of ground states, spin-Peierls, spin density wave, Peierls, superconductivity and magnetic-field induced spin density wave, etc...

Since it has been established from the beginning that this Institute was to be devoted essentially to the Physics of Low Dimensional Conductors, only one main course summarized the progress in chemistry and material preparation.

The number of lectures amounted to 60 hours of tutorial courses and 10 hours of workshops on more specialized topics. Bad weather helping, informal tutorial courses on theoretical backgrounds were organized on some of the afternoons and proved very successful. We were happy to see that in spite of a fairly heavy program, attendance was high till the end of the Institute. For this successful event we thank all the speakers who contributed to it as well as V.J.Emery and K.Bechgaard who helped in the organization of the scientific program. We also wish to thank our sponsors: NATO, the Hydro-Québec Research Institute, Alcan International, the Université de Sherbrooke and the Université de Paris-sud.

Sherbrooke and Orsay
June 1987

L.CARON and D.JEROME

CONTENTS

LOW-DIMENSIONAL CONDUCTORS AND SUPERCONDUCTORS: AN INTRODUCTION

J. Friedel

Université Paris Sud
91405 Orsay (France)

In this introductory talk, I wish to emphasise three points :

- The compounds studied in this school are complex. Many of them are new. A large fraction belongs to organic chemistry where they show behaviours which would have been undreamt of twenty years ago. The successes have come from an intimate collaboration of chemists and physicists to create and study new substances, and from the daily discussions of experimentalists with theoreticians to elaborate simple models.
- The low temperature phase changes characteristic of these compounds fall within a larger context the history of which will be shortly reviewed.
- This field gives examples of a process of successive superstructures which has been much used recently in studying complex organisations in condensed matter.

1 - SIMPLE MODELS FOR A COMPLEX FIELD

a - Simple models and their crossover

Ab initio computations can possibly help and describe the behaviour of a given compound in fixed physical conditions. They would be quite helpless to predict unknown behaviours or unknown compounds.

What are needed therefore are crude but simple models built on what one thinks are the dominant features of the problem, and neglecting irrelevant details. A saving grace of the fields is indeed that, in each case, the behaviour observed resembles rather closely behaviours well known in simpler solids.

But a characteristic of the field is its flexibility. Small changes in composition or in the physical conditions (temperature, pressure, magnetic field) can profoundly alter the properties, by altering the hierarchy in the features of the problem. New models have then to be developped, using other characteristics. The crossover between these different models are usually rather sharp.

This multiplicity of behaviours is a characteristic of the field which makes its study a priory very complex. But the study of these crossovers can help to understand what are in each case the main features. They can also hopefully be used practically as probe of small perturbations (radiation damages, weak magnetic fields), as amplifiers of signals by using often large non linearities in the response functions, or as memories by using hysteretic behaviours.

b - Intra-unit interactions

Because we are dealing with low-dimensional conductors, these models all start from simpler low-dimension units where interactions are strong, more weakly coupled together : chains or planes of atoms or molecules.

We know furthermore that, at least in a large domain of physical conditions, these units are able to carry a metallic current. This means that their valence electrons are delocalised on these units, which act each as a giant molecule. In these conditions, the kinetic energy of delocalisation, i.e. the band width w associated with the transfer integral $t_{//}$ between subunits in each unit is a leading parameter. It should in general be at least as strong as the correlation energies, as measured for instance by the energies to bring two electrons in close contact, on the same subunits (U) or on neighbouring subunits (v). The reasons why $w \gtrsim U$ and v is first that the subunits are in close contact, so that $t_{//}$ and w are relatively large. Second the subunits are either transitional metals or flat organic molecules with double bonds ; in both cases, the subunits are easily polarisable, thus electronic correlations are large within each subunit and reduce the Coulomb terms such as U, v. Indeed it is known that already for separate subunits (in a gas for d shells, in solutions for organic molecules), $U = I - A$ (I first ionisation potential, A electron affinity) is modest and indeed smaller than the corresponding values of w.

At low temperatures, the electron-electron interactions can however be responsible for small rearrangements. In the more usual cases, they can produce a spin modulation (or spin density wave SDW) with a wave length related to the Fermi wave length, or possibly a kind of superconductivity.

These possible rearrangements must be compared with other ones where electron-phonon interactions dominate : these can be a lattice modulation (or charge density wave CDW) with a wave length related to the Fermi wave length or possibly the classical BCS superconductivity.

The combined effects of electron-electron and electron-phonon interactions can lead, if they are of similar strengths, to new instabilities : in general to a doubling of the wave vector of the lattice modulation (4 k_M modulation) ; and, for specific integral numbers of carriers per subunit, to a dimerisation into singlet states (spin Peierls modulation).

The discussion between the four first types of instabilities can be made using a perturbation scheme which only involves couplings of electrons at the Fermi level. For chains, it takes a particularly simple form as a function of the various matrix elements g_i which couple the various types of Fermi electrons through a combination of electron-phonon and electron-electron interactions. The extension of such treatments to stronger couplings relies on scaling arguments which couple the various types of Fermi electron interactions. The extension of such treatments to stronger couplings relies on scaling arguments which are not necessarily predicting all the possible instabilities, as the Spin Peierls and 4 k_M cases show.

C - Inter-unit interactions

They are essential to explain the way the low-dimensional units assemble in real compounds. They also play a leading role in fixing the actual temperatures below which the distortions mentioned above transform from large scale dynamic fluctuations into real static phase changes.

When electron transfers between units are negligible, these interactions are dominated by the long range Coulomb (Madelung) terms which appear if there are static charge transfers between units of different natures. Further attractive terms are due to multipolar or dispersion forces. They are balanced by short range Coulomb overlap and eventually exchange interactions. The exact strength and consequences of these forces are relatively badly known, essentially because the units involved are highly polarisable, of low symmetry and not easily split into independent subunits. As a result, the local field corrections are very hard to estimate in a reasonable way and problems such as the amount of charge transfer in a given compound and its variation with pressure are not well understood at the present, anymore than the choice between different stacking possibilities of different units. A final difficulty is to explain in such cases why the electrons do not usually show any sign of localisation by disorder, as expected especially for 1d conductors. It has been argued, probably convincingly, that electron-phonon forward scattering helps delocalise the electrons in this case, at least in a large range of temperatures.

These problems are still at least partly present in the probably more usual case where the interunit electron transfer integral $t_\perp$ has sizeable effects. These can be classified into three regimes :

- If $t_\perp$, while being much smaller than $t_{//}$, is larger than any secondary feature of the intraunit physics, it will produce coherent 3d one-electron Bloch functions through the compound, with definite phase relations between the units, and produce a 'warping' of the Fermi surface without usually changing the topology it has for independent units (planes for 1d, cylinders for 2d units). It is with such Bloch functions and Fermi surface that one must look at the phase changes possibly produced by changes of physical conditions.

- If $t_\perp$ is smaller than the intraunit couplings, one must first build the fluctuations or phase changes due to these coupling, and then introduce the effect of $t_\perp$. As a result, $t_\perp$

transfers the electron-electron pairs or the electron-hole pairs involved in these intraunit couplings, with a lower probability, or order $t_\perp^2/\Delta$ is Δ is the coupling strength.

- Finally if $t_\perp$ is larger than these intraunit couplings but smaller than the intraunit relaxation rate $\nu_{//}$, the phase coherence between different units is lost. The electron motion between units is diffusive, with a rate $t_\perp^2/h\nu_{//}$. The Fermi surface looses its 3d warping, but is blurred over an energy range equal to the Dingle temperature $h\nu_{//}$. A possibility of localisation arises here too if $\nu_{//}$ is large enough.

2 - CHARGE AND SPIN DENSITY WAVES IN CONDUCTORS. A HISTORICAL PERSPECTIVE

These concepts emerged progressively, first in 3 then in lower dimensions.

a - 3 dimensions

The first mention of a relation between the atomic structure of a metallic conductor and the electronic structure of its valence electrons came when, in the early 30's, Hume Rothery noticed that the phase diagrams of Cu, Ag and Au base alloys nearly superimposed when plotted not in terms of atomic concentrations but of the average number of valence electrons per atom. This was then explained by Jones, noting that a special stability behaviour was expected for phases such that the Fermi sphere of free electrons just touch a number of Brillouin zone limits : the coherent scattering of the valence electrons by the atomic potentials opens a gap and thus lowers the energies of the occupied states if the Fermi level falls above the bottom of the gap. In 3 dimensions, the gap occurs at different energies in different directions of the reciprocal space, and the effect is maximum when the Fermi sphere just overlaps Brillouin zone limits, so that a large area of the sphere falls within the gap (Figure 1). The effect on the internal energy of the conductor is but slight, producing a small inflexion of its variation with increasing average valency. It can just be large enough to stabilise the corresponding phase near this optimum condition.

This analysis is at the root of all further analyses. It raised however two coupled questions which were only clarified in the early 60's by the work of Blandin and his coworkers.

- Jones only considered the sum ε of the one electron energies E for the N valence electrons. If $\delta n(E)$ is the charge of density n(E) due to the Brillouin zone scattering, the associated change in this total energy is given by :

$$\delta\varepsilon = \int^{E_M+\delta E_M} \big(n(E) + \delta n(E)\big) E dE - \int^{E_M} n(E) \quad dE.$$

with

$$N = \int^{E_M+\delta E_M} \big(n(E) + \delta n(E)\big) dE = \int^{E_M} n(E) \; dE$$

E_M is here the unperturbed Fermi level and $E_M + \delta E_n$ the Fermi level in the presence of the Brillouin zone effect. A development in small δn gives

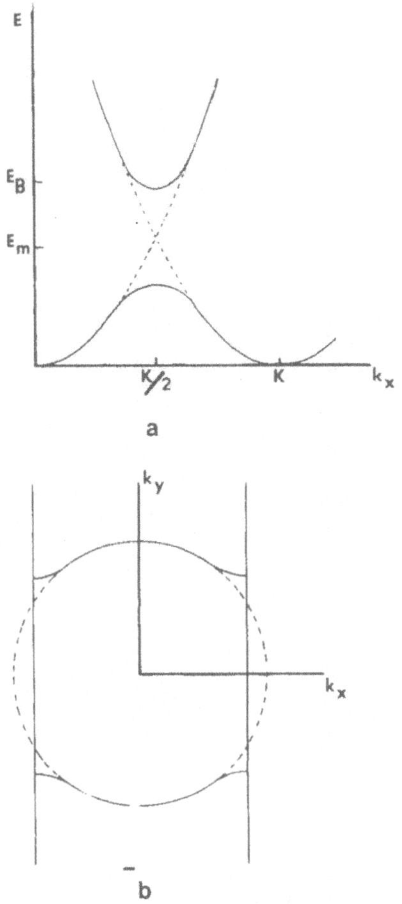

Fig. 1. Jones' explanation of Hume Rothery's phases.
a) $E(k_x)$; punctuated line: the kinematic approximation;
b) Fermi surface;

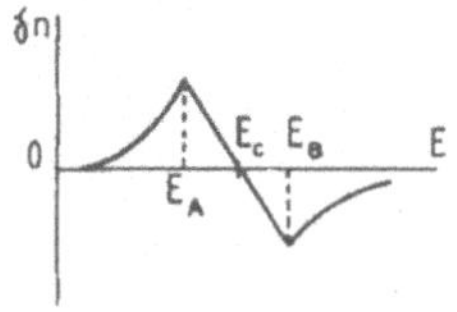

c

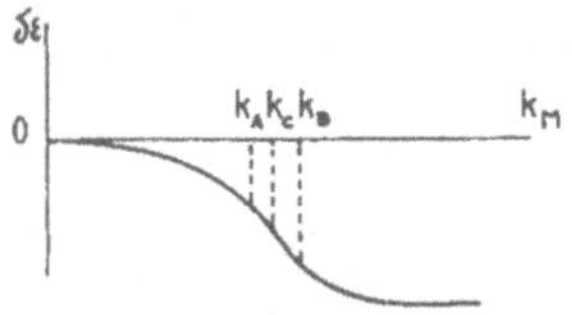

d

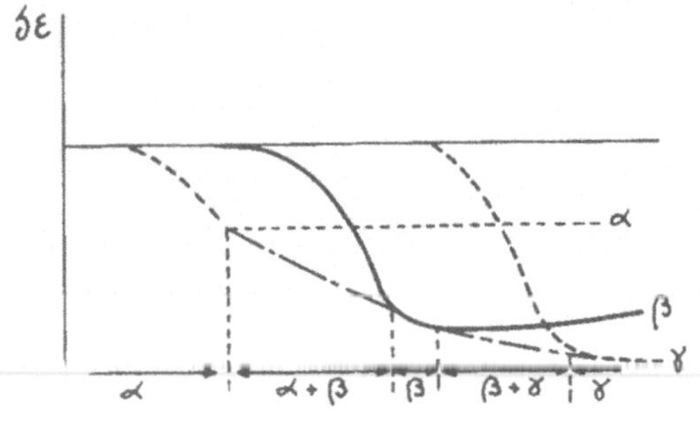

e

Fig. 1. c) change in density of states; d) change in energy; phases A, B, C.
e) possible phase diagram for three Hume Rothery

$$\frac{d^2\,\delta\varepsilon}{dE_M^2} \simeq -\,\delta n(E_M).$$

$\delta n(E)$ has the form pictured figure 1c : peak A is the van Hove anomaly when the perturbed Fermi surface just touches the Brillouin zone boundary in the first valence band ; peak B corresponds to new pockets of states just starting to be filled in the second valence band : E_A and E_B are thus the lowest values of the gap energies, figure 1a. With $E=\hbar^2k^2/2m$, one then deduces that $\delta\varepsilon(k_M)$ has the form pictured figure 1d, with an inflexion point of finite slope at k_c and two points at k_A and k_B where the curvature of $\delta\varepsilon(k_M)$ varies abruptly, with a continuous finite slope. If one considers the effects of different Brillouin zone boundaries corresponding to different atomic structures α, β, γ, the diagram of figure 1e that if they stabilise a phase such as β, it can only be over a narrow range of electronic concentrations in the neighbourood of k_B, figure 1d, thus the free electrons Fermi sphere just beyond touching the associated Brillouin zone boundary (figure 1b).

As $\delta n(E_M)$ is small, the changes in electronic entropy in such phase changes are small. When one compares different close packed structures of similar atomic volumes, the changes in vibrational entropy are also small. The phase boundaries are expected to be very little temperature dependent, a classical characteristic of Hume Rothery's phase diagrams.

- As in all Hartree analysis, Jones's treatment counts twice the Coulomb interactions between the valence electrons. This would have no importance if the electronic density remained unaffected by the Brillouin zone scattering. However this is not true. The same effect which leads to deviations from free electron behaviour in $E(\underset{\sim}{k})$ also produces a reinforcement or a weakening (depending on the sign of V_K, the matrix element that produces the gap) of the electronic density along interatomic bonds. To take this selfconsistency problem simply into account, one is lead to analyse the indirect interactions between atoms through the Fermi gas of valence electrons using the concept of its $\underset{\sim}{k}$ dependent dielectric constant $\varepsilon_{\underset{\sim}{k}}$, a method first used in the context by Kohn. This essentially assumes that one can develop the energy of the conductor in successive powers of the atomic potentiels. It is then clear that the first significant structure sensitive terms in the development are pair interactions v_{ij} between the naked charge ρ_i^o of one atom by the clothed potential V_j of another atom ; the clothing is computed as if the atom was an isolated impurity in the electron gas. Thus

$$w_{ij} = \int \rho_i^o\, v_j\, d\tau = \sum_{\underset{\sim}{k}} \rho_k^o \frac{v_{ik}^o}{\varepsilon_k}\, e^{i\underset{\sim}{k}\underset{\sim}{r}_{ij}}$$

The logarithmic anomaly of ε_k at $k = 2k_M$ related to the sharp Fermi cut off (figure 2) produces long range oscillations in $w_{ij}(r_{ij})$. Summation over the lattice produces a total coupling term w inversely proportional to ε_k, thus with a sharp increase in stability beyond Jones condition $K = 2\,k_M$ (figure 3).

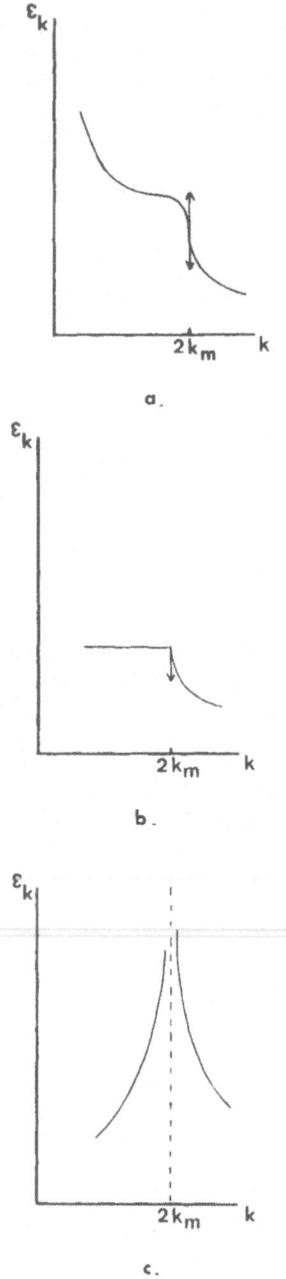

Fig. 2. k dependent dielectric constant of a conductor:

a) 3d; b) 2d; c) 1d.

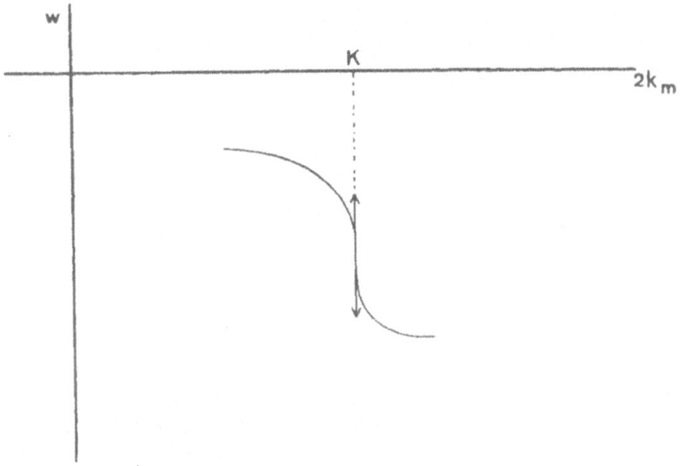

Fig. 3. Variation of the coupling term w with k_M near Jones' condition $K = 2\ k_M$.

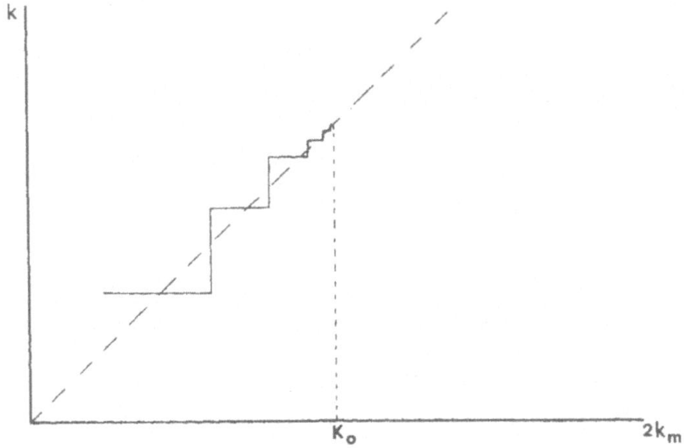

Fig. 4. Devil stair case for stable commensurate phases near to Jones' condition. $2\pi/k$ is the period of the stable phase.

- This use of a linear dielectric constant assumes that a non degenerate (one wave) perturbation scheeme is valid. This 'kinematic' approximation is obviously not valid near a Brillouin zone limit, where each Bloch state involves the degenerescence of at least two free waves. Blandin and Pick pointed out however that, in 3 dimensions, the errors involved in the use of the kinematic approximation compensated to give, to second order in the V_i, the correct result of the (two waves) dynamical approximation. Comparison of figures 1d and 3 shows that this approach is only valid for infinitely small gaps, as it obviously replaces the two anomalies at k_A and k_B (figure 1d) by a single one at K.

- In 2 or 1d, the use of a dielectric constant and the kinematic approximation is not valid, and the selfconsistency problem cannot be solved in this simple way.

Blandin also showed with Déplanté that if one considered a family of possible commensurate phases of increasing periods, built for instance by introducing increasing numbers of stacking faults in a simple close packed phase, one could predict that as one approached Jones' condition $2k_M = K$ for the initial close packed phase, the density of stable successive phases went to infinity (Figure 4). This was the earliest example of what Aubry will call a 'devil staircase' in the context of epitaxy.

Atomic modulations with continuously varying wave lengths, thus possibly uncommensurate with the underlying lattice, had be considered early by Kohn, who showed, using the dielectric constant approach, that the logarithmic anomaly of ε_k should produce a similar anomaly in the dispersion curves of the phonons at $k = 2k_M$. In agreement with experiment and contrary to later contensions by Overhauser, Kohn showed that this effect was in general small, especially for free electrons, and could not then produce static modulations at $k = 2k_M$. The name of 'charge density waves' coined by Overhauser has however subsisted for such static lattice modulations.

The concept of static antiferromagnetic magnetic modulation (similarly called spin density waves by Overhauser) progressed in parallel. After the discovery of antiferromagnetism by Néel, Slater was the first to point out that the magnetic superlattice due to the apparition of antiferromagnetism should produce new Brillouin zone limits which could help and stabilise the magnetic phase if the Fermi level was near enough to the Brilooouin zone limits. Using the concept of k dependent magnetic susceptibility χ_k, equivalent in this field to that of ε_k, Lomer extended the kinematic treatment to narrow bands and thus Fermi surfaces of any form. Because there is no problem of counting electron-electron interactions twice here, χ_k appears in the numerator of the energy change. From the fact that an antiferromagnetism with a reciprocal period $\underset{\sim}{K}_M$ couples the electronic states $\underset{\sim}{k}$ and $\underset{\sim}{k} + \underset{\sim}{K}_M$, Lomer also introduced the concept of nesting (Figure 5), i. e. the coupling is especially stable if the Fermi surface 'touches' its image translated by K_M over a large region, which is often far away from the Brillouin zone limits. The conditions of apparition of magnetism was a straight extension of Stoner's criterion for ferromagnetism.

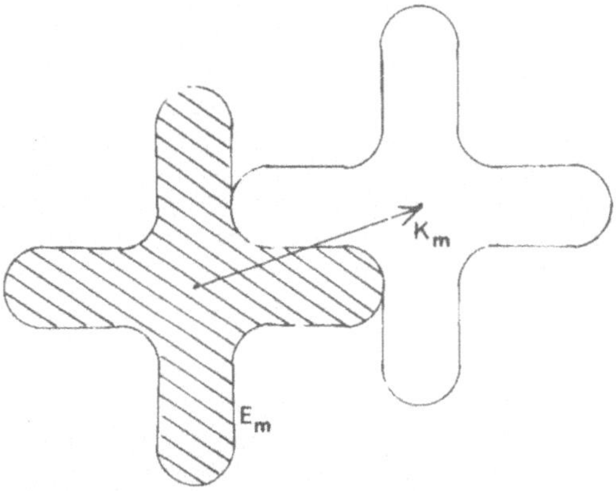

Fig. 5. Nesting condition.

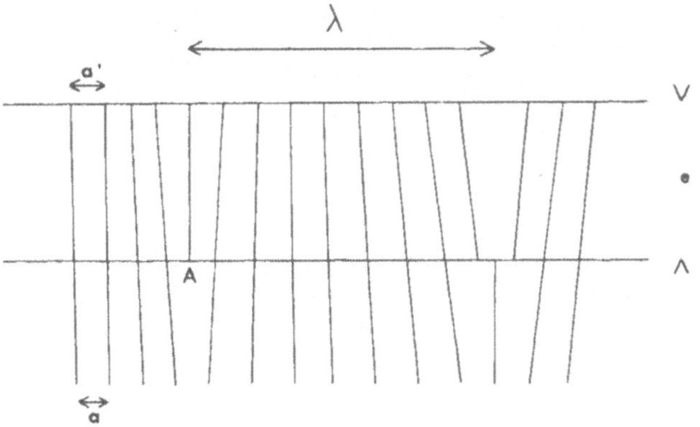

Fig. 6. Epitaxial dislocations.

This work followed shortly the first observations by neutrons of uncommensurate (helical) antiferromagnetic arrangements in Au_4Mn, in rare earth metals and in Chromium alloys.

For these magnetic modulations, the charge and lattice distortions are only second order effects, due to the spin orbit coupling term. The problem of charge selfconsistency does not therefore arise here.

b - 2 dimensions

It was known in the early 60's from the study of impurities in metals that long range oscillations decrease more slowly for lower dimensions and that, correlatively, the anomalies of ε_k at $k = 2\ k_M$ were more marked (Figure 2b, c).

It came then as no special surprise when essentially 2d lattice modulations were observed in transitional dichalcogenides. The main questions were to relate the nature of these modulations to the form of the Fermi surface, and to understand in what conditions this low temperature instability was preferred to supraconductivity. Also the role of the weak interactions between successive metallic planes was and remains poorly understood, especially in intercalate compounds where thick layers of extraneous matter can be inserted between these planes. However all analyses of the modulations are based on the use of ε_k and thus are not satisfactory from a quantitative point of view at low temperatures.

The fact that these modulations were mostly uncommensurate and occured on several crystallographically equivalant $\underset{\sim}{K}$ vectors at the same time could also be seen as a direct extension of the work of the 50's on epitaxial layers. It was indeed known by crystallographers in the 1920's that epitaxy of two crystals on close packed planes could be approximate, relative differences of lattice parameters or crystallographic angles of up to 15 % being allowed. This problem was studied in the 50's by Frank and van der Merwe who extended to 2d the 1d model of Frenkel and Kontorova of a chain of atoms united by springs of a natural lenght a' different from the parameter a of the lattice with which they interacted (Figure 6). For large enough differences $\Delta = \frac{a' - a}{a}$, the most stable state contains a network of epitaxial dislocations, with an average distance $\lambda = a/\Delta$, which can then be uncommensurate. In that case the epitaxial layer can glide freely on the substrate, by a translational motion which is the limit of the branch of 'phason' modes with frequency vanishing with k. It is indeed in such epitaxial layers that such phason modes were first observed by neutron scattering.

In most cases, the epitaxial dislocations also form a two dimensional network, so as to relax Δ in all directions of the plane. For two FCC crystals epitaxial on a closepacked plane for instance, depending the thickness of the layer and on λ and thus on Δ, one expects no dislocations, a family of parallel (split) dislocations or a network of three (split) types of dislocations (Figure 7). The splitting occurs because of the special structure of the FCC lattice. The network is the normal configuration for thick layers ; but the simple family can be more stable for atomically thin layers, such as the

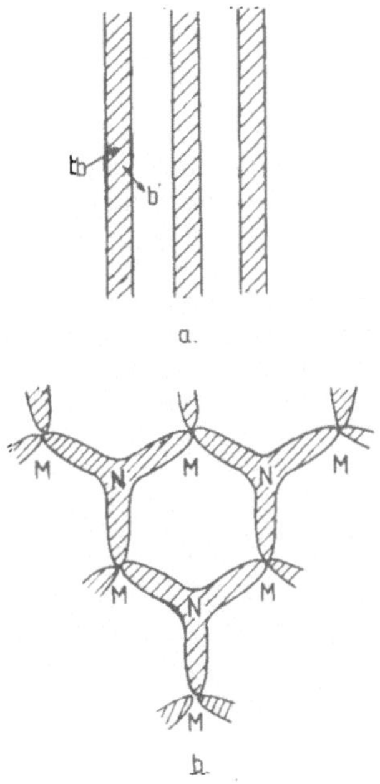

Fig. 7. One family or three-families of dislocations for epitaxy of two FCC crystals on a close packed plane.

(111) surface layer of Au on Au crystal.

c - <u>1 dimension</u>

Contrary to 2d, the 1d domain followed a rather independent course.

Peierls first showed that a 1d conductor becomes necessarily unstable with respect to a <u>lattice</u> modulation, in agreement with the infinite anomaly of ε_k at $k = 2k_M$ (Figure 2c). Furthermore he correctly developed in this case the dynamic (two waves) approximation which was also later used for BCS supraconductivity.

Fröhlich then pointed out that under an applied electric field, such lattice modulation could carry a current in parallel to the normal metallic current if the modulation was uncommensurate.

From an experimental point of view, dimerisation of unsaturated carbon chains (carotenoids) was known and understood early by chemists.

The way was then free for a direct interpretation of the first observations of charge density waves in (quasi) 1d compounds such as KCP, TTFTCNQ, $NbSe_3$ or $(CH)_x$.

Experimental and theoretical work developed in parallel on <u>spin</u> modulations, mostly an antiferromagnetic chains. The analysis of large thermal fluctuations in terms of solitons and their limitation by interchain couplings were early studied.

Three conceptual difficulties are still present today, sometimes behind the scene :

- Peierls treatment, usually followed as Jones' early treatment, merely adds one electron energies and is thus properly not selfconsistent for lattice distortions. As far as I know, only Barisic has looked at this problem, in a weak coupling limit which does not always apply.

- Localisation by disorder was predicted by Mott and studied by many authors mostly of the russian school. Its absence in a large domain of temperature can be attributed either to delocalisation by forward scattering on phonons, or by effective transverse escape of the electrons from chain to chain.

- The role of electrons-electron correlations within the chains or between chains was emphasized early for excitonic excitation of the carotenoids as well as for the 3d couplings of CDW in quasi one dimensional compounds.

3 - THE CONCEPT OF SUCCESSIVE SUPERSTRUCTURES

The field of low dimension conductors offers several examples of a modelling procedure which has been much used recently in many instances of (complex) condensed matter behaviours.

In the case of lattice modulations on organic chains for

instance, the basic unit is the chain built by a piling of equivalent molecules. These chains are packed into a d lattice structure. The 'lattice modulation' then introduces a new period along the chains, which can be uncommensurate with the basic 3d lattice.

At each level, one can define structural defects - broken bonds due to radiation damage on individual molecules, molecules of the wrong sort, chains of the wrong sort, dislocations in the lattice modulation. For many physical properties, when defining the new superstructure, the underlying structure can be assumed continuous and its periodicity neglected. However detailed physical properties depend on that periodicity neglected. Such are the special stability and friction of commensurate modulations or the lattice friction of dislocations.

It is clear that in this process, the stability of each successive superstructure is less than that of the structure on which it is based, and its size larger. Although conceptually endless, the procedure effectively terminates when the stability of an eventual new superstructure or its kinetic of formation are too low for it to be observed.

STRUCTURAL INSTABILITIES OF ONE-DIMENSIONAL CONDUCTORS

Jean Paul Pouget

Laboratoire de Physique des Solides, associé au CNRS
Université Paris-Sud
91405 Orsay (France)

INTRODUCTION

For 15 years, structural studies have played a role in the physics of one dimensional (1D) conductors. The reason is that most of these materials exhibit low temperature periodic lattice distortions (PLD) related to instabilities of the 1D electronic gas. The underlying physical mechanism was found by Peierls more than 30 years ago : a 1D metal, of ρ independent electrons per unit cell, is unstable,at T = OK, with respect to a periodic modulation of wave vector $2k_F$ ($\rho = 2 \times 2k_F$) opening a gap at $\pm k_F$ in the electronic structure (fig. 1a). In the Peierls-Fröhlich approach, the electron phonon coupling connects the intrachain displacive modulation and the electronic instability in the formation of a $2k_F$ charge density wave (CDW). But it is conceivable that other interactions of the electron gas, like its Coulomb coupling with orientational or positional (ionic) degrees of freedom, external to the chain, are able to stabilize a $2k_F$ lattice periodicity.

Because of the coupling between electronic and structural degrees of freedom, structural studies are able :
a) to probe the wide regime of thermal fluctuations above the Peierls transition, due to the ID nature of the electronic subsystem (see the lecture of Schulz),
b) to measure the wave vector of the CDW electronic instability with, in particular, its shifts from $2k_F$ to $4k_F$ (fig. 1) for strong enough electron electron interactions (see lectures of Barisic, Emery and Hirsch).

The format of these lectures is organized as follows. Part I will present the spatial and dynamical aspects of the $2k_F$ CDW structural instability. The $2k_F$ and $4k_F$ structural instabilities shown by the family of organic conductors will be analyzed in part II. Finally part III will describe more explicitly a new kind of structural instability seen in the ID conductors $(TMTSF)_2X$ and $(TMTTF)_2X$, when the anion X is non centrosymmetric, which leads to orientational anion ordering (A.O.) phase transitions.

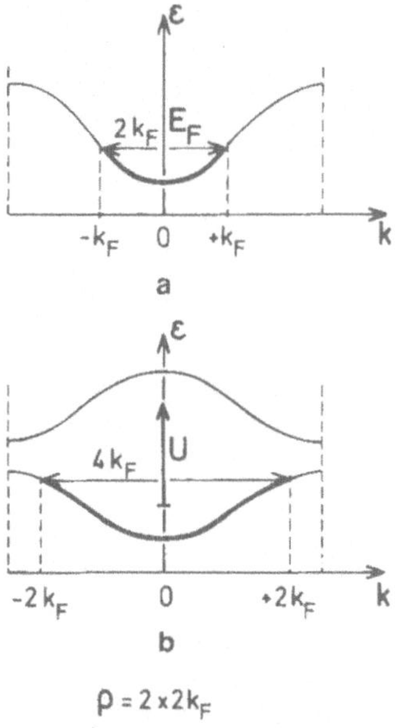

Fig. 1. Electronic structure of a 1D electron gas of (a) independent electrons and (b) strongly correlated electrons, showing respectively the $2k_F$ and $4k_F$ wave vectors of the electronic CDW instability. In b, the two Hubbard bands are separated by the intrasite Coulomb repulsion energy U.

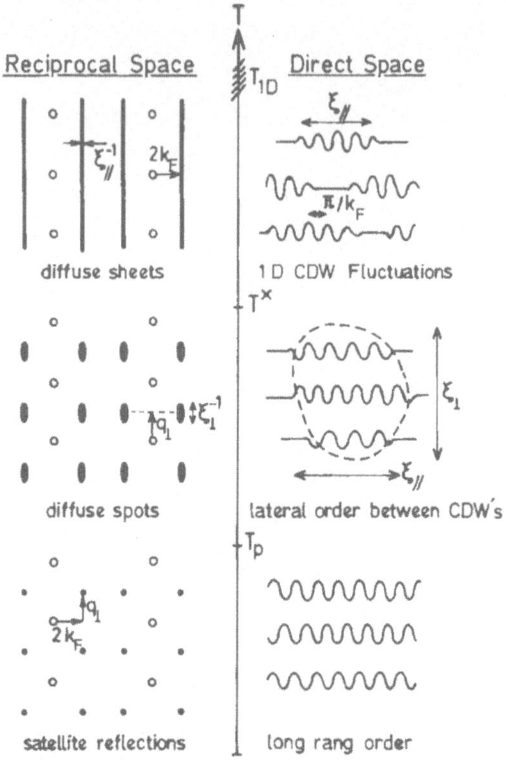

Fig. 2. Schematic representation of the diffraction (reciprocal space) from a 1D conductor at various temperatures. The corresponding CDW lattice is indicated (direct space)

I - $2k_F$ C.D.W. INSTABILITY OF 1 D CONDUCTORS

A - Spatial correlations

1) Dimensionality of the fluctuations

Let us for simplicity consider an array of identical chains showing a $2k_F$ CDW electronic instability, leading at low temperature (below T_p, the Peierls transition) to a P.L.D. Above T_p, the dimensionality of structural fluctuations (and that of the CDW's resulting from the linear coupling of the lattice and electronic degrees of freedom) can be best visualized by X-ray diffuse scattering experiments [1-3]. Figure 2 defines 3 temperatures, delimiting 3 important temperature ranges :

i) T_{1D} : onset of 1D structural fluctuations
ii) T_{3D}^{x} (T_{2D}^{x}) : cross over towards a regime of 3D (2D) fluctuations
iii) T_p : onset of the 3D long range order

Values of these temperatures for typical organic and inorganic 1D conductors showing a $2k_F$ CDW instability are given in Table I.

Table 1

Materials	T_{1D} (K)	T^x (K)	T_p (K)	atoms involved
TTF-TCNQ	150	60 (3D)	54	mostly TCNQ sublattice
TMTSF-DMTCNQ	225	$\sim$ 75 (3D)	42	mostly TMTSF sublattice
$NbSe_3$	$\gtrsim$ 300	?	144	type III chain
$K_{0,3}MoO_3$	?	$\gtrsim$ 300 (2D) $\sim$ 210 (3D)	183	$Mo_{10}O_{30}$ slab

2) X-ray diffuse scattering intensity

In the case of a displacive phase transition, the X-ray diffuse intensity at the wave vector $\vec{Q}$ ($\vec{Q} = \vec{G} + \vec{q}$, where $\vec{G}$ is a reciprocal lattice wave vector) can be expressed in the form[4] :

$$Id(\vec{Q}) = |Fd(\vec{Q})|^2 S(\vec{q}, t = 0), \qquad (1)$$

assuming the instability of a single lattice mode. In (1) $Fd(\vec{Q})$ is the structure factor of the unstable mode of generalized coordinate $A_{\vec{q}}$, and $S(\vec{q}, t = 0)$ is the instantaneous correlation function :

$$S(\vec{q}, t = 0) = \langle |A_{\vec{q}}|^2 \rangle \qquad (2)$$

In the classical limit ($k_B T > \hbar w_{2k_F}$, where $\hbar w_{2k_F}$ is the characteristic energy of the unstable lattice mode), the theorem of equipartition of the energy gives :

$$S(\vec{q}, t = 0) = k_B T \, \chi(\vec{q}), \qquad (3)$$

where $X(\vec{q})$ is the susceptibility associated with the order parameter, $A_{\vec{q}}$. To describe the approach of a phase transition, it is easiest to measure the extra diffuse scattering above the background, $\delta Id(\vec{Q})$, than the absolute value of the diffuse intensity, $Id(\vec{Q})$. The former quantity is directly related to the critical growth of pretransitional fluctuations. In the case of a linearily coupled electron phonon system, the extra scattering is proportional to the CDW response function $\chi_\rho(\vec{q})$[5,6] :

$$\delta \, Id(\vec{Q}) \propto k_B T \, \chi_\rho(\vec{q}) \qquad (4)$$

The anisotropy of $\delta \, Id(\vec{Q})$, shown in fig. 2, gives thus the anisotropy of $\chi_\rho(\vec{q})$. Treating the electron phonon coupling in the RPA approximation, $\chi_\rho(q)$ takes the form :

$$\chi_\rho(\vec{q}) = \frac{\chi_e(\vec{q})}{1-\lambda_{\vec{q}}\,\chi_e(\vec{q})} \quad , \tag{5}$$

where $\lambda_{\vec{q}}$ is the dimensionless electron phonon coupling constant, and $\chi_e(\vec{q})$ is the electronic polarizability, including the electron electron interactions. Depending on the strength and on the range of the electron electron interactions (See lectures of Barisic, Emery and Hirsch), $\chi_e(\vec{q})$, and thus $\chi_\rho(\vec{q})$, present a maximum for the wave vector component$(q_{//})$ $2k_F$ and/or $4k_F$ in chain direction.

In the regime of 1D fluctuations (i.e. between T_{1D} and T^x), assuming a $2k_F$ instability, $\chi_\rho(q_{//})$ has roughly a lorentzian dependence :

$$\chi_\rho(q_{//}) = \frac{\chi_\rho(2k_F)}{1 + \xi_{//}^2\ \delta q_{//}^2} \quad , \tag{6}$$

where $\delta q_{//} = q_{//} - 2k_F$. Because of 1D fluctuations, $\chi_\rho(q_{//})$ differs from the predictions of equation (5), especially in the vicinity of the mean field temperature of the lattice instability, T_p^{MF} defined by $1 = \lambda_q \chi_e(q_{//})$

From (4) the peak intensity of $\delta\, Id(\vec{Q})$ divided by k_BT gives a quantity proportional to the $2k_F$ (or $4k_F$) CDW response function χ_ρ. From (6), the half width at half maximum (H.W.H.M.) of $\delta I_d(\vec{Q})$ in chain direction gives (after a resolution correction) the inverse CDW correlation length, $\xi_{//}^{-1}$. Below T^x, $\xi_\perp^{-1}$ is obtained in a similar way from the H.W.H.M. of the diffuse spots in transverse directions. At T^x, $\xi_\perp$ amounts to one interchain distance in the directions coupled.

Figure 3 gives the temperature dependence of $\chi_{2k_F}^{-1}$, $\xi_{//}^{-1}$ and $\xi_\perp^{-1}$ for the $2k_F$ scattering of TMTSF-DMTCNQ. It defines clearly the 3 temperatures quoted in table I :

- $T_{1D} \sim 225$ K : the temperature at which the extra scattering δId begins to be observed under the form of diffuse sheets

- $T_{3D}^x \sim 75$ K, when $\xi_\perp$ amounts to the interstack distance (at this temperature, there is also a change in the temperature dependence of χ_ρ and $\xi_{//}$)

- The 2nd order Peierls transition $T_p = 42$ K, when $\xi_\perp^{-1}$, $\xi_{//}^{-1}$ and χ_ρ^{-1} vanishes (approximately like $\xi^{-1} \sim \sqrt{T-T_p}$ and $\chi^{-1} \sim (T - T_p)$ in figure 3). Similar measurements have been performed in $K_{0.3}MoO_3$ [5], TTF-TCNQ [7] and several other organic conductors [6].

3) 1 D CDW fluctuation regime

Above T^x, the temperature dependence of χ_ρ and $\xi_{//}$ can be roughly fitted with a power law of the temperature, which diverges, as expected for a 1D system, at 0K. The expression for the divergence depends on the salts; for example in TTF-TCNQ : $\chi_\rho(2k_F) \propto T^{-1.6}$, while in the Se analogue TSF-TCNQ : $\chi_\rho(2k_F) \propto \log \frac{350}{T}$. The strongest divergence of $\chi_\rho(2k_F)$ is observed in salts showing the $4k_F$ instability (i.e. materials with substantial Coulomb interactions). Usually the $4k_F$ scattering has a weaker divergence than the $2k_F$ scattering; in TTF-TCNQ : $\chi_\rho(4k_F) \propto T^{-0.7}$.

Electron electron interactions, g, (See lectures of Barisic, Emery, Hirsch) as well as the electron phonon coupling, λ, (See the lecture of

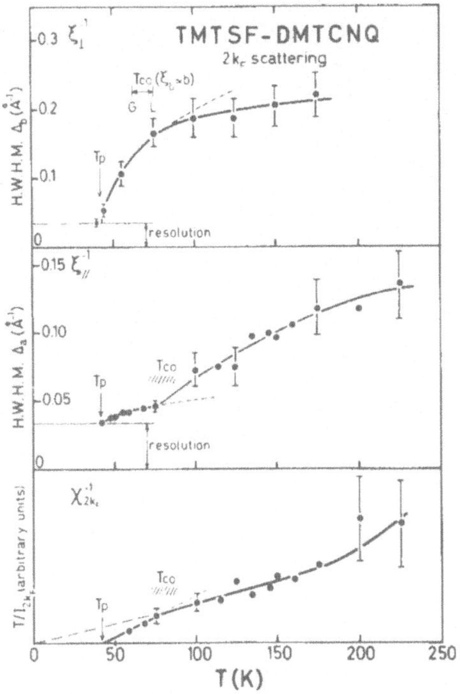

Fig. 3. Temperature dependence of the H.W.H.M. along b and a (related to $\xi_{\perp}^{-1}$ and $\xi_{//}^{-1}$) and of the inverse peak intensity divided by the temperature (related to χ_{ρ}^{-1}) of the $2k_F$ scattering of TMTSF-DMTCNQ. T_p and T^x (i.e. T_{Co}) are indicated.

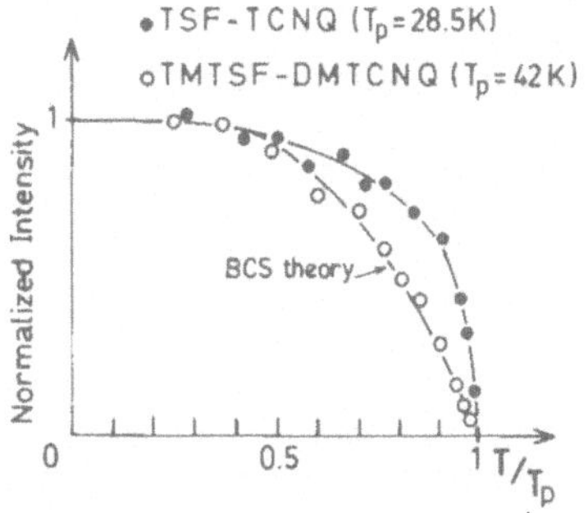

Fig. 4. Temperature dependence of the satellite intensity, is reduced scale, for TMTSF-DMTCNQ[3] and TSF-TCNQ[13]. The square of the BCS mean field order parameter is indicated.

Schulz) contribute at the divergence of χ_ρ and $\xi_{//}$. It has been argued that ID fluctuations are mostly due to electonr electron interactions (i.e. $g > \lambda$) in organic conductors like TTF-TCNQ[8] and mostly due to electron phonon coupling effects (i.e. $\lambda > g$) in inorganic conductors like $K_2Pt(CN)_4$ $0.3Br - xH_2O$(KCP), $NbSe_3$ or $K_{0.3}MoO_3$[9] (See the lecture of Barisić). This is only well above the mean field temperature of the lattice instability $T > T_p^{MF}$ (or $\lambda_{\vec{q}}\ \chi_e(\vec{q}) < 1$ in eq. (5)) that the extra diffuse scattering has an intensity proportional to the electronic polarizability $\chi_e(\vec{q})$. In this respect, TSF-TCNQ is especially interesting because above 50K, $\chi_\rho\ (2k_F)$ and $\xi_{//}^{-1}$ have roughly the temperature dependence expected from the electronic polarizability of a 1D Fermi gas of independent electrons[6]:

$$\chi_e(2k_F) = N(E_F)\ \log \frac{A}{k_B T} \tag{7}$$

$$\xi_{e//}^{-1} = 4.34\ \frac{k_B T}{\hbar\ v_F} \sqrt{\log \frac{A}{k_B T}} \tag{8}$$

Furthermore from the absolute value $\xi_{//}^{-1} = 0.06 \pm 0.015$ Å^{-1} measured at 150K, equation (8) gives a Fermi velocity $v_F = 1.3 \pm 0.3 10^5$ m/s, corresponding to a 1D tight binding (donor) band width $W_D = 0.5 \pm 0.1$ eV which agrees quite well with calculated[11] and optically measured[68] values : 0,6 eV[10] (In TSF-TCNQ, the $2k_F$ scattering originates mainly from the donor stack containing $\rho = 0.63$ hole per molecule).

4) Coupling between CDW's

The reduced temperature range of 3D fluctuations, $\Delta t = \frac{T^x}{T_p} - 1$, varies appreciably in the series of organic conductors. Table I gives Δt values from 0.1 (TTF-TCNQ) to 0.8 (TMTSF-DMTCNQ). For especially anisotropic compounds, a regime of 2D fluctuation can exist. This is the case of the blue bronzes $A_{0.3}MoO_3$ (A = K, Rb) between about 300 K and 210 K. CDW interchain coupling mechanisms are considered in the lecture of Schulz.

5) 3D long range order between CDW's

The P.L.D. is characterized by the formation of sharp satellite reflections. Three kinds of information can be generally extracted from X-ray, neutron or electron studies below T_p :

a) the wave vector (generally incommensurate) of the structural modulation,
b) the temperature dependence of the modulation,
c) the polarization of the atomic displacements.

Informations (a) and (b), concerning the phase diagram and the thermodynamics of the transition, are obtained by the measurement of few satellite reflections, while (c) requires a structural refinement from the collection of a great number of satellite reflections.

The $q_{//}$ component of the wave vector of the structural modulation characterizes the CDW instability of the 1D electron gas (figure 1). This component may vary appreciably in temperature ($K_{0.3}MoO_3$[5]) or under pressure (TTF-TCNQ[12]). The transverse components ($q_\perp$) are determined by the interchain coupling. In compounds like TTF-TCNQ, possessing two kinds of stacks with their own temperature of instability, $q_\perp$ (i.e. q_a) varies in temperature and pressure, leading at a very complex phase diagram[12]. A discussion of the phase diagram of TTF-TCNQ can be found in ref. 8.

For the simplest structural transitions, the satellite intensity is directly proportional to the square of the order parameter η (i.e. the amplitude of the distortion A_{2k_F}, which is related, via the electron phonon coupling constant, at the Peierls gap Δ). This is the case of TMTSF-DMTCNQ and TSF-TCNQ, considered in fig. 4. The behaviour of η^2 near T_p, is very different for these 2 salts. Surprisingly, the order parameter of TMTSF-DMTCNQ has roughly the BCS mean field temperature dependence[19], although the value of the Peierls gap 2Δ is two times larger than the mean field quantity : $3.52\ T_p$[14]. Similar observations have been made in $K_{0.3}MoO_3$ and $NbSe_3$. The mean field behaviour of η is not understood, owing to the large regime of 1D fluctuations. It could be due to weak fluctuations in the 3D CDW coupled regime, restoring a pseudo mean field behaviour. In this limit, deviations from the overall mean field law are also expected to occur in the 3D Ginzburg critical region near T_p.

B - Dynamics

Temporal fluctuations of the order parameter are mostly studied by neutron scattering technics. These experiments measure a quantity, the differential cross section, proportional to $S(\vec{q}, \omega)$, the Fourier transform

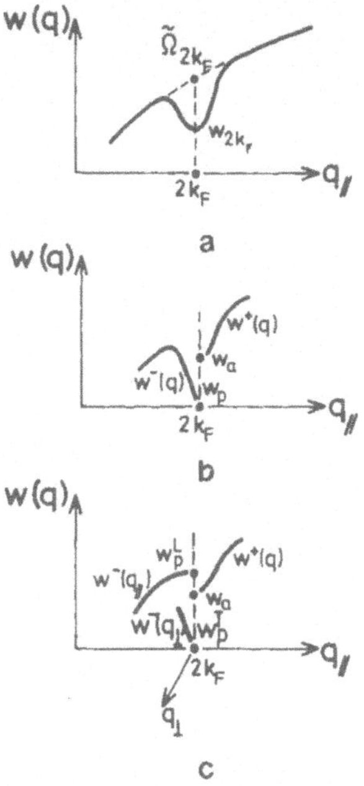

Fig. 5. Schematic representations for:

i) $T>T_p$ of the Kohn anomaly in a phonon branch (a);

$$\tilde{\Omega}^2_{2k_F} = \Omega^2_{2k_F}\,(1-\frac{\lambda}{2})$$

ii) $T<T_p$ of the phase [$\omega^-(q)$] and amplitude [$\omega^+(q)$] excitation branches for an incommensurate modulation (b : screened CDW; c : unscreened CDW). In the unscreened case, notice the splitting between longitudinal (along $q_{//}$) and transverse (along $q_\perp$) phase modes.

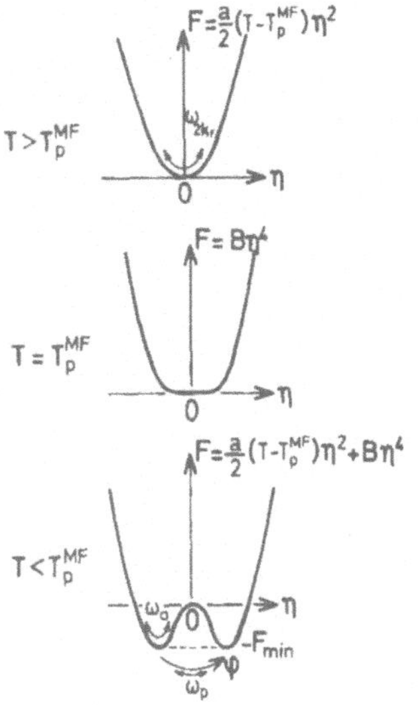

Fig. 6. Order parameter dependence of the free energy, F, in a mean field theory of the Peierls transition for $T>T_p^{MF}$, $T = T_p^{MF}$ and $T<T_p^{MF}$. The characteristic frequencies of the associated dynamics, above and below T_p^{MF}, related to the curvature of F near its minima, are indicated. η and φ are respectively the amplitude and the phase of the order parameter (case of a single q incommensurate modulation).

of the correlation function $<A_{\vec{q}}(0)A_{-\vec{q}}(t)>$. In the classical limit ($k_BT>\hbar\omega$), $S(\vec{q},\omega)$ is simply related to the generalized susceptibility $\chi(\vec{q},\omega)$, by :

$$S(\vec{q},\omega) = \frac{k_BT}{\hbar\omega}\,\mathrm{Im}\,\chi(\vec{q},\omega) \tag{9}$$

Using the Kramers Kronig relationship, it is easy to deduce, from (9), the X-ray diffuse scattering expression :

$$S(\vec{q},t=0) = \int S(\vec{q},\omega)d\omega = k_BT\,\mathrm{Re}\,\chi(\vec{q},\omega=0) \tag{3}$$

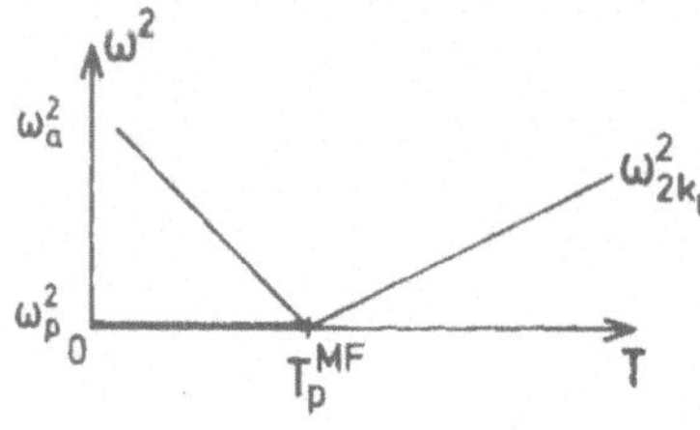

Fig. 7. Temperature dependence of $\omega^2_{2k_F}$, ω^2_a and ω^2_p, defined in figures 5(a, b) and 6, for a mean field theory of the Peierls transition.

1) Pretransitional fluctuations

As in any phase transition described by an order parameter, pretransitional fluctuations possess a dynamics above T_p. Because of the electron phonon coupling the dynamics is also controlled by the fluctuations of the 1D electron gas. In the limit of weak electron phonon coupling, temporal fluctuations are described by the softening of a particular phonon mode at a critical wave vector, assumed to be $2k_F$ in this section. More precisely, the bare phonon frequencies $\Omega(q)$ are reduced by the screening of 1D conduction electrons. By treating the electron phonon coupling in perturbation, the renormalized phonon frequencies $\omega(q)$ become :

$$\omega^2(\vec{q}) = \Omega^2(\vec{q})[1-\lambda\chi_e(\vec{q})] \tag{10}$$

with $\chi_e(\vec{q})$ showing a well defined peak around $2k_F$, resulting in a sharp anomaly in the phonon spectrum (fig. 5a). Such a Kohn anomaly has been observed in KCP [1, 24], $K_{0.3}MoO_3$ [15, 16] and to a less extent in TTF-TCNQ [1, 17]. In the harmonic approximation, $\omega^2_{2k_F}$ behaves in temperature as the curvature, $\chi^{-1}(2k_F)$, of the free energy near its minimum (fig. 6) [19] :

$$\omega^2_{2k_F} = a(T-T_p^{MF}) \tag{11}$$

The (mean field) temperature dependence of the softening is shown in fig.7 (T_p^{MF} is the mean field temperature of the lattice instability defined by the vanishing of (10)). This result can also be obtained from equations (9) and (3) using, for the soft mode, a response function having the ω dependence of a damped harmonic oscillator :

$$\chi(2k_F, \omega) = \frac{\omega^2_{2k_F}}{\omega^2_{2k_F} - 2\,i\Gamma_{2k_F}\,\omega - \omega^2}\,\chi(2k_F, 0) \tag{12}$$

and the Curie Weiss temperature dependence :

$$\chi(2k_F, 0) \propto (T - T_P^{MF})^{-1}.$$

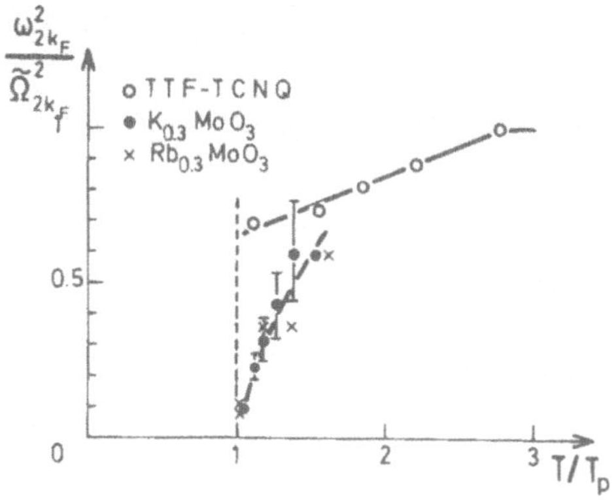

Fig. 8. Temperature dependence of $\omega^2_{2k_F}$, in reduced units, for TTF-TCNQ [17] and the blue bronzes $A_{0.3}MoO_3$ (A = K [18], Rb [18]).

From (12) one deduces :

$$S(2k_F, \omega) \propto k_B T \frac{2\Gamma_{2k_F}}{(\omega^2_{2k_F} - \omega^2)^2 + 4\Gamma^2_{2k_F}\omega^2} \qquad (13)$$

The quasi harmonic frequency $\omega 2k_F$ is usually obtained from a fit of the experimental cross section (minus the background) by expression (13) convoluted with the experimental resolution. Fig. 8 gives the temperature dependence of the quasiharmonic frequency, $\omega^2_{2k_F}$, obtained in quasi 1D conductors. It shows that at 10% of T_p, the softening is of 90% for the blue bronzes, but only of 30% for TTF-TCNQ. There is apparently an incomplete softening at T_p in the latter case.

Although the softening seems to be nearly complete in the blue bronzes the divergence of $\chi(2k_F)$ is achieved by the critical growth of another component, centered around $\omega \sim 0$, in $S(q, \omega)$ [15, 16], as shown by fig. 9. An elastic critical scattering has been also observed near T_p in TTF-TCNQ[49]. The formation of a central peak in $S(q, \omega)$ near the critical temperature is a common phenomenon for displacive structural phase transitions. The central peak can be viewed as an extra scattering coming from domains of the low temperature phase which are formed above T_p with a finite lifetime. In 1D conductors several mechanisms have been invoked to explain the formation of such domains [8, 21]. Theories are based either on an intrinsic origin (anharmonicity [8], electron electron interactions [8] or electron phonon coupling [20]) or on an extrinsic origin (formation of Friedel oscillations in the vicinity of defects [22]) for the central peak. Intrinsic mechanisms are also able, in the strong coupling limit, to give a purely relaxational pretransitional dynamics (i.e. without phonon softening).

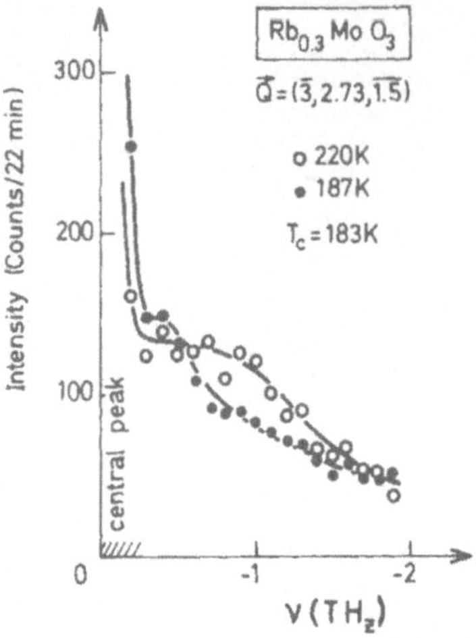

Fig. 9. $\nu(=\omega/2\pi)$ dependence of the neutron scattering differential cross section at $2k_F$ for $Rb_{0.3}MoO_3$ at 220K and 187K ($T_c = T_p$ = 183 K)[18]. Note the softening of the phonon response and the growth of an additional response near $\nu = 0$.

2) Phase and amplitude excitations

In the harmonic theory of an incommensurate Peierls transition, the decoupling of the amplitude and phase degrees of freedom, of the complex order parameter $\eta\, e^{i\varphi}$, occurs at $T_p (= T_p^{MF})$. Below T_p, associated fluctuations are described by two phonon branches called respectively ampliton and phason, ω^+ and ω^- in fig. 5b (see lectures of Bjelis and Schulz). For $q = 2k_F$, the free energy does not depend on φ. No restoring force can be opposed at the free sliding of the phase of the incommensurate modulation with respect to the average lattice, thus $\omega_- = 0$. Such is not the case for the amplitude fluctuations, which characteristic frequency remains finite below T_p and is, as $\omega^2_{2k_F}$ above T_p, related to $\chi^{-1}_{(2k_F)}$ (fig. 6) :

$$\omega_a^2 = 2\, a\, (T_P^{MF} - T) \tag{14}$$

In the mean field theory ω_a^2 increases linearily for decreasing T (fig. 7). In the vicinity of $2k_F$, the Golstone theorem applied to the phase degree of freedom predicts an acoustic like dispersion for the phason branch :

$$\omega^-(q) = v_\varphi\, |q - 2k_F| . \tag{15}$$

a and v_φ can be calculated from the microscopic theory of the 1D electron gas coupled to phonons (LRA theory - See ref (25) and lectures of Bjelis and Schulz).

In the case of an incomplete softening at T_p, the phase mode originates from the central peak and the amplitude mode is the continuation below T_p of the soft mode. These excitations have respectively diffusive and propagative dynamics [20, 21].

Amplitude and (pinned) phase modes at $2k_F$ can be observed respectively by Raman and infrared spectroscopies (see the lecture of Jandl). Neutron measurements are difficult to perform because of the narrowness of the Kohn anomaly in chain direction, but evidences for these modes have been reported for KCP [1, 24] and $K_{0.3}MoO_3$ [23]. Analysis of the data obtained show that the above theory is in fact oversimplified. Corrections should include :

i) the decoupling above T_p between amplitude and phase degrees of freedom, in a temperature range where, because of strong 1D fluctuations, the pseudo gap is well established. This is especially the case of KCP, for which $S(2k_F, \omega)$ does not change appreciably on heating above $T_p \sim 100K$ (KCP shows a well formed pseudo gap below room temperature). In connection with this observation, the amplitude mode frequency ($\nu_a \sim 1.3\ TH_Z$ for KCP; $\nu_a \sim 1.5\ TH_Z$ for $K_{0.3}MoO_3$) varies weakly in temperature, near T_p, contrasting with the mean field prediction of equation (14).

ii) the finite value of the phase mode frequency at $2k_F$. ω_p is different from zero because of the pinning of the CDW by defects (in KCP showing a substantial disorder due at the Br and H_2O : $\nu_p = \omega_p/2\pi \sim 0.6\ TH_Z$). Pinning due to Coulomb forces between CDW's built with oppositely charged carriers, in charge transfer salts like TTF-TCNQ, is also expected when the CDW are of comparable amplitude on unlike stacks [8, 25].

iii) the polar character of the phase mode. In that case, a longitudinal compression of the CDW involves charge redistribution (fig. 10a) and long range Coulomb forces, giving an optical frequency ($\omega_p^2 \neq 0$) at the longitudinal phason [25] (See the lectures of Barisic and Bjelis). Such charge redistribution does not occur for a transverse shear deformation of CDW lattice (fig. 10b). This preserves the acoustic like dispersion of the phase mode in $q_\perp$ directions. This effect gives the schematic representation of figure 5c which is apparently in conflict with that shown in fig. 5b for the $q_{//}$ direction. In fact, this analysis does not take into account the screening of long range Coulomb forces by electron-hole pairs created thermally accross the Peierls gap, 2Δ. Clearly, near T_p, fig. 5bis valid : there is screening of the Coulomb forces because the Fermi velocity of the quantum gas of quasi particles, v_F, is larger than v_φ. At low temperatures, $T < \Delta$, fig. 5c is valid because the (Maxwell Boltzman) velocity of the classical gaz of quasi particles, v_{MB} $(\sim\sqrt{\frac{T}{m_{eff}}} \sim v_F\sqrt{\frac{T}{\Delta}})$, is too slow, compared to v_φ, to screen the charges induced by the phase fluctuations. The cross over between the two regimes has been analyzed in ref(26) which also predicts an increase of v_φ on cooling. Screening effects occur for in chain wave vectors components such that $\delta q_{//}\lambda_{TF} < 1$, where λ_{TF} is the Thomas Fermi screening length. Partial screening effects could explain the apparent enhancement of the observed phason velocity ($v_\varphi \sim 10^5$ m/s around 100K in KCP [24, 27]; $v_\varphi = 2.1 \pm 0.3 10^4$ m/s at 175K in $K_{0.3}MoO_3$ [23]), with respect to the value expected from the LRA theory [25], using a reasonable value for the CDW effective mass.

II - $2k_F$ AND $4k_F$ STRUCTURAL INSTABILITIES IN 1D ORGANIC CONDUCTORS

Structural instabilities observed in organic conductors are usually connected with the $2k_F$ or $4k_F$ wave vector of the 1D electronic gas. These instabilities result either, via the electron phonon coupling, from the CDW instability of the electron gas or from the coupling of the electron gas

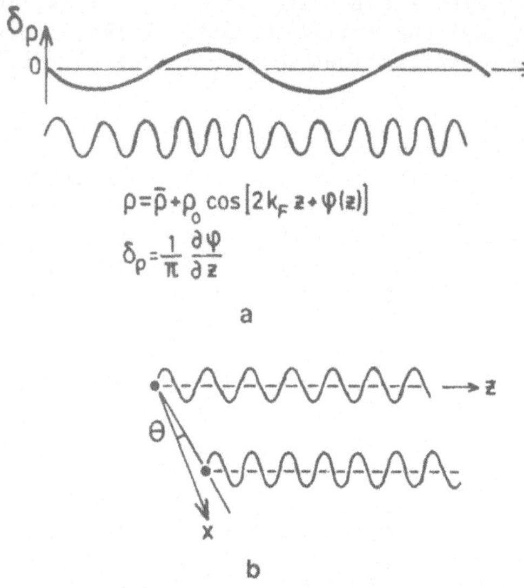

Fig. 10. a) longitudinal deformation of the phase $\varphi(z)$ of the CDW, $\rho(z)$, inducing a charge redistribution, $\delta\rho(z)$, with the wave length $2\pi/\delta q_{//}$.
b) transverse shift, θ, of the CDW lattice

with external degrees of freedom. A survey of the structural instabilities shown by ID organic conductors is the purpose of this part.

A - <u>Molecular array</u>

Although the 1D nature of the electron gas influences greatly the physics of organic conductors, real materials are in fact 3D. For example the molecular array determines the interstack coupling and thus the wave vector of the low temperature P.L.D. Neighbouring stacks provide an effective medium which influences also the intrastack interactions. Figure 11 represents some typical arrays observed in charge transfer organic conductors :

a : segregated donor (D) and acceptor (A) stacks of the same (in chain) periodicity, forming two metallic subsystems (TTF-TCNQ, NMP-TCNQ). The charge transfer from D to A, ρ, is generally incommensurate and comprised between 1/2 and 3/4 [5, 6]. Generally, ρ varies with the temperature, the pressure or upon alloying.

b - d : segregated donor (TMTSF, TTT) or acceptor (TCNQ) stacks, forming one metallic susbsystem. The full charge transfer occurs via a ionic sublattice composed either of anions ($X = ClO_4$, ReO_4; I_3) or cations (MEM, TEA). Depending of the stoichiometry of the salt, ρ is either commensurate (b and c) or incommensurate (d).

B - <u>k_F wave vectors</u>

For simplicity, let us use an independent electron description. Depending of the nature of the donor molecule and of the number of metallic

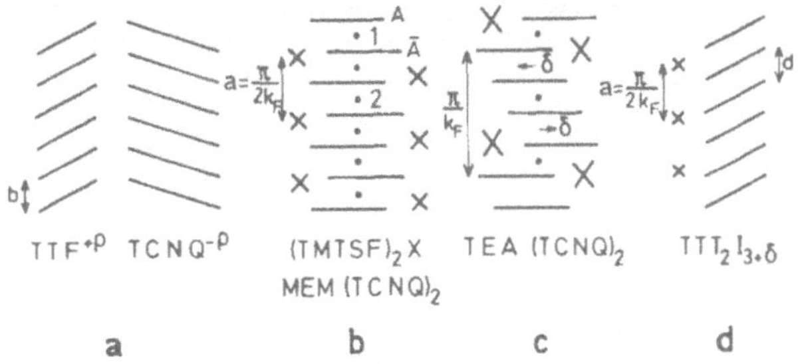

Fig. 11. Schematic representation of the molecular array of some charge transfer salts. The stacking direction is vertical. In (b),(c),(d) the anion and cation positions are shown by crosses. In (b) and (c) the dots indicate the intrastack inversion centres. In (b), A and $\bar{A}$ label the 2 (inversion related) molecules per unit cell. In (c), δ represents the shift of the centre of the inversion related dimers in a direction perpendicular to the stacking axis.

subsystems several band structures occur. For simplicity, we shall assume that there is only one conduction band per stack (i.e. the splitting between intramolecular energy levels is larger than the intrastack overlap integral).

In materials containing D and A metallic subsystems (fig. 11a) it is important to distinguish between closed shell donor (TTF) and unsaturated donor (NMP). In the first case, (fig. 12a), the number of electrons transferred to the acceptor is equal to the number of holes formed on the donor : k_F is the same for the D and A bands. Thus it is not straightforward to know on which stack a $2k_F$ or $4k_F$ instability develops. Structural analysis of the diffuse scattering or local measurements (EPR, NMR, infrared absorption etc..) are necessary for this assignment. For example, the structure factor analysis of the X-ray diffuse scattering from TSF-TCNQ [10] or HMTSF-TCNQ [28] shows that the observed $2k_F$ scattering mainly occurs on the donor stack. A concentration study of the diffuse scattering from the solid solution $(TTF)_{1-x}(TSF)_xTCNQ$ [6, 29] proves that the $4k_F$ scattering originates from the donor stack, while the $2k_F$ instability is driven by the TCNQ stack, in agreement with optical measurements (see the lecture of Jacobsen). In the second case, fig. 12b, the Fermi wave vector is different on the donor and the acceptor stacks. This occurs for NMP-TCNQ and the solid solutions $NMP_xPhen_{1-x}TCNQ$ and, in principle, there is no ambiguity to attribute the observed $2k_F$ or (and) $4k_F$ instabilities to the TCNQ stack [30].

The simplest situation occurs when there is only one metallic subsystem : the 1D instability occurs in this subsystem. Figure 12c shows the band structure of $(TMTSF)_2X$ in an extended Brillouin zone. However, in salts considered in fig. 11b, c, d, the periodicity of the ionic charge distribution, and the response of the 1D organic stack at its electrostatic potential have to be considered. For example, the gap Δ_G of $(TMTSF)_2X$

shown in fig. 12c, is believed to be mainly caused by the $(4k_F)$ ionic periodicity. Its value determines the strength of Umklapp electron electron interaction g_3 in quarter filled band ($\rho = 1/2$ systems [31] (See lectures of Bourbonnais and Emery).

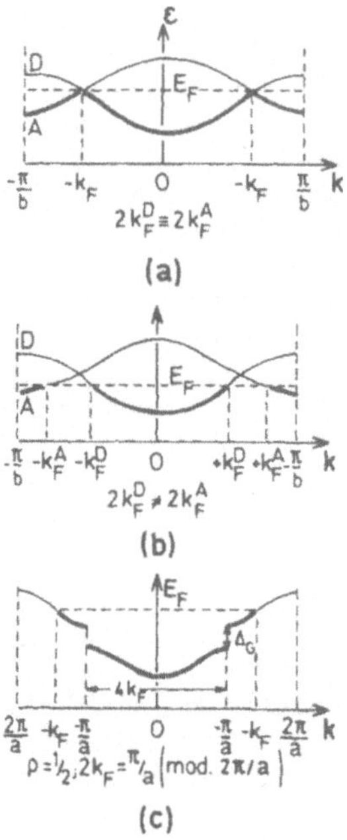

Fig. 12. One electron energy dispersion of the inverted band structure of charge transfer salts with two electrons to share per DA pair (a) and less than two electrons per DA pair (b). (c) gives, in an extended zone, the band structure of $(TMTSF)_2X$ or $(TMTTF)_2X$. Δ_G is the $(4k_F)$ band gap at the real Brillouin zone boundary.

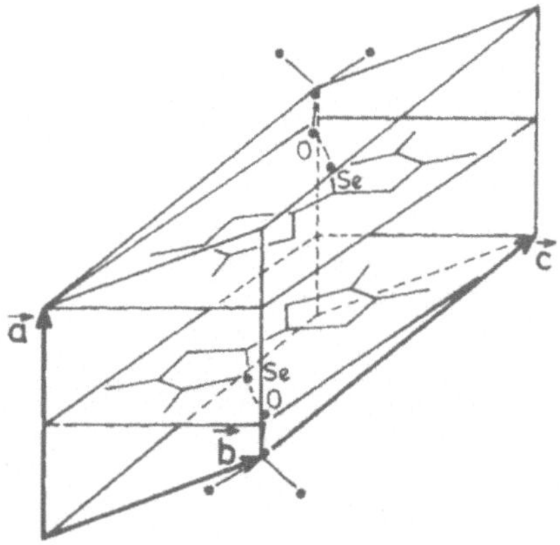

Fig. 13. Perspective view of the P $\bar{1}$ structure of $(TMTSF)_2X$ ($X = ClO_4$, ReO_4..) showing the 2 Se...O short contact distances per unit cell. Note that the anions are not situated in the molecular planes.
(From S. Ravy - thesis).

C - Periodicity of the ionic sublattice

This quantity can be calculated very simply from the knowledge of the structure and of the stoichiometry, x, of the charge transfer salt. Let us consider the salt $D^{+x}(A^-)_x$, where there is a full charge transfer from D to A. The average charge per D and A molecules is respectively $\rho^D = x$ and $\rho^A = 1$. The wave vectors associated to this charge distribution are respectively $4k_F^D = \frac{2\pi}{d}x$ and $4k_F^A = \frac{2\pi}{a}$, where d and a are the periodicities of D and A stacks. If r is the ratio of the number of chains A/D per unit cell (rd = xa), then, the (D) charges are transferred on r (A) stacks; This gives :

$$4k_F^D = r\ 4k_F^A \qquad (16)$$

Table II gives, for some salts, the wave vector ($4k_F$) of the 1st Fourier component of the ionic charge distribution and its relation with the Fermi wave vector of the metallic stack. Depending of the commensurate or incommensurate value of x (or the ratio d/a), the Fermi wave vector is or is not commensurate with the chain reciprocal periodicity.

Table II

Materials	x	r	relationship (16)
$(TMTSF)_2X$	$\frac{1}{2}$	1	$4k_F^D = \underline{4k_F^A}$
$MEM(TCNQ)_2$	2	1	$\underline{4k_F^D} = 4k_F^A$
$Qn(TCNQ)_2$	2	2	$\underline{4k_F^D} = 8k_F^A$
$TEA(TCNQ)_2$	2	1/2	$\underline{4k_F^D} = 2k_F^A$
$TTT_2(I_3)_{\frac{3+\delta}{3}}$	$\frac{3+\delta}{6}$	1	$4k_F^D = \underline{4k_F^A}$
$E_2P(I_3)_{0.53}$	0.53	1	$4k_F^D = \underline{4k_F^A}$
DIPS $\emptyset_4$ $(I_3)_{0.76}$	0.76	2	$2k_F^D = \underline{4k_F^A}$
$Per_2\{M(mnt)_2\}$ M=Pt,Pd,Au	$\frac{1}{2}$	$\frac{1}{2}$	$8k_F^D = \underline{4k_F^A}$

Let us begin with the commensurate salts shown in fig. 11b and c (per organic molecule $\rho = 1/2$ electron for $MEM(TCNQ)_2$ and $TEA(TCNQ)_2$ and $\rho = 1/2$ hole per organic molecule for $(TMTSF)_2X$). The ionic periodicity of fig.11b corresponds to the wave vector $4k_F$ of the electron gas, while the ionic periodicity of fig.11c corresponds to $2k_F$. Dimerization and tetramerization of these organic stacks stabilize respectively the $4k_F$ and $2k_F$ wave vectors. In $(TMTSF)_2X$ (and $MEM(TCNQ)_2$) the effect of the ionic potential on the organic stack is in fact relatively subtle :

i) the electrostatic potential viewed by the molecular sites is the same, because all the molecules are related by inversion symmetry ; (i.e. $<\psi_A^1|V|\psi_A^1> = <\psi_{\bar{A}}^1|V|\psi_{\bar{A}}^1>$ in the notations of figure 11b; V being the ionic potential, and $|\psi>$ being the H.O.M.O. of the TMTSF molecule).

ii) the electrostatic potential viewed by the bonds is different, because the structure has not a screw axis (2_1) symmetry (the anions are shifted from the TMTSF molecular planes in opposite directions as shown by fig. 13)(i.e. $<\psi_A^1|V|\psi_{\bar{A}}^1> \neq <\psi_A^2|V|\psi_{\bar{A}}^1>$). For strong enough V potential the interaction (ii) may induce a $4k_F$ hole localization on the bonds in near contact with pairs of anions. This ionic potential and the dimerization of the organic stack which my result, contribute at the $4k_F$ bank gap, Δ_G, shown in figure 12c.

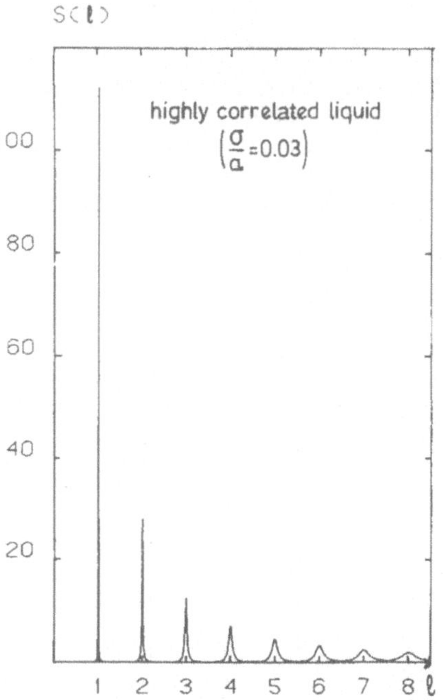

Fig. 14. 1D liquid like correlation function S(q, t=0) from I_3^- units in DIPS $\emptyset_4$ $I_{1.6}$ at room temperature (σ is the root mean square fluctuation of distance between 1st neighbouring anions, and a (see fig. 11d) is the average distance). Note the successive maxima for integer values of ℓ , the decrease of their intensity and the peak broadening for ℓ increasing. ℓ is expressed in $2\pi/a$ units (From P.A. Albouy - Thesis).

The charge distribution of the ionic sublattice provides an infinite set of inequivalent potential periodicities in the incommensurate case (fig. 11d), corresponding for the 1D electron gas, at ℓ $2k_F$ for DIPS $\emptyset_4$ $I_{1.6}$ and ℓ $4k_F$ for $TTT_2I_{3+\delta}$, (ℓ is an integer) . In these salts, mostly because of the incommensurability, there is no transverse order and no long range order in chain direction of the ionic sublattice at room temperature [32]. Thus at this temperature the organic sublattice experiences only local interactions from the iodine sublattice. The lack of long range order is manifested by the "damping" of the correlation function $S(q_{//}, t=0)$ for ℓ increasing (fig. 14). Long range periodic interactions occur only below the 1D liquid - 3D solid phase transition of the ionic sublattice ($\sim$ 180K for DIPS $\emptyset_4$ $I_{1.6}$; $\sim$ 170K for TTT_2 $I_{3+\delta}$).

D - Structural Phase transitions of organic conductors

In most of the two stack organic salts a $2k_F$ or $4k_F$ CDW instability leads to a low temperature phase transition provided that disorder effects do not prevent lateral coupling between CDW's (see the lecture of Zuppiroli). In salts of the TTF-TCNQ family, there is a sizeable regime of ID structural fluctuations, which result from the electron phonon coupling between lattice and electronic degrees of freedom. Divergence of these fluctuations leads below Tp ($\simeq$ 50K) to a displacive distortion of the organic stacks. The Peierls transition breaks the translational lattice symmetry, and the wave vector stabilized in the chain direction, $2k_F$ or $4k_F$, is generally incommensurate. This 2nd order phase transition corresponds, with a metal insulator transformation, to anomalies in electronic properties[33].

Structural transitions shown by one stack organic salts are more subtle, because in addition to the electron phonon coupling, the coupling of the electron gas with the ionic sublattice must be considered. In the linear approximation, the free energy of a CDW, ρ_q, coupled to an external electrostatic potential, of Fourier component V_q, is :

$$F = \frac{1}{2\chi_e(q)} |\rho_q|^2 - V_q \rho_{-q} , \tag{17}$$

where $\chi_e(q)$ is the q dependent electronic polarizability. By minimization of (17), it is easy to see that the external potential V_q induces a CDW, which amplitude depends on the electronic response function at the same wave vector :

$$\tilde{\rho}_q = \chi_e(q) V_q . \tag{18}$$

CDW response of the electronic gas is generally strong for the $2k_F$ and/or $4k_F$ wave vectors. In these cases, we are faced with the problem of a Peierls transition in an external field, conjugated to the CDW [71]. Thus, no symmetry breaking will occur in temperature. For a 2nd order Peierls transition, in absence of V, a continuous increase of ρ is obtained when $V \neq 0$ (fig. 15a). Only a change of slope $d\rho/dT$, at T_I, recalls the underlying critical transormation. If V is small enough, the 1st order nature of the underlying phase transition can be kept (fig. 15b). Figure 16 gives two examples of a continuous $2k_F$ transformation :

a) TEA$(TCNQ)_2$: simultaneous growth of the electrical gap, Δ, and of the dimer shift, δ, defined in fig. 11c. A change of slope is observed at $T_I \sim 210 \pm 10K$ [34]. The $2k_F$ potential is provided by the TEA sublattice (the well defined array of tetramers of TCNQ molecules contributes also at the $2k_F$ electrical gap),

b) DIPS $\phi_4 I_{1.6}$: onset of an electrical gap below the ($\sim$ 180K) 1D liquid - 3D solid phase transition of the iodine sublattice, when the $2k_F$ long range potential is established, and further growth of Δ_{eff} below about 140K[32,35].

A continuous $4k_F$ charge localization due to anion (χ) potential has been invoked in the $(TMTTF)_2X$ series from the observation below about 200K of an activated electrical conductivity [36,37] (see the lecture of Bourbonnais). Apparently, the 335 K 1st order structural phase transition of MEM$(TCNQ)_2$, also interpreted as being due to a $4k_F$ charge localization, occurs without breaking of the symmetry of the TCNQ stack (see the lecture of Kommandeur).
Real structural phase transitions with (translational) symmetry breaking, and related to sharp anomalies in electronic properties, have also been observed in one metallic chain salts. They correspond to :

-a $2k_F$ spin Peierls instability of chains of localized electrons : $(TMTTF)_2Pf_6$(T_{SP} ≃ 20K) and $MEM(TCNQ)_2$ (T_{SP} = 18K) - (see the lectures of Bourbonnais, Kommandeur and Schulz).

- a coupling of the electronic gas with degrees of freedom, external to the organic stack :

i) orientational degrees of freedom : case of non-symmetrical anions X in $(TMTSF)_2X$ and $(TMTTF)_2X$. The phase transition is attributed to anion ordering effects (See part III)

ii) positional degrees of freedom : case of I_3^- units in channels delimited by the DIPS $\emptyset_4$ or TTT host lattice. The phase transition is attributed to the 1D liquid - 3D solid transformation of the iodine sublattice[35].

iii) magnetic degrees of freedom : case of $M(mnt)_2$ stacks (M = Pt, Pd) of localized spins in Per_2 $M(mnt)_2$. The 28K phase transition, for M=Pd, corresponds at a $2k_F$ (spin Peierls ?) distortion of the magnetic chain[38,39].

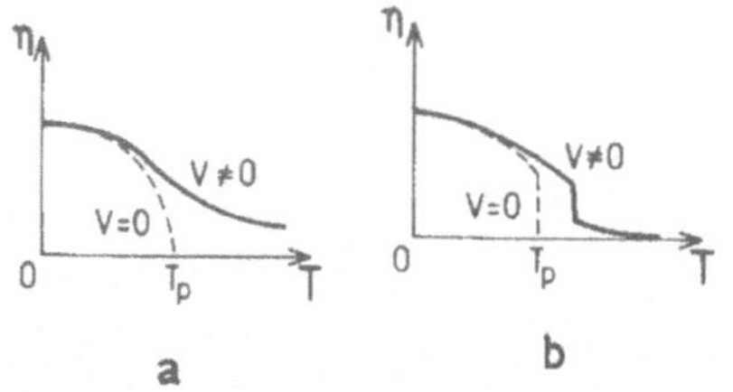

Fig. 15. Temperature dependence of the CDW order parameter η(≡ρ) in presence of an external potential V.(a) and (b) correspond respectively to underlying 2nd and 1st order phase transitions.

E - General view of intrastack instabilities in organic conductors

Table III gives a general classification of the organic conductors with respect to their $2k_F$ and $4k_F$ structural instabilities. Only salts showing an intrastack instability (i.e. not clearly associated at a coupling with external degrees of freedom) have been considered. Inside each block, substantial differences may exist between the salts. This is the case in the central column where the compounds have been classified from (TSF-TCNQ, HMTSF-TCNQ) to $DBTTF-TCNQCl_2$ on the structural manifestations of the increase of Coulomb interactions : increase in the power law divergence of $\chi(2k_F)$, presence of a $4k_F$ instability and of a $4k_F$ transition. Other differences exist among salts showing the $2k_F$ spin Peierls instability with respect to the anisotropy (1D versus 3D) of the pretransitional fluctuations[6].

TABLE III

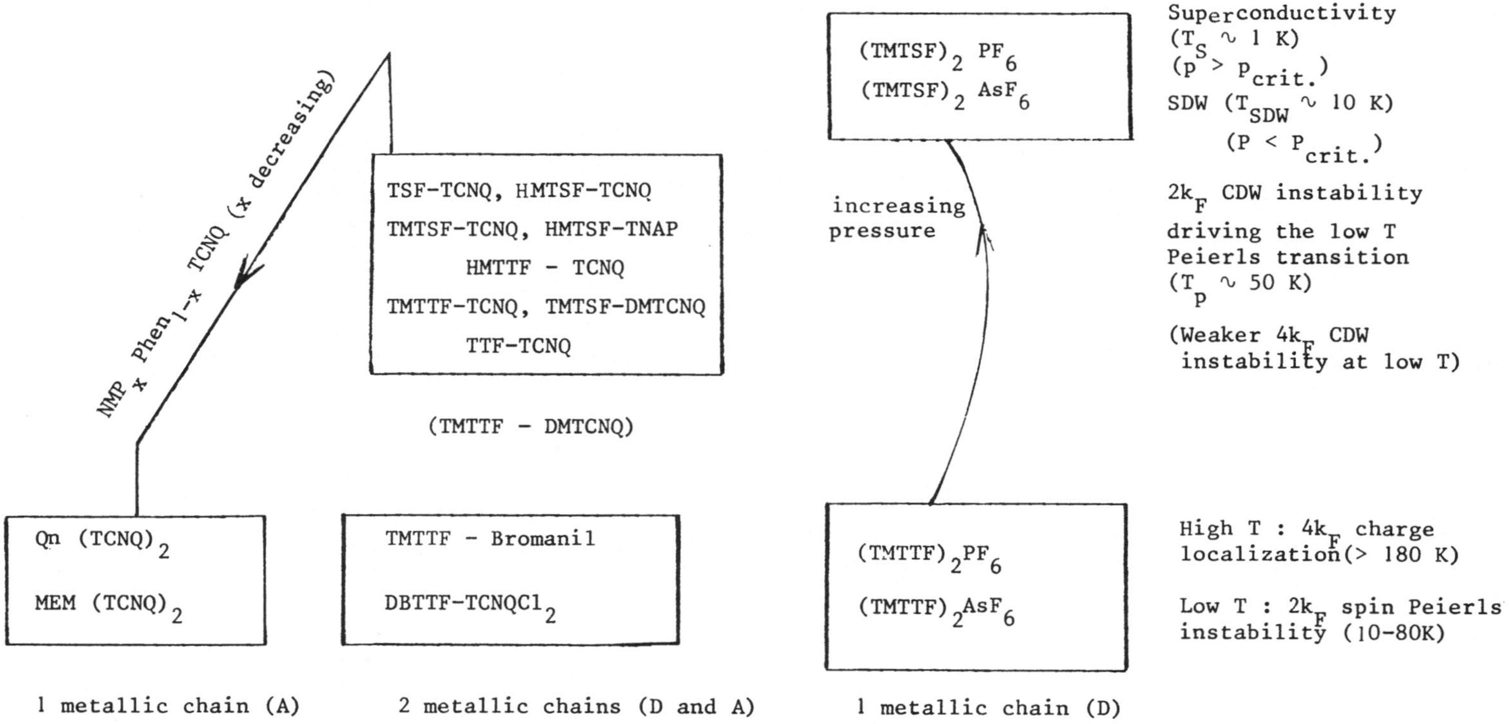

NMP x Phen1-x TCNQ (x decreasing)
TSF-TCNQ, HMTSF-TCNQ
TMTSF-TCNQ, HMTSF-TNAP
HMTTF - TCNQ
TMTTF-TCNQ, TMTSF-DMTCNQ
TTF-TCNQ
(TMTTF - DMTCNQ)
Qn (TCNQ)2
MEM (TCNQ)2
TMTTF - Bromanil
DBTTF-TCNQCl2
(TMTSF)2 PF6
(TMTSF)2 AsF6
increasing pressure
(TMTTF)2PF6
(TMTTF)2AsF6
Superconductivity
(TS ∿ 1 K)
(p > pcrit.)
SDW (TSDW ∿ 10 K)
(P < Pcrit.)
2kF CDW instability
driving the low T
Peierls transition
(Tp ∿ 50 K)
(Weaker 4kF CDW
instability at low T)
High T : 4kF charge
localization(> 180 K)
Low T : 2kF spin Peierls
instability (10-80K)
1 metallic chain (A)
2 metallic chains (D and A)
1 metallic chain (D)

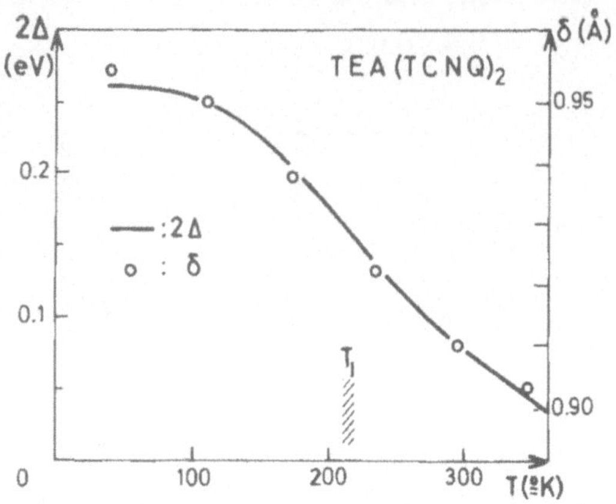

Fig. 16.(b) Temperature dependence of the electrical gap, Δeff, of DIPS $\emptyset_4$ $I_{1.6}$ (from ref. 32, 35)

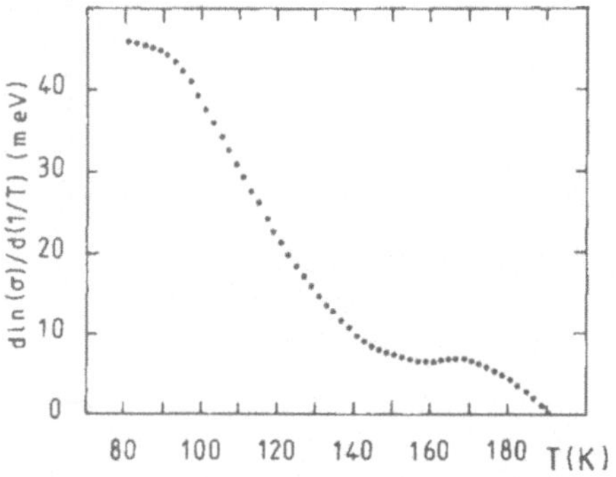

Fig. 16.(a) Temperature dependence of the full gap, 2Δ, and of the dimer shift ,δ, in TEA(TCNQ)$_2$ (adapted from ref. 34)

The different instabilities quoted in Table III correspond more likely to changes in the physical parameters characterizing the 1D electron gas and its coupling to the lattice. These changes can be triggered by external parameters as shown by :

- the pressure induced shift of the electronic ground state of $(TMTTF)_2PF_6$[55].

- the interchange between the relative importance of $2k_F$ and $4k_F$ instabilities, for x increasing in $NMP_xPhen_{1-x}TCNQ$, which can be associated with the emptying of the donor stack of mobile electrons[30].

III - ANION ORDERING PHASE TRANSITIONS OF $(TMTSF)_2X$ AND $(TMTTF)_2X$

A -Phase diagrams

At room temperature the Bechgaard salts crystallize in the triclinic space group $P\bar{1}$, with anions, X, situated at inversion centres of cavities delimited by the zig zag pattern of conducting chains (fig. 11b). The inversion symmetry implies a statistical disorder for non-centrosymmetrical anions (ClO_4^-, ReO_4^-, NO_3^-, SCN^-..), with at least two inverse orientations equally occupied (each orientation favors a short contact with one of the Se(S) atoms of the two inversions related TMTSF(TMTTF) molecules). This state of maximum configurational entropy is not stable at low temperature. Experimentally it is observed upon cooling, anion-ordering (A.O.) transitions leading to superstructures with either identical or alternate orientations of the anion along the 3 crystallographic axes. These superstructures are commonly characterized by the wave vector, $\vec{q}$, of the new reflections which appear below T_{AO} [3, 40].

The A.O. frequently plays a governing role in the nature of the electronic ground states [41, 42], as in the case of the tetrahedral anions ClO_4 and ReO_4 [33]. In $(TMTSF)_2ClO_4$, A.O. with $q_1=(0, \frac{1}{2}, 0)$ is stabilized by slow cooling below $T_{AO} = 24$ K. The metallic state is kept (only a slope anomaly is observed in the temperature dependence of the electrical resistivity at T_{AO} (fig. 17) and superconductivity occurs below 1.2 K. T_{AO} is nearly insensitive to pressure. At the ordering, the ClO_4 anion moves away from the organic molecule towards which it points in order to reduce the long Se..O distance with the other close TMTSF molecule. No sizeable distortion of the organic stack occurs at T_{AO} [43, 44]. In $(TMTSF)_2ReO_4$, the 1st order A.O. transition at $T_{AO} \sim 180$ K, with $\vec{q}_2 = (\frac{1}{2}, \frac{1}{2}, \frac{1}{2})$, leads to an insulating state ($\frac{1}{2} a^x \equiv 2k_F$, see fig. 12c). This transition involves also a distortion of the organic stack, synchronized with a shift of the ReO_4^- anion from the inversion centre, tending to preserve the short Se..O distance [45, 46]. However, the insulating state is suppressed under pressure, (above 12 Kbar), and superconductivity is recovered below 1.7 K. This behaviour is correlated with the stabilization of a new high pressure phase with $\vec{q}_3 = (0, \frac{1}{2}, \frac{1}{2})$, below $T_{AO} \sim 200$ K at 14 Kbar [47], also announced by a slope anomaly in the electrical resistivity (fig. 17) [48].

Competitions between A.O. have been observed in other salts. Let us just mention here the competition between the q_1 and q_2 A.O. in $(TMTSF)_2ClO_4$ alloyed with $(TMTTF)_2ClO_4$ [60] and $(TMTSF)_2ReO_4$ [61] and the competition between the $(\frac{1}{2}, \frac{1}{4}, 0)$ and $(\frac{1}{2}, \frac{1}{2}, \frac{1}{2})$ wave vectors in $(TMTSF)_2F_2PO_2$ [61].

B - Specific features of the A.O. transitions

The A.O. transitions differ from the (Peierls) structural phase transitions previously studied, by the observation, up to about 1.5 $T_{A.O.}$ of nearly isotropic pretransitional fluctuations. This is particularily true for the q_2 A.O. where nothing in the anisotropy and the temperature dependence of the correlation lengths recalls the 1D nature of the CDW instability of an electron gas, although the electrons are clearly involved in the q_2 metal insulator phase transition and that, T_{AO} is usually above the 1D-2D crossover temperature of the electron gas ($T_{x}^{e} = 10 - 100$ K according to the various estimates, the former number taking into account the electron-electron interactions. See the lecture of Bourbonnais) . In addition the q_2 A.O. transitions are of 1st order, which differs from the 2nd order Peierls transition, with typically $T_{A.O.}$ 2 or 3 times higher than T_P. Thus, it has been suggested that the A.O. transitions result from Coulomb interactions between the anions and the organic stacks [41, 42, 50].

Since in the ordered state, the ClO_4^- or ReO_4^- tetrahedra point towards a Se atom of one of the two inversion related TMTSF molecules (fig. 13), the order can be described by an Ising variable. With 2 possible orientations, I and II, for a given tetrahedron, the order parameter of an A.O. transition can be taken as $P_I - P_{II}$, P_i being the probability of the

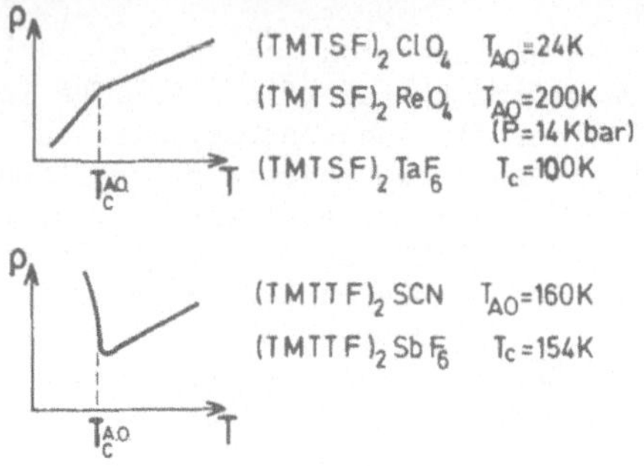

Fig. 17. Schematic drawing showing the two opposite behaviour of the electrical resistivity at the $\vec{q}_1$ A.O. of $(TMTSF)_2ClO_4$, the q_2A.O. of $(TMTSF)_2ReO_4$ under pressure and the $\vec{q}_2$ A.O. of $(TMTTF)_2SCN$. Similar anomalies are observed at T_c, in (TMTSF) TaF_6 and $(TMTTF)_2SbF_6$.

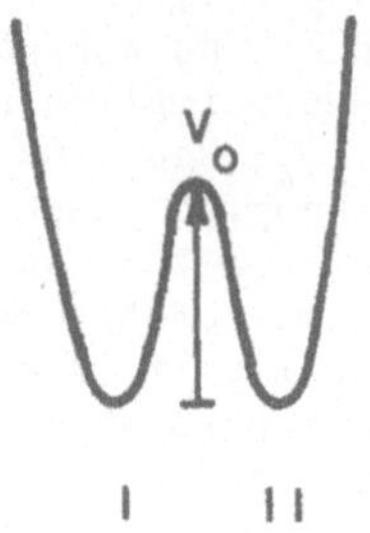

Fig. 18. Local potential experienced by a tetrahedron from its neighbourhood. I and II are the two inversion symmetry related orientations.

orientation i. Figure 18 gives a schematic representation of the local potential that a given tetrahedron has to overcome to change its orientation [51]. The dynamics associated to such an orientational order is relaxational. NMR T_2 measurements give evidence of very slow pretransitional dynamics for the A.O. transition of $(TMTSF)_2ClO_4$, with a potential height V_o estimated at 240K [52]. With V_o much larger that $k_B T_{A.O.}$ (=24K), kinetics has a substantial influence on the A.O. process. This was demonstrated by structural studies [43, 53, 54] showing that the fraction of ClO_4^- ordered and the size of ordered domains depend crucially on the cooling rate (fig. 19). Furthermore, fast cooling stabilizes a spin density wave state below 5K, instead of the superconducting state obtained for relaxed samples (see the lecture of Tomic). Thermal annealing effects occuring below $T_{A.O.}$, in initially quenched $(TMTSF)_2ClO_4$ samples,allow to study the specific features of the kinetics of the Ising Model. By its order parameter and the associated dynamics, the A.O. transition differs also of the displacive Peierls transition considered in part I.

C - Coupling between the anions

Figure 13 shows that the structure of $(TMTSF)_2X$ consists of $(\vec{a}, \vec{b})$ planes of anion separated by planes of organic molecules along $\vec{c}$. The Coulomb interaction between anions in the $(\vec{a}, \vec{b})$ plane is direct and probably also mediated by the electrons of the organic molecules. The interaction between planes of anions is mostly mediated by organic molecules. With respect to the directions of short Se.X contacts, mediated interactions in the $(\vec{a}, \vec{c} - \vec{b})$ plane are especially interesting[50]. However, due to the isotropy of pretransitional fluctuations, all these directions have to be equally considered in any model of the A.O. phase transition.

1) Electron anion coupling

The electrostatic coupling between an anion and an organic molecule depends on the charge distribution on the anion and of the Se..X (S..X) distance. The less electronegative the central atom of the anion, the more

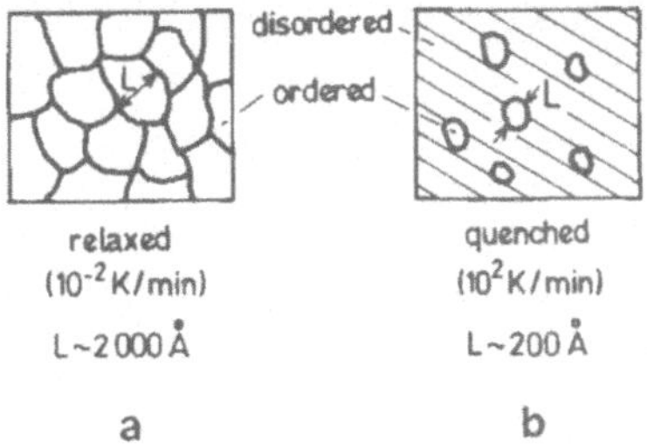

Fig. 19. Schematical representation of ordered regions in relaxed (a) and quenched (b) $(TMTSF)_2ClO_4$. The cooling rate at $T_{A.O.}$ is indicated, as well as the average size, L, of domains of ordered ClO_4^-.

charged are the outer atoms of the anion and the stronger is the electron anion coupling. The outer charge increases along the sequences :

- PF_6^-, AsF_6^-, SbF_6^-, TaF_6^- for fluoro octahedral anions, and

- ClO_4^-, FSO_3^-, $F_2PO_2^-$, ReO_4^- for oxo and oxofluoro tetrahedral anions.

In the case of ClO_4^-, there is already 0.44 electron per O atom, according to a recent quantum chemistry calculation. A correlation between the nature of the octahedral anion and the electrical resistivity has been noticed in the $(TMTTF)_2MF_6$ series [56]. It has also been observed [57] that, the critical temperature of the $\vec{q}_2$ A.O. increases when the electronegativity of the central atom of oxo and oxofluoro tetrahedral anions decreases in the $(TMTSF)_2X$ series.

These correlations indicate that the electron anion interaction influences the physical properties of the Bechgaard salts. The anion is expected to polarize the electronic charge of the molecule towards which it establishes a short contact [58]. For strong enough coupling, it has been argued that anions may trap the itinerant charges of the 1D electron gas [50]. The observation of large splitting of infrared active modes below $T_{A.O.}$ provides the best evidence for such a polarization effect [59]. As the interaction between anions is mediated, along certain directions, by the electrons of the organic molecule, the A.O. process must also depend on the metallic and molecular polarizabilities of the organic stack. The drastic change in the $\vec{q}_1$ A.O. observed in $(TMTSF)_2ClO_4$ alloyed with $(TMTTF)_2ClO_4$ emphasizes the important role of the organic medium in the A.O. process [60]. The metallic polarizability must be also considered, because the $2k_F$ and/or $4k_F$ in chain wave vector components are generally critical for the 1D CDW electronic response function. The distortion of the organic stack observed for the $\vec{q}_2$ (and probably $\vec{q}_3$) A.O. suggests, through the electron phonon coupling, a $2k_F$ ($4k_F$) response of the electron gas at the A.O. (but not necessarily a CDW electronic instability, as shown by the absence of a regime of 1D structural fluctuations above $T_{A.O.}$.

2) A.O. Superstructures

The A.O. superstructures can be classified [41] with respect to the periodicity of the short contacts experienced by a given organic stack from the two neighbouring anionic chains, situated respectively at the

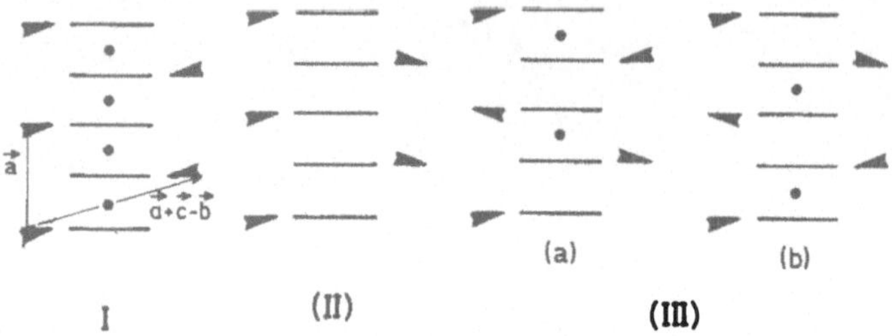

Fig. 20. Schematic drawing of the organic stack in the $(\vec{a}, \vec{c}-\vec{b})$ plane for various A.O. configurations. The arrow gives the direction of the short contact distance of the anion. Note that the anion contacts are not in the molecular planes.

origine and in $\vec{a} + \vec{c} - \vec{b}$ (See fig. 20). Assuming either uniform or alternate ordering along the 3 crystallographic axes, there are 4 different patterns of short and long contacts between the anions and the molecular stack in the $(\vec{a}, \vec{c} - \vec{b})$ plane. They belong to 3 classes[41, 64] :

α) class I : $\vec{q} = (0, \frac{1}{2}, 0)$ or $(0, 0, \frac{1}{2})$–uniform A.O.

Each stack is either in short or in long contact with the anions. All the inversion centres of the stack are kept at $T_{A.O.}$. As above $T_{A.O.}$ (see section C in part II), the molecular sites experience the same anionic potential, and there is a $4k_F$ bond potential. However there are tow inequivalent stacks, with respect to the strenth of the anionic potential. per unit cell. This is the case of the q_1–A.O. of $(TMTSF)_2ClO_4$, preserving the metallic state. The shift of the ClO_4^- anions below $T_{A.O.}$ tends to reduce the difference of anionic potential between the stacks of short and long contacts.

β) class II : $\vec{q} = (0, \frac{1}{2}, \frac{1}{2})$ or $(0, 0, 0)$–$4k_F$ A.O.

Each stack has its molecules successively in short and long contacts with the anions. All the inversion centres of the stack are suppressed at $T_{A.O.}$. The molecular sites now experience a $4k_F$ anionic potential. In principle this ordering increases the $4k_F$ gap Δ_ρ (fig. 12) and the Umklapp electron electron interaction g_3. The q_3 A.O. is observed in $(TMTSF)_2ReO_4$ under pressure[47] and at ambiant pressure in $(TMTTF)_2SCN$[62], with opposite temperature dependence of the electrical conductivity below $T_{A.O.}$ (fig. 17). In this respect the resemblance between electrical properties of $(TMTSF)_2ClO_4$ and $(TMTSF)_2ReO_4$ suggests that the g_3 interaction and $4k_F$ localization effects remain weak in the $(TMTSF)_2X$ series. However, the coincidence of $T_{A.O.}$ with the metal insulator phase transition of $(TMTTF)_2SCN$, confirms the importance of the g_3 interaction and $4k_F$ localization effects in the $(TMTTF)_2X$ series. Figure 17 shows also well defined electrical anomalies in $(TMTSF)_2TaF_6$ and $(TMTTF)_2SbF_6$ which recall that shown respectively by $(TMTSF)_2ReO_4$ (under pressure) and $(TMTTF)_2SCN$. They may be the signature of a structural phase transition at T_c. As no superstructure spots could be detected below T_c in these salts, the phase transition may result from a uniform ($\vec{q} = (0, \delta, 0)$) displacement of the anionic sublattice with respect to the organic stacks, caused by the strong electron anion coupling due to the TaF_6^- and SbF_6^- anions[56]. Effects of the class II A.O. on the electronic ground state, are also considered in ref. (63).

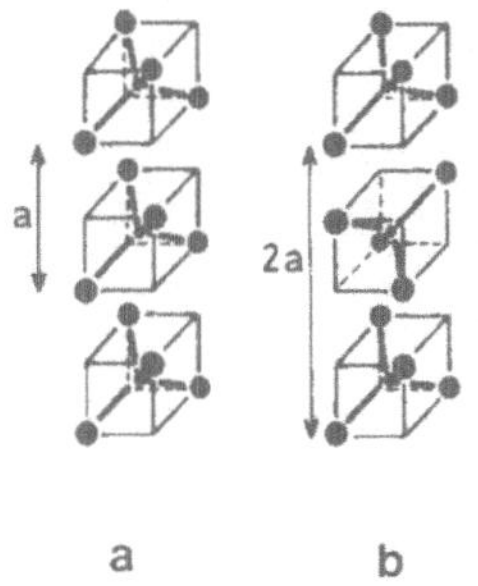

Fig. 21. Stacking of regular tetrahedra in uniform (a) and alternate (b) orientations, giving respectively a minimum and a maximum of Coulomb repulsion between charges.

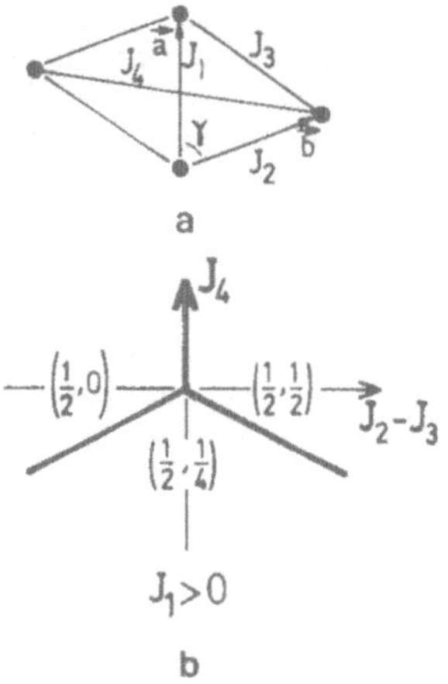

Fig. 22. (a) $(\vec a, \vec b)$ plane of anion
(b) (q_a, q_b) phase diagram as a function of the phenomenological interactions defined in (a) for $J_1 > 0$.

γ) Class III : a) $\vec q = (\frac{1}{2}, \frac{1}{2}, 0)$ or $(\frac{1}{2}, 0, \frac{1}{2})$
b) $\vec q = (\frac{1}{2}, \frac{1}{2}, \frac{1}{2})$ or $(\frac{1}{2}, 0, 0)$ } $-2k_F$ A.O.

Each stack can be viewed as composed of dimers of molecules successively in short and long contact with the anions. Only one inversion centre of the stack out of two is kept below $T_{A.O.}$. With respect to anionic positions, which do not coincide with the molecular planes (see fig. 13), configurations (a) and (b) are energetically different [70]. Experimentally the configuration (b) appears to be the most stable in salts with tetrahedral anions : the $\vec q_2$ A.O. is found at ambiant pressure in $(TMTSF)_2X$ with X = ReO_4, FSO_3, BF_4 and $(TMTTF)_2X$ with X = ReO_4, ClO_4, BF_4. The $\vec q_2$ A.O. is accompanied by a substantial distortion of the organic stack. In the $(TMTSF)_2X$ series, the $\vec q_2$ A.O. coincides with a metal insulator transition. In the $(TMTTF)_2X$ series the $\vec q_2$ A.O. occurs when the charge degrees of freedom are already localized (i.e. below the maximum of conductivity). An interesting compound is $(TMTSF)_2F_2PO_2$,[61] which stabilizes the $(\frac{1}{2}, \frac{1}{4}, 0)$ superstructure between 136.3K and 135.3K. In the A.O. hypothesis, the local environment of TMTSF stacks could be viewed as a mixture of the configurations (a) ($\vec q=(\frac{1}{2}, \frac{1}{2}, 0)$) and (b) ($\vec q_2=(\frac{1}{2},0, 0)$). Below 135.3K, the q_2 A.O. (configuration b) is stabilized. The $(\frac{1}{2}, 0, 0)$ A.O. is stabilized in $(TMTSF)_2NO_3$[49] and $(TMTTF)_2$ NO_3[40], with probably a rather long contact distance established between the NO_3 and the organic stack. The $(\frac{1}{2}, 0, 0)$ A.O. of $(TMTSF)_2NO_3$ is not accompanied by a metal insulator phase transition, because of the 2D nature of the Fermi surface around $T_{A.O.}$=40K, and the weak anionic $2k_F$ potential created on the organic stack.

3) Models of the A.O.

We shall consider here only A.O. of tetrahedra leading to a very rich variety of phase diagrams [47, 61]. By drawing regular tetrahedra on cubes, it is easy to see (fig. 21) that for a stacking of cubes on top of each other, the uniform orientation of tetrahedra (period a) corresponds, with respect to a disordered state, at a gain of electrostatic energy, while the alternate orientation (period 2a) produces a loss of electrostatic energy [49]. In real materials containing distorted tetrahedra, Madelung energy calculations in the a direction [65-66] confirm this observation. Thus the 1/2 a^x component of the $\vec{q}_2$ A.O. is certainly stabilized by the gain of energy of another degree of freedom, probably the electronic energy of the organic stack. In the $(TMTSF)_2X$ series, the large gas in the electronic energy spectrum ($2\Delta \sim$ 2000 K) occuring below $T_{A.0}$ has been interpreted either as a $2k_F$ Peierls like band gap [65] or as the trapping energy of holes on the dimers experiencing the strong anion potential [50]. A gap resulting from a CDW intrastack instability is unlikely because it should be announced by ID structural fluctuations above T_{AO}, which are not observed. In the $(TMTTF)_2X$ series, the nature of the gain of electronic energy below $T^S_{A.O.}$ ($\simeq T^{Se2}_{A.O.}$) is less clear. The electrons being partly localized, with the $4k_F$ wave vector, at temperatures higher than $T_{A.O.}$, it is reasonable to consider a spin Peierls like pairing energy gain.

The calculation of the A.O. in the b direction based on electrostatic coupling between tetrahedra must also include, because of the large deviation of γ from 90° (see fig. 22a), the interaction along $\vec{a}-\vec{b}$. With the uncertainties in the atomic positions and the shift of molecules below $T_{A.O.}$, the conclusions of such calculations are uncertain [66]. But one cannot exclude that the $\vec{q}_1$ A.O. of $(TMTSF)_2ClO_4$ in the $(\vec{a}, \vec{b})$ plane might be due to direct electrostatic interactions between tetrahedra. In any case, the $\vec{q}_1$ A.O. of $(TMTSF)_2ClO_4$ is stabilized by a very weak energy gain, Δ E. By writing the gain of free energy, with respect to the disordered state, under the form :

$$\Delta F = \Delta E - k_B T \ell n\ 2 \tag{17}$$

ΔE can be estimated at $\sim$ 1.5 meV from $T_{A.O.}$ = 24 K ($\Delta S = k_B \ell n\ 2$ is the change of entropy observed at the A.O. [67]). Models of the A.O. transition can be constructed using phenomenological interactions between Ising pseudo spin ($S_\ell = \pm 1$) :

$$H = \sum_{\vec{\ell},\, i} J_i\, S_{\vec{\ell}}\, S_{\vec{\ell} + \vec{r}i} \tag{18}$$

For example the $\vec{q}_1$ A.O. of $(TMTSF)_2ClO_4$ corresponds, in the $(\vec{a}, \vec{b})$ plane, at $J_1 < 0$ and $J_2 + J_3 > 0$ (see fig. 22a for the definition of $J_{\vec{\ell}}$ and $\vec{r}$) Depending of the values of J_i, competing interactions between various orders can be described. For example if a dominant "antiferromagnetic" interaction along $\vec{a}$ ($J_1 > 0$) is assumed in the $(\vec{a}, \vec{b})$ plane, the q_b order which is stabilized depends upon the sign and the value of the ratio J_4/J_2-J_3). In this respect, it is interesting to remark, in fig. 22b, that if there is a substantial compensation between J_2 and J_3 and if $J_4<0$, the (1/2, 1/4) order can be more stable than the (1/2, 1/2) order. The (1/2, 1/4, 0) superstructure is observed in a limited temperature range for $(TMTSF)_2F_2PO_2$. However, the calculation of its stability requires also a knowledge of the coupling between $(\vec{a}, \vec{b})$ planes via the organic stacks [70]. In this respect another interesting plane is that $(\vec{a}, \vec{c}-\vec{b})$ containing the two directions, $\vec{c}-\vec{b}$ and $\vec{a}+\vec{c}-\vec{b}$, along which the anions are connected by short contact with the molecular stack. This plane has been used to describe the competition between the $\vec{q}_1$ and $\vec{q}_2$ A.O. of $(TMTSF)_2ClO_4$ for a dominant "antiferromagnetic" interaction along the $\vec{a} + \vec{c} - \vec{b}$ direction [50]. It can be used also to discuss in a similar way the competition between the $\vec{q}_2$ and $\vec{q}_3$ A.O. of $(TMTSF)_2ReO_4$, for a dominant "ferromagnetic" interaction along $\vec{c} - \vec{b}$. However, because of the isotropy of the pretransitional fluctuations the phenomenological interactions in the $\vec{a}$, $\vec{b}$, $\vec{a} - \vec{b}$, $\vec{c} - \vec{b}$ and $\vec{a} + \vec{c} - \vec{b}$ directions must be considered on equal terms.

ACKNOWLEDGEMENTS

Most of the X Ray and Neutron experimental studies quoted in these lectures have been performed in collaboration with P.A. ALBOUY, R. COMES, C. ESCRIBE-FILIPPINI, B. HENNION, S. MEGTERT, R. MORET, A.H. MOUDDEN, S. RAVY, and G. SHIRANE. These studies have benefited from many discussions with S. BARISIC, K. BECHGAARD, C. BOURBONNAIS, V.J. EMERY, A.J. EPSTEIN, J. FRIEDEL, D. JEROME, S. KAGOSHIMA, C. NOGUERA, C. SCHLENKER, H.J. SCHULZ and K. YAMAJI.

REFERENCES

1. R. Comès and G. Shirane in "Highly Conducting One Dimensional Solids" edited by J.T. Devreese, R.P. Evrard and V.E. van Doren (Plenum 1979), p.17
2. S. Kagoshima in "Extended Linear Chain Compounds", volume 2, edited by J.S. Miller (Plenum 1982) p.303
3. R. Moret and J.P. Pouget in "Crystal Chemistry and Properties of Materials with quasi One dimensional Structures" edited by J. Rouxel (Reidel 1986) p. 87
4. J.P. Pouget in "Solid State Phase transformations in Metals and Alloys" (Editions de Physique - Orsay - 1978) p.523
5. J.P. Pouget, C. Noguera, A.H. Moudden and R. Moret, J. Physique **46**, 1731 (1985); erratum **47**, 145 (1986)
6. J.P. Pouget to be published in "Quasi One Dimensional Conductors" edited by E.M. Conwell (Academic Press)
7. S.K. Khanna, J.P. Pouget, R. Comès, A.F. Garito and A.J. Heeger, Phys. Rev. B 16, 1468 (1977)
8. S. Barisic and A. Bjelis in "Theoretical Aspects of Band Structures and Electronic Properties of Pseudo One Dimensional Solids" Edited by H. Kamimura (Reidel 1985) p.49
9. S. Barisic in "Electronic Properties of Inorganic Quasi One Dimensional Materials" Vol. I, Edited by P. Monceau (Reidel 1985) p.1
10. K. Yamaji, S. Megtert and R. Comès, J. Physique 42, 1327 (1981)
11. F. Herman, Physica Scripta 16, 303 (1977)
12. S. Megtert, R. Comès, C. Vettier, R. Pynn and A.F. Garito, Mol. Cryst. Liq. Cryst. 85, 159 (1982)
13. S. Kagoshima, T. Ishiguro, T.D. Schultz and Y. Tomkiewicz, Solid. State Commun. 28, 485 (1978)
14. C.S. Jacobsen, K. Mortensen, J.R. Andersen and K.Bechgaard, Phys. Rev. B18, 905 (1978)
15. M. Sato, M. Fujishita and S. Hoshino, J. Phys. C : Solid State Phys. 16 L 877 (1983)
16. J.P. Pouget, C. Escribe - Filippini, B. Hennion, R. Currat, A.H. Moudden, R. Moret, J. Marcus and C. Schlenker, Mol. Cryst. Liq. Cryst. 121, 111 (1985)
17. G. Shirane, S.M. Shapiro, R. Comès, A.F. Garito and A.J. Heeger, Phys. Rev. B14, 2325 (1976)
18. J.P. Pouget, C. Escribe-Filippini, B. Hennion, R. Currat and J. Marcus (unpublished results)
19. M.J. Rice and S. Strassler, Solid State Commun. 13, 125 (1973)
20. P.Y. Le Daeron and S. Aubry, Journal de Physique 44, C3-1573 (1983)
21. W. Dieterich, Adv. Phys. 25, 615 (1976)
22. L.J. Sham in ref. (1) p 227
23. C. Escribe-Filippini, J.P. Pouget, B. Hennion and M. Sato, Proceeding ICSM'86 to be published in Synthetic Metals

24. K. Carneiro, G. Shirane, S.A. Werner and S. Keiser, Phys. Rev. B13, 4258 (1976)
25. P.A. Lee, T.M. Rice and P.W. Anderson, Solid State Commun. 14, 703 (1974)
26. Y. Nakane and S. Takada, J. Phys. Soc. Japan 54, 977 (1985)
27. L.K. Hansen and K. Carneiro, Proceeding Yamada Conference XV, to be published in Physica B
28. K.Yamaji, J.P. Pouget, R. Comès and K. Bechgaard, Journal de Physique Colloque 44, C3 - 1321 (1983)
29. L. Forro, S. Bouffard and J.P. Pouget , J. Physique Lettres, 45, L543 (1984)
30. A.J. Epstein, J.S. Miller, J.P. Pouget and R. Comès, Phys. Rev. Lett. 47, 741 (1981)
31. S. Barisic and S. Brazovskii in "Recent Developments in Condensed Matter Physics" Vol. 1, edited by J.T. Devreese (Plenum Press 1981) p 327
32. P.A. Albouy, J.P. Pouget and H. Strezelecka, Phys. Rev. B (under press)
33. D. Jerome and H.J. Schulz, Adv. in Physics, 31, 299 (1982)
34. J.P. Farges, J. Physique 46, 465 (1985)
35. P.A. Albouy, P. Le Guennec, J.P. Pouget and C. Noguera, Proceeding ICSM'86 to be published in Synthetic Metals
36. C. Coulon, D. Delhaes, S. Flandrois, R. Lagnier, E. Bonjour and J.M. J.M. Fabre, J. Physique 43, 1059 (1982)
37. V.J. Emery, R. Bruinsma and S. Barisic Phys. Rev. Lett. 48, 1035 (1982)
38. R.T. Henriques, L. Alcacer, J.P. Pouget and D. Jérome, J. Phys. C : Solid State Phys. 17, 5197 (1984)
39. R.H. Henriques, L. Alcacer, D. Jérome, C. Bourbonnais and C. Weyl, J. Phys. C : Solid State Phys. 19, 4663 (1986)
40. R. Moret, J.P. Pouget, R. Comès and K. Bechgaard, J. Physique Colloque 44, C3-957 (1983)
41. V.J. Emery, J. Physique, Colloque, 44, C3-977 (1983)
42. L.P. Gor'kov, Sov. Phys. Usp 27, 809 (1984)
43. R. Moret, J.P. Pouget, R. Comès and K. Bechgaard, J. Physique, 16, 1521 (1985)
44. B. Gallois, A. Meresse, J. Gaultier and R. Moret, Mol. Cryst. Liq. Cryst. 131, 147 (1985)
45. R. Moret, J.P. Pouget, R. Comès and K. Bechgaard, Phys. Rev. Lett. 49 1008 (1982)
46. G. Rindorf, H. Soling and N. Thorup, Acta Cryst. C40, 1137 (1984)
47. R. Moret, S. Ravy, J.P. Pouget, R. Comès and K. Bechgaard, Phys. Rev. Lett. 57, 1915 (1986)
48. The coexistence of the $\vec{q}_2$ and $\vec{q}_3$ ordering is observed in $(TMTSF)_2ReO_4$ between 9.5 Kbar and 12.5 Kbar. In this intermediate pressure range, electronic properties show a complex behavior (see lecture of Tomic).
49. J.P. Pouget, R. Moret, R. Comès and K. Bechgaard, J. Physique Lett. 42, 543 (1981)
50. R. Bruinsma and V.J. Emery, J. Physique Colloque, 44, C3-1115 (1983)
51. By drawing a tetrahedron on a cube, it is easy to see that the inverse orientation of a tetrahedron can be simply achieved by a simple rotation around one of the axes of symmetry of the cube which is different from an axis of symmetry of the tetrahedron (R. Moret private communication)
52. M. Takigawa and G. Saito, J. Phys. Soc. Japan 55, 1233 (1986) and M. Takigawa, private communication
53. S. Kagoshima, T. Yasunaga, T. Ishiguro, M. Anzai and G. Saito, Solid State Commun. 46, 867 (1983)
54. S. Ravy, R. Moret, J.P. Pouget and R. Comès, Synth. Metals 13, 63 (1986)
55. F. Creuzet, D. Jérome and A. Moradpour, Mol. Cryst. Liq. Cryst. 119, 297 (1985)

56. R. Laversanne, C. Coulon, B. Gallois, J.P. Pouget and R. Moret J. Physique Lett 45, L 393 (1984)
57. F. Wudl, Acc. Chem. Res. 17, 227 (1984)
58. F. Wudl, J.A.C.S. 103, 7064 (1981)
59. R. Bozio, C. Pecile, J.C. Scott and E.M. Engler, Mol. Cryst. Liq. Cryst. 119, 211 (1985)
60. J.P. Pouget, R. Moret, R. Comès, G. Shirane, K. Bechgaard and J.M. Fabre, J. Physique Colloque 44, C3-969 (1983)
61. S. Ravy, R. Moret, J.P. Pouget and R. Comès, Proceeding Yamada Conference XV, to be published in Physica B
62. C. Coulon, A. Maaroufi, J. Amiell, E. Dupart, S. Flandrois, P. Delhaes, R. Moret, J.P. Pouget and J.P. Morand, Phys. Rev. B 26, 6322 (1982)
63. S. Brazovskii and V. Yakovenko, J. Phys. Lett. 46, L111 (1985)
64. S. Ravy, R. Moret, J.P. Pouget and R. Comès, Proceeding ICSM'86, to be published in Synthetic Metals
65. C.S. Jacobsen, H.J. Pedersen, K. Mortensen, G. Rindorf, N. Thorup, J.B. Torrance and K. Bechgaard, J. Phys. C : Solid State Phys. 15 2651 (1982)
66. S. Ravy private communication
67. F. Pesty, P. Garoche and A. Moradpour, Mol. Cryst. Liq. Cryst. 119 251 (1985)
68. C.S. Jacobsen, Mat. Fys. Medd. Dan. Vid. Selsk. 41, 251 (1985)
69. R. Comès, S.M. Shapiro, G. Shirane, A.F. Garito and A.J. Heeger, Phys. Rev. B 14, 2376 (1976)
70. The directions c-b and a-b+c are not equivalent: the anion-anion distance is slightly shorter along $\vec{c}-\vec{b}$ than along $\vec{a}-\vec{b}+\vec{c}$. In the pseudo spin model of eq (18), applied to the $(\vec{a},\vec{c}-\vec{b})$ "short contacts" plane for dominant "antiferromagnetic" interactions along a ($J_a>0$), the configuration (a) corresponds at $J_{c-b}>J_{a-b+c}$ and the configuration (b) at $J_{c-b}<J_{a-b+c}$. Configurations (a) and (b) are energetically equivalent for $J_{c-b}=J_{a-b-c}$. In that case, mixing of (a) and (b) configurations, like that occuring in the (1/2, 1/4, 0) phase of $(TMTSF)_2F_2PO_2$ (assuming an A.O.), can occur. The stability of this last phase depends also of the couplingsin the $(\vec{a},\vec{b})$ plane.
71. The Peierls transition in an external potential has been previously considered by :
M.J. Rice Phys. Rev. Lett. 37, 36 (1976)
L.K. Hansen and K. Carneiro, Solid State Comm. 49, 531 (1984)
The two order parameters considered in these papers are the total gap in the electronic spectrum, Δq, and the amplitude of the PLD, Aq (proportional to the CDW amplitude ρ_q, via the linear electron phonon coupling). Δq and Aq are no longer proportional :
$\Delta q = \lambda$ Aq + Vq.

BASIC IDEAS IN THE THEORY OF ORGANIC CONDUCTORS*

V. J. Emery

Department of Physics, Brookhaven National Laboratory
Upton, N. Y. 11973
and
IBM Zürich Research Laboratory, CH8803 Rüschlikon
Switzerland

INTRODUCTION

Organic conductors are charge-transfer compounds typically consisting of organic donor molecules such as TTF, TSeF, TMTTF, TMTSF, and acceptors which may be organic (TCNQ) or inorganic (AsF_6, PF_6, ClO_4, ReO_4). The structures of these molecules and the solids formed out of them are described in the articles of K. Bechgaard[1] and J. P. Pouget[2]. For present purposes, it is sufficient to know that the flat donors are arranged in stacks in which the spacing between molecules is much less than the distances between stacks. This structure leads to an anisotropy of physical properties such as the electrical conductivity, and to the possibility of regarding the motion of the charge carriers on the stacks as one-dimensional at sufficiently high temperatures.

A major goal of the field has been the discovery of organic superconductivity. Materials such as TTF-TCNQ showed a fairly high conductivity which increased as the temperature was lowered, but ultimately this was cut off by a structural phase transition to an insulating state[2]. This Peierls transition, caused by the interaction between the charge carriers and the molecular lattice, has itself become an important subject of study. Superconductivity was first achieved[3] in 1980 in $(TMTSF)_2PF_6$, and subsequently in the series of compounds obtained by replacing PF_6 with other anions such as AsF_6, ReO_4, ClO_4 etc., which differed from each other in their symmetry and in the electronegativity of the central atom. In the past two years, the introduction of $(ET)_2I_3$ and related compounds[1] has brought up to an eightfold increase in the transition temperature (to about 8K) and has added a new dimension to the study of organic superconductivity.

From a theoretical point of view, the central goals are to understand how the superconductors managed to avoid a Peierls transition, and to unravel the fundamental mechanism of superconductivity, the source of the pairing interaction between charge carriers. The aim of these lectures is to address these questions in a general way, after giving an elementary survey of the basic physics of organic conductors.

In the organic superconductors, each anion takes one electron from the organic stack, and the remaining holes, one to every two organic molecules, become the carriers of current. In making a model, the usual assumption is that each organic molecule has only one relevant spatial state which can accommodate at most two holes. Thus, the band is quarter-filled. This approximation is good when the energy of excitation to the second level is much larger than the kinetic and potential energies of the holes, and it will be adopted for much of the ensuing discussion. However, this condition is not extremely well satisfied in the superconductors, and consequently some of the effects of including a second level will be considered in the following chapter.

The commonly used model of organic conductors is known as the extended Hubbard model. It assumes one level per site and uses the tight-binding approximation in which the overlap between states on neighboring molecules is incorporated into "transfer" or "hopping" integrals t_a, t_b, t_c along the three crystallographic directions. The interaction between holes on the same site is denoted by U and between holes on first and second neighboring sites by V_1 and V_2. For the TMTSF compounds, the highly conducting (stacking) direction is along $\underline{a}$ and $t_a:t_b:t_c = 300:30:1$, with the longitudinal bandwidth $4t_a$ about 1.4 eV. The interactions are repulsive (reflecting the Coulomb force) and of about the same order of magnitude as the bandwidth. For the present, it will be assumed that the effects of lattice polarization (phonon exchange) have been incorporated into U, V_1 and V_2, although it is necessary ultimately to include the lattice degrees of freedom explicitly. It is also assumed that the polarization of a molecule has been incorporated as a negative contribution to U.

One final feature to be taken into account is the potential due to the anions. This has a period 2a and it may produce either a transfer of holes from one site to another[4] or, when there is anion ordering[5], a simple periodic potential. For ease of presentation, we shall assume the latter, although there is no difficulty in incorporating hole transfer into the theory.

When $t_b = t_c = 0$, the motion of the holes is purely one-dimensional, and the Fermi surface consists of two planes $k = \pm k_F$. In a single chain of length L with periodic boundary conditions, the allowed wave vectors are integral multiples of $2\pi/L$, and since there are two particles (of opposite spin) per state, N particles occupy a Fermi sea of length

$$2k_F = \frac{N}{2}\frac{2\pi}{L} = \pi/2a \tag{1}$$

where $L/N = 2a$ is the mean spacing between holes. When t_b, t_c are not zero but are much smaller than t_a, the transverse motion distorts the Fermi surface, but this effect is insignificant unless it exceeds the thermal broadening k_BT (where k_B is Boltzmann's constant). Thus, at sufficiently high temperatures, the system is effectively one-dimensional. In the simplest approximation[6], the crossover temperature T_x below which two- or three-dimensional effects become important is given by

$$T_x \sim t_b/\pi \tag{2}$$

from which $T_x \sim 100K$. A more elaborate estimate[7] implies that T_x is somewhat lower, although in practice the crossover cannot easily be evaluated from first principles and the best approach is to determine it from experiment. Below T_x, it is possible to introduce an effective Hamiltonian and to make use of the conventional methods of many-body theory to calculate the properties of the system[8]. The subsequent discussion will focus on the broad one-dimensional region where the physical effects are perhaps less familiar.

PHYSICS OF ONE-DIMENSIONAL CONDUCTORS

A great deal of effort has been expended in working out the properties of the extended Hubbard model for a one-dimensional system, using either the methods of quantum field theory[9,10] or numerical calculations[11]. But it is quite simple to get a feeling for what is going on by considering the strong coupling limit $U, V_1, V_2 \gg t_a$. Much of the argument does not depend on space dimension in an essential way but the peculiar feature of one-dimensional systems is that the conclusions are in qualitative and sometimes quantitative agreement with the more elaborate weak-coupling calculations, and therefore provide a reliable guide to what is going on.

For $t_a = 0$, the configuration is specified by the number $\rho_{n\pm}$ of holes with spin $\pm 1/2$ at site n. If the total charge density at n is denoted by $\rho_n \equiv \rho_{n+} + \rho_{n-}$, the Hamiltonian is given by

$$H_p = \sum_n \{U\rho_{n+}\rho_{n-} + V_1\rho_n\rho_{n+1} + V_2\rho_n\rho_{n+2} + (-1)^n\varepsilon\,\rho_n\} \qquad (3)$$

where ε is the strength of the anion potential. The objective now is to map out the ground states for various sets of values of the interactions and to estimate how they are modified when the hopping t_a is restored. Obviously this is too large a parameter space to describe completely so we shall merely illustrate what goes on for a few selected situations. Readers may then work out other examples for themselves.

In all cases, it will be assumed that $|U|$ is the largest energy in the problem.

$U < 0,\ \varepsilon = 0$

In this case, the electrons occupy molecular sites in pairs to take advantage of the on-site attraction. This is illustrated in Fig. 1a, where the crosses represent molecules and the arrows refer to up or down spin holes. The ground state is very degenerate because the energy does not depend upon which sites are occupied.

Charge-density wave states occur in an extreme form when there is an intersite repulsion V_n but still no hopping. To minimize the energy, the pairs are equally spaced, as shown in Fig. 1b for a half-filled band. Then the charge density varies with period 2L/N, so according to Eq. (1) the principal wave vector is $2k_F$. This is a possible ground state because the system is classical when $t_a = 0$. More realistically, when $t_a \neq 0$, the picture is not so static and there is a much smaller modulation of the charge density. But the wave vector remains at $2k_F$.

Singlet superconductivity can arise when hopping is included. The electron pairs are bosons, bound in a singlet state, and it is possible that they become superfluid (and hence superconducting since they are charged) at low enough temperatures. Actually this does not happen in a

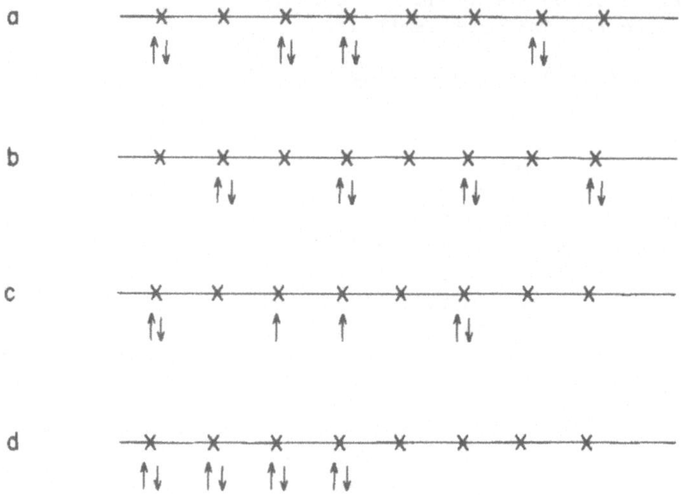

Fig. 1 Some configurations of holes for the strong coupling limit with $U < 0$. The arrows show the electron spin.

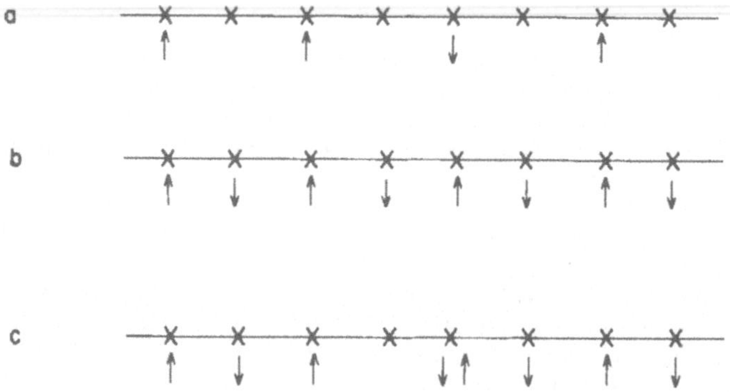

Fig. 2 Some static configurations for the strong coupling limit with $U > 0$. The arrows show the hole spin.

purely one-dimensional system since quantum mechanical fluctuations prevent superconductivity even at zero temperature. When hopping between chains is allowed, a transition does take place at a temperature T_c determined partly by the value of the perpendicular hopping amplitude. In general, T_c is not the same as the temperature $|U|/k_B$, which characterizes pair formation, a distinction which persists for weak coupling.

Triplet superconductivity will not occur because the electrons are bound into singlet pairs before long-range triplet correlations can build up.

The excited states are of two kinds. Charge-density wave excitations require the movement of pairs from site to site, and they can be phonons or plasmons according to whether the neutralizing background (ions or holes) moves in phase or out of phase with the electrons. However, in a spin-wave excitation, a spin is turned over and a pair must be broken since two electrons of the same spin cannot occupy the same site. This costs an energy $|U|$, which appears as a gap in the spin-wave spectrum. A static version of such an excitation is shown in Fig. 1c. In the same way, the Pauli susceptiblity χ is proportional to $\exp(-U/k_BT)$ at low temperatures since only thermally broken pairs can respond to a weak magnetic field. The spin-wave gap and exponential susceptibility are general features of one-dimensional systems with attractive interactions and they occur for weak coupling also.

$U > 0$, $\varepsilon = 0$, $V_2 = 0$

Now the electrons avoid the on-site repulsion and there are no doubly-occupied sites. The ground state is degenerate because the spin states have no effect on the energy.

Charge-density wave states occur again in an extreme form when there is an intersite repulsion V_2 but no hopping. In contrast to $U < 0$, however, single electrons rather than pairs are equally spaced so the period of the charge-density wave is halved and its wave vector is $4k_F$. This is shown in Fig. 2a for a quarter-filled band. (No particular significance is attached to the spin orientations at this stage.) Once again, hopping makes the charge-density wave weaker and less static. It also mixes in doubly occupied sites and restores the Fermi sea, which may lead to an additional $2k_F$ periodicity by the mechanism described for $U < 0$.

The spin-density wave instability is most clearly visualized for a half-filled band, which has exactly one electron per site so that only the spin degrees of freedom have to be considered. Virtual hopping produces an effective antiferromagnetic exchange interaction and the ground state has a modulation of the spin density, which is illustrated in Fig. 2b. The period is $2L/N$ and the wave vector $2k_F$. Actually, the picture in Fig. 2b should not be taken too seriously because quantum fluctuations wash out the spin density and prevent the establishment of long-range order for a one-dimensional system. However, if one spin were fixed, the system would show a weaker decaying $2k_F$ modulation in the density of neighboring spins.

Superconductivity may occur when the band is not half-filled, $t_b \neq 0$ and the interactions are attractive. The holes will form Cooper pairs with either singlet or triplet spin. Usually it requires a coupling to the lattice or to the collective modes of the carriers themselves in order to overcome the Coulomb repulsion and produce a net

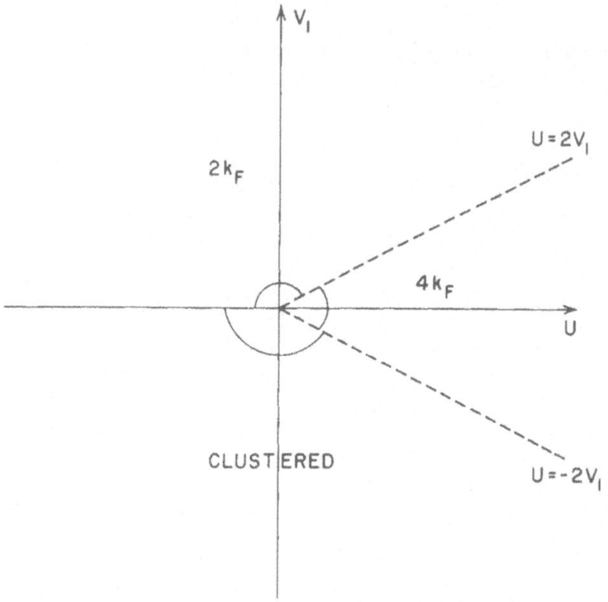

Fig. 3 Phase diagram for a half-filled band.

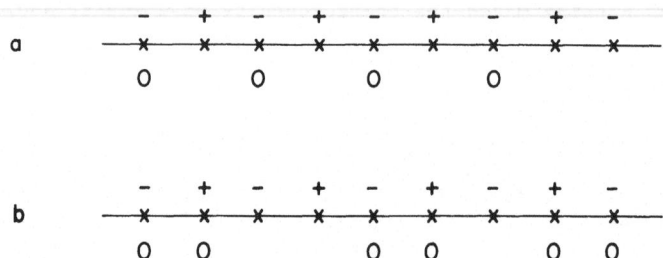

Fig. 4 Configurations for a quarter-filled band in the presence of an alternating external potential (denoted by +, -). The circles represent holes; the spin is not specified.

attraction. These mechanisms will be discussed later. Once again hopping between chains is required if there is to be a phase transition to a superconductive state.

The excited states are rather different in character from those described for $U < 0$. When the number of electrons is different from the number of sites, there is no difficulty in rearranging spins and charges to produce an excited state but, for a half-filled band, the charge-density wave excitations have an energy gap, for they require the double occupancy of a site as shown in Fig. 2c. The spin waves are known exactly for this case and have no gap.

Half-filled Band $\varepsilon = 0$, $V_2 = 0$

Here we combine features of the previous two cases, but fix the density at one hole per site. Let us compare the energies of the $2k_F$ and $4k_F$ lattices given in Figs. 1a and 2b. From Eq. 3 with $\varepsilon = 0$, $V_2 = 0$, the energy per hole is

$$2k_F \qquad E/N = U/2 \qquad (4)$$

$$4k_F \qquad E/N = V_1 \qquad (5)$$

The transition line from one state to another $U = 2V_1$ occurs where the energies cross. The other state which may enter the phase diagram has the pairs clustered on N/2 neighboring sites as shown in Fig. 1d. The energy of this "clustered" state is given by:

$$E/N = U/2 + 2V_1 \qquad (6)$$

which crosses the $4k_F$ energy at $U = -2V_1$ and the $2k_F$ energy at $V_1 = 0$. The "phase diagram" is shown in Fig. 3

When hopping is included then (a) the $2k_F$ state survives as a density wave but ultimately long-range order is destroyed. (b) In the $4k_F$ region spin-density wave correlations build-up and (c) the clustered phase gives way to superconductivity when t_a becomes sufficiently large. All of this follows from more elaborate calculations in the weak coupling limit[9,10], for which the phase boundaries remain the same. (Using Monte Carlo methods, Hirsch[11] has shown that there are small departures from the phase boundary of Fig. 3, for intermediate coupling.)

Quarter Filled Band

Here the anion potential is included and the assumption of one electron/two sites corresponds to the situation in the organic superconductors. When $U \gg V_1, V_2, \varepsilon$ there are two possible ground states, as shown in Fig. 4. The energies for the two states are given by

$$\begin{aligned} E_a &= V_2 - \varepsilon \\ E_b &= V_1/2 \end{aligned} \qquad (7)$$

which cross when $V_2 = \varepsilon + V_1/2$. As before 4(a) has principal Bragg vector equal to $4k_F$ and it is an extreme form of a Wigner crystal, a solid produced by Coulomb interactions. Note that it has the same period as the anion potential and therefore interacts strongly with it:

the positive holes are aligned in the position of lowest anion energy. On the other hand, Fig. 4b is another example of a solid with two particles per unit cell and lattice spacing equal to the period of a $2k_F$ density wave. This solid is free to move since its energy is independent of ε: the pairing correlations have suppressed the effects of the anion potential.

When hopping is included, the spatial order in Fig. 4a is weakened to a ($4k_F$) charge-density wave and is finally destroyed altogether when t_a is sufficiently large. Spin-density wave correlations appear as before.

In the other situation (b), if V is large, hopping mixes in a $4k_F$ modulation of the kinetic energy (bond order wave) and when $2t_a$ is greater than $2V_2-V_1-2\varepsilon$ (the energy to break a pair) the $2k_F$ order is destroyed[12]. Finally, when t_a is sufficiently large, quantum fluctuations remove the residual $4k_F$ bond order[12]. A proof of these statements and a detailed description of these transitions is beyond the scope of these simple arguments but may be found in Ref. 12. In second order, hopping moves a pair and it is reasonable to ask if the pairs Bose condense to form a superconducting state, particularly since their motion is not impeded by the external potential. However, a more detailed analysis shows that when $V_1 < 0$ and t_a is weak, it is favorable for the electrons to aggregate side-by-side at one end of the lattice. When t_a is large enough, superconducting fluctuations finally take over, but never with infinitely long range order, because of quantum fluctuations.

The situation is similar to that for the pairs in Fig. 1d except that the holes are on the same site in one case but on neighboring sites in the other.

CORRELATION FUNCTIONS FOR SPINLESS FERMIONS

The simple ideas set out in the preceding sections were intended to give a qualitative feeling for the physics of organic conductors. The whole picture is well supported by detailed analytical solutions of models in the strong and weak coupling limits[9,10], and by numerical calculations for intermediate coupling[11]. The analytical solutions involve methods of quantum field theory--mainly the use of boson representations of fermion operators and the quantum version of the renormalization group method. But some of the essential features of these developments may be conveyed by studying a model for which a rather simple solution may be obtained.

The possibility of making use of such an approach rests on the idea of "universality", which is borrowed from the field of critical phenomena[13]. The existence of phase transitions depends upon the long-distance behavior of correlation functions, and they in turn are determined by rather few features of the underlying models--symmetry and the number of space dimensions. Any model which has these features may be used to calculate the properties of interest. Actually the one-dimensional Fermi gas is not quite as "universal" as higher dimensional systems--there is one parameter which does depend on the details of the model. But this does not detract from the value of studying the simpler models, since it is often desirable to regard that one parameter as phenomenological.

For the present discussion, the fermion spin and the role of the molecular and anion lattices will be omitted. This will prevent us from considering all of the effects mentioned in the qualitative discussion of the preceding section, but it is sufficient to illustrate the essential ideas and to derive the power-law behavior of the density-density correlation functions.

Consider N fermion of mass M interacting via a potential V(r) which has a deep narrow well at interparticle separation r = d. If M is large enough, there are only small oscillations about the equilibrium separation d of neighboring fermions, and expanding V(r) to second order about the minimum, the Hamiltonian becomes

$$H = \sum_{n=1}^{N} \left\{ \frac{p_n^2}{2M} + \frac{Mc^2}{2d^2} (r_{n+1} - r_n - d)^2 \right\} \tag{8}$$

where r_n and p_n are the position and momentum of the nth fermion. The spring constant has been written in terms of the sound velocity c, which will appear in the subsequent calculations. It is convenient to rewrite

$$r_n = u_n + nd + \bar{r} \tag{9}$$

so that the interaction $(r_{n+1}-r_n-d)^2$ simplifies to $(u_{n+1} - u_n)^2$. The single parameter which characterizes the model is the dimensionless combination

$$\mu = \pi\hbar/Mcd \tag{10}$$

of the constants appearing in H. When μ is small, the system is quite classical and its properties are easy to evaluate. This limit is not of immediate interest for the electron gas problem but it is a very good approximation for a real physical system[14], the mercury chains in $Hg_{3-\delta}AsF_6$, and it provides a useful starting point for the present discussion. In the classical limit, thermal averages are separate gaussian integrals over the p_n and u_n and it is straightforward to show that

$$\langle (u_{n+\ell} - u_n)^2 \rangle = |\ell| \, \sigma^2 \tag{11}$$

where $\langle \ldots\ldots \rangle$ denotes thermal average and

$$\sigma^2 = k_B T d^2/Mc^2 \tag{12}$$

with T denoting the temperature and k_B Boltzmann's constant. Setting $\ell = 1$ in Eq. (11) shows that σ is the root mean square displacement from equilibrium separation, so the harmonic approximation (8) is valid when $\sigma \ll d$. Equation (11) also shows that, for $T \neq 0$, the electrons do not form a lattice, since deviations from equilibrium separation become progressively less correlated as the distance increases. Long range order does exist at zero temperature, where $\sigma = 0$, but this is a property of the classical system, and there is only quasi long-range order when quantum fluctuations are taken into account.

The nature of the electron distribution may be explored by evaluating the Fourier transform of the density-density correlation function, or structure factor:

$$S(Q) = \langle \rho_Q \rho_{-Q} \rangle \tag{13}$$

where

$$\rho_Q = N^{-1/2} \sum_{n=1}^{N} e^{iQr_n} \tag{14}$$

is the Fourier transform of the density. S(Q) may be measured by X-ray or neutron scattering. Using translational invariance and the properties of gaussian integrals, S(Q) becomes

$$S(Q) = \sum_{\ell} \exp[iQ\ell d - \frac{1}{2} Q^2 \langle u_{n+\ell} - u_n)^2\rangle] \tag{15}$$

Then, using Eq. (11) and evaluating the sum over ℓ,

$$S(Q) = \frac{\sinh(Q^2\sigma^2/2)}{\cosh(Q^2\sigma^2/2) - \cos Qd} \tag{16}$$

This function has a sequence of peaks centered at $Q = Q_m \equiv 2m\pi d$, and, in the neighborhood of the mth peak, setting

$$Q = Q_m + q \tag{17}$$

and assuming $qd \ll 1$, $\sigma^2 \ll d^2$, S(Q) becomes

$$S(Q_m + q) = \frac{2\kappa_m}{\kappa_m^2 + q^2} \tag{18}$$

where

$$\begin{aligned} \kappa_m &\equiv Q_m^2\,\sigma^2/2d \\ &= 2\pi^2 m^2 k_B T/Mc^2 d \end{aligned} \tag{19}$$

Thus the peaks are approximately Lorentzian, with widths κ_m that increase as m and T increase. When $T \to 0$, they all become delta functions, corresponding to the Bragg peaks of the ideal classical lattice. This form of structure factor shows that we are actually dealing with a fermion liquid, in which the short-range order extends to quite large distances. By contrast, in most liquids, the peaks are much broader, and have a considerable overlap.

The properties of the modes of oscillation may be explored by calculating the dynamical structure factor $S(Q,\omega)$, which describes the scattering of neutrons from the system[15]. This function is the Fourier transform of the time-dependent density correlation function, and it is given by an expression similar to Eq. (15), but with equal contributions from modes moving to the right and modes moving to the left. For small q, this amounts to making the replacements $\ell d \to x$ and $|x| \to \frac{1}{2}|x+ct| + \frac{1}{2}|x-ct|$ in the exponent. Then, Fourier transformation with respect to x and t gives (for $qd \ll 1$)

$$S(Q_m + q,\omega) = \frac{4\kappa_m^2/\pi dc}{[(q + \omega/c)^2 + \kappa_m^2]\,[(q - \omega/c)^2 + \kappa_m^2]} \tag{20}$$

This expression is a <u>product</u> of a function of $q + \omega/c$ and the same function of $q - \omega/c$, a form which is characteristic of one-dimensional systems. It is quite different from the dynamical structure factor[15] of a three-dimensional system, which consists of a sum of Lorentzians plus an elastic scattering term proportional to $\delta(\omega)$. The latter is derived from an expansion of the exponent in Eq. (15), but this is not permissible here, because the sum would diverge for large $|\ell|$. Equation (20) is an exact expression for $S(Q_n + q,\omega)$ when $qd \ll 1$, and it includes the effects of fluctuations to all orders. There is no elastic scattering because there is no true lattice. However, there are quite well-defined phonon peaks. The width κ_m is not a consequence of anharmonicity: it is an intrinsic property, related to the absence of long-range order. Indeed, κ_m is the inverse of the lattice coherence length, associated with the peak centered at Q_m, and the phonons are well defined if their wavelength is short compared $2\pi/\kappa_m$.

When the classical approximation is not applicable, the correlation functions have the same general form (15), since this is a property of harmonic oscillator Hamiltonians, but the exponent becomes more complicated than the expression (11). Nevertheless, the integrals may be evaluated in closed form[9] to give

$$G(Q_m + q,t) \equiv \langle\rho_Q(t)\rho_{-Q}\rangle\Big|_{Q = Q_m+q} = d^{-1} \int dx\, e^{iqx}\, \chi_m(x,t) \tag{21}$$

where

$$\chi_m(x,t) = F_m^+(x + ct)\, F_m^-(x - ct) \tag{22}$$

and

$$F_m^{\pm}(x) = \left[\frac{1}{x \pm i0}\; \frac{\pi x d k_B T/hc}{\sinh\left(\dfrac{\pi x k_B T}{\hbar c}\right)}\right]^{\mu_m} \tag{23}$$

with $\mu_m = m^2\mu$, where μ is defined in Eq. (10). When $h \to 0$

$$\chi_m(x,t) \to \exp\left\{\frac{-Q_m^2\sigma^2}{2d}\left[\,|x + ct| + |x - ct|\,\right]\right\} \tag{24}$$

the expression that led to the classical result (20). The same exponential occurs as a factor in the asymptotic form of the quantum mechanical correlation function. At the critical point $T = 0$, $\chi(x,t)$ is a simple power:

$$\chi_m(x,t) = \left(\frac{x^2 - c^2t^2}{d^2}\right)^{-\mu_m} \tag{25}$$

Note that the exponent depends on the parameters of the Hamiltonian since $\mu_m = m^2\mu$, with μ given by Eq. (10). The expression (25) for the density-density correlation function also is obtained for other models of the one-dimensional electron gas[9,10] the only difference is the dependence of μ on the parameters of the Hamiltonian. The essential feature is that the correlation functions are powers and that there is no long-range order. In principle the value of μ could be determined from experiment and used as the one quantity that characterizes the model.

The static structure factor S(Q) is equal to G(Q,0) and it may be evaluated in closed form from Eqs. (21), (22) and (23). In the classical limit, it agrees with Eq. (18), whereas for T = 0

$$S(Q_{m+q}) = \frac{1}{2}(1 - 2\mu_m)\sin\pi\mu_m\left(\frac{qd}{4}\right)^{2\mu_m-1} \tag{26}$$

Thus the δ-function peaks, to which Eq. (18) reduces as $T \to 0$, are replaced by power laws: true long-range order becomes what is known as quasi-long-range order when quantum effects are taken into account. Indeed, there will not be a divergence at all for large enough m, since $2\mu_m > 1$.

When $\mu > 1/8$ there is only one peak at $Q = 2\pi/d$ which is also equal to $2k_F$ for spinless fermions. This is the situation for the extreme quantum limit usually studied in the theory of the one-dimensional electron gas[9,10] and the expression (26) is the result usually obtained by the bosonization technique.

When the lattice and the fermion spin are restored, H is replaced by two Hamiltonians, one describing charge-density waves and the other spin-density waves. There are additional terms in H which depend on the cosine of the u_n and give what is known as the sine-Gordon Hamiltonian. The solution of the full problem is described in Refs. 9, 10, 16, where it is shown that the correlation functions relevant for spin density waves and superconductivity also have a power law behavior, and all may have an exponential decay related to the energy gaps described in Sec. 3. These results are used to obtain the phase diagrams and other properties as described in Sec. 3, and they constitute the modern theory of the one-dimensional electron gas.

References

* Work supported by Division of Materials Sciences U.S. Department of Energy, under Contract No. DE-AC02-76CH00016.

1. K. Bechgaard, this volume.
2. J. P. Pouget, this volume.
3. D. Jérome, A. Mazaud and M. Ribault, J. Phys. (Paris), Lett. 41, 95 (1980).
4. J. P. Pouget and S. Barišić, private communication.
5. R. Moret, S. Ravy, J. P. Pouget, R. Comès and K. Bechgaard, Phys. Rev. Lett. 57, 1915 (1986).
6. V. J. Emery, J. Phys. (Paris) Colloq. 44, C3-977 (1983).
7. C. Bourbonnais, this volume
8. V. J. Emery, R. Bruinsma and S. Barišić, Phys. Rev. Lett. 48, 1039 (1982).
9. V. J. Emery in Highly Conducting One-dimensional Solids, edited by J. T. DeVreese, R. P. Evrard and V. E. Van Doren (Plenum, New York 1979).
10. J. Solyom, Adv. Phys. 28, 201 (1979).
11. J. E. Hirsch, this volume
12. V. J. Emery and C. Noguera, to be published.
13. See for example, D. J. Amit, Field Theory, the Renormalization Group and Critical Phenomena (McGraw-Hill, New York, 1978).
14. V. J. Emery and J. D. Axe, Phys. Rev. Lett. 40, 1507 (1978).

15. G. L. Squires, Thermal Neutron Scattering (Cambridge University Press, Cambridge (1978)).
16. V. J. Emery, Proceedings of Kyoto Summer Institute 1979, Y. Nagaoka and S. Hikami, eds. (Publication Office, Progress of Theoretical Physics, 1979).

TRYING TO UNDERSTAND THE PROPERTIES OF TCNQ-SALTS FROM THEIR CRYSTAL STRUCTURES AND THEIR TRANSFER INTEGRALS

Sietse Oostra and Jan Kommandeur

Laboratory for Physical Chemistry
The University of Groningen
Nijenborgh 16
9747 AG Groningen, The Netherlands

I. INTRODUCTION

It has often been thought that the great advantage of organic conductors would be that their properties could be "tailored", i.e. the versatility of organic chemistry could be used to "tune" the properties of the compounds synthesized. Although the great variety of properties found in organic conductors and semi-conductors might lead one to doubt the application of this thought, some years ago, we nevertheless decided to further investigate it.

To this end we synthesized some twenty derivatives of morpholinium $(TCNQ)_2$, which differed only in the substitution on the nitrogen (by H, CH_3, C_2H_5 etc.) and the change of oxygen in the morpholinium ring into a sulphur. We then determined all the crystal structures and their electrical and magnetic properties. The great variety in these immediately showed that "tuning" them is out of the question.

Another matter is, whether we can understand the electrical and magnetic properties from the crystal structures. For this we need a theory, and we review the available theories in section 2. We also need numerical values for the parameters in the theory, such as the transfer integral t and the Hubbard U. Their estimation is discussed in section 3.

In sections 4 and 5 we discuss the application of theory to experiment for, respectively, the electrical and magnetic properties, while finally in section 6 we draw some conclusions.

2. THEORY

To interpret our experimental data we need <u>some</u> theory. For the TCNQ-salts there is little doubt that electron-electron repulsion plays a considerable role. If this were not so, all dimerized quarter filled bands such as most of the $M^+(TCNQ)_2^-$ compounds would be metals, and this is certainly not so. Electron-electron repulsion therefore must be included. Unfortunately analytical solutions for band systems with intermediate on-site electron-electron repulsion are not available, we have them only for the limit where the Hubbard U is either infinite or zero. Since we have excluded the latter limit we are forced to take the former. Electrons are then not allowed to doubly occupy any site. (Hubbard $U = \infty$). Since double occupancy is excluded the electrons behave like spinless fermions, which can move between sites, because of the transfer integrals t. The Hamiltonian then becomes:

$$H = t_1 \sum_m (C^+_{2m} \; C_{2m+1} + C_{2m+1} \; C_{2m}) +$$

$$t_2 \sum_m (C^+_{2m+1} \; C_{2m+2} + C^+_{2m+2} \; C_{2m+1}) ,$$

where we have ssumed the system to be dimerized and C^+_m and C_m have the usual meaning og electron creation and annihilation operators on site m.

It would seem reasonable to add a neartest neighbour repulsion term of the form $V \sum_m n_m n_{m+1}$, where V is the repulsion and n_m is the occupation number of site m, but then again we have no analytical solution. We therefore neglect this term.

We can add the external effect of the lattice. Because of the cations in general not every site will be at the same potential. Because of the symmetry of a dimerized system we can take the lattice potential to be symmetric around some arbitrary zero, i.e. $+\Delta$ and $-\Delta$. A large value for this potential will lead to a localization of the electron in the dimer on the site with the positive potential, which can sometimes be observed crystallographically. We therefore add the terms:

$$+\Delta \sum_m n_{2m} - \Delta \sum n_{2m+1} \quad \text{to the hamiltonian.}$$

Fourier Transforming to k-space and solving the secular determinant leads to the k-dependent energies:

$$E(k) = \pm \sqrt{t_1^2 + t_2^2 + \Delta^2 + 2t_1t_2 \cos k}$$

We thus have two bands with energies running from $E = - [(t_1 + t_2)^2 + \Delta^2]$

to $E = - [(t_1 - t_2)^2 + \Delta^2]$

and from $E = + [(t_1 - t_2)^2 + \Delta^2]$

to $E = + [(t_1 + t_2)^2 + \Delta^2]$

with a gap of $E_g = 2 [(t_1 - t_2)^2 + \Delta^2]$

For $\Delta = 0$ the charge at each site will be 1/2, for $\Delta \to \infty$ the charge on the sites will alternate between 0 and 1. The gap vanishes only for $t = 0$ and $\Delta = 0$, it is equal to 2Δ for $t_1 = t_2$. For intermediate values of Δ a complicated equation for its relation with the site density has been derived [18]. As it will turn out, the above treatment is useful to roughly interpret some of our conductivity data, probably because once correlation has been taken into account in a global way, it will at most affect the magnitude fo the conductivity, but not its temperature dependence.

This is not at all so for the magnetic susceptibility. It is critically dependent on correlation. In the limit of $U = 0$ we would expect a low temperature independent (Pauli) susceptibility for a dimerized quarter filled band, and in the limit $U = \infty$ a Curie susceptibility. Neither, how ever, is observed. We clearly have to deal with correlation more explicitly here.

Various efforts have been undertaken to do so [1] in a low temperature approximation. Applied to the systems of interest here, the results can be summarized as follows [2]:

We write the spin-hamiltonian as: $H = 2J \Sigma S_i S_{i+1}$, where J is the exchange energy and S are the spin operators. When $t_1 = t_2$, $\Delta = 0$ we have $J \simeq t^2 \mid (U + \sqrt{2})$, and in the dimer limit, where $t_1 >> t_2$, $\Delta = 0$ we have $J \simeq 1{,}25\, t_2^2 \mid (U + t_1)$. Since conductivity data show that $U > t_1, t_2$ we can generally take $J \simeq t_2^2 / U$, where t_2 is smaller of the two transfer integrals. This equation holds, even when we have a modest charge alternation due to $\Delta \neq 0$. J does, however, depend on Δ as was shown analytically, as well as numerically [2]. In the limit $\Delta \rightarrow J_0$, where J_0 is the exchange energy for $\Delta = 0$, J becomes very small, since the charges are localized and hardly "see" each other.

After having estimated J we need a calculation of the spin susceptibility (χ) of a linear chain of spins. Again, analytical solutions are lacking; but numerical calculations are available from Bonner and Fisher [3]. At high temperatures χ follows a Curie-Weiss Law ($\chi = 1/(T + J)$), it reaches a maximum at $T = 1.83\, J$ and it goes to a finite value at $T = 0$. Clearly, the latter is not allowed because of the third law and therefore a spin Peierls transition will have to come about, making the lattice tetrameric with two exchange integrals J_1 and J_2, and with

$$H = \sum_i (J_1 S_{2i} S_{2i+1} + J_2 S_{2i+1} S_{2i+2}$$

Numerical calculations of the spin susceptibility in this case have also been performed [4]. At high temperature we again have a Curie Weiss Law with a maximum at $T = 1.25\, J_1$ $(J_1 \geq J_2)$, and because there now is a gap in the spin excitation spectrum $\chi \rightarrow 0$ as $T \rightarrow 0$. The decrease of χ at low T is the steeper, the closer J_2 approaches J_1. We note here that the magnetic susceptibilities can also be estimated from numerical calculations on an eight site linear chain in the spirit of our work on the Peierls transition [5,6,7]. We can then use the full Hubbard hamiltonian, including intersite repulsion (V) and the lattice potential [2]. As long as $U, V > t$, kT, however, the results do not differ significantly from the ones obtained by Bonner et. al. [3,4]. This can easily be understood. At such low temperratures and with $U > t$ no electronic degrees of freedom are excited, and indeed the solutions of the Hubbard hamiltonian are equivalent to those of the Heisenberg hamiltonian.

To compare the theoretical results with experiment we need some estimates of the transfer integrals and of that elusive quantity, the Hubbard U. We treat their evaluation in the next section.

3. TRANSFER INTEGRALS AND THE HUBBARD U.

tn pure one electron theory the hamiltonian used is

$$H = t \sum_{i,\sigma} C^{+}_{i+1,\sigma} C_{i,\sigma},$$

where the transfer integral t is given by

$$t = < \phi^W_{i+1} |H(1)| \phi^W_i >,$$

with ϕ^W_i the Wannier function on site i and H(1) the one-electron operator, which incorporates interactions with all electrons except the band electrons in a Hartree-Fock manner. The Hubbard hamiltonian reads:

$$H_h = t \sum_{i,\sigma} c^+_{i+1,\sigma} + U \sum_i c^+_{i,\sigma} c_{i,\bar{\sigma}} c^+_{i,\sigma} c_{i,\bar{\sigma}}$$

where U is the on-site repulsion and again t is a transfer integral. Study of the derivation of the Hubbard hamiltonian [8] reveals that in this treatment a number of correlation and other terms are incoporated in t. Therefore, the t of tight-binding (one-electron theory) should numerically be very different from the one used in the Hubbard mode. Therefore, the transfer integrals used in band structure calculations [9,10,11], open to severe criticism themselves [12], cannot be used in a Hubbard treatment.

In view of all the uncertainties it appears best to rely on an expression derived from the Mulliken approximation [12]: $t = h_{ii} (K-1)S$, where S is the overlap of ϕ_{ii} and ϕ_{i+1}, K is a constant between 1 and 2 and h_{ii} is the orbital energy of the lowest unoccupied molecular orbital.

The transfer integrals are then proportional to S, and relative values can easily by obtained. A calibration for the 1:2 salts of TCNQ is obtained from the infrared work of Rice, Yartsew an Jacobson [13] and for the 1:1 salts from Hibma, Sawatzky and Kommandeur [14]. In this way reasonable values of t, which critically depend on the crystal structure can be obtained. This strong dependence is caused by the nodal structure of the orbitals; transfer integrals cannot be estimated by just "looking" at the crystal structure.

The value of the Hubbard U is even harder to estimate. It is affected by the correlation terms neglected [8], by the polarization of the surrounding lattice [15], and by the band filling [16]. Differences of about a factor of 10 between estimates can be obtained. As far as it plays a role in our interpretations we have taken values between 1 and 1.5 eV.

In the next section we make use of the calculations and we try to correlate our results on conductivity experiments with them.

4. THE ELECTRICAL CONDUCTIVITY

In view of the previous discussion we compare the conductivities and their activation energies with the calculated gaps in tables 1, 2 and 3. Table 1 shows the conductivities of the single chain compounds, where the TCNQ-ions are related by a crystal symmetry operation, i.e. $\Delta = 0$.

Taking the calibration for the transfer integral from the work by Hibma et al. [14], we find remarkably good agreement between the theoretical and the experimental values for the gap and the activation energies for the three compounds in question. If the calibration had been taken from the work of Rice, Yartsev and Jakobsen [13] the theoretical values would have been off by a factor of two, of course.

Table 1. Gaps and Activation energies for single chains, $\Delta = 0$

	(t_1-t_2) (meV)	Ea (meV)	$\sigma(\Omega^{-1}cm^{-1})$
MEM T_2 (>335 K)	20	10	40
DMM T_2	210	230	2×10^{-2}
DMTM T_2	240	260	14×10^{-3}

Table 2. Gaps and Activation energies for single chains, $\Delta \neq 0$

	$\rho \rightarrow$	Δ(meV)	Eg/2(meV)	Ea(meV)	$\sigma(\Omega^{-1}cm^{-1})$
MBTM T_2	.64	90	110	170	9
METM T_2	.67	120	120	80	4
HMM T_2	.69	130	140	170	.17
MEM T_2 (250 K)	.69	130	280	320	6×10^{-5}
DMM (tricl.)	186	280	290	290	6×10^{-5}

In table 2 we compare calculated and experimental values for the gaps of the compounds with single chains, but $\Delta \neq 0$. As outlined before Δ is calculated from the crystallographically observed charge distributions, together with the transfer integrals. The value of Δ thus derived is then inserted into the equation for the gap.

The agreement is much less satisfactory, probably because of the uncertainty involved in the determination of the charge distributions. Nevertheless, the values do not scatter too wildly.

Finally in table 3 we compare calcutated and experimental gaps for the two inequivalent chain compounds. Since now "indirect" transitions between the various chain bands can take place, we do not expect much agreement, although we would expect $E_a \leq 1/2E_g$, where E_g is the lowest value calculated. Even that is only born out in two of the three compounds.

It is worth pointing out the very small gap for DEM (T<272), compared to the theoretical estimates. Here, as was suggested before [18] the chains are at such different potentials that one chain has "leaked" electrons to the other, yielding a very small gap and a very low conductivity.

In conclusion it can be said that a reasonable theoretical estimate of the gap in the high U limit can be made, as long as crystal field complications do not interfere.

We can even try to estimate the actual conductivities with the calculated values for the gaps. We do need a σ_0 for the equation $\sigma = \sigma_0 \exp -Ea/kT$. We might take $\sigma_0 = \dfrac{n\, e^2\, \lambda_F}{m\, v_F}$, where n is the number of electrons per unit of volume, λ_F the scattering length for particles at the fermi surface in their mass and v_F the fermi velocity. With $n = 10^{21} cm^{-3}$ and $E_F = 0.3$ eV we obtain $\sigma_0 = 7.1 \times 10^{10} \lambda_F$. Assuming that each ion scatters the particles, we have $\lambda \simeq 3.5$ A, which leads to $\sigma_0 \simeq 300 \Omega^{-1} cm^{-1}$.

Table 3. Gaps and Activation energies for two chain compounds, $\Delta = 0$

	t_1	t_2(meV)	E_g(meV)	E_a(meV)	$\sigma(\neq^{-1}cm^{-1})$
MBM T_2	270	315	45	70	9
	293	315	22		
MPM T_2	400	225	90	50	112
	292	337	20		
DEM T_2	68	337	270	90	4×10^{-4}
	427	68	350		

Table 4. Theoretical and experimental conductivities

	$\sigma_{exp.}$ $(\Omega^{-1}cm^{-1})$	$\sigma_{theor.}$ $(\Omega^{-1}cm^{-1})$
MEM (>335)*	40	93
DMM	2×10^{-2}	3×10^{-2}
DMTM	1.4×10^{-2}	2×10^{-2}
MBTM	9	3.7
METM	4	2.5
HMM	.17	1.5
MEM (250 K)	6×10^{-5}	2×10^{-5}
MBM**	9	184
MPM**	.12	185
DEM**	4×10^{-3}	3×10^{-3}

* with σ_0 exp- Eg/2kT $(1 + \text{exp-Eg/2kT})^{-1}$ in view of small Eg

** Inequivalent chain compounds.

The values of the room temperature conductivities derived with this assumption and the theoretical values for the gaps are given in table 4.

Except for the inequivalent chain compounds the agreement between calculated and experimental values is surprisingly good and an a posteriori justification of our scattering assumtion.

We should not close this section without a remark about the socalled "steps" in the conductivity. It seems to be a general property of one-dimensional conductors that the conductivity on cooling changes discontinuously in a rather arbitrary way. Sometimes these steps are accompanied by the appearence of little cracks at the surface. The steps also appear on heating, by cycling the crystals usually makes them disappear.

It could be that the steps are due to the sudden coherent ordering of part of hte cations, to phase transitions occurring on single chains or bunches of them, or as due to macroscopic strain, induced by the highly anisotropic thermal expansion. One does not find much about these effects in the literature, but personal communications appear to bear out that for single chain compounds it is a generally observed effect.

The slopes of the σ vs. T curves are always retained after a step, and it is these slopes that have been presented in table 1, 2 and 3.

The spin susceptibility

Spin susceptibility was obtained by integrating the calibrated ESR signal. In general all compounds showed a Borner-Fisher-like [3] behavior, which is typical for antiferromagnetically coupled Heisenberg chains. At low temperatures all susceptibilities showed a tendency to fall to zero, be it that in some compounds a strong Curie tail appeared to mask this behavior. It would therefore seem that all these materials have a (spin) Peierls $2k_F$ transition [6].

All susceptibilities were fitted to a Bonner-Fisher plot, the results for the exchange constants are given in the various tables.

Since the Bonner-Fisher calculations were carried out for regular chains we first compare the results with a theoretical estimate $J = t^2/U$, using the average $t = 1/2(t_1+t_2)$ for the transfer integral and the value of U that best fits to the experimental data (U = 1.12 eV). The results in table 5 show that no reasonable consistency is obtained.

Table 5. Experimental and Theoretical exchange constants for dimerized compounds.

	experiment		theory				
	J from	J from	J	parameters used			
compound	χ_{RT} (K)[a,g]	T_{max} (K)		t_1	t_2	Δ	Q[f]
HBM T_2 *	75	60	-	-	-	-	-
HBTM T_2	(143)	-	-	-	-	-	0.50
METM T_2	80	80	211	0.178	0.177	0..34	0.67
HMM T_2	140	170	178	0.171	0.162	0.035	0.69
MBM T_2	63	55	A 173	0.169	0.148	0	0.5
			B 207	0.167	0.165	0	0.5
DMMT$_2$ (tricl.)	9	7[e]	53	0.145	0.111	0.088	0.86
MPM T_2	73	22	A 200	0.182	0.159	0	0.5
			B 136	0.209	0.127	0	0.5
DEM T_2	64	45/75[d]	A 13	0.190	0.039	0	0.5
			B 10	0.221	0.034	0	0.5
DMM T_2 (monocl.)	85/78[c]	-	33	0.182	0.062	0	0.5
DMTM T_2 (T>272 K)	42	-	16	0.179	0.043	0	0.5
DMTM T_2 (T<272 K)	81±10[b]	35	-	-	-	-	-
MEM T_2	53	53	26	-.213	0.057	0..53	0.69
MBTM T_2	81	80	197	0.172	0.166	0.024	0.64
MEM T_2 (348 K)	70	-	191	0.168	0.157	0	0.5

a error is about ± 5 K unless otherwise indicated
b from X at 268 K
c the second value is from Faraday balance measurements. The EPR value is probably more reliable
d two chains
e from a fit of X between 10 and 100 K
f charge of most negative TCNQ molecule
g the formula

$$J\chi/Ng^2\mu^2_\beta = \frac{0.25}{(T/J+1)} \left(1 - \frac{1}{1.056\,(T/J)^2 + 1.3}\right)$$

was used. It accurately fits the Bonner & F sher magnetic susceptibility above T = 1.0 J.

* T stands for TCNQ

Table 6. Experimental and Theoretical exchange constants for tetramers

compound	Experiment $J_1 \simeq T_{max}/1.25$	$\Delta E_{ST} \simeq (J_1-J_2)$	theory J_1	J_2	parameters used t_{12}	t_{23}	t_{34}	t_{41}
HEM T_2	-	435 K	449	52	0.187	0.193	0.187	0.103
EBM T_2	212 K	(200 K)*	209	167	0.155	0.167	0.155	0.150
EBTM T_2	234 K	215 K	226	2	0.212	0.127	0.212	0.024
$TEAT_2$ (110 K)			406	169	0.232	0.201	0.232	0.163
(173 K)		(229 K)*	338	147	0.217	0.181	0.217	0.153
(234 K)	240 K	(222 K)*	280	135	0.206	0.163	0.206	0.148
(295 K)		(205 K)*	218	121	0.192	0.142	0.192	0.142
(345 K)			193	111	0.181	0.133	0.181	0.137
MEM T_2 (5 K)	-	J_1=60 K J_2=46 K[1]	46	30	0.227		0.216	0.060

compound	lattice type	$\Delta_{A,B}$	Q_B
HEM T_2	ABBA	±0.087	0.68
EBM T_2	ABBA	0	0.50
EBTM T_2	ABBA	±0.105	0.67
$TEAT_2$ (110 K)	BAAB	±0.05	0.380
(173 K)	BAAB	±0.05	0.390
(234 K)	BAAB	±0.05	0.402
(295 K)	BAAB	±0.05	0.413
(345 K)	BAAB	±0.05	0.418
MEM T_2 (5 K)	ABAB	±0.053	0.68

(1) T stands for TCNQ
* numbers in parentheses are probably not reliable.

At least the exchange constants found experimentally are of the order of the ones calculated theoretically.

It will be clear that there is little if any agreement between theory and experiment. Theory predicts a much greater variation of J than is found experimentally. The theoretical J's are calculated with U = 1.4 eV, which as has been pointed out before is only a very rough estimate. Other values might have been chose, but they would not have affected the variation much. Clearly theory is in bad shape here as far as quantitative predictions are concerned. We fare somewhat better with the tetramerie compounds: the condensed data and theoretical results are collected in table 6.

It is worth noting the change by more than a factor of 2 in J for TEA $(TCNQ)_2$ between 110 K and 345 K, for which the crystal structure has been measured very accurately at different temperatures by Filhol [21]. This more or less points up the fact that it is rather hopeless to characterize the exchange in TCNQ compounds by measurements as a function of temperature. The conclusion of this chapter must be, that there is little or no relation between theory and experiment. Magnetic properties cannot be predicted from the chemical constitution but even knowledge of the crystal structure fails in this respect.

A special position is taken by DMM $(TCNQ)_2$. It has a very high Curie-type susceptibility which persists down to about 1.7 K, where probably a spin-Peierls transition occurs (22, 23, 24). In this compound the localization of the electron on one member of the dimer is so high due to the asymmetric crystal field that the exchange is extremely low. It is at least one effect that can be readily estimated from the crystal structure, which clearly shows the asymmetric ordering of the cations around the TCNQ chain [18].

6. CONCLUSIONS

One conclusion is firm. The chemical constitution of counter-ions does not predictably influence the electric and magnetic properties of TCNQ-compounds. The versatility of organic chemistry is of no use here. "Tuning" the properties in this manner is a fairytale.

Once the crystal structures are known, a better understanding can be reached. The electrical properties are dominated completely by the packing of the TCNQ-ions in the chain. The variation of the magnetic properties, however, cannot be connected to the crystal structure.

It may be worthwhile to enumerate some of the difficulties that face the "interpreter" of electric and magnetic properties. It is mainly a question of magnitudes. The Hubbard U seems to be of the order of 1 - 1.5 eV, 4t the total bandwidth of a regular chain is somewhere between 0.6 - 1 eV, not very far from U. The weaker characteristic interactions are exchange ($\simeq$ 0.02 eV) and the Debye temperature ($\simeq$ 100 K $\simeq$ 0.01 eV). The thermal expansion leads to variations in t, which are at least of the order of magnitude of J and Θ_D, while J itself may change by a factor of two through that effect. It seems that all interactions have to be brought into play simultaneously and that makes any theory very complicated and so far, not very successful.

REFERENCES

1. D.J. Klein & W.A. Seitz, Phys. Rev. B10, 3217 (1974)
2. S. Oostra, Ph. D. thesis, Groningen, 1985
3. J.C. Bonner & M.E. Fisher, Phys. Rev. 135, A640 (1964)
4. J.C. Bonner & H.W.J. Blote, Phys. Rev. B 25, 6956 (1982)
5. H.T. Jonkman, H.J. Zwinderman & J.Kommandeur, Journ. de Physique, 44, C3-1281 (1983)
6. S. Huizinga, J. Kommandeur, H. T. Jonkman & C. Haas, Phys. Rev. B 25, 1717 (1982)
7. These Proceedings, Chapter on Peierls transitions by J. Kommandeur
8. J. Hubbard, Proc. Roy. Soc. London Ser. A 276, 238 (1963)
9. A.J. Berlinsky, J.F. Carolan & L. Weiler, Solid State Commun, 15, 795 (1974)
10. F. Herman, Phys. Rev. B 16, 2453 (1977)
11. P.M. Grant, Phys. Rev. B 26, 6888 (1982)
12. S. van Smaalen & J. Kommandeur, Phys. Rev. B 31, 8056 (1985)
13. M.J. Rice, V.M. Yartsev & C.S. Jacobsen, Phys. Rev. B 21, 3437 (1980)
14. T. Hibma, G.A. Sawatzky & J. Kommandeur, Phys. Rev. B 15, 3959 (1977)
15. G.A. Sawatzky, P.I. Kuindersma & J. Kommandeur, Solid State Commun. 17, 569 (1975)

16. S. Mazumdar & A.N. Block, Phys. Rev. Lett. 50, 207 (1983)
17. B. van Bodegom, Ph. D. thesis University of Groningen (1985)
18. R.J.J. Visser, Ph. D. thesis University of Groningen (1985)
19. S. van Smaalen, J.L. de Boer, C. Haas & J. Kommandeur, Phys. Rev. B 31, 3496 (1985)
20. See f.i. A.J. Dekker, Solid State Physics, (Prentice Hall, 1957) p.283
21. A. Filhol & M. Thomas, Acta Cryst. B40, 44 (1984)
22. W.H. Korving, T.W. Hijmans, H.B. Brom, S. Oostra, G.A. Sawatzky & J. Kommandeur, J. de Physique 44, C-1425 (1983)
23. G.J. Kramer, H.B. Brom, Mol. Cryst. Liq. Cryst. 120, 153 (1984)
24. T.W. Higmans, Ph. D. thesis University of Leiden (1985)

MONTE CARLO SIMULATION OF MODELS FOR LOW-DIMENSIONAL CONDUCTORS

J. E. Hirsch

Department of Physics
University of California, San Diego
La Jolla, CA 92093

ABSTRACT

Monte Carlo simulations are a useful tool to study properties of interacting quantum systems beyond perturbative regimens. We give here an overview of the application of Monte Carlo techniques to the study of collective effects in models for low-dimensional conductors. Topics discussed in detail include the phase diagram of the half-filled extended Hubbard model, $2p_F$ and $4p_F$ instabilities, and the effect of Coulomb interactions and quantum fluctuations in electron-phonon systems. We conclude with a discussion of future directions.

I. INTRODUCTION

A theoretical understanding of the physics of low-dimensional conductors often involves sorting out delicate interplays between competing interactions. Although powerful theoretical techniques[1,2] have been very successful in understanding collective phenomena in these systems in weak and strong coupling regimes, there remains a wide region of parameter space where the validity of these analytic techniques remains doubtful. In addition, although it is often possible to obtain analytically the asymptotic power-law divergence of some quantity, both the way this asymptotic region is approached, as well as the amplitude of the divergence are not easily obtained analytically.

In this paper we discuss a quantum Monte Carlo technique that has been developed and applied to low-dimensional systems during the past 5 years.[3,4] This technique is most efficient precisely in the parameter regimes where analytic techniques like weak or strong coupling perturbative theory are most likely to fail, thus providing an important complement to these. Also, as we will illustrate, simulation results are sometimes useful to interpret the predictions of the analytic calculations. By making contact with the analytic calculations in limiting cases and interpolating in-between, this technique allows one to predict the properties of the model over the entire parameter range. The outcome of these calculations are thermodynamic properties and correlation functions that one can more or less directly compare with experiments; by using finite-size scaling techniques one can also obtain the exponents of various power-law divergences to compare with analytic predictions.

In this paper we will review some of the results obtained by quantum Monte Carlo simulations that are relevant to low-dimensional conductors. Some of the results discussed exist in the literature and some are new. The purpose of this paper is to provide an up-to-date review of existing results, and point out possible future directions.

As is well known, the extended Hubbard model is though to be appropriate to describe many physical phenomena in low-dimensional conductors. The Monte Carlo method discussed here is particularly suited for such a tight-binding Hamiltonian defined in real space. After reviewing the main features of the numerical technique in Section II, we discuss in Section III two transitions that occur in the extended Hubbard model in the half-filled band sector: the charge-density-wave (CDW) to spin-density-wave (SDW) transition and the condensation transition. In Section IV we discuss results for $2p_F$ and $4p_F$ instabilities in non-half-filled and half-filled band sectors. In Section V we discuss electron-phonon interactions and in particular the effect of Coulomb interactions and quantum fluctuations on the Peierls instability. Section VI gives an overview of other results that have been obtained by Monte Carlo simulations on low-dimensional systems, particularly on a model for excitonic superconductivity. We conclude in Section VII with a short discussion.

II. FORMALISM

The partition function is written as:

$$Z = \mathrm{Tr}\, e^{-\beta H} = \sum_{\{i_1 \ldots i_{2L}\}} \langle i_1 | e^{-\Delta\tau H_1} | i_2 \rangle \langle i_2 | e^{-\Delta\tau H_2} | i_3 \rangle \langle i_3 | e^{-\Delta\tau H_1} | i_4 \rangle \cdots \langle i_{2L} | e^{-\Delta\tau H_2} | i_1 \rangle \quad (1)$$

$$H = H_1 + H_2 \quad (2)$$

To obtain eq. (1), the temperature axis was divided into L slices of size $\Delta\tau = \beta/L$ and the approximation

$$e^{-\Delta\tau H} = e^{-\Delta\tau H_1} e^{-\Delta\tau H_2} + O(\Delta\tau^2 [H_1, H_2]) \quad (3)$$

was made. The error in eq. (3) can be made negligible for sufficiently small $\Delta\tau$. For the fermion part of the Hamiltonian, it is useful to choose H_1 and H_2 as sums of non-overlapping cell Hamiltonians which can be easily diagonalized. For phonon degrees of freedom, it is convenient to choose for H_1 and H_2 kinetic and potential energy parts respectively.

The intermediate states $|i_\ell\rangle$ are chosen to be diagonal in the fermion occupation number representation and in the lattice displacement coordinates. The sum over intermediate states is performed using a Metropolis Monte Carlo algorithm choosing only transitions that respect the conservation laws implicit in the Hamiltonian, i.e. local particle number and spin conservation (world line algorithm). For details on the methodology, see Refs. 3 to 5.

III. ELECTRONIC TRANSITIONS IN THE HALF-FILLED HUBBARD MODEL

We consider the extended Hubbard Hamiltonian

$$H = -t \sum_{i,\sigma} (c^{\dagger}_{i\sigma} c_{i+1,\sigma} + h.c.) + U \sum_i n_{i\uparrow} n_{i\downarrow} + V \sum_i n_i n_{i+1} \qquad (4)$$

in the half-filled band sector. We will discuss two-phase transitions that occur in this model as a function of U and V:

a) CDW-SDW transition

Figure 1 shows Monte Carlo cycles for the charge density wave order parameter[6]:

$$m = \sum_i (-1)^i n_i \qquad (5)$$

for three values of U as a function of V. They clearly show that a phase transition to a CDW state occurs around $V = U/2$. The fact that the cycles show larger hysteresis as U increases indicates that the transition becomes first order for large U.

It is easy to see in strong coupling ($U/t \to \infty$, $V/t \to \infty$) that the CDW-SDW transition occurs at $V = U/2$, by balancing the energies of both states. In weak coupling, renormalization-group (RG) predicts a transition when the backscattering $g_1 = 0$, which again corresponds to $U = 2V$. However, as we will see there are small deviations from this relation for intermediate values of U. To obtain accurate answers, we consider the CDW structure factor:

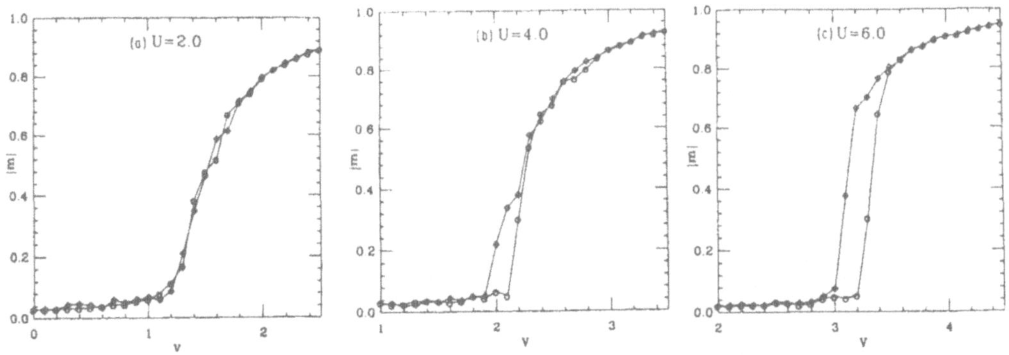

Fig. 1. Absolute value of CDW order parameter vs V for U = 2, 4, 6 on a 32 × 32 lattice. At each point 100 sweeps through the lattice were performed, and V was changed in steps of 0.1. The circles (diamonds) were obtained with increasing (decreasing) V.

$$S(q) = \frac{1}{N} \sum_{i,j} e^{iq(R_i - R_j)} \langle n_i n_j \rangle \qquad (6)$$

and the SDW zero-frequency susceptibility:

$$\chi(q) = \frac{1}{N} \int_0^\beta d\tau \sum_{i,j} e^{iq(R_i - R_j)} \langle (n_{i\uparrow}(\tau) - n_{i\downarrow}(\tau))(n_{j\uparrow}(0) - n_{j\downarrow}(0)) \rangle \qquad (7)$$

By increasing the lattice size and inverse temperature β by the same factor one expects $S(q = \pi)$ to diverge linearly in the CDW phase and $\chi(q = \pi)$ to diverge linearly in the SDW phase. Figure 2 shows these correlation functions plotted versus lattice size for $U = 2$. It is clear that the transition occurs around $V = 1.15$. Similar calculations for $U = 3, 4, 6$, and 8 yield critical values of V of 1.675, 2.163, 3.158 and 4,131, respectively.

To investigate the character of the transition (whether it is first order or continuous), it is useful to look at histograms of the CDW order parameter.[6,7] For small U, these histograms show a peak at $m = 0$ for small V which moves continuously to $m \neq 0$ as V increases beyond its critical value; for large V, the peak at $m = 0$ becomes smaller while a new peak starts forming at a finite m as V is increased, signaling a first order transition. This is illustrated in Fig. 3 for the case $U = 6$. The tricritical point, where the transition changes from continuous to first order, is found to be around $U \sim 3$.[6]

As discussed by Haldane,[8] it is possible to predict the changeover from continuous to first order transition from weak coupling as the point where Umklapp scattering of electrons with parallel spins becomes relevant. It can also be understood from strong coupling as the point where the surface tension of droplets of the opposite phase vanishes.[6] These arguments, however, do not provide reliable quantitative estimates for the tricritical point. Similarly, although strong coupling perturbative theory does indicate that the transition line should deviate from $U = 2V$ towards the CDW

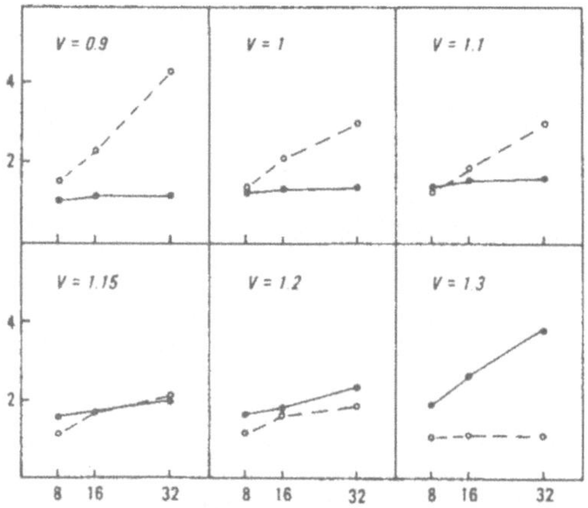

Fig. 2. CDW structure factor (solid line) and SDW susceptibility (dashed line) vs lattice size for $U = 2$ and several values of V.

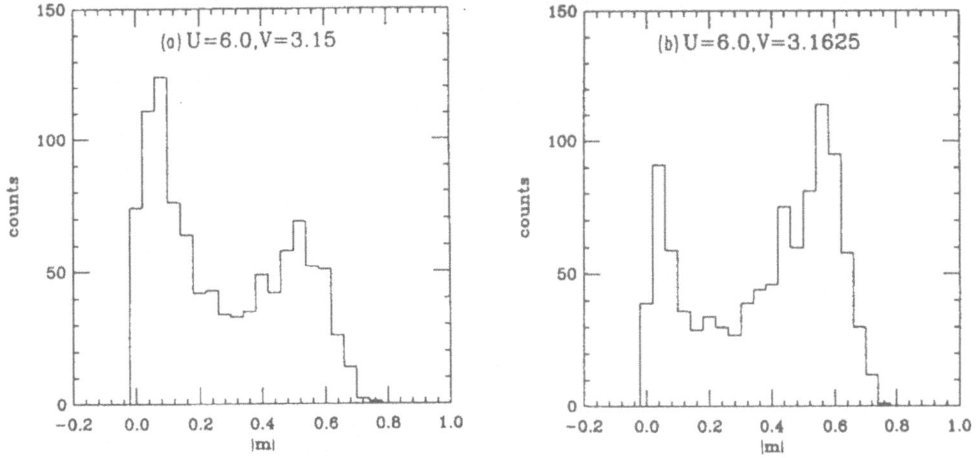

Fig. 3. Histograms of the absolute value of the CDW order parameter for $U = 6$ and two values of V on a 32×32 lattice.

phase, only simulations can go all the way between strong and weak coupling and provide quantitative answers.

b) Condensation transition

As V becomes negative and large, a first order transition occurs where the electrons condense. This is seen, for example, in the CDW structure factor at $q = 0$. As Fig. 4 shows, for $U = 0$, it increases abruptly as V is decreased below -1.15. One can obtain strong coupling estimates for this phase boundary both for $U \gg 0$ and $U \ll 0$. We have found that for $|U| \geqslant 4$ the numerical results join the strong coupling predictions. These results are discussed in Ref. 9.

Figure 5 shows the phase diagram of the half-filled extended Hubbard model that emerges from these calculations. The transition line between

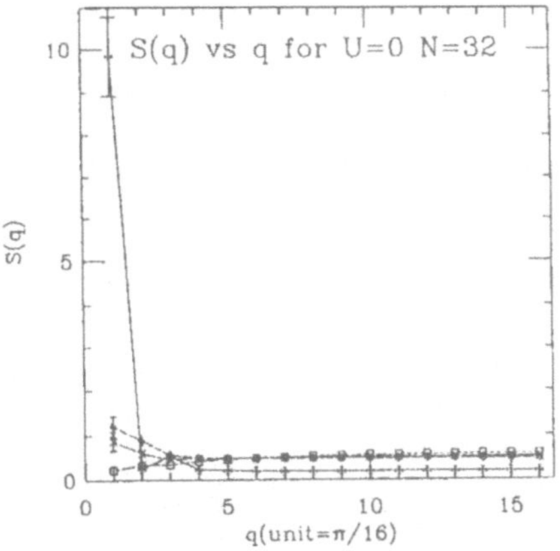

Fig. 4. Structure factor S(q) for a 32-site ring at $U = 0$ and $V = -1.0$ (dotted line and square), $V = -1.1$ (dotted-dashed line and cross), $V = -1.15$ (dashed line and diamond), and $V = -1.2$ (solid line and fancy diamond). The sudden jump at $q = Q_{min}$ for $V = -1.2$ indicates that the transition is first order.

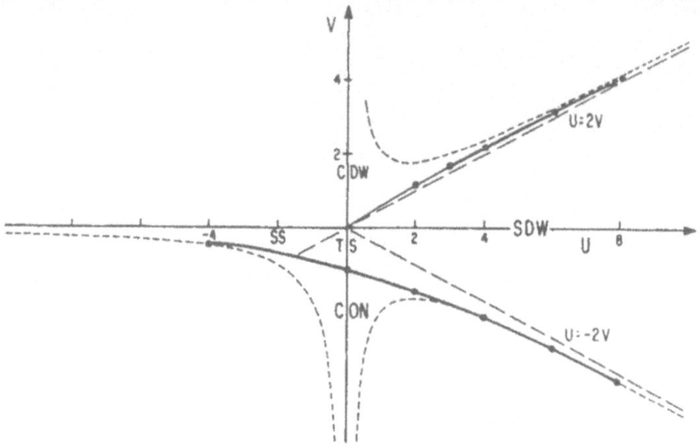

Fig. 5. Ground state phase diagram of the one-dimensional extended Hubbard model in the half-filled band sector. The CDW-SDW transition is continuous for small U and first-order for large U ($U \geqslant 3$), and deviates slightly from the $U = 2V$ line toward the CDW phase. The condensation transition is always first-order. The short-dashed lines are results from various strong coupling expansions.

SS (singlet superconducting) and CDW phases ($V = 0$, $U < 0$) can be obtained from symmetry arguments, so that all the phase boundaries separating phases with long-range order are well established for arbitrary coupling regimes. The dashed lines delimiting singlet and triplet superconducting phases are renormalization group results which should be checked beyond weak coupling by Monte Carlo simulations.

IV. $2p_F$ AND $4p_F$ INSTABILITIES

As is well known, a one-dimensional non-interacting electron system exhibits a $2p_F$ (Peierls) instability. In the presence of strong on-site repulsion, one expects also a $4p_F$ instability, and both instabilities are observed in various charge-transfer compounds by diffuse X-ray scattering experiments.[10] Monte Carlo simulations can shed light on the interplay between $2p_F$ and $4p_F$ instabilities as a function of the Coulomb interaction and band filling.[11, 12] It was found in Ref. 11 that the charge susceptibility

$$N(q) = \frac{1}{N}\int_0^\beta d\tau \sum_{i,j} e^{iq(R_i - R_j)} \langle (n_{i\uparrow}(\tau) + n_{i\downarrow}(\tau))(n_{j\uparrow}(0) + n_{j\downarrow}(0))\rangle \qquad (8)$$

at $q = 4p_F$ is small in the presence of on-site repulsion only and is in fact non-divergent for any $U < \infty$. As shown in Fig. 6, it only diverges logarithmically for $U = \infty$. This indicates that in systems where $4p_F$ peaks are observed, longer-range Coulomb repulsion is important. In particular, Monte Carlo simulations for the 1/4-filled band case show a strong $4p_F$ response when a nearest neighbor repulsion V exists,[11] and it is easy to see that as $U \to \infty$, one will obtain a transition to a $4p_F$ CDW state for $V \geqslant 2t$.

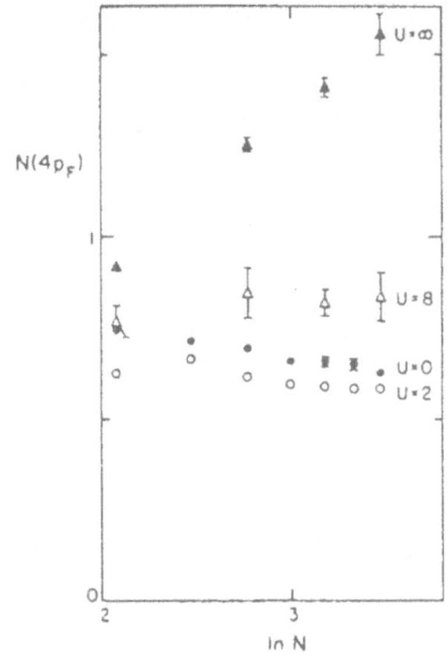

Fig. 6. $N(4p_F)$ vs ln N for various sized lattices with the number of τ slices $L = N$. With $\Delta\tau = 0.25$ one has $\ln N = \ln(4t/kT)$ and one can view the figure as showing how $N(4p_F)$ scales with $\ln(4t/kT)$. For $U = \infty$ we see that $N(4p_F) \sim \ln(4t/kT)$, for finite U these results indicate that $\tilde{N}(4p_F)$ does not diverge as $T \to 0$.

Concering the $2p_F$ instability, as emphasized in Ref. 12, it is important to distinguish between the site CDW susceptibility eq. (8) and the bond CDW susceptibility:

$$\tilde{N}(q) = \frac{1}{N}\int_0^\beta d\tau \sum_{\substack{i,j\\ \sigma}} e^{iq(R_i - R_j)} \langle (c^\dagger_{i+1,\sigma}(\tau)\, c_{i\sigma}(\tau) + h.c.)(c^\dagger_{j+1\sigma}(\tau)\, c_{j\sigma}(\tau) + h.c.)\rangle \tag{9}$$

Although renormalization-group calculations do not distinguish between both susceptibilities, one finds in Monte Carlo simulations that their behavior can be quite different.[12] As an example, Fig. 7 shows N(q) and $\tilde{N}(q)$ for $U = 4$ in the 1/4-filled band case. U has completely suppressed the site $2p_F$ CDW peak while the bond $2p_F$ peak is still large at low temperatures. The reason for this difference is that $\tilde{N}$ is sensitive to the spin-Peierls instability for large U while N is not. Diffuse X-ray scattering experiments will measure N or $\tilde{N}$ depending on whether the phonons that are softening are intra-molecular or inter-molecular modes, respectively.

For non-half-filled band cases, Monte Carlo simulations show that an on-site repulsion suppresses both the $2p_F$ charge and charge transfer peaks.[12] This result, in apparent disagreement with asymptotic RG predictions, can, however, be reconciled with these predictions by numerically integrating the RG equations, as discussed in Ref. 11. For the half-filled-band case, N and $\tilde{N}$ show drastically different behavior. As seen in Fig. 8, U rapidly suppresses $N(2p_F)$ while $\tilde{N}(2p_F)$ is first enhanced by a small U and only suppressed for larger values of U. In the presence

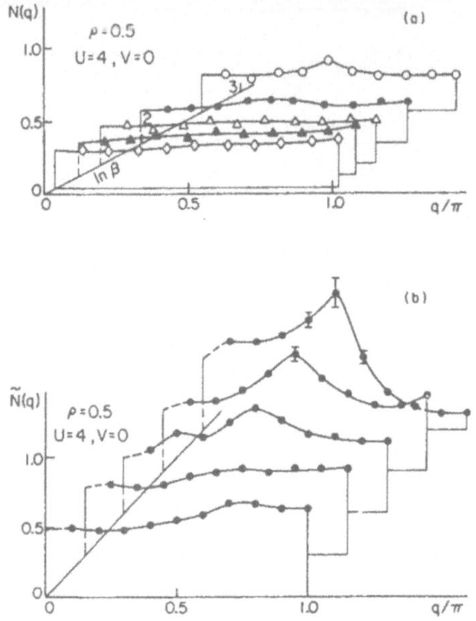

Fig. 7. (a) Charge-density susceptibility N(q) for the $\frac{1}{4}$-filled ρ = 0.5 Hubbard model with U = 4. (b) Charge-transfer susceptibility Ñ(q) for ρ = 0.5 and U = 4.

of a small nearest-neighbor repulsion V, Fig. 8 shows that both $N(2p_F)$ and $\tilde{N}(2p_F)$ are enhanced. These results imply that in a coupled electron-phonon system the Peierls distortion will first increase as U is turned on if the coupling is to inter-molecular modes, and decrease if the coupling is to intra-molecular modes. This is indeed found in numerical simulations of electron-phonon systems as discussed in the next section.

It is interesting to study the behavior of $N(2p_F)$ and $\tilde{N}(2p_F)$ as the temperature is lowered and the lattice size increased. As Fig. 9 shows, the divergence in $\tilde{N}(2p_F)$ turns immediately from logarithmic to linear as U is turned on. For large U the amplitude starts to be suppressed but the linear divergence remains. In fact, as $U \to \infty$, $\tilde{N}(2p_F)$ should diverge in the same way as the four-spin susceptibility that signals the spin-Peierls instability in the equivalent Heisenberg model, which is known to diverge linearly from theoretical arguments[13] and was confirmed by Monte Carlo simulations.[14]

In contrast, the site CDW is exponentially suppressed as U is turned on, as seen in Fig. 9. It is interesting to note that RG calculations that don't distinguish between site and bond CDW's, in fact had predicted a linear divergence of the charge susceptibility in the half-filled band case for U = 0. However, because of the gap in the charge-density excitations

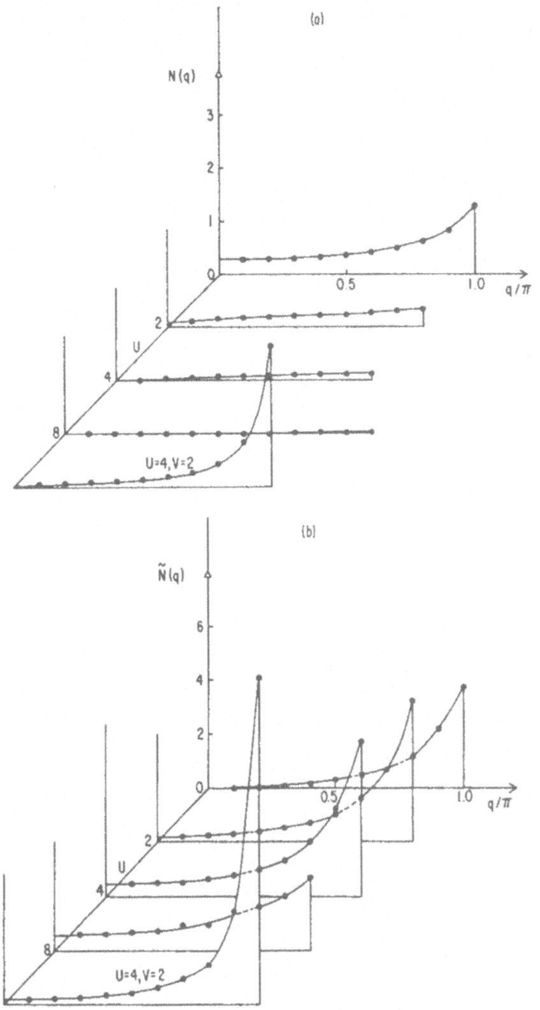

Fig. 8. $\rho = 1$, $\beta = 7.25$. (a) N(q) and (b) $\tilde{N}(q)$ for U = 0, 2, 4, 8 with V = 0 and for U = 4, V = 2.

that develops for U = 0 it had been argued that in fact the CDW susceptibility should be exponentially suppressed.[15] As we find in our simulations, and can also be argued on theoretical grounds,[16] it is the bond CDW that follows the RG predictions and the site CDW is exponentially suppressed due to the gap. Thus the Monte Carlo simulations can be of help in interpreting results of analytic calculations.

It is also of interest to study the behavior of $N(2p_F)$ and $\tilde{N}(2p_F)$ versus U for different temperatures, as shown in Fig. 10. $N(2p_F)$ is rapidly suppressed with U and becomes temperature independent because of the finite gap. $\tilde{N}(2p_F)$ shows a maximum at low temperatures at U ~ bandwidth which shifts to smaller U as T is increased. These results suggest[17] that in the coupled electron-phonon system the maximum dimerization will occur for U ~ bandwidth for small electron-phonon coupling constant λ and shift to smaller U as λ is increased. This behavior is indeed found in Monte Carlo and finite size calculations[18,19] as well as in variational calculations.[20]

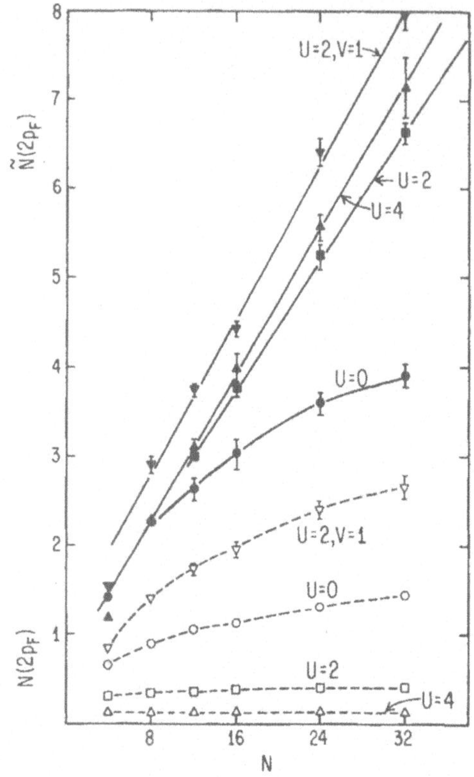

Fig. 9. $\tilde{N}(2p_F)$ (solid symbols, full lines) and $N(2p_F)$ (open symbols, dashed lines) versus lattice size N (with $\beta = N/4$) for U = 0, U = 2, U = 4, and U = 2, V = 1.

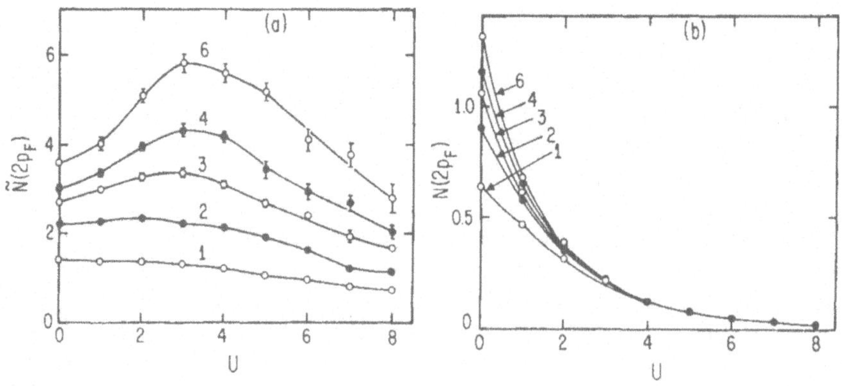

Fig. 10. (a) $\tilde{N}(2p_F)$ and (b) $N(2p_F)$ versus U for β = 1, 2, 3, 4 and 6. Lattice sizes are $N = 4\beta$.

V. ELECTRON-PHONON SYSTEMS

A variety of simulations on electron-phonon models have been performed to study the Peierls instability beyond mean-field theory and the connection between Peierls and spin-Peierls transitions. Two electron-phonon Hamiltonians were studied: The Su-Schrieffer-Heeger (SSH) model:

$$H = \sum_i \left(\frac{p_i^2}{2M} + \frac{1}{2} K(q_{i+1} - q_i)^2 \right) + \sum_{i,\sigma} (t - \lambda(q_{i+1} - q_i))(c_{i\sigma}^\dagger c_{i+1\sigma} + h.c.) \tag{10}$$

describing inter-molecular vibrations and the molecular crystal model for intra-molecular vibrations:

$$H = \sum_i \left(\frac{p_i^2}{2M} + \frac{1}{2} K q_i^2 \right) - t \sum_{i,\sigma} (c_{i\sigma}^\dagger c_{i+1\sigma} + h.c.) + \lambda \sum_{i,\sigma} q_i n_{i\sigma} \tag{11}$$

both with and without added Coulomb interaction terms:

$$H_1 = U \sum_i n_{i\uparrow} n_{i\downarrow} + V \sum_i n_i n_{i+1} \quad . \tag{12}$$

In the limit of infinite ionic mass, mean field theory provides an exact solution for these models in the absence of Coulomb interactions. For the half-filled band case, the systems undergo a Peierls transition to a dimerized state for arbitrary electron-phonon coupling λ. The same is true for 1/4-filled band systems in the limit $U \to \infty$ (since they are equivalent to non-interacting spinless fermions).

For finite ionic mass, one might expect that the long-range order would be destroyed if the $2p_F$ phonon frequency is larger than the dimerization gap. It was found by simulations, however, that the Peierls state is stable even for $M \to 0$ in the half-filled band case.[21] In the quarter-filled band case with large U (half-filled band spinless) quantum fluctuations do destroy the dimerized state for λ smaller than a critical value λ_c which is a function of the phonon frequency.[21]

A question that has generated great interest is the effect of Coulomb interactions on the Peierls state in the SSH model. Hartree-Fock calculations predict that the dimerization is independent of U up to a critical value U_c, where a transition to a non-dimerized spin-density-wave state occurs.[22,23] Perturbation theory in U, however, predicted that the dimerization is increased when U is turned on.[24,23] As we discussed in the previous section, the behavior of the charge transfer susceptibility $N(2p_F)$ as a function of U suggests that the dimerization will be first enhanced and then suppressed with U. This is found in simulations of the SSH model, as shown in Fig. 11.

For large U, the electronic model will become equivalent to an anti-ferromagnetic Heisenberg model coupled to inter-molecular phonons that

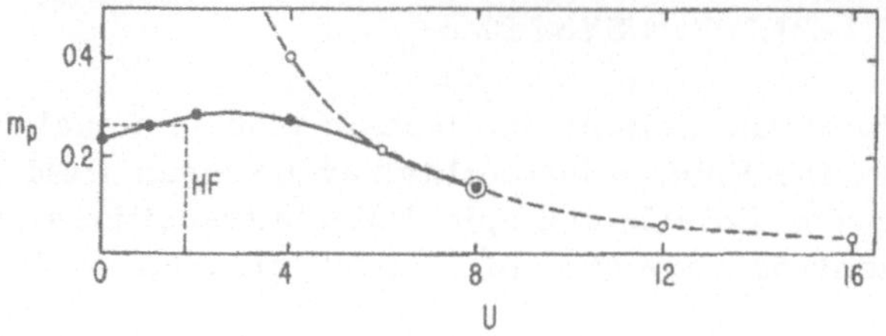

Fig. 11. Lattice distortion m_p versus U for SSH model, t = 1, $\lambda = 0.25$, K = 0.25, $\omega = 2(K/M)^{1/2} = 0.066$. The dashed line labeled HF is the Hartree-Fock prediction. The open circles are obtained from simulations of the equivalent Heisenberg Hamiltonian.

modify the effective exchange coupling:

$$H_{eff} = 2 \sum_i J_{i,i+1} \vec{S}_i \cdot \vec{S}_{i+1} \tag{13}$$

$$J_{i,i+1} = \frac{4t^2}{U} \left(1 - \frac{2\lambda}{t} (q_{i+1} - q_i) \right) , \tag{14}$$

which will undergo a spin-Peierls transition. This model can also be directly simulated by our Monte Carlo technique. Figure 12 shows the spin-spin correlation function in the SSH model for several values of U and in the corresponding limiting spin model. It can be seen that for U = 8 the correlations in the SSH model are already very close to those in the spin model.

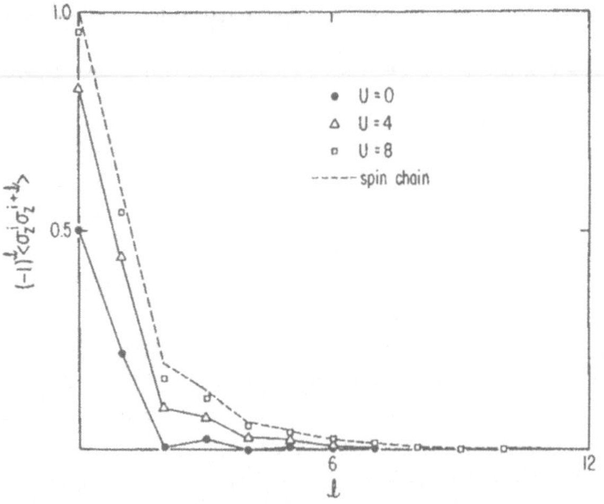

Fig. 12. Staggered spin-spin correlation function versus distance. The dashed line shows the results obtained for the equivalent spin chain as $U \rightarrow \infty$.

Since in the Monte Carlo simulation a smaller $\Delta\tau$ (and correspondingly more time slices) has to be taken when U increases, it becomes difficult to simulate the SSH model directly for large U. As Fig. 11 shows, however, the simulation results for the dimerization from the SSH model match the simulation results for the strong coupling Hamiltonian Eq. (13) for $U=6$ and $U=8$. Similarly, the correlation functions were found to match in Fig. 12. Thus, the combination of the results for the SSH Hamiltonian and the antiferromagnetic Heisenberg chain Hamiltonian Eq. (13) provide a complete description of the Peierls-spin-Peierls transition from $U=0$ to $U=\infty$. No analytic method is known that can provide such a global description, although a variational calculation[20] appears to be rather successful up to $U \sim 4t$.

To summarize, our results show that there is a continuous cross-over between Peierls and spin-Peierls regimes, i.e. a "generalized Peierls transition" encompassing both regimes. Because the charge transfer susceptibility $\tilde{N}(2p_F)$ changes from logarithmically divergent to linearly divergent for any non-zero U, strictly speaking one is in the spin-Peierls regime for any non-zero U. Investigation of this cross-over for other band fillings should be of interest.

The effect of a nearest-neighbor repulsion was also studied[18] and it was found that it enhances the dimerization further for $V \leqslant U/2$. The divergence of $\tilde{N}(2p_F)$ for $U > 0$ and non-zero V is also found to be linear up to $V \leqslant U/2$.[25] For $V > U/2$, a transition to a site CDW occurs.

Independent numerical studies on the question of the effect of U and V on the Peierls distortion have been performed by Mazumdar and Dixit[19] (exact diagonalization) and Campbell, de Grand and Mazumdar[26] (a different Monte Carlo method), with results that are in agreement with the ones discussed here. The method discussed here, however, appears to be the best suited to study electron-phonon systems in the presence of phonon dynamics. Other Monte Carlo work on the SSH model has dealt with properties of solitons and we refer the reader to refs. 26 and 27 for a detailed discussion.

For the molecular crystal model, Eq. (10), the effect of Coulomb interactions on the Peierls distortion has also been studied by simulations.[7] The results are essentially in agreement in the mean-field predictions and show that U rapidly suppresses the CDW while V enhances it. The transition changes from continuous to first order in a similar way as in the purely electronic model (Sec. II). In general, it appears that the intermolecular Peierls distortion is more robust than the intra-molecular one since it is less affected by finite phonon frequency and on-site Coulomb repulsion, so that we expect it to be the dominant one in most systems.

VI. OTHER SIMULATION STUDIES

A detailed study of a model for a one-dimensional excitonic superconductor of the type proposed by Little[28] has been performed.[29]. The results show that such a model is very unlikely to display dominant superconducting correlations. The main reason, which actually can be found both in weak and strong coupling perturbation theory,[29] is that <u>retardation</u>

(i.e. finite exciton frequency) rapidly suppresses pairing while it enhances CDW formation. Our simulations for a wide range of parameters show that the dominance of CDW in one dimension is inescapable except for extreme parameters. Simulations for a two-dimensional model show much more encouraging results since the CDW instability disappears in the absence of nesting. In addition, if the Fermi energy is close to the logarithmic singularity in the density of states, superconducting pairing is strongly enhanced.[30]

The effect of randomness in one-dimensional systems is an area that remains wide open for numerical simulations. Recently, simulation studies of a random antiferromagnetic Heisenberg chain were performed,[14,31] which confirmed the expected power-law divergence of the susceptibility induced by randomness that had been found in renormalization-group studies.[32] Studies of the effect of randomness on the Peierls instability and in pairing correlations are in progress.

Finally, it is important to go beyond the single-chain problem and study systems of coupled chains. The first step in that direction has been a recent study of the CDW transition in a coupled chain system of spinless fermions.[33] It is simple from a technical point of view to study several chains if the interactions between chains are only density-density or spin-spin interactions, but difficulties with the technique discussed here appear in the presence of single particle or pair transfer between chains. In that case, other simulation techniques have to be used.[34] Nevertheless, there are many interesting questions, for example, regarding $2p_F$ and $4p_F$ instabilities in weakly coupled chains that could be addressed using the method discussed here.

VII. CONCLUSIONS AND PROSPECTS

As we have tried to show in this review, quantum Monte Carlo simulations can provide answers on properties of models for low-dimensional conductors that cannot be obtained by analytic techniques. The field is still in its early stages, and although several different systems have been studied there is still much to be learned by more detailed studies of those systems and on more realistic models for low-dimensional conductors.

Concerning the extended Hubbard model, more detailed studies need to be performed on the dependence of correlation exponents on coupling constants for different band fillings. Recent analytic predictions[35] on the behavior of these exponents for arbitrary values of U should be checked numerically, and extended to study the effect of V. The addition of next-nearest-neighbor interactions should bring in interesting new phenomena[36] and it is straightforward to include these couplings in the simulation program. For electron-phonon systems, there is a wide region in the parameter space of band filling, Coulomb interactions and phonon frequency that remains unexplored.

As discussed earlier, models including randomness in the couplings should be studied since in many of the materials of interest, this is thought to be an important factor. Although it is straightforward to include random couplings in the simulation program, the questions one wants to address should be very well defined since these simulations are likely to be very time consuming. With increasing computational power available, this area is likely to become important in the future.

Finally, one should attempt to study more realistic models in quasi-one-dimensional conductors by simulating coupled chains with electron-electron and electron-phonon interactions and possibly even randomness. This appears to be within reach of present algorithms. We believe it is likely that numerical simulations will become an increasingly important tool in the study of collective phenomena in low-dimensional quantum systems in the future.

ACKNOWLEDGMENTS

I am grateful to D. J. Scalapino for collaborations and stimulating discussions on many of the problems discussed here. This work was supported by the National Science Foundation under grant No. NSF-DMR-85-17756. I am grateful to Exxon Corporation, Cray Research and Ridge Computers for their support.

REFERENCES

1. V. J. Emery, in "Highly Conducting One-Dimensional Solids," ed. by J. Devreese, R. Evrand and V. van Doren (Plenum, New York, 1979), and references therein.
2. J. Solyom, Adv. Phys. 21, 201 (1979) and references therein.
3. J. E. Hirsch, D. Scalapino, R. L. Sugar and R. Blankenbeckler, Phys. Rev. B26, 5033 (1982).
4. J. E. Hirsch, in "Monte Carlo Methods in Quantum Problems," ed. by M. H. Kalos (D. Reidel Publ. Co., Dordrecht, 1984) contains a review of some of the early applications of the method.
5. R. L. Sugar, in Ref. 4.
6. J. E. Hirsch, Phys. Rev. Lett. 53, 2327 (1984).
7. J. E. Hirsch, Phys. Rev. B31, 6022 (1985).
8. F. D. M. Haldane, private communication.
9. H. Q. Lin and J. E. Hirsch, Phys. Rev. B33, 8155 (1986).
10. J. P. Pouget, these proceedings.
11. J. E. Hirsch and D. J. Scalapino, Phys. Rev. B27, 7169 (1983).
12. J. E. Hirsch and D. J. Scalapino, Phys. Rev. B29, 5554 (1984).
13. M. C. Cross and D. S. Fisher, Phys. Rev. B19, 402 (1979).
14. J. E. Hirsch and R. Kariotis, Phys. Rev. B32, 7320 (1985).
15. Ref. 1, Appendix C.
16. H. J. Schultz, private communication.
17. C. Bourbonnaise and J. Voit, private communication.
18. J. E. Hirsch, Phys. Rev. Lett. 51, 296 (1983).
19. S. Mazumdar and S. N. Dixit, Phys. Rev. Lett. 51, 292 (1983).
20. D. Baeriswyl and K. Maki, Phys. Rev. B28, 2065 (1983).
21. E. Fradkin and J. E. Hirsch, Phys. Rev. B27, 1680 (1983); Phys. Rev. B27, 4302 (1983).
22. K. R. Subbaswamy and M. Grabowski, Phys. Rev. B24, 2168 (1981).

23. S. Kivelson and D. Heim, Phys. Rev. B27, 4278 (1982).
24. P. Horsch, Phys. Rev. B24, 7351 (1981).
25. J. E. Hirsch, unpublished.
26. D. W. Campbell, T. A. de Grand and S. Mazumdar, Phys. Rev. Lett. 52, 1717 (1984); Jour. Stat. Phys. 43, 803 (1986).
27. J. E. Hirsch and M. Grabowski, Phys. Rev. Lett. 52, 1713 (1984).
28. W. A. Little, Phys. Rev. 134, A1416 (1964).
29. J. E. Hirsch and D. J. Scalapino, Phys. Rev. B32, 117 (1985).
30. J. E. Hirsch and D. J. Scalapino, Phys. Rev. Lett. 56, 2732 (1986).
31. H. B. Schüttler, D. J. Scalapino and P. M. Grant, Phys. Rev. B (to be published).
32. C. Dasgupta and S. Ma, Phys. Rev. B22, 1305 (1980); J. E. Hirsch and J. V. Jose, Phys. Rev. B22, 5339 (1980).
33. D. J. Scalapino, R. L. Sugar and W. D. Toussaint, Phys. Rev. B (to be published).
34. R. Blankenbecler, D. J. Scalapino and R. L. Sugar, Phys. Rev. B24, 2278 (1981).
35. H. J. Schultz, private communication.
36. V. J. Emery, private communication.

THE PEIERLS TRANSITION SPLIT BY CORRELATION:

THE ELECTRONIC ($4K_F$) AND THE SPIN ($2K_F$) TRANSITION

Jan Kommandeur

Laboratory for Physical Chemistry
University of Groningen
Nijenborgh 16
9747 AG Groningen, the Netherlands

INTRODUCTION

Many years ago, R.E. Peierls in his book on the quantum theory of solids [2] proposed that a one-dimensional metallic system would at some temperature show an inherent thermodynamic instability. Peierls considered a half-filled band, one electron on every atomic (or molecular) site, and he showed that going to lower temperatures one would expect an initially equidistant one-dimensional lattice to dimerize.

Probably because the idea was only publised in a textbook, the transition was rediscovered a number of time [2,3], but it took until the advent of one-dimensional conductors, before clear experimental evidence for such phase transitions could be presented [4]. Even then it was not realized that one had a Peierls transition at hand, but as the understanding of the field grew, the Peierls transition and its application to the one-dimensional organic conductors found its proper place.

A real problem in the beginning of the field was the fact that the half-filled band systems one worked with were always semi-conductors, at temperatures above the phase-transition, they were not metals as Peierls had supposed. Therefore, Beni and Pincus [5] proposed that one-dimensional spin systems, with a spin on each site would show the same behavior: at high temperature one would have a regular chain at low temperature it would be dimerized. Things looked fine now: metals would of course constitute the completely delocalized mode, spin systems would be the completely localized model, it seemed safe to assume that the transition would occur in all systems with one electron per site (half filled bands), independent of their localization.

In the metallic cases, the band is filled up to k_F, the Fermi wave vector. In the half-filled band case $k_F = \pi/2a$, where a is the lattice parameter. A dimerization has wavelength $\lambda = 2a$, the associated momentum is $k = 2\pi/\lambda = \pi/a$, i.e. the distortion has the momentum $2k_F$. The Peierls transition therefore became called the $2k_F$-transition.

As is common in science the beautifully constructed apple cart for the delocalized and localized models was upset by the first reports of a $4k_F$ diffuse scattering, as observed in systems that were not half-filled bands [6]. Usually, a phase-transition has a "precursor", i.e. a dynamic instability, which has the same momentum as the distortion of the lower temperature phase. This dynamic instability, which leads to diffuse scattering can be observed at temperatures well above the temperature of the actual phase transition. This is particularly so in one-dimensional systems, since a well-known theorem of Landau and Lifshitz [7] states that in one dimension no phase transitions can occur. In the TTF/TCNQ case, which was the material in which Pouget et al. [6] first observed the $4k_F$ instability, the $4k_F$-transition indeed never comes about, the $2k_F$-transition is the one that establishes itself at lower temperatures.

When we consider a quarter-filled band, $k_F = \pi/4a$, $2k_F = \pi/2a$ with $\lambda = 4a$ and $4k_F = \pi/a$ with $\lambda = 2a$. If the two instabilities lead to static deformations, we would expect them to amount to a dimerization ($4k_F$) and a tetramerization ($2k_F$) of the one-dimensional lattice. Such a relatively simple case in found in MEM(TCNQ)$_2$, where MEM stands for methyl-ethyl-morpholinium. We will use this material as an experimental illustration of the $4k_F$ and $2k_F$ phenomena.

After shortly reviewing the experimental evidence for the two instabilities, we will discuss their theoretical interpretation. We will argue that the Peierls transition is split by the effects of correlation, into an electronic ($4k_F$) and a spin ($2k_F$) part. We will bolster up this assertion by reviewing the results of a computer calculation on finite ring systems. Then we will discuss how a Peierls transition may be avoided, if such is desirable and finally we will try to derive some pleasure from the "inverse Peierls transition".

METHYL-ETHYL-MORPHOLINIUM (TCNQ)$_2$

There are three crystal phases of this compound. At T > 335 K it is an (almost) regular chain. In the range 335 K > T > 19 K it is a dimerized chain [8], and at T < 19 K it is a very slightly tetramerized chain [9]. The crystal structures of the chains are illustred in figures 1a and 1b. In the 335 K ($4k_F$) transition, the conductivity shows a sharp drop (see fig. 2), the magnetic susceptibility remaining unaffected.

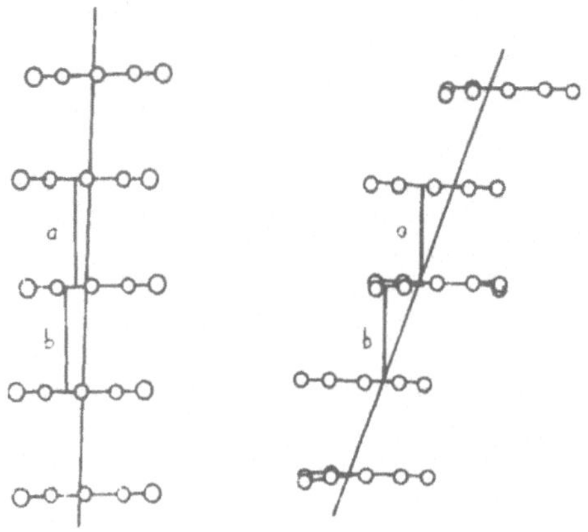

Fig. 1a. High temperature (left) and low temperature (right) structure of the TCNQ chain in MEM (TCNQ)$_2$.

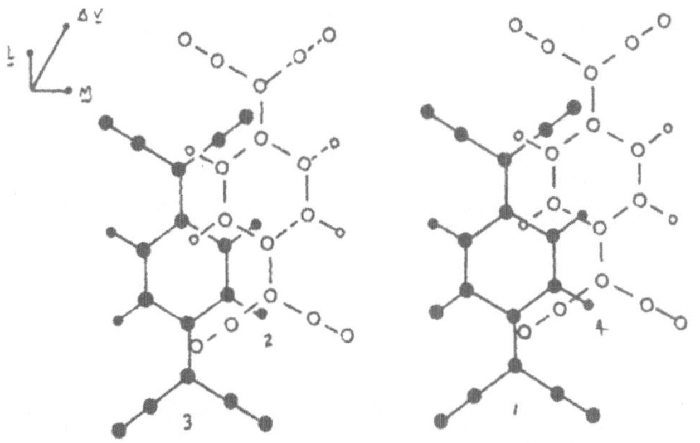

Fig. 1b. Close inspection shown that the overlaps between the dimers in MEM $(TCNQ)_2$ differ at temperatures below 19 K.

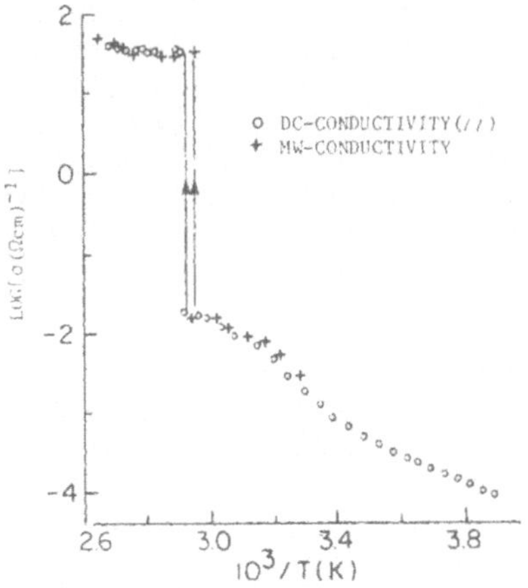

Fig. 2. The DC and micro wave conductivity of MEM $(TCNQ)_2$ as a function of temperature.

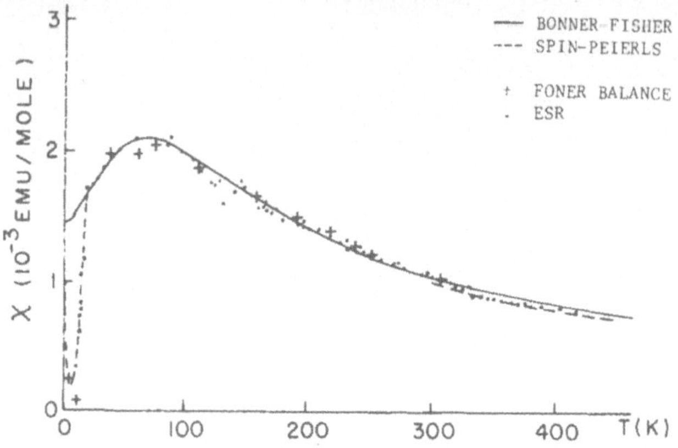

Fig. 3. The magnetic spin susceptibility of MEM $(TCNQ)_2$ as a function of temperature. Note the sharp drop at 19 K.

In the 19 K ($2k_F$) transition the magnetic susceptibility shows a sharp drop (10) (see fig. 3). How can we explain these phenomena ?

INTERPRETATION OF THE TWO TRANSITIONS

If there were no electron-electron interaction, there would only be a $2k_F$-transition [1], which for a quarter-filled band would amount to a tetramerization. Four sites would share tow electrons, which means that every site would be partially doubly occupied. Electron-electron interaction leads to avoidance of such double occupancy. Therefore, it leads to the sharing of one electron by two sites. The bond formed between these two sites can be compared to the one-electron bond in H_2^+. Since the sites are now bounded in pairs, the system dimerizes ($4k_F$). At lower temperature, when smaller energetic considerations play a role, the unpaired spins on the dimers prefer to form singlet states by pairing, because these singlet states have lower energy (by 2J, the exchange energy). Thus, the dimers will dimerize and form a tetramerized lattice ($2k_F$).

In the dimerization the electrons lose (part of) their distributive entropy over the whole chain, in the tetramerization they lose their spin entropy. At high temperatures the entropy plays a dominating role and the system is regular and a metal. At in-between temperatures it gives up the distributive entropy, but retains the spin entropy, at very low temperatures it gives up the spin entropy. Depending on the magnitudes of these quantities, we therefore have two phase transitions, and if the transitions themselves do not come about two diffuse scatterings: $4k_F$ and $2k_F$.

Since in the $4k_F$ transition the electronic degree of freedom is lost, the metal will turn into a semiconductor as is observed, and since in the $2k_F$-transition the spins are lost and therefore the magnetic susceptibility disappears, it seems appropriate to call the $4k_F$-transition the <u>electronic</u> Peierls-transition and the $2k_F$-transition the <u>spin</u> Peierls transition. The separation of these two is caused by the electron-electron correlation.

How big should the electron-electron correlation be to cause this splitting ? This is the point at which numerical work became necessary. We discuss it in the next section.

Electrion electron interactions can be taken into account by the use of the Hubbard hamiltonian [11]. Unfortunately, there are no analytical solutions of this hamiltonian for an infinite chain with more than two electrons. Therefore, recourse was sought in numerical calculations on a finite (ring) system. First on a ring with 4 sites [12], but the same results were obtained on an 8 membered ring [13]. The procedure was simple but required a lot of computer time. the input parameters were a transfer integral t, exponentially dependent on the distance between the sites; an energy U which represented the energy of a doubly occupied state and a repulsive term of the form Br^{-12}, where r was the distance between the sites.

The ring was allowed to distort according to the $4k_F$ and $(2k_F + 4k_F)$ distortions (see fig. 4). At points along the path the 1860 electronic states E_i were calculated, from these the electronic partition function $Z_{el} = \Sigma_1^i \exp -E_i/kT$ and the free energy $F = kT\ln Z_{el} + E_{rep}$, where E_{rep} represents the repulsive term. We thus obtain for each temperature $F = F(\xi, \eta)$, where ξ and η represent the $4k_F$ and $2k_F$ distortion parameters, and its minimum can be found. Sharp phase transition temperatures were obtained as a function of the repulsive energy U. The results are displayed in fig. 5, which also compares with the results of the 4 site calculation. They both lead to the same conclusion: at U = 4t, where the repulsive energy equals the total bendwidth of the metallic system, the $(2k_F + 4k_F)$ transition is split into the $4k_F$ and $2k_F$ part.

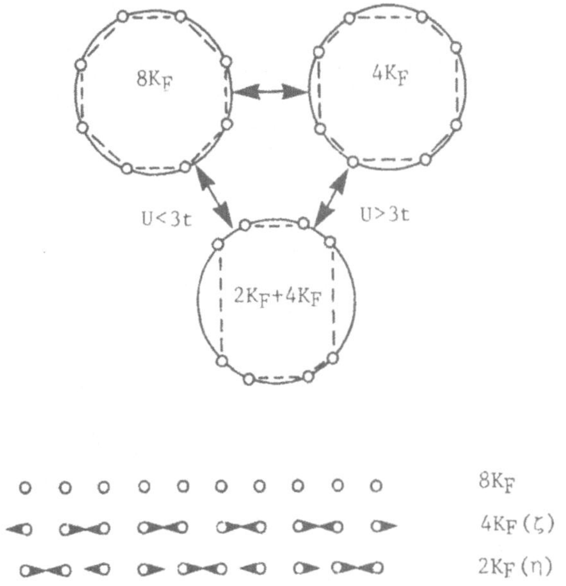

Fig. 4. Conformations of the 8-membered ring in the regular ($8k_F$), the dimerized ($4k_F$) and the tetramerized ($2k_F$) phase with the corresponding lattice deformations in a one-dimensional chain.

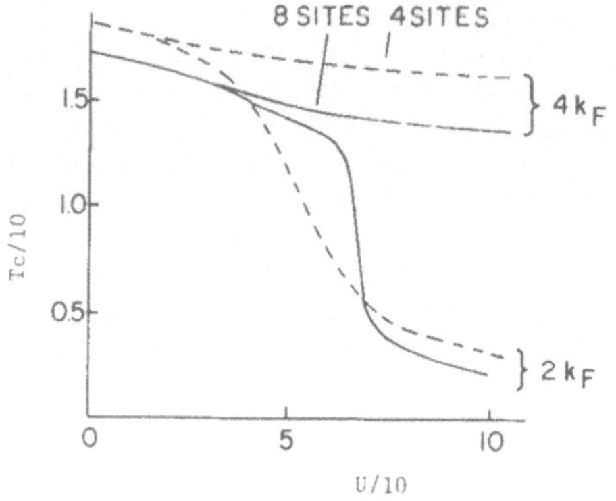

Fig. 5. Reduced phase transition temperatures as a function of the reduced correlation energy for calculations with 4 and 8 sites.

We note here that the phase with the dimerized dimers has indeed the symmetry ($2k_F + 4k_F$). In the 4 site calculation we also found a purely $2k_F$ phase [12], which occurs by itself only at U = 0 corresponds to the clas-classical Peierls case.

ONLY A FREE CHAIN LEADS ITS OWN LIFE

Of course, TCNQ chains as most other one dimensional chains are embedded in a crystal lattice. In almost all considerations of the Peierls transition the effect of the embsedding lattice is completely neglected, or it is taken as the dominant cause for a distortion.

In MEM$(TCNQ)_2$ we are in the lucky situation that there is a little extra space in the lattice as shown by the considerable dynamics of the MEM ions [14]. This may well be the reason that the $2k_F$ and $4k_F$ phase transitions are so nicely expressed in this compound. In any case, the dynamics and thus the "free" space can be removed by substituting thiomorpholinium for morpholinium, in other words replacing an O by S[15]. Within experimental error the volume of the unit cell and the lattice parameters do not change over the full range of the substitution, even though the space required for S is about 15 A^3 larger than that for O.

The $4k_F$-transition temperature as monitored by the drop in conductivity, does change from 335 K in the pure oxygen crystal to 230 K in the crystal containing less than 20 % of the sulphur compound (see fig. 6.) At higher concentrations it appears to be absent. At the same time the dynamics of the MEM ions disappears.

All this is strongly indicative of the $4k_F$ Peierls transition, which comes largely about through lateral motion (see fig. 1a), being impeded by the lack of free space in the lattice. Packing the crystals hard may be a way of preventing Peierls transitions, at least in TCNQ compounds !

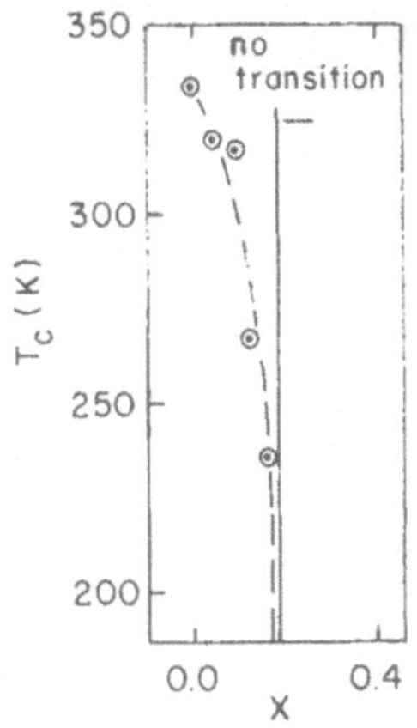

Fig. 6. The $4k_F$-transition temperature as a function of the molar fraction x of thiomorpholinium in MEM $(TCNQ)_2$.

THE "SLREIEP" TRANSITION

After all this serious discussion of the Peierls transition it is fun to consider its inverse: the Slreiep transiton. Fig. 7 shows the behavior of the conductivity as a function of temperature in dimethylthimorpholinium $(TCNQ)_2$ [16]. It is a semiconductor at high, but a metal at low temperature ! This is clearly the upside down world.

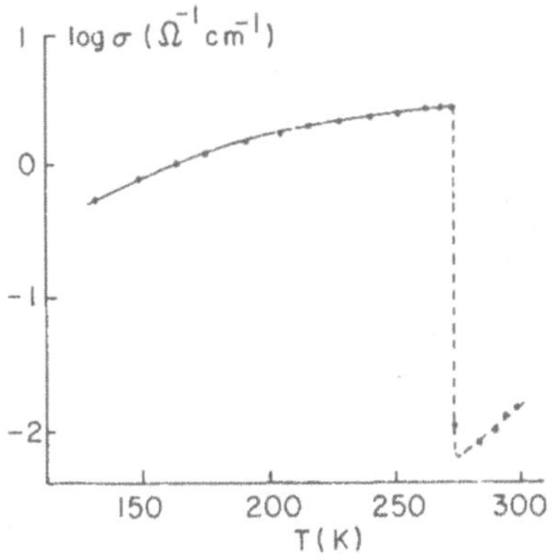

Fig. 7. The conductivity as a function of temperature in DMTM $(TCNQ)_2$.

The explanation appears to be the following. At high temperature the two different TCNQ chains are crystalographically connected by a symmetry element. This means each gets half of the available electrons, each of them will be exactly quarter-filled.

At the phase transition (at T ≃ 272 K) the crystal deforms and the two chains become inequivalent in the low temperature phase. They will not be at the same crystal potential any more and in all likelyhood charge will flow from one to the other. They are now ($\frac{1}{4}$ + δ) or ($\frac{1}{4}$ - δ) filled. The dimerization does not impede the flow of the excess electrons and therefore the system exhibits metallic behavior. Depending on δ, one would expect another, possibly incommensurate phase transition at much lower temperature, but such has as yet not be found.

SUMMARY

It can be reliably argued that the $4k_F$ transition or its dynamic equivalent is due to an electron correlation splitting of the ($2k_F$ + $4k_F$) transition. We find the $4k_F$ component at higher and the $2k_F$ at lower temperatures.

The effect of the crystal field was discussed. In MEM$(TCNQ)_2$ it is clear that the loosely packed crystal lets the $4k_F$ transition, involving lateral motion, occur. It can be impeded by filling up the free space. The crystal field may also lead to a slight disproportionation of two different TCNQ chains, which gives rise to the inverse Peierls (or Slreiep) transition.

REFERENCES

1. R.E. Peierls, Quantum Theory of Solids (Oxford University Press, London, 1955).
2. D. Alder and H. Brooks, Phys. Rev. 155, 826 (1967).
3. J.J. Hallers and G. Vertogen, Phys. Rev. Lett. 28, 563 (1972).
4. J.G. Vegter, T. Hibma and J. Kommandeur, Chem. Phys. Lett. 3, 427 (1969).
5. G. Beni and P. Pincus, J. Chem. Phys. 57, 3531 (1972).
6. J.P. Pouget, S.K. Khanna, F. Denoyer, R. Comes, A.F. Garito and A.J. Heeger, Phys. Rev. Lett. 37, 437 (1976).
7. L.D. Landau and H.M. Lifshitz, Statistical Physics (Pergamon Press, Oxford, 1959) Ch. XI.
8. A. Bosch and B. Van Bodegom, Acta Cryst. Sect B 33, 3013 (1977).
9. R.J.J. Visser, S. Oostra, C. Vettier and J. Voiron, Phys. Rev. B 28, 2074 (1983).
10. S. Huizinga, J. Kommandeur, G.A. Sawatzky, B.T. Thole, K. Kopinga, W.M. de Jonge and J. Roos, Phys. Rev. B 19, 4723 (1979).
11. J. Hubbard, Proc. Roy. Soc. London A276, 238 (1963) and ibid. A281, 401 (1964).
12. S. Huizinga, J. Kommandeur, H.T. Jonkman and C. Haas, Phys. Rev. B 25, 1717 (1982).
13. H.T. Jonkman, H.J. Zwinderman and J. Kommandeur, J. de Physique, Tome 44, C3-1281 (1983).
14. S. Huizinga, Ph. D. Thesis, Croningen, 1980.
15. S. Van Smaalen, J.L. de Boer, J. Kommandeur and G.J. Kramer, Mol. Cryst. Liq. Cryst. 120, 173 (1985).
16. R.J.J. Visser, S. Van Smaalen, J.L. de Boer and A. Vos, Mol. Cryst. Liq. Cryst. 120, 167 (1985).

THE CROSSOVER FROM ONE TO THREE DIMENSIONS: PEIERLS AND SPIN-PEIERLS INSTABILITIES

H.J. Schulz

Laboratoire de Physique des Solides
Université Paris-Sud
91405 Orsay, France

1 INTRODUCTION

Compared to three-dimensional metals, quasi-one-dimensional conductors present two fundamental and apparently conflicting peculiarities: on the one hand, a one-dimensional electron gas is unstable against the formation of charge- or spin-density waves [1,2,3], in addition to the usual Cooper pairing instability [4] (which is present in any dimension); on the other hand, strong thermodynamic fluctuations forbid symmetry breaking phase transitions in a strictly one-dimensional system [5], and therefore in order to observe symmetry breaking a (weak) three-dimensional coupling (which is always present in a real quasi-one-dimensional system) is absolutely essential. Experimentally, both a variety of different phase transitions and strong fluctuation effects are often observed [6,7]. In order to understand these results, one needs to know the one-dimensional effects and the way a system crosses over from one-dimensional properties at high temperature to three-dimensional behaviour in the vicinity of its phase transition. The present paper is an attempt to clarify this crossover using the relatively simple example of the Peierls instability. In the following chapter, the mean-field theory of the Peierls transition is described. Chap. 3 demonstrates how one-dimensional fluctuation effects destroy the symmetry-breaking found in the mean-field approximation. In chap. 4 the establishment of three-dimensional order in a system of coupled chains (quasi-one-dimensional system) is discussed. In chap. 5 analogous considerations for the spin-Peierls transition occuring in strongly correlated systems are made. Some of these points have been discussed previously in papers by Barišić [8] and Dieterich [9]. Throughout the paper, units are chosen so that $\hbar = k_B = 1$.

2 MEAN-FIELD THEORY OF THE PEIERLS TRANSITION

One of the fundamental peculiarities of a one-dimensional electron gas is its low-temperature instability against the spontaneous modulation of the charge-density ("charge-density wave", CDW) and of the lattice periodicity (Peierls instability [1,2]). The basic mechanism is easy to understand. Consider a one-dimensional metal, all states with $k \leq k_F$ being occupied (fig.1). A modulation of the underlying lattice with wavevector $2k_F$ via Bragg reflection mixes states close to k_F with states close to $-k_F$, and in particular leads to the appearance of a gap $2\mid\Delta\mid$ in the single-particle spectrum at the Fermi level. Consequently, the energy of all the occupied states is lowered, and the total energy gain of the electrons is proportional to $-\mid\Delta\mid^2 \ln\mid\Delta\mid$. On the other hand, the elastic energy of the lattice increases by an amount

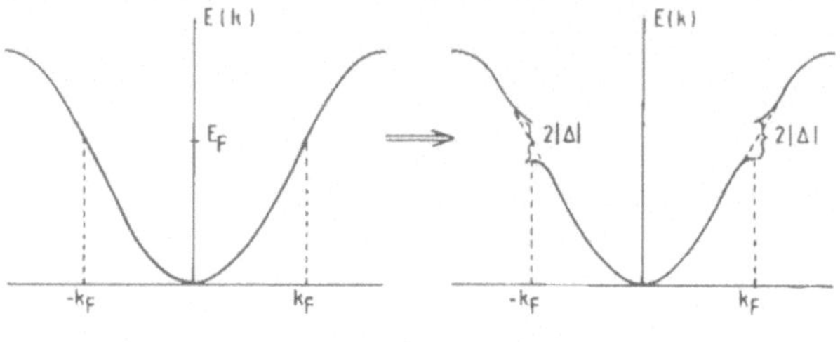

Figure 1: Electron dispersion of the undistorted (left) and of the Peierls modulated (right) one-dimensional electron-phonon system. States with $|k| \leq k_F$ gain energy by the distortion.

proportional to $|\Delta|^2$. However, due to the logarithmic factor in the electronic contribution, the modulated state with a nonzero Δ is always energetically favourable.

To make these arguments more quantitative, consider the standard microscopic model of a one-dimensional electron gas interacting with a deformable lattice. The Hamiltonian is

$$H = \sum_{k\sigma} \varepsilon_k a^\dagger_{k\sigma} a_{k\sigma} + \sum_q \omega_q b^\dagger_q b_q + \frac{g}{\sqrt{L}} \sum_{kq\sigma} a^\dagger_{k+q,\sigma} a_{k\sigma} (b_q + b^\dagger_{-q}) \quad . \tag{2.1}$$

Here $a_{k\sigma}$ is the destruction operator for an electron in Bloch state k with spin σ and energy ε_k ($\varepsilon_{k_F} = 0$), b_q is the destruction operator for a phonon of wavenumber q and energy ω_q, and L is the length of the system. The first two terms describe the non-interacting electron and phonon systems respectively. In the third (interaction-) term, the electron-phonon coupling constant generally depends both on k and on q, however in the following electrons with $|k| \approx k_F$ and phonons with $|q| \approx 2k_F$ are most important, so that the wavenumber dependence of g can be neglected.

For coupling to optical or intramolecular phonons the lattice displacement field u is given in terms of the phonon operators as

$$u(x) = \sum_q \frac{1}{\sqrt{2L\rho\omega_q}} (b_q + b^\dagger_{-q}) e^{iqx} \quad , \tag{2.2}$$

where ρ is the ionic mass density. In the case of acoustic phonons, the electrons are coupled to the gradient of the displacement field, which leads only to minor changes in what follows. A static lattice modulation of finite amplitude and wavenumber $2k_F$ is described by $\langle b_{2k_F} \rangle = \langle b^\dagger_{-2k_F} \rangle \propto \sqrt{L}$, i.e. phonon modes with wavenumber $\pm 2k_F$ are macroscopically occupied. We now introduce the complex "order parameter"

$$\Delta = |\Delta| e^{i\phi} = \frac{2g\langle b_{2k_F} \rangle}{\sqrt{L}} \tag{2.3}$$

and neglect for the moment fluctuations of the lattice displacement field around its mean value ("mean-field approximation"). Then the total ground state energy is minimized by

$$|\Delta| = 2E_F e^{-1/\lambda} \ , \ \lambda = \frac{2g^2}{\pi v_F \omega_{2k_F}} \quad , \tag{2.4}$$

where E_F is the Fermi energy measured from the bottom of the band, and for simplicity the electron dispersion has been linearized around k_F ($\varepsilon_k = v_F(|k| - k_F)$). The energy gap between the highest occupied and the lowest unoccupied state is $2|\Delta|$ (fig.1). The modulation of the charge-density is given in terms of the order parameter as

$$n(x) = n_0 - \frac{2|\Delta|}{\pi v_F \lambda} \cos(2k_F x + \phi) \quad , \tag{2.5}$$

where n_0 is the average electron density. Similarly, one finds for the lattice modulation

$$\langle u(x)\rangle = \sqrt{\frac{2}{\rho\omega_{2k_F}}}\frac{|\Delta|}{g}\cos(2k_F x + \phi) \ . \tag{2.6}$$

From eqs. (2.5) and (2.6) one sees that the phase of the order parameter determines the position of the modulation with respect to the underlying lattice. On the other hand, the energy of the system depends only on $|\Delta|$, not on ϕ. This is so because we have here implicitly assumed that the bandfilling is incommensurate, i.e. the reciprocal lattice vector G is not a simple multiple of $2k_F$, and consequently the position of the charge-density wave is not fixed with respect to the lattice.

It is fairly straightforward to generalize the mean-field calculation to finite temperatures [10]. One finds that the modulated state described above is stable below a critical temperature T_c given by

$$T_c = 1.13 E_F e^{-1/\lambda} \ . \tag{2.7}$$

At T_c a second order phase transition into the ordered state occurs, characterized by $\Delta \neq 0$. At the transition there is a jump ΔC of the specific heat given by

$$\Delta C = \frac{12}{7\varsigma(3)} C_n(T_c) = 1.43 C_n(T_c) \ , \tag{2.8}$$

where $C_n(T) = (2\pi T/3v_F)L$ is the normal-state electronic specific heat, and $\varsigma(x)$ is Riemann's zeta function ($\varsigma(3) = 1.202...$). The condensation energy, i.e. the energy difference between the ground state and the undistorted metallic state ($\Delta = 0$) at $T = 0$ can be expressed in terms of T_c and ΔC:

$$E_c = \frac{L}{2\pi v_F} |\Delta(T=0)|^2 = \frac{9e^{-2\gamma}}{7\varsigma(3)} T_c \Delta C = 0.34 T_c \Delta C \ , \tag{2.9}$$

where $\gamma = 0.5772..$ is Euler's constant. From eqs. (2.4) and (2.7) one obtains

$$2|\Delta| = 3.52 T_c \tag{2.10}$$

One should note that the numerical factors in eqs. (2.4) and (2.7) are not universal but rather depend on details of the bandstructure ε_k. On the other hand, the ratios (2.8), (2.9), and (2.10) are independent of such details, and moreover can be experimentally tested. These ratios are, however, modified by Coulomb type interactions between electrons (which are neglected in our model (2.1), see the contribution of S. Barišić in this volume), and by the fluctuation effects to be discussed in the next chapter.

Due to the gap in the excitation spectrum, the single-particle conductivity is thermally activated and vanishes at zero temperature. In an ideal incommensurate system dc conductivity by a collective motion of the charge-density wave is possible [2]. However, in most cases this mechanism is suppressed either by commensurability effects or by impurity pinning [11,12], and the conductivity then is indeed semiconductor-like, at least for small electric fields. Similarly, the spin susceptibility is activated and vanishes at $T = 0$.

3 FLUCTUATION EFFECTS IN ONE DIMENSION

As is well known, in one dimension no symmetry breaking phase transitions can occur at nonzero temperatures [5]. The results of the preceding chapter are in obvious contradiction with this, due to the use of the mean-field approximation. This deficiency can be cured taking into account fluctuation effects, which will be done in the following first in the low-temperature domain and then in the (wide) vicinity of the mean-field transition temperature T_c^{MF} (as calculated in the preceding chapter).

3.1 Low temperatures

In order to assess the importance of fluctuation effects the exited states of the system have to be investigated. Most types of excitations like single-particle excitations accross the gap, solitons [14], polarons [15], or oscillations of the amplitude of the order parameter [11] have a finite excitation energy and therefore can be neglected at sufficiently low temperatures. However, as already noted above, a spatially constant shift of the phase of the order parameter, $\phi \to \phi + \phi_0$, does not cost energy at all, and consequently one expects very little energy cost if ϕ_0 varies slowly in space. Moreover, one may notice from eqs. (2.5) and (2.6) that a CDW moving with constant velocity v can be represented by $\phi = -2k_F vt$, and therefore that a nonzero value of $\dot{\phi}$ leads to a nonzero kinetic energy. More precisely, the excitations of the phase of the order parameter can be described by an effective Hamiltonian [12,13]

$$H_\phi = \int dx \left\{ \frac{1}{2\rho}\Pi_\phi^2(x) + \frac{c}{2}\left(\frac{\partial\phi}{\partial x}\right)^2 \right\} , \tag{3.1}$$

$$\rho = \frac{1}{2\pi v_F}\left(1 + \frac{4 \mid \Delta \mid^2}{\lambda\omega_{2k_F}^2}\right) = \frac{1}{2\pi v_F}\frac{m^*}{m} , \; c = \frac{v_F}{2\pi} . \tag{3.2}$$

In (3.1) $\Pi_\phi = \rho\dot{\phi}$ is the momentum density conjugate to ϕ. A moving CDW implies motion both of the electrons and of the ions, and consequently the effective mass density ρ contains both a contribution from the ions (the second term in parentheses in eq. (3.2)) and from the electrons. $m = 1/v_F$ is the electronic effective mass, and m^* is called the "CDW effective mass". A slow variation of the phase of the order parameter in space is equivalent to a variation of k_F. Consequently, the coefficient c contains only parameters of the electronic system.

The effective Hamiltonian (3.1) can be used to evaluate the effect of the fluctuations of the phase on different physical quantities. The excited states of (3.1) are long-wavelength oscillations of ϕ, called phasons, with a sound-like dispersion relation [11]

$$\omega_\phi(q) = v_\phi \mid q \mid = v_F\sqrt{\frac{m}{m^*}} \mid q \mid . \tag{3.3}$$

Due to the large effective mass m^*, v_ϕ usually is much smaller than the Fermi velocity v_F. We now consider the effect of the phason mode on the correlation function of the ionic displacement which determines static and dynamic structure factors measured in scattering experiments. Allowing for spatial and temporal variations of the phase, but neglecting the (small) fluctuations of the amplitude of the order parameter, one finds in analogy to eq. (2.6):

$$\begin{aligned} \langle u(x,t)u(0,0)\rangle &= \frac{1}{\pi v_F \omega_{2k_F}^2 \rho} Re\left\{e^{2ik_F x}\langle\Delta(x,t)\Delta^*(0,0)\rangle\right\} \\ &= \frac{\mid \Delta \mid^2}{\pi v_F \omega_{2k_F}^2 \rho} Re\left\{e^{2ik_F x}\langle e^{i\phi(x,t)}e^{-i\phi(0,0)}\rangle\right\} . \end{aligned} \tag{3.4}$$

The average over ϕ can be easily evaluated:

$$\langle e^{i\phi(x,t)}e^{-i\phi(0,0)}\rangle = \left(\frac{\alpha\pi T}{v_\phi}\right)^\eta \{\sinh[\pi T(x - v_\phi t)/v_\phi]\sinh[\pi T(x + v_\phi t)/v_\phi]\}^{\eta/2} \tag{3.5}$$

$$= \left(\frac{\alpha^2}{x^2 - v_\phi^2 t^2}\right)^{\eta/2} \quad (T = 0) , \tag{3.6}$$

where $\eta = 1/(2\pi\sqrt{c\rho}) = \sqrt{m/m^*}$ is usually considerably smaller than unity, and α is a short-range cutoff. Let us first consider the zero-temperature limit, eq. (3.6). Then only quantum fluctuations are present. In the "classical limit" $\rho \to \infty$ one has $\eta = 0$. The correlation function (3.6) goes to a constant value at large distances, i.e. there is long-range order, and from the Fourier transform of (3.4) one finds true Bragg peaks in the static structure factor at

$\pm 2k_F$. Quantum effects (i.e. zero-point oscillations of the phase) start to play a role for any finite value of ρ. One than has $\eta > 0$, and the correlation function falls of algebraically to zero, i.e. there is no long-range order even at $T = 0$. The Bragg peaks disappear, however as long as $\eta < 1$ there are still power-law divergences of the static structure factor around $Q = \pm 2k_F$. Similarly, in the dynamic structure factor there are power-law singularities around $\omega = \pm v_\phi q$.

At nonzero temperature, thermal fluctuations are important, and the long-distance (or long-time) behavior of the correlation function (3.5) changes drastically: the algebraic decay is replaced by an exponential law

$$\langle e^{i\phi(x,0)} e^{-i\phi(0,0)} \rangle \approx e^{-|x|/\xi(T)} \;, \; \xi(T) = 2c/T \; . \tag{3.7}$$

One should note that the correlation length ξ is independent of the importance of quantum effects (i.e. independent of ρ). At any finite temperature there is no long-range order and no symmetry breaking, in agreement with the general remark at the beginning of the chapter. The present argument should make clear that the long-wavelength excitations of the order parameter itself are actually rsponsible for the destruction of long-range order. It might appear that there is an inconsistency in the reasoning used here in that one assumes long-range order to exist to derive the effective Hamiltonian (3.1), which then is used to demonstrate the absence of long-range order. However, to derive H_ϕ it is actually suffcient to assume the existence of a well-defined absolute value of Δ, together with slow variations (on atomic scales) of ϕ [12,13,16]. These conditions can always be satisfied at low temperatures.

Due to the exponential decay (3.7), the singularities in structure factors are replaced by peaks of finite height and width proportional to $\xi(T)^{-1}$. At sufficiently low temperatures the correlation length becomes very large, and then the differences between the "real" behavior of the (in any case hypothetical) one-dimensional system and the mean-field description of the last chapter become very small.

We have considered here an incommensurate system, where the broken-symmetry groundstate has a continuous degeneracy ($\phi \rightarrow \phi + \phi_0$ with arbitrary ϕ_0), and consequently there are excited states of arbitrary low energy (3.3). In a commensurate situation (e.g. a half- or quarter-filled band) there is only a finite number of degenerate ground states, and instead of phase fluctuations domain wall excitations (with finite energy) destroy the long-range order. In that case the correlation length diverges exponentially at low temperatures ($\xi(T) \approx \exp(E_0/T)$), and at $T = 0$ there always is long-range order.

3.2 The vicinity of T_c^{MF}

The standard way to describe properties in the vicinity of the critical temperature is to use a Ginzburg-Landau type approach [8,9,17]: the free energy of the system for a fixed configuration $\Delta(x)$ is expanded in powers of $\Delta(x)$ and its derivatives, and one has (cf. fig. 2)

$$F\{\Delta(x)\} = F(0) + \int dx \left\{ a \mid \Delta \mid^2 + b \mid \Delta \mid^4 + c \left| \frac{\partial \Delta}{\partial x} \right|^2 \right\} \; . \tag{3.8}$$

For the moment, we shall consider b and c as arbitrary positive constants, and for a assume the form

$$a = a'(T/T_c^{MF} - 1) \; , \tag{3.9}$$

where T_c^{MF} is the mean-field transition temperature. Explicit expressions for the coefficients appropriate for the Peierls transition will be given below. In the temperature range considered quantum effects are of minor importance, and consequently Δ can be regarded as a classical quantity in (3.8).

In the mean-field approximation one simply minimizes $F(\Delta)$, with the result

$$\begin{aligned} \mid \Delta \mid^2 &= 0 \quad (T > T_c^{MF}) \\ &= -a/2b \, (T < T_c^{MF}) \; . \end{aligned} \tag{3.10}$$

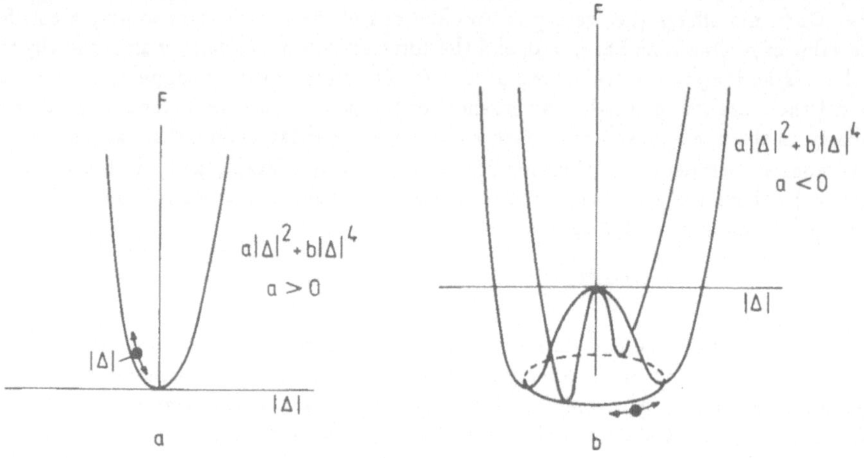

Figure 2: Shape of the free-energy functional, eq. (3.8): (a) $T > T_c^{MF}$ $(a > 0)$, the amplitude of the order parameter (heavy dot) fluctuates around $|\Delta| = 0$; (b) $T < T_c^{MF}$ $(a < 0)$, the phase of the order parameter "rotates" in the minimum of $F(\Delta)$, at $|\Delta| \approx const.$.

The specific heat jump at the transition is

$$\Delta C = L \frac{a'^2}{2bT_c^{MF}} \quad . \tag{3.11}$$

As already discussed, the mean-field approximation is insufficient in one dimension. To obtain a satifactory description of the thermodynamics, the partition function

$$Z(T) = \int \mathcal{D}\Delta \exp[-F(\Delta)/T] \tag{3.12}$$

has to be evaluated. Here the integration over $\mathcal{D}\Delta$ means integration over all possible configurations of $\Delta(x)$. In general, this is a quite complicated task, due to the presence of the fourth order term in F.

For the subsequent discussion it is useful to go over to Fourier-transformed variables:

$$\Delta(x) = \frac{1}{\sqrt{L}} \sum_k \Delta_k e^{ikx} \quad . \tag{3.13}$$

One should note that $\Delta(x)$ is complex and therefore $\Delta_k \neq \Delta^*_{-k}$, i.e. Δ_k and Δ_{-k} are different degrees of freedom. In terms of the Δ_k $F(\Delta)$ can be rewritten as

$$F\{\Delta\} = F(0) + \sum_k a_k \mid \Delta_k \mid^2 + \frac{b}{L} \sum_{k_1,k_2,k_3} \Delta^*_{k_1} \Delta^*_{k_2} \Delta_{k_3} \Delta_{k_1+k_2-k_3} \quad , \tag{3.14}$$

with $a_k = a + ck^2$. The contribution of the order parameter to the static structure factor is

$$S(k \pm 2k_F) \approx \langle \mid \Delta_k \mid^2 \rangle \quad . \tag{3.15}$$

As a first approximation, one may neglect the troublesome fourth-order term (the Gaussian approximation). In microscopic calculations this is equivalent to the random-phase approximation (RPA). One then has

$$\langle \mid \Delta_k \mid^2 \rangle = T/a_k \quad , \tag{3.16}$$

and the correlation function of the order parameter is given by

$$\langle \Delta(x) \Delta^*(0) \rangle = \frac{1}{2\pi} \int dk \langle \mid \Delta_k \mid^2 \rangle e^{ikx} = \langle \mid \Delta \mid^2 \rangle e^{-|x|/\xi(T)} \quad , \tag{3.17}$$

$$\xi(T) = \sqrt{c/a} = \sqrt{c/a'} (T/T_c^{MF} - 1)^{-1/2} \quad . \tag{3.18}$$

The Fourier transform of the correlation function (3.17) gives the static structure factor, which in the present case is a Lorentzian centered at $\pm 2k_F$ of width $1/\xi(T)$ (cf. eq. (3.15)). This peak represents the thermal fluctuations into an ordered state which are present even above T_c^{MF}. The average size of the ordered regions is given by the correlation length $\xi(T)$.

The problems associated with the Gaussian approximation become apparent looking at the **local** fluctuations of the order parameter

$$\langle|\Delta|^2\rangle = \frac{1}{2\pi}\int dk \langle|\Delta_k|^2\rangle = \frac{T}{2\sqrt{ac}} = \frac{T}{2\sqrt{a'c}}(T/T_c^{MF}-1)^{-1/2} \ . \tag{3.19}$$

This expression diverges at T_c^{MF}, and consequently the neglect of the fourth-order term cannot be justified in the vicinity of T_c^{MF} (note that in three dimensions $\langle|\Delta|^2\rangle$ remains finite at a second-order phase transition). To estimate the region of validity of the Gaussian approximation we calculate the lowest-order corrections due to a nonzero value of b. These corrections are given by the "Hartree-Fock" [18] replacement $|\Delta|^4 \rightarrow 4\langle|\Delta|^2\rangle|\Delta|^2$ or equivalently

$$a \rightarrow a + \Sigma = a + 4b\langle|\Delta|^2\rangle \tag{3.20}$$

in the free energy functional (3.8). Clearly, close to T_c^{MF} the "correction" Σ becomes much more important than the bare coefficient a. The relative importance of a and Σ then determines the validity of the Gaussian approximation: the approximation fails if $\Sigma \approx a$. This occurs at a temperature T_δ above T_c^{MF} (Δt if measured in units of T_c^{MF}) given by:

$$\Delta t = T_\delta/T_c^{MF} - 1 = 2\left(\frac{bT_c^{MF}}{a'^{3/2}c^{1/2}}\right)^{2/3} , \tag{3.21}$$

and the Gaussian approximation fails if

$$t - 1 = T/T_c^{MF} - 1 < \Delta t \ . \tag{3.22}$$

Specifically, for small b the Gaussian approximation is valid over a large temperature range **above** T_c^{MF}. The fluctuation corrections considered here are related to fluctuations of the amplitude of the order parameter around its most probable value $|\Delta| = 0$.

One may note that, taken literally, eq. (3.20) indicates the absence of a phase transition at any nonzero temperature: a transition would necessitate $a + \Sigma = 0$, implying (cf. eq. (3.19)) $\langle|\Delta|^2\rangle = \infty$, in obvious contradiction with (3.20). Nevertheless, the argument is not very satisfactory as we have considered here only the lowest-order corrections. To convince oneself of the absence of long-range order at any nonzero temperature one should rather rely on the reasoning presented in the preceding section.

Fortunately, in one dimension the static properties of models like eq. (3.8) can be calculated exactly [19]. In fig. 3 the exact result for the inverse correlation length is shown, for different values of the parameter Δt. For small values of Δt, the exact result is very similar to the Gaussian approximation over most of the temperature range, and even though $\xi(T)$ does not diverge at T_c^{MF}, there is obviously a marked change in behaviour around that temperature. Analogous anomalies are also found in other physical quantities like the specific heat [19]. With increasing Δt these anomalies are progressively wiped out, and vanish nearly completely for $\Delta t \approx 1$. Comparing to the exact solution one also observes that the Hartree-Fock approximation gives a reliable description of the system above T_c^{MF}, even for large Δt [20].

For a Peierls instability, the coefficients in the free energy functional (3.8) can be obtained [17] from the microscopic model (2.1):

$$a = \frac{1}{\pi v_F}\ln(T/T_c^{MF}) \approx \frac{1}{\pi v_F}(T/T_c^{MF}-1) \ , \tag{3.23}$$

$$b = \frac{1}{\pi v_F}\frac{7\zeta(3)}{16\pi^2 T^2} \ , \tag{3.24}$$

$$c = v_F^2 b \ , \tag{3.25}$$

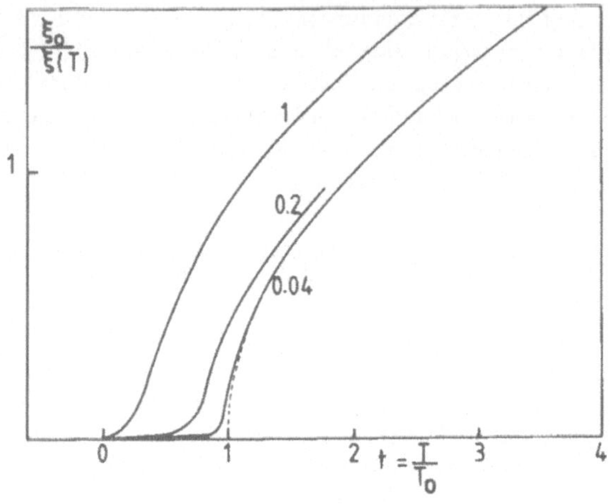

Figure 3: The inverse correlation length as a function of reduced temperature $t = T/T_c^{MF}$, for various values of the parameter Δt, as obtained from the results of [19]. a', b and c are assumed to be temperature independent, $a(T) = a'(T/T_c^{MF} - 1)$, and $\xi_0 = \sqrt{c/a'}$. The dashed line is the Gaussian approximation, eq. (3.17).

and from this one obtains (setting approximately $T = T_c^{MF}$)

$$\Delta t = 1.6 \ . \tag{3.26}$$

This large value of Δt shows that there is a large temperature range above T_c^{MF} where the Gaussian approximation (or, in more microscopic terms, the random-phase approximation) gives bad results and does not provide an adequate microscopic description. Within the framework of the Ginzburg-Landau model, eq. (3.8), the effects of the quartic terms have to be taken into account. However, if fourth-order terms are important, one may wonder wether higher-order terms, omitted in (3.8), are not equally important. To answer this question, a systematic high-temperature expansion of correlation functions etc. would be necessary.

3.3 Summary

In this chapter we have seen how one-dimensional fluctuation effects destroy the long-range ordered CDW state. The physical properties of the system can often be described starting from the mean-field (or RPA) solution and taking the fluctuation corrections into account.

At low temperature, the fluctuations of the phase of the order parameter dominate (fig. 2b), and the correlation length becomes very large. Even though there is no true long-range order, the properties of the system are very similar to those obtained in the mean-field approximation [13]: the spin susceptibility is strongly reduced with respect to its free-electron value, and there is a marked minimum in the density of states at the Fermi level ("pseudogap"), a remnant of the true single-particle gap found in chap. 2. Similarly, there are narrow Lorentzian peaks in the static structure factor instead of the Bragg (i.e. δ-function) peaks typical of true long-range order, and a well-defined phason mode exists in the dynamic structure factor for $| q - 2k_F | > 1/\xi(T)$. Corrections to the phase-fluctuation-dominated behavior are typically of order $\exp(- | \Delta(T) | /T)$, and consequently can be neglected for $T \leq T_c^{MF}/2$ (note that the coefficients of the Ginzburg-Landau free-energy (3.8) are strongly modified if $T \leq | \Delta(T) |$, and eqs. (3.23) to (3.25) are certainly inappropriate in such a situation).

In the vicinity of T_c^{MF} (and more generally above T_c^{MF}) amplitude fluctuations are important. Even well above T_c^{MF} the Gaussian approximation (or RPA) fails and interactions between the fluctuation modes (e.g. the fourth-order term in (3.8)) have to be taken into account (cf. eq (3.26)). In this temperature region there is a slow decrease of the spin susceptibility and of the density of states with decreasing temperature. The fluctuations of the order parameter give rise to Lorentzian peaks around $\pm 2k_F$, however there is no phason-like mode in the dynamics [9].

4 COUPLED CHAINS

Obviously, in any real quasi-one-dimensional material the individual conducting chains are not completely isolated from each other (as assumed in the last chapter). There is rather some finite coupling between them, which, amongst other things, makes symmetry-breaking phase transitions possible. This situation will be described in the following two sections.

4.1 The vicinity of T_c^{MF}

The Ginzburg-Landau model used in section 3.2 can be generalized so as to include interchain coupling; the functional for a system of coupled chains then is

$$F\{\Delta_i(x)\} = \sum_i F_1\{\Delta_i(x)\} + 2\lambda_c \sum_{\langle i,j\rangle} \int dx \, Re[\Delta_i^*(x)\Delta_j(x)] \ , \tag{4.1}$$

where F_1 is the functional for an isolated chain, eq. (3.8), i, j label the chains, and $\langle i, j\rangle$ indicates summation over nearest-neighbor pairs in a square lattice of parallel chains (each nearest-neighbor pair being counted once). Going over to Fourier-transformed variables as in eq. (3.13) the form (3.14) is recovered, with

$$a_k = a + ck_x^2 + 2\lambda_c[\cos(k_y d) + \cos(k_z d)] \ , \tag{4.2}$$

where d is the distance between chains, and in addition one has to make the trivial replacement $L \rightarrow V$ in the fourth-order term.

In most cases one has $\lambda > 0$, so that (4.1) is minimized if the sign of the order parameter alternates between adjacent chains. In order to eliminate this fast oscillation define a modified order parameter

$$\Delta(\vec{r}) = (-1)^{(y_i+z_i)/d}\Delta_i(x) \ , \ \vec{r} = (x, y_i, z_i) \ , \tag{4.3}$$

where (y_i, z_i) is the position of chain i in the transverse direction. If $\Delta(\vec{r})$ now varies only slowly in the transverse directions (on the scale of d), the interchain coupling term in (4.1) may be approximated by a gradient and the sum transformed into an integral:

$$F\{\Delta(\vec{r})\} = F(0) + \frac{1}{d^2}\int d^3r \left\{ a \mid \Delta \mid^2 + b \mid \Delta \mid^4 + c\left|\frac{\partial \Delta}{\partial x}\right|^2 + c_\perp |\nabla_\perp \Delta(\vec{r})|^2 \right\} \ , \tag{4.4}$$

where $c_\perp = \lambda_c d^2$, $\nabla_\perp$ is the gradient with respect to the transverse coordinates y, z, and a small correction of order λ_c to a has been neglected. In the Fourier-transformed variables one now has

$$a_k = a + ck_x^2 + c_\perp(k_y^2 + k_z^2) \ . \tag{4.5}$$

In the case of the Peierls instability, the principal mechanisms of coupling between adjacent chains are:

1. the Coulomb interaction between CDW's on adjacent chains [11,21]. Because of the repulsive nature of the Coulomb interaction, maxima of the electronic charge on adjacent chains repell each other, so that the CDWs on adjacent chains are out of phase ($\lambda_c > 0$).

2. tunneling of electrons between adjacent chains, described by a transfer integral $t_\perp$. Due to the nesting of opposite sides of the Fermi surface Δ again alternates between adjacent chains [22]. One finds

$$c_\perp = 2\left(\frac{t_\perp d}{v_F}\right)^2 c \ . \tag{4.6}$$

3. the dependence of the bare phonon frequency on the transverse components of the wavevector which comes from the interaction between ions on adjacent chains. In this case

$$c_\perp \approx \frac{\partial^2 \omega_q}{\partial^2 q_\perp} . \tag{4.7}$$

More generally, the microscopic expression for a_k is

$$a_k = \frac{2g^2(2k_F + k)}{\omega_{2k_F+k}} + \Pi(2k_F + k) , \tag{4.8}$$

where $\Pi(k)$ is the static electronic polarisability at wavevector k. The second derivatives of a_k at its minimum then determine the coefficients $c, c_\perp$. If necessary, in expressions like (3.16) the full form (4.8) can be used instead of the approximations (4.2) or (4.5), but in general the differences are minor.

Due to the coupling between chains a nonzero correlation length in the transverse directions exists. From (4.4) one finds in the Gaussian approximation

$$\langle \Delta^*(\vec{r})\Delta(0)\rangle \approx \frac{1}{\sqrt{(x/\xi_\parallel)^2 + (r_\perp/\xi_\perp)^2}} \exp\left[-\sqrt{(x/\xi_\parallel)^2 + (r_\perp/\xi_\perp)^2}\right] , \tag{4.9}$$

with

$$\xi_\parallel = \sqrt{c/a} \,, \; \xi_\perp = \sqrt{c_\perp/a} . \tag{4.10}$$

In the structure factor (cf. eq. (3.15)) the correlations between ajacent chains manifest themselves by a Lorentzian peak of transverse width $1/\xi_\perp$, situated at $\vec{k} = (2k_F, \pi/d, \pi/d)$. Sufficiently close to T_c^{MF} $\xi_\perp$ is large, and therefore the peak is well-defined and the gradient approximation in the transverse direction (eq. (4.4)) is well justified. With increasing temperature $\xi_\perp$ decreases, and at a "crossover temperature" T_x one has $\xi_\perp(T_x) = d$, i.e. adjacent chains are only weakly correlated. In this situation the transverse width of the peak in the elastic structure factor becomes comparable to the size of the Brillouin zone $(2\pi/d)$, and with further increase of temperature any dependence of $S(\vec{k})$ on the transverse components of $\vec{k}$ disappears quickly, i.e. adjacent chains are essentially uncorrelated and behave one-dimensionally as far as phenomena related to the Peierls instability are concerned. In this situation the gradient approximation is obviously inadequate, and rather the full form (4.4) should be used.

Considering interchain tunneling as the main coupling mechanism (cf. eq. (4.6)), one finds

$$\Delta t_x = T_x/T_c^{MF} - 1 = 0.1(t_\perp/T_c^{MF})^2 . \tag{4.11}$$

Experimentally one finds, for example for $TTF - TCNQ$ [23], $\Delta t_x \approx 0.1$ (using the true transition temperature instead of T_c^{MF}).

As in the purely one-dimensional case, there are corrections to the Gaussian approximation due to the presence of the fourth-order term in (4.4). Often the crossover to three-dimensional fluctuations occurs at temperatures much lower than T_δ (i.e. $\Delta t_x \ll \Delta t$, cf. eqs. (3.21) and (4.11)), and then the importance of the fourth-order term is determined by the purely one-dimensional criterion (3.21). Moreover, there is a shift of the critical temperature from T_c^{MF} to lower values. To lowest order in b T_c is given by

$$a(T_c) + 4b\langle|\Delta|^2\rangle_{T_c} = a(T_c) + \frac{1}{\pi}\frac{T_c b \Lambda_\perp}{c}\frac{\xi_\parallel}{\xi_\perp} = 0 . \tag{4.12}$$

Here the quadratic Brillouin zone of the transverse momenta has been replaced by a circle of diameter $\Lambda_\perp = 2\sqrt{\pi}/d$, so that the total number of degrees of freedom is conserved. This can be shown numerically to be a very good approximation. In the chain direction no momentum cutoff is needed. If the corrections to T_c^{MF} are important, the true T_c instead of T_c^{MF} has to be used in (4.11).

Another deviation from the Gaussian approximation occurs due to critical fluctuation effects close to T_c, well known from isotropic three-dimensional phase transitions. As close

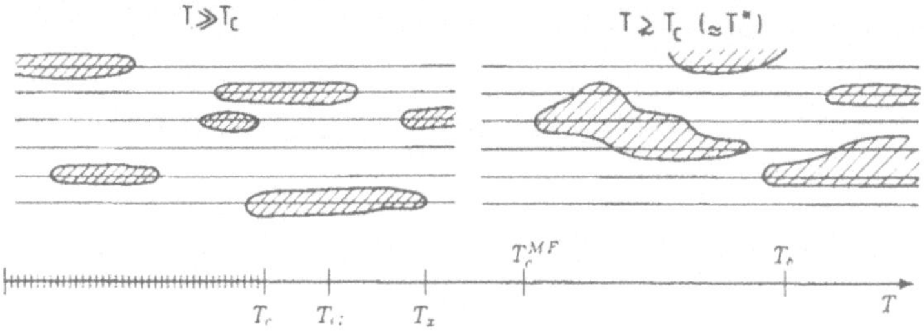

Figure 4: Qualitative shape of the ordered regions (fluctuations, hatched areas) above and slightly below $T_x (\equiv T^*)$. Fluctuations below T_c can be visualized taking the hatched areas as disordered. The location of the diffent characteristic temperatures discussed in the text is shown schematically on the temperature scale at the bottom of the figure. The hatched part of the scale represents the ordered region. In general, there is no fixed relation between T_c^{MF} and T_G, T_x. In all cases one has $T_x > T_G > T_c, T_0 > T_c^{MF} > T_c$.

to T_c the transverse correlation length is large, these effects are insensitive to the quasi-one-dimensional structure, and the critical exponents are expected to be those of a regular three-dimensional system. The temperature region around T_c where these effects are expected to be important is given by the "Ginzburg critical width" [24] which in the present case can be evaluated as

$$\Delta t_G = T_G/T_c - 1 = \frac{4T_c^2 b^2 d^2}{\pi^4 c^3 a'} \left(\frac{\xi_\parallel}{\xi_\perp}\right)^4 . \tag{4.13}$$

Note the occurrence of the anisotropy of the correlation length to the fourth power. It should also be pointed out that in eqs. (4.12) and (4.13) all the coefficients can be temperature dependent, as already found in the preceding chapter. The different characteristic temperature introduced here are shown in figure 4, together with a schematic representation of the order parameter fluctuations.

Using the coefficients derived for the Peierls instability, from eq. (4.12) one obtains

$$\ln(T_c/T_c^{MF}) = -2\sqrt{\pi}\frac{T_c d}{v_F}\frac{\xi_\parallel}{\xi_\perp} . \tag{4.14}$$

Values appropriate for $TTF-TCNQ$ are [6,23] $\xi_\parallel/\xi_\perp \approx 15, T_c = 53K, v_F/b \approx 2000K, d \approx 3b$, where b is the lattice constant in the chain direction, and then one has $T_c \approx T_c^{MF}/10$. This almost certainly overestimates fluctuation effects, due to the neglect of higher order corrections and the use of coefficients which are incorrect well below T_c^{MF}, however large corrections to the mean-field critical temperature are certainly to be expected.

Similarly, for the Ginzburg critical width one has

$$\Delta t_G = \frac{64}{7\zeta(3)}\left(\frac{T_c d \xi_\parallel}{v_F \xi_\perp}\right)^4 \tag{4.15}$$

Due to the fourth power of the correlation length anisotropy, this number is not necessarily small, and critical effects may thus well be observable.

For comparison one may note that in a three-dimensional superconductor one has (d is the lattice constant)

$$T_c^{MF} - T_c \approx T_c^{MF}\left(\frac{T_c d}{v_F}\right)^2 \approx 10^{-6} T_c^{MF} , \quad \Delta t_G \approx \left(\frac{T_c d}{v_F}\right)^4 \approx 10^{-12} , \tag{4.16}$$

i.e. all corrections to the standard mean-field-RPA treatment are completely negligible.

4.2 Low temperatures

As discussed in section 3.1, at low temperatures only fluctuations of the phase of the order parameter are important. A coupling term as in (4.1) then is proportional to $\cos(\phi_i - \phi_j)$, and in analogy to (4.1) one finds

$$H\{\phi_i(x)\} = \sum_i H_1\{\phi_i(x)\} + 2\lambda_c \sum_{\langle i,j\rangle} \int dx \cos(\phi_i(x) - \phi_j(x)) \ , \qquad (4.17)$$

where H_1 is the Hamiltonian of the individual chains, eq. (3.1). Even though the coupling mechanisms are the same as those discussed in the last section, the values of the coupling parameter λ_c are different from those used in (4.1). In addition, the long-range part of the Coulomb interaction should also be included [25], but this interaction will be neglected here. Treating now the interchain coupling in the mean-field approximation and the one-dimensional effects exactly [26], a new effective Ginzburg-Landau functional for the order parameter $\Psi(\vec{r}) = 4\lambda_c\langle e^{i\phi}\rangle$ can be derived:

$$F\{\Psi(\vec{r})\} = \frac{1}{d^2}\int d^3r \left\{ a \mid \Psi \mid^2 + b \mid \Psi \mid^4 + c\left|\frac{\partial \Psi}{\partial x}\right|^2 + c_\perp |\nabla_\perp \Psi(\vec{r})|^2 \right\} \ . \qquad (4.18)$$

Neglecting quantum effects, i.e. in the realistic case $m^* \gg m$, the coefficients are

$$\begin{aligned} a &= -\frac{4\gamma}{T^2} + \frac{1}{4\lambda_c}, \quad b = \frac{28\gamma^3}{T^6}, \quad c = \frac{16\gamma^3}{T^4}, \\ c_\perp &= \frac{d^2}{16\lambda_c}, \quad \gamma = \frac{v_F}{2\pi} \ . \end{aligned} \qquad (4.19)$$

The form of the coefficients including quantum effects is given in [27]. The "mean-field" critical temperature of the model (4.18) is given by the zero of the coefficient a:

$$T_c^{(0)} = 4\sqrt{\gamma\lambda_c} \ , \qquad (4.20)$$

i.e. it is essentially determined by the strength of the interchain coupling. Including quantum effects one finds $T_c^{(0)} \propto \lambda_c^{1/(2-\eta)}$. From scaling arguments [28] one can conclude that this type of power-law is indeed correct for $\lambda_c \to 0$.

One can now discuss corrections to the mean-field result (4.20) along the same lines as in the preceding section [27]. The downward shift of the critical temperature due to the b-term is found to be

$$T_c^{(0)} - T_c \approx 0.2 T_c^{(0)} \ , \qquad (4.21)$$

and the crossover temperature and the Ginzburg critical width are

$$\Delta t_x \approx 0.25 \ , \ \Delta t_G \approx 0.1 \ . \qquad (4.22)$$

In the limit of weak interchain coupling the true critical temperature thus is much smaller than the mean-field result of the second chapter (and vanishes for $\lambda_c \to 0$). As shown by the relative smallness of the corrections (4.21), (4.22), a consistent description is possible in the weak-coupling limit treating the interchain coupling in the mean-field approximation.

4.3 Summary

In this chapter we have introduced and discussed the different characteristic temperatures associated with a phase transition in a quasi-one-dimensional system (cf. fig. 4, see also ref. 29). At temperatures above the crossover temperature T_x, purely one-dimensional fluctuation behaviour occurs. Between T_x and the Ginzburg temperature T_G fluctuations are three-dimensional and Gaussian, with a renormalized critical temperature. Finally, between T_G and T_x one has three-dimensional critical fluctuations. Both at low temperatures, where phase fluctuation dominate, and in the vicinity of T_c^{MF} (and above it), where the amplitude

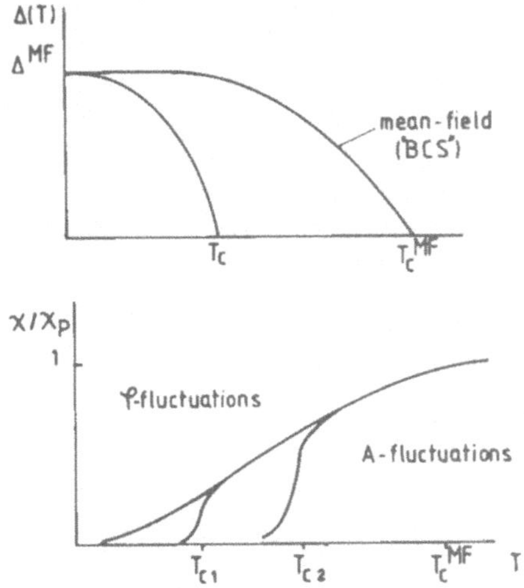

Figure 5: Above: schematic temperature dependence of the Peierls order parameter from the mean-field solution ("BCS-curve"), and in a real situation with $T_c \ll T_c^{MF}$. Below: schematic temperature dependence of the spin succeptibility. Even in the presence of strong fluctuation effects ($T_{c1} \ll T_c^{MF}$) χ drops sharply only below the critical temperature.

fluctuations are important, a satisfactory (semi-) quantitative discussion can be given. In many real cases, however, one will be in an intermediate situation, (e.g. there is no well-defined phason mode above T_c) where much less precise information is available. Experimentally, the crossover behaviour is most directly observable in scattering experiments, but can also be seen in magnetic and transport measurements.

One important consequence of the discussion of the two preceding sections is that one generally expects the observed critical temperature of the Peierls transition to be much below the calculated mean-field value T_c^{MF} (unless the coupling between adjacent chains is quite strong). T_c^{MF} then merely sets a temperature scale where corrections due to CDW fluctuations become apparent in different physical quantities (shown schematically in fig. 5 for the spin susceptibility). In general it will be difficult to determine T_c^{MF} dirctly from experimental results. For example, fig. 3 shows that (for $\Delta t = 1$, approximately appropriate for the Peierls instability) above T_c^{MF} the true correlation length can be well approximated by the replacement $T_c^{MF} \rightarrow T_c^{MF}/3$ in the Gaussian (RPA) result.

A sequence of characteristic temperatures similar to the one discussed here occurs also below T_c, in the reverse order. Of course, in that case one is concerned with fluctuations around the static mean value of the order parameter. Well below the critical temperature thermal fluctuations disappear. Moreover, as discussed in section 3.1, quantum effects are usually very weak ($m^* \gg m$). Consequently, the low temperature properties of a three-dimensionnally ordered state are well described by the mean-field results. In particular one expects $\Delta(T=0) \approx \Delta^{MF}(T=0)$, and therefore

$$2 \mid \Delta(T=0) \mid \gg 3.52 T_c \ , \qquad (4.23)$$

in agreement with experimental results, where the numerical factor in (4.23) is usually of the order of 10.

The discussion in this and the preceding chapter is largely based on a Ginzburg-Landau like approach, treating the order parameter as slowly varying in time compared to the electronic motion. This approximation is valid if phonon-mediated exchange interactions can be neglected, i.e. if $\omega_{2k_F} < 2\pi T_c^{MF}$ [8,30], or equivalently $m^* \gg m$. In the opposite case (phonon exchange important) a more microscopic treatment of the one-dimensional electron gas, based e.g. on the "g-ology" picture, is necessary [30,31]. The same remark applies to electronic instabilities other than the Peierls instability, like spin-density-formation or superconductivity. In these cases a Ginzburg-Landau functional like eq. (3.8) can formally be derived. However, a systematic expansion around the mean-field solution is usually not possible, one reason being that the collective modes have velocity v_F and therefore cannot be separated from the single-particle motion. In cases where the spin degrees of freedom are frozen out at low temperature the Hamiltonian (4.17) still provides a good starting point for the discussion of the crossover from one-dimensional to three-dimensional behaviour [27]. A discussion for more general situations is given by C. Bourbonnais in this volume.

5 THE SPIN-PEIERLS TRANSITION

The Peierls transition discussed in the preceding chapters leads from a metallic high-temperature state to a low-temperature CDW state with thermally activated conductivity and spin susceptibility. The low-energy excitations responsible for the conductivity and spin susceptibility of the metallic state are lost **simultaneously** at the transition temperature T_c. In an incommensurate system, this situation is not qualitatively changed by the presence of Coulomb interactions (even though the values of various numerical constants may be strongly affected). The situation may be quite different in a commensurate situation, where charge localisation (equivalent to an energy gap in the spectrum of long-wavelength charge fluctuations) due to Coulomb repulsion between electrons may occur, whereas the spin degrees of freedom remain massless.

To be specific, let us consider the one-dimensional Hubbard model:

$$H = t\sum_i (a^+_{i,s}a_{i+1,s} + a^+_{i+1,s}a_{i,s}) + U\sum_i n_{i,\uparrow}n_{i,\downarrow} \quad , \tag{5.1}$$

where $a^+_{i,s}$ is the creation operator for a fermion on site i with spin s, $n_{i,s} = a^+_{i,s}a_{i,s}$, and U parametrizes the on-site Coulomb repulsion. If there is one particle per site (half-filled band) there is a gap Δ_ρ in the charge excitation spectrum [32,44]. For $T \ll \Delta_\rho$ the charge excitations then are frozen out, i.e. the charge is localized and one has a thermally activated conductivity. On the other hand, there is no gap in the spin excitations and the spin susceptibility remains therefore finite. In this situation the low energy excitations can be described by an antiferromagnetic spin-chain Hamiltonian

$$H_{spin} = \sum_i J\vec{S}_i \cdot \vec{S}_{i+1} \quad , \tag{5.2}$$

where the $\vec{S}_i$ are spin-1/2 operators ($\vec{S}_i^2 = 3/4$), and J is an effective exchange constant, $J = 2t^2/U$ for large U [33]. The low-temperature spin susceptibility of (5.2) is $\chi(T=0) = (g\mu_B/\pi)^2/J$ [34].

On a deformable lattice, the exchange integral J depends on the lattice spacing. The spin system then can gain energy by dimerizing [35,36]. The phase transition leading to the nonmagnetic (singlet) dimerized state ($\chi(T=0) = 0$) is called "spin-Peierls" transition, in analogy to the instability of a one-dimensional metal discussed previously. The phase transition occurs at a temperature well below Δ_ρ, i.e. charge and spin degrees of freedom freeze out at very **different** temperatures in this scenario. It should be pointed out that the type of broken symmetry is the same for both the Peierls and spin-Peierls instabilities; there is a continuous crossover from Peierls to spin-Peierls behaviour if the electron-electron repulsion is increased [30] (provided that no other instability like spin-density wave formation occurs).

In a dimerized state one has to replace $J \rightarrow J[1 + (-1)^i\delta]$ in (5.2), where $\delta = cu$, u is the dimerization amplitude and $c = (1/J)\partial J/\partial u$. As a function of δ, the ground state energy per site then is given by [36] (up to logarithmic corrections [38])

$$E_0(\delta) = J[1/4 - \ln 2 - A\delta^{4/3}] \quad , \tag{5.3}$$

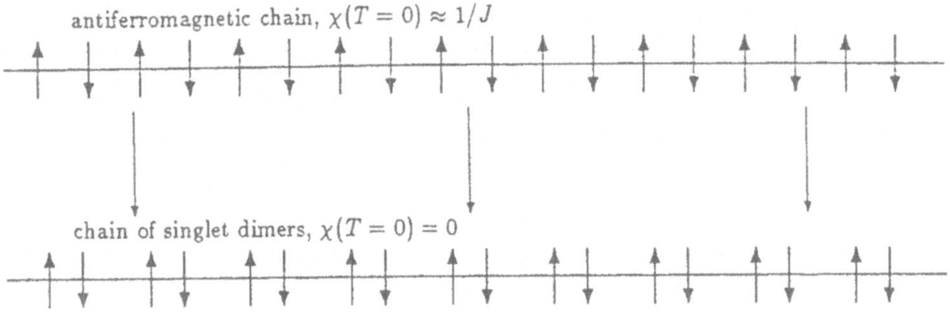

Figure 6: Spin-Peierls transition

where the value for $\delta = 0$ is exact [39]. One should note that for small δ the present spin system gains energy much faster than the free-electron system of chap. 2 (4/3-power law versus $\Delta^2 \ln \Delta$). For nonzero δ there is a (singlet-triplet) gap in the spectrum [36]:

$$\Delta_\sigma = BJ\delta^{2/3} \ . \tag{5.4}$$

Numerical calculations [37] give $A \approx 0.3, B \approx 2$. Adding an elastic lattice energy to (5.3), the total energy of the dimerized state is

$$E_{tot}(\delta) = E_0(\delta) + \frac{K\delta^2}{2c^2} \ . \tag{5.5}$$

This expression is minimized by

$$\delta = \left(\frac{4AJc^2}{3K}\right)^{3/2} , \quad \Delta_\sigma = \frac{4ABJ^2c^2}{3K} \approx 0.8\frac{J^2c^2}{K} \ . \tag{5.6}$$

The dependence of the gap on the electron-phonon coupling constant c found here is very different from the one found in chap. 2. Generalizing these calculations to finite temperatures, the critical temperature is given by [36]

$$T_{sp}^{MF} = 0.26\frac{J^2c^2}{K} \ . \tag{5.7}$$

This leads to a relation analogous to (2.10):

$$\Delta_\sigma \approx 3.1 T_{sp}^{MF} \ . \tag{5.8}$$

Relations analogous to (2.8) and (2.9) can also be found.

The considerations of the last paragraph are based on the mean-field approximation, i.e. spatial variations of the order parameter δ (or u) are neglected. The typical one-dimensional fluctuation effects can be taken into account in the same way as in chap. 3. However, we are here considering a commensurate (dimerized) system, and therefore the order parameter is real (an incommensurate spin-Peierls transition occurs in a nonzero magnetic field). This has some important consequences. In particular, there aren't any long-wavelength low-energy phase fluctuations. The low-temperature properties are rather dominated by thermally excited diffusing domain walls between the two degenerate ground states. This leads to a correlation length diverging exponentially for $T \to 0$ (compare eq. (3.7)). Moreover, at $T = 0$ quantum effects are of minor importance and in particular correlation functions like (3.4) decay to a nonzero constant at large distances, i.e. there is true long-range order at $T = 0$. At higher temperatures (T_{sp}^{MF} and above) a Ginzburg-Landau description, based on a functional like (3.8) with real order parameter can be used. From the microscopic expressions for the coefficients [36] one finds $\Delta t \approx 1$, i.e. fluctuation effects are expected to become important well above T_{sp}^{MF}, as in the standard Peierls case.

The differences in the low-temperature properties between the case discussed here and chap. 3 are due to the different type of broken symmetry considered, rather than to the difference between Peierls and spin-Peierls instabilities. In the present case the **discrete** translation symmetry by one lattice constant is broken by the dimerization, and quite generally the lower critical dimension (i.e. the dimension at and below which fluctuation effects prevent symmetry breaking) for breaking of discrete symmetries is $d_{lc} = 1$, i.e. the present one-dimensional case is just marginal. On the other hand, the broken symmetry of an incommensurate Peierls state is a **continuous** translation symmetry (the ground state energy is invariant under $\phi \to \phi + \phi_0$ with arbitrary ϕ_0), and for continuous symmetries one has $d_{lc} = 2$.

The coupling between adjacent chains which is inevitably present in any real quasi-one-dimensional substance will generally make a spin-Peierls transition at a nonzero temperature possible, as discussed in chap. 4 for the Peierls case. If the coupling between chains is sufficiently strong, the mean-field results should be valid, without any sizeable one-dimensional fluctuation effects. However, because $\Delta t \approx 1$, in order to avoid the one-dimensional effects altogether, the crossover temperature from one- to three-dimensional fluctuations also has to be rather large, i.e. one needs $\Delta t_x \approx 1$ (cf. eq. (4.11)). This needs interchain couplings of the order of the coupling on a single chain, and in such a situation an approach starting from isolated chains may not be appropriate at all. For weaker interchain coupling the transition temperature decreases and the one-dimensional fluctuation effects of the spin-Peierls type become more and more important. For example, the spin susceptibility should start to decrease slowly well above the transition temperature (as shown in fig. 5 for the Peierls transition). Such an effect has been observed in $(TMTTF)_2PF_6$ [40]. For very weak interchain coupling the transition temperature vanishes as $T_{sp} \propto 1/\ln \lambda_c$ [26]. The difference with the square-root law (4.20) is due to the discrete symmetry considered here and the consequent exponential divergence of the correlation length for $T \to 0$.

Our discussion of one-dimensional fluctuation effects is based on a Ginzburg-Landau type approach. As already discussed in section 4.3, this is a reasonable approach as long as the phonon dynamics can be considered as slow compared to the electronic degrees of freedom, e.g. as long as $\omega_{2k_F} < 2\pi T_{sp}^{MF}$. However, in many cases this condition is badly violated. In particular, in spin-chain systems one typically has $J \approx 100K$, i.e. the phonon frequency is of the order of the exchange integral, and therefore much larger than T_{sp}^{MF}. Phonon exchange then gives rise to a quasi-instantaneous interaction between localized spins, leading to the renormalization $J \to J_{eff} < J$, but otherwise no drastic one-dimensional fluctuation effects occur [41]. In particular, the spin-susceptibility is determined by J_{eff}, and remains finite for $T \to 0$ (contrary to the adiabatic case). A phase transition into a three-dimensionally ordered state will be uniquely driven by the interchain coupling. Treating the interchain coupling in the mean-field approximation, one finds $T_{sp} \propto \lambda_c$, and thermodynamic relations found in the mean-field calculation, like (5.8), are again valid. As in the case discussed in section 4.2, fluctuation corrections are small, essentially because of the three-dimensional nature of the coupling. The nonadiabatic nature of the phonons probably plays an important role in insulating spin systems like $TTFCuBDT$ or $MEM(TCNQ)_2$, where measurements of the magnetic susceptibility indicate the absence of any spin-Peierls precursor effects (no decrease below the Bonner-Fisher curve) even in the close vicinity of the transition temperature [42]. In the nonadiabatic case one also has to keep in mind that there will generally be a competition between spin-Peierls and antiferromagnetic ordering. The antiferromagnetic ordering (Néel) temperature depends on the interchain exchange interaction $J_\perp$ as $T_{AF} \propto J_\perp$ (provided that $J_\perp \ll J$, and in general the phase with the highest critical temperature will be realized. The thermodynamic relations at the antiferromagnetic transition can be obtained using the results of ref. [27], using the boson representation of spin operators [43], and one obtains for example $\Delta C/C_n(T_{AF}) = 2.5$.

6 CONCLUSION

In this paper I have tried to explain the different characteristic temperatures determining the physical properties of a quasi-one-dimensional system in the vicinity of a phase transition. Emphasis has been put on the fluctuation effects typical of such systems. In the cases of the Peierls or spin-Peierls instabilities these fluctuation effects are experimentally accessible in scattering experiments or magnetic measurements (see e.g. the contributions by J.P. Pouget, C. Berthier, and C. Bourbonnais in this volume). In the framework of the models discussed here at least semiquantitative predictions can be made for the behaviour both at low temperatures and in the vicinity of the mean-field transition temperature (and above it). Most real substances, however, exhibit a phase transition in a intermediate temperature region, where less reliable results are available.

The discussion given here was restricted to cases where the order parameter dynamics is slow compared to typical electronic times, so that the distinction between order parameter fluctuations and single particle excitations is well defined. This is the physical situation typical of the Peierls and possibly the spin-Peierls instabilities, where the large ionic mass makes the order parameter dynamics slow. However, in superconducting or antiferromagnetic instabilities (and in the spin-Peierls instability in the case $\omega_{2k_F} \approx J$) order parameter and single particle excitations cannot be separated. In this case one has to start from the picture of the one-dimensional interacting electron gas ('g-ology') [44]. In cases where the spin degrees of freedom are frozen out (or if the interchain coupling is dominated by a direct coupling between order parameters) crossover temperatures and thermodynamic relations can be obtained following the approach of refs. 27 and 45.

Finally, it should be pointed out that the present article is only concerned with the crossover from one- to three-dimensional behaviour of correlations of different type of order, i.e. with the onset of some type of order with decreasing temperature. Quite different temperature scales may appear if single-particle properties are considered. For example, in the presence of a transverse hopping integral $t_\perp$ single-particle properties will be anisotropic three-dimensional below $T \approx t_\perp$, whereas the crossover to three-dimensional CDW fluctuations, given by eq. (4.11), will generally happen at a quite different temperature.

Bibliography

1 R.E. Peierls, "Quantum Theory of Solids" (Oxford University Press, London, 1955), p. 108.

2 H. Fröhlich, Proc. Roy. Soc. London A **223**, 296 (1954).

3 J.C. Slater, Phys. Rev. **82**, 538 (1951); J. des Cloiseaux, J. Phys. (Paris) **20**, 607 (1959); A.W. Overhauser, Phys. Rev. Lett. **4**, 462 (1960).

4 J. Bardeen, L.N. Cooper, and J.R. Schrieffer, Phys. Rev. **108** 1175 (1957).

5 L.D. Landau and E.M. Lifshitz, "Statistical Physiscs" (Pergamon, London, 1959), p. 482; P.C. Hohenberg, Phys. Rev. **158**, 383 (1967); N.D. Mermin and H. Wagner, Phys. Rev. Lett. **17**, 1133 (1966); N.D. Mermin, Phys. Rev. **176**, 250 (1968).

6 D. Jérome and H.J. Schulz, Adv. Phys. **31**, 299 (1982), and references therein.

7 Recent Conference Proceedings are: Molec. Cryst. Liq. Cryst. **117-121**, and Proceedings of the Yamada Conference on One-Dimensional Conductors, Lake Kawaguchi, Japan, 1986, Physica B, to appear.

8 S. Barišić, in "Organic Conductors and Semiconductors", ed. by L. Pál et al. (Springer, Berlin, 1976), p. 85.

9 W. Dieterich, Adv. Phys. **25**, 615 (1976).

10 C.G. Kuper, Proc. Roy. Soc. London A **227**, 214 (1955); D. Allender, J.W. Bray, and J. Bardeen, Phys. Rev. B **9**, 119 (1974).

11 P.A. Lee, T.M. Rice, and P.W. Anderson, Sol. St. Comm. **14**, 703 (1974).

12 H. Fukuyama, J. Phys. Soc. Jpn. **41**, 513 (1976); H. Fukuyama and P.A. Lee, Phys. Rev. B **17**, 535 (1978).

13 S.A. Brazovskii and I.E. Dzyaloshinskii, Sov. Phys. JETP **44**, 1233 (1976).
14 M.J. Rice, A.R. Bishop, J.A. Krumhansl, and S.A. Trullinger, Phys. Rev. Lett. **36**, 342 (1976); S.A. Brazovskii, JETP Lett. **28**, 606 (1978), Sov. Phys. JETP **51**, 342 (1980); W.P. Su, J.R. Schrieffer, and A.J. Heeger, Phys. Rev. Lett. **42**, 1698 (1979); H. Takayama, Y.R. Lin-Liu, and K. Maki, Phys. Rev. B **21**, 2388 (1980).
15 D.K. Campbell and A.R. Bishop, Phys. Rev. B **24**, 4859 (1981); S.A. Brazovskii and N.N. Kirova, JETP Lett. **33**, 4 (1981).
16 K. Takano, Progr. Theor. Phys. **68** 1 (1982).
17 P.A. Lee, T.M. Rice, and P.W. Anderson, Phys. Rev. Lett. **31**, 462 (1973); K. Levin, S.L. Cunningham, and D.L. Mills, Phys. Rev. B **10**, 3832 (1974).
18 J.R. Tucker and B.I. Halperin, Phys. Rev. B **3**, 3768 (1970).
19 D.J. Scalapino, M. Sears, and R.A. Ferrell, Phys. Rev. B **6**, 3409 (1972).
20 D. Jérome and H.J. Schulz, in "Extended Linear Chain Compounds" vol. 2, ed. by J.S. Miller (Plenum, New York, 1982) p. 159.
21 K. Saub, S. Barišić, and J. Friedel, Phys. Lett. A, **56**, 302 (1976).
22 B. Horowitz, H. Gutfreund, and M. Weger, Phys. Rev. B **12**, 3174 (1975).
23 S.K. Khanna, J.P. Pouget, R. Comès, A.F. Garito, and A.J. Heeger, Phys. Rev. B **16**, 1468 (1977).
24 V.L. Ginzburg, Sov. Phys. Solid State **2**, 1824 (1960); A.P. Levanyuk, Sov. Phys. JETP **36**, 571 (1959); D.J. Amit, J. Phys. C **7**, 3369 (1974).
25 D.J. Bergman, T.M. Rice, and P.A. Lee, Phys. Rev. B **15**, 1706 (1977); P.A. Lee and H. Fukuyama, ibid. **17**, 542 (1978).
26 D.J. Scalapino, Y. Imry, and P. Pincus, Phys. Rev. B **11**, 2042 (1975).
27 H.J. Schulz and C. Bourbonnais, Phys. Rev. B **27**, 5856 (1983).
28 S. Barišić and K. Uzelac, J. Phys. (Paris) **36**, 1267 (1976).
29 N. Menyhard, J. Phys. C **11**, 2207 (1978); ibid. **12** 1297 (1979).
30 L.G. Caron and C. Bourbonnais, Phys. Rev. B **29**, 4230 (1984).
31 J. Voit and H.J. Schulz, Phys. Rev. B **34**, 7429 (1986).
32 E.H. Lieb and F.Y. Wu, Phys. Rev. Lett. **20**, 1445 (1968).
33 V.J. Emery, Phys. Rev. B **14**, 2989 (1976).
34 R.B. Griffiths, Phys. Rev. **133**, A768 (1964).
35 E. Pytte, Phys. Rev. B **10**, 2309 (1974).
36 M.C. Cross and D.S. Fisher, Phys. Rev. B **19**, 402 (1979).
37 J.C. Bonner and H.W.J. Blöte, Phys. Rev. B **25**, 6959 (1982); Z.G. Soos, S. Kuwajima, and J.E. Mihalick, Phys. Rev. B **32**, 3124 (1985); K. Okamoto, H. Nishimori, and Y. Taguchi, J. Phys. Soc. Jpn. **55**, 1458 (1986); G. Spronken, B. Fourcade, and Y. Lépine, Phys. Rev. B **33**, 1886 (1986).
38 J.L. Black and V.J. Emery, Phys. Rev. B **23**, 429 (1981).
39 L. Hulthén, Ark. Mat. Astron. Fys. **26A**, 1 (1938); C.N. Yang and C.P. Yang, Phys. Rev. **150**, 321, 327 (1967).
40 J.P. Pouget, R. Moret, R. Comès, K. Bechgaard, J.M. Fabre, and L. Giral, Molec. Cryst. Liq. Cryst. **79**, 129 (1982); F. Creuzet, C. Bourbonnais, L.G. Caron, D. Jérome, and K. Bechgaard, to appear in Synthetic Metals.
41 H. Fukuyama, in the Proceedings of the Yamada Conference on One-Dimensional Conductors, Lake Kawaguchi, Japan, 1986, Physica B, to appear.
42 J.W. Bray, L.V. Interrante, I.S. Jacobs, and J.C. Bonner, in "Extended Linear Chain Compounds" vol. 3, ed. by J.S. Miller (Plenum, New York, 1983), p. 353, and references therein.
43 A. Luther and I. Peschel, Phys. Rev. B **12**, 3908 (1975).
44 J. Sólyom, Adv. Phys. **28**, 209 (1979); V.J. Emery, in "Highly Conducting One-dimensional Solids", ed. by J.T. Devreese et al. (Plenum, New York, 1979), p. 247.
45 H.J. Schulz, J. Phys. (Paris) **44**, C3-903 (1983).

AN OVERVIEW OF ORGANIC SOLIDS: THE RELATION BETWEEN THEIR ELECTRONIC, OPTICAL, MAGNETIC AND STRUCTURAL PROPERTIES

Jerry B. Torrance

IBM Research
Almaden Research Center
650 Harry Road
San Jose, California 95120-6099

INTRODUCTION

Most of the effort and interest in organic solids has focussed on the study of as many detailed physical properties as possible of a very few materials, *e.g.*, TTF-TCNQ and $TMTSF_2\ ClO_4$. While a tremendous amount has been learned[1] about each of these material systems, we still have a very poor understanding of what are the important differences between them that are responsible for such large differences in their properties. A complimentary approach to understanding these materials is to study a few basic properties of as many different organic solids as possible and attempt to understand what they have in common and what is the origin of their differences. In this chapter we shall attempt to describe what we can learn from this latter approach. It is important to have the proper expectation level for what we can hope to learn from a comparison of a wide variety of materials having very different structures and very different properties. As with other solids, we cannot expect to be able to predict which structure or stoichiometry two molecules will adapt when forming a solid, although we might hope to rationalize some aspects of this in hindsight. Similarly, only the most basic properties can be understood, *i.e.*, only those properties common to many materials that are determined by the largest energies in these solids, for example, the Madelung energy, the ionization potential, the bandwidth, the intra-molecular Coulomb repulsion, the energy of the lowest intra-molecular exciton, and the electron-phonon coupling. With this approach we can more clearly get evidence as to which of these energies are the largest. In the complimentary approach, the detailed properties of a specific material are often dominated by competition between smaller energies. These details often appear to be "understood," even though incorrect assumptions have been made about the relative magnitude of the largest energies.

Organic charge transfer solids are molecular solids, containing as structural units the individual molecules from which the solid was formed. (This is in contrast with conducting polymers.) The electronic, magnetic, optical and structural properties that are of interest in these solids are not molecular properties, but are associated with intermolecular interactions. The most basic unit exhibiting these interactions is the dimer. Our approach will emphasize the molecular nature of these materials by first considering the properties of two molecules and studying the effects when they overlap to form an interacting dimer. The stacks of the solid will then be viewed as formed by stacking these dimers and studying the effects of their overlap. We shall see that in this approach there is a strong, intimate inter-relation between the basic electronic, magnetic, optical and structural properties. This common inter-relation gives rise to many common features found in different types of structures, and provides a view of the unifying features of organic solids. We shall start by considering the electronic properties of dimers. Next, we shall examine the structural, optical and magnetic properties of dimers and then stacks. For an early, more rigorous discussion of a similar approach, see the review by Soos.[2]

DIMERS

Let us first consider the dimer formed from a pair of adjacent radical anions, $A^{\bar{\cdot}}$, *i.e.*, $A_1^{\bar{\cdot}} A_2^{\bar{\cdot}}$. As shown schematically in Fig. 1, the electronic states of this dimer may be viewed from two opposite limits: the Heitler-London Limit and the Molecular Orbital Limit. The former limit applies when the overlap between the molecules within the dimer is small, in which case the ground state consists of the unpaired electrons each localized on one molecule (left side of Fig. 1). The lowest excited state is a charge transfer state, corresponding to exciting one of the electrons over to the adjacent radical, thus forming A^oA^{2-}. This excited state is higher than the ground state by an energy U, the Coulomb correlation energy or the disproportionation energy. This electron repulsion energy tends to keep each of the electrons localized on one molecule.

The other extreme limit applies when the kinetic energy associated with the electrons moving back and forth between molecules is much stronger than U. In this case one can use Molecular Orbital theory and form two orbitals for these electrons: a lower, bonding orbital and a higher, antibonding one. The energy difference between these orbitals is 2t, where t is the transfer (or resonance) integral, which is a measure of the strength of the overlap and the kinetic energy. The lowest state corresponds to putting both electrons into the lowest orbital (right of Fig. 1). In the next excited state, one electron is in the lowest and one in the highest orbital. This state lies at an energy 2t above the ground state. The highest energy state has both electrons in the antibonding orbital (right of Fig. 1). In this limit the electrons are free and non-interacting. They are independently delocalized over both molecules of the dimer and are often on one of the molecules "at the same time," since the repulsion between them has been neglected.

In practice, neither of these two extreme limits is realistic: in the cases of interest there is usually some overlap, so the electrons are not completely localized. Similarly, there is usually some Coulomb repulsion between the electrons, so that they are not noninteracting. That is, $t \neq 0$ and $U \neq 0$. The energies for the region in between these two extreme limits are sketched in Fig. 1. In organic charge transfer compounds, there is not a strong chemical bond between adjacent molecules; there is only a relatively weak charge transfer interaction which causes the intermolecular separation to be somewhat less than the van der Waals distance. Experiment and theory indicate that 2t is usually in the range between 0.1 and 0.5, with a maximum of ~1.0 eV for fluoroanthene$_2$ PF_6. Experimental and theoretical

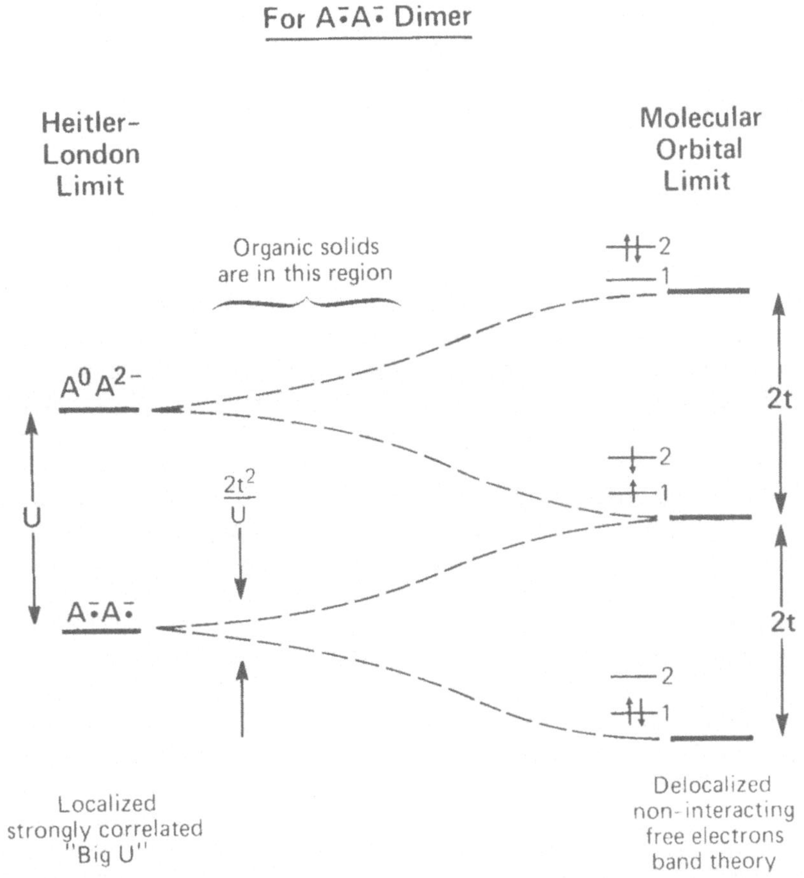

Figure 1. Energy levels for states of a A ·̄ A ·̄ dimer in the two extreme limits: Heitler-London and Molecular Orbital. Dashed lines represent schematic solution for the intermediate regime.

estimates for U (and the charge transfer band energy) range between 0.5 and 2.0 eV. From these numbers we can see that in general neither the $U/2t \gg 1$ nor the $U/2t \ll 1$ limits are realistic. Our materials lie in the difficult region of intermediate coupling. Nevertheless, we will discuss these materials in the vocabulary and the theory of strong correlations (Big U) for the following reasons: (1) the intermediate region is extremely difficult and if we want physical insight into the problem we must choose one of the two limits; (2) the numbers are such that $U/2t > 1$ is usually valid (although this is a controversial point); (3) this limit is pedagogically the clearest way to illustrate many of features of the physical properties of these materials,[3,4,5] *i.e.*, one can understand more features and obtain more physical insight from this limit than from the $U/2t \ll 1$ limit. Thus, we shall consider the $U/2t \gg 1$ limit, always recognizing that the materials we are measuring are certainly not in such an extreme limit, but hopefully the insight gained will nevertheless be applicable in the intermediate region.

STRUCTURAL PROPERTIES

We will now consider a variety of dimers formed from planar aromatic π-donors(D) and π-acceptors (A) from the point of view of the strongly correlated limit, with weak overlap. Consider first the dimer formed from a donor molecule adjacent to an acceptor, thus forming a DA pair. This dimer can be nominally neutral (D^oA^o) or nominally ionic ($D \overset{+}{\bullet} A \overset{-}{\bullet}$). The energies of these mixed or hetero-molecular dimers are shown schematically in Fig. 2. The lowest excited state of D^oA^o is $D \overset{+}{\bullet} A \overset{-}{\bullet}$, *i.e.*, where an electron is excited from D^o to A^o. Correspondingly, the lowest excited state of $D \overset{+}{\bullet} A \overset{-}{\bullet}$ is D^oA^o, where an electron is excited from $A \overset{-}{\bullet}$ to $D \overset{+}{\bullet}$. These dimers can be stacked on top of each other, to form Mixed Stacks of alternating donor and acceptor molecules (as indicated in Fig. 2). There are a vast number of Neutral Mixed Stack organic solids. Anthracene-PMDA and the Neutral phase of TTF-chloranil are two prominent examples. There are only a few Ionic Mixed Stack solids, including TMPD-TCNQ and TMPD-chloranil.

In addition to these hetero-dimers, there are two important types of homo-molecular dimers: the fully ionic $A \overset{-}{\bullet} A \overset{-}{\bullet}$ and the mixed valence $A \overset{-}{\bullet} A^o$ dimer (Fig. 2). These dimers can also be stacked on top of each other, to form Segregated Stacks, as shown in Fig. 2. There are also dimers of the type $D \overset{+}{\bullet} D \overset{+}{\bullet}$ and $M \bullet M \bullet$ ($M\bullet$ is a neutral free radical) as well as $D \overset{+}{\bullet} D^o$ and $M \bullet M^+$ which are equivalent to $A \overset{-}{\bullet} A \overset{-}{\bullet}$ and $A \overset{-}{\bullet} A^o$, respectively. (There are also possible dimers of the type D^oD^o, A^oA^o, $D \overset{+}{\bullet} A^o$, $D^oA \overset{-}{\bullet}$, $D^{2+}A^{2-}$, *etc.*, but these are not as relevant to the charge transfer materials as the four main types shown in Fig. 2.) Examples of fully Ionic Segregated Stacks include K-TCNQ, TMPD-I, and HMTTF-TCNQF$_4$. Segregated stacks can be mixed valence for two reasons: because of a non-1:1 stoichiometry, such as in MEM-TCNQ$_2$, TTF-Br$_{0.79}$ and TMTSF$_2$ PF$_6$, or because there is incomplete transfer of charge from donor to acceptor, as in TTF-TCNQ and HMTTF-TCNQ.

The effect of including the intermolecular overlap between the two molecules of the dimer causes a configuration interaction between the ground state and the excited charge transfer state. That is, the transfer integral mixes the excited state into the ground state. This interaction/mixing results in a decrease in the energy of the lowest state (Fig. 2). For example, the overlap between D^o and A^o in D^oA^o

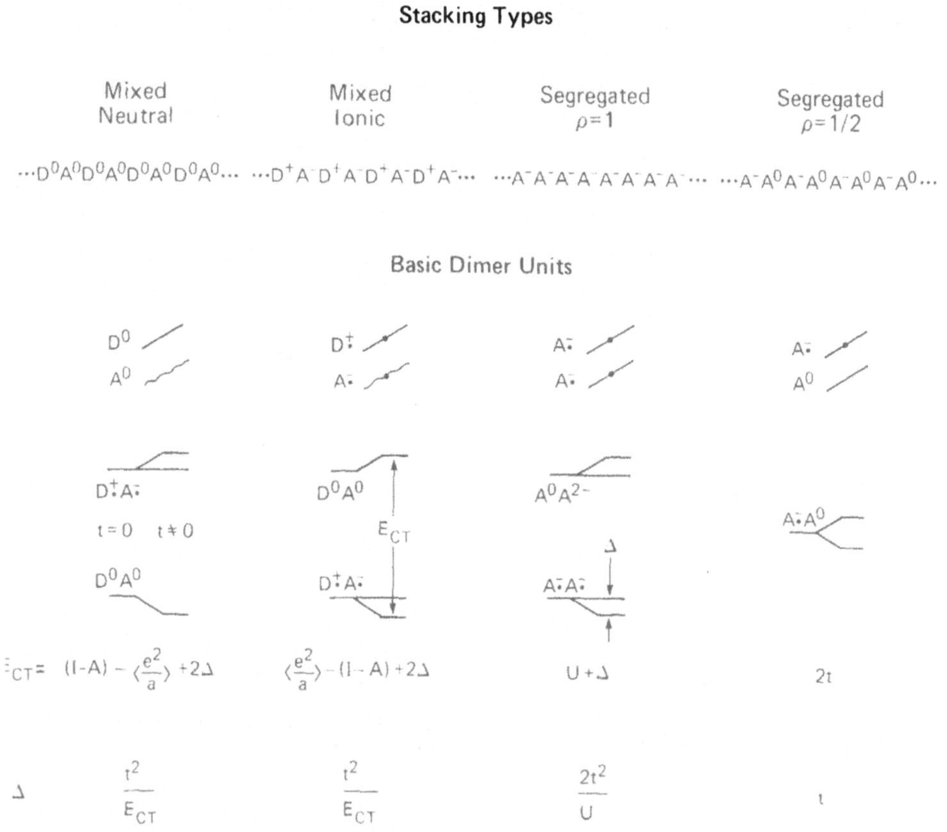

Figure 2. Four principal types of stacks in organic solids and four corresponding types of dimers, together with their energy levels, the charge transfer band energy (E_{CT}), and the stabilization energy (Δ).

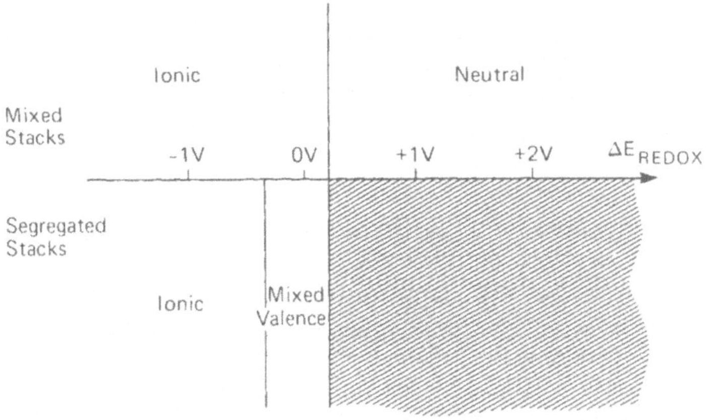

Figure 3. The different phases of 1:1 D:A Mixed and Segregated stacking solids as a function of ΔE_{REDOX} (Schematic).

mixes some of the ionic $D^{\dot{+}} A^{\dot{-}}$ state into D^oA^o, with two consequences: first, the charge (q) on the molecules in the $D^{+q}A^{-q}$ dimer is increased from zero, and secondly, the energy of the dimer is lowered by an energy t^2/E_{CT}, where E_{CT}, is the energy of the $D^{\dot{+}} A^{\dot{-}}$ state relative to D^oA^o. That is, the D^oA^o state is stabilized by the delocalization of the electrons induced by the overlap. Similarly, the ground $D^{\dot{+}} A^{\dot{-}}$ state is stabilized by an energy t^2/E_{CT} because of mixing in some D^oA^o character, caused by the off-diagonal matrix element of t between $D^{\dot{+}} A^{\dot{-}}$ and D^oA^o. For the homo-molecular dimers, the energy of the ground state $A^{\dot{-}} A^{\dot{-}}$ is decreased by $2t^2/U$ due to the mixing of both $A^{2-}A^o$ and A^oA^{2-}. The two states $A^{\dot{-}} A^o$ and $A^oA^{\dot{-}}$ are degenerate and hence the effect of the matrix element t is to split this degeneracy by an energy 2t (Fig. 2). Using the range of 2t between 0.1 and 1 eV with the range of ΔE between 0.5 and 2 eV gives a range in the stabilization energy, $\Delta \sim t^2/\Delta E$, between 15K and 6,000K.

It is this charge transfer stabilization energy, Δ, that is responsible for the very common observation that D^oA^o dimers are readily formed in solution. More remarkably, dimers of $A^{\dot{-}} A^{\dot{-}}$ and $D^{\dot{+}} D^{\dot{+}}$ are also observed in solution, in spite of the Coulomb repulsion between the charges. In solids, this same energy is responsible for the formation of stacks, in which each pair of neighboring molecules are held together in the same way as a dimer. Thus, many examples are found of each of the four stacking types of Fig. 2. Note that the dimers of D^oD^o (and A^oA^o) would not be stabilized very well, because the relevant excited charge transfer states D^+D^- (and A^+A^-) are generally very high in energy (hence, $t^2/\Delta E$ is small). For this reason, dimers of the type D^oD^o and A^oA^o are not commonly observed. For the same reason, there are no examples of Segregated Stacks of neutral molecules (...$D^oD^oD^oD^oD^o$...) and (...$A^oA^oA^oA^oA^o$...); they always form Mixed Stacks (...$D^oA^oD^oA^oD^oA^o$...) (See discussion of Fig. 3 below.)

What determines which stacking type will occur for a given D and A? This is of course an extremely complex problem and there is certainly no simple answer that can accurately explain all of the observations. Nevertheless, an oversimplified picture has been given for 1:1 compounds that surprisingly well accounts for many features of the observed behavior.[6,7] The presumption is that the principal variable is ΔE_{REDOX}, *i.e.*, the difference between the electrochemical oxidation potential of the donor and the reduction potential of the acceptor. The results are schematically shown in Fig. 3. For Mixed Stack solids with high ΔE_{REDOX} (*i.e.*, bad D and A), Neutral Mixed Stacks are formed, whereas excellent D and A (low ΔE_{REDOX}) form Ionic Mixed Stacks. Futhermore, a relatively sharp boundary is found between them, as expected from the simplest ($t = 0$) theory.[8] Some neutral materials near this boundary can actually be induced to transform from Neutral to Ionic under pressure[9] or at low temperature,[10] notably TTF-chloranil. For Segregated Stacks, excellent D/A combinations form fully Ionic stacks ($\rho = 1$, where ρ is the average number of unpaired electrons per molecule). Since Segregated Stacks of neutral donors (or acceptors) are not well stabilized (see above), Segregated Stacks are not formed by combinations of poor donors and acceptors: they must form Mixed Stacks. It has been noted how the vast majority of charge transfer compounds form Mixed Stacks. This fact is largely associated with the fact that ΔE_{REDOX} for this vast majority is too high to have the possibility of forming Segregated Stacks (Fig. 3).[11]

For a range of intermediate values of ΔE_{REDOX}, Segregated Stacks of Mixed Valence molecules can be formed in which the degree of charge transfer from donor to acceptor is not complete. All of the known highly conducting 1:1 charge transfer compounds fall in this region.[6,7,12] Of course, Mixed Valence ($\rho < 1$) Segregated Stacks may also be formed if the D:A stoichiometry is not 1:1 or for cation and anion radical salts with non-1:1 stoichiometry.

OPTICAL PROPERTIES: CHARGE TRANSFER ABSORPTION

Perhaps the most striking indication of intermolecular interactions in dimers is in their optical absorption spectrum: there is often a new absorption band observed that is not present in the spectrum of the monomers. When a donor is mixed in solution with an acceptor, D^oA^o dimers are often formed. When a solution of radical cations (or anions) is concentrated and cooled, $D\overset{+}{\bullet}D\overset{+}{\bullet}$ (or $A\overset{-}{\bullet}A\overset{-}{\bullet}$) dimers may be formed. In these cases, the extra absorption observed corresponds to transferring an electron from one molecule in the dimer to the other. In the stacks of a solid, there is a similar extra absorption with a similar origin. Note that for each of the four types of dimers considered in Fig. 2, the lowest excited state corresponds to such a charge transfer excitation. The overlap between molecules (the transfer integral) mixes ground and excited states, thereby giving oscillator strength to this charge transfer (C.T.) absorption.

Expressions for the energies of the C.T. absorption, E_{CT}, are given near the bottom of Fig. 2 for each of the four dimer types. For a D^oA^o dimer, the energy is proportional to (I-A), the ionization potential of the donor minus the electron affinity of the acceptor. For hundreds of D-A dimers in solution, experimental agreement with this simple relationship has been well documented,[13] using either calculated values for (I-A) or using ΔE_{REDOX}. In the solid state a test of this simple relationship for a variety of D^oA^o solids is shown[9] in Fig. 4. Here the energy of the charge-transfer band in the solid is plotted *versus* an electrochemical measurement of the isolated molecules in solution. The agreement is surprisingly good, with the data having a slope of 1. For the few examples we have of D^+A^- solids (solid rectangles), the agreement with the expected behavior[8] (straight line in figure) is not very good, suggesting this simple model is too simple.

For ($A\overset{-}{\bullet}A\overset{-}{\bullet}$) and ($D\overset{+}{\bullet}D\overset{+}{\bullet}$) dimers, the energy of the charge transfer band is given by U, the disproportionation energy. Values of U obtained from a variety of dimers in solution are given in Fig. 5. Note that $U \sim 1.5 \pm 0.3$ for the donors, acceptors and free radicals used in most charge transfer solids. Compared with variations of the charge transfer band energies E_{CT} for D^oA^o materials (factor of ~4), with variations in the value of 2t (factor of ~10), and with variations in t^2/E_{CT} (factor of ~400), the value of U is relatively constant at ~1.5 eV. In fully Ionic Segregated Stack solids, an absorption is observed at an energy somewhat lower than the value of U in solution.

In Fig. 6, we compare[14] some spectra for $TTF\overset{+}{\bullet}$ in solution and in the solid.[15] The short dashed curve is for isolated $TTF\overset{+}{\bullet}$ molecules in solution, showing the intra-molecular absorption for $TTF\overset{+}{\bullet}$. The solid curve above it is for $(TTF\overset{+}{\bullet})_2$ dimers in solution, showing a new excitation at $U \sim 1.7$ eV, in addition to the peaks of the $TTF\overset{+}{\bullet}$ monomer (blue shifted as expected). In the monovalent

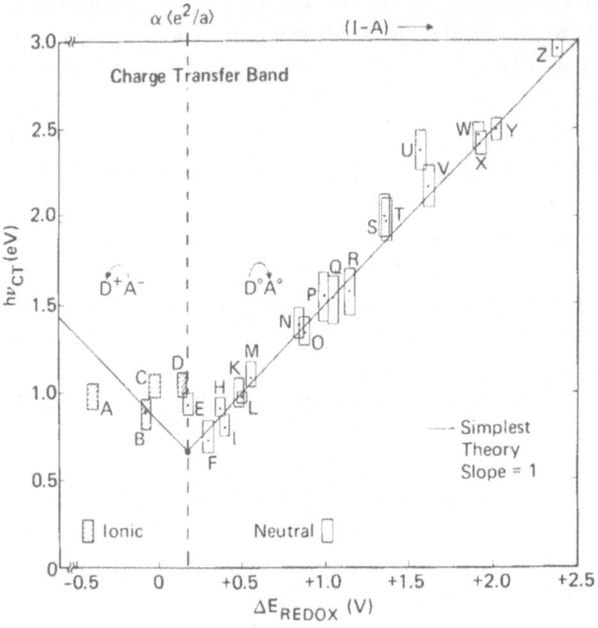

Figure 4. E_{CT} for a variety of Mixed Stack Solids *versus* ΔE_{REDOX}.

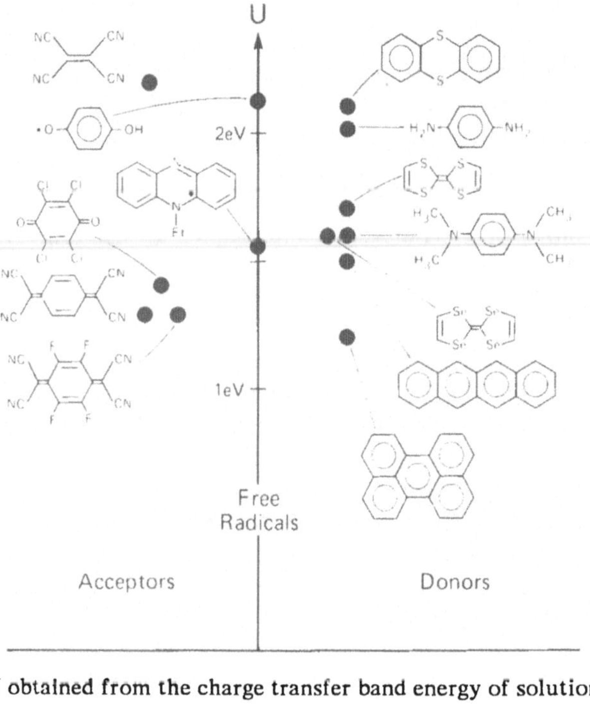

Figure 5. U obtained from the charge transfer band energy of solution dimers.

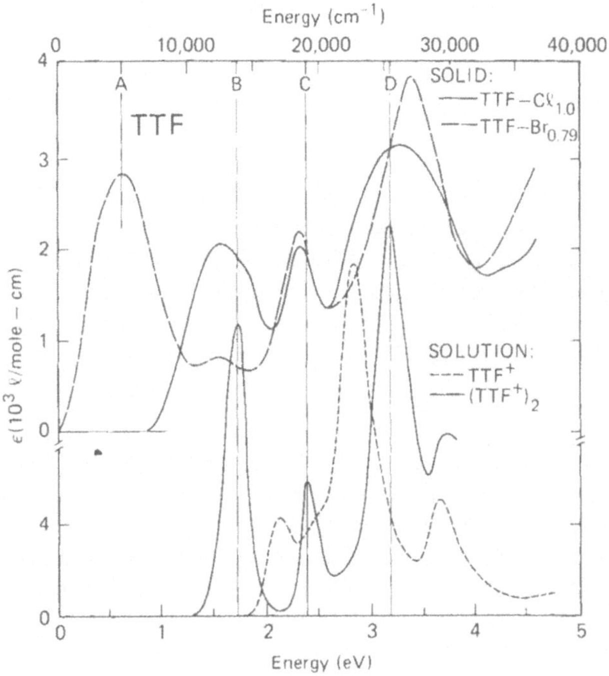

Figure 6. Optical absorption of TTF $\dot{+}$ molecules, dimers, as well as fully ionic (ρ = 1) and Mixed Valence Segregated Stacks.

TTF-Cl salt, the solid shows a spectrum almost identical with that of the solution dimer. In Fig. 7, similar spectra are shown[3,16] for the radical anion TCNQ$^{\bar{\bullet}}$. The lowest spectrum is that of dimers of TCNQ$^{\bar{\bullet}}$ in solution, showing the absorption peak at 1.3 eV, corresponding to U. The absorption peaks above this energy are intra-molecular excitations of TCNQ$^{\bar{\bullet}}$. In each of the solids K-TCNQ, N-methylacridinium-TCNQ, and HMTTF-$TCNQF_4$, there is an absorption peak assigned to the charge transfer excitation at U ~ 1 eV, which is somewhat lower in energy than that observed in solution.

For a Mixed Valence dimer (*e.g.*, A$^{\bar{\bullet}}$ A^{o}), the charge transfer band should occur much lower in energy than for the other cases, since the energy 2t (0.1-1.0 eV) is in the infrared, whereas U(~1.5 eV) and ΔE_{CT} (0.7-3.0 eV) are in the UV-VIS range. Probably for this reason, this mixed valence C.T. absorption is not often observed in dimers. As we imagine stacking these mixed valence dimers to form a Mixed Valence stack, the energy of the charge transfer excitation in this simple picture decreases toward zero, corresponding to the simple expectation that such a stack will be a metal. Because of a number of complicating factors, however, the mixed valence charge transfer band occurs at a finite frequency, typically 0.3-0.6 eV (but this energy is *not* a measure of 2t or U). Examples of spectra of Mixed Valence stacks are shown in Figs. 6 and 7. The spectrum of TTF-$Br_{0.79}$ shows the broad low energy mixed valence absorption peaking near 0.6 eV, in addition to a reduced absorption intensity at U ~ 1.5 eV. With ρ = 0.79 radicals per site on average, radicals will often be adjacent to other radicals as well as neutrals, *i.e.*, they will be involved in monovalent dimers as well as mixed valence dimers. In Fig. 7, the spectra of TTF-TCNQ and HMTTF-TCNQ show the low frequency absorption of their Mixed Valence stacks that is related to their high conductivity.

As a historical note, it was earlier imagined[17] that the highly conducting organic solids were conducting because the polarizability of the organic molecules on neighboring stacks screened (decreased) U sufficiently that U << 2t. It is clear now that high conductivity in organic solids is correlated with Mixed Valence stacks[6,7] and not high polarizability. The difference of 10^6 in room temperature conductivity[16] between the two isostructural compounds HMTTF-TCNQ and HMTTF-$TCNQF_4$ is caused by a difference in the degree of charge transfer (the polarizability of all the molecules being essentially identical).[16] The point is that these "highly polarizable" molecules are not polarizable enough to substantially decrease U. For example, the optical dielectric constant for TTF-TCNQ is only 2.4. (The issue of how much the conductivity of mixed valence compounds effectively decreases U is still a subject of controversy.)

The intensity or oscillator strength of these charge transfer absorptions is another aspect of the effect of 2t on the Heitler-London states of the dimer. Corresponding to the shifts in the energy of the states in Fig. 2, there is also a mixing of the wavefunctions caused by 2t. One result of this admixture is for the nominally D^oA^o ground state to become $D^{+q}A^{-q}$ (as the excited D^+A^- state is mixed in). Another consequence is that the oscillator strength of the C.T. transition is increased (it is zero if 2t = 0). In fact, the relation between these two consequence can be tested: from a measurement of the oscillator strength, one can obtain a value of 2t, from which a value of q can be calculated. At three different temperatures in TTF-chloranil, three different values of q thus obtained[18] are in excellent agreement

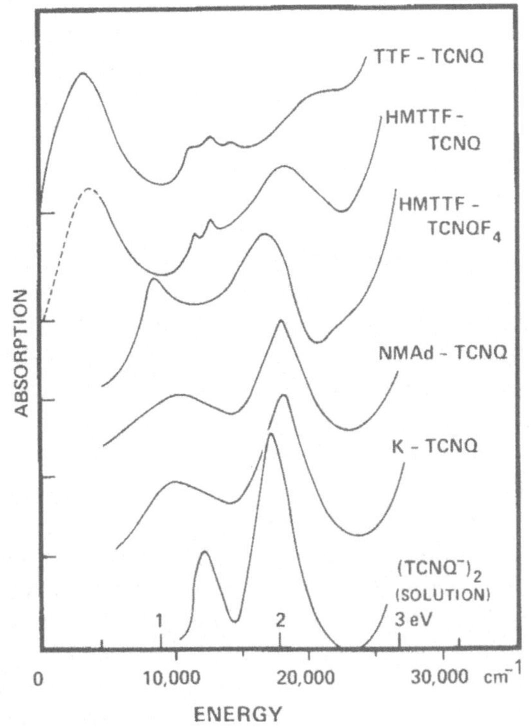

Figure 7. Optical absorption of TCNQ $\bar{\cdot}$ in dimers in solution and in fully Ionic (ρ = 1) and Mixed Valence Segregated Stacks.

with values of q obtained independently by studying infrared vibrational frequencies.[19]

MAGNETIC PROPERTIES

Let us start by reexamining the energy levels of a $A^{\bar{\bullet}} A^{\bar{\bullet}}$ dimer, as shown in Fig. 8, but now including the effects of the spin of the unpaired electrons. In the ground state, the two unpaired spins may be in either a Singlet or Triplet configuration. If there were no overlap between the molecules, there would be no spin-spin interaction (other than weak dipole-dipole) and the Singlet and Triplet states would be degenerate. As discussed throughout this chapter, the effect of the overlap (2t) is to cause a configuration interaction between the ground and excited charge transfer states. Since this excited state ($A^{2-}A^o$) is usually a Singlet state and since the interaction is spin conserving, 2t will couple the excited Singlet state to only the ground Singlet state, lowering its energy by $2t^2/U$ (as discussed earlier). Thus, including the spin dependent aspects of the problem, we see that the same charge transfer stabilization we have been discussing is also a magnetic interaction, since the Singlet state is stabilized while the Triplet is not. The energy difference between Singlet and Triplet is a spin-spin interaction, or an exchange interaction, J, which is equal to Δ, the charge transfer stabilization energy (Fig. 2). A similar discussion also applies to the cases of $D^{\dot{+}} D^{\dot{+}}$ and $D^{\dot{+}} A^{\bar{\bullet}}$ dimers, with similar conclusions. In the case of the neutral D^oA^o dimer, there are no spins, while for the mixed valence $A^{\bar{\bullet}} A^o$ there is only one unpaired spins.

Let us now extend the discussion to stacks formed from dimers, using Fig. 9. In a Neutral Mixed Stack, there are no unpaired electrons. In an Ionic Mixed Stack, the spins will have an antiferromagnetic interaction between them along the stack, since the interaction between each of the pairs (dimers) favors an antiparallel alignment of the unpaired electronic spins. This is also the case for a fully Ionic Segregated Stack, as shown schematically in Fig. 9. The case of a Mixed Valence Segregated Stack is somewhat more subtle. The electrons in this stack can conduct and should be imagined as being somewhat delocalized along the stack. During the time that they are not adjacent to another spin, there is no interaction, but when they are adjacent they may be viewed as forming an ionic dimer and hence, the Singlet state is stabilized. For the case when the electrons have a repulsion only when they are on the same molecule (Hubbard model), the above hand waving arguments[3] are supported by calculations in the "big U" limit. Coll[20] has shown for a segregated stack with ρ electrons per molecule that the spins have an antiferromagnetic interaction between them given by

$$J = \frac{2t^2}{U}\rho\left(1 - \frac{\sin 2\pi\rho}{2\pi\rho}\right).$$

Thus, the spins in a Mixed Valence Segregated stack should have an antiferromagnetic interaction between them, but the magnitude of this interaction is reduced from the $\rho = 1$ value of $J = (2t^2/U)$. For $\rho = 1/2$, the exchange interaction is reduced by a factor of two. The magnetic susceptibility per molecule is also reduced by a factor of ρ, since there are only ρ spins per molecule.

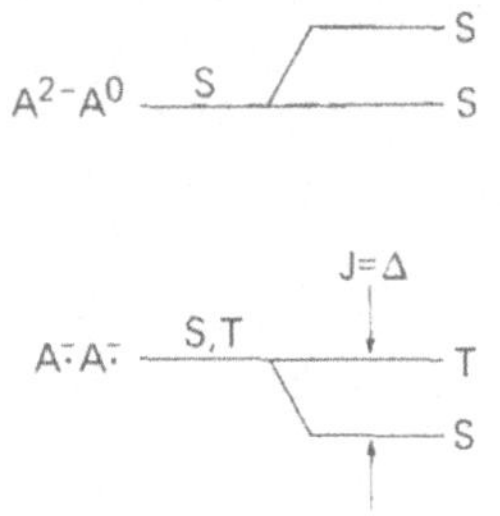

Figure 8. Energy levels of $A^{\bar{\cdot}}\,A^{\bar{\cdot}}$ dimer, including spin.

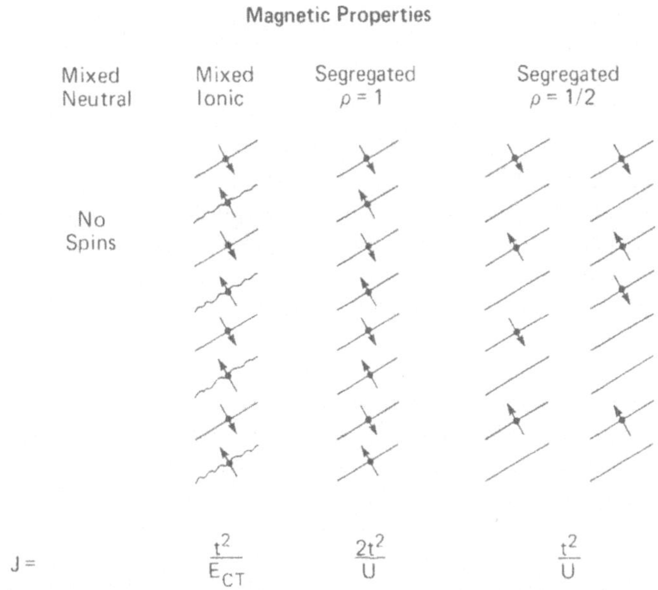

Figure 9. Schematic of three types of stacks of spins with antiferromagnetic interactions, with expression for exchange interaction J.

Thus, we conclude that the three types of stacks are magnetically very similar: they are stacks of spins that are coupled with each other by an antiferromagnetic exchange interaction J (Fig. 9). The fact that in Mixed Stacks there are both donors and acceptors does not give rise to any unusual magnetic property. Similarly, a metallic Mixed Valence Segregated Stack has similar magnetic properties to those of insulating $\rho = 1$ stacks and Mixed Stacks (as long as the coupling between stacks can be neglected). Furthermore, the Ionic Mixed Stack salts have $E_{CT} \sim 1$ eV (Fig. 4) and the segregated stacks have $U \sim 1$ eV, so that E_{CT} will be quite similar. Thus, for Mixed Stacks, for Segregated Stacks with $\rho = 1$, and for Segregated Stacks with $\rho = 1/2$, the exchange J is approximately $t^2/1$ eV (neglecting factors of 2). For this reason, variations in the magnetic susceptibility will be caused by variations in t (the overlap) and not by differences in the structure or variations in ρ, U, or E_{CT}.

For such a stack of spins with antiferromagnetic interactions, the magnetic susceptibility is expected to follow the Bonner-Fisher susceptibility[21,22] shown as the dashed line in Fig. 10. The value of the exchange interaction, J and the number of spins, ρ, per molecule are the only parameters. At low temperature, $\chi(T)$ scales like ρ/J and the temperature scales like J. So, for a value of J twice as large, χ at low temperatures would be half as large, but the temperature at which $\chi(T)$ peaks and then decreases would be twice as high. At high temperatures (*i.e.*, $T >> J$), χ approaches the Curie-Weiss susceptibility ($\chi = C/(T + J)$). With a very few notable exceptions, the magnetic susceptibility of organic charge transfer compounds is similar to that of TMPD-I shown[23] in Fig. 10. It does not follow the expected Bonner-Fisher susceptibility, especially at lowest temperatures: rather than approaching a finite value at low temperatures, $\chi(T)$ is observed to exponentially decrease toward zero.

In Figs. 11-13 we plot log $\chi(T)$ *versus* log T for a wide variety of representative Mixed Stack (Fig. 11), Segregated $\rho = 1$ Stack (Fig. 12), and Segregated Mixed Valence (Fig. 13) compounds, listed in Table I. The fact that all the $\chi(T)$ data are smaller than the Curie χ indicates that the interactions between spins are antiferromagnetic in all of these compounds. With the notable exceptions of compounds T and U (to be discussed later), the magnetic susceptibility of these compounds approaches zero at low temperatures, contrary to the expectation of Bonner and Fisher for a uniform antiferromagnetic stack. The critical word here is "uniform", because in a number of examples of Segregated $\rho = 1$ stacks, a distortion has been observed which results in a dimerization of the stacks. Such a dimerization, takes the uniform stack into a stack of dimers, which will cause $\chi(T)$ to go exponentially to zero at low T. Recently, such a dimerization has been inferred from infrared measurements in Mixed Stack compounds.[24] In Mixed Valence compounds, a lattice distortion at $2k_F$ has been observed in a number of systems which also causes $\chi(T)$ to be thermally activated. In Figs. (11-13) the black dots indicate where a phase transition has been found. In Table I, the temperatures of these transitions are listed, followed by a (D) if a distortion was observed at that transition.

Thus, the general behavior of $\chi(T)$ that is similar for all the compounds in Fig. (11-13) (except T and U) (even for those where no one has looked for a distortion) is that at low T, $\chi(T)$ vanishes because of a phase transition which distorts their stacks of antiferromagnetically coupled spins.[25] What is the origin of this phase transition? Does $\chi(T)$ become activated below the transition as an innocent consequence of the distortion which itself is caused by other factors which

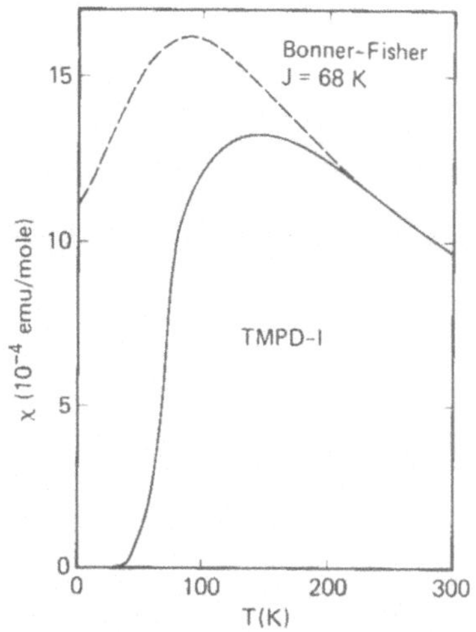

Figure 10. Measured magnetic susceptibility of TMPD-I compared with that expected for a uniform antiferromagnetic stack with J = 68K (J. B. Torrance, unpublished work).

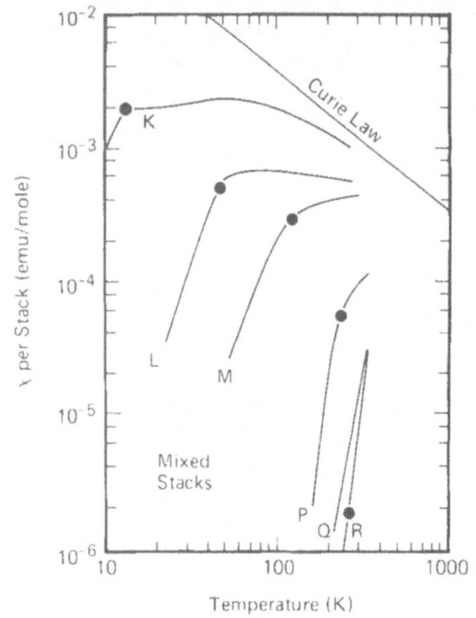

Figure 11. Log χ *versus* log T for a representative series of Mixed Stack solids from Table I.

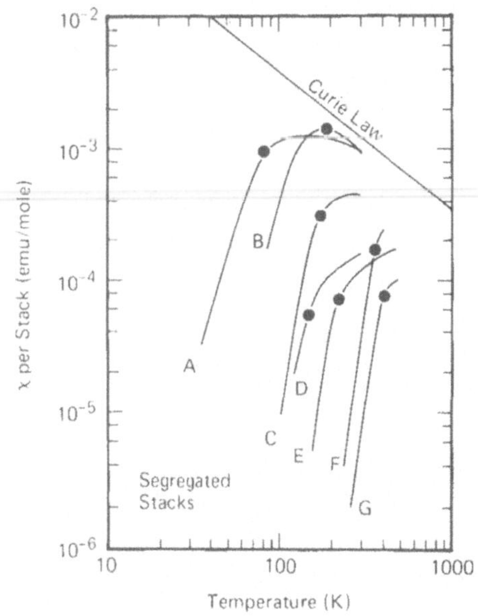

Figure 12. Log χ *versus* log T for a representative series of $\rho = 1$ Segregated Stack solids from Table I.

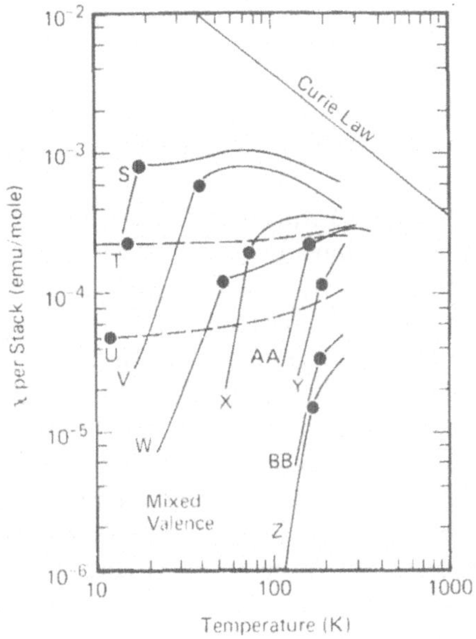

Figure 13. Log χ *versus* log T for a representative series of Mixed Valence ($\rho < 1$) Segregated Stack solids from Table I.

drive the phase transition? Or, is the phase transition itself driven by the magnetic properties of these antiferromagnetic stacks? It is known that one-dimensional antiferromagnetic stacks of $s = 1/2$ spins are potentially unstable to a Spin-Peierls phase transition,[26–29] which would distort the stacks and cause $\chi(T)$ to be activated.

There are two general arguments which strongly indicate that these phase transitions are driven by the magnetic interactions.[25] First, the phase transition temperature, T_0, appears to roughly scale with the antiferromagnetic exchange interaction J (which is determined by the overlap). For example, compounds G, D, and A in Fig. 12 have progressively lower T_0 and higher χ at lower T (indicating lower J). The overlap between molecules in the stacks of G(K-TCNQ), D($TCNQF_4$ stack in HMTTF-$TCNQF_4$) and A(TMPD-I) show progressively weaker overlap. In the Mixed Stack compounds of Fig. 11, T_0 is progressively lower for R, P, L, and K. Correspondingly, the overlap between TMPD and chloranil in R is very strong, weaker between TMPD and TCNQ in P, and quite poor between TTF and bromanil in compound L. Furthermore, in compound K the overlap between TTF and CuBDT is quite good, but χ is high and T_0 low, in an apparent contradiction to this trend. This is because the $CuBDT^-$ anions have no unpaired spins and the magnetic exchange interaction is between TTF^+ molecules on *different* stacks.[27] Thus, the relevant overlap is very poor, consistent with the high χ and low T_0.

In order to obtain more quantitative estimates for the values of J, we can try to fit the $\chi(T)$ data for these compounds to a Bonner-Fisher $\chi(T)$. From Fig. 10, it is clear that a fit is expected only for T well above T_0, where the observed $\chi(T)$ has the weak temperature dependence expected. For the compounds in Table I with low values of T_0, an estimate of J can thus be made and is listed in Table I. For those compounds with a high T_0, the distortion has decreased $\chi(T)$ below the Bonner-Fisher $\chi(T)$ for the stack if it were undistorted. From the maximum observed value of χ, we can only obtain a upper limit on an estimate for J. These values are listed in Table I, along with the calculated ratio T_0/J. It is seen that T_0 varies by a factor of 30, but the ratio of T_0/J appears to cluster in the $T_0/J \sim 0.6 \pm 0.3$ range. (In addition, to the exceptions U and T, the phase transition in TMPD-ClO_4 (B) is known to have strong involvement of the ClO_4 groups, which might account for its anomalously high value of T_0.)

The second evidence that this phase transition is determined by the magnetic properties comes from the fact that the relationship of T_0 to J is so similar for the wide range of materials from three classes of stacks. An alternative origin of the driving force for the transition might involve the Fermi surface of a metal, a special aspect of the structure (*e.g.*, Mixed stacks, or both donor and acceptor stacks) or some aspect of the vibrational properties of these materials. The fact that similar behavior and is found for metals, semiconductors and insulators, for Segregated and Mixed Stacks, for charge transfer salts, radical anion salts, and radical cation salts suggests that the phenomenon is related to something they have in common. They have two things in common: stacks of radicals with similar vibrational properties and stacks of antiferromagnetically coupled spins. The qualitative correlation of T_0 with the overlap would be consistent with both possibilities. The semi-quantitative correlation with T_0 and J favors the magnetic driving force. (Note that T_0 and t do not correlate as well). A definitive example is TTF-CuBDT, where the overlap along the stack would favor a high T_0 whereas the overlap between stacks which determines J favors a low T, as observed. Furthermore, the fact that the wave vector

TABLE I.

	Material	(Ref.)	Structure	T_0 (K)	J(K)	T_0/J
A	TMPD-I	(a,b)	Segregated	80	68	1.2
B	TMPD-ClO_4	(c)	"	187(D)	45	4.2
C	TMPD-Br	(a)	"	175	~230	~0.8
D	Acceptor stack in HMTTF – $TCNQF_4$	(d)	"	150(D)	<670	>0.2
E	Cs-TCNQ	(e)	"	210(D)	<575	>0.4
F	TEA-HCTMCP	(f)	"	364(D)	<440	>0.8
G	K-TCNQ	(e)	"	395(D)	<910	>0.4
K	TTF-CuBDT	(g)	Mixed	12(D)	38	0.32
L	TTF-bromanil	(h)	"	56(D)	~190	~0.3
M	M_2P-$TCNQF_4$	(i)	"	122	~250	~0.5
P	TMPD-TCNQ	(b)	"	210	<790	>0.3
Q	TMPD-bromanil	(j)	"		<2400	
R	TMPD-chloranil	(j)	"	240	<2600	>0.1
S	MEM-$TCNQ_2$	(k)	Seg. M.V.	18(D)	53	0.34
T	$TMTTF_2$-SCN	(l)	"	<9	~180	<0.05
U	$TMTSF_2PF_6$	(m)	"	<12	≤435	<0.03
V	DBTTF-$TCNQCl_2$	(n)	"	38(D)	80	0.48
W	TTF-TCNQ	(p)	"	52(D)	~185	~0.3
X	TMTTF-bromanil	(q)	"	75(D)	~110	~0.7
Y	TMTTF-chloranil	(q)	"	190(D)	<220	>0.9
Z	TTF-$Br_{0.79}$	(r)	"	170	<1800	>0.1
AA	$ET_3(ClO_4)_2$	(s)	"	171	~270	~0.6
BB	$fluoranthene_2PF_6$	(t)	"	190	~950	~0.2

REFERENCES

a. J. Tanaka, M. Inoue, M. Mizuno, and K. Horai, *Bull. Chem. Soc. Japan* **43**, 1998 (1970).

b. J. B. Torrance (unpublished results).

c. K. Okumura, *J. Phys. Soc. Japan* **18**, 69 (1963).

d. J. B. Torrance, Y. Tomkiewicz, R. Bozio, C. Pecile, C. R. Wolfe, and K. Bechgaard, *Phys. Rev. B*. **26**, 2267 (1982).

e. J. G. Vegter and J. Kommandeur, *Mol. Cryst. Liq. Cryst.* **30**, 11 (1975).

f. S. C. Abrahams, et al., *Phys. Rev.* **B29**, 1258 (1984).

g. I. S. Jacobs, et al., *Phys. Rev.* **B14**, 3036 (1976).

h. J. B. Torrance, J. J. Mayerle, V. Y. Lee, R. Bozio and C. Pecile, *Solid State Commun.* **38**, 1165 (1981).

i. Z. G. Soos, H. J. Keller, K. Ludolf, J. Queckbörner, D. Wehe, and S. Flandrois, *J. Chem. Phys.* **74**, 5287 (1981).

j. Y. Sato, M. Kinoshita, M. Sano, and H. Akamatu, *Bull. Chem. Soc. Japan* **43**, 2370 (1970).

k. S. Huizimga, et al., *Phys. Rev.* **B19**, 4723 (1979).

l. C. Coulon, *J. de Phys. C3* **44**, 885 (1983).

m. J. C. Scott, H. J. Pedersen, and K. Bechgaard, *Phys. Rev. Letters* **45**, 2125 (1980).

n. C. S. Jacobsen, H. J. Pedersen, K. Mortensen, and K. Bechgaard, *J. Phys. C* **13**, 3411 (1980).

p. J. B. Torrance, Y. Tomkiewicz and B. D. Silverman, *Phys. Rev. B* **15**, 4738 (1977).

q. A. Girlando, C. Pecile and J. B. Torrance, *Solid State Commun.* **54**, 753 (1985).

r. P. M. Chaikin, et al., *Phys. Rev. B* **22**, 5599 (1980).

s. S. S. P. Parkin, M. Miljak and J. R. Cooper, (preprint).

t. E. Dormann and U. Köbler, *Solid State Commun.* **54**, 1003 (1985).

of the distortion changes as the wave vector ($2\,k_F$) of the magnetic excitations is strong evidence of the magnetic nature of the transition. Any suggestion that these phase transitions involve only the lattice requires a postulated driving force which is unknown and must be specified.

Thus, we conclude that almost all organic charge transfer compounds at high temperature have stacks of antiferromagnetically coupled spins. As the temperature is lowered, a Spin-Peierls-like transition occurs, below which $\chi(T)$ becomes thermally activated because of a $2\,k_F$ lattice distortion. The temperature of this phase transition scales approximately with the magnitude of the exchange interaction J. The fact that T_0/J is so large (~0.3-0.9) indicates that this transition involves extremely strong spin-phonon coupling.

The primary exceptions to the generally observed behavior are the Bechgaard salts, exemplified here by $TMTTF_2SCN$ (Curve T in Fig. 13) and $TMTSF_2PF_6$ (Curve U). From their value of J, one would expect a Spin-Peierls-like transition in the vicinity of 50-150K for T and 125-375 for U. Such a transition is clearly not found. Rather, these materials exhibit three dimensional antiferromagnetic ordering at 9 and 12K, respectively. Why and how is the Spin-Peierls-like phase transition surpressed? Presumably by interstack interactions, but the situation is not at all clear.

CONCLUSION

Many of the basic properties of organic solids can be described by starting from a model of localized electrons on dimers and considering the effects of the transfer integral. This effect is to mix the excited charge-transfer state into the ground state. This approach is applied to the four types of dimers and then extended to the corresponding four types of organic stacks formed by stacking these dimers. It is shown how this admixture gives rise to the following interrelated effects: charge transfer stabilization, magnetic susceptibility and optical absorption. This approach emphasizes the interrelation of these properties and their similarity in a wide variety of different materials from each of the four broad types of organic charge transfer solids. These stacks are shown to have the same type of stabilization, the same type of antiferromagnetic exchange interaction, related charge transfer absorption, and a Spin-Peierls-like distortion.

Many of these features are illustrated in a comparison of the two compounds[16] HMTTF-TCNQ and HMTTF-$TCNQF_4$. The major effect of substituting 4 fluorenes on TCNQ to increase its electron affinity, but not its size. Therefore, it is not surprising that ρ is higher for the $TCNQF_4$ salt, nor that their crystal structures are almost identical. In fact, $\rho = 1$ for the $TCNQF_4$, while $\rho = 0.72$ for the TCNQ salt. This difference gives rise to a 10^6 difference in their conductivity at 300K and a related low frequency electronic absorption in the TCNQ salt. Nevertheless, the values of χ/ρ are virtually the same at high temperature. At low temperature there is a Spin-Peierls-like transition on both the donor and acceptor stacks of each material, with T_0 of the acceptor being higher in both compounds. Thus, the magnetic susceptibility behavior and the structure are more general properties, only weakly affected by changes in ρ, which cause tremendous changes in the conductivity.

REFERENCES

1. See, for example, Proceedings of International Conference on Low Dimensional Compounds, Abano Terme, Italy; *Mol. Cryst. Liq. Cryst.* **117-121** (1985); Proceedings of International Conferences on Synthetic Metals, Kyoto (1986), Synthetic Metals (in press).
2. Z. G. Soos, *Annual Rev. Phys. Chem* **25**, 121 (1974).
3. J. B. Torrance in *Chemistry and Physics of One-Dimensional Metals*, Ed. by H. J. Keller (Plenum, New York, 1977), p. 137-166.
4. J. B. Torrance and B. D. Silverman, *Phys. Rev. B* **15**, 788 (1977).
5. J. B. Torrance, Y. Tomkiewicz and B. D. Silverman, *Phys. Rev. B* **15**, 4738 (1977).
6. J. B. Torrance, *Accts. Chem. Res.* **12**, 79 (1979).
7. J. B. Torrance, *Proc. of Symp.* on "Celebration of 30 Years of Mulliken's Charge Transfer Theory," *Mol. Cryst. Liq. Cryst.* **126**, 55 (1985).
8. H. M. McConnell, B. M. Hoffman and R. M. Metzger, *Proc. Nat. Acad. Sci. U.S.A.* **53**, 46 (1965).
9. J. B. Torrance, J. E. Vazquez, J. J. Mayerle and V. Y. Lee, *Phys. Rev. Lett.* **46**, 253 (1981).
10. J. B. Torrance, A. Girlando, J. J. Mayerle, J. I. Crowley, V. Y. Lee, P. Batail and S. J. La Placa, *Phys. Rev. Lett.* **47**, 1747 (1981).
11. J. B. Torrance in *Molecular Metals*, Ed. by W. E. Hatfield (Plenum, New York, 1979), p. 7.
12. G. Saito and J. P. Ferraris, *Bull. Chem. Soc. Japan* **53**, 2141 (1980).
13. G. Briegleb, *Elektronen-Donator-Acceptor-Komplex* (Springer, Berlin, 1961).
14. J. B. Torrance, B. A. Scott, B. Welber, F. B. Kaufman and P. E. Seiden, *Phys. Rev. B* **19**, 730 (1979).
15. D. B. Tanner, in "Extended Linear Chain Compounds," Ed. by J. S. Miller, (Plenum, New York, 1982), Vol. 3, p. 205.
16. J. B. Torrance, J. J. Mayerle, K. Bechgaard, B. D. Silverman and Y. Tomkiewicz, *Phys. Rev. B* **22**, 4960 (1980).
17. A. F. Garito and A. J. Heeger, *Accts. Chem. Res.* **7**, 232 (1974).
18. C. S. Jacobsen and J. B. Torrance, *J. Chem. Phys.* **78**, 112 (1983).
19. A. Girlando, F. Marzola, C. Pecile and J. B. Torrance, *J. Chem. Phys.* **79**, 1075 (1983).
20. C. F. Coll, *Phys. Rev. B* **9**, 2150 (1974).
21. J. C. Bonner and M. E. Fisher, *Phys. Rev.* **135**, A640 (1964).
22. For a simple analytic fit to the Bonner-Fisher susceptibility, see footnote 60 of Ref. 5.
23. J. B. Torrance (unpublished work).
24. A. Girlando, C. Pecile and J. B. Torrance, *Solid State Commun.* **54**, 753 (1985).
25. J. B. Torrance, J. J. Mayerle and J. I. Crowley, *Bull. Am. Phys. Soc.* **23**, 425 (1978).
26. E. Pytte, *Phys. Rev. B* **10**, 4637 (1974).
27. J. W. Bray, L. V. Interrante, I. S. Jacobs and J. C. Bonner, in *Extended Linear Chain Compounds*, Vol. III, Ed., J. S. Miller (Plenum, 1983) p. 353.
28. S. Huizinga, *et al.*, *Phys. Rev.* **B19**, 4723 (1979).
29. H. Fukuyama, *Proc. Int. Conf. Synthetic Metals* Kyoto, 1986. *Synthetic Metals* (in press).

ANTIFERROMAGNETIC RESONANCE AMONG ORGANIC CONDUCTORS

C. Coulon and R. Laversanne

Centre de Recherche Paul Pascal, C.N.R.S.
Doamine Universitaire de Bordeaux I
33405 Talence Cedex, France

INTRODUCTION

Since the discovery of the first organic antiferromagnet, many other magnetic organic compounds have been discovered[1]. They all present related structural organizations derived from that of the well known Bechgaard salts.

One way to probe the long range magnetic order is the study of the static magnetic susceptibility. However, this technique requires a sample of a few milligrams, not often available as a single crystal. A more sensitive experiment is the antiferromagnetic resonance (AFMR) which can be easily performed on organic compounds at least when no selenium is present in the material[2]. Moreover, this method gives directly the characteristics of the anisotropy, including the position of the magnetic axes.

We first give in this paper a brief review of the AFMR theory. In a second part, we present typical AFMR results recently obtained in the BEDT series and we discuss briefly the characteristics of the observed AF ground state.

AFMR APPLIED TO ORGANIC COMPOUNDS

The AFMR theory, due to Nagamiya[3], is a generalization of the electron spin resonance (ESR) when long range magnetic order is present. In this case, a given spin experiences both the external field and internal fields (exchange and anisotropy) which vanish with the order parameter. The frequencies of the corresponding collective excitations of the spins are strongly dependent on both the orientation and the magnitude of the external field. As an example, we present in Fig. 1a the dependence of the resonance frequencies when the field is applied along the easy direction (along which the anisotropy energy is minimum) in the zero T limit. Three parameters should be introduced to describe this curve :

- $r = 1 - \chi_{//}/\chi_{\perp}$ ($r = 1$ in the limit of Fig. 1 since $\chi_{\perp}$ $(T = 0) = 0$)
- Ω_- and Ω_+ the zero field resonance frequencies related to the magnetic energies through the relation[3,4].

$$\hbar\,\Omega_- = (\mu / \mu_B)\sqrt{2JW_{ei}} \qquad \hbar\,\Omega_+ = (\mu / \mu_B)\sqrt{2JW_{eh}}$$

where μ is the local magnetic moment. μ_B is the Bohr magneton. J is the exchange energy. W_{ei} and W_{eh} are the difference of anisotropy energy between easy and intermediate and between easy and hard directions respectively.

Working with a conventional spectrometer, the frequency is fixed and for a given direction of the field H, the magnitude of H is scanned to search for a resonance. For example, two modes are obtained along the easy axis when the experimental frequency ω_0 is lower than Ω_- (see the dashed line in Fig. 1a). Starting from this situation and rotating the sample around a given axis, rotation patterns are generated giving the resonance fields as a function of the rotation angle[4]. Such patterns are given in Fig. 1b and 1c for rotations around simple magnetic axes.

Starting from one single crystal of an organic compound, it is easy to rotate the sample about simple crystal axes. Because of the low symmetry of these materials, this may not correspond to simple magnetic directions and the rotation axes should be determined from the fit of the experimental data.

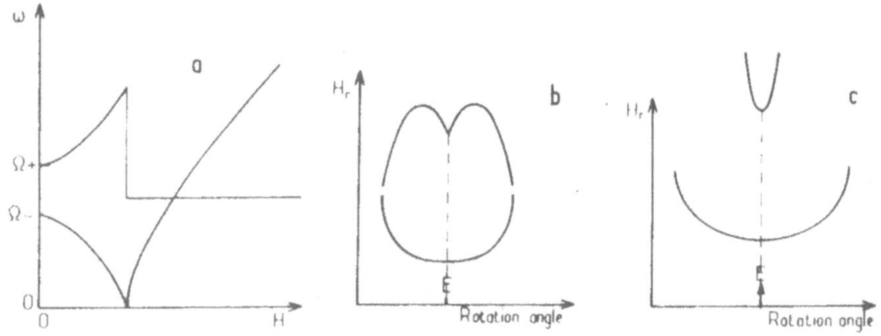

Fig. 1 : (a) Frequency versus field when H is applied along the easy axis The dashed line gives the experimental frequency.
(b) Correpondng rotation patterns in the easy intermediate plane
(c) In the easy-hard plane.

EXAMPLE OF $(BEDT)_2ICl_2$

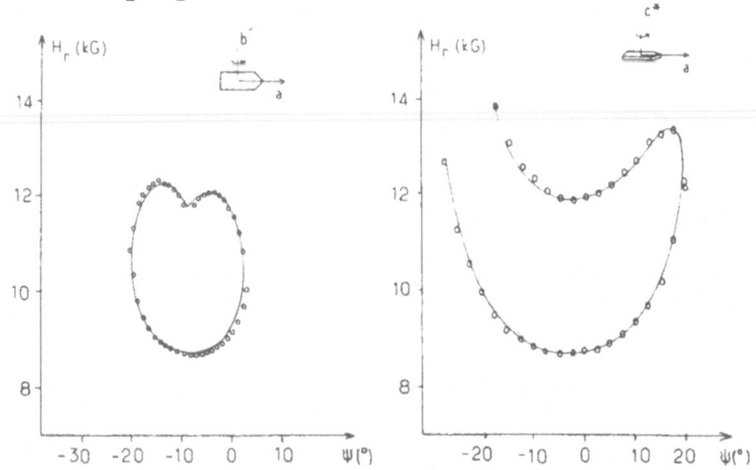

Fig. 2 : AFMR rotation patterns for $(BEDT)_2ICl_2$ at 4.5 K
(a) rotation around b', (b) rotation around c*
The continuous line gives the fit with Ω_+ = 13kGauss, Ω_- = 11kGauss. The polar angles of the rotation axes in the magnetic frame are : (a)θ = 85.5°, ψ = 95°,
(b)θ = 15°, ψ = -60°.

Many organic antiferromagnets have been investigated through AFMR[2]. Typical experimental results recently obtained for $(BEDT\text{-}TTF)_2ICl_2$ which presents an AF ground state[5] below 20 K are given in Fig. 2a and 2b. Closely related results are found for the $AuCl_2$ salt[6]. These experimental rotation patterns are close to the theoretical patterns given in Fig. 1a and 1b. However the asymmetry of the figures indicates a misorientation of the magnetic axes relative to the rotation axes b' and c*. The fits shown as the continuous lines in Fig. a and b give the position of the magnetic axes and the values of Ω_- and Ω_+ (see figure 2).

The crystal structure of $(BEDT)_2ICl_2$, schematized in Fig. 3, can be described in terms of weakly interacting dimers of BEDT molecules[5]. Consequently, the electrical conductivity of this material is very low and its magnetic susceptibility is characteristic of a chain of spins[5] (due to the stoichiometry, the charge on each dimer is one electron).

To modelize the AF ground state, two extreme descriptions are possible. The first one refers to the band structure of the material resulting from transfer integrals between neighboring molecules. These energies are defined in Fig. 3. The resulting electronic energy is :

$$\varepsilon(\vec{k}) = 2\, t_- \cos \vec{k}\vec{a} \pm \left| t + t_b\, e^{ikb} + t_\delta e^{i\vec{k}\vec{b} - i\vec{k}\vec{a}} + t_+ e^{-i\vec{k}\vec{a}} \right|$$

where t is the intradimer transfer integral, larger than the interdimer transfer integrals. When an approximate expression of $\varepsilon(\vec{k})$ is used to find the shape of the Fermi surface the corresponding optimal nesting wave vector can be derived[7]. In the present case one obtains :

$$q_{nest} = \frac{\pi}{a} + \frac{2\phi}{a} \quad \text{with } tg\phi = \left(t_\delta - \frac{t_b}{t}\, t_+\right) \Big/ \left(t_+ + 2t_- \sqrt{1 + \frac{t_b^2}{t^2}} + \frac{t_b t_\delta}{t}\right)$$

One important point is the possibility of obtaining a negative value of t_- as first mentioned in the TMTXF series[8]. The competition between t_+ and t_- leads to $q_{nest} = 0$ when $t_+ \approx -2t_-$. More generally q_{nest} is incommensurate and different from π/a.

The low conductivity of $(BEDT)_2ICl_2$ suggests that a localized picture is more appropriate. In this case, the transfer integrals introduced above are used to determine the exchange energies between neighboring dimers. Due to the competition between t_+ and t_-, we have shown that the exchange in the direction may become ferromagnetic in presence of electron-electron interactions[9]. The resulting wave vector of the AF ordering is then $q_a = 0$. Although the effect of the competition between transfer integrals is still to reduce the transverse component of the AF wave vector, the actual of q_a differs in the two limits and it would be useful to measure the AF wave vector

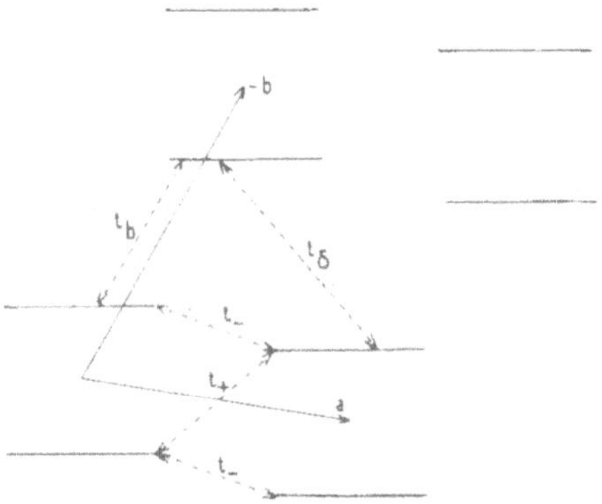

Fig. 3 : Schematization of the structure of $(BEDT)_2ICl_2$

to discriminate between the two descriptions. However the good agreement between AFMR results and the modelization of the anisotropy in the localized limit is in favor of this description[6,9].

In conclusion, AFMR is a powerful technique to study the magnetic behavior of organic conductors. Applied to $(BEDT)_2ICl_2$ it confirms the occurence of an AF ground state below 20 K and shows the strong similarity existing between this compound and other Bechgaard salts. In particular, the occurence of an AF phase implies repulsive electron-electron interactions which are probably also present for other members of the BEDT series including those for which a superconducting state is observed.

ACKNOWLEDGEMENTS

We thank M. Abderraba and L. Ducasse for fruitful discussions.

REFERENCES

1. For a recent review, see Proceedings of the ICSM Conf. Kyoto 1986.
2. C. Coulon, R. Laversanne and J. Amiell, Antiferromagnetic resonance : a probe for various organic conductors, Proceedings of the Yamada Conf. XV, Lake Kawaguchi 1986.
3. T. Nagamiya, Theory of antiferromagnetic resonance in $CuCl_2$, $2H_2O$, Adv. Phys. 4, 1 (1954).
4. C. Coulon, J.C. Scott and R. Laversanne, Antiferromagnetic resonance in TMTTF salts, Phys. Rev. B33, 6235 (1986).
5. T.J. Emge, H.H. Wang, P.C.W. Leung, P.R. Rust, J.D. Cook, P.L. Jackson, K.D. Carlson, J.M. Williams, M. Whangbo, E.L. Venturini, J.E. Schirber, L.J. Azevedo, J.R. Ferraro, New cation- anion interaction motifs, electronic band structure and electrical behavior in $\beta(ET)_2X$, J. Amer. Chem. Soc. 108, 695 (1986).
6. C. Coulon, R. Laversanne, J. Amiell and P. Delhaes, Antiferromagnetic resonance in the BEDTTTF series, submitted to J. Phys. C.
7. K. Yamaji, Pressure and anion dependence of the SDW phase transition in the Bechgaard salts, J. Phys. Soc. Jpn 55, 860 (1986).
8a. P.M. Grant, Band-structure parameters of a series of tetramethyltetraselenafulvalene $(TMTSF)_2X$ compounds, Phys. Rev. B26, 6888 (1986).
8b. L. Ducasse, A. Abderraba, B. Gallois, Temperature dependence of the Fermi surface topology in the $(TMTSF)_2X$ and $(TMTTF)_2X$ families, J. Phys. C 10, L947 (1986).
9. R. Laversanne, C. Coulon and J. Amiell, Magnetic properties of $(DIMET)_2$ SbF_6 : quantitative discussion of the antiferromagnetic behaviour, Europhys. Lett. 2, 401 (1986).

ENERGY SCALE IN ORGANIC CONDUCTORS

AND THE PROBLEM OF SUPERCONDUCTIVITY IN THE BECHGAARD SALTS

L.G. Caron

Centre de recherche en physique du solide
Faculté des sciences, Université de Sherbrooke
Sherbrooke, Qc, Canada J1K 2R1.

ENERGY SCALE

There are various electronic energy or frequency scales in all solids. The high energy limit is set by core atomic orbitals with energies typically 10^3 eV corresponding to temperatures of 10^7 K (I shall use the temperature T scale from now on equivalently for energies or frequencies thereby implying that $k_B=\hbar=1$). The atomic or molecular valence orbitals which are more relevant to the chemical processes have energies in the range 10^4-10^5 K. The conduction processes, which are our prime concern, occur within the bands resulting from single-electron tunneling processes between orbitals of neighboring molecules. The energy scale for this motion is set by the tunneling amplitudes $t_\parallel$ (along the stacks) and $t_{\perp i}$ (in both perpendicular directions). One has $t_\parallel \sim 10^3$ K[1-3], $t_{\perp 1}$ can be 10 K[1] in the highly anisotropic TTF-TCNQ salts for instance[1], 100 K in the Bechgaard salts[2] ($TMTSF_2$-X), or close to 10^3 K in the BEDT-TTF superconductors[3] which are not very anisotropic, and $t_{\perp 2}$ is a few K[1-3]. The phonon energies θ are typically 100 K. Finally the strength of the Coulomb interaction, which I shall bluntly characterize by a single parameter U (the Hubbard contact interaction), is of the order of 10^4 K[4]. The scale is shown in Fig. 1.

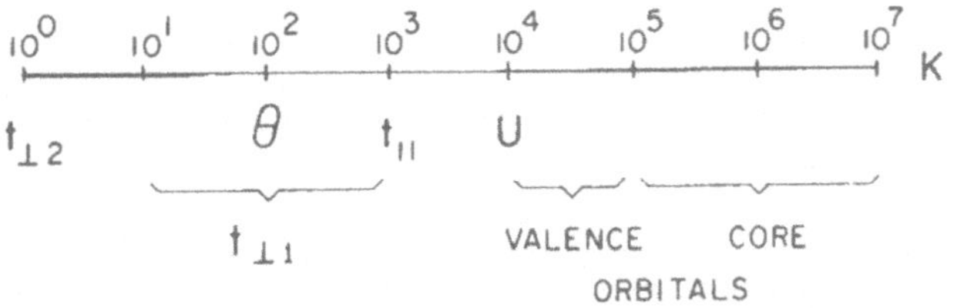

Figure 1. Energy scale in organic conductors.

EFFECTIVE HAMILTONIAN

The interesting Physics in the organics occurs between 1K and 300 K. $t_{\perp 2}$ is therefore not a relevant variable since the large thermal fluctuations average its effects to zero. The higher frequency (quantum) dynamics are surely relevant but their effect is independent of temperature. One usually tries to integrate them into parameters of an effective Hamiltonian which describes the system in the frequency range of interest. There are various ways of doing this. Perturbation theory is a standard approach[5] in which the lower frequency processes have a perturbing influence on the high frequency behavior. Renormalization group (RG) is another, though more modern, way in which one gradually integrates out the high frequency degrees of freedom of the system[6]. In the case at hand, the high frequency information in the orbitals results in an intermediate frequency (10^3 K) single-band effective Hamiltonian, of the molecular crystal or Su-Schrieffer-Heeger (SSH) type[7,8], or possibly a two-band model[9]. At lower frequencies still (10-100 K), the RG can efficiently integrate out these intermediate frequencies and yield an effective low-frequency Hamiltonian with renormalized parameters or even new ones that were generated by the RG[6,10]. For the quarter filled band situations we will be interested in, the RG analysis splits the Coulomb interaction into two dominant parameters g_1 and g_2, the backward and forward processes respectively[6]. In the one-dimensional (1D) regime, g_1 is rapidly screened out[6]

$$g_1 \approx U/[1+(U/\pi v_F)\ln(t_\parallel/T)] \quad , \tag{1}$$

while g_2 decreases slightly as T is lowered (g_1-$2g_2$ is approximately constant). Here v_F is the Fermi velocity.

APPLICATION TO $(TMTTF)_2$-X AND $(TMTSF)_2$-X

Coulomb Interaction

The strength of the bare Coulomb interaction parameter U can be estimated from static magnetic susceptibility data using the order of magnitude random phase approximation (RPA) formula[11]

$$\chi_S \approx \mu_B^2\ \chi°/[1-(U/2)\chi°] \quad , \tag{2}$$

where $\chi° = (\pi v_F)^{-1}$ is the non-interacting electron Pauli response function (these conducting salts have delocalized electrons and the non-interacting limit is a reasonable starting point) and μ_B is the Bohr magneton. One finds from the experimantal results[12] and $t_\parallel \approx 2000$ K that $\chi_S/\mu_B^2 \approx 2\chi°$ implying that $U/(\pi v_F) \approx 1$. This corresponds roughly to $U \approx 4t_\parallel$, a typical value.

Electron-Phonon Coupling

The electron-acoustic phonon coupling[8] $\lambda \propto t_\parallel$ results in effective electron-electron interactions[13] for a quarter filled band with value

$$g_{1ac} = -8\lambda^2/\kappa \ ; \qquad g_{2ac} \approx 0 \quad , \tag{3a}$$

whereas the electron-optical phonon coupling leads to[13]

$$g_{1op} = g_{2op} = -\lambda^2/\kappa \equiv g_{op} \quad , \tag{3b}$$

where κ is the effective phonon spring constant. These interactions have a frequency cutoff θ (retardation)[8b]. The mean-field spin-Peierls transition in $(TMTTF)_2$-PF_6, which can be seen by X-rays[14] and nuclear magnetic resonance[15] around 50 K, occurs at the pole of the effective RPA coupling[13]

$$(g_{lac})_{eff} \approx g_{lac}/[1+g_{lac}X(2k_F)] \quad , \tag{4}$$

where $X(2k_F)$ is the charge response function of the SSH model at the wave vector $2k_F$ corresponding to the natural 1D instability of the electron gas[8]. In this salt, the charge degrees of freedom are frozen for $T < 200$ K[16] and thus[10] $X(2k_F) \sim 200/(\pi v_F T)$ which, from Eq. 4, gives $g_{lac} \approx -0.25\pi v_F$. This value is not expected to be very different in the Bechgaard salts (the TMTSF family). $t_\parallel$ and thus λ may increase[2] but this is compensated by a larger κ resulting from the reduced intermolecular distance.

Nesting Model

In the traditional nesting model[9,17], the quasi-1D solid becomes two-dimensional (2D) when $T \leq T_X \approx t_{\perp 1}/\pi$. The thermal coherence time $(\pi T)^{-1}$ is then larger than the transverse hopping time $t_{\perp 1}^{-1}$ and the transverse motion becomes coherent. Nesting means that the Fermi surface at $-k_F$ maps fairly well onto the Fermi surface at $+k_F$ with a proper translation (see Fig. 2 of ref. 18). The RPA treatment in this 2D regime leads to the following phase transition temperatures[6,13,16,19]

$$T_N \approx T_X \exp[-2\pi v_F/g_2^X] \quad , \tag{5a}$$

$$T_P \approx T_X \exp[-2\pi v_F/(g_2^X-2g_1^X-2g_{lac}^X-g_{op}^X)] \quad , \tag{5b}$$

$$T_S \approx T_X \exp[2\pi v_F/(g_2^X+g_1^X+g_{lac}^X+2g_{op}^X)] \quad , \tag{5c}$$

where T_N, T_P, and T_S are the Néel antiferromagnetic (AFM), the Peierls and the singlet superconductivity transition temperatures respectively. g_i^X is the renormalized value of g_i at T_X. The g_{iac} and g_{iop} interactions are not appreciably renormalized if T_X is far from a lattice modulation transition (see Eq. 4). The expressions (5) are only valid when the arguments in the exponentials are negative. There are no transition temperatures otherwise. There is thus from (5) a possible competition for long-range order (LRO). It can be seen that coupling to the acoustic phonons favors the Peierls state more than the superconducting one and vice versa for the coupling to optical phonons. On the other hand, the Coulomb interaction favors the AFM state.

$(TMTSF)_2ClO_4$

In quenched $(TMTSF)_2ClO_4$, AFM LRO sets in at $T_N \sim 4$ K[20]. With $T_X \sim 80$ K[17], Eq. (5a) gives $g_2^X \sim 0.7\pi v_F$ which is close to the estimated bare value of U. From Eq. 1 and using $t_\parallel \sim 2000$ K[2] we estimate that $g_1^X \sim 0.25\pi v_F$. Since there is no Peierls transition, this means from (5b) that

$$g_{lac}^X+g_{op}^X \geq -0.25\pi v_F \tag{6a}$$

which is compatible with the estimate deduced earlier for g_{lac}. This implies that the coupling to optical phonons is not very important at low temperature. Indeed, considering that $g_{lac}^X < 0$, one has

$$g_{op}^X > -0.25\pi v_F \quad . \tag{6b}$$

When slowly cooled, this salt becomes superconducting at $T_S \sim 1$ K[20]. AFM disappears because of a transition in the anions which destroys the good nesting conditions[17]. Eq. (5c) would imply that

$$g^{x}_{1ac}+2g^{x}_{op} \approx -1.4\pi v_F \qquad (6c)$$

which is incompatible with the non-existence of a Peierls transition in the quenched state. This can be shown by substituting (6a) into (6c) to give $g^{x}_{op} < -1.15\pi v_F$ which is both unphysical and inconsistent with (6b). This brings into focus the problem of explaining superconductivity in the Bechgaard salts with the traditional approach. Emery[9] is investigating spin-exchange and two-band mechanisms of superconductivity while Bourbonnais *et al.*[10] are proposing an alternate mechanism within a modified RG theory incorporating the effect of $t_{\perp 1}$.

REFERENCES

1. S. Shitzkovsky, M. Weger, and H. Gutfreund, J. Physique 39, 711 (1978).
2. C.S. Jacobsen, this volume; P. Grant, J. Physique Colloque C3, 846 (1983); D. Jérome, in *Physics and Chemistry of Electrons and Ions in Condensed Matter*, Proc. of the NATO ASI, Cambridge, England (Reidel, Netherlands, 1984).
3. P. Grant (unpublished); F. Creuzet, C. Bourbonnais, D. Jérome, D. Schweitzer, and H.J. Keller, Europhysics Lett. 1, 467 (1986).
4. J. Torrance, this volume; M. Mehring, this volume.
5. See for example P. Löwdin, J. Chem. Phys. 19, 1396 (1951).
6. a) C. Bourbonnais, Mol. Cryst. Liq. Cryst. 119, 11 (1985); b) J. Solyom, Adv. Phys. 28, 201 (1979).
7. D.J. Scalapino and R.L. Sugar, Phys. Rev. B24, 4295 (1981); T. Holstein, Ann. Phys. (NY) 8, 325 (1959); W.P. Su, J.R. Schrieffer, and A.J. Heeger, Phys. Rev. B22, 2099 (1980).
8. a) J.E. Hirsch, this volume; b) S. Barisic, this volume.
9. V.J. Emery, this volume.
10. C. Bourbonnais and L.G. Caron, in Proc. of Yamada Conference on Physics and Chemistry of quasi-1D conductors, Lake Kawaguchi, Japan (1986), to appear in Physica B; also in Proc. of the International Conference on Science and Technology of Synthetic Metals, Kyoto, Japan (1986), to be published in Synthetic Metals; C. Bourbonnais, this volume.
11. See J. Bardeen, in *Highly Conducting One-Dimensional Solids*, edited by J.T. Devreese, R.P. Evrard, and V.E. vanDoren (Plenum, New York, 1979).
12. C. Coulon, J. Physique Colloque C3, 885 (1983).
13. L.G. Caron and C. Bourbonnais, Phys. Rev. B29, 4230 (1984).
14. J.P. Pouget, R. Moret, R. Comes, K. Bechgaard, J.M. Fabre, and L. Giral, Mol. Cryst. Liq. Cryst. 79, 129 (1982).
15. F. Creuzet, C. Bourbonnais, L.G. Caron, D. Jérome, and K. Bechgaard, in Proc. of the ICSM, Kyoto (ref. 10).
16. V.J. Emery, R. Bruisma, and S. Barisic, Phys. Rev. Lett. 48, 1039 (1982).
17. V.J. Emery, J. Physique Colloque C3, 977 (1983).
18. G. Montambeaux, this volume.
19. D. Jérome and H.J. Schulz, Adv. Phys. 31, 299 (1982).
20. T. Takahashi, D. Jérome, and K. Bechgaard, J. Physique Colloque C3, 805 (1983).

MECHANISMS OF ORGANIC SUPERCONDUCTIVITY*

V. J. Emery

Department of Physics, Brookhaven National Laboratory
Upton, N. Y. 11973
and
IBM Zürich Research Laboratory, CH8803 Rüschlikon
Switzerland

In applying the ideas of the preceding chapter to organic superconductors, it is necessary to keep in mind three essential experimental facts: (a) there is a crossover from spin-density wave order to superconductivity as the pressure is increased (b) there is no $2k_F$ charge-density wave and (c) the spin susceptibility is not activated. Using the general discussion of the previous chapter, it follows that $U > 0$ and V_1 (or an appropriate combination of V_n) is negative at high pressure. This conclusion is not restricted to the one-dimensional region but it does assume that the pairing involves electrons on the same stack. However, in the general context of organic conductors, it is not easy to imagine that V_1 can be as attractive as the experiments seem to require[1]. There is ample evidence of quite strong repulsive interactions in organic conductors, and in the conventional BCS picture of superconductivity this has to be overcome by the retarded electron-phonon interaction, to produce pairing. For the latter, an electron or hole distorts the lattice in its immediate neighborhood, and a second electron or hole feels the distortion as an effective attractive coupling to the first. The Coulomb repulsion can be overcome because the distortion takes some time to decay, and is active after the first electron or hole has moved away. But this "retardation" is not so effective in one dimension because two holes moving in opposite directions cannot help but come close to each other and feel the strong Coulomb repulsion. This conclusion is borne out by detailed calculation and applies to forward scattering from phonons, which is most important for producing superconductivity[2].

Of course, the superconducting transition takes place well below the crossover temperature T_x, where the behavior of the system becomes truly three-dimensional, but this does not make the problem go away. A recent X-ray experiment[3] on $(TMTSF)_2ReO_4$ shows an anion ordering transition above about 8 kbar, with a transition temperature that increases from 150K as the pressure increases. Thus it encompasses the one-dimensional region and possibly the superconductive phase[4]. The point is that the wave vector (0,1/2,1/2), associated with the ordered state[3] gives rise to a periodic potential of the kind shown in Fig. 4 of the preceding chapter[1], yet the electrical conductivity continues to increase as the temperature decreases and finally there is superconductivity. This behavior should be contrasted with that of $(TMTTF)_2$ SCN which has a similar anion ordering at 160K but enters a magnetic state[5] below 7K. Thus it appears that, in $(TMTSF)_2ReO_4$ at high pressure, correlations in the motion of the holes suppress the effects of the anion potential in the one-dimensional region. This too requires some kind of attractive interaction, if a $2k_F$ density instability is to be avoided.

We shall consider two ways of getting around these difficulties. The first is that the pairing involves electrons on different organic stacks[1], and it fits well with a particular source of the pairing force--the exchange of spin-density fluctuations[6]. The other takes account of the second band of hole states[7] which may be exceptionally important in the organic superconductors. Both make use of additional degrees of freedom to overcome the kinematic restrictions of a one-dimensional system.

PAIRING ON DIFFERENT STACKS

The simplest way to weaken the effects of the Coulomb repulsion is to have the members of a Cooper pair lie on neighboring stacks in the b-direction. Then the gap function has the form[1]

$$\Delta \sim \cos k_b \tag{1}$$

which vanishes for $k_b = \pm\pi/2$. The existence of zeros in Δ is a general feature of off-chain pairing and it can be tested experimentally. At first sight, the jump in the specific heat at the superconducting transition might be useful, since it is given by

$$\frac{\Delta C}{C} = 0.95 \tag{2}$$

for Eq. (1), compared to 1.43 for a constant gap[1]. However, the measured value[8] is 1.67, so $\Delta C/C$ probably is modified by strong-coupling effects[9], as in liquid He^3. Another consequence of zeros is that the specific heat[8] and sound attenuation[10] will not decrease exponentially with temperature in the ordered state. In this regard the evidence is not unequivocal at the present time. In particular, the electronic specific heat[8] is obtained by subtracting a lattice contribution from the experimental values, but this could be contaminated by an anomalous contribution from spin fluctuations.

What is the origin of the interaction responsible for pairing on different stacks? It is natural to think of the electron-phonon mechanism once again because there is no known counterexample in solid superconductors. But this would involve phonon modulation of the transverse hopping integral t_b which is only .035 eV and would need some kind of enhancement to produce a transition temperature as high as 1K. However, there is another mechanism--the exchange of spin fluctuations--which is quite plausible for organic superconductors and leads naturally to the idea of pairing between electrons on different stacks.

SPIN-DENSITY FLUCTUATIONS

The generic phase diagram for the organic superconductors has a spin-density wave state at low pressure giving way to a superconducting phase at high pressure and low temperature. Near to the spin-density wave state, there are long-lived spin fluctuations which may provide a pairing mechanism[6] in much the same way as the phonons do--an electron or hole produces a polarization of the spin density in its neighborhood, and another electron or hole interacts with it. For the non-extended Hubbard model ($V_1 = V_2 = 0$), the effective interaction may be written in terms of the functions

$$\phi_3(\underline{q}) = \frac{U^2 \chi^0(\underline{q},\omega)}{1 - U^2[\chi^0(\underline{q},\omega)]^2} \tag{3}$$

$$\phi_1(\underline{q}) = \left[2U\,\chi^0(\underline{q},\omega) + 1\right]\phi_3(\underline{q}) \tag{4}$$

where $\chi^0(\underline{q},\omega)$ is the dynamic spin correlation function for free fermions. Then, for the scattering of a pair of holes from momenta $(\underline{p},-\underline{p})$ to $(\underline{k},-\underline{k})$, the singlet and triplet scattering amplitudes are given by[6,11]

$$V_s(\underline{p},\underline{k}) = \frac{1}{2}\left[\phi_1(\underline{p}-\underline{k}) + \phi_1(\underline{p}+\underline{k})\right] + 2U \tag{5}$$

$$V_t(\underline{p},\underline{k}) = \frac{1}{2}\left[\phi_3(\underline{p}+\underline{k}) - \phi_3(\underline{p}-\underline{k})\right] \tag{6}$$

The terms involving $\phi_1(\underline{q})$ and $\phi_3(\underline{q})$ represent the effects of spin fluctuations and, in the simplest theory, the transition to an ordered spin-density wave state occurs when the denominators in Eqs. (3) and (4) vanish for $\omega = 0$ and $\underline{q} = \underline{q}_0$. Near to the transition, $\phi_n(\underline{q})$ has a peak at $\underline{q} = \underline{q}_0$ and, from Eq. (4), $\phi_1(\underline{q}_0) \approx 3\phi_3(\underline{q}_0)$ so the fluctuations are more important for singlet states. Since $U > 0$ and $\chi^0(\underline{q},0) > 0$, $\phi_n(\underline{q}_0)$ is positive. The forward scattering amplitude, for which $\underline{p} = \underline{k}$, is a measure of the average interaction and, according to Eqs. (5) and (6) this is repulsive for singlet states but can be attractive ($\underline{q}_0 \approx 0$) or repulsive ($\underline{q}_0 \approx 2p$) for triplet states. At least the $\underline{a}$-component of $\underline{q}_0$ is equal to $2k_F$, so it appears that the exchange of spin-density wave fluctuations gives rise to a generally repulsive interaction. However, since $\underline{q}_0 \neq 0$, the effective interaction oscillates in space and is attractive in some regions but repulsive in others. If the pair wave function is able to confine the relative motion of the holes in such a way as to obtain a net attraction, superconductivity may result[6]. The precise way in which this comes about depends on $\underline{q}_0$: for example, if $\underline{q}_0 = (2k_F,\pi/b,0)$ in the organic superconductors, then the interaction is attractive for electrons on neighboring stacks in the (a,b) plane and the gap would be given by Eq. (1). Such a gap vanishes when $\underline{p} = \pm\,\underline{q}_0/2$ and therefore strong repulsive forward or backward scattering ($\underline{p} = \pm\,\underline{k} = \pm\,\underline{q}_0/2$) is avoided.

It is not easy to evaluate the transition temperature produced by this mechanism, taking account of all relevant effects[6] as well as the competition with other repulsive interactions. For this reason, it has not yet been established that spin-fluctuation exchange really is a likely mechanism for organic superconductivity but, as a minimum, it could help to determine the shape of the gap, and force interchain pairing.

A similar idea has been proposed for heavy fermion superconductors[12].

TWO BAND MODEL

Although the exchange of spin fluctuations may play a role or even be the primary reason for superconductivity, it cannot be the whole story, for it is a low-temperature mechanism which cannot account for the problems raised by the anion ordering transition[3] alluded to above. Furthermore, experiment might rule out the anisotropy of the gap, forcing us to look at other possibilities. One candidate is the inclusion of a second band.

The energy to move a hole to an excited state on an isolated molecule is about 1 eV. This is the same order as the longitudinal bandwidth $4t_a$ and the interactions which appear in the Hubbard model. In the solid, the second level will also be broadened into a band by overlap between neighboring molecules and, although we have no knowledge of the relevant bandwidth, it is presumably of the same order as $4t_a$. Thus the neglect of the second band is not really justified in the organic superconductors as it may well come close to the Fermi level or even pass below it. Although there is no consensus on the band structure, one calculation[13] does indeed find that the bands overlap. It is therefore reasonable to explore some of the consequences, and to seek experimental evidence of overlapping bands.

If the two bands have Fermi wave vectors k_A and k_B, then the fact that all holes have to be accommodated into the two Fermi seas gives

$$k_F = k_A + k_B \tag{7}$$

where k_F is the Fermi wave vector for a single band. It follows that the principal Bragg vector ($4k_F$) of the anion potential or the (0,1/2,1/2) anion order[3] is $4k_A + 4k_B$ which can only couple to the holes via a complicated (and less relevant) joint umklapp process involving the two bands. Thus one possibility is that the bands do not overlap at low pressure (where the anions aid the spin-density wave ordering) but they do so at high pressure, where the bands are broader, and this helps eliminate the sensitivity to the (0,1/2,1/2) anion order, remove the spin-density wave state and produce superconductivity. On the other hand, for anion order with wave vector $2k_F = 2k_A + 2k_B$, the ordinary "$4k_F$", or umklapp processes, described in the previous chapter are effective--they merely involve one hole from each band instead of one hole of each spin. Thus, in this case, the conductivity would be affected for a two-band model just as for a single band.

A first step in establishing a two-band theory of the one-dimensional electron gas[7] simplified the problem by considering spinless fermions. It is then possible to think of a two-fluid system (A and B) in much the same way as we considered the up-spin and down-spin fluids in the previous chapter. The important difference is that the interactions, and particularly the Fermi velocities, are not symmetric under interchange of A and B. The outcome is that superconductivity can be produced by purely repulsive interactions[7], although this competes with a state in which the densities of A and B fermions are modulated and out of phase (the analog of a spin-density wave). Consequently, the existence of overlapping bands might remove the principal objections to having a significant pairing of holes on the same stack.

Evidently there is much more to do in pursuing this model--incorporating spin, studying the low-temperature, three-dimensional region and considering the possibility that the bands come close but do not overlap. These and other options should be explored in seeking to understand the mechanism of organic superconductivity.

References

* Work supported by Division of Materials Sciences U.S. Department of Energy, under Contract No. DE-AC02-76CH00016.

1. V. J. Emery, J. Phys. (Paris), Colloq. 44, C3-977 (1983).
2. I thank J. Voigt and S. Barišić for discussion of this point.
3. R. Moret, S. Ravy, J. P. Pouget, R. Comès and K. Bechgaard, Phys. Rev. Lett. 57, 1915 (1986).
4. The experiment[4] did not actually extend into the superconducting phase and, in principle another structural transition could intervene.
5. C. Coulon et. al., Phys. Rev. B26, 6322 (1982).
6. V. J. Emery, Syn. Met. 13, 21 (1986).
7. K. A. Muttalib and V. J. Emery, Phys. Rev. Lett. 57, 1370 (1986).
8. P. Garoche, R. Brusetti, D. Jérome and K. Bechgaard, J. Phys. (Paris) Lett. 43, 147 (1982).
9. A. J. Leggett, J. C. Wheatley, Rev. Mod. Phys. 47, 331 (1975).
10. C. J. Pethick and D. Pines, Phys. Rev. Lett. 57, 118 (1986).
11. M. T. Béal-Monod, C. Bourbonnais and V. J. Emery, Phys. Rev. B34 (1986). Note that here we use the even (odd) part of the singlet (triplet) interaction.
12. D. J. Scalapino, E. Loh, Jr. and J. E. Hirsch, Phys. Rev. B34 (1986); K. Miyake, S. Schmitt-Rink and C. M. Varma, Phys. Rev. B34 (1986). See also Ref. 11.
13. C. Minot and S. G. Louie, Phys. Rev. B26, 4793 (1986).

SUPERCONDUCTIVITY OF β-(BEDT-TTF)$_2$I$_3$

F. Creuzet, G. Creuzet, B. Hamzic* and D. Jérome

Laboratoire de Physique des Solides (associé au CNRS)
Université de Paris-Sud, 91405 Orsay (France)

INTRODUCTION

β-(BEDT-TTF)$_2$I$_3$ is the first ambient pressure sulfur gased organic superconductor [1,2], exhibiting two different superconductiong phases. If the sample is cooled under ambient pressure, a pure low-T_c state (called β-L) is observed, with T_c around 1.2 K [1,2]. Cooling under pressure first decreases T_c for moderate pressures but a high-T_c phase (called β-H) with T_c around 7.2 K is observed for pressures above a critical one, originally estimated to be around 1.3 kbars [3,4], but which is more probably around 500 bars [5,6]. Finally, this high T_c is also reduced with further increase of pressure.

In an attempt to obtain the β-H phase under ambient pressure, the low temperature conductivity was studied either after temperature cycling [7] or after the release of a high pressure at ambient temperature [8]. In both cases a "two-step" incomplete transition was obtained around 7-8 K. This can be understood within the simple picture of a mixture of small amount of β-L phase.

The crystallographic study reveals the existence of an incommensurate distortion, upon cooling at 1 bar, below a critical temperature estimated around 200 K bu neutron scattering experiments [9] and around 175 K by X-ray diffusive scattering experiments [10].

PREPARATION AND STUDY OF THE PURE HIGH-T_c PHASE AT 1 BAR

a) Conductivity

The low temperature conductivity curve displayed in Figure 1 was obtained after a particular cooling process as shown in the inset. A pure hydrostatic (helium gas) pressure of 1.3 kbars is applied at room temperature, followed by a cooling down to around 34 K under monitored constant pressure, depressurisation at this temperature and finally cooling down to 1.5 K. The four-probe measurements give a sharp and complete superconducting transition with T_c = 8.1 K evaluated at the centre of the transition. The upper critical field H_{c_2} is typically around 3 T at 1.5 K

---

* Permanent address : Institute of Physics of the University, P.O. Box 304 41001 Zagreb, Yugoslavia.

(more details can be found in reference 11). At this stage, we assumed that using our pressure cycling process we obtained a pure β-H phase, and we suggested that the β-H and β-L phases correspond respectively to the structure stable at room temperature and to the modulated superstructure. Later on, extended neutron scattering experiments under a large number of cycling conditions [6] have confirmed this point. Finally, the metastable character of the β-H phase was shown by annealing up to room temperature and quantitatively studied by AC susceptibility and $\underline{c}^*$-resistivity as discussed below. It must be noted that similar conductivity results have been obtained simultaneously and independently by the Russian group [5].

b) AC susceptibility

The superconductivity detection was obtained in this experiment measuring small shifts in resonance frequency of a Robinson oscillator (details can be found in ref. 12). Figure 2 shows that the two states obtained under pressure (a) or after the pressure cycling process (b) give the same signature, which reinforce the assumption that both cases correspond to the β-H phase. Different annealing experiments at 1 bar showed that the β-H state remains stable as long as the annealing temperature does not exceed 125 K (after this process the transition temperature is not modified but the amplitude of the signal is reduced) while it disappears completely if the annealing temperature is above 131 K (see figure 2). As a consequence, it appears that two high temperature regions are concerned with the stabilization of the β-H phase : 175-180 K [13] where the crystallographic transition to the β-L phase takes place upon cooling, and 125-130 K which is the upper limit of the stability of the β-H phase on heating. Extended annealing experiments measuring the ab-conductivity [14] give very similar results for all these treatments. In addition, it has been shown in this AC susceptibility study that during the pressure cycling process the depressurization can be done at any temperature below 110 K [12].

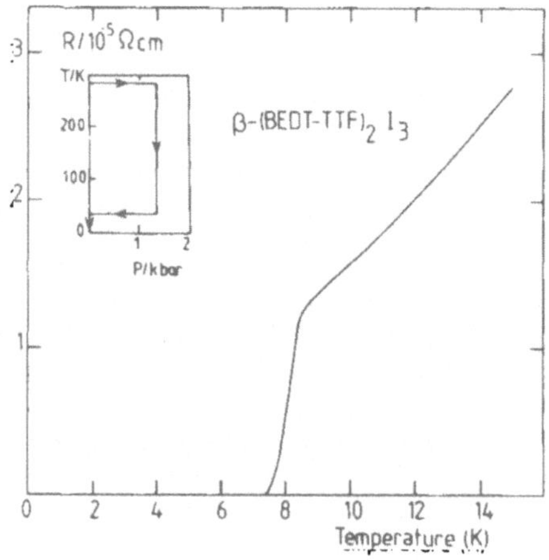

Figure 1

Superconducting transition at 8.1 K in β-$(BEDT\text{-}TTF)_2I_3$ under 1 bar after sample cooling following the T-P path displayed in the inset.

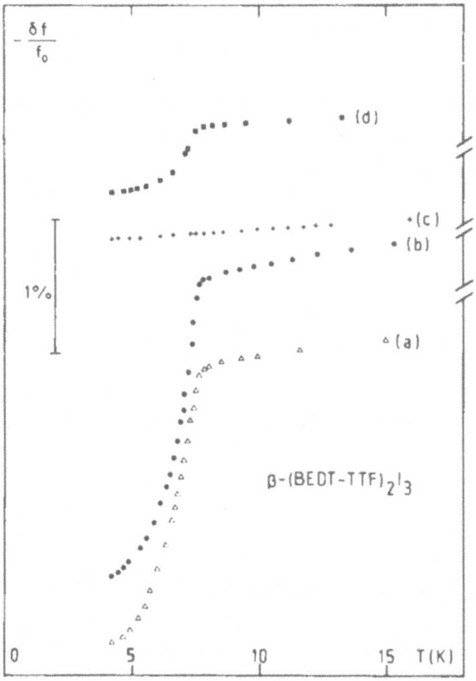

Figure 2

AC susceptibility signal in β-$(BEDT\text{-}TTF)_2I_3$, under 1.6 kbars(a), after pressure release at low temperature (b), after annealing above 131 K (c) and after annealing at 125 K (d).

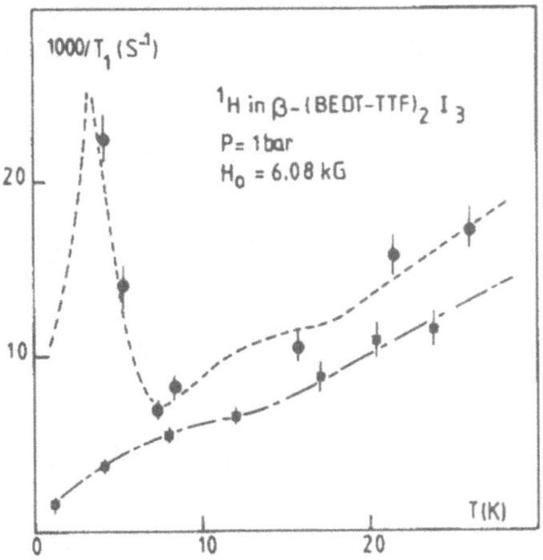

Figure 3

^1H spin-lattice relaxation rate in β-$(BEDT\text{-}TTF)_2I_3$ under 1 bar in both states β-L (■) and β-H (•).

c) NMR

^{1}H spin-lattice relaxation rate was measured under pressure of 1.6 kbars and under ambient pressure in both β-H (obtained by pressure cycling process described above) and β-L state (after annealing at high temperature). Detailed results can be found in reference 15. We want to emphasize the clear anomaly around 8 K for the β-H phase (data are closely related to those obtained under pressure which are interpreted as the occurrence of the SC transition) compared with the flat behaviour in the β-L phase (see figure 3).

METASTABILITY AND HIGH-TEMPERATURE TRANSITIONS

We performed an extended study of c^*-conductivity from room temperature down to low temperature under a large variety of cyling conditions. Under ambient pressure, the β-H to β-L transition is identified by a clear anomaly around 181 K leading to a S-shape behaviour as reported before by the Zagreb group [16]. This anomaly is characterized by a cusp in the resistivity derivative, which is smoothed out by pressure but not displaced in temperature (more details can be found in reference 17).

The most important result of this c^*-conductivity study is displayed on figure 4. First we performed a pressure cycling cooling as described before and we observed at low temperature and ambient pressure the same superconductive transition as in the previous ab-conductivity study. During this process the sample is always in the β-H state. Upon warming we first observed at $T_L \sim 132$ K, which is exactly the annealing limit deduced from AC susceptibility, a sharp increase of the resistivity. Later on, we observed at $T_H \sim 181$ K the same anomaly as for the β-H to β-L transition upon cooling. It is then clear that T_L corresponds to the β-H to β-L transition and T_H to the β-L to β-H one. In addition we checked the metastable character of the β-H phase by temperature explorations around T_c, after the transition, where we still remain on the β-L curve as shown in the inset.

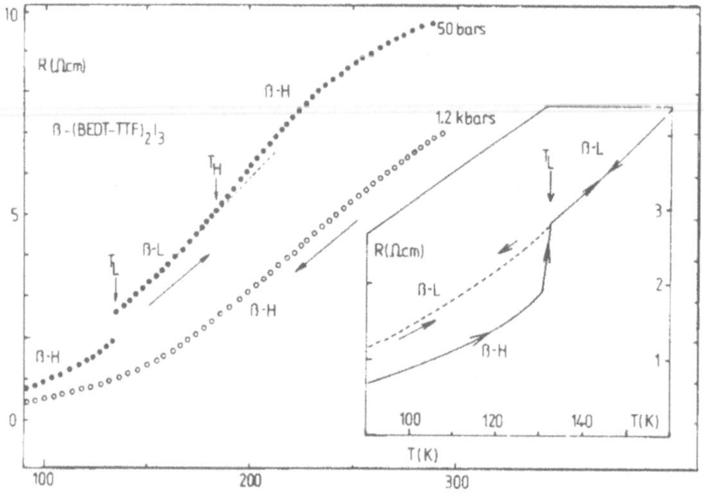

Figure 4

High temperature c^*-resistivity of β-$(BEDT\text{-}TTF)_2I_3$ under typical T-P cycling (see text). Clear evidence is given for the β-H → β-L → β-H transitions upon warming and for the metastable character of the β-H state by temperature explorations around 130 K displayed in the inset.

CONCLUSION

The β-H phase (T_c = 8.1 K) of β-$(BEDT-TTF)_2I_3$ can be prepared at low temperature when the sample is cooled under pressure (> 500 bars) and depressurized at temperature below 110 K. This is deduced from ab-conductivity, AC susceptibility, NMR and c*-conductivity. The β-H state has the high temperature state crystallographic structure. Cooling under pressure locks the system in the β-H state which is metastable at low temperature and under ambient pressure. The maximum annealing temperature of the β-H state is around 132 K, leading to the β-H → β-L → β-H succession upon warming.

REFERENCES

1. E.B. Yagubskii, I.F. Schegolev, V.N. Laukhin, P.A. Kononovich, M.V. Kartsovnik, A.V. Zvarykina and L.I. Buravov, JETP Lett. 39, 12 (1984).
2. G.W. Crabtree, K.D. Carlson, L.N. Hall, P.T. Copps, H.H. Wang, T.J. Emge M.A. Beno and J.M. Williams, Phys. Rev. B30, 2958 (1984).
3. V.N. Laukhin, E.E. Kostyuchenko, Yu.V. Susko, I.F. Schegolev and E.B. Yagubskii, JETP Lett. 41, 81 (1985).
4. K. Murata, M. Tokumoto, H. Anzai, H. Bando, G. Saito, K. Kajimura and T. Ishiguro, J. Phys. Soc. Jap. 54, 1236 (1985).
5. V.B. Ginodman, A.V. Gudenko, I.I. Zasavitskii and E.B. Yagubskii, JETP Lett. 42, 472 (1985).
6. A.J. Schultz, M.A. Beno, H.H. Wang and J.M. Williams, Phys. Rev. B33, 7823 (1986).
7. I.F. Schegolev, E.B. Yagubskii and V.N. Laukhin, Mol. Cryst. Liq. Cryst. 126, 365 (1985) ; V.B. Ginodman, A.V. Gudenko and L.N. Zherikhina, JETP Lett. 41, 49 (1985).
8. T. Tokumoto, K. Murata, H. Bando, H. Anzai, G. Saito, K. Kajimura and T. Ishiguro, Solid State Commun. 54, 1031 (1985).
9. T.J. Emge, P.C.W. Leung, M.A. Beno, A.J. Schultz, H.H. Wang, L.M. Sowa and J.M. Williams, Phys. Rev. B30, 6780 (1984).
10. S. Ravy, R. Moret and J.P. Pouget, private communication.
11. F. Creuzet, G. Creuzet, D. Jérome, D. Schweitzer and H.J. Keller, J. Physique Lett. 46, L1079 (1985).
12. F. Creuzet, D. Jérome, D. Schweitzer and H.J. Keller, Europhys. Lett. 1, 461 (1986).
13. The first estimation of 200 K from neutron scattering experiments seems to be too high because i) X-ray results from ref. 10 with identical characteristics for the modulated superstructure give a transition temperature of 175 K and ii) this estimation was checked in Orsay using different techniques like thermal expansion (B. Hamzic, G. Creuzet, D. Schweitzer and H.J. Keller, to be published in Solid State Comm.) or c*-resistivity (see text).
14. B. Hamzic and G. Creuzet, to be published.
15. F. Creuzet, C. Bourbonnais, D. Jérome, D. Schweitzer and H.J. Keller, Europhy. Lett. 1, 467 (1986).
16. J.R. Cooper, L. Forro and D. Schweitzer, ICSM'86, to be published in Synthetic Metals.
17. B. Hamzic, G. Creuzet and C. Lenoir, to be published in Europhys.Lett.

ELECTRONIC CORRELATIONS IN ORGANIC CONDUCTORS AND SUPERCONDUCTORS

Claude Bourbonnais

Laboratoire de Physique des Solides
Université de Paris-Sud,
Orsay 91405, France

INTRODUCTION

In the extensive experimental study of the sulphur $(TMTTF)_2X$ and the selenide $(TMTSF)_2X$ (the Bechgaard salts) series of organic conductors with the anions $X=PF_6$, AsF_6, Br, ClO_4, ..., it soon became apparent that there is unity in their properties and therefore in the complex mechanisms at the origin of their rich variety of phase transitions.[1-3] At ambient pressure, depending on the anion X or the organic stack, either a spin-Peierls (SP) or an Antiferromagnetic (AF) or finally a superconducting phase transition is observed.[1,4] At this pressure, the SP state is only stabilized[5] for the sulphur compounds $(TMTTF)_2PF_6$ and $(TMTTF)_2AsF_6$ whereas the superconducting phase has been detected exclusively for the slow cooled Bechgaard salt $(TMTSF)_2ClO_4$.[4] The AF state however, is observed in the same conditions for many different compounds of both series.

It is the effect of pressure that shows most clearly the close similarity between both series. In this respect, the case of $(TMTTF)_2PF_6$ is particularly illustrative. Indeed, at 13 kbar of pressure, the SP state is found to be no longer stable and it is an AF state that is observed,[6] which appeared to be quite similar to the one taking place for other sulphur compounds at ambient pressure.[7] When further pressure is applied, the characteristics of the AF state are essentially the same as those commonly seen in AF $(TMTSF)_2X$ compounds.[1-4,8] For the Bechgaard salt family, the application of a finite critical pressure can lead to the observation of superconducting long range order.[4] This has led to infer that at sufficiently high pressure, the sulphur series should show superconductivity. According to Barisic and Brazovskii[9] and Emery et al.[2], the origin of antiferromagnetism is a consequence of the electronic properties of the isolated organic chains which are purely one-dimensionnal(1D) systems. This is essentially due to the existence of dominant repulsive electron-electron interactions along the chains and sizeable Umklapp scattering effects. The 1D electron gas theory[10] predicts that this will favor AF correlations. In fact, the origin of Umklapp terms for the interactions comes mainly from the anion X potential which induces a $4k_F$ bond modulation of the electronic charge along the organic stack. This in turn produces a slight but observable $4k_F$ dimerization of the stack. Thus, it will lead to the production of a small $4k_F$ gap, $\Delta(4k_F)$, in the electronic spectrum. The band will be effectively half-filled at low temperature which otherwise would be 3/4 filled[4]. The strength of the electron-electron Umklapp process, commonly called g_3 in the "g-ology" decomposition of the direct electron-electron interaction, is directly proportional to $\Delta(4k_F)$.[9] When the g_3 coupling is a relevant parameter, the 1D electron gas theory predicts the existence of a non-zero charge degree of freedom gap Δ_ρ. If the couplings are repulsive, this gives rise to strong 1D AF correlations at $T \ll \Delta_\rho$. Emery et al.[2] claimed that the resistivity minimum generally observed in the sulphur series at high temperature is a clear manifestation of Δ_ρ.[1] As the amplitude of Δ_ρ is directly connected to the one of dimerization, the argument was further supported by the observation of a rapid decrease of Δ_ρ under pressure.[8] Actually, Δ_ρ is no longer seen at very high pressure where the AF characteristics of the sulphur series become quite similar to those of the selenide series. This is why a g_3 coupling was said to be a necessary condition for the occurrence of superconductivity[2]. In this lecture, we will show how the scaling hypothesis, well known in the theory of critical phenomena, can be applied to existing nuclear relaxation data and thus allows a clear distinction of the various microscopic energy scales involved for every phase transition of both series of conductors. The importance, the nature and the dimensionality of correlations precursor to each type of phase transition are specified. The above mentionned importance of 1D correlation effects is confirmed and further characterized. In addition, the non-trivial evolution of the electron-phonon coupling and especially of the interchain AF exchange mechanism under pressure are discussed in some details. Finally, we show how under some favorable conditions the interchain exchange coupling together with the AF paramagnons observed as precursors of the superconducting transition can lead to an attractive pairing mechanism for superconductivity.

Magnetic properties: One Dimensional Part

For all AF compounds studied[1,13] in the $(TMTTF)_2X$ series with X = SbF_6, Br, SCN...., the Pauli susceptibility χ_P decreases monotonically with temperature for the entire paramagnetic regime. The value of χ_P is enhanced compared to its free electron gas prediction however, indicating the presence of repulsive interactions. There is no visible[1,13] change of χ_P at T_ρ where the resistivity minimum takes place($T_\rho \simeq \Delta_\rho/\pi < 230K$). Therefore, only charge degrees of freedom are affected at T_ρ and spin degrees of freedom remain gapless. In a "g-ology" model, the low temperature enhancement of χ_P is given by[11] $\chi_P(T\to 0)/\chi_P^o \simeq [1-g_1/2\pi v_F]^{-1}$. g_1 is the electron-electron backward scattering interaction and χ_P^o is the bare Pauli susceptibility. The sizeable enhancement generally found in sulphur series, imposes to locate the interaction g_1 in the repulsive sector with $g_1 \simeq \pi v_F$.[14] The Umklapp scattering term g_3 is proportional to the dimerization gap:[9] $g_3 \approx g_1 \Delta(4k_F)/E_F \ll g_1$. If $\Delta(4k_F)$ is non zero, this supports a singular $4k_F$ bond response of the hole gas to the anion potential. From the microscopic theory, this is possible if the inequality $(2g_2-g_1)/\pi v_F \geq 1$ prevails in the repulsive sector.[10] Thus this imposes $g_2 \gtrsim \pi v_F$, for the forward scattering part of the coupling. These couplings sastisfy the condition $g_1-2g_2<|g_3|$ for the relevance of Umklapp scattering and a non-zero correlation gap $\Delta_\rho \ll E_F$ for conductivity.[2] This condition also imposes precise predictions for the 1D AF correlation function. Indeed, according to ref.10, this important quantity has a power law profile in 1D:

$$\chi_{1D}(q,\omega\ T) \ \alpha \ [MAX(v_F q,\omega,T)/E_o]^{-2+\gamma_\rho+1/\gamma_\sigma}\ , \qquad (1)$$

where q is the wavevector taken from $2k_F$, ω is the frequency and E_o is a cut-off energy scale ($0 < E_o \leq E_F$) for the charge (γ_ρ) and the spin (γ_σ) degrees of freedom parts of the exponent. In general, γ_ρ and γ_σ are non-universal functions of the bare g's. At $T<T_\rho$, charge degrees of freedom are frozen and then $\gamma_\rho \to 0$. On the other hand, if we assume that the magnetic anisotropy is small compared to the bare direct interaction energy, we can write[10] $\gamma_\sigma \simeq 1$ and this leads to:

$$\chi_{1D}(q,\omega,T) \ \alpha \ [MAX(v_F q,\omega,T)/T_\rho]^{-1}\ . \qquad (2)$$

This power law singularity is the consequence of strong 1D AF correlations along the chains. As the true phase transition temperature observed $T_N (\leq 20K)$[1,6-8,12] is much smaller than T_ρ, the regime of 1D AF precursors is described by $\chi_{1D}(2k_F)$ and should be very large in temperature.

Magnetic Properties: Interchain Coupling

It is well known that a purely 1D system cannot sustain long range ordering at finite temperature. The observation of a finite T_N in organic conductors indicates that interchain coupling must be taken into account. Owing to a finite overlap of electronic wavefunctions between chains, the most important source of transverse coupling comes from the interchain single electron tunneling (IST) along the b* direction.[4] In $(TMTTF)_2X$ compounds at ambient pressure,[15] its bare amplitude $t_{\perp b} \simeq 100K$, is much smaller than the one along the high conducting axis namely, $t_{\perp b}/t_a \simeq 1/30$. Neglecting the effect of interactions, a coherent transverse motion of holes will only occur for $T < t_{\perp b}$. However, from the considerations given in the previous section, the apparition of a 1D correlation gap Δ_ρ at a much higher energy $\pi T_\rho \gg t_{\perp b}$, will completly freeze IST motion since each hole is bound to an electron with an energy of the order of Δ_ρ. Nevertheless, there is an effective transverse motion but it will be two-particle or electron-hole like. This gives rise to the so-called AF interchain exchange (IEX) mechanism. In second order of perturbation theory(see also below),[16] the IEX amplitude is given by the expression $J_b \simeq \pi v_F (\bar{t}_{\perp b}/\Delta_\rho)^2$ where $\bar{t}_{\perp b}$ is[17] the renormalized IST amplitude. This leads therefore to consider in addition to the g's at $T=T_\rho$, a transverse AF coupling of the form: $H_\perp = \int \sum_{\langle i,j\rangle} J_b \vec{S}_i(x) \cdot \vec{S}_j(x) dx$. Here $\vec{S}_i(x)$ is the three component spin operator at position x along the chain i. There is a similar IEX coupling along the c* direction: $J_c \simeq \pi v_F (\bar{t}_{\perp c}/\Delta_\rho)^2$, which is much smaller than J_b since:[4,15] $t_{\perp c} \ll t_{\perp b}$. Nevertheless, J_c is necessary to induce a true long range ordering at finite temperature.

It is through the IEX mechanism that AF correlations of a single chain can be propagated coherently in the tranverse direction. This is well illustrated through a random phase approximation (RPA) for the total AF susceptibility:

$$\begin{aligned}\chi_{3D}(\vec{q},\omega,T) &= cst + \chi_{1D} + J(\vec{q}_\perp)\chi_{1D}^2 + \ldots, \\ &= cst + \chi_{1D}(q,\omega,T)\ [1-J(\vec{q}_\perp)\chi_{1D}(q,\omega,T)]^{-1}, \quad (3)\end{aligned}$$

where $J(\vec{q}_\perp) = 2J_b \cos(q_{\perp b} d_\perp) + 2J_c \cos(q_{\perp c} d_\perp)$ is the Fourier transform of the total IEX coupling. Here, we have assumed for simplicity an underlying orthorombic tight binding band structure which leads in presence of Δ_ρ to a square lattice of weakly coupled AF chains. One must note that in (3), the contribution to $\chi_{1D}(cst)$ comes from the temperature range $T \leq (>) T_\rho$ where Δ_ρ is relevant(irrelevant). Long range order is predicted when the Stoner factor becomes singular that is, at $T_N \simeq T_\rho [\bar{t}_{\perp b}/\Delta_\rho]^{2/\gamma}$ for $\vec{q}_o = (2k_F, \pi/d_\perp, \pi/d_\perp)$. For $(TMTTF)_2X$ compounds below T_ρ, $\gamma = 2-\gamma_\rho - 1/\gamma_\sigma \simeq 1$(see (2)). Near T_N however, the critical behavior is of the form:

$$\chi_{3D}(\vec{q}_o, T) \sim |T-T_N|^{-\mathring{\gamma}} , \tag{4}$$

where from (3), $\mathring{\gamma}=1$ in RPA. The dimensionality crossover temperature T_x2 for AF correlations occurs when the pole of (3) becomes perceptible. This is only possible at $T_x2 \sim T_N$. The subscript 2 emphasizes the fact that a two-particle propagation is involved in the crossover. The non-classical critical width is known to be small[17] for such a model so that a mean-field approach for the critical indices near T_N is justified. It is clear from the above expression for T_N and T_x2 that $T_\rho \gg T_N, T_x2$. Thus, the 1D domain of AF correlations should dominate the electronic properties.

Nuclear Relaxation Analysis: General Scaling Formulation

In order to have a direct experimental evidence for the predicted importance of AF precursors, one needs an experimental probe that is sensitive to staggered correlations and especially to 1D $2k_F$ fluctuations. Neutron scattering experiment is potentially suited for this task but no successfull attempts have been obtained so far. A more easily measurable quantity that can give priceless details about the microscopic nature of correlations in organic conductors is the nuclear relaxation rate obtained by NMR.[3,18-20] This quantity symbolized by T_1^{-1}, probes the local variations of the magnetic hyperfine field at a NMR active nucleus. These variations are produced by the electronic spin fluctuations. The hole spin along the stack can be coupled to three different types of nucleus: the protons (^1H) of the methyl groups situated at both ends of the TMTTF and TMTSF molecules, the carbons(^{13}C) and in the case of TMTSF molecules only, to the seleniums (^{77}Se). The local nature of the coupling between these nucleus and the conduction holes makes $T_1{}^{-1}$ sensitive to correlations of all wavelengths. This can be put explicitly using the well known result derived by Moriya:[20]

$$T_1^{-1} = 2\gamma_N^2 |A|^2 T \int d^d q \, \chi''(\vec{q}, \omega)/\omega \qquad (k_B=1) , \tag{5}$$

where $|A|$ is the hyperfine matrix element, γ_N is the gyromagnetic ratio of the nucleus and χ'' is the imaginary part of the retarded susceptibility. The external frequency ω

coincides with the Larmor frequency of the nuclear spin. For an electronic system with repulsive electron-electron interactions and AF correlations, one expects that the staggered part of (5) will dominate the relaxation. As we have already mentionned however, the Pauli susceptibility, though being a non-critical quantity, is enhanced by Coulomb repulsion. Thus, uniform correlations can give an important contribution to T_1^{-1} outside the critical domain. It appears then natural to consider the following decomposition:

$$T_1^{-1} = T_1^{-1}(\vec{q} \simeq 0) + T_1^{-1}(\vec{q} \simeq \vec{q}_o) . \tag{6}$$

For the uniform part, we can write:[18]

$$T_1^{-1}(\vec{q} \approx 0) = cst\ T\chi_P^2(q=0,T) . \tag{7}$$

This expression accounts only for the 1D coherent uniform correlations and neglects spin diffusive effects[18]. As we will see, the relevance of (7) can be assessed by checking the linear shape of T_1^{-1} versus the quantity $T\chi_P^2$, obtained via a separate experiment. For the derivation of the staggered part of the relaxation, we shall be making use of the so-called "extended scaling hypothesis" (ESH) which is well known to be useful in the theory of critical phenomena with crossover features.[21,22] This is particularly interesting here, since the AF phase transition in organic conductors is precisely a many energy scale problem. This supports the existence of a variety of crossovers in the whole range of temperature. In the model discussed before, for example, Umklapp processes induce a non zero Δ_ρ and this introduces a crossover in the exponent γ of χ(see eqs (1-2)). Under the influence of the IEX coupling at low temperature, this is followed by another characteristic energy $T_{x}2$, announcing the change $\gamma \to \mathring{\gamma}$ in the critical indices of χ. For the stagerred part, the ESH states that the three quantities $\int d^dq$, χ'' and ω in (5), scale with the AF correlation length ξ to some characteristic power (exponent). In presence of crossovers, each different energy scale will contribute to the relaxation in the same way except for a possible change in the exponent. In the usual way,[22] two energy scales or two sets of exponents are connected trough a crossover exponent ϕ. Applying ESH, it was then possible to derive the general expression:[21]

$$T_1^{-1}(\vec{q} \simeq \vec{q}_o) = cst\ T\ [\xi_1(T_{h1})]^{\gamma_o/\nu_o + z_o - 1} \ldots$$

$$\times\ [\xi_{h_\perp}(T_x)]^{\gamma_n/\nu_n + z_n - 1}\ [\xi_d(T)]^{\mathring{\gamma}/\mathring{\nu} + \mathring{z} - d} , \tag{8}$$

for the stagerred part of T_1 near the critical point. h_i, for $i=1,\ldots n$, is a small perturbative parameter that produces the change $(\gamma_{i-1},\nu_{i-1},z_{i-1}) \longrightarrow T_{hi} \sim hi^{1/\phi_i} \longrightarrow (\gamma_i,\nu_i,z_i)$, in the 1D AF correlation function (γ), the correlation length (ν) and the dynamical ($z;\omega \sim \xi^{-z}$) exponents at the crossover temperature T_{hi} of the 1D domain. At each scale, $\xi_i(T) \sim [T/T_{hi-1}]^{-\nu_i}$. As far as the 1D▸d dimensionality crossover is concerned, it is induced by $h_\perp$ at the temperature $T_x \sim h_\perp^{1/\phi_\perp}$, where $(\gamma_n,\nu_n,z_n) \rightarrow (\mathring{\gamma},\mathring{\nu},\mathring{z})$, and below which, $\xi_d \sim |T-T_N|^{-\mathring{\nu}}$. It is clear from the general expression that all energy or length scales of the critical AF cluster contribute in the same way to T_1^{-1}. In fact, all the details of the AF correlation length at smaller scales are contained in the coefficients of this expression. Consequently, from the direct computation of $\gamma_i(\mathring{\gamma})$, $\nu_i(\mathring{\nu})$, $z_i(\mathring{z})$ and $T_{hi}(T_x)$ for all $i(h_\perp)$ in a particular model, T_1^{-1} [$2k_F$ $(\mathring{q}_o)$] can be readily given. Conversely, by looking at the data, the nature and the dimensionality of correlations can be investigated. This confrontation is certainly unique and is probably crucial to assess the degree of relevance of any theoretical models for organic conductors.

Ambient Pressure Antiferromagnetism

In the context of the IEX model for the description of the AF state in $(TMTTF)_2X$ at low pressure, we are in a position to make clear predictions for the low temperature profile of the nuclear relaxation. As we have already mentionned before for the paramagnetic regime, the important energy is that of the resistivity minimum temperature induced by Umklapp scattering (see figure 1a). Hence, this allows to write $h_1 = g_3$, and from the 1D electron gas theory,[10] $T_\rho \sim g_3^{1/\phi}$ with $\phi \simeq 1$, in absence of magnetic anisotropy. Furthermore, $v_F q$, ω and T always enter in the same way in various correlation functions. This in turn implies that $\nu_o = \nu_1 = z_o = z_1 = 1$, which is exact in 1D. On the other hand, for γ_o and γ_1, we respectively get $\gamma_o = 2 - \gamma_\sigma - 1/\gamma_\rho$, $E_o = E_F$ and $\gamma_1 = 1$, $E_o = T_\rho$. Under the influence of the IEX coupling, $h_\perp = J_b$ and the 1D to 3D two-particle like crossover is achieved at $T_x 2 \sim J_b^{1/\phi_\perp}$, where from (3), $\phi_\perp = \gamma_1$. In RPA at $T < T_x 2$, we take the well known mean field exponents: $\mathring{\gamma}=1$, $\mathring{\nu}=1/2$ and $\mathring{z}= 2$ for AF critical effects.[21] Collecting all the results, we get for the staggered part, the following three distinct regimes of relaxation:

$$T_1^{-1}(2k_F) = \text{cst}\ T(T/E_F)^{-\gamma_o} \quad (T_\rho < T < E_F)\ , \tag{9a}$$

$$T_1^{-1}(2k_F) = \text{cst}\ T(T_\rho/E_F)^{-\gamma_o}\,(T/T_\rho)^{-1} \quad (T_N \ll T < T_\rho)\ , \tag{9b}$$

$$T_1^{-1}(\mathring{q} \approx \mathring{q}_o) = \text{cst}\ T(T_\rho/E_F)^{-\gamma_o}\,(T_x 2/T_\rho)^{-1}\,|T-T_N|^{-1/2} \quad (d=3,\ |T-T_N|/T_N \ll 1)\ , \tag{9c}$$

It is clear from these expressions that AF precursors to T_N will introduce strong deviations to a Korringa type of relaxation ($[T_1T]^{-1} \sim$ cst). One must note that since $\chi_P(q=0,T)$ is already enhanced at very high temperature while $\chi_{1D}(2k_F)$ is not, the uniform contribution $T_1^{-1}(q=0)$ will emerge somewhere in order to dominate the relaxation in the high T paramagnetic regime. However, the temperature domain where the staggered part prevails can be quite large.

There is much experimental evidence of AF long range order in various sulphur compounds. At ambient pressure in $(TMTTF)_2Br$, $(TMTTF)_2SCN$ and $(TMTTF)_2SbF_6$ compounds for example, the anisotropy of magnetic susceptibility with field orientation, the spin-flop transition,[13] the divergence of the EPR linewidth at T_N[1] and the existence of antiferromagnetic resonance below T_N[23] all confirm the existence of a spin modulation. The inhomogeneous broadening of the NMR linewidth at T_N is also consistent with a magnetic ordering. However, as we have seen, the information given by the nuclear relaxation rate is not restricted to the determination of the nature of the ground state. This is well illustrated by the ^{1}H T_1^{-1} data of Creuzet et al.[24] in fig. 1b for the $(TMTTF)_2Br$ salt at 1bar. The divergence of T_1^{-1} at $T_N \simeq 14K$ is a clear indication of a magnetic long range order. A power law of the form (9c) has been confirmed from the log-log plot of these data for $T_N<T<T_x{}^2 \approx 20K$. The classical exponent 1/2 is only valid for a 3D AF. This implies that the 2D AF critical region for which $T_1^{-1} \sim |T-T_N|^{-1}$, is quite small and cannot be detected. In the paramagnetic regime, T_1^{-1} is clearly temperature independent which is in remarkable agreement with (9b). This result strongly supports the existence of strong 1D AF correlations and the relevance of Umklapp processes.[2,3] It also stresses that magnetic anisotropy is quite small ($\gamma_\sigma=1$), which is also compatible with the relatively low value of the spin-flop field ($\leq$ 9kG) in these compounds.[23] A similar profile of T_1^{-1} for $(TMTTF)_2SCN$ has also been observed.[25] A major inconvenience in dealing with proton relaxation in organic conductors comes from the difficulty of probing the electronic contribution in the high temperature domain due to the occurence of resonant methyl group rotation effects on the relaxation. In $(TMTTF)_2Br$ for example, at H=10.6kOe, these effects emerge at $T \geq 35K$ and completely mask the electronic parts (8a), (8b) or (7).[24] It appears therefore preferable to probe another type of nucleus which can be free from any of these extrinsic effects. This

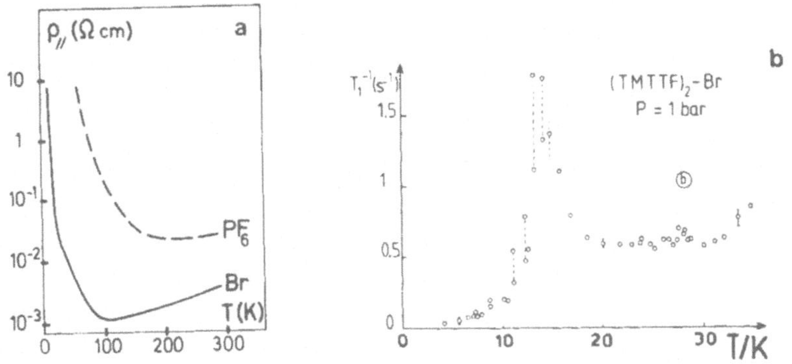

Fig. 1. (a) Resistivity vs temperature data for the two sulphur compounds $(TMTTF)_2PF_6$ (dashed line) and $(TMTTF)_2Br$ (continuous line). After Coulon et al., in ref. 1; (b) Proton nuclear relaxation vs temperature of $(TMTTF)_2Br$ at ambiant pressure. After Creuzet et al., ref. 24.

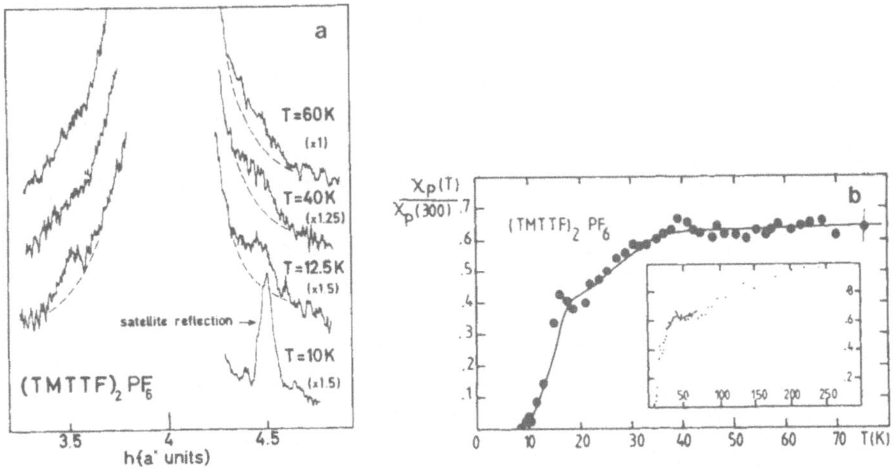

Fig. 2. (a) Microdensimeter reading of X-Ray patterns from $(TMTTF)_2PF_6$ below 60K at 1bar. After Pouget et al., ref. 26; (b) EPR susceptibility data vs temperature of $(TMTTF)_2PF_6$ (samples of figure 3) at 1bar. In the insert, the data up to 250K. After Creuzet et al., ref. 31.

will be illustrated next with the ^{13}C NMR analysis of the SP and AF transitions in sulphur compounds.

$(TMTTF)_2PF_6$ and $(TMTTF)_2AsF_6$: Two Spin-Peierls Systems

The magnetic susceptibility experiments made on $(TMTTF)_2PF_6$ and $(TMTTF)_2AsF_6$ at ambient pressure have revealed the existence of a non-magnetic ground state below the critical temperature 15K(PF_6) and 11K(AsF_6).[7,13] For both compounds, the susceptibility is isotropic with respect to field orientation and goes to zero as $T \to 0$.[12] On the other hand, X-ray experiments of Pouget et al.,[26] on both compounds have shown the presence of 1D $2k_F$ diffused scattering sheets below 60K(PF_6) and 40K(AsF_6) indicating a 1D $2k_F$ phonon softening of the organic stack. As shown in figure 2a, these diffused sheets condense into Bragg spots below T_{SP} where there is a static 3D distortion. Taking into account the $4k_F$ dimerization of the stack, the $2k_F$ distortion leads to the "dimerization of the dimerized chain"(see the lectures of J. Kommandeur). The lattice precursors of the transition and the static distortion are at first sight similar to those of a classical quasi-1D Peierls transition.[4] However, both compounds are characterized by a resistivity minimum that occurs at a temperature much higher than the one of structural fluctuations ($T_\rho \simeq 220K$). Accordingly, the 1D $2k_F$ phonon softening occurs when there are strong 1D AF correlations and a Mott-Hubbard localization of holes over a distance of the order of v_F/T_ρ.[27] In such a case, the distortion should be rather called a spin-Peierls transition. One should emphasize that the quasi-1D character of the SP transition discussed here is rather unique compared to the well known SP systems like TTF-CuBdT and $MEM(TCNQ)_2$.[28] Actually, the present systems are metallic at ambient temperature and the electronic system produces its own charge localization and AF fluctuations due to Umklapp scattering. This contrasts with previous systems which are localized at ambient temperature.[28] In addition here, the coupled electron-phonon system produces a large domain of 1D lattice fluctuations which are generally not detected in other compounds above the true transition.[28] In order to give a correct description of this transition, one must incorporate the effect of electron-phonon interaction to the model discussed above. In a "g-ology" picture, the electron-acoustic phonon interaction leads to an effective attractive backward ($g_{1ph} < 0$) and repulsive Umklapp ($g_{3ph} > 0$) electron-electron scattering terms. In the adiabatic limit and small electron-phonon coupling constant, g_{1ph} and g_{3ph} are strongly renormalized by the presence of purely electronic correlations.

In mean-field approximation, we can write:[27,29]

$$g'_{1ph}-g'_{3ph} = (g_{1ph}-g_{3ph})\bar{N}^{-}(T)\,[1-(g_{1ph}-g_{3ph})N^{-}(T)]^{-1}, \quad (10)$$

where $N^{-}(T)$ is the total 1D $2k_F$ bond charge density wave or the charge transfer correlation function of the 1D electron gas.[30] $\bar{N}^{-}(T)$ is the corresponding auxilliary quantity defined as $\bar{N}^{-}(T) = -\pi v_F\, \partial N^{-}(T)/\partial \ln(E_F/T)$. At $T<T_\rho$, $\bar{N}^{-}(T) = \bar{N}^{-}(T_\rho/E_F)(T/T_\rho)^{-1}$, which means that the $2k_F$ charge transfer fluctuations have the same power law behaviour as $\chi_{1D}(2k_F,T)$. In fact, AF exchange between dimers automatically implies a coherent charge transfer of two particles. It is via the coupling to these electronic correlations[29] that the 1D $2k_F$ acoustic phonon softening can occur. The characteristic energy for such a softening is of the order of the 1D mean field transition temperature T^o_{SP} which determines the simple pole of(10) namely,

$$T^o_{SP} \approx \bar{N}^{-}(T_\rho)(g_{1ph}-g_{3ph})T_\rho\ , \quad (11)$$

where $T^o_{SP} \ll T_\rho$. The occurrence of 1D $2k_F$ lattice fluctuations will damp significantly the growth of AF correlations. This is the pseudo-gap effect for the spin degrees of freedom and for which, $T^o_{SP} \sim h_2^{1/\phi_2}$ will act as another energy scale. The small parameter is now $h_2 = g_{1ph}-g_{3ph}$, and $\phi_2 = 1$. As shown by the EPR data Creuzet et al.[31], in figure 2b, the occurrence of the SP pseudo-gap clearly affects the Pauli susceptibility at $T^o_{SP} \simeq 40K$ The spin degrees of freedom reduction is around 35% at the true spin-Peierls transition temperature observed at $T_{SP} \simeq 20K$ for the sample used in this experiment. The reduction of χ_P is only partial in the pseudo-gap domain so that the remaining spin degrees of freedom can still give rise to AF collective effects. This means that the singularity in $\chi_{1D}(2k_F,T)$ is not completely suppressed at T^o_{SP}. From the bosonization approach[32] to the study of correlation functions, the quantity γ_σ in (1) is directly related to the velocity v_σ and the compressibility K_σ of the spin excitations as $\gamma_\sigma = 2\pi K_\sigma v_\sigma$. Actually, K_σ is the derivative of the total spin with respect to the magnetic field which coincides with the definition of χ_P ($\chi_P = 1/2\Pi v_\sigma$, at $T \to 0$). From figure 2b, χ_P is no longer temperature independent below 40K and this naturally suggests to use the temperature dependent exponent:

$$\gamma_2(T) = 2 - \gamma_\sigma(T)^{-1} = 2 - \gamma_\sigma(T^o_{SP})^{-1}\chi_P(T^o_{SP})/\chi_P(T)\ , \quad (12)$$

for $\chi_{1D}(2k_F)$ in the $T < T^o_{SP}$ regime. Here, $\gamma_\sigma(T^o_{SP})=1$. In this semi-phenomenological approach, the ratio $\chi_P(T)/\chi_P(T^o_{SP})$ can be taken directly from the EPR data of figure 2b. From (3) and

(12), the AF critical temperature becomes $T_N \sim J_b^{1/\gamma_2(T)} \sim T_x 2$ and will be reduced in presence of a 1D SP lattice softening. Roughly speaking, if $T_{SP} > (<) T_N$ then, a SP(AF) transition is likely to occur. This criteria does not take into account the competition between different types of 3D correlations whenever $T^o_{SP} \simeq T_{SP} \simeq T_N$. In order to get a real SP phase transition, one must specify the type of interchain coupling for the stabilization of SP long range ordering. It is generally believed that a modulation of the IEX coupling by acoustic phonons propagates the SP order in the transverse direction.[28] From the work in references 16 and 29 however, the 1D $2k_F$ bond correlations can also be propagated in the transverse direction via an electron-hole tunneling mechanism. Actually, there is an effective interchain bond wave coupling with an amplitude $V_{SP} \alpha (\vec{t}_{\perp b}/\Delta_\rho)^2$, which is of the same order of magnitude as the IEX coupling. Consequently, this coupling should be dominant in the adiabatic limit. The existence of a pseudo-gap means that a one component SP Landau order parameter is a meaningful physical quantity.[27] When the interchain coupling is small compared to the intrachain combinaison $g_{1ph}-g_{3ph}$, it allows the use of a generalized Landau-Ginzburg approach to the SP transition. This leads to the characteristic formula:[33]

$$T_{SP} = T^o_{SP}/\ln|g_{1ph}-g_{3ph}/V_{SP}| \,. \tag{13}$$

From EPR data, $T^o_{SP} \simeq 40K$ and $T_{SP} \simeq 20K$, which leads to an anisotropy ratio $|g_{1ph}-g_{3ph}/V_{SP}| \sim 10$. Below T_{SP}, the Pauli susceptibility is clearly activated. The magnetic activation energy is found to satisfy $\Delta_M \simeq 3.5\ T_{SP}$. A non-zero Δ_M is the result of a static lattice distortion. As far as the nuclear relaxation rate is concerned, it is clear that it will be also affected by the SP pseudo-gap. Indeed, from (8) and (12) we will get:

$$T_1^{-1}(2k_F) = \text{cst } T\ (T_\rho/E_F)^{-\gamma_0}(T^o_{SP}/T_\rho)^{-1}(T/T^o_{SP})^{-\gamma_2(T)} . \tag{14}$$

We therefore expect a damping of T_1^{-1} below T^o_{SP}. This prediction is in remarkable agreement with the nice data of figure 3 obtained by Creuzet et al.[31] on enriched ^{13}C $(TMTTF)_2PF_6$ single crystals. From the data, the plateau of relaxation seen below 65K agrees with the one predicted in (9b). It is followed by a sudden drop of T_1^{-1} at 40K, in agreement with (14) and EPR data. The reduction of relaxation at T_{SP}=20K is about 25%. From (14), we will get $\gamma_2(T_{SP}) \simeq .6$, which corresponds roughly to the observed 35% reduction of χ_P in figure 2b. Like χ_P,[31] T_1^{-1} is activated with the same value of Δ_M. An activated behavior for the relaxation is reminescent of the one seen in TTF-CuBdt and predicted by Erhenfreund et al.[34] for a dimerized Heisenberg chain. In this work however, the activation energy observed for T_1^{-1} should be twice the one appearing in the magnetic susceptibility, in contrast with the present case. The physical reason for this disperancy is not known so far.

In the high temperature domain T≥65K, we immediately see that T_1^{-1} is no longer temperature independent, but rather increases with a slight upward curvature. This clearly disagrees with the expected profiles (9a) and (9b). As already mentionned however, the expressions in (9) are only concerned with the staggered part of the relaxation and neglect the uniform part (7), which is almost linear in T and must take over somewhere in the high temperature range. Indeed, if we plot the T_1^{-1} data vs the product $\chi_P(T)^2T$, taken from EPR data of figure 2b, we can easily check from figure 4 that the decomposition

$$T_1^{-1} = \text{cst}\ \chi_P(T)^2T \ + T_1^{-1}(2k_F) \quad , \tag{15}$$

where $T_1^{-1}(2k_F)$=cst, is remarkably followed in the whole temperature range studied above the SP pseudo-gap. Therefore, the change of regime observed at 65K must not be identified with another energy scale for the 1D $2k_F$ correlations but it simply coincides with the dominance of uniform correlations for the relaxation.

Effect of Pressure

The lattice constants of organic conductors decrease under the application of hydrostatic pressure.[4] Consequently, the $4k_F$ dimerization is gradually suppressed and this leads to a reduction of the amplitude of Umklapp scattering term g_3 under pressure. This implies in turn that the characteristic energy $T_\rho \propto g_3$ for the activation of charge degrees of freedom also decreases under pressure. From the resistivity data of Creuzet et al.[8], and Parkin et al.[1], on $(TMTTF)_2PF_6$ and $(TMTTF)_2Br$, T_ρ decreases rather rapidly under pressure. At 15 kbar in $(TMTTF)_2PF_6$ for example, $T_\rho \simeq$ 50K and in $(TMTTF)_2Br$ at 5 kbar, T_ρ is no longer seen. This behaviour of T_ρ will have strong consequences on the stability of the SP phase since according to (11-13), $T_{SP} \propto T_{SP}^o \propto T_\rho$. The reduction of the 1D domain of SP fluctuations will favor AF correlations (see eq. (12)). Thus, there will be some critical pressure range where the criteria $T_N < T_{SP}$, for the stability of the SP phase, will be no longer satisfied. When the pressure is such that $T_N \simeq T_{SP}$, AF critical fluctuations become dominant and a SP → AF crossover is likely to taking place. Using a somewhat similar criteria,

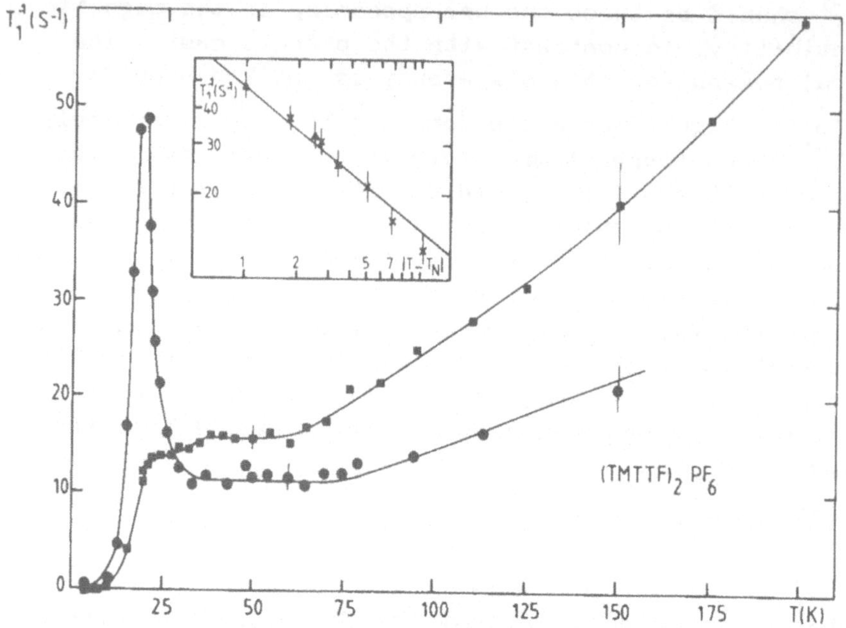

Fig. 3. ^{13}C T_1^{-1} vs temperature data of $(TMTTF)_2PF_6$ single crystal at P=1bar (squares) and 13kbar (circles). In the insert, a log-log plot of the 13kbar data on both sides of $T_N \simeq 17K$. After Creuzet et al., ref. 31.

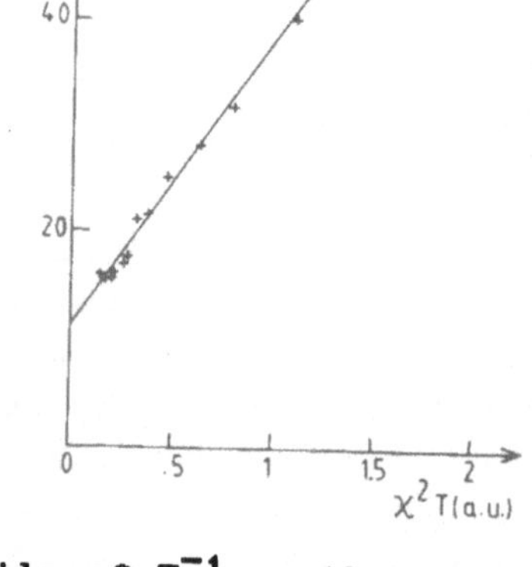

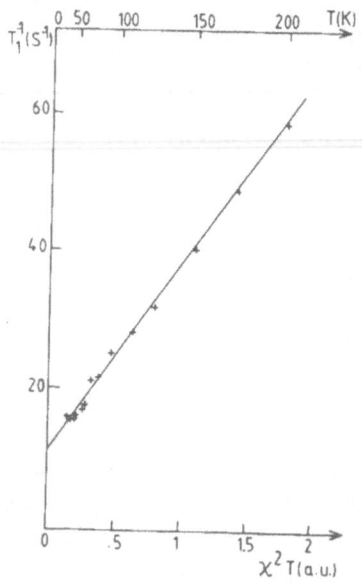

Fig. 4. Linear profile of T_1^{-1} vs the quantity $\chi(q=0,T)^2T$ taken from the EPR data of figure 2 (b).

Caron et al.[29] claimed that the crossover should occur near 10 kbar. A more rigorous treatement that would take into account critical effects of both types of ordering near the crossover is not expected to change this value very much. This order of magnitude analysis is consistent with ^{1}H NMR data of Creuzet et al.[6], on $(TMTTF)_2PF_6$ at 13kbar which clearly showed a T_1^{-1} divergence at $T_N \simeq 17K$. However, in order to avoid the methyl group rotation contribution to the relaxation in the paramagnetic domain, the experiment was repeated on ^{13}C enriched samples and the results are shown in figure 3. In the insert of this figure, the T_1^{-1} divergence at T_N 17K is found to follow quite well the 3D classical profile predicted in (9c) for both sides of the transition. The error on the exponent is less than 10%. The superiority of probing the carbons in this NMR experiment is remarkably illustrated by the presence of a plateau of 1D AF relaxation of the form (9b) outside the critical domain and up to $75K \simeq T_\rho$. In the higher temperature range, the uniform part (7) of the relaxation becomes in turn dominant. Compared to the P=1bar data, the smaller slope of T_1^{-1} versus T in this range of temperature can be assigned to an increase of v_F($\simeq$ 2% /kbar, see ref.35) which mainly decreases the density of states at the Fermi level and the ratio $g_1/2\pi v_F$ under pressure.

From the results presented in this section, it is clear that $(TMTTF)_2PF_6$ at 13kbar (fig.3) and $(TMTTF)_2Br$ at 1bar (fig. 1b) have similar properties. This illustrates how pressure can be equivalent to anion substitution. According to Emery et al.[2], this should be true for all symmetrical anions and to some extent to non-symmetrical ones.

When higher pressure is applied to $(TMTTF)_2PF_6$ and $(TMTTF)_2Br$ compounds, resistivity data show that T_ρ is further decreased.[8] For the PF_6(Br) compound, T_ρ is no longer clearly seen around 20kbar(5kbar) where $T_{x2} \simeq T_\rho$. An interesting feature of the above model that have been predicted by Caron et al.[29] is that if J_b increases with pressure thus, T_N also increases and reaches a maximum when T_ρ is no longer relevant and finally, it will decrease monotonically at higher pressure. There is some indication of a maximum of T_N for $(TMTTF)_2Br$ between 1bar and 13kbar[8,24](see also the lecture of C.Coulon). Additional experimental works are needed in order to confirm this result. At sufficiently high pressure namely, when T_ρ is irrelevant, one is confronted with the delicate problem of the absence of a small parameter for J_b, since $\bar{t}_{\perp b}/\Delta_\rho$ is no longer well defined. At this point, one can therefore ask the question: when T_ρ is no longer a relevant quantity, what becomes the mechanism at the origin of the AF long range ordering ? Actually, there are two possibilities: i) the IEX mechanism is still active but somewhat differently than before and can stabilize the AF order(see below); ii) as predicted by usual perturbation methods,[36] the IEX coupling disappears with T_ρ and it is only with good 2D or 3D nesting of the Fermi surface, taking place only below the single particle dimensionality crossover T_x1, that AF order can be achieved.[2,3] This will be called the NAF mechanism in the following.

This question about the mechanism behind the stabilization of antiferromagnetism is not only relevant for the sulphur series under presssure but also for the selenide series for which there is no T_ρ above all AF transitions observed.[4] Furthermore, the Neel temperatures observed for sulphur compounds under high pressure are known to reach the values commonly seen in the selenide series at P≥1 bar (≤12K).[4,8] Finally, for both series, a finite critical pressure P_c is sufficient to destroy AF long range ordering. For the selenide compounds at $P \geq P_c$, superconducting long range order is seen at 1K or so.[4] For the sulphur compound $(TMTTF)_2Br$, Parkin et al.[1], have reported the appearance of superconductivity at 24kbar. This result, though not reproduced afterward,[8] gives yet strong indications for both series to be considered as a whole for their theoretical interpretation.[2]

PROPERTIES OF $(TMTSF)_2X$ COMPOUNDS

Antiferromagnetism and Precursors to Organic Superconductivity

In this section, we will analyze the NMR data obtained for the $(TMTSF)_2X$ series in order to support the existence of either the IEX or the NAF mechanism for antiferromagnetism. As we will see, this problem is not only relevant to magnetism but it is also fundamental to the puzzling proximity between magnetism and superconductivity in the phase diagram. Here again, a nuclear relaxation rate analysis appears to be quite useful. Actually, from the general expression (8), it is rather easy to show that the relaxation profile precursor to the 3D AF critical domain is quite different depending if it is either the IEX or the NAF mechanism that is involved in the transition. From the arguments that will be presented in the next section and the work of reference 16, the IEX coupling is still active for $T > T_x1 > T_\rho$, if Coulomb interactions are non-zero. Therefore, if we assume for this particular situation that scaling exists at least approximatively in the 1D region and in the 3D critical domain, we can write from (8):

$$T_1^{-1}(2k_F) = cst T\,(T/E_F)^{-\gamma_o} \qquad (T \gg T_N)\ , \qquad (16a)$$
$$T_1^{-1}(\vec{q}_o) = cst\ |T-T_N/T_N|^{-1/2} \qquad (|T-T_N/T_N| \ll 1)\ . \qquad (16b)$$

In absence of T_ρ, deviations to scaling are expected when the the critical domain is approached from above and this leads to an effective increase of γ_o with temperature.[16] It follows then, that strong relaxation enhancement must be expected outside the critical domain. For the NAF model[2,37] the critical temperature is given by $T_N = T_x1\ \exp(-2\pi v_F/\lambda)$ where $\lambda = \lambda(T_x1)$ is the effective AF coupling at T_x1. Therefore, as the weak coupling condition $\lambda < 2\pi v_F$ is expected to be satisfied in $(TMTSF)_2X$ compounds, one has $T_N \ll T_x1$ that is, 2D or 3D Fermi liquid features should take place in a sizeable temperature domain. From (8), it follows that the staggered part of the relaxation for the NAF mechanism is separated into three distinct parts:[36]

$$T_1^{-1}(2k_F) = cst\ T\,(T/E_F)^{-\gamma_o} \qquad (T \geq T_x1)\ , \qquad (17a)$$
$$T_1^{-1}(\vec{q}_o) = cst\ T\,(T_x1/E_F)^{-\gamma_o} \qquad (T_N < T < T_x1)\ , \qquad (17b)$$
$$T_1^{-1}(\vec{q}_o) = cst\ |T-T_N/T_N|^{-1/2} \qquad (|T-T_N/T_N| \ll 1)\ . \qquad (17c)$$

Aside from the same type of critical profile near T_N for both mechanisms, only the NAF model is characterized by a Korringa type of relaxation ($1/T_1T$ = cst) outside the critical width namely, where 2D Fermi liquid properties prevail. The Korringa constant is renormalized however,[19] compared to the bare one by the 1D power law factor $(T_x1/E_F)^{-\gamma_o}$.

For the low temperature domain we are interested in, the Pauli susceptibility χ_P is constant in temperature (T<80K).[38] We therefore expected according to (7), that the uniform contribution to the relaxation will be also Korringa like. Without an accurate knowledge of the amplitude of the latter, one must always keep in mind its possible emergence in the paramagnetic temperature domain.

Electronic correlations of the entire paramagnetic regime in $(TMTSF)_2X$ conductors are best investigated by performing the ^{77}Se nuclear relaxation. A NMR experiment of this kind has been done in Orsay by F. Creuzet[39] on $(TMTSF)_2PF_6$ single crystals below 50K at four different pressures: P=1bar, 5.5, 8 and 11kbar namely, below and above the critical pressure $P_c \simeq 6$ kbar.[4] The data are shown in figure 5. A singularity in T_1^{-1} is observed at $T_N \simeq 12K$ and 8.7K for P=1bar and 5.5kbar repectively. As shown in fig. 6, a log-log plot of the data in the critical domain (T<20K) shows unambigously that T_1^{-1} follows the 3D critical profile predicted for both models. The error on the 1/2 exponent is estimated to be less than 10%. This exponent

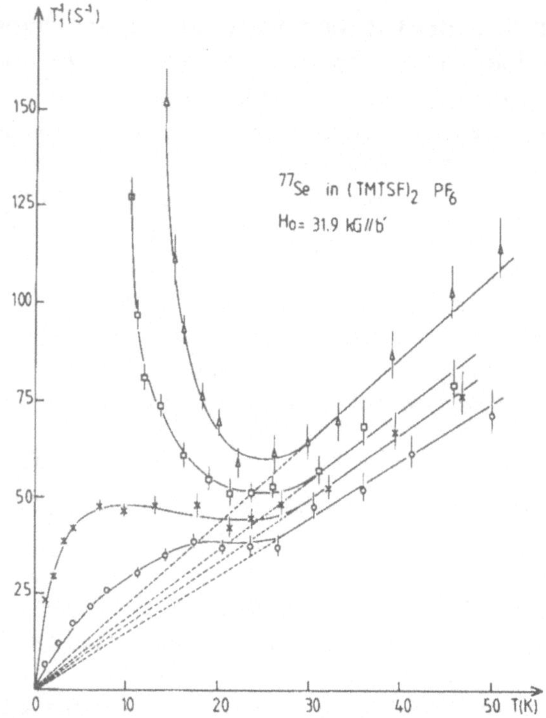

Fig. 5. ^{77}Se relaxation rate data of $(TMTSF)_2PF_6$ vs temperature at P=1bar. (triangles), 5.5kbar (squares), 8kbar (crosses) and 11kbar (circles). After Creuzet et al., ref. 39.

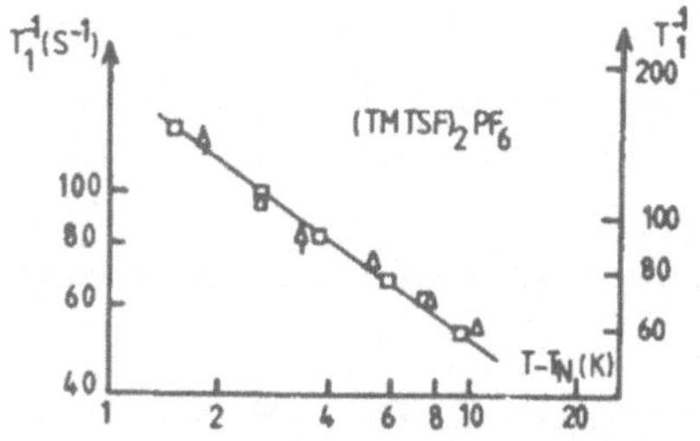

Fig. 6. Log-log plot of the ^{77}Se T_1^{-1} data of fig. 5 at 1bar (triangles, rigth scale) and 5.5kbar (squares, left scale). The classical exponent 1/2 is found with an error estimated to be less than 10%.

agrees with recent results obtained on protons.[37] In contrast to a NMR experiment made on protons however, the inhomogeneous broadening of the ^{77}Se NMR line is so important as we approach T_N, that the intensity of the line drops to zero[40] at T_N and the relaxation cannot be measured in the condensed phase.

At first sight, the appearance of a Korringa type of relaxation at T>30K for both pressures fits well with the prediction of the NAF model (see eqs (15b,c)). This interpretation would imply that at 50K, the system is already a 2D Fermi liquid (T_x1 >50K). By looking at the $P \geq 8kbar > P_c$ data however, the NAF model is unable to explain the strong enhancement that still takes place below 30K. Indeed, at 8kbar for example, $(TMTSF)_2PF_6$ is metallic down to 1K or so where it becomes a superconductor with no sign of "residual" 3D AF order above T_c.[4] This is clearly supported by the absence of decrease in the NMR intensity at T≤30K.[39] This is also consistent with other experiments in this range of temperature which show metallic behavior.[4,38] Actually, residual 2D or 3D AF critical correlations lead only to logarithmic corrections to the Korringa type of relaxation in the NAF model.[39] These corrections are obviously too weak to account for the observed enhancement which persists down to 8K(12K) or so at 8kbar(11kbar). At T<8K(12K), a significative drop of T_1^{-1}, almost linear in temperature, reappears. This sequence of anomalies in the relaxation is not peculiar to the PF_6 compound at $P \geq P_c$, but it seems to be a rather universal precursor to organic superconductivity in the Bechgaard salts. Indeed, the same type of sequence has been observed[19,41] in slow cooled $(TMTSF)_2ClO_4$ samples at $P \geq 1bar > P_c$ and also in $(TMTSF)_2FSO_3$ at $P=10kbar > P_c$. Some of the data are reproduced in figure 7. For the perchlorate, the enhancement is independent of the orientation and the amplitude of the applied magnetic field for H < 32kG. At 64 kG//b' however, a slight down shift in temperature of the sequence has been observed.[19]

For the PF_6 and ClO_4 salts,[38] the Pauli susceptibility is observed to be temperature independent below 50K. This confirms that staggered correlations are involved in the relaxation enhancement. The interpretation given in the context of the ClO_4 salt in ref.19 for such an enhancement was based on the presence of 1D $2k_F$ AF correlations for $8K \leq T \leq 25K$. It was claimed that if the electronic system is still dominated by 1D effects then, the expression (14a) with $\gamma \simeq 1$ can easily account for the observed shape of T_1^{-1}. Its sudden drop at $T \simeq 8K$ was in turn interpreted as the manifestation of the single particle dimensionality crossover temperature T_x1 below which, according to (15b), $T_1^{-1} \alpha T$, in qualitative agreement with the observation(see also figure 8). The same interpretation was applied[39] to the PF_6 salt at P=8kbar($T_x 1 \simeq 8K$) and P=11kbar ($T_x 1 \simeq 12K$). Therefore, if this interpretation is correct, this would imply that the AF phase transition observed below P_c in $(TMTSF)_2PF_6$ salt is not induced by the NAF mechanism, but it is still driven by the IEX coupling(see below).[39] In the work of ref. 19, no microscopic explanation was given to the temperature (25K) where the enhancement suddenly emerges in the ClO_4 compound. The effect of anion ordering that takes place at T≃24K in this salt must be discarded as a possible mechanism for it. Indeed, for the PF_6 salt, we have the same sequence of anomalies and there is no anion ordering at 30K.[4] It is worthwhile to note however, that the change of profile observed here is clearly similar to the one already discussed for the $(TMTTF)_2PF_6$ compound near 65K at ambient pressure (see figure 3). It was shown that it is the uniform part of the relaxation (7) that becomes dominant and completly masks the 1D staggered contribution. Creuzet et al.,[39] argued that the same must be true at 30K(25K) for the $(TMTSF)_2PF_6$ ($(TMTSF)_2ClO_4$) salt. This interpretation is consistent with a constant χ_P in this temperature domain.[38] It agrees also with the field dependence analysis of the ^1H relaxation performed by Stein et al.,[42] which shows that the ratio $T_1^{-1}(q \simeq 0)/T_1^{-1}(2k_F)$ is around 3 at ambient temperature for both compounds, making highly probable the interchange between both types of relaxation at much lower temperature.

The effect of pressure on these relaxation anomalies is also of interest. As shown by Creuzet et al.,[35] on $(TMTSF)_2ClO_4$, the temperature at which the anomaly appears, increases with pressure and reaches 50K at 10kbar. This experimental fact alone, strongly disfavors the 2D or 3D interpretation of the enhancement since 3D AF ordering effects are generally suppressed and pushed down to lower temperature in selenide compounds under pressure.[4] This is not so however, for the temperature at which the uniform contribution to the relaxation is supposed to take over. Actually, the uniform and the staggered parts of the relaxation are related to different types of correlations and combinaison of microscopic couplings[10] and they will act as essentially independent quantities under pressure. In $(TMTSF)_2PF_6$ for example, the onset of the enhancement levels off at 30K between 1bar and 11kbar whereas for $(TMTTF)_2PF_6$, it increases from 65K to 75K or so for the same change of pressure (see figure 3).

The temperature of the restoration of a "linear" regime at

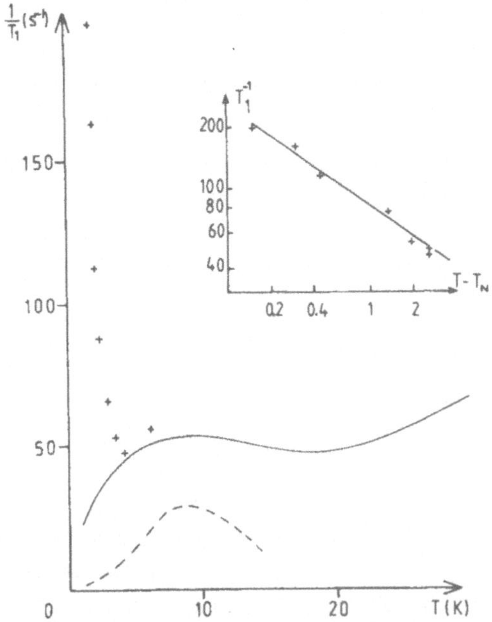

Fig. 7. ^{77}Se T_1^{-1} data vs T of slow cooled $(TMTSF)_2ClO_4$ single crystals at 1bar and H < 32kG//c* (full line), and at 750bar and H = 73.9 kG//c* (crosses); and of $(TMTSF)_2FSO_3$ at 10 kbar and H = 55.3kG//b' (dashed line). In the insert, the log-log plot of the critical relaxation for the high field phase of the perchlorate compound. After Creuzet et al., in refs. 19 and 35.

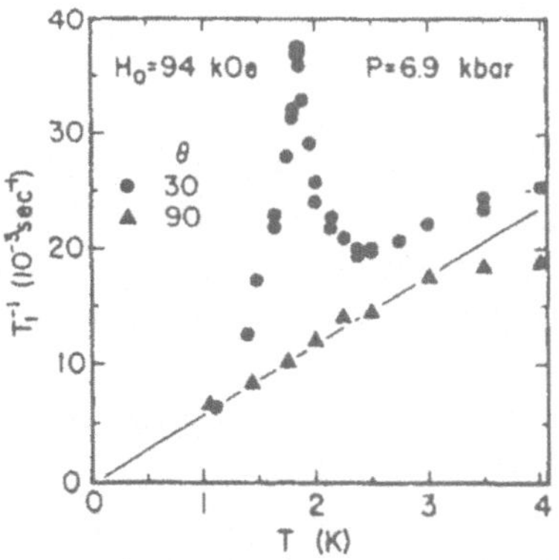

Fig. 8. ^{1}H T_1^{-1} vs T of $(TMTSF)_2PF_6$ at 6.9kbar and for H ≃ 80kG//c* (black circles) and H = 94kG//b'. After Azevedo et al., ref. 51.

T_x1 also evolves under pressure. From figure 5 and also the data of ref.35, T_x1 increases under pressure. Before analyzing the pressure effect on T_x1 however, one must try to understand the physical origin for the apparent irrelevance of T_x1 below P_c and its rather low value just above P_c. The first part of this problem is connected to the origin of the IEX coupling below P_c and will be discussed in the next section. For the second part, Emery[43] predicted that the temperature scale at which the single particle motion undergoes a dimensionality crossover is in lowest order, $T_x^o1 \simeq t_{\perp b}/\pi$. Under favorable conditions, this temperature is also connected with the dimensionality crossover of the nesting vector namely, from $2k_F$ in 1D to $(2k_F, Q_b(, Q_c))$ in 2D(3D). Here, $Q_b(Q_c) \neq 0$ depends on the details of the Fermi surface. From various experimental probes[4] together with band calculations,[15] one gets from this result $T_x^o1 \simeq 50.....80K$. This predicted range of values is obviously much higher than the one deduced above from the relaxation data. The explanation given in ref. 19 for the reduction of T_x1, was based on the effects of 1D electronic correlations on the transverse motion of electrons. Actually, many-body effects affect the efficiency of interchain tunneling thereby decreasing $t_\perp$[7,44]. Quite recently, Caron and the author[45] calculated the effect of the tranverse extension of the correlation cloud aroud each electron. They found that it further reduces the value of T_x1, for which they derive the expression:

$$T_x1 \simeq [\,(1-11\theta)/1-\theta]^{1/2}\ \bar{T}_x1 \,. \qquad (18)$$

$\theta \geq 0$ is the interaction dependent 1D exponent for the decrease of the density of states at the Fermi level due to electronic correlations.[46] $\bar{T}_x1 = T_x^o1\ (t_{\perp b}/E_F)^{\theta/1-\theta}$ is the crossover temperature of refs 17 and 19, which neglects the transverse extension of the correlation cloud. From (18), this effect is quite important however, since it can completly suppress T_x1 even for relatively weak coupling ($\theta \geq 1/11$). The low temperature emergence of T_x1 deduced by NMR would suggest that we are not too far from the critical value of θ. As pressure increases $t_{\perp b,c}$ and v_F, which in turn decreases the ratio $g_1/\pi v_F$ and θ, T_x1 will then increase under pressure. This is clearly seen for $(TMTSF)_2PF_6$ by comparing the 8kbar and the 11kbar relaxation data of figure 5. In the perchlorate compound, the crossover temperature reaches the value of 25K or so under 10kbar.[35] One must note that the increase of the Fermi velocity under pressure can mainly account[35] for the decrease of the slope of $T_1^{-1}(q{=}0)$ at $T \geq 30K$ in Fig. 5, since $T_1\ (q{=}0)$ is proportional to the square of the Fermi energy and the enhancement depends on the ratio $g_1/2\pi v_F$.

THE INTERCHAIN ANTIFERROMAGNETIC EXCHANGE IN WEAK COUPLING

In this section, we would like to give qualitative arguments that support the existence of the IEX mechanism in the metallic state of $(TMTTF)_2X$ and $(TMTSF)_2X$ compounds. As already mentionned in a previous section: when there is no small paramater of the form $\bar{t}_\perp/\Delta_\rho$, standard perturbation[36] approaches have failed to predict a non-zero IEX amplitude. As demonstrated in ref.16 however, the IEX coupling should always be present at $T \geq T_x 1$, in spite of the absence of Δ_ρ in a metallic state. Actually, it is well known that local Coulomb repulsion always generate locally correlated pairs (electron-hole and electron-electron) of particles.[10] Since the pairs have finite correlation length and time, a coherent tunneling of these pairs from chain to chain via IST should be possible. Very simple arguments support this view.[16] In fact, in the absence of gaps and for $T > T_x 1$, both particles forming any type of locally correlated pairs can hop to neighboring chain within the thermal coherence time scale $\tau_{th} \simeq 1/\pi T$. They can each do so through the effective interchain tunneling $\bar{t}_\perp$ which is compatible with (18). The probability for a coherent hopping is thus $(\tau_{th}/\tau_\perp)^2 = (\bar{t}_\perp/\pi T)^2$. On the other hand, the local nature of the pair is assured by the local nature of the Coulomb interaction itself. In fact, the probability of finding a locally correlated pair of particles is $\tilde{g}' = g'/\pi v_F$ for each chain where $1/\pi v_F$ is the density of state at the Fermi level and $\tilde{g}'$ is the strenght of the effective electron-electron coupling associated with the type of correlated pair involved in the process. This leads to a total tunneling matrix element proportional to $\tilde{g}'^2(\bar{t}_\perp/\pi T)^2$. This result, though qualitative, has been confirmed rigourously using the renormalization group method.[16] For the IEX coupling for example, we have $J_b \propto (\tilde{g}'_2 + \tilde{g}'_3)^2 (\bar{t}_{\perp b}/\pi T)^2$ in weak coupling which is now a temperature dependent quantity. The same arguments can be used in the strong coupling case where $g' \simeq \mp\pi v_F$, namely when there is a gap (Δ) in the spin(-) or charge(+) excitations. In this case, there is a pair binding coherence time $\tau_{gap} \simeq 1/\Delta$ that replaces $\tau_{th} (>\tau_{gap})$. For the IEX amplitude, the pair tunneling matrix element reduces to the standard form that is,[37] $J_b \propto (\bar{t}_{\perp b}/\Delta_\rho)^2$. In the context of organic conductors, this result is quite interesting since it demonstrates that the IEX mechanism is non-zero and continuous at $\pi T \simeq \Delta_\rho$. In a recent work,[14] Caron and the author used this result to show that T_N in $(TMTTF)_2X$ compounds is continuous when Δ_ρ is no longer seen with pressure. The rapid drop of T_N at low (high) pressure in $(TMTSF)_2X$ ($(TMTTF)_2X$) series near P_c was interpreted as two-dimensional effects on the IEX mechanism. In fact, as the pressure increases there is some characteristic temperature where $\bar{t}_\perp/\pi T \simeq 1$ and is in turn no longer a small paramater. This temperature is precisely T_x1 given in (18). Thus, the relevance of the IEX mechanism for the AF ordering is cut-off by the single particle dimensiolity crossover, consistently with the fact that T_x1 becomes only observable at $P>P_c$.

THE FIELD-INDUCED TRANSITION AS SEEN BY NMR

The complete disparition of 3D AF long range ordering for $T < T_x1$ is generally believed to be due to the absence of good 2D or 3D nesting conditions for the Fermi surface, which makes the NAF mechanism ineffective.[47] A very nice theoretical support for the absence of perfect nesting at $P>P_c$ in $(TMTSF)_2X$ salts, was given by the Gorkov and Lebed interpretation for the occurrence of a succession of AF phase transitions[48] observed when the applied magnetic field along the c* direction is greater than some treshold value $H_T \simeq 40$kG.[49] According to Gorkov and Lebed, the electronic motion is "unidimensionalized" under a c* tranverse magnetic field and this improves nesting conditions. The argument was further refined by Heritier et al.,[50] who show that the field induced NAF mechanism is in fact quantized.

The first evidence of the AF character of the field-induced transition was given by Azevedo et al.,[51] using the ^{1}H NMR probe on $(TMTSF)_2PF_6$ at 6.9 kbar. The relaxation data obtained are reproduced in figure 8. At H≃80kG//c* (black circles), there is a clear-cut divergence of T_1^{-1} at $T_N(H) \simeq 1.8$K, showing the growth of 3D AF order. From this figure, it is also interesting to note that when H//b', there is no phase transition at all and the relaxation is Korringa like below 4K ($T_1^{-1} \alpha$ T). The high temperature part of this linear profile is also seen outside the critical domain of the field induced transition. This experimental fact is important since it allows a clear distinction between the AF correlations leading to the critical effects in T_1^{-1} near $T_N(H)$ and those taking place near 8K (see figures 5 and 7) which were associated to 1D effects. Other groups[35,41,52] have also reported a similar upturn of the relaxation above $T_N(H)$. The ^{77}Se data of Creuzet et al.,[35] in figure 7 for $(TMTSF)_2ClO_4$ under 72kG//c*, clearly show a 3D classical divergence of the form (17c)(see the log-log plot in the insert). Outside the critical domain, there is an upturn of T_1^{-1}, just before reaching the plateau of relaxation seen in absence of transition. Thus, from the profiles predicted for the NAF and the IEX mechanisms, only the NAF prediction in (17) seems to be consistent with the observed relaxation profiles for field-induced transitions. This interpretation then agrees with the Gorkov and Lebed model.

It is generally observed[49,50] that $T_N(H)$ increases under field (see the lectures of Chaikin and Ribault). Therefore, one can conjecture that a NAF to IEX crossover can be induced under sufficiently high tranverse field. In fact, when the temperature of onset of critical fluctuations, which obviously increases with $T_N(H)$, will be of the order of or greater than T_x1 then the IEX coupling should in turn drive the transition.

ON THE NATURE OF THE SUPERCONDUCTING PAIRING

As we have just discussed, the assumption of bad nesting conditions that are responsible for the disappearance of the AF sate at $P \geq P_c$ is remarkably supported by the existence of field-induced phases. Although the disappearance of magnetism is obviously a necessary condition for superconductivity to take place, deviations from perfect nesting cannot explain the origin of the superconducting pairing. The behaviour under high field and the presence of strong AF precursors above T_x1 supported the existence of dominant repulsive interactions at low temperature and thus, lead to infer that phonons might not play a significant role in the origin of superconducting pairing (see the comment of Caron). However, the close proximity between antiferromagnetism and superconductivity in the phase diagram have led to consider a possible participation of antiferromagnetic correlations to the pairing interaction.[14,16,52,53] There are two distinct approaches to this problem. One possibility, that has been first discussed by Emery[52] and also worked out by Beal-Monod et al.[53], is to look at the effective electron-electron interaction in presence of 2D or 3D AF correlations that are produced *below* T_x1 by weak deviations to perfect nesting. Another possibility that has been considered recently by Caron and the author,[14,16] was to look at the pairing in presence of AF correlations that are produced *above* T_x1 and that are triggered by the IEX mechanism. Only the latter approach will be discussed here (see the lecture of V. Emery for the first approach). Note that the AF correlations involved in this approach are precisely those seen by NMR (see figures 5 and 7). If we assume that there are sizeable deviations from perfect 2D or

3D nesting at T_x1 then, only the superconducting channel can develop long range ordering. Actually, if we take into account the effect of all intra and interchain couplings, the gap equation for the existence of singlet superconducting long range order below T_x1 reads:

$$\Delta(k_\perp) = -1/2 \sum_{k'} \Delta(k'_\perp)/\pi v_F \quad [(g'_1(T_x1)+g'_2(T_x1)) + V_S -3/2 J'_b(k_\perp+k'_\perp)] \; th(\varepsilon_k/T_c) \; . \tag{19}$$

Here $k_\perp+k'_\perp$ is the tranverse momentum transfer of the IEX coupling and ε_k is the electron energy close to the Fermi surface. V_S is the effective interchain Josephson coupling which is always *attractive* at T_x1 ($\tilde{V}_S \simeq -2(\tilde{g}'_1(T_x1) + \tilde{g}'_2(T_x1))^2$) and *which exists even for repulsive g's !*(see the section on the interchain coupling).[16] This term can screen a sizeable part of the effective local pairing repulsion at T_x1. From (19) however, one notes the remarkakle fact that the effective IEX amplitude J'_b at the crossover gives a non-zero contribution to the singlet pairing matrix element. In the neighborhood but not nescessarly too near an AF critical domain($T_x1 \geq T_x2$), the quantity $J'_b(k_\perp+k'_\perp) \simeq J'_b \cos(k_\perp+k'_\perp)d_\perp$ can be sufficiently large to screen completly the repulsion and gives an attractive effective pairing. Physically, this IEX pairing-induced mechanism can be visualized as an interchain AF paramagnon exchange. Indeed, the IEX coupling is equivalent to consider that a $+k_F$ single electron on a given chain, emits a 1D $2k_F$ AF paramagnon and in the process, the electron is backscattered to $-k_F$ while the emitted paramagnon has a finite probabilty via $t_\perp$, to tunnel coherently to a neighboring chain where a $-k_F$ electron can absorb it and be in turn backscattered at $+k_F$. Because the paramagnon emission and absorption take place on different chains, an attraction between electrons that participate in the process is possible. The non-local character of the IEX pairing in the tranverse direction is showed by its dependence on $k_\perp+k'_\perp$ in (19). In presence of a net attraction, the superconducting gap $\Delta(k_\perp)$ will be anisotropic with zeros on the Fermi surface. From numerical renormalization group calculations, Caron and the author have shown[14] that starting with weak coupling values for the bare g's and typical values for t_a and $t_{\perp b}$, one can justify the low value of T_x1 observed. Furthermore, the use of renormalized couplings at T_x1 along with the gap equation (19) in absence of good nesting conditions, lead to a finite superconducting temperature of the rigth order of magnitude. T_c is also found to be strongly pressure dependent, in qualitative agreement with the observation.[4]

It is worth noting that it is the existence of interchain pair tunneling processes in absence of gaps that makes this pairing mechanism possible. In the model discussed here, it is clear that these couplings play a central role for the whole phase diagram of both series of compounds.

ACKNOWLEDGMENTS

I would like to thank Prof. L. G. Caron and Dr F. Creuzet for their close collaboration in the development of the ideas presented in these lectures. I also benefit of several important discussions with the Profs. D. Jerome, S. Barisic, V. Emery, J. Friedel, M.T. Beal-Monod, H. Schulz and P. Stein.

References

1. C. Coulon, P. Delhaes, S. Flandrois, R. Lagnier, E. Bonjour and J.M. Fabre, J. Physique 43, 1059(1982); S. Parkin, F. Creuzet, M. Ribault, D. Jerome, K. Bechgaard and J. M. Fabre, Mol. Cryst. Liq. Cryst. 79, 249(1982).
2. V. Emery, R. Bruisma and S. Barisic, Phys. Rev. lett. 48, 1039(1982).
3. C. Bourbonnais, F. Creuzet and L. G. Caron, J.M.M.M. 54-57, 1249(1986).
4. D. Jerome and H. J. Schulz, Adv. Phys. 31, 299(1982).
5. J. P. Pouget, Chemica Scripta, 17, 85(1981); A. Maaroufi, S. Flandrois, G. Fillion and J. P. Morand, Mol. Cryst. Liq. Cryst. 119, 311(1985).
6. F. Creuzet, D. Jerome and A. Moradpour, Mol. Cryst. Liq. Cryst. 119, 287(1985).
7. T. Takahashi, F. Creuzet, D. Jerome and J. Fabre, J. Physique(Colloque) 44, C3-1095(1983).
8. F. Creuzet, S. S. P. Parkin, D. Jerome and J. Fabre, J. Physique(Colloque) 44, C3-1099(1983).
9. S. Barisic and S. Brazovskii, in: "Recent developpements in condensed matter physics", J. T. Devreese ed., Vol. 1, Plenum, N.Y. (1981).
10. J. Solyom, Adv. Phys. 28, 201(1979); V. J. Emery, in: "Highly conducting one-dimensional solids, J. T. Devreese, R. P. Evrard and V. E. Van Doren eds, Plenum, New York, (1979).
11. P. A. Lee, T. M. Rice and R. A. Klemm, Phys. Rev. B15, 2984(1977).
12. B. Liautard, S. Peytavin, G. Brun and M. Maurin, J. Physique(Colloque) 44, C3-951(1983).
13. R. Laversanne, C. Coulon, B. Gallois, J. P. Pouget and R. Moret, J. Phys. Lett. 45, L393(1984); A. Maaroufi, S. Flandrois, C. Coulon, P. Delhaes, J. P. Morand and G. Fillion, J. Physique(Colloque) 44, C3-1091(1983); R. Laversanne, J. Amiell, C. Coulon, C. Carrigou-lagrange and P. Delhaes, Mol. Cryst. Liq. Cryst. 119, 317(1985).
14. L. G. Caron and C. Bourbonnais in Proc. of Yamada Conference in Physics and Chemistry of Quasi-1D Conductors, Lake Kawaguchi, Japan (1986), to appear in Physica B.
15. P. M. Grant, J. Physique(Colloque) 44, C3-847(1983); L. Ducasse, M. Abderraba, J. Hoarau, M. Pesquer, B. Gallois and J. Gaultier, J. Phys. C: Solid State Phys. 19, 3805(1986).
16. C. Bourbonnais and L. G. Caron in ref. 14 and submitted for publication.
17. C. Bourbonnais, Mol. Cryst. Liq. Cryst. 119, 11(1985) and Ph.D thesis, Universite de Sherbrooke (1985), unpublished.

18. G. Soda, D. Jerome, M. Weger, J. Alizon, J. Gallice, H. Robert, J. M. Fabre and L. Giral, J. Physique 38, 931(1977).
19. C. Bourbonnais, F. Creuzet, D. Jerome and A. Moradpour, J. Phys. Lett. 45, L755(1984).
20. T. Moriya, J. Phys. Soc. of Jpn. 18, 516(1963).
21. C. Bourbonnais, in Proc. of the International Conference on Science and Technology of Synthetic metals, Kyoto, Japan (1986), to appear in Synthetic Metals.
22. P. Pfeuty, D. Jasnow and M. E. Fisher, Phys. Rev. B10, 2088(1974); B. I. Halperin and P. C. Hohenberg, Phys Rev. 177, 952(1969).
23. J. B. Torrance, J. Physique(Colloque) 44, C3-799(1983) and references therein; C. Coulon, J. C. Scott and R. Laversanne, Mol. Cryst. Liq. Cryst. 119, 307(1985).
24. F. Creuzet, T. Takahashi, D. Jerome and J. M. Fabre, J. Phys. Lett. 43, L755(1982).
25. M. Peo, J. C. Scott and E. M. Engler, Phys. Rev. B30, 3639(1984).
26. J. P. Pouget, R. Moret, R. Comes, K. Bechgaard, J. M. Fabre and L. Giral, Mol. Cryst. Liq. Cryst. 79, 129(1982); J. P. Pouget, unpublished results.
27. C. Bourbonnais and L. G. Caron, Mol. Cryst. Liq. Cryst. 119, 287(1985); L. G. Caron and C. Bourbonnais, Phys. Rev. B29, 4230(1984).
28. J. D. Bray, L. V. Interrante, I. S. Jacols and C. B. Bonner, in: "Extended Linear Compounds," J. S. Miller, ed., Plenum press, N. Y. (1982).
29. L. G. Caron, C. Bourbonnais, F. Creuzet and D. Jerome, in ref. 21.
30. M. Kimura, Prog. Theor. Phys. 53, 955(1975).
31. F. Creuzet, C. Bourbonnais, L. G. Caron, D. Jerome and K. Bechgaard, in ref. 21.
32. K. B. Efetov, Sov. Phys. JETP 43, 1221(1976).
33. D. J. Scalapino, Y. Imry and P. Pincus, Phys. Rev. B11, 2042(1975).
34. E. Ehrenfreund and L. J. Scott, Phys. Rev. B16, 1870(1977).
35. F. Creuzet, D. Jerome, C. Bourbonnais and A. Moradpour, J. Phys. C: Solid State Phys. 18, L821(1985).
36. Y. A. Firsov, V. N. Prigodin and Chr. Seidel, Physics Reports 126, 245(1985); S. Brazovskii and V. Yakovenko, J. Phys. Lett. 46, L111(1985).
37. C. Bourbonnais, P. Stein, D. Jerome and A. Moradpour, Phys. Rev. B33, 7608(1986).
38. M. Miljak, J. R. Cooper and K. Bechgaard, J. Physique(Colloque) 44, C3-893(1983); M. Miljak and J. R. Cooper, Mol. Cryst. Liq. Cryst. 119, 141(1985); L. Forro, J. R. Cooper, B. Rothaemer, J. S. Schilling, M. Weger and K. Bechgaard, Sol. State Comm. 60, 11(1986).
39. F. Creuzet, C. Bourbonnais, L. G. Caron, D. Jerome and A. Moradpour, in ref. 21 and to be submitted.
40. A. Andrieux, D. Jerome and K. Bechgaard, J. Phys. Lett., 46, L87(1981).
41. T. Takahashi, D. Jerome and K. Bechgaard, J. Physique 45, 445(1984); M. Takigawa and G. Saito, J. Phys. Soc. of Jpn 55, 1233(1986).
42. P. C. Stein, A. Moradpour and D. Jerome, J. Phys. Lett. 46, L241(1985).
43. V. J. Emery, J. Physique(Colloque) 44 C3-977(1983).
44. V. N. Prigodin and Y. A. Firsov Sov. Phys. JETP 49, 813(1979).
45. L. G. Caron and C. Bourbonnais, submitted for publication.
46. Y. Suzumura, Prog. Theor. Phys. 63, 51(1980).
47. K. Yamaji, J. Phys. Soc. of Jpn 51, 2787(1982); 52, 1361(1983).
48. L. P. Gorkov and A. G. Lebed, J. Phys. Lett. 45, L433(1984).

49. J. F. Kwak, Phys. Rev. B28, 3277(1983); K. Kajimura, M. Tokumoto, K. Murata, T. Ukachi, H. Anzai, T. Ishiguro and G. Saito, J. Physique(Colloque) 44, C3-1059(1983); M. Ribault, D. Jerome, T. Tuchendler, C. Weyl and K. Bechgaard, J.Phys. Lett. 44, L953(1983); F. Pesty, P. Garoche and K. Bechgaard, Phys. Rev. Lett. 55, 2495(1985); P. M. Chaikin, Mu-Yong Choi, J. F. Kwak, J. S. Brooks, K. P. Martin, M. J. Naughton and E. M. Engler, Phys. Rev. Lett. 51, 2333(1983).
50. M. Heritier, G. Montambaux and P. Ledderer, J. Phys. Lett. 45, L943(1984).
51. L. J. Azevedo, J. E. Schirber and E. M. Engler, Physica 108B, 1183(1981).
52. V. J. Emery, Synthetic Metals 13, 21(1986).
53. M. T. Beal-Monod, C. Bourbonnais and V. J. Emery, Phys. Rev. B34 (Nov. 1986).

ELECTRON TRANSPORT IN $(FA)_2X$ -TYPE ORGANIC CONDUCTORS

SPECIAL MAGNETIC RESONANCE TECHNIQUES.

M. Mehring

2. Physikalisches Institut der Universität Stuttgart
D-7000 Stuttgart 80/Fed.Rep.Germany

1. STRUCTURE AND BANDSTRUCTURE

Fluoroanthene radical cation salts of the $(FA)_2X$-type, where X stands for all possible counter ions like AsF_6, SbF_6, PF_6 etc. show a similar structure as the Bechgaard salts $(TMTSF)_2X$ [1,2].

In contrast to the Bechgaard salts, fluoranthenyl (FA) is a pure hydrocarbon, where the highly one-dimensional conduction is achieved by stacking the molecules alternatively with an average separation of 3.3 Å. Strong overlap of the p_z-orbitals occurs, leading to a tight binding bandstructure as shown in Fig.1. Note, that the repeat unit along the conduction axis (a) with a lattice constant a= 3.33 + 3.28 = 6.6 Å contains two FA molecules. This dimerization exists already at room temperature[1] and corresponds to the $4k_F$ transition caused by Coulomb repulsion. It opens a gap whoose width $4\Delta_{\|}$ is currently unknown.

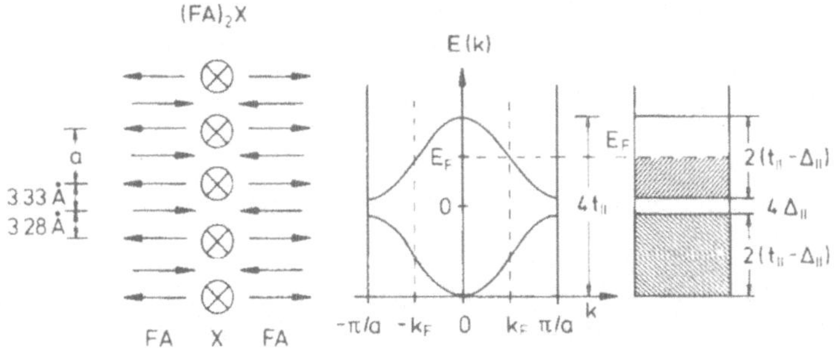

Fig.1 Left: Stack structure of $(FA)_2X$ with unit cell spacing a. Right: Corresponding tight binding bandstructure. The $4k_F$ splitting which exists already at room temperature is indicated. A $2k_F$ transition occurs at about 190 K, opening a gap at the Fermi level.

Since one electron is transferred to the anion per dimer one can speak of a 3/4 filled electron band or 1/4 filled hole band. The corresponding Fermi wave vector $k_F=\pi/(2a)$ is shown in Fig. 1. The lower part of the band is filled with two electrons and the upper part with one electron per dimer. The total bandwidth amounts to $4t_{\|}$ where $t_{\|}$ is the intermolecular transfer integral. If we restrict ourselves to the half filled upper band we arrive at the following parameters: Fermi energy $E_F = t_{\|}$ with Fermi velocity $v_F = \hbar^{-1}(dE/dk)_{k_F} = \hbar^{-1}t_{\|}a$ and the density of states per spin $N(E_F)=(\pi t_{\|})^{-1}$.

From reflectance measurements[3] it is known that $(FA)_2X$-type conductors have a very large bandwidth as compared with TTF-TCNQ and $(TMTSF)_2X$ type crystals. We adopt here the value $t_{\|}=0.5$ eV [3]. Although the measured value of the magnetic susceptibility[4] $\chi_S=0.94\ 10^{-4}$ emu/mole^{-1} is smaller than in TTF- TCNQ [5,6], there may still be an enhancement due to Coulomb correlations also in $(FA)_2X$ as is advocated here. Following the argument of Jerome and cowor- kers [5,6], we can express the observed magnetic susceptibility χ_S as

$$\chi_S = \frac{\chi_P}{1-\alpha} \;; \quad \text{where } \chi_P = \frac{1}{2}\,(g\mu_B)^2\, N(E_F) \tag{1}$$

and

$$\alpha = U\, N(E_F) = U/(\pi t_{\|}) \tag{2}$$

The Hubbard U which takes care of the Coulomb correlations can be estimated from the measured susceptibility as $\alpha \simeq 0.56$. At the end of this contribution we will discuss an alternative determination of α.

At about 190K a phase transition takes place[1] which was not only observed in X-ray scattering[1] and susceptibility measurements[4] but also in numerous magnetic resonance experiments. It shows all the signatures of a $2k_F$-Peierls transition with a gap opening of about 0.12 eV. In this contribution we will restrict ourselves, however, to the metallic region ($190K \le T \le 300K$) only.

2.0 SPIN TRANSPORT IN A MAGNETIC FIELD GRADIENT

A number of ESR experiments (continuous wave (cw) and pulsed) have been performed on $(FA)_2X$ type organic conductors[7-14]. Most of these are concerned with electron spin relaxation (T_1,T_2) and linewidth measurements[7,11,14]. These type of measurements do allow the determination of τ_S (longitudinal phonon scattering time) and $\tau_\perp$ (transverse hopping time) by applying a model for one-dimensional motion and the knowledge of the process governing the electron spin relaxation behaviour. This information is therefore rather indirect and relies on several assumptions. There is, however, a method, namely the "magnetic field gradient method" which allows the direct observation of spin motion without any adjustable parameter[11-13]. This technique has been used in NMR a lot, but was applied in ESR to dertermine electron transport in organic conductors only recently[11-13]. In the course of some electron spin echo experiments[9,10] the gradient effect was also noticed, but not utilized.

Fig.2 shows the pulse sequences which have been used to determine electron spin diffusion in a magnetic field gradient[11-13]. A $(\pi/2)$-τ- (π) spin echo sequence is always applied at the Larmor frequency ($\omega_e/2\pi$=9GHz) of the electron spins. The electron spin echo appears at $t_E = 2\tau$ and would decay without diffusion as $\exp(-t_E/T_2)$, where T_2 is the characteristic decay time (neglecting special non-exponential decay processes here). An echo can only be formed if the ESR spectrum is inhomogeneous. In $(FA)_2X$ organic conductors the ESR line is homogeneous and a magnetic field gradient G_z (in z- direction) has to be applied to make it inhomogeneous. The application of the π-pulse inverts the spins and therefore lets time run backwards. Another way of looking at it is that the inverted spins see a gradient G_z whose sign

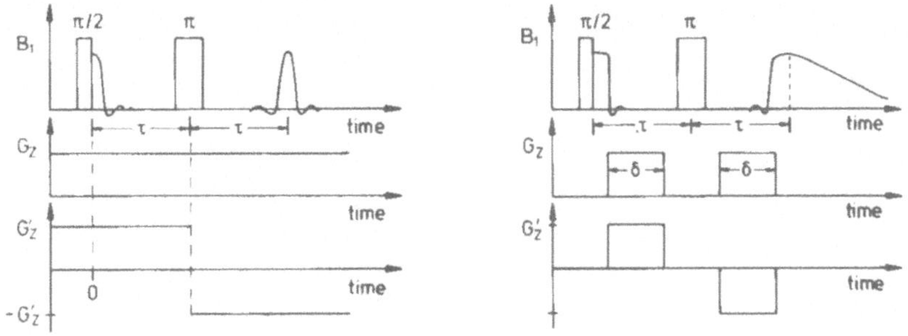

Fig.2 Two different magnetic field gradient spin echo techniques for spin diffusion measurements. Left: Constant gradient. Right:Pulsed gradient.

has changed. In fact we have performed in our laboratory an experiment recently, where instead of the π-pulse only the gradient was inverted at time τ creating an echo at time $t_E=2\tau$[15]. If there is no motion of the spins, the phase change $\Delta\phi=\gamma G_z z\tau$ of a spin (where γ is the gyromagnetic ratio of the electron) at position z during time τ is exactly cancelled at time 2τ. Only when random spin transport occurs the spin visits different positions z which cause a random fluctuation of local fields, which is not refocussed at $t_E=2\tau$. It leads therefore to a loss of coherence as

$$G(2\tau) = \exp(-2\tau/T_2)\ \exp\left(-\frac{2}{3}\ \gamma^2\ G_Z^2\ D_Z\ \tau^3\right) \tag{3}$$

with only the D_z component of the diffusion tensor $\tilde{D}$ contributing to the decay as

$$D_z = D_{\|}\cos^2\beta + D_{x\perp}\cos^2\alpha\,\sin^2\beta + D_{y\perp}\sin^2\alpha\,\sin^2\beta \tag{4}$$

where $D_{\|}$ is the diffusion constant parallel to the highly conducting axis a (corresponding to β=o) and $D_{x\perp}$ and $D_{y\perp}$ are the corresponding perpendicular components. Note, that no unknown parameters besides D_z are contained in the extra decay function. The gradient G_z can be experimentally determined to a high degree of accuracy, i.e. the diffusion constant can be determined directly in absolute units[11-13]. Although it is adventageous to use a single crystal to determine $D_{\|}$ and $D_{\perp}$ separately, at least $D_{\|}$ can be determined in a powder quite accurately too[11].

The constant gradient technique (Fig.2 left) is limited, because large gradients during the microwave pulses affect spin excitation and can lead to wrong results. It is therefore adventageous to apply pulsed field gradients as is shown in Fig.2 (right)[12,13]. Fig.3 shows some results which were obtained with this technique. Note the large anisotropy of the diffusion tensor. Even larger anisotropies ($D_{\|}/D_{\perp}\simeq1000$) have been observed recently[13]. The following values are characteristic for $(FA)_2X$ organic conductors: $D_{\|}=1.6cm^2s^{-1}$ and $D_{\perp}\simeq10^{-3}cm^2s^{-1}$. Note, that $D_{\|}$ and $D_{\perp}$ are practically temperature independent in the metallic regime ($190K\leq T\leq300K$) and only $D_{\|}$ decreases slightly but not dramatic below the phase transition ($T_c\simeq190K$)[13]. This leads to the conclusion that the drastic reduction in conductivity and susceptibility below T_c is not caused by a drastic change in the transport parameters.

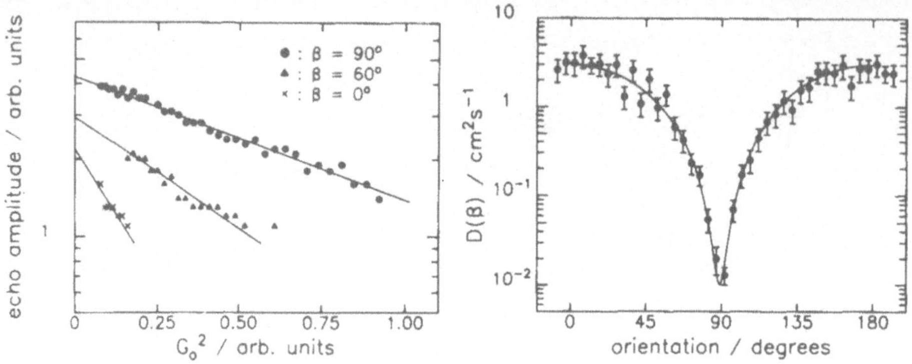

Fig.3 Left: Conduction electron spin echo decay curves versus gradient squared for different orientations of the crystal. β=0 corresponds to B_0 parallel to the 1D conduction axis. Right: Spin diffusion coefficient D versus crystal orientation with respect to the magnetic field gradient.

These data allow us to derive the following conduction electron parameters:

$D_{\parallel} = v_F{}^2\tau_s = 1.6\ cm^2s^{-1};\qquad \tau_s = 0.64\ 10^{-15}\ s\quad$ for $v_F = 5\cdot10^5\ cm^2s^{-1}$

$D_{\perp} = b^2/\tau_{\perp} = 10^{-3}cm^2s^{-1};\qquad \tau_{\perp} = 1.6\cdot10^{-11}\ s\quad$ with $b = 10A$

Similar values were obtained from other measurements[3,8-10]. Following Soda et.al.[5] the transverse tunneling integral $t_{\perp}=\hbar/(\tau_s\tau_{\perp})^{1/2}$=8meV can be obtained and a classification according to (a) dimensionality $\delta=t_{\perp}/E_F$ =1.6 10^{-2} (with E_F=0.5eV) and (b) disorder $\kappa=\hbar/(E_F.\tau_s)$=2.1 is possible. This allows to label the $(FA)_2X$ organic conductors as highly one-dimensional, highly disordered organic metals. It might be speculated, that strong $4k_F$ fluctuations in the metallic regime are causing the disorder.

3.0 ^{1}H AND ^{13}C KNIGHT SHIFTS

The Knight shift (named after its discoverer), observed in metals, is a shift of the NMR resonance frequency by an amount $\Delta\omega = K\omega_n$, with

$$K = \chi_s\ a/(\hbar\gamma_e\gamma_n) \qquad (5)$$

where ω_n is the nuclear Larmor frequency, γ_e and γ_n are gyromagnetic ratios of the electrons and nuclei respectively, χ_s is the spin susceptibility and a is the hyperfine interaction. The effect is caused by the rapid motion of the conduction electrons passing by the nuclei. A spin up electron shifts the resonance of the nucleus to +a/2 and a spin down electron to -a/2. If both spin directions would be equally probable K would be zero. By the virtue of the magnetic field B_0 a slight unbalance of spin up and spin down electron expressed by the spin susceptibility χ_s results in the net shift K. For known χ_s the Knight shift therefore allows to determine the hyperfine interaction a between conduction electrons and the nuclear spins of the sample. The importance of Knight shift measurements for the determination of the molecular orbitals of the conduction band (CBMO) in organic conductors has been demonstrated recently[17-19] and will be disscused in the following.

In solids, proton lines are usually broad (typical 50kHz) and ^{13}C lines (natural abundance 1.1%) are usually weak and broad if not special techniques are applied[16]. In order to obtain highly resolved proton spectra in solids, a special multiple-pulse NMR technique, namely the Waugh, Huber, Haeberlen sequence (see ref.16) has to be applied. There are numerous other sequences (like MREV-8 and BR-24 used here) derived from this mother sequence. The essential aspect of these techniques is that they average the dipole-dipole interaction, which causes the dominant part of the broadening to zero, but retaining the shift interaction, however scaled. In ^{13}C spectroscopy first the ^{13}C spins have to be polarized via the proton spin bath and must be decoupled from the protons consecutively to give a highly resolved spectrum according to Pines, Gibby and Waugh (see ref. 16).

Both sequences are scetched in Fig.4 (left). The top sequence is applied to protons, where all pulses are 90°-pulses with phases ±x, ±y in the rotating frame. The cycle is repeated several hundred times. The sequence at the bottom creates a spin polarization of the S spins (here ^{13}C) by contact with the abundant I spins (here ^{1}H). All fields are in the rotating frame. After the B_{1S} field (^{13}C) has been turned off, a free induction decay (FID) results with the proton spins decoupled by the presence of the B_{1I} decoupling field. After Fourier transform of the FID the ^{13}C spectrum results[16]. In order to erase all anisotropic interactions in addition magic angle sample spinning (MAS) was applied in combination with both pulse schemes.

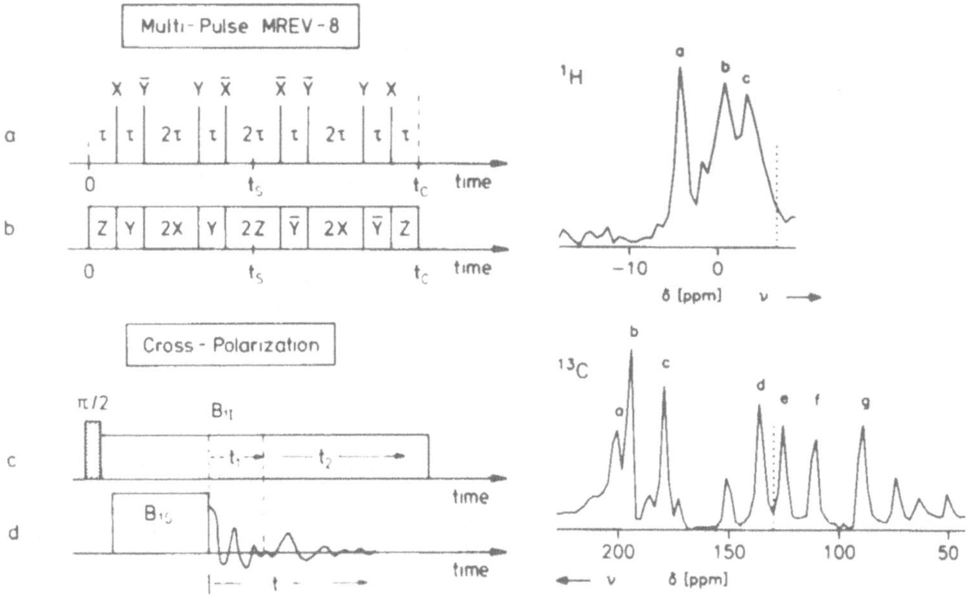

Fig.4 Top: Multi-pulse sequence (left) and corresponding high resolution ^{1}H NMR Spectrum. Bottom: Cross-polarization pulse scheme (left) and corresponding ^{13}C NMR spectrum. The labelling of the lines is identical to the one in the Table.

The highly resolved ^1H and ^{13}C spectra in $(FA)_2X$ are shown in Fig.4 (right). The protons of diamagnetic FA in the solid state show a single line at the position of the dotted line. The lines labelled a,b,c are Knight shifted from the diamagnetic resonance frequency (about 180MHz)[18,19]. The corresponding ^{13}C spectrum is shown at the bottom and contains a larger number of Knight shifted lines as is expected for the 9 different carbons of the FA-molecule (see Fig.5)[17]. The assignment of the lines is by no means a trivial task and has been performed with the help of two-dimensional spectroscopy[16] to be discussed elsewhere[19]. The result of this analysis is shown in the Table. In order to relate the locally resolved hyperfine interactions a_j to the spin density at position ρ_j on the molecule a connecting expression between a_j and ρ_j must be used.

Table: Experimental and theoretical local spin densities ρ_j per $(FA)_2$ dimer and the corresponding hyperfine interactions (in 10^6 rad s^{-1}; 1G = 17.6 10^6 rad s^{-1}) and Knight shifts (in ppm).

Pos	^1H line	^1H K_j	a_j	ρ_j exp	ρ_j th.	^{13}C line	K_j (^{13}C) th.	exp.
1	-	-	-	-	-0.012	g	-44.4	-42.4
2	-	-	-	-	-0.011	g	-45.4	-40.4
3	a	-10.3	-32.8	0.075	0.066	a	50.3	76.0
4	c	- 3.0	- 9.5	0.021	0.024	e	-16.5	- 3.4
5	b	- 5.2	-16.6	0.038	0.025	f	0.7	- 8.6
6	-	-	-	-	0.065	b	40.8	57.0
7	-	-	-	-	-0.011	g	-44.7	-49.6
8	a	-10.3	-32.8	0.075	0.067	c	49.4	57.0
9	c	- 3.0	- 9.5	0.021	0.026	d	-10.9	8.1

The following relations were applied in this analysis:

^1H: $a_j/2\pi = Q\,\rho_j$ with $Q = 70$ MHz

and

^{13}C (2 neighbours): $a_j/2\pi = 99\text{ MHz}\,\rho_j - 39\text{ MHz}\,(\rho_1+\rho_2)$

^{13}C (3 neighbours): $a_j/2\pi = 85\text{ MHz}\,\rho_j - 39\text{ MHz}\,(\rho_1+\rho_2+\rho_3)$

where $\rho_{1,2,3}$ is the spin density of the neighbours of the carbon atom. The experimental values for ρ_j at the different nuclear sites j of the FA-molecule compare fairly well with a McLachlan-type theory[19,20] performed for a $(FA)_2$ dimer. In a sense these spin densities are the expectation values $|C_j|^2$ of the conduction band MO (CBMO) at position j as is scetched in Fig.5. In fact a tight binding band calculation does not alter drastically the values derived for the dimer, i.e. the molecular properties are only slightly changed by the band formation. Here we have discussed only the isotropic hyperfine interaction, we did observe, however, also anisotropic Knight shift, which will be discussed elsewhere. The nature of the ^1H and ^{13}C Knight shift in these compounds is, of course, entirely due to corepolarization. Any contribution due to molecular orbital paramagnetism is included in the chemical shift and has been subtracted. There might be, however, an (so far unknown) orbital contribution caused by the narrow conduction band, as was pointed out by C. Berthier at this meeting. I propose that it can be determined by observing the NMR shift under saturating ESR conditions. Similar experiments as discussed here have been performed by Bernier et.al.[20] and the Heidelberg group around Schweitzer

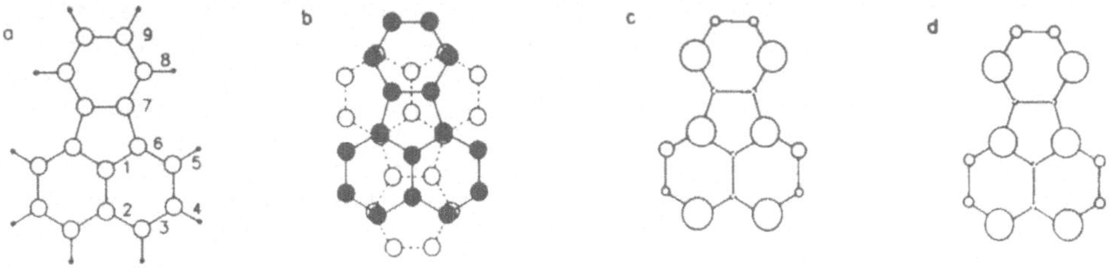

Fig.5 a) Numbering of carbon and proton positions on the FA molecule,
b) Eclipsed stacking of the FA molecules in the organic conductor,
c) Experimental spin density distribution according to the Table,
d) Theoretical spin density distribution (see Table and Text). The magnitude of the spin density is represented by the diameter of the circles.

Haeberlen on different systems. Hyperfine interactions of conduction electrons are also nicely demonstrated by the Overhauser shift experiments of the Bayreuth group[22].

Finally we discuss briefly the implications of ^{13}C spin lattice relaxation T_1 which has been measured in $(FA)_2$ X crystals[17]. Since the T_1 relaxation is caused by the same random fluctuations of the hyperfine field whose average value results in the Knight shift, there is a rigid connection between both, called the Korringa relation[5,6] (usually expressed in a slightly different form)

$$K^2\, T_1\, T\, C_o\, S_K = 1 \qquad \text{with } C_o = \frac{4\pi k_B}{\hbar}\left(\frac{\gamma n}{\gamma e}\right)^2 \tag{5}$$

where T is the temperature and

$$S_K = \left\{\frac{1}{2}\left(\frac{\tau_\perp}{\tau_S}\right)^{\frac{1}{2}}\left[\frac{3}{5}\,\varepsilon\, j(\omega_n) + \left(1+\frac{7}{5}\varepsilon\right) j(\omega_e)\right]\kappa_o(\alpha) + \frac{1}{2}\,(1+2\varepsilon)\kappa_{2k_F}(\alpha)\right\} \tag{6}$$

is a Korringa scaling factor which takes account of the low dimensionality of the crystal and of Coulomb·correlation. Here we have also considered dipole contributions d to the hyperfine interaction ($\varepsilon=d^2/a^2$)[17]. For $\varepsilon=0$ Eq.(6) reduces to the expression of Soda et.al.[5],with

$$j(\omega) = \left[\frac{(1+\omega^2\tau_\perp{}^2)^{\frac{1}{2}}+1}{2(1+\omega^2\tau_\perp{}^2)}\right]^{\frac{1}{2}} \tag{7}$$

$$\text{with} \quad \kappa_0(\alpha) = (1-\alpha)^{1/2} \qquad \text{and} \qquad \kappa_{2k_F}(\alpha) = \frac{(1-\alpha)^2}{[1-\alpha F(2k_F)]^2}$$

where $F(2k_F)=1/2\,\ln(4.56\, T_F/T)$ is the Lindhard function and T_F the Fermi temperature. All parameters in S_K (Eq.6) besides α are precisely known. The correlation times $\tau_\perp$ and τ_S were determined as discussed in section 2.0.

The experimentally observed average value of S_K for ^{13}C nuclei[17] at 45MHz in $(FA)_2X$ is $S_K \simeq 80$. The derivation of α from Eq.(6) is quite straightforward and leads to α=0.4-0.42 depending only weakly on ε. With $U/4t_{\|} = (\pi/4)\alpha$ this leads to $U/4t_{\|}$=0.33 a value somewhat smaller than the corresponding values for TTF- TCNQ[5,6] ($F:U/4t_{\|}$=0,75; $Q:U/4t_{\|}$=0.81). This does not mean, however, that Coulomb correlation is weak in $(FA)_2X$ organic conductors. On the contrary, due to the large bandwidth the Hubbard U for $(FA)_2X$ is with U=0.7 eV substantially larger, than the values U_F=0.49 eV and U_Q=0.32 eV observed in TTF- TCNQ[5,6]. Note, that the susceptibility discussed in section 2 leads to a similar but slightly larger value of α.

Acknowledgments: I would like to acknowledge continuous discussions with my co-workers A. Grupp, M. Helmle, D. Köngeter and G.G. Maresch on the subject. This work has benefitted greatly from the cooperation with J. v. Schütz and numerous discussions with E. Dormann. The Stiftung Volkswagenwerk and the Deutsche Forschungsgemeinschaft have supported this project financially.

REFERENCES

1. V. Enkelmann, B.S. Morra, Ch. Kröhnke, G. Wegner and J. Heinze, Chem.Phys. 66, 303 (1982)

2. D. Jerome and H.J. Schulz, Adv.Phys. 31, 299 (1982)

3. H.P. Geserich, B. Koch, W. Ruppel, R.W. Wilckens, D. Schweitzer, V. Enkelmann, G. Wegner, G. Wieners and H.J. Keller, J.Physique 44, C3-1461 (1983)

4. E. Dormann and E. Köbler, Sol.State Commun. 54, 1003 (1985)

5. G. Soda, D. Jerome, M. Weger, J. Alizon, J. Gallice, H. Robert, J.M. Fabre and L. Giral, J. Physique 38, 931 (1977)

6. T. Takahashi, D. Jerome, F. Masin, J.M. Fabre and L. Giral, J.Phys. C: Solid State Phys. 17, 3777 (1984)

7. H. Eichele, M. Schwoerer, C. Kröhnke and G. Wegner, Chem.Phys.Lett. 77, 311 (1981)

8. W. Stöcklein, B. Bail, E. Dormann, G. Sachs and M. Schwoerer, J.Physique 44, C3/1413 (1983)

9. W. Stöcklein, B. Bail, M. Schwoerer, D. Singel and J. Schmidt in: Electronic Excitations and Interaction Processes in Organic Molecular Aggregates, P. 228 Springer Verlag (1983)

10. J. Sigg, Th. Prisner, K.P. Dinse, H. Brunner, D. Schweitzer and K.H. Hausser, Phys.Rev. B 27, 5366 (1983)

11. G.G. Maresch, A. Grupp, M. Mehring, J.U. v. Schütz and H.C. Wolf, Chem.Phys. 85, 333 (1984)

12. G.G. Maresch, A. Grupp, M. Mehring, J.U. v. Schütz and H.C. Wolf, J. Physique 46, 461 (1985)

13. G.G. Maresch, A. Grupp, M. Mehring and J.U. v. Schütz, Synth.Metals (in press)

14. G. Sachs, W. Stöcklein, G. Denninger, M. Schwoerer and E. Dormann, (to be published)

15. G.G. Maresch, M. Mehring and J.U. v. Schütz (to be published)

16. M. Mehring, Principles of High Resolution NMR in Solids (Springer Verlag 1983)

17. M. Mehring and J. Spengler, Phys.Rev.Lett. 53, 2441 (1984)

18. M. Mehring, M. Helmle, D. Köngeter, G.G. Maresch and S. Demuth, Proceedings of the ICSM86, Kyoto/Japan (1986)

19. S. Demuth, M. Helmle, D. Köngeter and M. Mehring (to be published)

20. P. Bernier, M. Andenaert, R.J. Schweizer, P.C. Stein, D. Jerome, K. Bechgaard and A. Moradpur, J.Physique Lett. (in press)

21. J. Wieland, doctoral thesis, Univ. Heidelberg 1986

22. W. Stöcklein and G.Denninger, Mol.Cryst.Liq.Cryst. 136, 335 (1986)

^{1}H-NMR STUDIES OF SDW PROPERTIES IN $(TMTSF)_2PF_6$

Toshihiro Takahashi

Department of Physics
Gakushuin University
Mejiro 1-5-1, Toshima-ku, Tokyo 171, Japan

INTRODUCTION

The SDW instability is an important phenomenon characteristic of the low dimensional electron system, together with another type of instability; charge density waves (CDW). The conducting organic compounds $(TMTSF)_2X$ and $(TMTTF)_2X$ have been found to have an antiferromagnetic SDW ground state at ambient pressure. It is believed that this is a common feature of this family with centrosymmetric anions X.[1,2] However, we had little information about the characteristics of SDW's in these materials: even the fundamental parameters such as the SDW amplitude and the wave number had not yet been determined, mainly because of the lack of available probes to detect SDW's. Recently, we have succeeded in determining these parameters in a typical material $(TMTSF)_2PF_6$ by a precise analysis of ^{1}H-NMR lineshape. In this lecture, we describe how a conventional NMR technique can determine the SDW properties and discuss some of the main results. For the details, you can refer the publications.[1,2]

PRINCIPLE OF LINESHAPE ANALYSIS

It has been observed that the proton linewidth in a polycrystalline sample is extremely broadened in the SDW state. The broadening is clearly caused by the onset of a local magnetic field at nuclear sites due to the SDW condensation. This local field should contain all information about the static nature of SDW's and a single crystal analysis should be quite fruitful.

The single-crystal ^{1}H-NMR absorption line in the metallic (paramagnetic) state is already complex and strongly angular dependent. (The lineshape in the metallic state is not shown here because of space limitations. See ref.1.) However, the results are well explained by considering the nuclear dipolar coupling among three protons on the same methyl group. A theoretical lineshape for an isolated methyl is a triplet, as shown schematically in Fig.1(a).[3] Fig.2(a) shows the observed satellite positions in the metallic state (at 77 K) compared with theoretical calculations. The agreement is fairly good.

In the SDW state, the observed absorption line is much broader than in the metallic state. The distinguishable satellites are plotted in Fig.2(b), and the absorption derivative at different field orientations are shown in Fig.3(a). One can show that the modification of lineshape

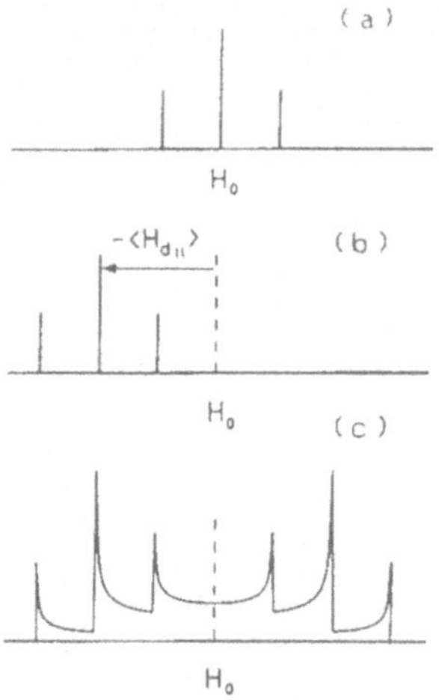

Fig. 1. (a) Schematic lineshape of ^1H in an isolated methyl group. (b) shows the effect of excess local field. $\langle H_{d//}\rangle$ is the component of the local field along the external field. (c) The expected lineshape in the case of incommensurate sinusoidal SDW's.

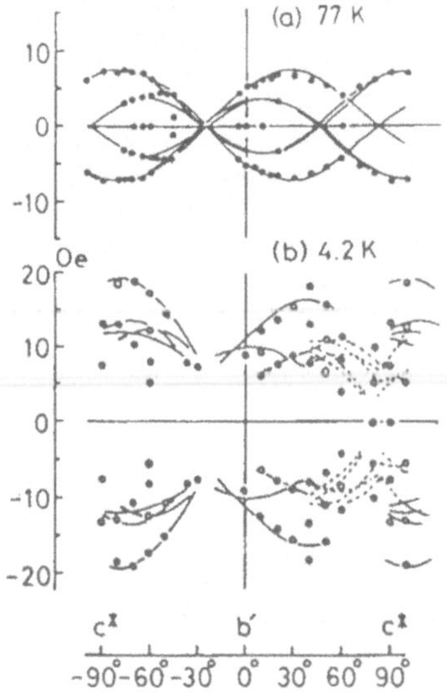

Fig. 2. Angular dependence of satellite positions (a) at 77 K and (b) at 4.2 K, which were defined as the steepest descent points of absorption derivatives. The curves show the calculated positions. In (b), the solid and broken curves correspond to the cases of electron spins along the a-axis and in the b'c*-plane, respectively.

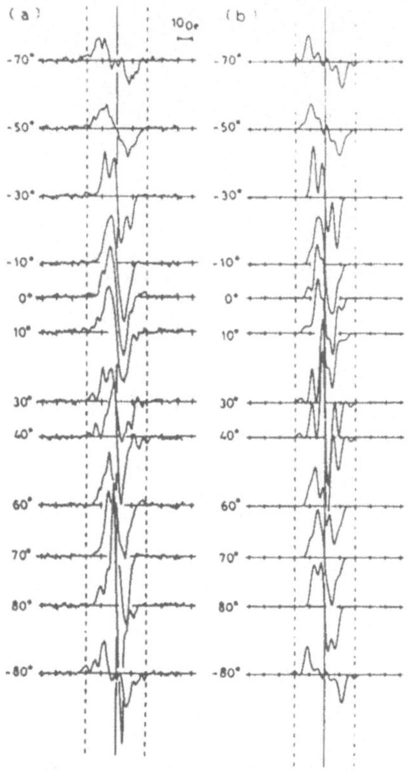

Fig. 3. (a) ^1H-NMR absorption derivative in $(TMTSF)_2PF_6$ at 4.2 K. The external field (1.17 T) was perpendicular to the a-axis and the angle was measured from the b'-axis. (b) The calculated lineshape, using the parameters given in the text.

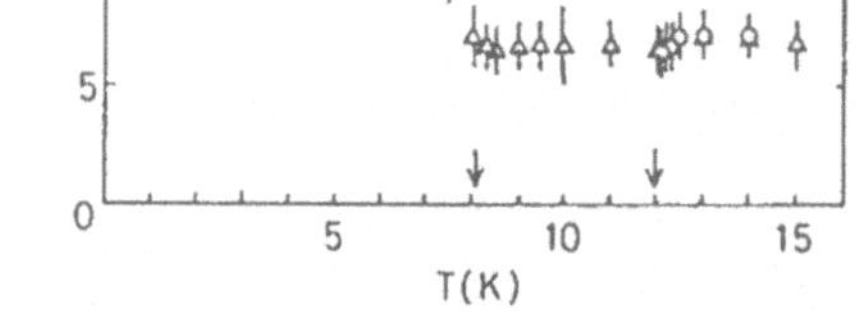

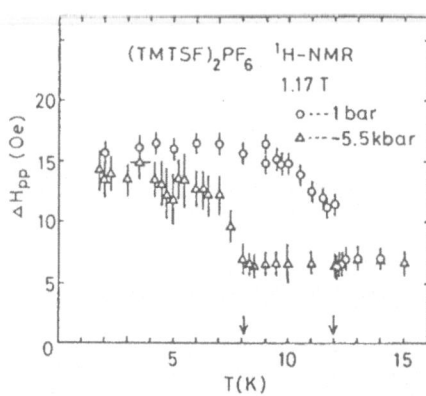

Fig. 4. Temperature dependence of the SDW satellite peaks at ambient pressure and at 5.5 kbar. Above the transition, The peak-to-peak width of absorption derivative is plotted (closed symbols).

(broadening and new structures) is caused by the component of the local magnetic field at nuclear sites along the external field. (See Fig.1(b),(c)) Now, our problem is to calculate the local field on the basis of a proper model of SDW's and to find the characteristic parameters which can reproduce the observed lineshape.

The local field at proton sites originates from two different mechanisms; the intra-atomic contact hyperfine coupling and the inter-atomic magnetic dipolar coupling. In order to neglect the former contribution, we have chosen a resonance field of 1.17 T (about twice as large as the spin-flip field) in the present experiment. In this condition, the isotropic hyperfine field is perpendicular to the external field and is not responsible for the modification of the observed lineshape. This is a big advantage of the present analysis, since the calculation of the dipolar field is straightforward, without any ambiguous parameters.

It is worthy to note that the proton lineshape is very sensitive to the SDW wave number. Since the proton nuclei are located at the four corners of the molecule,[4] the inter-molecular contribution is expected to be comparable to the intra-molecular ones. Thus, the local dipole field, as a vector sum of all contributions from neighbouring atomic sites, should be sensitive to the relative phase difference of SDW's among neighbouring molecular sites, which are determined by the wave number.

SIMULATION OF ^{1}H-NMR ABSORPTION LINESHAPE

Since we have reliable estimates of i) the spin distribution within the crystal[5] and ii) the orientation of the magnetic moment for a given external field orientation,[6] the adjustable parameters are only three; one for the SDW amplitude and two for the 'transverse' components of the wave number. We have tried to reproduce the observed highly anisotropic lineshape and found the best values for the SDW wave number are (0.5, 0.24±0.03, -0.06±0.20) and the amplitude at 4 K, 8%μ_B/molecule. The calculated satellite positions and lineshapes are shown in Fig.2(b) and Fig.3(b), respectively. The wave number is definitely incommensurate with the lattice periodicity. All trials to fit the observation with a commensurate wave number failed.

The observed and calculated angular dependence of the satellite positions indicates the existence of a spin-flip region when the external field is rotated in the b'c*-plane. It is suggested that the easy- and hard-axes are very near the shorter and longer molecular axes, respectively, contrary to previous expectations.

DISCUSSIONS

The SDW wave number

The SDW wave number was found to have an incommensurate value near a commensurate (1/2, 1/4, 0). This is the first determination of the SDW wave number in this family of material. The results agree with those of an independent analysis by Delrieu et al.[7] within the experimental errors. Moreover, the wave number is very near the optimal nesting vector obtained by the molecular orbital calculations. A recent calculation by Ducasse et al. has given (0.5, 0.3, ?).[8] This may be an evidence that the SDW condensation is caused by the Fermi surface instability.[9]

The SDW amplitude

The SDW amplitude of 8%μ_B at low temperatures is somewhat larger than the mean field estimate of 2 or 3%μ_B by a factor 3 or 4.[9] It is not clear whether this discrepancy is too large or not. Recently Yamaji informed us that a more elaborate model including an attractive interaction via electron-phonon coupling can remove this discrepancy.

Fig.4 shows the temperature dependence of the SDW amplitude, which is evaluated from the SDW satellites position at -25 (the external field parallel to the shorter molecular axis). A systematic deviation from the BCS curve can be seen. More interesting feature is the pressure dependence; by applying the pressure the transition temperature decreases as already known, but the SDW amplitude at absolute zero does not change much. This behaviour has been predicted by Yamaji,[9] and may again be a support of the nesting model for SDW condensation.

CONCLUDING REMARKS

We have shown that the precise analysis of NMR lineshape can determine the SDW amplitude and the wave number. This is the only way to obtain such information on the SDW properties at present. The evidence so far obtained seems to be favorable to the nesting model. The application to the other materials and systematic comparison should be important to clarify the mechanism of the SDW condensation.

The relaxation measurements are also very fruitful, especially to understand the dynamic properties. Several important results in the paramagnetic state are discussed in the lecture by C. Bourbonnais. The results in the SDW state have been published elsewhere.[2] There has been observed an anomalous behaviour of $^1H\text{-}T_1^{-1}$ below 4 K, suggesting a new phase transition with the onset of a small gap of about 10 K. The observed anomaly should be related to those reported by Ulmet et al.;[10] they found an oscillation of magnetoresistance in this material below 10 K, which suddenly vanished below about 4 K. It was found that the SDW properties do not exhibit appreciable change at the transition. More information is required for this problem.

ACKNOWLEDGMENTS

This work was carried out in collaboration with Y. Maniwa, H. Kawamura, K. Murata and G. Saito. We thank K. Yamaji for valuable discussions and comments. This work was partially supported by Grant-in-Aid for Special Project Research from the Ministry of Education, Science and Culture.

REFERENCES

1. T. Takahashi, Y. Maniwa, H. Kawamura and G. Saito, J. Phys. Soc. Jpn. 55 (1986) 1364. Other references are therein.
2. T. Takahashi, Y. Maniwa, H. Kawamura and G. Saito, Proc. Yamada Conf. XV on Physics and Chemistry of Quasi One-Dimensional Conductors, Lake Kawaguchi, May 1986, Physica B, ed. S. Tanaka, in press.
3. F. Apaydin and S. Clough, J. Phys. C 1 (1968) 932.
4. N. Thorup, G. Rindorf and H. Soling, Acta Cryst. B 37 (1981) 1236.
5. R.M. Metzger, J. Chem. Phys. 75 (1981) 482.
6. J.B. Torrance, J. Physique Colloq. 3 (1983) C3-799.
7. J.M. Delrieu, M. Roger, Z. Toffano, A. Moradpour and K. Bechgaard, J. de Physique 47 (1986) 839.
8. L. Ducasse, M. Abderrabba and B. Galois, preprint.
9. K. Yamaji, J. Phys. Soc. Jpn. 51 (1982) 2787; Superconductivity in Magnetic and Exotic Materials, eds. T. Matsubara and K. Kotani (Springer Verlag, 1984) p. 149.
10. J.P. Ulmet, P. Auban, A. Khmou, S. Askenazy and A. Moradpour, J. Physique Lett. 46 (1985) L-535.

EXPERIMENTAL STUDIES OF MAGNETIC FIELD INDUCED PHASE TRANSITIONS

Michel Ribault

Laboratoire de Physique des Solides-Bât.510
Université de Paris-Sud
91 405 Orsay Cedex, France

INTRODUCTION

In the following I shall discuss properties of electrical conducting materials in which the electronic behaviour is well described by electronic energy bands. In a magnetic field (H), the Lorentz force (F) which is perpendicular to the electronic velocity (v) bends the electron trajectory ($\vec{F} \propto \vec{v} \times \vec{H}$). If the relaxation time τ is long enough then the electronic motion, perpendicularly to the magnetic field, becomes periodic with the pulsation ω. This corresponds to the usual relation $\omega_c \tau > 1$. The corresponding energies are quantized in $\hbar\omega$ unit. The energy quantum has to be compared with the relaxation processes connected to the temperature, the purity and the crystallographic quality of the sample. At low temperature, in high quality single crystals of pure material, the effect of the energy quantization can be easily detected by magneto-transport or magnetostatic measurements. But usually the quantization of the energy of conduction electrons only gives rise to oscillations of the free energy of the system and not a discontinuity of any of its derivatives : there are no phase transitions associated with this quantization. The Shubnikov-de Haas effect in magneto-transport and de Haas-van Alphen effect in magneto-static measurements are then signatures of this condensation of the electronic states.

Field induced phase transitions may be considered in materials with anisotropic electronic properties. If the electronic velocity along the $\vec{H}$ direction is zero then the electronic energy has only a quantized distribution. In the energy balance which determines the occurrence of a phase transition this quantization may be of importance. In graphite, when H is perpendicular to the conducting layers, an electronic phase transition has been observed in very high magnetic field[1]. A further increase of the anisotropy of the material allows the observation of such a phase transition at moderate magnetic fields (in the few Tesla range) and the observation of a cascade of phase transitions. That has been observed in the so-called Bechgaard salts (salts of the di-tetramethyl tetraselenafulvalenium : $(TMTSF)_2X$[2]).

In the following, at first, I concentrate on one of these salts, the $(TMTSF)_2C\ell O_4$, with which most of the experiments have been done. I will show you that electronic phase transitions are well identified, then that they have specific characteristics connected to the energy quantization.

In this chapter it will appear that the specificities of this material, very useful in a first approach, do not allow to fully reveal the richness of the physical situation. The results on the $(TMTSF)_2PF_6$, discussed after, will be encouragements to try harder.

Among the properties of the Bechgaard salts let me now summarize those of importance for us.

THE BECHGAARD SALTS

It is a large family of materials initiated by K. Bechgaard[2]. Single crystals are obtained through an electrochemical synthesis. The planar molecules of TMTSF are stacked in the a direction, sheets of molecules are in the (a, b) plane and in the c direction are alternate sheets of molecules and anions. Connected to this crystallographic structure, the electrical properties are very anisotropic. The conductivities (σ_i) along the axes (i) ($\sigma_c/\sigma_b/\sigma_a$) are in the ratio (1/300/30000); at room temperature $\sigma_a \simeq 5$ S. These materials are quasi-one-dimensional electrical conductors. A review of their properties and of the physics in one dimension can be found in ref.3. The electronic band structure has been calculated by Grant[4]. The corresponding Fermi surface (F.S.) is composed of two warped planes, the warping in the c direction is very small ($\lesssim$ 1 meV).

At room temperature, the electrical conductivity has a metallic behaviour. For salts with centrosymetric anions, this behaviour remains down to 12 K. At this temperature a transition to an antiferromagnetic state occurs. This transition to a spin density wave (S D W) state is relevant to the physics of quasi-one-dimensional materials. The coupling of electronic states initiated by the spin modulation reduces the number of free electrical carriers. In the present case the reduction goes to zero; below 12 K the material is an insulator.

In the Fermi surface picture the coupling of the electronic states can be visualized through the nesting of the two sheets of the Fermi surface : the nesting vector is the translation vector corresponding to the modulation wave length[3,5]. The perfect nesting in which all the states at the Fermi level are coupled is the picture of the transition to an insulating state. In the energy balance which determines the occurrence of a phase transition the perfect nesting is not necessarily achieved. Then it can vary either versus temperature (as in TTF-TCNQ) or versus magnetic field (as in the Bechgaard salts). Under pressure the transition temperature decreases. For pressures above 7-9 kbar, at low temperature, the ground state is a superconducting state.

For noncentrosymmetric anions the crystallographic structure is not stable at low temperature[5]. In the $(TMTSF)_2ClO_4$ the anion ordering occurs at 24 K[6]. In the ordered phase the ground state is a superconducting state[7]. This ordering is supposed to have the same effect as pressure : it reduces the energy gain which can be achieved by the nesting of the Fermi surface.

Electronic band structure calculations have been performed for $(TMTSF)_2ClO_4$ in the ordered phase and for $(TMTSF)_2PF_6$ under pressure, at low temperature[8]. In both cases the Fermi surface is composed of warped planes. The superconducting ground state is destroyed by magnetic fields lower than a few Tesla and then the other electronic instabilities must be considered[9]. The magnetic field effect has to be compared with transition temperatures of the order of a few Kelvins and with deviation from perfect nesting corresponding to energies in the meV range.

In the energy scale :

$1 \text{ meV} = 1.6\times10^{-22}\text{J} \qquad 1 \text{ K} \rightarrow k_BT = 1.38\times10^{-23}\text{J}$

and for free electrons

$$1\,\mathrm{T} \rightarrow 2\,\mu_B H = \hbar\omega_c = 1.85 \times 10^{-23}\ \mathrm{J}$$

Therefore, in the Bechgaard salts, all the energies under consideration are of the same order of magnitude : transitions between competing ground states can occur. As explained in the introduction, magnetic field effects are to be stronger for H perpendicular to conducting planes; this magnetic field direction H // c^* will be our reference in all the following discussions.

FIELD INDUCED PHASE TRANSITION IN $(TMTSF)_2ClO_4$

The threshold field

A magnetic field applied in the c^* direction suppresses superconductivity as soon as $H > 0.1$ T. Then, the open F.S., composed of two warped planes is the relevant reference to understand the following measurements: Hall voltage[10], electronic specific heat[11], sound velocity[12].

At a magnetic field which is temperature dependent, the Hall voltage changes sign and increases rapidly (Fig. 1.b). All the physical properties have anomalies : the open F S is no more relevant. This type of transition has been detected, at first, by Kwak et al in $(TMTSF)_2PF_6$[13] and a short time later in $(TMTSF)_2ClO_4$[14] it has also been reported in $(TMTSF)_2ReO_4$[15] and can be identified in $(TMTSF)_2AsF_6$[16]. The angular dependence of the threshold field[13,17] clearly shows that the transition is due to orbital effects : at fixed temperature the threshold field varies as $1/\cos\theta$ where θ is the angle between H and c^*. The magnetic field is more efficient when it is perpendicular to the conducting planes.

In NMR measurements performed in a fixed magnetic field on ^{77}Se, the transition to a magnetic state is identified by i) a peak of the relaxation process (T_1^{-1}) at the transition temperature, ii) a broadening of the line width coupled with decreasing of the intensity of the resonance below the transition temperature[18]. Thus the state induced by the magnetic field has the same nature as the insulating state of the Bechgaard salts with centrosymmetric anions and so it is tempting to make both similar; but we must remember the differences. In $(TMTSF)_2ClO_4$, at 7.3 T, in the field induced

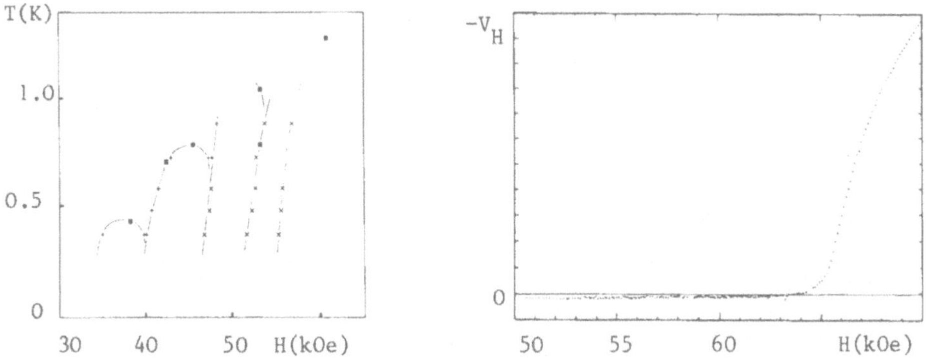

Fig. 1 a. Phase diagram of $(TMTSF)_2ClO_4$ as deduced from specific heat measurement (20).
b. Hall voltage versus magnetic field in $(TMTSF)_2ClO_4$. Note the change in sign and the strong increase of V_H above 65 kOe (10).

state (F.I.S.), the electrical resistance decreases with temperature[19]: it is a metallic not a semiconducting behaviour. The reduction of the number of carriers, as determined by Hall voltage measurements, is by orders of magnitude lower in F.I.S. than at the 12 K transition. The similarities are due to the fact that the same interactions are efficient in both cases but the magnetic field is the new ingredient that allows them to reveal. The magnetic field introduces specificities such as periodicity in (1/H), degeneracy of the energy levels in H. As a result, the variation of the threshold field versus temperature is not monotonous[20](Fig. 1.a) An approach that takes into account only the degeneracy of the energy levels[21] draws a monotonous line through the oscillating behaviour. The theoretical approach[22] developed in these proceedings by G. Montambaux leads to a more satisfactory description but the expressions are not easy to exactly compare with the experimental results.

The calorimetric measurements[11,20,23] prove that the transition on the threshold field (the Kwak transition) is a second order phase transition : there is a jump of the electronic specific heat (C_e) without any variation of the entropy of the system.

As referenced to the electronic specific heat on the high temperature side of the transition (γT_c), the jump (Δc_e) is magnetic field dependent. The ratio ($\Delta C_e/\gamma T_c$) varies from less than 1.4 to more than 4.3. Therefore the system cannot be analysed simply in the weak coupling limit.

In the technique developed by Naughton et al[24] it is possible to measure on the same sample both the electrical resistance and the magnetization. The Kwak transition is identified by resistive measurement; then the magnetization is analysed. The magnetization is a first derivative of the free energy therefore a first order phase transition is a jump of magnetization. A second order phase transition is a change in the slope of the magnetization. At low temperature, magnetization[25] and specific heat measurements are in agreement : the Kwak transition is a second order phase transition. At temperature higher than 3 K their analyses disagree : second order phase transition in a strong coupling system for specific heat, first order phase transition for magnetization. Up to now I think that the second order phase transition is more likely because the magnetic transition at 4.2 K[26] appears very broad; no latent heat of transition has been detected in this temperature range[27].

In the Kwak transition there is a nesting of the Fermi surface. The number of free carriers is strongly reduced. This reduction is by 68% at 0.1 K and at least 98% at 1.3 K[10]. From these results it is obvious that, at low temperature, the Kwak transition is not the only transition and that a cascade of transitions can be expected.

Cascade of transitions at low temperature

The first evidence for such a cascade was obtained from magnetoresistance measurements[28]. The experimental results do not fit with the usual analysis of orbital effects, in metals, at low temperature : the oscillations of the magnetoresistance are neither periodic in 1/H nor obtained at a magnetic field independent of the temperature.

The Hall voltage V_H (Fig. 2.b) increases by steps[10,29]. The magnetization[25] M strongly jumps on these steps and has a diamagnetic behaviour between them. The mean magnetization is increasing with magnetic field(Fig.2.b) and there are anomalies of the specific heat[20] on the strong increase of magnetization. The conjonction of these three sets of experiments clearly established the nature of the cascade of transition. They are first order phase transitions between magnetic states and they involve a strong reduction of the number of carriers. The diamagnetic behaviour between transitions has to be connected with the importance of the orbital effects.

The difference between the present results and measurements in standard metals are the following : magnetic and calorimetric effects correspond to peaks, not to oscillations around a constant value. In the same way steps in the Hall voltage are very different from small oscillations connected to Shubnikov de Haas effect. But the differences appear stronger in the temperature variation and it is what leads, at first, to the suggestion that they were phase transitions[10,29]. The anomalies do not die down consistently in the whole magnetic field range. Only the anomalies at the lowest magnetic field are modified, they are reduced while keeping the same shape.

The jumps in magnetization (M) are connected to jumps in entropy (S) via the Clapeyron relation $\Delta S = -\Delta M(dH_T/dT)$[25]. Direct measurements of the latent heat of transition have been performed above 1.2 K[27]. In both cases the thermodynamic analysis is in agreement with the Hall voltage measurements : the initial number of conduction electrons is strongly reduced above 10 T. Each method of analysis has its shortcomings. On one hand, in the thermodynamic analyses, the entropy variations within each phase and the entropy variations due to hysteresis are neglected. On the other hand, the discussion of Hall voltage measurements is based on a very simple model in which V_H varies with the inverse of the number of carriers. Nevertheless, the order of magnitude must be correct : less than 1/1000 of the low magnetic field free carriers remains at 10 T.

The agreement between experimental results is now well established in the magnetic field range down to 5 T. Between the threshold field and 5 T, the calorimetric measurements bring out a great number of transitions that may be connected to changes in slope of the Hall voltage[31]. To compare experimental results we must keep in mind that calorimetric measurements are sensitive to excited states. The electronic specific heat versus magnetic field, at fixed temperature, can show anomalies in situations where nothing appears in magneto-transport. Moreover the specific heat connected to the density of carriers is more sensitive than magneto-transport effects at low magnetic field when the number of carriers is fairly high. Before discussing the specificities of the phases connected to the quantization of the orbital effects we can draw the first conclusion. Above a threshold magnetic field characterized by a second order phase transition a series of first order phase transitions lead to a nearly total shrinking of the Fermi surface.

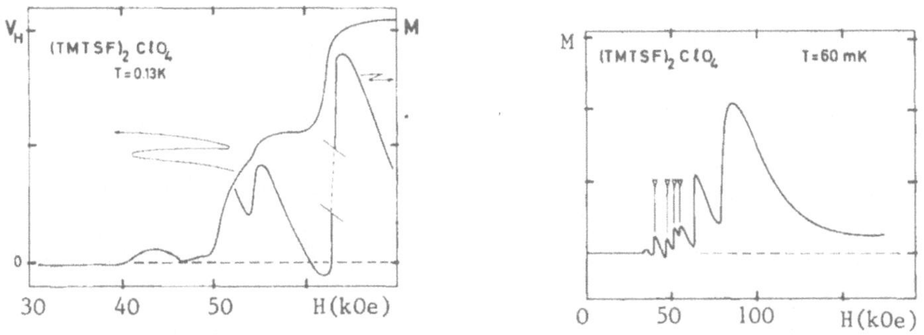

Fig. 2 a. Hall voltage and magnetization versus magnetic field in $(TMTSF)_2Cl O_4$. There are jumps in magnetization on the steps of the Hall voltage (after 10 and 25).
b. Magnetization at 60 mK in $(TMTSF)_2Cl O_4$ (25). Triangles are referred to anomalies in calorimetric measurements (20).

Field induced phases and quantization

The Hall voltage variation (Fig. 2.a, Fig. 3.a) versus magnetic field is characterized not only by steps but also by plateaus between steps. Therefore the results are reminiscent of the quantum Hall effect (Q H E). Let us compare.

First we know that steps are associated with phase transitions in F.I.S. and with standard orbital quantization in Q.H.E.. As a result, steps are periodic in (1/H) in the last case only. Second, in Q H E the plateaus of V_H result of quantization of the Hall conductivity : $\sigma_{xy} = n\ (e^2/h)$. Plateaus of V_H are as $(1/n)(h/e^2)$. The magnitude of V_H on the plateaus is not temperature dependent. In F.I.S. plateaus are observed in the same way as in Q H E but the magnitude of the Hall voltage on the plateaus varies with temperature[10]. In the simple model we use, we consider only one type of carriers. If N is the number of carriers $V_H \propto H/N$. The observation of a plateau in the Hall voltage means that H/N is a constant. In Q H E the variation of N with H is obtained in shifting electrons from localized to delocalized states; in F I S states it is suggested[19] and theoretically explained[22] that the S D W wave length varies with magnetic field. The driving force is the quantization of the electronic states : the electronic energy has sharp minima for completely filled energy levels and the energy level degeneracy varies with H. We obtain steps in going from one set of completely filled energy levels to another one and plateaus in adjusting the number of carriers to maintain the level filling[30]. In this approach, the lattice periodicity enters only in determining the initial electronic band structure. The continuous adjustement of the S D W wave length implies that there is no commensurability between the crystal and the S D W, therefore no superstructure. The unit cell is still the zero magnetic field unit cell, not a much bigger one as would be the case in the approach developed by P. Chaikin[33].

If measurements connected with Q H E and the series of F.I.S. compare, that must be only far from the transitions. In magneto-transport measurements, as an example, we must discuss only results corresponding to saturated values of the Hall voltage. The magnetic field and temperature range

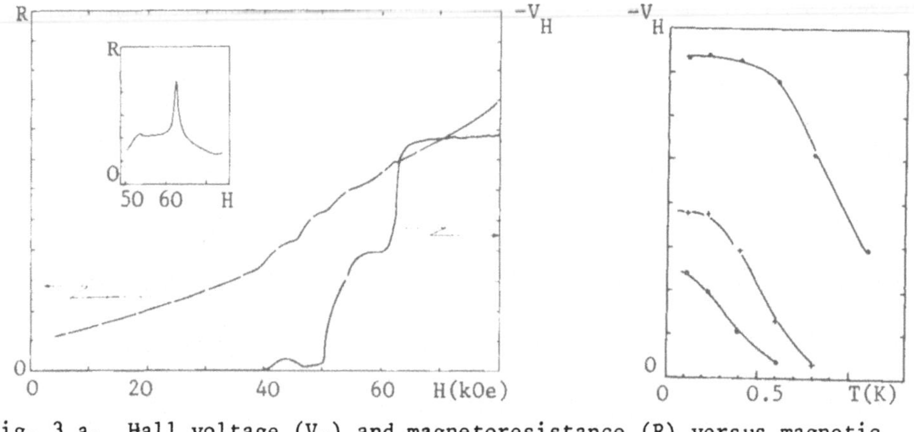

Fig. 3 a. Hall voltage (V_H) and magnetoresistance (R) versus magnetic field in $(TMTSF)_2ClO_4$ at 0.11 K. Note the peak in R on the step in V_H, at 62 kOe. In the insert the same peak on a different sample (10,19).

b. Hall voltage on the plateau of V_H, versus temperature, in $(TMTSF)_2ClO_4$ (., +, o) at 67, 60 and 54 kOe respectively.

deduced from this analysis can be used to discuss the other results. To improve magnetic field and temperature range, the best way is improving the sample quality. It is well known that measurements involving orbital effects are much more sensitive than others to diffusion processes : the phase coherence is cancelled even by small angle scattering. Therefore if the first step is to select and fit a high quality sample the second is to control the cooling down process through the 24 K anion ordering transition[5] . On Fig. (3.b) are plotted the Hall voltage amplitudes on the 6.7 T, 6.0 T and 5.4 T plateaus. At the lowest temperature they scale as 1, 1/2, 1/3 and, from a statistic on our measurements in Orsay, this corresponds, for the Hall conductivity σ_{xy}, to 6, 12, 18 times (e^2/h). On the steps of the Hall voltage are peaks in magneto-resistance (Fig.3.a). From sample to sample the peaks can be more or less difficult to identify. On Fig.3.a are two examples. The main points are i) in both cases the amplitude of the peak corresponds to the same variation of resistivity and ii) even in the case of the sharp peak, the resistance does not go to zero at the plateau. This illustrates the problem we have to compare exact values in these materials : first we never know perfectly the effective width and thickness of the sample. This explains the uncertainty we have on the exact value of the Hall conductivity. For this reason it is better to determine the ratio of the Hall conductivities from one plateau to the other rather than calculate their exact value. Second the samples are not perfect single crystals. On measuring the magneto-resistance, we measure not only the longitudinal conductivity in the high conductivity plane but also contributions from grain boundaries and conductivity in the less conducting direction. Strong peaks in resistivity can be visualized, zero resistance cannot be. In decreasing temperature a saturation value corresponding to resistive bridges is obtained. This situation explains the experimental problem and why the sharp peak in Fig.3.a was obtained in a sample with a resistivity ratio $[R(T = 300\ K, H=0)/R(R\to 0,H\to 0)] \simeq 2000$ whereas usually the resistivity ratio is in the range of hundreds.

Thus, in the comparison between F I S and Q H E, there are not any definitive arguments against the similitude of the ground states. Most of the experiments have been performed in poor quality samples at a too high temperature. Even in the best samples at the lowest temperature, the physical limitations prevent us from observing the expected behaviours. The

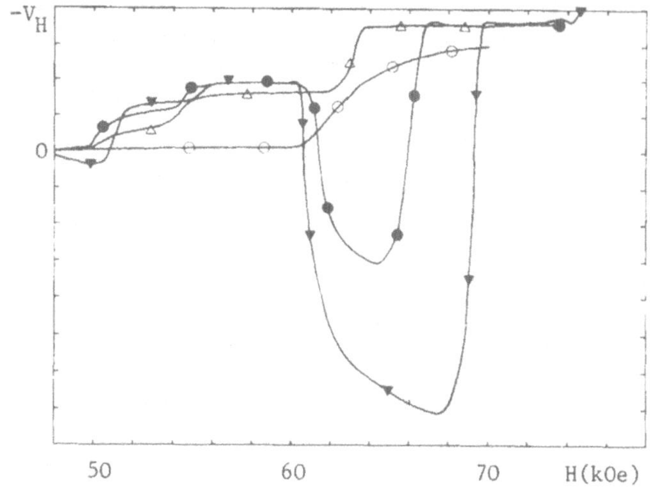

Fig.4. Hall voltage in $(TMTSF)_2ClO_4$ for various cooling rate between 30 K at 4.2 K(o,Δ ,●,▼) respectively 2×10^{-3}, 1.6×10^{-3}, 0.7×10^{-3} and 10^{-4}K/s (31).

existence of plateaus in the magnetic field variation of the Hall voltage and the integer values of the ratios between their amplitudes is the only thing that we can assert with certainty. If the quantization of the Hall voltage in this material is very stimulating, the occurrence of a new set of transitions in improved quality samples is still surprising and may be the beginning of new things.

Phase transition and cooling rate

In Fig.4 you can see the effects of a reduction of the cooling rate in the remperature range below the ordering temperature of the $C\ell O_4^-$ anions[31]. The Hall voltage reaches a saturation value on the 5.8 T plateau and seems to reach a saturation value on the 5.4 T plateau. Moreover new features appear, on one hand, between 4.8 and 5.1 T, and on the other hand, between 6 and 6.8 T. In these magnetic field ranges the Hall voltage changes sign, takes a high value, and varies versus magnetic field. At the same time magneto resistance shows anomalies : the carrier mobility (defined as the ratio of the Hall voltage to the magneto resistance voltage times H) is at least three times lower in the (6-6.8) T range than on neighbouring plateaus. The magnetization has an extra jump[33] connected to the change in sign, thus it can be identified as a phase transition. Nevertheless the magnetic field inducing the transition does not vary versus temperature on one side of this anomaly. At the lowest cooling rate, on the electron like plateaus, the Hall conductivity varies as 2. 6, 12, 18 times (e^2/h) whereas on the hole like variation, it reaches 3 (e^2/h).

The extra transition may be connected to another S D W wave vector. In such a case, either the saturation value of the Hall voltage is not reached or there is not only one type of carriers and the minority carriers are not fully quantized states. These new S D W wave vectors can connect parts of the F S more sensitive to anion disorder and thus are efficient only in the best ordered samples. But in both cases we cannot explain why the magnetic field inducing the transition (H_T) is temperature independent. An alternative explanation[30] links the extra transition with tunneling effects. In such a process there are not any entropy variations so H_T can be temperature independent. Tunneling implies that the real orbit is equivalent to the initial one run along several times. Thus we can understand that it needs a larger mean free path. But the path is not really, let us say, m times the same orbit but m orbits linked in the easy tunneling direction. In this way there is a gap between the observation of plateaus and steps , and the occurrence of extra transitions. For the last one a correlation length of the anion order several times larger than the longitudinal orbital length is needed. In such an explanation we are still within the zero magnetic field unit cell : the tunneling does not modify the adjustment process. Thus, some problems remain : why is there a change in sign ? Why is there such an amplitude and not a smaller one ? If during the adjustment process you allow a locking of the S D W wave length to the lattice parameter then you have a superstructure. You have to describe the physical properties with a much smaller Brillouin zone. Tunneling processes can develop in the new extended Brillouin zones. Then you can shift from electron orbits to hole orbits of nearly the same area : the change in sign and the amplitude problems do not exit any more. But we have now introduced a new ingredient in the problem : the real lattice. This may be even more important than the equivalence between F I S and Q H E[33].

The measurements in $(TMTSF)_2C\ell O_4$ at ambiant pressure are easier than in $(TMTSF)_2PF_6$ under pressure. The control of the anion ordering allows us to select the phase transitions and thus to break up its component parts : energy quantization and magnetic lattice-crystallographic lattice interactions. To test the validity of this approach we now discuss measurements in a material without any anion ordering effect, the $(TMTSF)_2PF_6$.

On Fig.5 are compared versus 1/H the Hall voltage in $(TMTSF)_2PF_6$ at 10.5 kbar (PF_6 in the following) and $(TMTSF)_2C\ell O_4$ at ambiant pressure ($C\ell O_4$ in the following). In PF_6 a series of plateaus cannot be clearly identified in the 15 to 18 $(MOe)^{-1}$ range because they do not improve in decreasing temperature. The basic period, a minimum + two plateaus, is the same in both compounds up to 14 $(MOe)^{-1}$. The periodicities are, within 10%, the same, the amplitudes of V_H are of the same order of magnitude. In PF_6, hysteresis can be detected[32], the magneto-resistance exhibits peaks on changes in V_H. Thus these oscillations of V_H in PF_6 are due to phase transitions in the same way as they are in $C\ell O_4$. The specificities of PF_6 must be connected to the absence of any perturbation potential induced by the residual anion disorder. For this reason the effects of the electronic energy quantization in a periodic lattice can develop fully. As soon as they are detectable, effects are periodic in 1/H. Neighbouring states have nearly the same amplitude coefficient. Therefore the temperature variation of the transition fields must be small, detectable only near the threshold field. By magneto-transport measurements these variations cannot be distinguished from thermal variations of the threshold field.

But the most exciting difference is connected to the temperature variation of V_H in the 15 to 18 $(MOe)^{-1}$ range; not only don't the V_H plateaus improve in decreasing temperature but also a depression develops. There is an anomaly in magnetoresistance associated with this depression. We have attributed these connected features to a new phase developing in this field range, at low temperature. A careful analysis shows that at the lowest temperature two other new phases can be identified. The domain of existence of these new phases is different from the domain of existence of the phases studied previously. In $C\ell O_4$ phases exist up to a temperature where they are rubbed out at the threshold field (Fig.1.a). These new phases cannot be identified to such a high temperature. As a consequence the phase diagram has no longer the aspect of (Fig.1.a) but looks like a family tree. These new states correspond to what is called substates by M. Héritier[30]. They are a direct consequence of the interaction of two periodicities resulting from the electronic quantization in a crystal. The superstructure, leading to the definition of a huge unit cell and the discussion of the Thouless-Streda[34] approach is certainly to be considered (see P. Chaikin (33)).

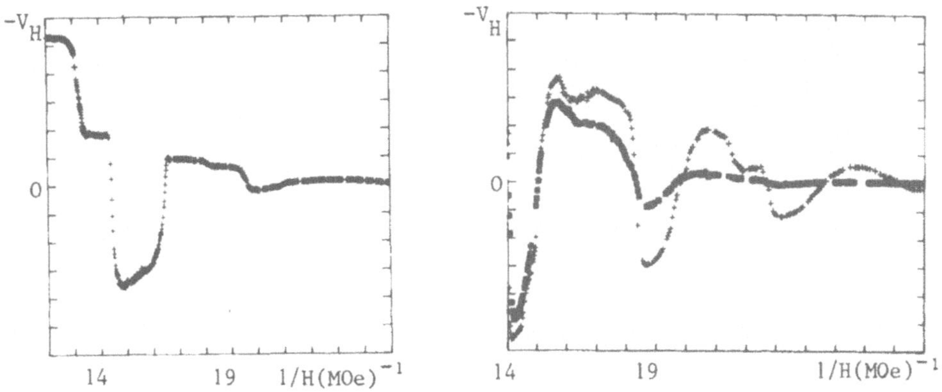

Fig.5. Hall voltage versus the inverse of the magnetic field.
a. in $(TMTSF)_2C\ell O_4$ at 0.15 K (cooling rate 10^{-4} K:s)
b. in $(TMTSF)_2PF_6$ at 0.15 K(+) and 0.39 K (□), (after 31,32).

CONCLUSION

As a result of the orbital motion of electrons in a magnetic field, it is possible to induce phase transitions in an anisotropic material if the mean free path of the electron is larger than the orbit. Within each phase the ground state is similar to the ground state inducing quantum Hall effect. If the mean free path is much larger than the orbit it is then possible to observe effects connected to interactions between orbital periodicity and crystal periodicity.

ACKNOWLEDGEMENTS

I am indebted to my colleagues in Orsay for many controversial discussions and clarifications. This contribution reflects my understanding of the subject at the end of the A S I, many thanks to all participants and mainly to the directors of the A S I and to my bar mates.

REFERENCES

1. Y. Iye, P.M. Tedrow, G. Tunp, M. Shayegan, M. Dresselhaus, G. Dresselhaus, A. Furukawa and S. Tanuma, High-Magnetic-field electronic phase transition in graphite observed by magnetoresistance anomaly, Phys. Rev. B 25 : 5478 (1982)
2. K. Bechgaard, C.S. Jacobsen, K. Mortensen, H.J. Pedersen and N. Thorup, The properties of five highly conducting salts : $(TMTSF)_2X$, X=PF_6^-, AsF_6^-, AsF_6^-, BF_4^- and NO_3^- derived from tetramethyltetraselenafulvalene (TMTSF), Sol. St. Comm. 33 : 1119(1980)
K. Bechgaard, These proceedings
3. D. Jérome and H.J. Schulz, Organic conductors and superconductors, Advances in Physics, 31 : 299 (1982)
4. P.M. Grant, Band-structure parameters of a series of $(TMTSF)_2X$, Phys. Rev. B, 26 : 6888 (1982) - J. Phys. (Paris), 44 : 847 (1983)
5. J.P. Pouget, These proceedings
6. J.P. Pouget, G. Shirane, K. Bechgaard and J.M. Fabre, X-ray evidence of a structural phase transition in $(TMTSF)_2C\ell O_4$, pristine and slightly doped, Phys. Rev. B, 27 : 5203 (1983)
7. K. Bechgaard, K. Carneiro, M. Olsen, F. Rasmusen and C.S. Jacobsen, Zero-Pressure organic superconductor : $(TMTSF)_2C\ell O_4$, Phys. Rev. Lett. 46, 852 (1981)
8. L. Ducasse, M. Abderraba, B. Gallois, Temperature dependence of the Fermi surface topolography in the $(TMTSF)_2X$ and $(TMTSF)_2X$ families, J. Phys. C, 18 : L 947 (1985)
L. Ducasse, M. Abderraba, J. Hoarran, M. Pesquer, B. Gallois and J. Gaultier, Temperature dependence of the transfer integrals in the $(TMTSF)_2X$ and $(TMTSF)_2X$ families, J. Phys. C, 19 : 3805 (1986)
9. G. Montambaux, These proceedings
10. M. Ribault, D. Jérome, J. Tuchendler, C. Weyl and K. Bechgaard, Low field and anomalous high-field Hall effect in $(TMTSF)_2C\ell O_4$, J. Physique (Paris), 44, 953 (1983)
11. P. Garoche, R. Brusetti, D. Jérome and K. Bechgaard, Specific heat measurements of organic superconductivity in $(TMTSF)_2C\ell O_4$, J. Phys. Lett. 43 : L 147 (1982)
12. P.M. Chaikin, M.Y. Choi, J.F.Kwak, J.S. Brooks, K.P. Martin, M.J. Naughton, E.M. Engler, R.L. Greene, $(TMTSF)_2C\ell O_4$ in high magnetic fields, Mol. Cryst. Liq. Cryst., 119 : 79 (1985)
13. J.F. Kwak, J.E. Schirber, R.L. Greene and E.M. Engler, Magnetic quantum oscillations in $(TMTSF)_2PF_6$, Phys.Rev. Lett.,46 : 1296(1981)

14. J.F. Kwak, J.E. Schirber, R.L. Greene and E.M. Engler, Magnetotransport in $(TMTSF)_2PF_6$ and $(TMTSF)_2Cl O_4$ under pressure, Mol. Cryst. Liq. Cryst., 79 : 111 (1982)
15. H. Schwenk, S.S.P. Parkin, R. Schumaker, R.L. Greene and D. Schweitzer Magnetic field induced transition and quantum oscillations in $(TMTSF)_2ReO_4$, Phys. Rev. Lett., 56 : 667 (1986)
16. R. Brusetti, M. Ribault, D. Jérome and K. Bechgaard, Insulating, conducting states of $(TMTSF)_2AsF_6$ under pressure and magnetic field, J. Phys. C, 43 : 801 (1982)
17. L.J. Azevedo, J.E. Schirber, R.L. Greene and E.M. Engler, High field phase transition in $(TMTSF)_2PF_6$, Physica, 108 B : 1183 (1981)
18. T. Takahashi, D. Jérome and K. Bechgaard, A magnetic state in the organic superconductor $(TMTSF)_2Cl O_4$: cooling rate and strong magnetic field effects, Journal Phys. (Paris) 44 : 805 (1983)
19. M. Ribault, J. Cooper, D. Jerome, D. Mailly, A. Maoradpour and K. Bechgaard Quantum Hall effect and Fermi surface instabilities in $(TMTSF)_2Cl O_4$, J. Phys. (Paris), 45 : L 935 (1984)
20. F. Pesty, P. Garoche and K. Bechgaard, Cascade of fields induced phase transitions in the organic metal $(TMTSF)_2Cl O_4$, Phys. Rev. Lett., 55 : 2495 (1985)
21. P. Chaikin, Magnetic-field-induced transition in quasi two dimensional systems, Phys. Rev. B, 31 : 4770 (1985)
22. M. Héritier, G. Montambaux and P. Lederer, Stability of the spin density wave phases in $(TMTSF)_2Cl O_4$: quantized nesting effect, J. Phys. Lett. (Paris), 45 : L 943 (1984)
23. M. Ribault, F. Pesty, L. Brossard, B. Piveteau, P. Garoche, J.Cooper, S. Tomic, A. Moradpour and K. Bechgaard, Field induced phase transitions in the Bechgaard salts, Proceedings of the Yamada, Conference (1986)
24. M.J. Naughton, J.S. Brooks, L.Y. Chiang, R.V. Chamberlin and P.M.Chaikin to be published
25. M.J. Naughton, J.S. Brooks, L.Y. Chiang, R.V. Chamberlin and P.M.Chaikin Magnetization study of the field induced transition in $(TMTSF)_2Cl O_4$, Phys. Rev. Lett., 55 : 969 (1985)
26. P.M. Chaikin, E.J. Mele, L.Y. Chiang, R.V.Chamberlin, M.J.Naughton, J.S. Brooks, On the Kwak transition : field induced states iN 2D organic conductors, Synt. Met. 13 : 45 (1986)
27. B. Piveteau, J.R. Cooper and D. Jérome, Differential thermal analysis of the field induced phase transitions of $(TMTSF)_2Cl O_4$ above 1.2 K, submitted July 1986
28. K. Kajimura, H. Tokumoto, M. Tokumoto, K. Murata, T. Ukachi, H. Auzai, T. Ishiguro and G. Saito, Magnetoresistance of $(TMTSF)_2Cl O_4$, J. Phys. (Paris) 44 : C 3 1029 (1983)
29. P.M. Chaikin, M.Y. Choi, J.F. Kwak, J.S. Brooks, K.P. Martin, M.J. Naughton, E.M. Engler, R.L. Greene, $(TMTSF)_2Cl O_4$ in high magnetic field, Phys. Rev. Lett. 51 : 2333 (1983)
30. M. Héritier, These proceedings
31. M. Ribault, Electronic states below 5 K in $(TMTSF)_2Cl O_4$, Mol.Cryst. Liq. Cryst., 119 : 91 (1985)
32. B. Piveteau, L Brossard, F. Creuzet, D. Jérome, R.C. Lacoe, A. Moradpour and M. Ribault, Hall effect study of the field induced instabilities in $(TMTSF)_2PF_6$ under pressure, J. Phys. C, 19: 4483(1986)
33. P.M. Chaikin, These Proceedings
34. P. Streda, Quantized Hall effect in a 2D periodic potential, J.Phys.C: L 1299 (1982)
35. B. Horovitz, H. Gutfreund and M. Weger, Phys. Rev. B, 12 : 3174 (1979)

MAGNETIC FIELD INDUCED TRANSITIONS IN ORGANIC CONDUCTORS: EXPERIMENTS

X. Yan[a], R. V. Chamberlin[a], L. Y. Chiang[c]
M. J. Naughton[a,b], J. S. Brooks[b], and P. M. Chaikin[a,c]

[a] Dept. of Physics, Univ. of Penn. Phila PA 19104
[b] Dept. of Physics, Boston University, Boston, MA 02215
[c] Exxon Research, Rt. 22 East, Annandale, NJ 08801

INTRODUCTION

The Bechgaard salts, $(TMTSF)_2X$ (where X is ClO_4, PF_6, ReO_4 etc.), are best known for being the first organic superconductors, and exhibiting a rich phase diagram with spin density wave (SDW) and charge density wave formation, and many unusual anion ordering transitions as pressure, temperature and composition are varied[1]. However, their most interesting and to date unique properties result in the presence of an moderate magnetic field. The PF_6 salt has the prototype behavior. At zero magnetic field, the material is metallic at room temperature and remains metallic until it undergoes a transition at ~12K to a SDW insulator. With application of pressure the SDW transition temperature is reduced until at a critical pressure P_c ~6kbar, the metallic state is stable to zero temperature. When the metallic state persists below ~ 1K there is a superconducting transition with T_c ~ 1K, which in turn is reduced as pressure is increased. Both SDW and superconducting transitions have of course been observed in other materials.

Sitting at a pressure slightly above P_c we now start increasing the magnetic field from zero. At very low fields ~200 Gauss, we destroy the superconductivity, as expected. As the field is further increased above ~ 30,000 Gauss (3T), we reach a phase boundary which appears to take us back into the lower pressure SDW state. But the new state is semimetallic, rather than insulating. Moreover, the initial transition is followed at higher field by a whole series of transitions taking us further toward an insulating state, but never quite reaching it. The same effects occur in the ClO_4 salt where P_c~0, and the ReO_4 salt with P_c again ~6kbar.

To understand why these results are so remarkable, we must remember that the Bechgaard salts are variously regarded as being quasi-one or quasi-two dimensional. The magnetic field induced transitions are observed depend only on the component of the field which is perpendicular to the highly conducting two dimensional plane. This strongly suggests that the effects are orbital in nature rather than related to the electron spins, where the effects should be largely isotropic. Orbitally induced transitions are very rare and if anything might be related to orbital quantization and the formation of Landau levels. However, the Bechgaard salts are quasi-one dimensional in the sense that the Fermi surface is separated into two nonintersecting warped sheets. With the field aligned parallel to these sheets there are no closed orbits and we would expect no Landau quantization.

In this paper we will describe the questions that needed to be addressed experimentally and how we went about trying to answer them. Some of the experimental techniques and problems are discussed. Finally some of the more recent questions and experiments involving the three dimensionality of the electron motion, the presence of Schubnikov-deHaas (SdH) oscillations and additional transitions at high field are presented.

<u>What did we need to know?</u>

Up until about five years ago, it was expected that the magnetotransport of the Bechgaard salts would be relatively uninteresting. The measured mobilities were exceedingly high, at least 10^5 cm^2/volt-sec below 4K, but the open orbit Fermi surface should have presented no pronounced structure in magnetotransport and no Landau quantization. The discovery by Kwak et al.[2] of SdH like oscillations in the resistance of $(TMTSF)_2PF_6$ (under sufficient pressure to barely make it metallic) came as a complete surprise. Furthermore the observed oscillations did not start gradually from low field, but sprang up once a threshold magnetic field had been exceeded. Within a year of this remarkable discovery, several groups had evidence from NMR measurements[3] that the threshold (or Kwak) field H_k was the onset of a transition to some sort of magnetic state, possibly a SDW. Two questions which immediately arose were whether the orbits remained open up until the transition, and whether the carrier density was changing with magnetic field before or during the transition.

The obvious experiment to try was the Hall effect for the number density, and this along with other experiments to determine whether the orbits were open or closed. As we shall see, the Hall effect measurements were startling. They indicated not only a loss of carriers at the threshold field, but a whole cascade of phase transitions at higher field. In addition they showed steplike structure reminiscent of the quantum Hall effect (QHE). Soon after these measurements a number of theories presented us with an explanation for the cascade of field induced transitions in terms of the instability of two dimensional (2D) Fermi surfaces in the presence of a magnetic field.[4-8]

The experimental questions which arose then involved a determination of whether the transitions were thermodynamic, and of what order (1st or 2nd), what effects the three dimensionality of the samples would have on the 2D instability, and the persistent question of the relation of the these effects to the usual QHE. The experiments that were required were further transport studies, specific heat and magnetization measurements, and studies at higher field were the theories all suggested we should find an insulating state. In the experiments which followed there were additional surprises. A new set of SdH oscillations with an unusual temperature dependence was found. The existence of sign changes and fine structure were observed in the Hall resistance, as cleaner samples and lower temperatures were used, and a new set of transitions were observed at the highest fields available.

SOME EXPERIMENTAL TECHNIQUES

<u>Transport Measurements</u>

Transport measurements have traditionally been the most common experiments performed on the organic conductors. The techniques are straight forward but experimentalists just entering the field should be aware of some cautions which result from the characteristics of the samples. A typical lead configuration for conductivity measurements is shown in figure 1(a). For ohmic, homogeneous materials the resistance is the longitudinal voltage difference divided by the current, $((V_2-V_1)/I$ or $(V_4-V_3)/I))$, and the Hall resistance is the transverse voltage difference divided by the current in the presence of a magnetic field perpendicular to the plane, $((V_3-V_1)/I$ or $(V_4-V_2)/I))$. However, if the contacts are not exactly aligned parallel and perpendicular to the current lines in the sample, there will be a component of the longitudinal resistance in the Hall measurement and vice versa. This is especially true in the context of the Bechgaard salts, where inhomogeneities are augmented by thermal cracking of the samples and the current flow is often very distorted. In this case it is best to use the symmetry properties of the conductivity tensor with respect to field direction to determine the magnetotransport coefficients. The resistance is the even part and the Hall resistance is the odd part according to $\rho_{xx}(H)=(V_{\pm1,2}(H+) + V_{\pm1,2}(H-))/2I_{\pm}(A/L)$ and $\rho_{xy}(H)=(V_{\pm1,3}(H+) - V_{\pm1,3}(H-))/2I_{\pm}(A/L)$. The $\pm$ indicates that the averages with respect to current direction must also be taken. The experiments are often difficult in that determining the Hall resistance at high field often requires subtraction of large numbers to obtain small differences.

Transport measurements and particularly the Hall resistance was crucial to unraveling the properties of the Bechgaard salts in high magnetic field. Some longitudinal and Hall resitance data for the ClO_4 salt are shown in figure 2. Both groups that initially performed these measurements[9,10] were immediately struck with implications of the Hall resistance. In most materials the Hall resistance is a direct measure of the carrier density. $\rho_{xy}=R_H H=H/ne$, where H is the applied field, n the electron density and e the elementary charge. Thus if the density were constant, ρ_{xy} vs. H should be a straight line through the origin. Even for the quantum Hall effect ρ_{xy} oscillates in a staircase manner about this straight line. As might have been expected the Hall resistance increases markedly for $H>H_K$, indicating a substantial (>90%) loss of carriers at the transition. However, there is no way to draw a line through the data and the origin above H_K. Thus we were all led to the conclusion that each step corresponded to an additional loss of carriers and hence to the idea that the Kwak transition was just the first of a series of transitions which occurred as the field was increased.

The second striking feature of ρ_{xy} is the similarity of the plateaus with what is seen in the QHE. It was immediately noticed, however, that there were also substantial dissimilarities. The present plateaus are temperature dependent, not in the precise ratio 1/n, not equally spaced in

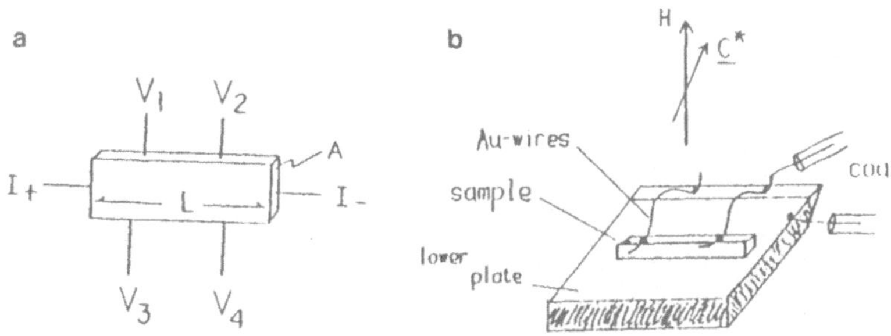

Figure 1(a). Typical lead configuration for Hall and magnetoresistance measurements on $(TMTSF)_2X$ samples. (b) schematic of capacitance magnetometer used, with sample cantilevered above a conducting plate.

1/H, and ρ_{xx} did not approach zero at the plateaus in ρ_{xy}. It should be noted that it is not easy to miss the QHE. Normally, resistance and Hall measurements in these sample are complicated by geometry, anisotropy, and the lead alignment. But for the QHE, independent of these factors, each of the two edges of the sample between the current leads is an equipotential. Any voltage difference between the edges should yield $(h/ne^2)I$ where I is the current, and any voltage difference on the same edge should give zero when one is on a Hall plateau. The experimental results on the Bechgaard salts have not shown these effects. Thus if we have some form of the QHE, it is not the same as in previously studied systems. Furthermore, both theory and experiment indicate that the thermopower is zero in both transverse and longitudinal directions at QHE plateaus. Thus the thermopower cannot miss a real QHE even if you don't know where your leads are. Thermopower experiments are easy to do on the Bechgaard salts, but they have also shown that the conventional QHE is not present.[11]

After the Hall measurements and the subsequent theories showing the instability of a two dimensional open orbit Fermi surface in a magnetic field to form a SDW, we wanted to see whether the Fermi surface had in fact remained open up to the Kwak field. The Hall resistance below H_K was consistent with calculations for an open orbit with reasonable parameters, but did not prove it. Sound velocity measurements on the other hand have a characteristic H^2 dependence for closed orbits and H for open orbits. Using a "vibrating reed" technique the sound velocity for the Bechgaard salts is easily made and in fact showed the open orbit linear dependence right up to H_K.[12]

Magnetization Measurements

To see whether the transitions were thermodynamic and of what order, one can do specific heat or magnetization measurements. Specific heat is most sensitive to entropy changes, whereas magnetization is most sensitive to energy changes. We will concentrate on the magnetization measurements,[13] where novel techniques[14] were required because of the high magnetic fields, small samples with small susceptibilities, and low temperatures. A schematic of the apparatus is shown above in figure 1(b).

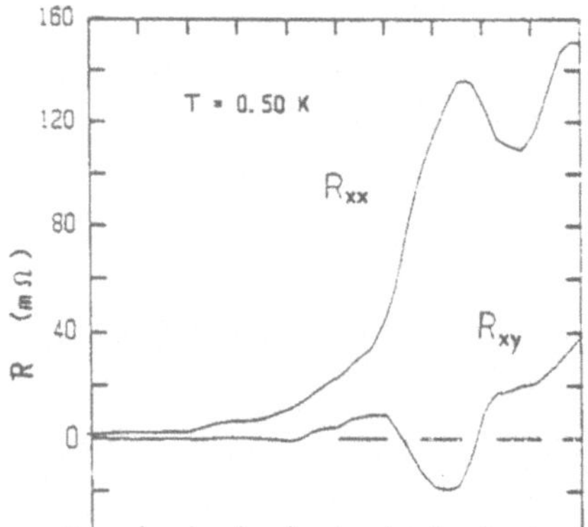

Figure 2. Typical Hall and magnetoresistance data for X=ClO_4, with threshold field at 4T.

Basically a sample is mounted with two or four fine gold leads (as for a transport measurement which in fact can be made simultaneously). The sample is suspended over one or two electrodes. A sample-electrode pair serves as a capacitor with which very small displacements can be measured. In the presence of a magnetic field with a small gradient, there is both a force and a torque on the sample, $F=M\partial H/\partial z \propto H^2$, $\tau=M\times H \propto \Delta\chi H^2$, where χ is the susceptibility and $\Delta\chi$ is the anisotropy in χ. The torque and force are resisted by the restoring force from the wires supporting the sample. The torque is measured by the difference in displacements obtained by the two capacitors while the force is related to the sum. The apparatus can be calibrated in several ways. Flipping the apparatus "upside-down" reverses the force due to gravity, and measuring the displacement even without knowing the mass gives a calibration for M in emu/gm. Using a capacitive coupling drive, the resonance of the sample-wire system gives the "spring constant" of the wires, and there are many other in situ methods.

The sensitivity of the apparatus is controlled by the restoring force from the wires. The displacement of the end of a rod (wire) of length L, radius r and Young's modulus E is $\delta y=4FL^3/(3E\pi^4)$, where F is the force applied to the end. With 12 μm diameter wires, ~.5 cm long, $\partial H/\partial z \sim 0.1$ Tesla/cm (easily obtained by putting the apparatus above or below the center of a conventional solenoid) and a capacitance bridge which can detect $\Delta C/C \sim 10^{-5}$ the sensitivity is $\delta M \sim 10^{-12}$ Joule/Tesla (10^{-9} emu), which is comparable to the sensitivity of a superconducting magnetometer (SQUID). However, the real advantage of the present system is that it can be used to measure χ independent of the mass of the sample. The support wires can be adjusted so that the addition of the sample causes a small displacement (say .01cm) at the free end. Since we can detect $\delta y \sim 10^{-8}$cm the sensitivity is $\sim 10^{-6}mg=\Delta\chi mH^2 \partial \ln H/\partial z$. With $\partial \ln H/\partial z \sim .1$ cm^{-1} and H=1 Tesla we can measure 10^{-10} emu/gm independent of sample mass.

Since this is a summer school it is worth mentioning the one problem with this apparatus. There is invariably a torque on any object in a magnetic field (even with isotropic χ) which is not shaped like a sphere. This leads to a background quadratic term which must be subtracted from the capacitance measurement. For example some "raw" data from the study of the ClO_4 salt at 0.70K is shown in figure 3(a). The background to be subtracted (dashed line) is determined from the low field measurements or from raising the temperature so that no transitions are observable in the region of interest. After subtraction of this background term, we can get magnetization measurements such as those shown in figure 3(b).

The magnetization shows a series of paramagnetic jumps which correspond to the fields at which the transitions are seen in the transport measurements. The curve resembles that which would be expected from conventional deHaas van Alphen oscillations in a two dimensional electron gas near zero temperature, if the Fermi energy rather than the electron density is fixed. In the present results the magnetization jumps are nearly discontinuous even at finite temperature as can be seen from the blow-up of the 2nd jump at 225mK. The fact that the magnetization jumps, coupled with the observation that the fields for these jumps are temperature dependent, shows that the transitions are first order. This follows from the magnetic version of the Clausius-Clapeyron equation: $\Delta S=-\Delta M(dH/dt)$. With a finite ΔM and a finite dH/dt, there is a discontinuity in the entropy S which serves to define a first order transition. Note that for a second order transition the characteristic behavior is a change in slope rather than a discontinuity in M vs H. At low temperature, where there are several transitions as the field is increased, we cannot tell whether the the initial Kwak transition is first or second order. However, all of the subsequent transitions appear as first order. At high temperatures T>4K the magnetization is smeared, but appears much more as a jump in magnetization rather than a slope change. We have therefore suggested that the single transition observed in this temperature range is first order. Further evidence for this being a first order transition comes from the fact that using the Clausius-Clapeyron equation yields an entropy jump which accounts for virtually all of the electronic entropy of the system.

Subsequent studies of the Hall resistance have indicated that when samples are sufficiently "clean" as for example the ClO_4 salt after extremely slow cooling through the anion ordering transition at 24K, there are additional steps in the Hall measurement. In particular the Ribault anomaly appears as a drastic sign change in the Hall resistance. It has been suggested that this is not a thermodynamic transition but as can be seen in figure 4, the magnetization measurement clearly shows that there is an additional magnetization jump for slowly cooled samples, at the field originally found by Ribault.[10]

MORE RECENT RESULTS

<u>But it's three dimensional</u>

All of the theories which describe the field induced transitions in the organic conductors relate to the instability of the open orbit two dimensional Fermi surface in the presence of a perpendicular magnetic field. If the system is truly two dimensional, all of the theories suggest that there should be a transition temperature for any finite magnetic field, or conversely that as the temperature is lowered we should continue to observe additional transitions at lower field. Previously it was shown[5] that the experimental data for the Kwak transition, the first transition in going from the metallic to the SDW phase, follows a curve

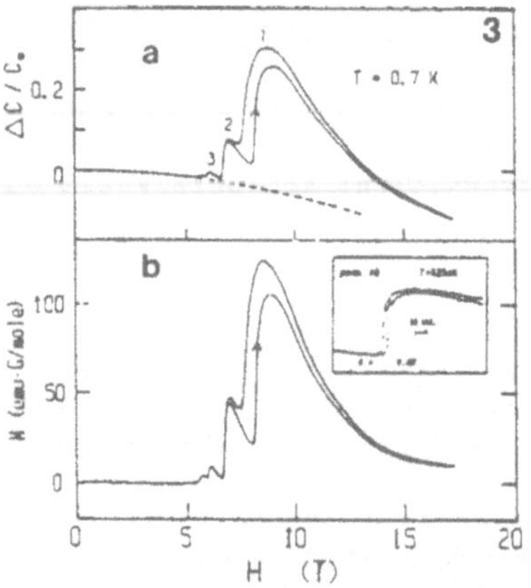

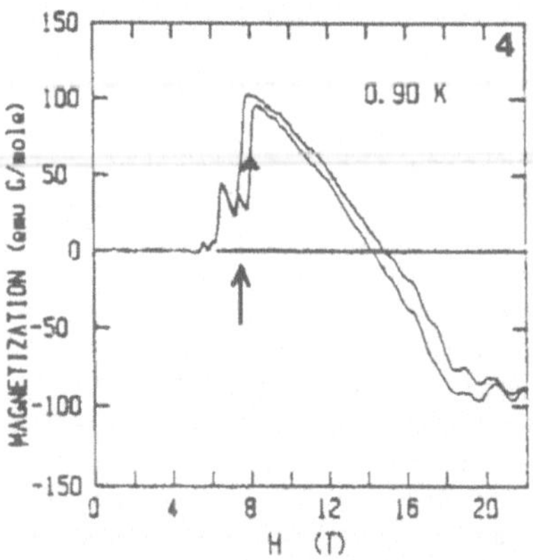

Figure 3.[left] (a) raw capacitance response of magnetometer for $(TMTSF)_2ClO_4$ at T=0.70K, with SDW phases labeled. Dashed line is low field background. A typical capacitance value is C_0= 1 pF. (b) Magnetization at 0.70 K obtained from the data in (a), with background removed. inset: raw data of sharp jump #2 at 225 mK.
Figure 4.[right] Magnetization at 0.90K in very relaxed state, where jump at 7T corresponds to negative Hall step.

described by $\ln(T_K) \sim 1/H$, and the spacing of the subphase transitions in magnetic field is very roughly $1/H$. Thus lowering the temperature from 100mK to 8mK should reduce H_k form ~3 Tesla to ~2.5 Tesla and introduce an additional 5 transitions. We therefore performed transport measurements[15] on the ClO_4 salt down to 8mK. We found that although the transitions sharpened up considerably there were no new transitions to be found below 2.95 Tesla. This transition could be observed at all temperatures below ~ 100mK. Three dimensionality must be playing a role.

Although these materials are highly anisotropic, the bandwidth in the third, c direction is not negligible, with $4t_c$ about 4K in energy. Since the observed transition temperatures are on the scale of 1K we would expect that the SDW gaps which are created should be of comparable magnitude. With this argument all of the transitions should have been destroyed by the three dimensionality. The riddle is resolved by using an old trick that we learned a long time ago in the one dimensional business. In one dimension the Fermi surface is always unstable to a distortion with wavevector $2k_f$. If we make the system somewhat two dimensional by adding a transverse bandwidth $4t_b$, then we can enhance the chance of an instability by making a distortion of $(2k_f,\pi/b)$. The effect is to better nest the opposing sides of the Fermi surface reducing the effect of the transverse bandwidth from t_b to $\sim t_b^2/E_f$. In the present case the system must be using the same trick along the c direction, choosing a wavevector of π/c. The effect is to reduce the third direction effective bandwidth to $4t_c^2/E_f$ or ~70mK. That is why there are no additional transitions below this temperature. The three dimensioanlity plays only a small role and only at the lowest temperatures.

SdH oscillations at high magnetic fields

Along with the series of phase transitions which are observed in the region from 3-8 Tesla at low temperature, there are other structures seen in various measurements as a function of field. Even in the earliest transport measurements there were reports[9] of oscillations at fields above ~ 14 Tesla which corresponded to a very different period $\Delta(1/H) \sim 1/(260$ Tesla) than observed for the transitions, $\Delta(1/H) \sim 1/(20$Tesla). These oscillations looked very much like conventional SdH oscillations, but disappeared at low temperature. Moreover, SdH oscillations result from quantization of closed orbits and as we have argued above, there are none in the present samples. Recently, we have focused some of our experimental effort on studying these oscillations. In figure 5 we show the magnetoresistance of a sample of the ClO_4 salt at two temperatures. At 6.5K, H_k is above 15T, and there are no transitions in the region shown. The oscillations are evident down to around 7T. At 1.5K where the transitions are present, the oscillations seem to have been wiped out. However, a blow-up of the low field region with subtraction of a background, shown in the insert, indicates that they are present both above and below the field induced transitions. The logarithmic derivative of the magnetoresistance in figure 6 shows that the oscillations go right through the transitions. The same effect was previously reported for the ReO_4 salt.[16] Note however, that below the transitions the amplitude increases with decreasing temperature, while above the transitions the behavior is complicated.

Recently, it has been proposed[17] that the oscillations are the result of the form of the wavefunction for an open orbit metal in a magnetic field. (The wavefunctions are Mathieu functions instead of simple plane waves.) The scattering matrix elements are then magnetic field dependent with a basic frequency given by $4t_b/\hbar\omega_c$, $\omega_c \sim eHk_fb/mc$. Such an explanation would imply that the oscillations are not visible in thermodynamic experiments. We therefore looked at the magnetization at high field, with some results shown in figure 7. It is clear that the "SdH" oscillations are present at high field. Moreover, they are extremely sharp and look very much like the two dimensional electron gas. (In three dimensions we would have a sinewave like oscillation instead of the observed sawtooth.)

Some additional remarks on the oscillations. They are much more apparent in very slowed cooled clean samples. The oscillations are much more readily observed in magnetoresistance than in Hall resistance where it is not yet certain that they show up at all. The oscillations above ~15 Tesla increase in amplitude and sharpen up as temperature is decreased. They are thermodynamic and are still not understood.

New Transitions at the Highest Fields

Our magnetization measurements indicate that there is pronounced hysteresis above 25T, and in some measurements we have observed a jump in the magnetization at these fields, aside from the oscillations. This suggests that there is yet another transition. High field magnetotransport measurements shown in figure 8 also point toward an additional transition. We have observed the resistance jump, typically 5-100 times the value at 20T, in a number of samples, starting at ~25T.[18] Other groups have recently detected the onset of this transition as well.[19,20] With such a large resistance change, it is difficult to do the subtraction required for a Hall measurement with much accuracy. The problem is additionally complicated by the fact that the hybrid magnet (superconducting outside a Bitter solenoid) is not reversible and the sample must be flipped instead. The measured Hall resistance is shown in the insert. What we can say is that ρ_{xy} changes by less than 50% while ρ_{xx} changes by up to two orders of magnitude.

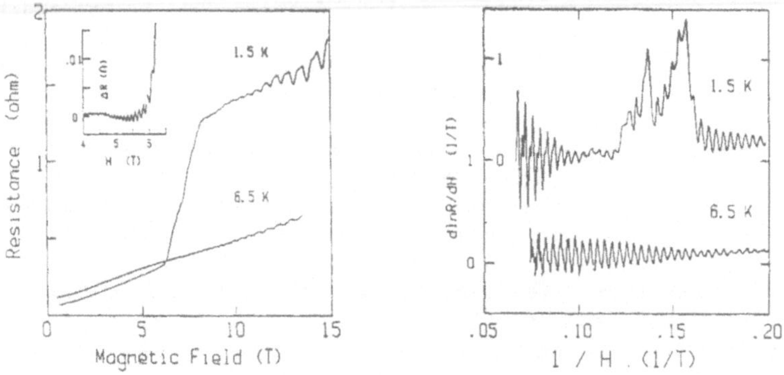

Figure 5.[left] Magnetoresistance at 1.5K and 6.5K, showing SdH oscillations appearing BELOW the threshold field for the SDW transitions. Inset: expanded view of 1.5K data, with H_K~6T. Cooling rate ~ 7 mK/min below 30K. Figure 6.[right] Logarithmic derivative of data in figure 5 against inverse field, showing SdH oscillations prevailing throughout the entire magnetic field range.

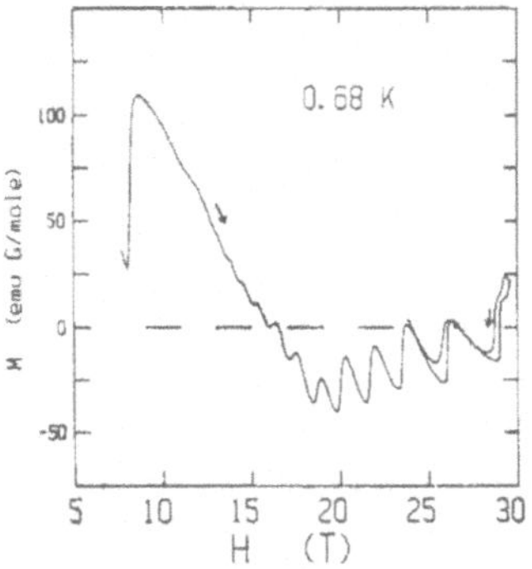

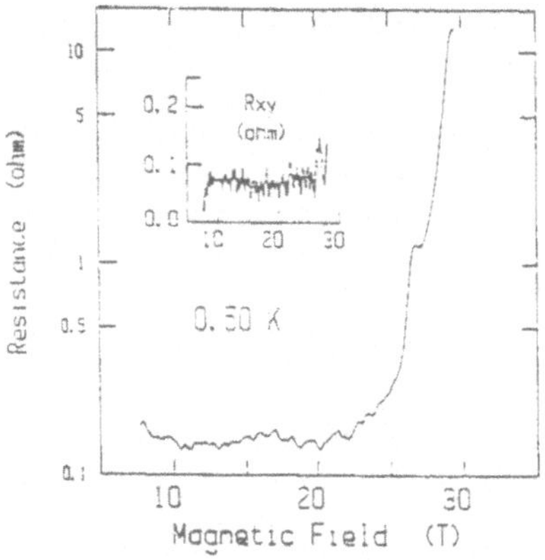

Figure 7.[left] **Magnetization at 0.68K in very relaxed state. The jump at 8T is the last SDW transition (#1). The dHvA oscillations are evident above 10T, becoming pronounced above 15T, with increased hysteresis above 25T. The field position at which M changes sign is sensitive to the background term, and is difficult to determine accurately.**

Figure 8.[right] High field magnetoresistance showing 'insulating' transition above 25T, and the SdH oscillations above 10T. Note log R-scale. Inset: Hall effect in same field range.

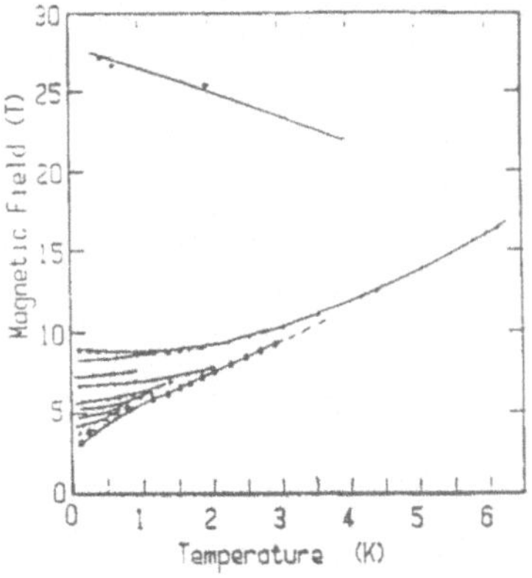

Figure 9. Magnetic field - temperature phase diagram for $(TMTSF)_2ClO_4$ in the very relaxed state.

Most of the theories of the field induced transition predict that the high field state of the system is an insulator. In trying to fit the experiments they suggest that the final transition should be at ~8T where indeed the last of the low field series of transitons is found. However, the system remains semimetallic and not insulating (or QHE like in resistance) above the 8T transition all the way to 25T. This final transition may be to the insulating state. Note that the behavior of the Hall resistance and the magnetization at the 25T transition is quite different from the low field transitions. Furthermore, the transition temperature vs magnetic field, shown in the phase diagram in figure 9, has the opposite slope to the other transitions. We therefore believe that he origin of this transition must involve different physics than the earlier ones.

There have been suggestions by a group of us,[21] and by other authors[20] that there is a transition at ~ 15 Tesla. In our more recent work we find that the evidence for a transition in this field range is weak. There is a slight change in slope of the magnetotransport coefficients, and depending on the sample, this is generally the regime where the SdH oscillations become appreciable, but trying to find a unique characteristic to define the transition field or temperature has proven elusive, and there is certainly no thermodynamic transition in this region from the magnetization measurements, unless we associate a transition with each SdH oscillation.

References

1. see recent review articles:
   (a) R.L. Greene and P.M. Chaikin, Physica, **126B**, 431 (1984).
   (b) J. Freidel and D. Jerome, Cont. Phys., **23**, 583 (1982).
2. J.F. Kwak, et al. Phys. Rev. Lett. **46**, 1296 (1981) and Mol. Cryst. Liq. Cryst. **79**, 121 (1981).
3. L.J. Azevedo, J.E.Schirber, J.C.Scott, Phys.Rev.Lett. **49**, 826 (1982) T. Takahashi, D.Jerome, K.Bechgaard, J.de Phys.Lett. **43**, L565 (1983).
4. L.P. Gor'kov, A.G. Lebed, J. de Phys. Lett., **45**, L433 (1984).
5. P.M. Chaikin, Phys. Rev. **B31**, 4770 (1985).
6. K.Yamaji, J.Phys.Soc.Jap. **54**, 1034 (1985), Syn.Metals, **13**, 29 (1986).
7. M.Heritier, G.Montambaux, P.Lederer, J.de Phys.Lett. **45**, L943 (1984), G.Montambaux, M.Heritier, P.Lederer, Phys.Rev. Lett. **55**, 2078 (1985).
8. K. Maki, and A. Virosztek, L. Chen, K. Maki, preprints.
9. P.M. Chaikin, M.Y. Choi, J.F. Kwak, J.S. Brooks, K.P. Martin, M. J. Naughton, E.M. Engler, R.L. Greene, Phys. Rev. Lett. **51**, 2333 (1983).
10. M. Ribault, D. Jerome, J.Tuchendler, C.Weyl, K.Bechgaard, J. de Phys. Lett., **44**, L953 (1983), M. Ribault, et al., J. de Phys. Lett. **45**, L935 (1984), and M. Ribault, Mol. Cryst. Liq. Cryst., **119**, 91 (1985).
11. P.M. Chaikin, et al., Synthetic Metals, **13**, 45 (1986).
12. et al. and M.J. Naughton , Mol. Cryst. Liq. Cryst., **119**, 27 (1985).
13. M.J. Naughton, J.S. Brooks, L.Y. Chiang, R.V. Chamberlin, P.M. Chaikin, Phys. Rev. Lett., **55**, 969 (1985).
14. J.S. Brooks, M.J. Naughton, Y.P. Ma, P.M. Chaikin, R.V. Chamberlin, Rev. Sci. Instr., to be published
15. D.C.Johnston, L.Y.Chiang, P.M.Chaikin, Bull. Am.Phys.Soc. **31** (1986).
16. H. Schwenk, S.S. Parkin, R. Schumaker, R.L. Greene, D. Schweitzer, Phys. Rev. Lett., **56**, 667 (1986).
17. K. Yamaji, J. Phys. Soc. Jap., **55**, 1424 (1986).
18. P.M. Chaikin, et al. Proc. XV Yamada Conf. on Phys. and Chem. of Quasi-One Dim'l Conductors, Japan, to be published.
19. T. Osada, N. Miura, G. Saito, ibid.
20. J.P. Ulmet, A.Khmou, P.Auban, L.Bachere, Sol.St.Comm. **58**, 753 (1986).
21. J.S. Brooks, M.J. Naughton, R.V. Chamberlin, P.M. Chaikin, J. Mag. Mag. Mat., **54-57**, 637 (1986).

EFFECTS OF INCOMMENSURATE POTENTIALS: LOW-DIMENSIONAL CONDUCTORS IN HIGH MAGNETIC FIELDS

P. M. Chaikin[a], M. Ya. Azbel[b], and P. Bak[c]

[a] Dept. of Physics, Univ. of Penn. Phila PA 19104 and Exxon Research, Rt. 22 East, Annandale, NJ 08801
[b] University of Tel Aviv, Tel Aviv, Israel
[c] Brookhaven National Lab, Upton, New York, 11973

INTRODUCTION

The purpose of the present lecture is to relate the experiments and theories on the magnetic field induced transitions (MFIT) in the Bechgaard salts, to the wider problem of electrons which are subjected to both a strong magnetic field and a periodic potential. The general problem is a quite old and intriguing one whose experimental consequences have yet to be exposited but for which we may have a unique oportunity with the organic conductors. The problem is one of commensurability. The magnetic field requires a quantization of areas, while the periodic potential requires the translation symmetry of the lattice. Only at special fields are the two conditions compatable and the energy spectrum takes a most remarkable form.

These effects are most pronounced in two dimensional systems in a perpendicular magnetic field, the relevant case for the Bechgaard salt studies. There are very few solutions for the eigenstates or eigenvalues for systems with incommensurate or quasi-periodic potentials but one of the finest examples is the case of a tight binding square lattice in a magnetic field. The result is a complex self-similar spectrum which has served as a paradigm for many incommensurate problems and whose implications for experiments such as the quantum Hall effect has only been brought to our attention over the past 10 years.

The actual system of interest for the organic conductors is quite different from the square lattice in that the tight binding bands are very anisotropic. As a result there are none of the closed orbits at the Fermi energy which are vital to the square lattice spectrum. However, the introduction of a magnetic field still yields a problem with incommensurate potentials and produces a very different, but also non trivial, excitation spectrum with a series of oscillating gaps and bands. The nature of this spectrum is vital to an understanding of the series of MFIT. Moreover, it connects the present problem with many others which occur in nature.

MAGNETIC FIELDS AND PERIODIC POTENTIALS: THE BASIC PROBLEM

We can easily and directly write the exact (nonrelativistic) equation for a two dimensional electron in a periodic potential with an applied magnetic field.

$$(1/2m)(-i\hbar\nabla-eA/c)^2\phi+V(x,y)\phi=E\phi, \quad V(x+na,y+mb)=V(x,y) \tag{1}$$

where A is the vector potential, n and m are integers, and a and b are primitive lattice vectors. We now need to choose a gauge. Throughout this lecture we will stick with the Landau gauge.

$$A=(0,Hx) \tag{2}$$

Why can't we just go ahead and solve 1 exactly, or at least do a well defined perturbation theory if the field is small? Accessible magnetic fields in the laboratory are ~10 Tesla, with the highest achievable fields presently at ~100 Tesla. On an atomic scale such fields are quite weak compared to electronic energies: the bandwidth or E_f. The answer lies in eqs. 1 and 2. For an infinite system the vector potential grows without bound for any H and is never a simple perturbation. The best we could hope for is that A will appear as a phase factor (as does the momentum), and that its effect on the energy will be bounded. However, in introducing the phase factor we have to worry about the associated wavelength and how it compares with the periodicity of the potential V.

To see how the two periodicities play against one another let's take another approach. Look at the wavefunction. When we turn on a magnetic field there is an additional phase shift:

$$\phi\to\phi e^{i\oint(eA/\hbar c).dl} \tag{3}$$

proportional to the line integral of the vector potential. In order for the wavefunction to be single valued, the phase change on any closed path must be 2π times an integer.

$$\oint(eA/\hbar c).dl=2\pi\times\text{integer} \tag{4}$$

The line integral of A around a closed loop is equal to the magnetic flux through the loop which is therefore quantized.

$$H\times(\text{area})=nhc/e, \quad hc/e=\Phi_0\sim4\times10^{-7}\text{gauss-cm}^2 \tag{5}$$

Φ_0 is called the flux quantum. Now what about the periodic potential? From Bloch's theorem it requires a wavefunction of the form:

$$\phi(r+T)=\phi(r)e^{ik.T} \quad T=na+mb \tag{6}$$

which always returns to the same phase when returning to the same point after a walk around the lattice. Suppose we apply a magnetic field such that $H\times a\times b=\Phi$ where Φ is the flux through a unit cell. If $\Phi=\Phi_0$, then going around a unit cell, or any number of unit cells, results in no phase shift. Everything is OK, there is no modification of the Bloch states or the energies. However, if Φ is equal to $\Phi_0/2$, then going around a single unit cell we have $\phi(0)=-\phi(0)$ which doesn't work. To make everything work we have to go around two unit cells. The translational symmetry of the lattice potential has been broken, we now have to take cells two at a time or rather define a new unit cell which is twice as big as the old one. This is OK, but as always when we double the unit cell we have to split each band into two bands. It is evident how to proceed to the general case. If the $\Phi/\Phi_0=p/q$, a rational number, where p and q are integers then we take groups of q cells as our new unit cell. This insures that the new unit cell contains an integral number (p) of flux quanta. Since we need q old cells for each new one, any old band splits into q new ones. This is the key to much of what comes later.

The way that most of us learned to treat magnetic fields and periodic potentials is via the Landau-Peierl's substitution[1]. Drop the vector potential from eq. 1 and it can readily be solved for the bandstructure. The dispersion is $E(k_x,k_y)$. If we interpret the wavevector as momentum, $k\to p/\hbar$, then we know how to deal with momentum in a magnetic field, $p/\hbar\to(i\nabla-eA/\hbar c)$ and we can turn the dispersion relation into an operator equation for the energy and states.

$$E(i\partial/\partial x, i\partial/\partial y-eHx/\hbar c)\phi=\varepsilon\phi \tag{7}$$

where we have again chosen the Landau gauge. This approximation was first written down by Peierls in the 30's and a great deal of work has gone into its justification[2]. Basically it is okay as long as we are not concerned about interband mixing, which is the case for weak fields and large bandgaps. The more drastic approximation that we learn is the semi-classical approximation[3]. In this case we interpret the wavevector as the momentum and the velocity, $v_k=\nabla_k E(k)$ literally as the classical derivative of the position r. The Lorentz force then gives $\hbar\dot{k}=ev\times H/c$ and if we have closed orbits we use the old quantum theory to obtain

$$\oint p.dr = 2\pi(n+\gamma) \Rightarrow k \text{ space area}=(n+\gamma)eH/hc=(n+\gamma)H/\Phi_0 \tag{8}$$

Thus the fundamental quantization condition for closed orbits is in terms of the area in k space. In real space the required area is that which yields a flux quanta in the applied field. This approximation, although intuitive and easy, neglects both interband mixing and the tunneling between the localized orbits which it produces. Nontheless, it tends to be the way in which most of us visualize the formation of quantum levels in a magnetic field.

In the absence of a periodic potential, that is for free electrons, either method above gives the same results. If we drop the $V(x,y)$ term from equation 1 above, we see that its only position dependence occurs in the squared term as $eA/c=eHx/c$. The equation can be rewritten as displaced harmonic oscillator with an effective spring constant of $(1/m)(eH/c)^2$. The discrete equally spaced levels of the quantized oscillator are given by $E_n=(n+1/2)\hbar\omega_c$ with the cyclotron frequency $\omega_c=eH/mc$. The splitting of these Landau levels is linear in the applied field. Moreover, the degeneracy of the levels is simply the number of them that can sit without overlapping on the sample. Since each level corresponds to an area Φ_0/H, the number per unit area is $H/\Phi_0=eH/hc=g$ where g is the degeneracy/unit area.

There are several nice things that come out of this simple picture. The Hall resistance is $\rho_{xy}=R_H H$ where R_H is the Hall coefficient and is proportional to the reciprocal of the k space area below the Fermi energy. The k space area is just the number of states and if we are restricted to less than the Fermi energy this is the total number of occupied states. $R_H=1/nec$ where n is the number of carriers. If the orbits are quantized then the number of carriers is the number of Landau levels below E_f, j, times the degeneracy of each level, g. Putting this all together, we find that the Hall resistance is $\rho_{xy}=h/je^2$ or the Hall conductance $\sigma_{xy}=je^2/h$, when we have only filled levels. Thus if for some reason (such as having localized states between the Landau levels) the Fermi level lies in the gap between Landau levels, the Hall conductance is quantized! This is one part of the Quantum Hall Effect. The other part is to realize that in the presence of crossed electric and magnetic fields, the electrons travel in the direction perpendicular to both E and

H. Thus the current is perpendicular to E and there is no dissipation. The longitudinal resistance ρ_{xx} and the conductance σ_{xx} are both zero.

The simple analysis of free electrons in a magnetic field is also very instructive in seeing what happens to band electrons in the semi-classical approximation. We now have a scale for the magnetic energy, $\hbar\omega_c$, which for band electrons becomes $\hbar\omega_c=eH/m_c c$ where m_c is the effective mass for a closed cyclotron orbit and depends on the bandstructure. Moreover, we now have a way of estimating self consistently how bad the semiclassical approximation is. The formation of bands and complicated orbits is largely governed by the presence of band gaps at the boundaries of the Brillouin zones (B.Z.). The gaps are responsible for pinching off regions of the Fermi surface and creating closed orbits. If the gaps are always large compared to the magnetic energy then the separate orbits never see one another, there is no tunneling. If the magnetic energy is large compared to the gaps then the gaps effectively no longer exist, we have magnetic breakdown and the electrons follow paths directly through the small gaps until they encounter one which has a larger energy than $\hbar\omega_c$. To quantify this phenomena we note that the probability of tunneling between two orbits which are separated by a gap E_g is $P\sim\exp(-E_g^2/\hbar\omega_c E_f)$ [4] in the limit when $\hbar\omega_c \ll E_g$. The picture which we will find very useful in understanding the complex spectra will simply envolve the intuitive semi-classical quantization combined with tunneling between the levels. Note however, that the MFIT typically involves gaps of order the transition temperature ~5K at 10 Tesla, while $\hbar\omega_c$ is ~10 K at this field. For these transitions we are always in the strong breakdown region and the orbits and energies must be quite different from those given by semi-classical arguments.

TIGHT BINDING SQUARE LATTICE IN A FIELD

Conceptually the first problem one might treat to see the combined effects of a periodic potential and magnetic fields, is the case of a single tight binding band, separated by large energies from the other bands so that we do not have to worry about interband admixture. The dispersion relation is

$$E(k_x,k_y)=2t(\cos(k_x a)+\cos(k_y a)) \tag{9}$$

where t is the transfer integral. With the Landau-Peierls substitution, eq. 7, we are left with the operator equation or effective hamiltonian:

$$2t[\cos(ia\partial/\partial x)+\cos\{(i\partial/\partial y-eHx/\hbar c)a\}]\phi=E\phi \tag{10}$$

Since y occurs only in the derivative in the second cosine above we know that the solution in the y direction can be written as a plane wave, $\phi=e^{ik_y y}\phi(x)$. The $\partial/\partial x$ operation is understood by expanding the cosine as: $\cos(ia\partial/\partial x)\phi(x)=(1/2)[\exp(a\partial/\partial x)+\exp(-a\partial/\partial x)]\phi(x)=(1/2)[\phi(x+a)+\phi(x-a)]$, where we recognize the exponential operators as the Taylor series expansion for a finite displacement. We are then left with a difference equation rather than a differential equation.

$$t[\phi(x+a)+\phi(x-a)+2\cos(k_y-eHx/\hbar c)\phi(x)=E\phi(x) \tag{11}$$

This is a one dimensional equation, but there are two basic periodicities, a and $\hbar c/eH$. The equation can be solved by a continued fraction expansion[5], or by the transfer matrix method[6]. With the substitutions $\beta=k_y a$, $x=ma$, $y=na$, $E/t=\varepsilon$, and $\alpha=a^2H/(hc/a)=\Phi/\Phi_0$ we have:

$$\phi(m+1)+\phi(m-1)+2\cos(2\pi m\alpha-\beta)\phi(m)=e\phi(m) \tag{12}$$

By definition a quasi-periodic function has two or more irrationally related periods (i.e. incommensurate periods). Eq.12 has such a form with periods 1 and α.If $\alpha= 1$, i.e. one flux quanta per unit cell, then we have a single band with the continuous variable β, just as we would for $\alpha=0$. If $\alpha=p/q$, then the equations close after q cells, which is the new periodicity. There are q bands. The problem was originally solved by Azbel[5], but the results did not receive wide attention until the spectrum was graphically presented in the work of Hofstadter[6] who used a computer to solve the transfer matrices. The remarkable spectrum of allowed energy states is shown in figure 1. The figure itself is striking. We can find regions of the figure reproduced on a smaller scale throughout. The pattern has a self-similarity. There are bands and gaps at all energy scales. Some parts we can understand immediately. The interval shown ranges from $0<\Phi/\Phi_0<1$. The pattern repeats with period unity. In the lower left hand corner are a set of lines whose slopes increase with the integers, 1,2,3 etc. These are just the Landau levels with $E_n=(n+1/2)\hbar\omega_c$ coming from the quadratic bottom of the band. Now go to $\Phi/\Phi_0=1/2$. There are two bands (which happen to meet at the center where the original band had no curvature). For $\Phi/\Phi_0=2/3$ there are 3 bands separated by 2 gaps, for $\Phi/\Phi_0=3/5$ 5 bands and 4 gaps etc. But note how unusual this is. For for $\Phi/\Phi_0=1/2$ we have seen there are two bands. Now change the field by a very small amount say to 499,999/1,000,000, there are then 1 million bands separated by a similar number of gaps. We are always infinitessimally close to having an infinite number of gaps in the spectrum.

What difference can these small gaps make? Normally we would expect that the smallest gaps have the smallest effects on any physically measurable quantity. However, Thouless and coworkers[7] decided to look at the Hall conductance. As it turns out the equations above are self-dual. For our purposes this means that the spectrum shown in fig. 1 not only describes the situation where we have a single band split by magnetic field, but also corresponds to the situation where we have a single Landau level split by a weak periodic potential. In the latter case the

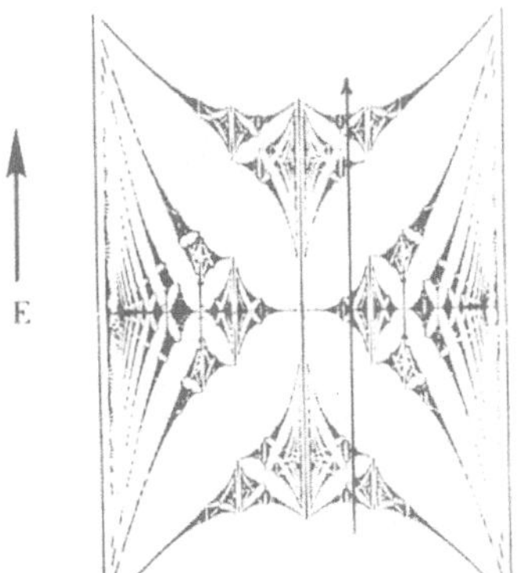

Fig. 1. Energy spectrum of an electron in a square lattice and a mangetic field. X-axis is field in flux quanta per unit cell. Line is path followed in figure 2.

lower axis becomes Φ_0/Φ. So we are going to look at a single landau level with a periodic potential and see what happens as we go from below to above it by changing the energy. We know that traversing a single Landau level we must gain a quantum of conductance e^2/h in the σ_{xy} [8,9]. What happens if we go part way though and stop in one of the gaps. In figure 2 we show the result from ref. 7 of being able to see some of the smaller gaps, which is accomplished in this case by traversing the Landau level at different values of Φ_0/Φ near the irrational value, $\tau=(sqrt(5)-1)/2\sim.618$, taking rational approximates. For 3/5 we encounter 4 gaps, for 8/13, 12 gaps etc. This is similar to what might be expected if one could see the smaller gaps by decreasing the temperature or making a cleaner sample.

What is remarkable in figure 2 is that one goes from $\sigma_{xy}=0$ below to $\sigma_{xy}=1$ above the level by way of large excursions of either sign (but always in integral units of e^2/h [7,8,9]). In fact the smallest gaps correspond to the largest values of σ_{xy}. If one could traverse the Landau level with infinite resolution at an incommensurate value of Φ_0/Φ then the Hall conductance would be completely chaotic varying from $-\infty$ to $+\infty$. With finite resolution, say for a reasonably clean sample, we might only see σ_{xy} as indicated by dashed line, while for a cleaner sample we might be able to pick up σ_{xy} as shown on the dotted line. One might like to associate a similar set of curves with the dramatic Hall changes and fine structure seen the experiments by Ribault[10,11] on the Bechgaard salts, as cleaner samples or lower temperatures are attained.

How can we understand such a complex spectrum and behavior of the Hall conductance. Let's see what we can do simply with the ideas of semi-classical quantization and magnetic breakdown or tunneling[12]. Start with the individual atoms in the system and their discrete levels. The way we get a tightbinding dispersion is to assume that the electrons are essentially localized on atomic sites, with the atomic wavefunctions, but with a small probability of tunneling to a neighboring atom. Each atomic level then gives rise to a band whose dispersion relation is $E(k)=E_a-2t(\cos(k_xa)+\cos(k_ya))$ where t is the transfer integral mixing near neighbor degenerate levels (fig. 3a). We turn on the magnetic field and quantize the k space areas. The band splits into discrete Landau levels (fig. 3b). In semi-classics this is where we stop. The part we have left out is the tunneling between these Landau levels. Pick one of the Landau levels in a particular B.Z. There are degenerate levels centered in all the neighboring B.Z.'s in k space, and there is a certain tunneling

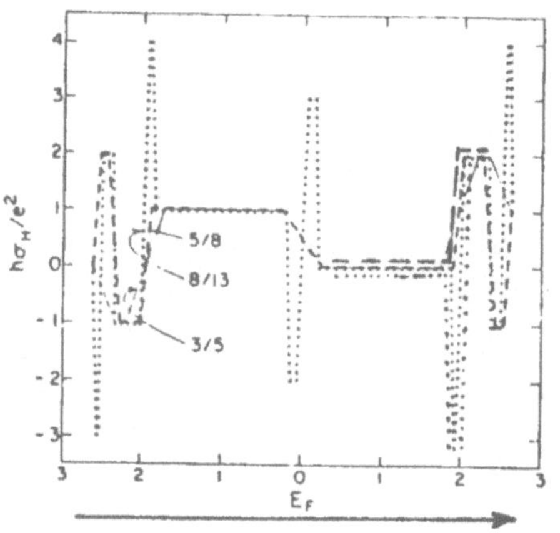

Fig. 2. Hall conductance as a function of energy in traversing fig. 1., sampling different gaps. Dashed line - 'clean' sample, dotted line - 'cleaner' sample.

probability or transfer integral between the states (fig. 3c). We have seen this problem before, it is just the situation which led to the tight binding band for each atomic level in fig.3a. The result is that each discrete Landau level becomes a tight binding Landau band. But the problem is not over, we must now quantize the new bands in the magnetic field. Each band becomes a set of levels. Since we never have an exact solution in this semiclassical model we must continue the iteration process to approach the correct solution (fig 3d) at least as long as we have an irrational value of Φ/Φ_0. The self-similarity and infinite series of gaps on all scales results from these continued iterations. So much for the spectrum.

Now what about the strange behavior of the Hall resistance. Suppose we can only see gaps to a minimum size because of finite temperature, scattering, electric fields or the commensurability of Φ/Φ_0'. Let's say the iteration before our cutoff the orbits looked like those shown in fig.4. Transfer between these levels would produce a continuous band, but if we require from commensurability or scattering a finite size cell then the new levels are discrete. They are orbits which circulate around the original orbits and then ocasionally hop to an adjacent cell until they complete a loop in the new cell. Two of these orbits are illustrated. Note that they can orbit the larger cell in opposite directions, corresponding to electron-like and hole like orbits. Moreover, the k space area determines the size of σ_{xy} and these orbits correspond to large values. The energy of these orbits is almost the same, differing only by the small tunneling term, but the σ_{xy}'s are vastly different. If

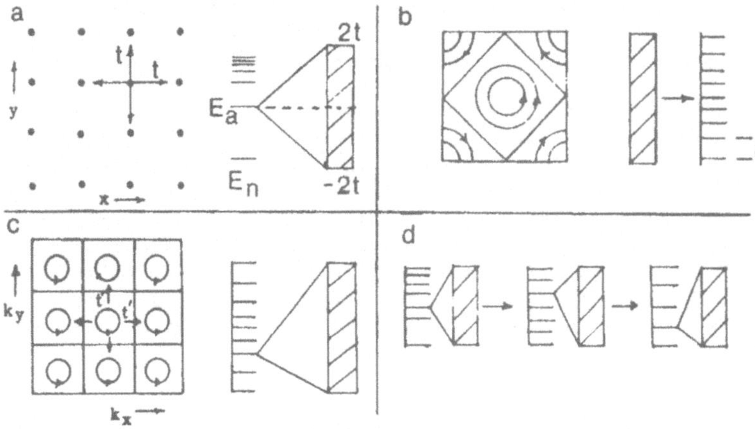

Fig. 3.a)Atomic levels develop into a band. b)Band splits into Landau levels in a magnetic field. c)Orbits break down to form Landau bands. d)Levels->Bands->Levels etc. approproximate solution assymptotically.

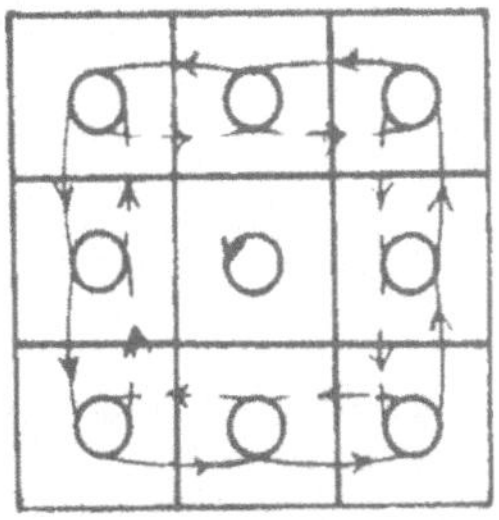

Fig. 4. Two possible orbits for a 3x3 new unit cell. Solid is electronlike and dashed is holelike.

the electrons scattered before they completed the orbits shown, then the two levels would collapse into one, with the original small σ_{xy}. The situation is further complicated by the fact that the bands which arise from different levels may overlap, causing 'wild' variation of the sign and magnitude of σ_{xy}.

OPEN ORBITS AND ORGANIC CONDUCTORS

In the organic conductors the above treatment is not directly applicable. The bands are tight binding but not 'square' or isotropic. There are nothing but open orbits. Taking the dispersion relation as $E(k)=\hbar^2 k_x^2/2m-2t_b\cos(k_y b)$ and applying the Landau-Peierls substitution we we have:

$$-(\hbar^2/2m)(\partial^2\phi/\partial x^2)-2t_b\cos(k_y b-eHxb/\hbar c)\phi=\varepsilon\phi \quad (13)$$

k_y occurs only in the argument of the cosine and can be eliminated by simply displacing x. We are left with a simple one dimensional equation for an electron in a sinusoidal potential. The solution has gaps for all $k_x=2\pi/\lambda$, $\lambda=hc/eHb$, which are sizable for $\varepsilon<2t_b$ and decay exponentially above[13]. We have no quasi-periodicity and no complexity of the spectrum. However, we do have something unusual. The magnetic field has changed the two dimensional dispersion into one dimensional[14,15]. To see physically what is going on it is instructive to look at the open orbit Fermi surface illustrated in fig. 5a. In the presence of a magnetic field an electron moves on the Fermi surface as shown. When it reaches the B. Z. edge it merely reappears on the other side. The velocity is always perpendicular to the Fermi surface and from this we can obtain the real space motion which is illustrated in fig. 5b. It is limited along y but infinite along x. This is as good a definition of one dimensional as any.

We know that one dimensional systems are unstable against CDW and SDW transitions. The essence of the MFIT in the Bechgaard salts is precisely the fact that the magnetic field has turned a two dimensional problem into a one dimensional one[14-16]. The usual way one might treat such a transition is to introduce a distortion (or spin distortion) into the potential calculate the effect on the electronic spectrum and compare the energy reduction with the cost of making the distortion. For free electrons in one dimension this corresponds to solving $-\partial^2\phi/\partial x^2-\Delta\cos(qx)\phi=\varepsilon\phi$, where q is the distortion wavevector and Δ is its amplitude. The result is a gap 2Δ in the spectrum at $k_x=q/2$. If we try the same approach in the present case we must solve:

$$-(\hbar^2/2m)(\partial^2\phi/\partial x^2)-2t_b\cos(2\pi x/\lambda)\phi-\Delta\cos(q.r)=\varepsilon\phi \quad (14)$$

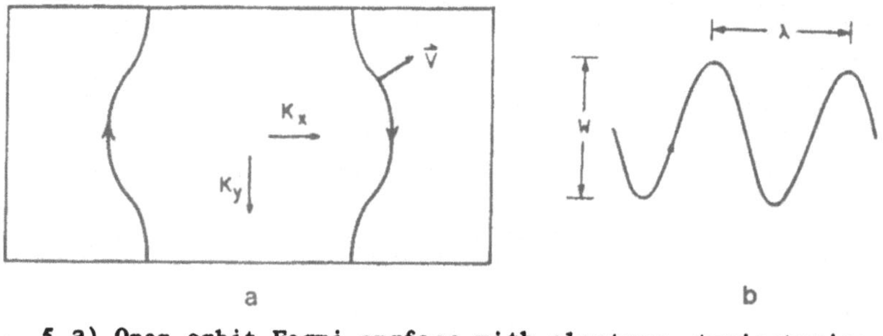

Fig. 5 a) Open orbit Fermi surface with electron trajectories. b) Real space electron trajectory.

This is not so easy. We have recovered an equation where we have a quasi-periodicity in $2\pi/\lambda$ and q. Naively, one might expect that there is little effect of the gaps at $\varepsilon<2t_b$ which we have already seen, and the gap at E_f from $q\sim 2k_f$. But incommensurate potentials are full of surprizes. The actual spectrum consists of a series of gaps over a width $4t_b$ about E_f, all of which oscillate with the magnetic field as illustrated in fig. 6[17].

If we place one of the gaps at the Fermi energy by choosing a certain value of q, then we can lower the electronic energy. This is essentially responsible for the field induced SDW transition. If we did not know about the presence of the other gaps, the fact that the gap at E_f oscillates as the magnetic field is changed would force the system back to the normal state periodically[14]. However, when the gap at $q/2=k_f$ vanishes the gap at $q/2(+/-)2\pi/\lambda$ is at a maximum. Thus instead of returning to the normal state it is merely necessary to change the distortion wavevector to $q=2k_f+2\pi/\lambda$ or any of the other values where the gaps are large[17]. Of course the actual situation is more complicated because q may take on different values in the y direction as well as along x[16,18-20].

Intrinsic to the problem of the MFIT is the interplay of the generally incommensurate wavelengths $2k_f$ and $2\pi/\lambda$. As a consequence we must always have a series of gaps rather than the single gap found in the usual SDW problem, and along with these gaps we must have a series of transitions which take advantage of the most favorable gaps.

Quasiperiodic potentials almost always have some feature which seems unusual or not intuitive. In the present case the limit $\lambda\to\infty$ is weird. We would expect the effect of the slow potential to disappear. What happens instead is that the number of gaps and bands increases with λ while the bandwidths decrease exponentially. The spectrum tends to a set of discrete values over a region $4t_b$ wide about E_f which is entirely filled by the gaps[17,21]. As far as transport is concerned there is one big gap, at the wavevector given by the small potential, i.e. $\sim k_f$, but with the large amplitude given by the large potential with the small wavevector $2\pi/\lambda\sim 0$. To understand how this arises let us look at a semiclassical treatment of eq. 14. In semi-classics we are supposed to treat the fast potentials first. The potential $\Delta\cos(qx)$ produces a gap of size 2Δ at E_f. We then modulate this picture (fig. 7) by the slow potential a $\cos(2\pi/\lambda)$ (much as we treat band bending for a semi-conductor). The states in the new potential well must be quantized leading to discrete levels. But it is always possible to tunnel through the gap region, and this again makes each level into a band. As the λ gets bigger the wells become wider,

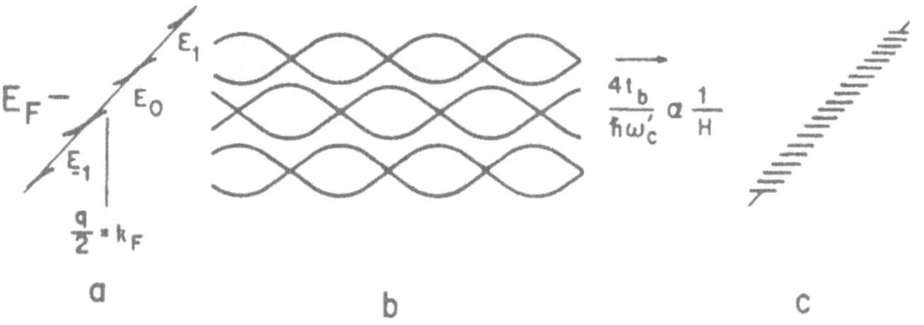

Fig. 6. a)Bands and gaps near E_f (from solution to eq. 14), b) oscillations as H is changed. c)Spectrum as $\lambda\to\infty$.

implying algebraically more levels, but the tunneling decreases exponentially since the tunneling distance is increasing.

We have investigated two different cases for electrons in quasi-periodic potentials, the square tight binding lattice and the open orbit case for the MFIT. We have seen that the latter case goes a long way in explaining the experiments on the field induced transitions. However, the experiments also have some unexplained observations, such as the Ribault anomalies and the fine structure in the Hall resistance at very low temperatures, which are similar to the expected behavior in the first case we looked at. What has been left out from our treatment of the field induced transitions? In the dispersion relation for the open orbit case all of the present theoretical models left out the periodicity along the a direction, they took Ak_x or Bk_x^2 rather than the tight binding form $2t_a\cos(k_xa)$. We know that in usual charge or spin density waves that including the periodicity of the lattice can strongly effect the nature of the transitions. In particular, if the distortion wavevector is commensurate with the lattice, we gain additional energy and the gap and transition temperature are enhanced. Thus in the case of the MFIT we should really consider three periods, π/k_f, λ and a, especially since $2a{\sim}\pi/k_f$. We will have to wait detailed calculations to see whether the more complex spectrum and additional structure arises from this commensurability effect of the MFIT with the lattice.

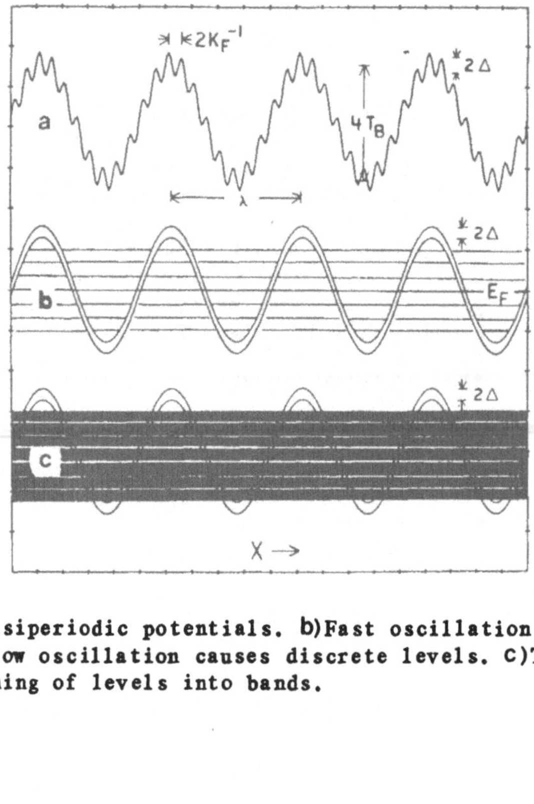

Fig. 7. a)Quasiperiodic potentials. b)Fast oscillation produces small gap, slow oscillation causes discrete levels. c)Tunneling allows broadening of levels into bands.

REFERENCES

1. R. E. Peierls, Z. Phys. 80, 763 (1933).
2. L. Onsager, Philos. Mag. 43, 1006 (1952), G. H. Wannier, Rev. Mod. Phys. 34 645 (1962). E. I. Blount, Phys. Rev. 126 (1962).
3. K. von Klitzing, G. Dorde and M. Pepper, Phys. Rev. Lett. 45, 494, (1980), D. C. Tsui, H. L. Stormer, and A. C. Gossard, Phys. Rev. B25, 1405 (1982)
4. M. H. Cohen and L. M. Falicov, Phys. Rev. Lett. 7, 231 (1961).

5. M. Ya Azbel, Zh. Eksp. Teor. Fiz. 46, 939 (1964) [Sov. Phys. JETP 19, 634 (1964)].
6. D. R. Hofstadter, Phys. Rev. B14, 2239 (1976).
7. D. J. Thouless, M. Kohmoto, M. P. Nightingale, and M. den Nijs, Phys. Rev. Lett. 49, 405 (1982), D. J. Thouless, Physics Reports, 110, 279 (1984).
8. R. B. Laughlin, Phys. Rev. B23, 5632 (1981).
9. B. I. Halperin, Phys. Rev. B25, 2185 (1982).
10. M. Ribault, Mol. Cryst. Liq. Cryst. 119, 91 (1985).
11. M. Ribault, in this volume.
12. This discription is largely due to M. Ya. Azbel.
13. P. M. Chaikin, T. Holstein, and M. Ya. Azbel, Phil. Mag. B48, 457 (1983).
14. L. P. Gorkov and A. G. Lebed, J. Phys. (Paris), Lett. 45, L433 (1984).
15. P. M. Chaikin, Phys. Rev. B31, 4770 (1985).
16. M. Heritier, G. Montambaux, and P. Lederer, J. Physique Lett. 45, (1984) L-943, G. Montambaux, M. Herritier and P. Lederer, Phys. Rev. Lett. 55, (1985) 2078.
17. M. Ya Azbel, Per Bak and P. M. Chaikin, Phys. Lett. A117, 92 (1986).
18. K. Yamaji, J. Phys. Soc. Japan 54, 1034 (1985), K. Yamaji, Synthetic Metals, 13, 29 (1986).
19. A. Virosztek, L. Chen and K. Maki, to be published in Phys. Rev., K. Maki, to be published in Phys Rev.
20. See articles by M. Herritier and G. Montambaux in this volume.
21. M. Ya Azbel, Per Bak and P. M. Chaikin, Phys. Rev. A34, 1392, (1986).

SUSCEPTIBILITY AND INSTABILITY OF THE QUASI-ONE-DIMENSIONAL ELECTRON GAS UNDER MAGNETIC FIELD

Gilles Montambaux

Laboratoire de Physique des Solides (associé au CNRS)
Université Paris-Sud
91405 Orsay, France

I - INTRODUCTION

It is well known that the magnetic field has dramatic effects on the dynamics of 2D electron gas. In the case of closed electronic orbits, the field effect leads to the quantization of orbits in Landau levels and thus to remarkable consequences on the transport properties such as the Hall effect [1]. The discovery of the so-called quantized Hall effect has been a stimulation for studying the physics of the 2D electron gas under magnetic field. However, a vast majority of papers so far have dealt with 2D isotropic systems. The 2D anisotropic electron gas, such as can be found experimentally in weakly coupled chain systems, has specific properties under magnetic field which deserve more attention.

Prominent among them is the possibility of open Fermi surfaces, which results in a qualitative change of the quasiclassical motion of the electron wave packet. The occurence of quasi parallel sheets of the Fermi surface leads to the instability of the metallic phase versus formation of Charge Density Wave (CDW) or Spin Density Wave (SDW) phases. The description of such an instability implies the study of the susceptibility χ_o of the non interacting electron gas. In zero field, it is given by the well-known expression

$$\chi_o(\vec{Q}) = \sum_{\vec{k}} \frac{f(\varepsilon_{\vec{k}+\vec{Q}}) - f(\varepsilon_{\vec{k}})}{\varepsilon_{\vec{k}} - \varepsilon_{\vec{k}+\vec{Q}}} \qquad (I.1)$$

where $\varepsilon(\vec{k})$ is the energy of a state with given wave vector $\vec{k}$, $f(\varepsilon)$ is the Fermi filling factor. The structure of $\chi_o(\vec{Q})$ strongly depends on the geometry of the Fermi surface. In mean field, the instability criterion is given by the Stoner identity : $1 - \lambda\chi_o$, where λ is the molecular field constant.

Particularly interesting in that case is the interplay between this instability and the orbital effect induced by the magnetic field. We are now going to study this interplay.

The field effects observed in the 3D anisotropic compounds of the $(TMTSF)_2X$ family have motivated this study so that we will keep in mind the physical situation in these so-called Bechgaard salts. In some compounds of this family, the metallic phase, stable in low field is destroyed by a moderate field of a few teslas applied along the c^* direction, which induces a succession of SDW phases[2]. Understanding of such a field-induced instability requires the study of the susceptibility χ_o in a magnetic field. This has been done in a first step by Gor'kov and Lebed[3]. But we have shown the importance of orbital quantization to understand the nature of the instability and the profound physical origin of the field-induced phases[4-8].

These compounds are triclinic and present a strong anisotropy of the electronic properties[9]. In a first approximation, they can be described with a hierarchy of three transfer integrals along the three crystal axis[10-12]. For our purpose, we replace the triclinic symmetry by an orthorhombic symmetry so that we postulate the following form of the electronic energy :

$$\varepsilon(\vec{k}) = -2t_a \cos k_x a - 2t_b \cos k_y b - 2t_c \cos k_z c \qquad \text{I.2}$$

with $t_a \gg t_b \gg t_c$. Typically, in Bechgaard salts, $t_a \simeq 3000$ K, $t_b \simeq 300$ K and $t_c \simeq 10$ K. The band is 3/4 filled. In these conditions, the Fermi surface is open and consists of two sheets.

Throughout this paper, we are dealing with a 2D electron gas. We will discuss the role of the third direction in the last section.

II - NEW PHYSICS INTRODUCED BY THE MAGNETIC FIELD

We postulate the following form of the energy

$$\varepsilon(\vec{k}) = \varepsilon_{/\!/}(k_{/\!/}) + \varepsilon_\perp(k_\perp) \qquad \text{II.1}$$

where $k_{/\!/}$ and $k_\perp$ are the components of the 2D wave vector. $k_{/\!/}$ indicates the direction $\vec{a}$ of highest conductivity, along the chains. The essential ingredient in our physical problem is an open Fermi surface and the periodicity of the dispersion relation in the transverse direction

$$\varepsilon_\perp(k_\perp + 2\pi/b) = \varepsilon_\perp(k_\perp) \qquad \text{II.2}$$

b being the distance between 1D chains.

Under magnetic field applied along the direction perpendicular to the 2D gas, the substitution $\vec{p} \to \vec{p} - e\vec{A}$ leads to the Schrödinger equation :

$$\mathcal{H}\phi = [\,\varepsilon_{/\!/}(\tilde{k}_{/\!/}) + \varepsilon_\perp(\tilde{k}_\perp - eHx/\hbar)]\phi = E\phi \qquad \text{II.3}$$

where the Landau gauge $\vec{A} = (0, Hx, 0)$ has been used (k is a number and $\tilde{k}$ an operator).

The fundamental effect of the magnetic field is to transform the periodicity in the transverse (b) direction into a <u>new periodicity along the parallel direction</u> (a). This new periodicity has the wavelength[4-6, 13] :

$$\lambda = 2\pi\, x_o = \frac{2\pi}{G} = \frac{2\pi\hbar}{eHb} \qquad \text{II.4}$$

We are going to show that all the new properties introduced by the field are intimately connected with this new periodicity.

Another important point due to the field is that <u>the wave vector $k_{/\!/}$ is no longer a good quantum number</u> :

$$[\mathcal{H}, \tilde{k}_\perp] = \frac{eH}{i\hbar}\,\frac{\partial\varepsilon_\perp}{\partial k_\perp} \neq 0 \qquad \text{II.5}$$

The eigenstates of the energy are no longer eigenstates of the wave vector. $k_\perp$ is a good quantum number : $[\mathcal{H}, \tilde{k}_\perp] = 0$. Obviously, the energy does not depend on $k_\perp$. Thus eq. II.3 describes a <u>one-dimensional problem</u> of a particle in a <u>periodic potential</u>, with wave-vector $G = eHb/\hbar = 1/x_o$ [13].

Let us now explore the consequences of these statements on the structure of the electronic wave function. To do this, we shall use a peculiar form of the dispersion relation which leads to simple analytical results. This form is obtained when the dispersion relation along the chains is linearized around the Fermi level. Let us choose the following form ($\hbar = 1$) :

$$\mathcal{H}\phi_E = [\mu + v(|k_{/\!/}| - k_F) + t_\perp(k_\perp b - eHbx)]\phi_E = E\,\phi_E \qquad \text{II.6}$$

μ being the Fermi energy and v the Fermi velocity. $t_\perp$ is an even function with period 2π, with $< t_\perp > = 0$.

The eigenfunction of Schrödinger equation II.6 writes for $k > 0$ (for $k < 0$, $v \to -v$) :

$$\phi_E^R(k_\perp,x) = \exp i\{kx + k_\perp y + \frac{x_o}{v} T_\perp(p - x/x_o) - \frac{x_o}{v} T_\perp(p)\} \qquad \text{II.7}$$

where $T_\perp(p) = \int_o^p t_\perp(p')\,dp'$, $p = k_\perp b$ and $E = \mu + v(|k| - k_F) = \varepsilon_{/\!/}(k)$

The eigenfunctions have the following Fourier expansion (Bloch theorem) :

$$\phi_E^R(k_\perp,x) = e^{ikx}\, u(k_\perp,x) = e^{ikx} \sum_n \gamma_n(k_\perp)\, e^{-inGx} \qquad \text{II.8}$$

where

$$\gamma_n(k_\perp) = \beta_n \exp\{- i\frac{x_o}{v} T_\perp(k_\perp b) + ink_\perp b\} \qquad \text{II.9}$$

$$\beta_n = \frac{1}{2\pi}\int_o^{2\pi} \exp\{i\frac{x_o}{v}\; T_\perp(u) - inu\} \qquad \text{II.10}$$

The amplitudes γ_n obey the sum rules

$$\sum_n \gamma_n = 1\,, \qquad \sum_n \gamma_n^2 = 1\,, \qquad \sum_{n\neq m} \gamma_n\,\gamma_m = 0 \qquad \text{II.11}$$

The projection of a state $|\phi_E^R>$ over an eigenstate of the wave vector $|k\ , k >$ can be easily deduced

$$|\ \phi_E^R > \ = \sum < k_{/\!/},\ k_\perp\ |\ \phi_E^R > |\ k_{/\!/},\ k_\perp > \qquad \text{II.12}$$

with

$$< k_{/\!/},\ k_\perp\ |\ \phi_E^R > \ = \ \sum_n \gamma_n(k_\perp)\ \ \delta\,[\,E - \ \varepsilon_{/\!/}(k_{/\!/} + nG)\,] \qquad \text{II.13}$$

This form II.13 has to be compared to the zero field structure :

$$< k_{/\!/},\ k_\perp|\ \phi_E > = \delta\ [\,E - \varepsilon(\vec{k})\,] \qquad \text{II.14}$$

The 2D dispersion relation is replaced by a series of "1D dispersion relations" with a probability amplitude γ_n. We can think of a pseudo-dispersion relation as shown on fig. 1. The successive sheets have weight γ_n and are separated by $G = 1/x_o$. This structure has consequences on the behaviour of the metallic phase, for example on the magnetoresistance[14]. In this paper, we concentrate on the consequences on the structure of χ_o. It is well known that, in the 1D case, it exhibits a logarithmic singularity due to perfect nesting of the Fermi surface. Now, in our effective 1D problem under field, we expect successive logarithmic singularities for a discrete set of wave vectors $\vec{Q} + n\vec{G}$ (fig. 1).

III - CALCULATION OF THE SUSCEPTIBILITY $\chi_o(\vec{Q})$ $\quad \vec{Q} = (\vec{Q}_{/\!/},\ \vec{Q}_\perp)$

The usual derivation of the susceptibility results straightforwardly from a first order perturbation expansion :

$$\chi_o(\vec{Q}) = \sum_{EE'}\ |\ < \phi_E\ |e^{i\vec{Q}\vec{r}}\ |\ \phi_{E'} >|^2\ \frac{f(E') - f(E)}{E - E'} \qquad \text{III.1}$$

where the perturbation at wave vector $\vec{Q}$ couples eigenstates of the non perturbed hamiltonian. In a first step we are now going to study the structure of $\chi_o(\vec{Q})$ in zero field, how it is connected to the geometry of the Fermi surface, in order to see later on how this structure is modified by the magnetic field.

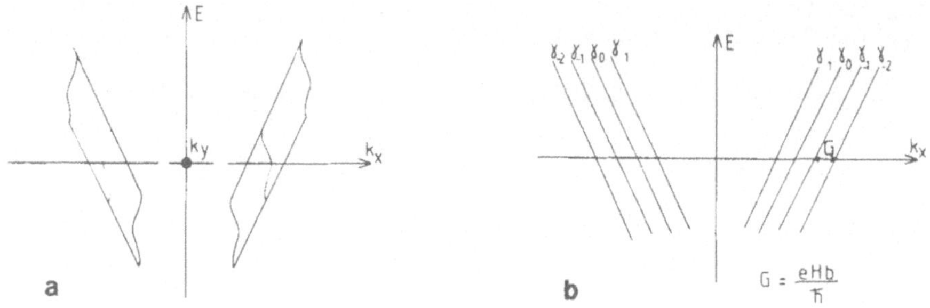

Fig. 1 (a) Dispersion relation $E(\vec{k})$ of a 2D anisotropic electron gas in zero field. The susceptibility depends on the nesting properties at the Fermi level.

(b) Under magnetic field the structure of the dispersion relation is qualitatively different. It consists in a discrete set of 1D spectra with a probability amplitude γ_n. Nesting is perfect whenever $Q_{\parallel}=2k_F+neHb/\hbar$. The amplitude of χ_o depends on the sequence of products $\gamma_p\gamma_{p'}$ as revealed in formula III.3.

A - ZERO MAGNETIC FIELD

In this case $\langle\phi_E|\vec{k}\rangle=\delta[E-\varepsilon(\vec{k})]$, so that the usual expression I.1 is recovered. Let us now study the susceptibility in some peculiar cases.

a) $t_\perp(p) = 0$

In this trivial case, the two sheets of the Fermi surface are strictly planar. Thus, all the states $|\vec{k}\rangle$ located on one sheet are coupled to the states $|\vec{k} + 2k_F\rangle$ located on the other sheet. There is a perfect nesting of the Fermi surface. This causes a logarithmic divergence of χ_o. As a result, the metallic phase is unstable below a critical temperature given by the Stoner criterion : $T_c = E_o \exp(-1/U)$ where $U = N(o)\lambda = \lambda/2\pi bv$. In fact, in this strictly 1D case, mean field does not apply and there is no SDW or CDW long range order (except at zero temperature of quantum fluctuations are neglected)

b) $t_\perp(p) = -2t_b \cos p$

In this simplest case, the Fermi surface has a sinusoïdal warping and the vector of components $(2k_F, \pi/b)$ perfectly nests one sheet onto the other one (fig. 2) .

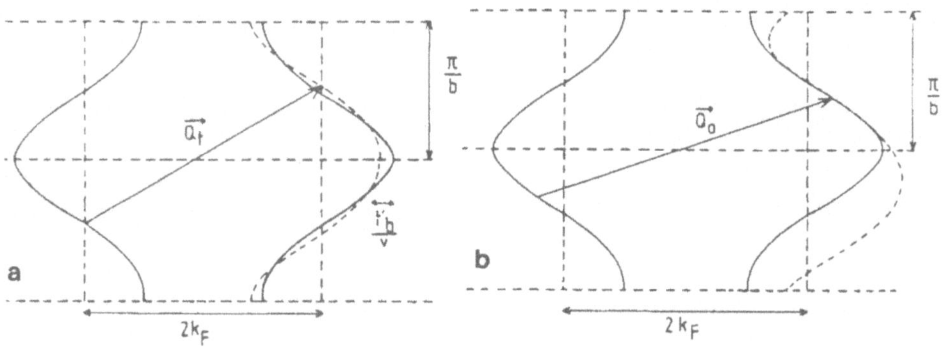

Fig. 2 : Fermi surface in the case $t_\perp(p) = -2t_b \cos p - 2t'_b \cos 2p$. When $t'_b \neq 0$, the vector $\vec{Q}_t$ does not perfectly nest the Fermi surface (a). The best nesting vector $\vec{Q}_o$ connects the inflexion points (b).

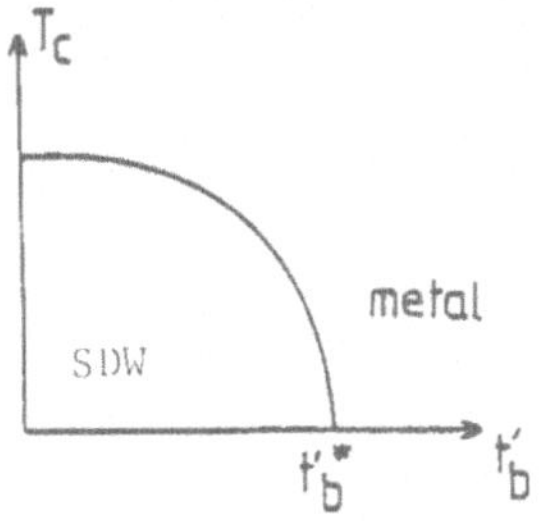

Fig. 3 : Schematic variation of the SDW critical temperature versus t'_b, deviation from perfect nesting.

But now, because of 2D couplings (or 3D in the real situation), the mean field result of a) is valid below a cross-over temperature T_X.

c) $t_\perp(p) = -2t_b \cos p - 2t'_b \cos 2p$

This dispersion can be obtained by a linearization of the tight binding relation I.2. t'_b is a correcting term due to the linearization :

$$t'_b = -\frac{\cos k_F a}{4 \sin^2 k_F a}\,\frac{t_b^2}{t_a} \qquad \text{III.2}$$

This energy is of order of 10 K in Bechgaard salts. Now, due to this t'_b term, neither the vector $Q_t = (2k_F, \pi/b)$ nor any other one exactly nest one sheet onto another (fig. 2). There is imperfect nesting of the Fermi surface. As a result, the susceptibility has the structure shown on fig.4b. The main point is that χ_o no longer diverges, at any wave vector. At 0 K, the best nesting vector i. e. that for which χ_o is maximum, is no longer $(2k_F, \pi/b)$ but the one which connects the inflexion points of the Fermi surface.

Now, the existence of a SDW phase depends crucially on the energy scale t'_b. If t'_b is larger than a critical value $t'^*_b(\lambda)$ which depends on interactions, the metallic phase remains stable, even at zero temperature because Stoner criterion cannot be achieved (fig. 3). This situation occurs in Bechgaard salts under given experimental conditions[2], where there is no SDW. Now when the magnetic field is applied, it can restore a SDW. To understand the origin of this restoration and the original structure of the field-induced phase we need now calculate the susceptibility under magnetic field.

B - FINITE MAGNETIC FIELD $(Q_{/\!/} = 2k_F + q_{/\!/})$

Now we calculate χ_o from eq. III.1. Using the expression II.7 for the wave function, it follows straightforwardly :

$$\chi_o(\vec{Q}) = \sum_N I_N^2 \sum_k \frac{f[\varepsilon_{/\!/}(k)] - f[\varepsilon_{/\!/}(k+Q_{/\!/}-NG)]}{\varepsilon_{/\!/}(k+Q_{/\!/}-NG) - \varepsilon_{/\!/}(k)} \qquad \text{III.3}$$

where $I_N^2 = \sum_{n,k_\perp} \gamma_n(k_\perp)\,\gamma_{-n-N}(k_\perp + Q_\perp)$ III.4

I_N can be written as a function of $T_\perp$

$$I_N = \frac{1}{2\pi}\int_0^{2\pi} \exp\{i\,\frac{x_o}{v}\,T_\perp(u) + i\,\frac{x_o}{v}\,T_\perp(u+Q_\perp b) + iN(u + \frac{Q_\perp b}{2})\}du \qquad \text{III.5}$$

These coefficients obey the sum rules similar to eqs. II.11

Eq. III.3 shows that $\chi_o(\vec{Q}, H, T)$ exhibits a series of logarithmic divergences of amplitude I_N^2 at quantized values of $q_{/\!/} = N/x_o = NG$. At T = 0 K,

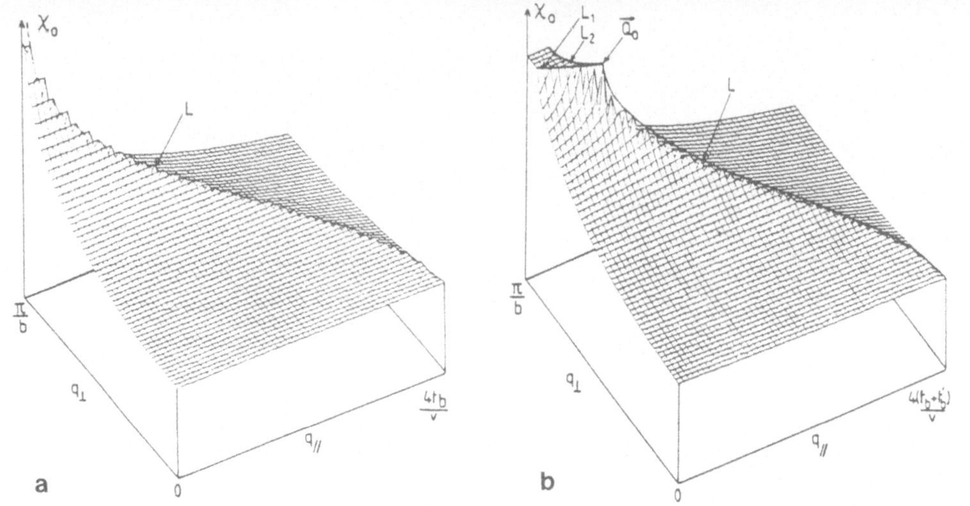

Fig. 4 : Susceptibility $\chi_o(\vec{Q})$ of the non interacting electron gas versus the two components of the nesting vector $\vec{Q} = (2k_F + q_{\parallel}, q_{\perp})$ at zero temperature. a) $t_b' = 0$, there is a logarithmic divergence of χ_o at $\vec{Q}_t = (2k_F, \pi/b)$. b) $t_b' \neq 0$, due to imperfect nesting there is no divergence of χ_o. $\vec{Q}_o$ connects the inflexion points of the Fermi surface. The lines L, L_1, L_2 describe the sliding of the two sheets of the Fermi surface one onto another.

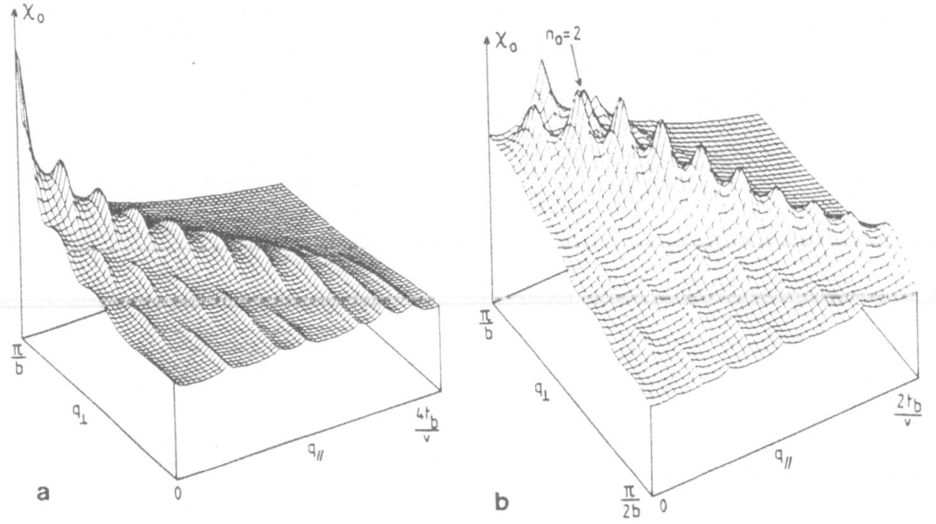

Fig. 5 : Susceptibility under magnetic field. a) $t_b' = 0$, the best nesting vector is the zero field one. b) $t_b' \neq 0$, the absolute maximum varies with the magnetic field. In this case, it is labelled by $n_o = 2$.

the susceptibility takes the particularly simple form :

$$\chi_o(\vec{Q}, T = 0) = \sum_N I_N^2 \ln \left| \frac{2k_F}{Q_{\parallel} - 2k_F - NG} \right| \qquad \text{III.6}$$

We have called this a quantized nesting condition. Let us now look at the details of this structure in some special cases.

a) $t_\perp(p) = -2t_b \cos p$

In this case $I_N = J_N\left(\frac{4t_b}{eHvb}\cos\frac{Q_\perp b}{2}\right)$. χ_o exhibits a series of peaks, the abscissa of which are quantized (fig. 5a) : $q_\parallel = neHb/\hbar$ but, the absolute maximum lies always at the vector $(2k_F, \pi/b)$.

b) $t_\perp(p) = -2t_b \cos p - 2t_b' \cos 2p$

More interesting is this case of imperfect nesting. In this case the magnetic field restores logarithmic divergences which were absent in zero field[4-6]. The best nesting vector corresponds to the largest of these peaks. In particular, it varies with the field, so that

$$q_\parallel = \frac{n_o eHb}{\hbar} \qquad \text{III.7}$$

This fundamental variation is at the origin of all properties described in this paper and the following one. The susceptibility has been computed in this case. The coefficients I_N are[15]

$$I_N = \langle \cos(z \sin u + z' \sin 2u - iNu)\rangle = \sum_{\ell=-\infty}^{\infty} J_\ell(z')\, J_{N-2\ell}(z) \qquad \text{III.8}$$

with $z = \frac{4t_b}{eHvb}\cos\frac{Q_\perp b}{2}$ $\qquad z' = \frac{2t_b'}{eHvb}\cos Q_\perp b$

The structure of χ_o is shown in fig. 5b.

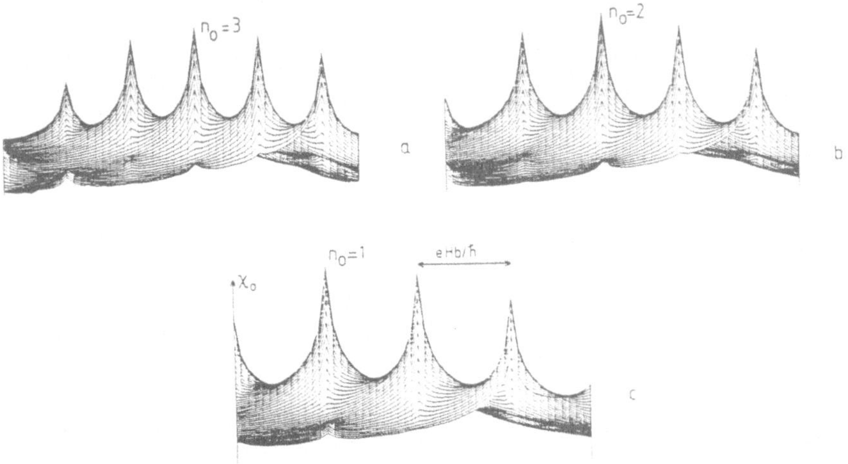

Fig. 6 : Projections of χ_o along the direction $q_\parallel$, on the $Q_\perp = \pi/b$ plane. The different ridges of $\chi_o(Q,H,T)$ are now superimposed on the same plane and the envelope shows the main series of peaks. On figure a, the abscissa of the absolute maximum is characterized by $n_o = 3$. It increases linearly with the field until the peak $n_o = 2$ becomes in turn the absolute maximum (b). At higher field, the peak labelled by $n_o = 1$ is the main one (c). This behaviour is the origin for the cascade of SDW phases.

At a given field, there is a main series of maxima. Their positions $\vec{Q}(H)$ are close to zero field continuous line of maxima which corresponds to the condition that the two sheets of the Fermi surface are tangent. This maxima deviate from this line when the field increases. This series is shown by arrows and labelled by quantum number n in fig. 5 b. In this figure, the absolute maximum is labelled by $n_o = 2$. When the field is varied, each of these peaks become in turn the absolute maximum as shown in fig. 6 [5,6]. As a result $\chi_o(H, T)$ the absolute maximum of the susceptibility, is made of a series of sheets, each of them being described by successive values of the quantum number n_o (fig. 7a). The wave vector jumps at the transition field between each segment (fig. 7b).

The sequence of critical fields that separates the successive sheets of the susceptibility and for which the wave vector jumps, has the form [6,7] : $H_n = H_f/(n + \gamma)$ where $\gamma \simeq 3$ for $t_b'/t_b = 1/10$. H_f is related to $q_{\parallel}$ in zero field : $H_f = \hbar q_{\parallel}/eb$. It can also be related to the size of the pocket of carriers defined in zero field, by the area between one sheet of the Fermi surface and the other one translated by $\vec{Q}$

$$A_o = \int_o^{2\pi} \left(q_{\parallel} - t_{\perp}(p) - t_{\perp}(p - Q_{\perp}b)\right) \frac{dp}{b} = \frac{2\pi q_{\parallel}}{b}$$

so that $H_f = \dfrac{A_o \hbar}{2\pi e}$ III.9

The best nesting vector varies between its zero field value and its value if t_b' was zero. Qualitatively the effect of the field is to suppress

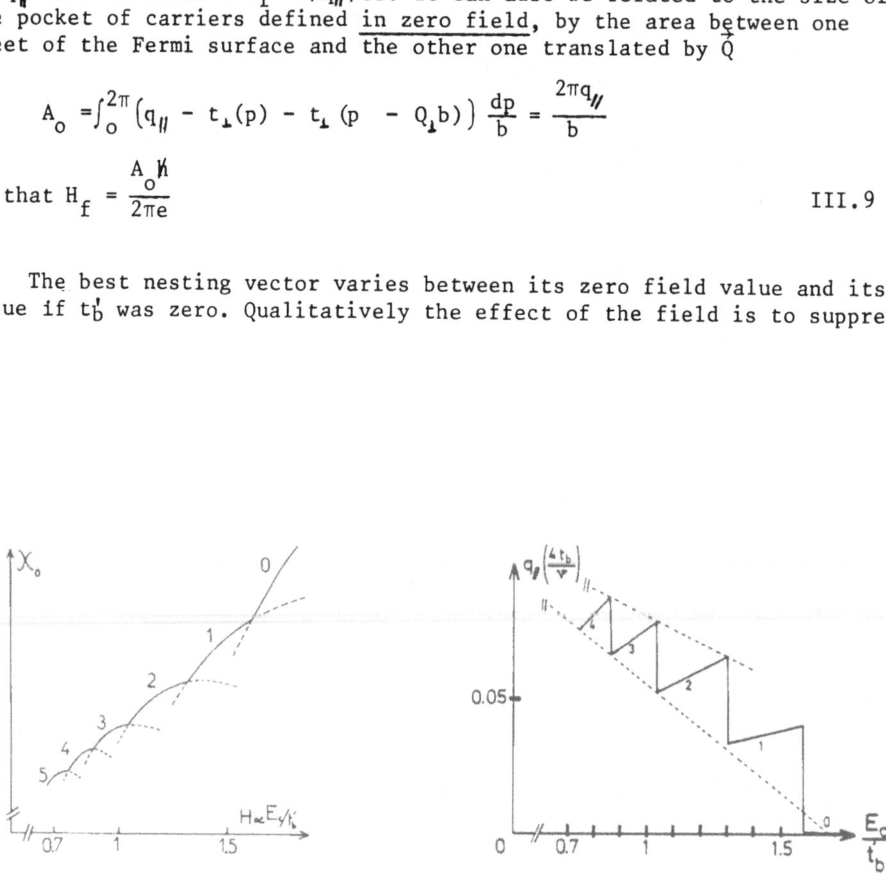

a b

Fig. 7 : a) Absolute maximum of $\chi_o(Q, H, T)$ as a function of the field H (scale in units of E_c/t_b'). It is a succession of segments characterized by successive values of the quantum number n_o. At the transition fields between different values of n_o, the wave vector jumps discontinuously.
b) Evolution with the field (E_c=eHvb) of the longitudinal component of the best nesting vector, deduced from the model. $q_{\parallel}$ varies linearly with H in each subphase. As soon as $n_o > 2$, critical fields vary roughly as $H_o = H_f/(n_o + \gamma)$ so that extreme values of $q_{\parallel}$ align along two straight lines (dotted lines). $t_b/t_b' = 10$.

the effect of deviations from perfect nesting. This is done for a field of order $H \gtrsim t'_b/evb$.

There is in the problem a characteristic area which is the area included between one sheet of the Fermi surface and the other sheet translated by Q. When χ_o is maximum, this area is given by :

$$A(H) = \frac{2\pi}{b} q_{/\!/}(H) = \frac{neH}{h} 4\pi^2 \qquad \text{III.10}$$

It contains an integer number of a given degeneracy which is nothing but the degeneracy of a Landau level. This remark will have all its signification in the comprehension of the ordered phase[8]. Now, when the field increases this area increases linearly at the expense of the nesting properties (fig. 8). There is a critical field above which nesting is so bad that it is better to decrease the quantum number by one unit to get back to a better nesting condition. Whence the jump of the wave vector and the particular structure of the susceptibility

IV - INSTABILITY OF THE METALLIC PHASE. TRANSITION LINE. 3D COMPOUNDS

The structure of the transition line can be easily deduced from [6,7]

$$1 - \max \ [\lambda(\vec{Q})\ \chi_o(\vec{Q}, H, T] = 0 \qquad \text{IV.1}$$

Here, we assume $\lambda(\vec{Q})=\lambda$. One obtains the transition line between the metallic phase and a series of subphases which are labelled by successive values of the quantum number n_o (see fig. 1 of the following paper[8]). These phases are described with a decreasing number N of carriers. Along the transition line $N(H)=neH/\hbar$ but the frequency of the phase transitions, which is given by H_f, is related to the zero field nesting properties. The subphase are separated by first order transitions due to the discontinuity of the wave vector.

Consider now the case where there is a dispersion in the third direction c, along the magnetic field. The linearized dispersion relation writes :

$$\varepsilon(\vec{k}) = \mu + v(|k_x|-k_F)-2t_b\cos k_y b-2t'_b\cos 2k_y b-2t_c\cos k_z c-2t'_c\cos 2k_z c \qquad \text{IV.2}$$

Consider first the case where t'_c is 0. There is a perfect nesting of the Fermi surface along the c direction with wave vector $q_z=\pi/c$. As a result, the susceptibility $\chi_o(q_{/\!/}, q_\perp, \pi/c, H, T)$ is exactly the same as the one $\chi_o(q_{/\!/}, q_\perp, H, T)$ of the 2D system. The instability properties of the electron gas with perfect nesting along c is identical to these of the 2D gas. Now if t'_c is non zero,

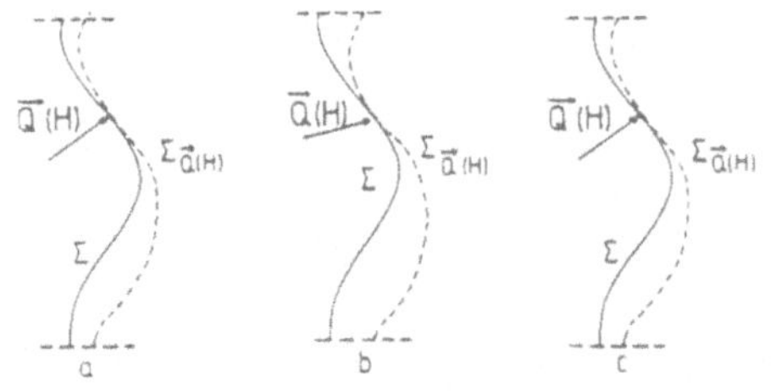

Fig. 8 : Mechanism of quantized nesting (semi-classical). The area A included between on sheet of the Fermi surface Σ and the other one translated by $\vec{Q}$, $\Sigma_{\vec{Q}}$, is quantized and increases linearly with the field as $A = 2\pi$ neH/h and thus the nesting vector varies with the field (a→b). When nesting becomes too bad, n decreases by one unit, to bet back to a better nesting situation (b→c).

maxima of susceptibility are cut off at $\sim \ln \max(T,t'_c)$[6]. As a result there is now a threshold field H_t for the appearance of the SDW cascade of subphases[3,6]. H_t is defined as $T_c(H_t) = t'_c$.

V - CONCLUSION

We have given a complete study of the instability of a 2D electron gas with open Fermi surface under magnetic field. We have shown that the field induces a set of logarithmic divergences of the susceptibility for a discrete series of wave vectors. This induces a cascade of SDW subphases, the original structure of which is studied in the following paper[8] and is strongly reminiscent of the spectrum of the metallic phase shown on fig. 1b. We have used a very simple model in which the wave vector of the instability always corresponds to the same sign of the quantum number N. However some interesting features could be model dependent. First, a refined dispersion relation can lead to alternance of positive and negative values of N[6,17]. Secondly, we have assumed a non $\vec{Q}$ dependent interaction parameter $\lambda(\vec{Q})=\lambda$. An appropriate slow $\lambda(\vec{Q})$ variation, combined or not with the first effect, should lead to a non trivial succession of N values and could also favor change in sign of the quantum number. This could explain the change in sign observed for the Hall effect in Bechgaard salts[2]. Another possible explanation is investigated in the following paper[8]. A strong $\lambda(\vec{Q})$ variation could also favor a totally different nesting vector, as the longitudinal one proposed by some authors[3,13]. But we think that the main features of the observed cascade of subphases, the frequency of transitions and the threshold field, are well accounted for by this best nesting vector (at least for $H \lesssim 8T$).

If some specific aspects remain to be understood and could be model dependant, the essential features of this physical problem are very general[7] and based on the essential idea of a competition between <u>nesting</u> properties of the Fermi surface and <u>quantization</u> of electronic motion under magnetic field. This quantization also alters the superconducting response function[17]. We expect such a cascade of phase transitions not to be specific of Bechgaard salts and should be observed provided simple conditions are fulfilled between the characteristic energies of the problem E_c, t'_b, $\hbar/\tau$, t'_c. It is not specific of SDW ordering and should be also observed in CDW system (with a slight modification due to the Zeeman term).

P. Lederer, D. Poilblanc and M. Héritier are gratefully aknowledged for a critical reading of the manuscript.

REFERENCES

1. K. von Klitzing, G. Dorda and M. Pepper, Phys.Rev.Lett. 45, 494 (1980)
2. P.M. Chaikin, this conference, M. Ribault, this conference, and ref.
3. L.P. Gor'kov and A.G. Lebed, J. Physique Lett. 45, L 433 (1984)
4. M. Héritier, G. Montambaux and P. Lederer, J. Physique Lett. 45, L943 (1984)
5. G. Montambaux, M. Héritier & P. Lederer, Phys. Rev. Lett. 55, 2078 (1985)
6. G. Montambaux, Thesis, Université Paris-Sud, Orsay (1985)
7. G. Montambaux, M. Héritier & P. Lederer, J. Phys. C 19, L293 (1986)
8. M. Héritier, this conference ; D. Poilblanc, M. Héritier, G. Montambaux & P. Lederer, J. Phys. C 19, L 321 (1986) and to appear
9. D. Jérome and H. J. Schulz, Adv. Phys. 31, 299 (1982) and refs. therein
10. P.M. Grant, J. Physique Colloq. 44, C3 847 (1983)
11. K. Yamaji, J. Phys. Soc. Japan, 51, 2787 (1982)
12. J.M. Delrieu, M. Roger, Z. Toffano, A. Moradpour & K. Bechgaard, J. Physique, 47, 838 (1986)
13. P.M. Chaikin, Phys. Rev. B 31, 4770 (1985)
14. K. Yamaji, J. Phys. Soc. Japan, 55, 1424 (1986)
15. L. Chen and K. Maki, preprint (1986)
16. K Yamaji, Syn. Met. 13, 29 (1986)
17. G. Montambaux, to be published

FIELD-INDUCED QUANTIZED MAGNETIC ORDERING IN QUASI-ONE-DIMENSIONAL CONDUCTORS

Michel Héritier

Laboratoire de Physique des Solides (associé au CNRS)
Université Paris-Sud
91405 Orsay, France

I - INTRODUCTION

The purpose of this paper is to give a theoretical interpretation of the novel properties of Bechgaard salts in strong magnetic field[1]. These organic compounds are considered as the first examples of a whole class of materials exhibiting the same magnetic phenomena. The only conditions to belong to this class are, interactions favouring electron-hole pairing and a strong anisotropy of the electron band structure, more precisely the hierarchy in the electron transfer integrals along the three crystal axes : $t_a >> t_b >> t_c$.

The main features that a theory should account for are :

1) The SDW instability of the metal in a moderate field of a few Teslas[2]

2) The structure of the metal-SDW transition line[3,4].

3) The phase diagram below the transition line, the cascade of first order transitions, the thermodynamics of the ordered phase[5,6], the succession of steps and plateaux in the Hall effect[5,7]

4) The new features[1] observed in very well ordered samples of $(TMTSF)_2ClO_4$ [8] and in $(TMTSF)_2PF_6$ [9] under pressure, in which new phase transitions as well as a complex behaviour of the Hall effect appear.

A susceptibility analysis[3] was sufficient to account for the first two points, as discussed in the preceding paper[10], but a detailed description of the ordered phase is necessary to understand the last two ones, which we are discussing in the following.

These metals are strongly anisotropic and can be considered as one-dimensional above a cross over temperature T_{co}. However, the effects discussed here occur at much lower temperature. The Fermi liquid approach which we shall use is therefore valid.

II - THE INTEGER QUANTIZED SDW PHASES

a) The Gor'kov equations

We consider a simple model, with an orthorombic crystal structure, and an electron dispersion relation linearized in the longitudinal a-direction :

$$\varepsilon(\vec{k}) = v(|k_x| - k_F) + \varepsilon_\perp(\vec{k})$$

$$\begin{aligned} \varepsilon_\perp(k) = &- 2t_b \cos \vec{k}\,\vec{b} - 2t_c \cos \vec{k}.\vec{c} \\ &- 2t'_b \cos 2\vec{k}.\vec{b} - 2t'_c \cos 2\vec{k}.\vec{c} \end{aligned} \tag{1}$$

t'_b and t'_c describe the deviations from perfect nesting of the Fermi surface. First, we neglect $t'_c \ll t'_b$, so the third direction does not play any role.

We define Matsubara Green's functions :

$$G_\uparrow(x,x') = - \langle T_\tau (\psi_{1\uparrow}(x,\tau)\, \psi^+_{1\uparrow}(x',\tau')\rangle$$
$$F_\uparrow(x,x') = - \langle T_\tau (\psi_{2\downarrow}(x,\tau)\, \psi^+_{1\uparrow}(x',\tau')\, e^{i\vec{Q}.\vec{r}}\rangle$$

where $\psi_1(k_\perp, x)$ and $\psi_2(k_\perp, x)$ are the field operators describing an electron on the right and on the left sheets of the Fermi surface, written in the mixed representation (in the direct space in the longitudinal direction, in the reciprocal space in the transverse direction) $\vec{Q}$ is the magnetic superstructure wave vector. The self-consistent SDW potential couples the mean field equations of motion for G and F, which read :

$$\begin{aligned} &\left(i\mu_n - \mu_B H + iv\frac{\partial}{\partial x} + vk_F - \varepsilon_\perp(p - \frac{x}{x_o})\right) G + \Delta F = \delta(x-x') \\ &\left(i\omega_n + \mu_B H - iv\frac{\partial}{\partial x} - v.k_F - vq_{/\!/} - \varepsilon_\perp(p - Q\,b - \frac{x}{x})\right) F + \Delta\, G = 0 \end{aligned} \tag{2}$$

We have set $\vec{k}.\vec{b} = p$, $eHb = 1/x_o$, $\vec{Q} = (2k_F + q_{/\!/}, Q)$.
The field H is parallel to z. The ω_n are the Matsubara frequency. The order parameter Δ must be solution of the self-consistency condition :

$$\Delta^* = \lambda\, T \sum_{\omega_n} \langle F(x,x)\rangle_p \tag{3}$$

where λ is the molecular field constant, T the temperature and $\langle ...\rangle$ means the average over p. It is possible, by a transformation on the phases of G and G and F, to define new Green's functions g and f, which obey simpler equations :

$$\begin{aligned} &(i\omega_n + iv\frac{\partial}{\partial x})g + \tilde{\Delta}(x) f = \delta(x-x') \\ &(i\omega_n - iv\frac{\partial}{\partial x} - vq_{/\!/}) f + \tilde{\Delta}^*(x)\; g = 0 \end{aligned} \tag{4}$$

where the complexity of the original Green's function Fourier spectrum has been incorporated in the effective potential $\tilde{\Delta}(x)$, defined by its Fourier series

$$\tilde{\Delta} = \Delta \sum_n a_n(p)\, I_n(z, z') \exp - in \frac{x}{x_o} \tag{5}$$

where $I_n(z, z')$ is defined by equation III-8 of the preceding paper[6,10] and $a_n(p) = \exp(-iz \sin p - i\, z' \sin 2p + inp)$. The effective potential $\tilde{\Delta}$ is now periodic, with a wavelength $2\pi x_o$ imposed by the field. The transverse degrees of freedom appear only as phase factors of the Fourier components : perfect nesting alone c and orbital effect of the field have made the problem one-dimensional.

b) The quasiparticle spectrum

We obtain the quasiparticle spectrum by diagonalizing the following matrix

$$\begin{vmatrix} iv\dfrac{d}{dx} & \tilde{\Delta}(x) \\ \tilde{\Delta}^*(x) & - iv\dfrac{d}{dx} - v_{q_{/\!/}} \end{vmatrix} \tag{6}$$

Below the critical line, Δ is non zero and the orbits are closed. The effective potential couples not only states of wave vector $\vec{k}$ and $\vec{k} + \vec{Q}$, but also $\vec{k}$ and $\vec{k} + Q + \frac{n}{x_o} \frac{\vec{a}}{|a|}$. This is reflected in the quasiparticle spectrum, in which a series of gaps of width $|\Delta I n|$ are opened (fig. 1). The closed orbits imply a Landau quantization of the orbital motion. These gaps should be considered as an effect of the SDW ordering and of orbital quantization.

c) The thermodynamic properties

We make an hypothesis of weak coupling limit $\lambda N(E_F) \ll 1$, in which the gaps are small enough to be considered as independent.

We consider quantized values of $q = N/x_o$, because this is the condition for the opening of the gap ΔI_N at the Fermi level. As in the mechanism for the Peierls instability in one-dimensional conductors, we have a one dimensional spectrum and the Fermi level is in the middle of one of the gaps opened in this spectrum. The self-consistency condition (3) can be written in the form :

$$1/(\lambda N(E_F) I_N^2) = f(\Delta) = f_o(\Delta) + f_1(\Delta) \quad (7)$$

in which f_o is the contribution of the main gap ΔI_N, and f_1 is the correction due to the presence of the other gaps. A diagrammatic expansion of the gap function $f(\Delta)$ can be obtained. The second order diagram of f_o in Δ^2 diverges logarithmically as $T \to 0$ for quantized values of q , because of the opening of a gap at the Fermi level, which gives a finite critical temperature T_{CN}, whatever small λ may be :

$$1/(\lambda N(E_F) I_N^2) = \text{Log} \frac{\pi T_{CN}}{2E_o \gamma} + \sum_{n \neq 0} \frac{I_{N-n}^2}{I_N^2} \text{Log} \frac{2x_o E_o}{|n| v} \quad (8)$$

which is the BCS expression, corrected by the last term, due to the auxiliary gaps. E_o is a cut off energy. To fourth order, we obtain the free energy F_N of the subphase N, compared to that of the metallic phase F_o, near T_c :

$$F_N - F_o = - \frac{4\pi^2}{7\zeta(3)} \times \frac{N(E_F)}{1+\varepsilon_N} (T_c - T)^2 \quad (9)$$

This is the BCS expression corrected by the factor $(1 + \varepsilon_N)^{-1}$, where ε_N, due to the auxiliary gaps, has a complicated expression in terms of the $I_{n's}$, oscillates with the field and is in general not small.

We expect, in increasing field a first order transition from the

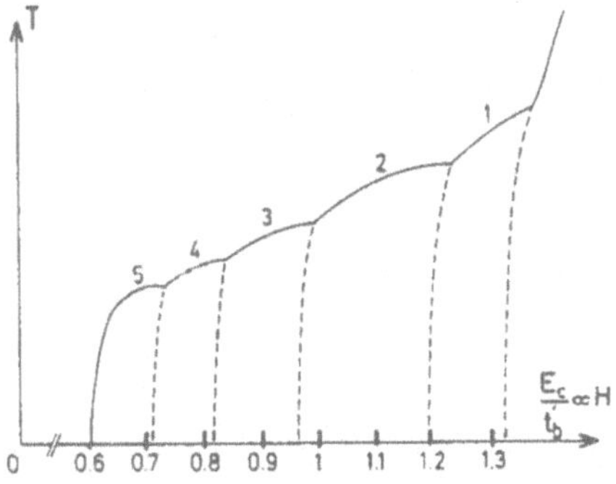

Figure 1 : Phase diagram with integer SDW sub-phases

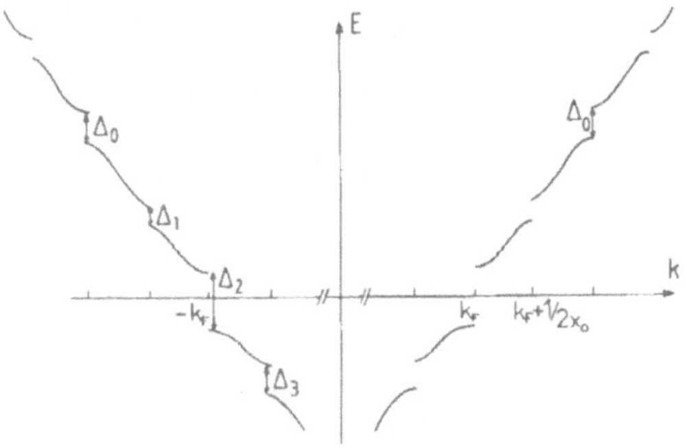

Figure 2 : Quasiparticle spectrum in the phase N=2

phase N to the phase N-1. In the BCS approximation, $\varepsilon_N = 0$, which corresponds to a transition line independent of temperature in the T,H) plane and to a vanishing latent heat. The finite value of ε_N makes the slope of the transition line finite. The frequency of the cascade of first order transitions is given by the field dependence of the I_N's. The reciprocal of the transition fields are roughly periodic, with a characteristic frequency proportional to t_b'.

We have considered quantized values of q because they make diverge the non interacting susceptibility (or the 2nd order term of f_o) as $T \to 0$. However, at finite T, f_o remains finite. Therefore, we must take into account the correction f_1, which makes q slightly departs from N/x_o by a quantity :

$$\delta q_{/\!/} = \frac{8}{7\zeta(3)} \sum_{n\neq 0} \frac{1}{n} \frac{I_{N+n}^2}{I_N^2} \left(\frac{\pi T x_o}{v_F} \right)^2 \tag{10}$$

This correction goes to zero as T 0. A wave vector dependent exchange interaction can also cause a deviation from exact quantization, but still much smaller at $T \sim T_{CN}$. At $T \to 0$, we can prove that the wave vector is such that the Fermi level stays inside the Landau gap, because of the logarithmic slope singularity in the energy when the Fermi level merges into the Landau band.

In conclusion, it is possible to label each SDW phase by a quantum number N. Because of the wave vector quantization, the Fermi level lies in a Landau gap. At zero temperature, the correct description of these quantized phases are in terms of Landau bands which are either completely filled or completely empty. These bands result from the broadening of Landau levels due to quantum tunnelling between localized Landau orbits. The quantum number N is the number of Landau bands between the Fermi level and the gap I_o, i. e. the number of completely filled electron or hole Landau bands in the semi-classical "carrier pocket".

d) Quantized Hall effect

The energy scale t_c which characterizes the electron dispersion in the direction along which the field is applied has vanished from the problem. We are left with an effective two-dimensional electronic structure. We assume the presence of impurities to pin the density wave. Then the electrons are submitted to a self-consistent periodic potential generated by the SDW. In each SDW phase, the Fermi level lies in a gap of extended states. Since perfect nesting in the third dimension makes the problem effectively strictly two-dimensional, the gauge arguments given by Laughlin[11] apply. Therefore, the Hall conductivity per atomic layer is exactly quantized, at zero temperature, according to the values $\sigma_{xy} = n\, e^2/h$. It can be proved rigorously that n = N in the phase defined by the quantum number N, by using the Streda expression[12] of the quantum part of the Hall current (which is the only non vanishing part when the Fermi level lies in a gap) properly modified to account for the field dependence of the self consistent potential generated by the SDW[7].

In strong field, in the phase N = 0, the Hall conductivity vanishes. The integrated density of states below the gap ΔI_o gives a zero contribution to the Hall current. This result remains true in any field : although the number of these electron states depends on the field through q (H) in the physical system, their total contribution to σ_{xy} is zero. Only the states between the gap ΔI_o and ΔI_N participate to the Hall current. At fixed chemical potential, the total number of electrons remains constant because of two opposite effects : (i) the increase of the cyclotron energy (i. e. the energy difference between the center of two successive Landau bands) ;

(ii) the increase of the wave vector Q (H). These effects cancell exactly because of the quantization condition on Q (H). However, although the number of electrons remains constant, σ_{xy} which is related only to the first effect does not vanish. This result becomes clear when one thinks of the semi-classical picture : only the small quantized "pockets" of carriers (electrons or holes) determined by the position of the Fermi level (above or under respectively) with respect to the gap ΔI_o, contribute to the Hall conductivity. The states below ΔI_o can be considered as playing the role of an effective "reservoir" to fill in completely the N Landau bands[3].

Therefore, the Hall conductivity is independent of the field and strictly quantized at T = 0 in each phase N and undergoes discontinuities in increasing field at the first order transitions $N \to N - 1$. The behaviour is similar to the usual quantized Hall effect, but the steps here are related to first order transitions.

However, in real systems, deviation from perfect nesting along c, t_c', which has been neglected avove, introduces a cross-over to a three-dimensional behaviour when $T \sim t_c'$. As the temperature is decreased below this cross-over, the longitudinal conductivity σ_{xx}, as well as the slopes of the Hall effect plateaux stop decreasing and never vanish : the quantization of the Hall effect is only approximate. The metallic phase is very far from the conditions of observation of the quantized Hall effect, but SDW ordering allows to fulfill approximately these conditions : (i) the 3D cross-over $t_c' \sim 0.01$ K is orders of magnitude smaller than in the metal ($t_c \sim 10$ K) because of nesting properties ; (ii) the number of carriers in the ordered phase per atomic layer $\sim 10^{11}$ cm^{-2} is also much smaller because of nesting (iii) the Fermi level is pinned in a gap of extended states because of the magnetic ordering.

This theory seems to account for the main features of the experimental phase diagram of $(TMTSF)_2ClO_4$ as determined by Hall effect, magnetization and specific heat experiments, and for the positive Hall effect plateaux observed in the same salt.

III - RATIONAL QUANTIZATION OF NESTING

However, some improvements of the theory are now clearly needed to interpret new experimental data. Hall effect and magnetoresistance studies in the PF_6 salts[9] have revealed a much more complex phase diagram, exhibiting a number of unexpected transition lines, almost independent of temperature and closely spaced in the (T, H) plane. The phase diagram of well relaxed $(TMTSF)_2ClO_4$ seems also to be richer than first suspected[8]. The negative Hall effect[8, 9] in these two salts is also unexpected. The interpretation of these new data requires to take into account new ingredients in the theoretical model.

In the ordered phase, two different periodicities appear in the longitudinal direction (in addition to the lattice periodicity a): one, $x_\sigma = 1/(eHb)$, is due to the orbital effect of the field on the phase of the electron wave function. The other is due to the period of the SDW self-consistent potential. In fact, because of nesting properties the relevant periodicity is not $2\pi/Q$, but rather $1\pi/q$. These two periodicities open complex families of gaps[13] in the quasiparticle spectrum, which have not been completely taken into account in the model using a linearized dispersion relation of the metal[2-6].

We now consider a model in which the metal dispersion relation preserves the simplification of the linearization near the Fermi level but includes the periodicity along the longitudinal direction. (The case of a half-filled band considered here could be extended to any rational filling of the band).

$$\varepsilon(\vec{k}) = \varepsilon\,(k_x) + \varepsilon\,(\vec{k}\,)$$

where $\varepsilon\,(\vec{k}\,)$ is still given by equation (1), but

$$\varepsilon\,(k_x) = v\left(\,|k_x - 4r\,k_F| \;-\; k_F\right) \qquad (11)$$

for $(4r - 2)k_F < k_x < (4r + 2)k_F$.

This dispersion relation introduces a new periodicity along x, namely $2\pi/(4k_F) \equiv a$ and therefore new commensurability effects, which we shall completely neglect, because $x_o \sim 2\pi/q_{/\!/} >> a$. The important effect of the periodic dispersion relation, which we want to discuss in the following is that the periodic potential $\Delta(x)$ can now open a gap near the Fermi level, not only to first order as in the linear dispersion model, but to any odd order of perturbation.

The infinite set of linear branches given in (11) is described by the field operators $\psi_i(x)$ and $\psi_i^+(x)$. We consider a self consistent potential of the form

$$V = \sum_i \Delta_2(x)\,\psi_{2i+1}(x)\,\psi_{2i}^+(x)\,e^{-i\vec{Q}_2.\vec{r}} + \Delta_1(x)\,\psi_{2i+2}(x)\,\psi_{2i+1}^+(x)\,e^{-i\vec{Q}_1.\vec{r}} + \text{h.c.} \qquad (12)$$

We define new Matsubara Green's functions

$$G_{2i+2p+1,2i} = -\,\langle T_\tau\,\psi_{2i+2p+1}(x)\,\psi_{2i}(x')\rangle e^{-i\left((p+1)\vec{Q}_2.\vec{r}+p\vec{Q}_1.\vec{r}\right)}$$

which obey equations of motion similar to (2), but in which $G_{2i+2p-1,2i}$ is coupled to $G_{2i+2p+1,2i}$ by Δ_2 and to $G_{2i+2p-1,2i}$ by Δ_1. The order parameters $\Delta_2(x)$ and $\Delta_1(x)$ are determined self-consistently by the condition :

$$\Delta_2(x) \;=\; \lambda\,T\sum_{\omega_n} \langle\, G_{2i+1,2i}(x,\,x,\,\omega_n,\,p)\,\rangle_p \qquad (13)$$

and a similar condition for $\Delta_1(x)$. As in the non-periodic model, a diagrammatic expansion of these self consistency conditions can be obtained[14]. New diagrams appear because now an electron can be scattered on the right neighbouring branch by Δ_2 or on the left one by Δ_1. A typical sixth-order diagram is shown in figure 3. Diagrams which diverge as $T \to 0$ can be obtained to $2m^{th}$ order for "rational" wave vectors of the form $Q_i = 2k_F + s/(mx_o)$ where s and m are integers. Such diagrams represent physical processes which open a gap at the Fermi level. Indeed, for such wave vectors the self consistent potential generated by SDW ordering and Landau quantization will open a gap at the Fermi level to m^{th} order in perturbation, provided m is an odd integer, since two opposite sheets of the Fermi surface must be coupled. For integer wave vectors, the quasiparticle spectrum is made of Landau bands separated by Landau gaps at wave vectors $n/2x_o$. For fractional wave vectors, the SDW potential splits each Landau band into m sub-bands. If the widths of these subgaps is large enough, they can stabilize new SDW sub-phases, corresponding to fractionaly quantized wave-vectors. This mechanism is the quantum version of a semi-classical mechanism proposed independently by Friedel[15] : by tunnelling between m semiclassical pockets, we obtain a large pocket, the quantization of which implies a quantization of mq , and therefore a fractional quantization of q .

Consider now the simplest case, namely $Q_1 = Q_2 = 1/(3x_o)$. For simplicity, we make the assumption that in the field range considered only $I_o(H)$, $I_1(H)$ and $I_{-1}(H)$ are non negligible and that $\varepsilon = I_1/I_o \simeq \varepsilon' = I_{-1}/I_o << 1$.

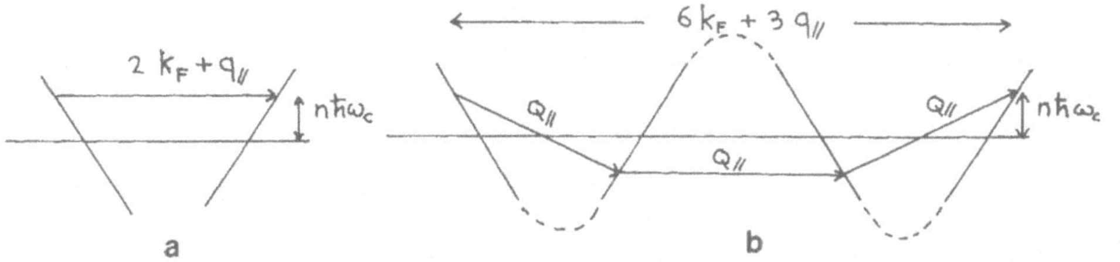

Figure 3 : a) linearized dispersion : SDW ordering opens a gap at the Fermi level if $vq/2 = n\hbar\omega_c$. b) periodic dispersion : to third order a gap is opened at the Fermi level if $3vq/2 = n\hbar\omega_c$.

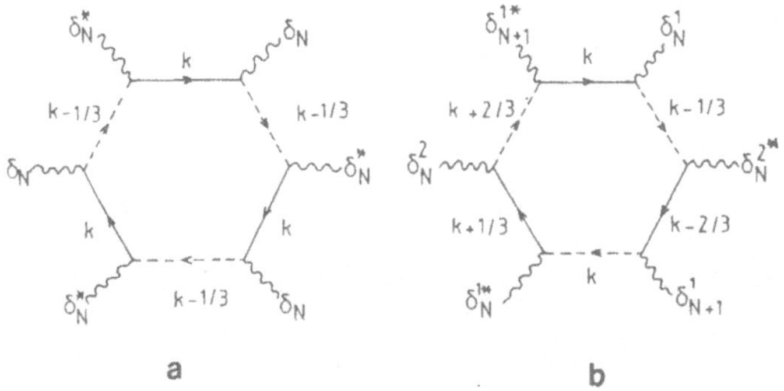

Figure 4 : typical sixth-order diagrams in the free energy. a) in the linearized dispersion model. b) in the periodic model. The latter corresponds to the physical process schematized in figure 3, which opens a gap at the Fermi level. This leads to a logarithmic divergence as $T \to 0$.

Then $\Delta_1 \simeq \Delta_2$, and the Landau expansion of the gap equation can be written[14] :

$$\frac{1}{\lambda N(E_F) I_o^2} = \text{Log}\frac{6E_o}{\omega_c} - \alpha\left(\frac{3\Delta I_o}{\omega_c}\right)^2 + \varepsilon^2\left(\frac{3\Delta I_o}{\omega_c}\right)^4 \text{Log}\frac{\omega_c}{3\pi T} + O\left(\left(\frac{3\Delta I_o}{\omega_c}\right)^6\right) \quad (14)$$

where $\omega_c = v_F/(2x_o)$

The first two terms are convergent as $T \to 0$ because the wave vector is not an integer multiple of $1/x_o$. However the third term, which is due to the diagram shown in figure 4b diverges, because of the opening of a gap at the Fermi level, as schematized in figure 3b). The result is a first order transition, at low temperature, to a large value of the order parameter. However, the transition temperature is lower than the critical temperature of the integer phase N = 0 : therefore, when the temperature is decreased, we have, first, a second order transition to the phase N = 0. However, it is possible that the "fractional phase", unfavoured at high T by the large value of Δ, becomes more stable at low T. To settle this point, the Landau expansion is not sufficient because of the large value of the order parameter. Rough interpolations of the gap function $f(\Delta)$ between the small Δ limit and the large Δ limit show that $f(\Delta)$ exhibits a maximum near $I_o \Delta \omega_c/3$, allowing the possibility of a fractional phase with negative energy for large enough coupling parameter $\lambda > \lambda_c$. Then, rough estimates of the energy difference between the phases $q = 0$ and $q = 1/(3x_o)$ seem to indicate that the fractional phase might be more stable for large enough value of $\varepsilon(H) = I_1/I_o$.

The same kind of arguments can be repeated qualitatively for any fractional wave vector of the form $q = s/(mx_o)$, where m is an odd integer. Furthermore, for such a wave vector, each Landau band is splitted in m subbands by the modulation of the order parameter. Now, we can imagine that a new modulation defined by $q x_o = s/(m + s'/m')$ can split again each subband into m' sub-sub-bands. The new gaps opened in the quasi-particle spectrum could, then, stabilize new sub-sub-phases. The phase diagram would exhibit a self-similar structure, with an infinite set of nested sub-phases. However, the higher order of the rational number defining a sub-phase, the sharper sould be the definition of the wave vector and the longer the electron mean free path. In Friedel's semiclassical picture, it is clear that, while an integer phase can be defined if $\omega_c\tau > 1$, where τ is the electron relaxation time, a fractional sub-phase s/m requires $\omega_c\tau > m$, a sub-sub-phase $\omega_c\tau > mm'$, and so on. It is plausible that, in real system, disorder effects wash out most of the rational phases and that only a few of them can exist in practice. A typical phase diagram based on these conjectures is shown in figure 5. The transition lines between fractional phases occur at almost saturated values of the order parameter and are therefore almost independent of temperature. In our simple model for the Fermi surface, the I_N's are larger for positive values of the quantum number. Therefore, the stable integer phases are expected to correspond to $N > 0$. In that case, the gap I_o opened by the SDW periodicity alone is above the Fermi level, which corresponds to a hole pocket and a positive Hall effect plateau.

In a fractional phase, the situation is less simple. Consider the case $q x_o = 1/3$. To third order, the SDW modulation alone opens a gap which is $\hbar\omega_c$ above the Fermi level and a gap which is $\hbar\omega_c$ below the Fermi level (and therefore a Landau gap is opened at the Fermi level)(see figure 6). The first gap yields a hole pocket while the second yields an electron one. At finite temperature, or in presence of disorder, there is a partial compensation

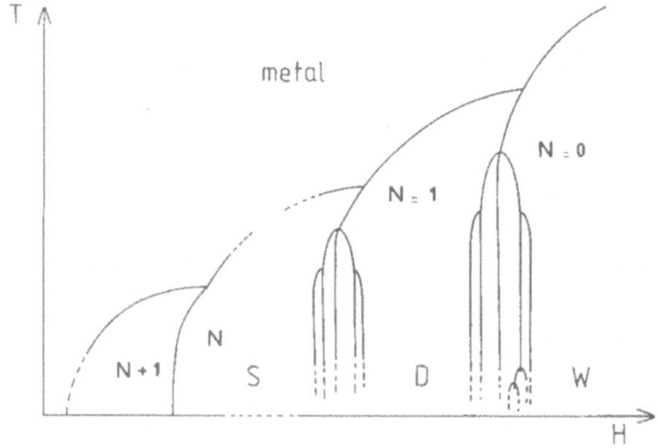

Figure 5 : Conjecture for the phase diagram : for large enough coupling parameter, fractional phases might appear in addition to the integer ones.

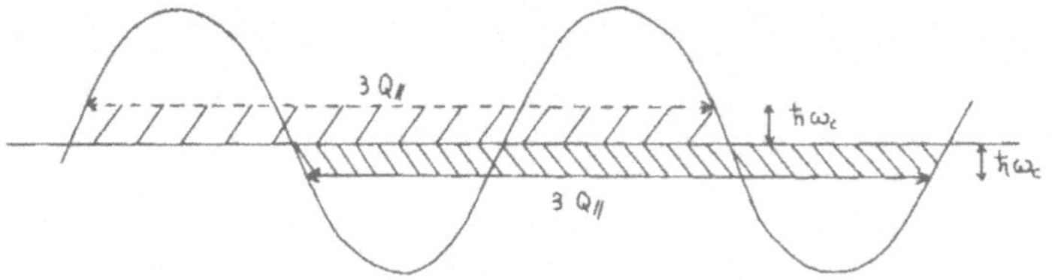

Figure 6 : For q x_o = 1/3, electron and hole pockets partially compensate at finite temperature.

between the two kinds of carriers. Since I(H) is in general larger than $I_{-1}(H)$, the gap below the Fermi level is larger than the gap above the Fermi level ; therefore the majority carriers are electrons and the Hall effect is negative. Since the ratio I_1/I_{-1} varies with the field, the carrier compensation also varies. For that reason we expect a field dependant Hall effect and not a plateau, in contradistinction with the integer phase case. Similar arguments lead to a field dependent negative Hall effect for $q_{/\!/} x_o$ = 1/3, 1/7,... and a positive one for $q_{/\!/} x_o$ = 1/5, 1/9,... Such an effect might explain the negative anomalies observed in well relaxed $(TMTSF)_2ClO_4$ and in $(TMTSF)_2PF_6$.

IV - CONCLUSION

It is now clear that the field-induced SDW phases in Bechgaard salts are the joint effect of nesting properties of the Fermi surface and orbital effect of the field. The latter induces a new periodicity along the chain direction $2\pi\ x_o$, much larger than a, the interatomic distance. SDW ordering yields another one, which because of nesting, reduces to $2\pi/q_{/\!/}$. These two wavelengths are comparable for fields of the order of a few Teslas. Quantized Nesting, which is responsible for the novel phase diagram observed experimentally, is related to the commensurability of these two periods : equivalently, the flux Φ of the magnetic field through the unit cell area $S(q_{/\!/}) = b \times 2\pi/q_{/\!/}$ defined by the two-dimensional periodic lattice experienced by an electron in the ordered phase (b in the transverse crystal periodicity) is commensurate with the flux quantum Φ_o :

$$\Phi_o/\Phi = \nu$$

In usual crystal lattices, this ratio is extremely large for available fields and no sizeable commensurability effect is expected. The fact that ν can be of the order of 1 for this self-consistent periodic potential is due to the very large value of $S(q_{/\!/})$ related to nesting properties. Because π/k_F is commensurate with a, rational and not only integer values of ν must be considered. It is tempting to apply these ideas to well-relaxed $(TMTSF)_2ClO_4$ and to $(TMTSF)_2PF_6$, where the centrosymmetric anions do not induce a disordered potential. Even in an integer phase, or even if disorder washes out the fractional phases, rational quantization have probably important effects on the fluctuation spectrum and therefore on various physical properties.

Nevertheless, the phase diagram and the positive Hall effect plateaux of moderately relaxed $(TMTSF)_2ClO_4$ seem well described by the stabilization of integer field-induced SDW phases. Fortunately, anion disorder preserves an apparent simplicity which allowed us to be conscious of quantized nesting effects. It is tempting to ascribe the new complexities observed in the phase diagram and the negative and field dependent anomalies of the Hall effect occurring in well relaxed samples to new fractional phases which become stable when the electron mean free path is large enough. In this picture, the PF_6 salt should be considered as very well ordered, even compared to the best relaxed perchlorate samples. This property allows to display the full complexity of the phase diagram, as for example, in the Hall effect data. Another explanation, (which does not exclude the first) would be that small Fermi surface deviations, very sensitive to disorder, are able to stabilize nearly defenerate subphases with negative N[10].

REFERENCES

1. For an experimental review, see P. M. Chaikin, this issue, and M. Ribault, ibid.
2. L. P. Gor'kov and A. G. Lebed, J. Physique Lett. 45, L 433 (1984) ; P. M. Chaikin, Phys. Rev. B 31, 4770 (1985)
3. M. Héritier, G. Montambaux and P. Lederer, J. Physique Lett. 45, L 943 (1984) ; G. Montambaux, M. Héritier and P. Lederer, Phys. Rev. Lett. 55, 2078 (1985)
4. K. Yamaji, Syn. Met. 13, 29 (1986)
5. D. Poilblanc, M. Héritier, G. Montambaux and P. Lederer, J. Phys. C 19, L 293 (1986)
6. A. Virosztek, L. Chen and K. Maki, Phys. Rev. B 34, 3371 (1986)
7. D. Poilblanc, G. Montambaux, M. Héritier and P. Lederer, to appear
8. M. Ribault, Mol. Cryst. Liq. Cryst. 119, 91 (1985) ; G. Faini, F. Pesty and P. Garoche, to be published
9. B. Piveteau, L. Brossard, F. Creuzet, D. Jérome, R. C. Lacoe, A. Moradpour and M. Ribault, J. Phys. C 19, 4483 (1986)
10. G. Montambaux, this issue
11. R. B. Laughlin, Phys. Rev. B 23, 5632 (1981)
12. P. Streda, J. Phys. C 15, L 717 (1982)
13. D. R. Hofstadler, Phys. Rev. B 14, 2239 (1976)
14. D. Poilblanc et al. to be published
15. J. Friedel (private communication).
16. P. M. Chaikin, this issue.

INFRARED PROPERTIES OF ORGANIC CONDUCTORS

C.S. Jacobsen

Physics Laboratory III
Technical University of Denmark
2800 Lyngby, Denmark

INTRODUCTION

The scope of these lectures is to give an introduction to the method of infrared spectroscopy applied to organic conductors. The interesting excitations, signatures of interactions and instabilities, and plasma oscillations, all occur in the infrared to near infrared range corresponding to photon energies up to about 1.5-2 eV. However, most of what is said is also valid for many of the inorganic, wider band low-dimensional conductors, although the interesting range here extends to higher frequencies.

The wide band unpolarized optical absorption of organic conductors typically looks like that shown in Fig. 1 for a series of TCNQ-salts of varying conductivity. There are always a number of high lying strong absorption bands corresponding to intramolecular excitations. The bands of main interest to us are those involving transfer of charge carriers between molecules. These are the lowest lying, primarily the A-band, which connect to low frequency, where the fascinating conductivity phenomena takes place.

In the following we shall first give a tutorial description of the basic concepts necessary to discuss these charge transfer bands. We proceed stepwise by first treating one- and two-dimensional conductors with non-interacting electrons, and next considering the effects of electron-electron and electron-phonon interactions. The effect of instabilities will be examplified by the Peierls-Fröhlich type. Other lectures deal more specifically with other types of instabilities as well.

After this theoretical part, we include a section on experimental techniques. The second part of the lectures deals with discussions of a number of illuminating examples of actual experimental data.

Finally we note that the spectroscopist often uses wavenumber (number of waves per cm in vacuum) as a frequency measure. 1,000 cm^{-1} ~ 124 meV ~ 1440 K ~ 1.89 x 10^{14} rad/sec.

For an isotropic medium the infrared (or optical) properties are determined by the complex, transverse dielectric function, $\tilde{\varepsilon}(\omega)$[1]. A plane electromagnetic wave propagates with a complex wave vector, $\tilde{q}$, given by

$$\tilde{q}^2 = (\omega^2/c^2)\tilde{\varepsilon}(\omega) \tag{1}$$

This equation, together with the appropriate boundary conditions on the fields, allows one to relate measurable optical quantities, like power reflectance, or absorption coefficient, to $\tilde{\varepsilon}(\omega)$. The reflectance at normal incidence, $R(\omega)$, is for example given by

$$R(\omega) = \left| \frac{\sqrt{\tilde{\varepsilon}} - 1}{\sqrt{\tilde{\varepsilon}} + 1} \right|^2. \tag{2}$$

$\tilde{\varepsilon}(\omega)$ has in general an imaginary part, $\varepsilon_2(\omega)$, from currents in phase with the electric field. For conductors $\varepsilon_2 \to \infty$ as $\omega \to 0^+$. Hence it is useful to introduce a frequency dependent conductivity, $\sigma(\omega)$, as:

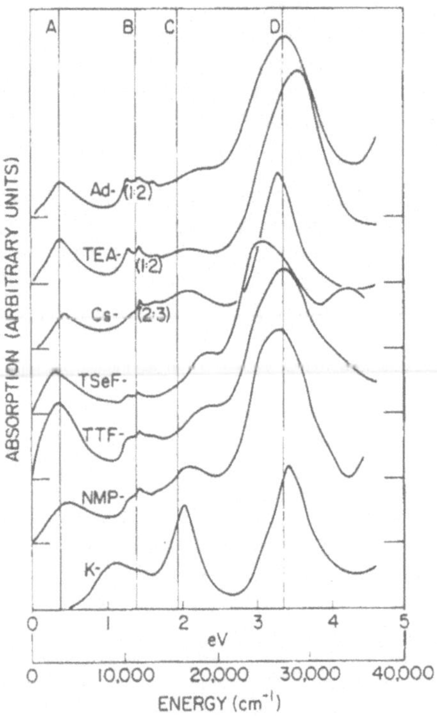

Fig. 1. Powder absorption spectra of a number of TCNQ-compounds, normalized per mole TCNQ. Reproduced after J.B. Torrance, B.A. Scott, and F.B. Kaufman, Solid State Commun. 17:1369 (1975). Copyright, Pergamon Journals Ltd.

$$\tilde{\varepsilon}(\omega) = \varepsilon_1(\omega) + i\varepsilon_2(\omega) = \varepsilon(\omega) + i\frac{\sigma(\omega)}{\varepsilon_o\omega} \tag{3}$$

ε_o is the vacuum permittivity and $\sigma(\omega\to 0)$ is the true dc conductivity.

Low-dimensional conductors are of course highly anisotropic, also optically. Thus ε and σ are tensorial. In general ε and σ will have different principal axes and these axes may rotate with frequency. However, in most of the experimental systems, at least one direction tends to be a common principal axis over wide frequency ranges. This is particularly true for the stacking axis of organic conductors in the infrared range. For the rest of the chapter we shall assume that we are dealing with such axes.

One-Electron Model

For conducting substances local field effects may be assumed negligible, and the first step is to apply the random phase approximation, where $\tilde{\varepsilon}$ may be written[1]

$$\tilde{\varepsilon}(\omega) = 1 + \Delta\tilde{\varepsilon}_{intraband} + \Delta\tilde{\varepsilon}_{interband} \tag{4}$$

with

$$\Delta\varepsilon_{intraband} \cong -\frac{e^2}{\varepsilon_o\omega^2}\left(\sum_{\bar{k}}\frac{1}{\hbar^2}\frac{\partial^2\varepsilon_{\bar{k}}}{\partial k_e^2}f(\varepsilon_{\bar{k}})\right) \tag{5}$$

and

$$\Delta\varepsilon_{interband} \cong \frac{2e^2}{\varepsilon_o m}\left(\sum_{\bar{k},n,n'}\frac{|P^{\bar{k}}_{n'n}|^2/\hbar\omega_{n'n}(\bar{k})}{\omega^2_{n'n}(\bar{k}) - \omega^2}f(\varepsilon_{\bar{k},n})\right). \tag{6}$$

Eq. (5) is only included if one or more bands are partially filled (metal). $\varepsilon_{\bar{k}}$ are the one-electron energies, and $\bar{k}$ is running over the first Brillouin zone. $f(\varepsilon_{\bar{k}})$ is the Fermi-Dirac occupation number. $\partial^2\varepsilon_{\bar{k}}/\partial k_e^2$ is to be calculated along the direction of the electric field. Eq. (5) may be written in the simple form

$$\Delta\varepsilon_{intraband} = -\frac{\omega_p^2}{\omega^2}, \quad \omega_p^2 = \frac{Ne^2}{\varepsilon_o}\left(\frac{1}{m^*_{opt}}\right), \tag{7}$$

where ω_p is the plasma frequency, N is the density of carriers in the band, and $(m^*_{opt})^{-1}$ is the inverse effective mass in the field direction averaged over occupied states according to Eq. (5).

In Eq. (6) n,n' are band indices, $\hbar\omega_{n'n}(\bar{k}) = (\varepsilon_{\bar{k},n'} - \varepsilon_{\bar{k},n})$ and $P^{\bar{k}}_{nn'}$ is a matrix element connecting Bloch states:

$$P^{\bar{k}}_{n'n} = \frac{1}{\Omega}\int_{\text{unit cell}} u^*_{\bar{k},n}(\bar{e}\cdot\bar{p})u_{\bar{k},n'}\,d^3r. \tag{8}$$

$\bar{e}\cdot\bar{p}$ is the momentum operator along the field and Ω is the unit cell volume. Note that momentum conservation requires a virtually unchanged $\bar{k}$-vector in the optical transitions (direct optical transition).

Assuming we have a metal where all $\omega_{n'n} \gg \omega_p$, Eq. (6) yields a nearly frequency independent contribution, ε_c, to ε for $\omega \lesssim \omega_p$. Then at infrared frequencies:

$$\tilde{\varepsilon}(\omega) = \varepsilon_c - \omega_p^2/\omega^2. \qquad (9)$$

So far dissipation has been neglected. Eq. (9) corresponds to $\sigma(\omega)$ being a δ-function at zero frequency, as can be seen by inserting in the general, causality implied Kramers-Kronig relations[1] connecting ε and σ

$$\varepsilon(\omega) = 1 + (2/\varepsilon_0\pi)P\int_0^\infty \frac{\sigma(\omega')}{\omega'^2-\omega^2}d\omega' \qquad (10)$$

$$\sigma(\omega) = \sigma_0 - (2\varepsilon_0\omega^2/\pi)\ P\int_0^\infty \frac{\varepsilon(\omega')-1}{\omega'^2-\omega^2}\ d\omega'. \qquad (11)$$

The simplest way to introduce dissipation is to relax the momentum conservation by introducing an elastic scattering mechanism. In that case[2] near Drude behaviour is found:

$$\tilde{\varepsilon}(\omega) = \varepsilon_c - \frac{\omega_p^2}{\omega(\omega+i\gamma)}\ . \qquad (12)$$

Here γ is a relaxation rate corresponding to finite dc-conductivity and finite IR absorption. The Drude expression is frequently used to describe the low frequency optical properties of metals. A numerical example is shown in Fig. 2. Note the characteristic plasma edge in reflectance

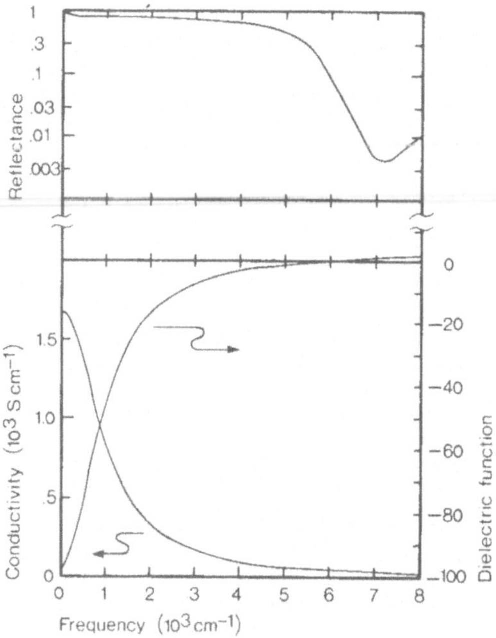

Fig. 2. Optical properties of the Drude mode. Parameters: ε_c=3, ω_p=10^4cm^{-1}, γ=10^3 cm^{-1}.

near $\omega_p/\sqrt{\varepsilon_c}$, where ε crosses zero while σ is small. As will be recalled this identifies the frequency of longitudinal, long wavelength oscillations of the conduction electrons, i.e. the plasmons. The plasmons are not usually excitable in normal incidence experiments, but lead to the edge structure in reflectance. Below it there is total reflection, above it electromagnetic waves can propagate.

Before applying the general theory to low-dimensional organic conductors, we add a few remarks on sum rules. At very high frequencies, $\omega \gg$ all $\omega_{nn'}$, the electrons can be considered free. Then from Eq. (7) it follows that

$$\varepsilon(\omega) \cong 1 - \frac{ne^2}{\varepsilon_0 m\omega^2} \tag{13}$$

where n is the total electron density and m is the free electron mass. By comparing Eq. (10) and (13) we further have

$$\int_0^\infty \sigma(\omega)d\omega = \frac{\pi}{2}\frac{ne^2}{m}, \tag{14}$$

proportional to the number of electrons in the system. This is basically the Thomas-Reiche-Kuhn sum rule for solids. If intraband transitions are well separated from interband transitions, it is meaningful to introduce a partial sum rule based on Eqs. (7) and (14):

$$\int_0^\omega \sigma(\omega')d\omega' = \frac{\pi}{2}\frac{n_{band}^{eff}(\omega)e^2}{m^*_{opt}}. \tag{15}$$

Here $n_{band}^{eff}(\omega)$ can be interpreted as the effective number of electrons participating in transitions up to frequency ω. We can see that the area below the σ(ω) curve in a natural way corresponds to the optical oscillator strength.

<u>The one-dimensional</u> single-stack conductor based on a single, non-degenerate orbital with near-neighbor overlap integral, t, is described by the tight-binding Hamiltonian

$$H_t = -\sum_{i,\sigma} t(c^+_{i,\sigma}c_{i+1,\sigma} + c^+_{i+1,\sigma}c_{i,\sigma}). \tag{16}$$

$c^+_{i,\sigma}$ creates an electron of spin projection σ on site i. In momentum space this may be written

$$H_t = -\sum_{k,\sigma}(2t\cos kd)c^+_{k,\sigma}c_{k,\sigma} \tag{17}$$

where d is the molecular repeat distance. With ρ electrons per site the band is filled to $\pm k_F = \pm\pi\rho/2d$ (assuming t>0). The low-frequency dielectric constant is within the theory described above (Eqs. (4,5)):

$$\tilde{\varepsilon}(\omega) = \varepsilon_c - \frac{\omega_p^2}{\omega(\omega+i\gamma)}, \quad \omega_p^2 = \frac{4td^2e^2\sin(\pi\rho/2)}{\pi\varepsilon_0\hbar^2 V_m}. \tag{18}$$

Here V_m is the crystal volume per molecule. For a double-stack conductor or a system with weakly dimerised stacks, t is to be considered an average value.

For the two-dimensional case we restrict ourselves to the orthorhombic lattice with one type of active molecule and two significant transfer integrals, t_1 and t_2. Thus the band structure is now

$$E(k_1,k_2) = -2t_1\cos(k_1d_1) - 2t_2\cos(k_2d_2) \tag{19}$$

With no loss of generality we may assume $t_1<t_2$, t_1 and $t_2>0$, and $\varepsilon_F<0$. Then the Fermi surface is closed if $-2t_1 -2t_2 \leqq \varepsilon_F \leqq -2t_1 + 2t_2$. It is open if $-2t_1 + 2t_2 < \varepsilon_F \leqq 0$. The procedure is now first to determine ε_F for the appropriate band filling and next calculate ω_{pi} from Eq. (7). This must in general be done numerically. A useful case is $\rho = 0.5$, where the normalized results are shown in Fig. 3. It is noteworthy that the optical anisotropy is approximately proportional to t_1/t_2 and not to $\sqrt{t_1/t_2}$ as might have been expected from Eq. (18).

Finally, a few remarks on the range of applicability of the above results are in order. From the way they are derived, it follows that they mainly describe the real part of $\tilde{\varepsilon}$ in the plasmon range. Absorption is only included through a phenomenological relaxation rate. There is no a priori reason for expecting Drude behavior throughout the infrared[4], and such behavior is indeed not usually found. But provided the absorption is centered well below the plasmon range and is insignificant in this range, the equations give useful relations between plasma edge position and band structure parameters.

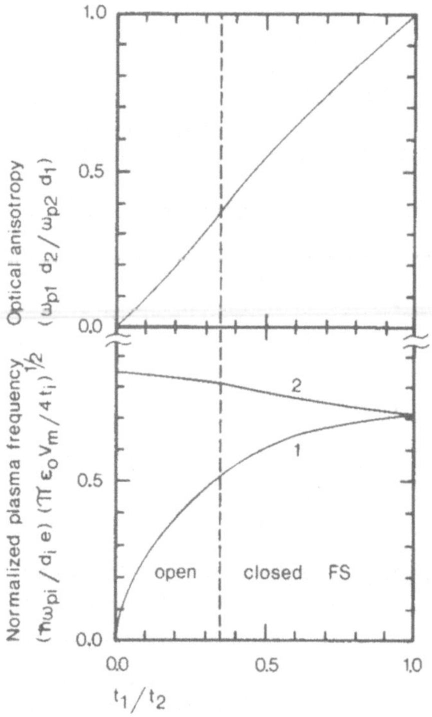

Fig. 3. Anisotropic plasma behavior in 2D orthorhombic and quarter-filled tight-binding model.

Interactions

The actual size and distribution of the oscillator strength (absorption) is not only determined by the transfer integrals, but also by the microscopic interactions in the crystal.

The short range electron-electron Coulomb interaction is after much debate now generally agreed to be an important factor in organic conductors. It is usually described by the extended Hubbard model, which adds to the Hamiltonian H_t the terms:

$$H_{int} = U \sum_i n_{i\uparrow} n_{i\downarrow} + V \sum_{i,\sigma,\sigma'} n_{i,\sigma} n_{i+1,\sigma'} . \qquad (20)$$

Here $n_{i\sigma} = c^+_{i\sigma} c_{i\sigma}$ is an occupation number for site i, so U is the extra energy cost of having two carriers on the same site, V correspondingly on neighbor sites. More distant interactions may be added, but are presumably not important.[4,5]

The localization tendency induced by finite Coulomb interaction acts to reduce the overall low frequency absorption. This can be seen from the sum rule[6],

$$\int_{intra} \sigma(\omega) d\omega = -(\pi e^2 d^2 / 2\varepsilon_o) \langle H_t \rangle , \qquad (21)$$

which includes all transitions generated from the original nondegenerate orbital with the Hamiltonian $H_t + H_{int}$. The brackets indicate the ground state expectation value.

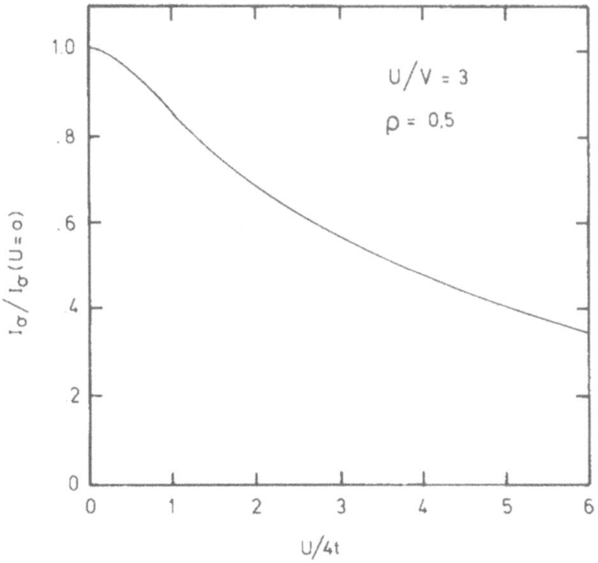

Fig. 4. Relative reduction in oscillator strength for a Hubbard tetramer with two electrons as function of U/4t with U=3V fixed.

In the absence of interactions the variational principle states that Eq. (21) has its maximum value (given by $(\varepsilon_o\pi/2)\omega_p^2$). Localization of the wavefunctions in the presence of correlations increases the kinetic energy (which is a negative number) and thus induces a reduction in oscillator strength. The magnitude of the reduction may be illustrated by a few examples. For a pair of molecules with two electrons the oscillator strength is proportional to[7] $(1 + (U/4t)^2)^{-\frac{1}{2}}$. For the infinite chain calculations are non-trivial[8], but it may be argued that studies on fairly small systems provide good guidelines[8,9]. In Fig. 4 we show the results for two electrons on four sites, asuming[5] V = U/3. For accepted values of U ≃ 1-1.5 eV the reduction for 4t = 0.5 eV is 30-40%.

Since the plasma edge at the zero-crossing of $\varepsilon(\omega)$ is associated with the existence of plasmons, i.e. the long wavelength density oscillations, the frequency of which in turn must be rather insensitive to the values of U and V, it seems that the zero-crossing itself will depend only weakly on the short range interactions. Considerations of the Kramers-Kronig relations then imply that some of the (reduced) oscillator strength must shift up in frequency. Finite chain calculations[6] indeed show that a number of correlation bands situated at for example V, U-V, U,... appear in addition to the Drude like band associated with the dc conductivity. Physically the correlation bands correspond to charge transfer processes creating doubly occupied sites, occupied neighbor sites, etc. Calculations also show[6,10] that most of the oscillator strength remains in the low frequency region.

One special case should be stressed: For $\rho = 1$ (one carrier per molecule) Mott insulator-like states are always found[11]. Then a gap of order U is formed and in realistic cases no sharp plasma edge is anticipated.

The electron-phonon coupling has several sources. The most important are usually considered to be (1) modulation of the transfer integral by external modes (involving translations or rotations of the molecules) and (2) modulation of the orbital energy by internal modes (molecular vibrations). The latter have frequencies spanning the infrared range (up to ~ 3,000cm^{-1}).

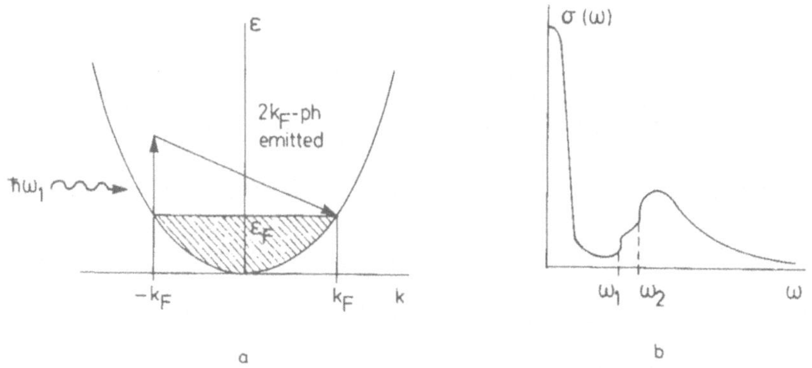

Fig. 5. Holstein absorption ($T \ll \theta_D$). (a) Schematic absorption process. (b) $\sigma(\omega)$ for rather strongly coupled metal.

For conductors with uniform chains the situation is much as in ordinary metals. The main deviations from the Drude spectrum occur at low temperature where the electron-phonon coupling does not contribute to the dc resistivity. Then $\sigma(\omega)$ consists of a narrow Drude-like contribution centered at dc plus an infrared part which turns on at the phonon frequencies. This part is due to the Holstein process[12,13] of scattering electrons by phonon emission and is sketched in Fig. 5. At high temperature both phonon absorption and emission contributes to the dc scattering rate, and a single Drude absorption is found with a relaxation rate

$$\gamma(T) = 2\pi\lambda k_B/\hbar\ (T + \theta_D/2) \tag{22}$$

where θ_D is of order the Debye temperature. This expression, given by Hopfield[14], assumes that k_BT is large compared to all important phonon frequencies, a condition not fulfilled for the vibrational modes in the molecular conductors.

When the molecular chains are not uniform, the electron-molecular vibration (emv) coupling plays a special role[15]. For the nondegenerate level only the fully symmetric (A_g) modes couple linearly to the electrons. The A_g-modes are not infrared active (but Raman active). However, when the chain is distorted, an external field can couple to the charge oscillations, which resonates strongly near the A_g-mode frequencies. This phenomenon often gives rise to spectacular effects in the infrared spectra.

Finally it is noted that both the Holstein model and the various emv coupling models conserve oscillator strength: The electron-phonon coupling primarily redistributes the absorption.

Instabilities

Other lectures at this school deal in more detail with the different low-dimensional instabilities. Here we shall just as an example describe the expected effect[16] of the $2k_F$-Peierls-Fröhlich instability on the infrared spectrum.

Disregarding the pure effects of interactions as discussed above, the quasi-onedimensional conductor should display a Drude-like spectrum at high temperatures. Below the three-dimensional ordering temperature of the Peierls transition, T_c, a true gap opens in the excitation spectrum, i.e. $\sigma(\omega)$ should show an onset at some frequency $\omega_g = 2\Delta/\hbar$, corresponding to the excitation of electron-hole pairs across the Peierls gap. Additionally, the electric field can couple to phase oscillations of the charge-density-wave (CDW), which is pinned to the lattice below T_c. This effect produces a low intensity, far infrared absorption band. A phenomenological model for $\tilde{\varepsilon}(\omega)$ may be written:

$$\tilde{\varepsilon}(\omega) = \varepsilon_c + \frac{\Omega_p^2}{\omega_F^2 - \omega^2 - i\omega\gamma_F} + \frac{\omega_p^2 - \Omega_p^2}{\omega_o^2 - \omega^2 - i\omega\gamma}\ , \tag{23}$$

where $\Omega_p^2 = (m^*_{opt}/M)\omega_p^2$ with[16]

$$\frac{M}{m^*_{opt}} = 1 + \frac{1}{\lambda}\left(\frac{2\Delta}{\hbar\omega_{2k_F}}\right)^2 . \tag{24}$$

Here ω_{2k_F} is the unperturbed phonon frequency at $q = 2k_F$ (one important phonon) and λ is a dimensionless electron-phonon coupling constant related to the bare coupling constant, g, by $\lambda = g^2N(0)/\hbar\omega_{2k_F}$, where N(0) is the metallic density-of-states at E_F. Since typically $\lambda \le 1$ and $\hbar\omega_{2k_F} << 2\Delta$, $\Omega_p^2 << \omega_p^2$. In essence the effective mass of the CDW is big because its motion involves displacements of the molecules, not only the electrons. Hence the oscillator strength of the pinned CDW (Fröhlich model) is small. However, since the pinning frequency, ω_F, may be quite small, it can give rise to a very high low-frequency dielectric constant.

At $T \gtrsim T_c$, fluctuation effects may be important, i.e. depinned CDWs with short range order may exist and contribute to the dc conductivity. Then the Fröhlich mode is expected to move to zero frequency ($\omega_F \to 0$) and the Peierls gap is somewhat smeared. The temperature dependence of $\sigma(\omega)$ is sketched in Fig. 6.

In the real molecular crystal, many phonon modes participate in the CDW formation. In that case, in addition to the Fröhlich mode, a number of infrared absorption lines appear near the unperturbed mode frequencies.[17,18] For the A_g-modes this may be viewed as a special case of infrared activation, now caused by the CDW broken symmetry. However, it has been stressed[17,18] that all modes contribute to the stability of the CDW, although its dynamical properties are mostly determined by the lowest lying phonons.

A simplified (T=0) expression for $\tilde{\varepsilon}(\omega)$ in the multiphonon model has been given by Rice[18]. The model calculation in Fig. 7 illustrates the importance of the emv coupling for typical parameter values. Notice that a mode with $\hbar\omega_{ph}>2\Delta$ produces an indentation in $\sigma(\omega)$ (Fano antiresonance).

EXPERIMENTAL METHODS

The highly opaque materials in question are usually investigated by means of near normal incidence, polarized reflectance measurements. If the power reflectance, $R(\omega)$, is known at all frequencies, then the phase shift on reflection, $\theta(\omega)$, can be calculated by a Kramers-Kronig transformation[1]:

$$\theta(\omega) = \frac{\omega}{\pi} P \int_0^\infty \frac{\ln R(\omega')}{\omega^2 - \omega'^2} d\omega' , \qquad (25)$$

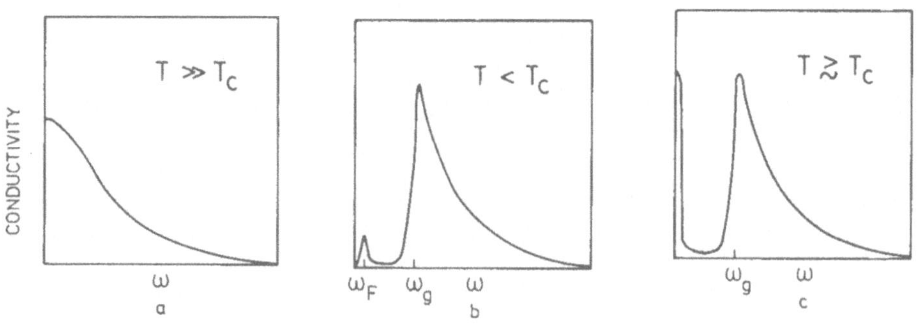

Fig. 6. Sketch of the frequency dependent conductivity of the Peierls-Fröhlich conductor. (a) For $T>>T_c$ (the three-dimensional transition temperature. (b) $T<<T_c$. (c) $T\gtrsim T_c$.

and then

$$\tilde{\varepsilon}(\omega) = \left(\frac{1 + \sqrt{R(\omega)}\ e^{i\theta(\omega)}}{1 - \sqrt{R(\omega)}\ e^{i\theta(\omega)}}\right)^2, \qquad (26)$$

In Eq. (25) suitable extrapolations for the ranges not covered must be adopted. This is particularly important when sum rule calculations are to be performed.

The low frequency extrapolation is not so critical provided the data covers nearly the entire range of interest. However, it is adjusted to make the result agree with independent information on low frequency properties (e.g. dc conductivity and microwave dielectric constant). A frequently adopted low frequency extrapolation for conductors is the Hagen-Rubens form $R = 1-A\sqrt{\omega}$, which follows the Drude spectrum for $\omega \to 0$.

At high frequencies the measured data should cover a range well beyond the frequencies of main interest. Even so the result of especially sum rule calculations will to some extent depend on the choice of high frequency extrapolation. Therefore the latter is chosen so that Eq. (25) agrees with independently measured values of $\theta(\omega)$ at a few frequencies near the plasmon frequency. $\theta(\omega)$ may, especially for quasi-one-dimensional systems, be measured by simple ellipsometric methods. A good method has been described by Young and Walker[19].

Another experimental difficulty consists in determining the absolute value of R with sufficient accuracy. Although a constant factor on $R(\omega)$ does not influence the phase calculation, Eq. (25), it clearly influences $\tilde{\varepsilon}(\omega)$. When reasonably sized, good optical quality crystals are available, there is no problem. For smaller crystals with imperfect faces, we have found that reproducible values can be obtained by first comparing the reflected intensity to that from a reference mirror, and next evaporating a gold film onto the crystal and repeating the measurement. For low frequencies (the far infrared), where R is often close to unity, it is in

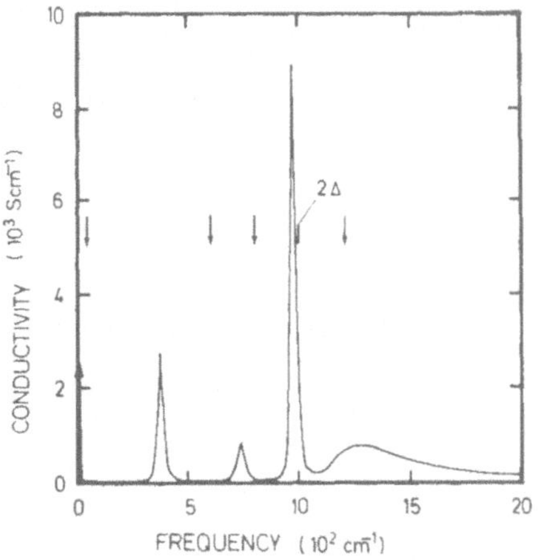

Fig. 7. Example illustrating the multiphonon-model. ω_p=6000 cm^{-1}, 2Δ=1000 cm^{-1}. Phonons at 40, 600, 800, and 1200 cm^{-1} with dimensionless coupling constants .1,.1,.05, and .1.

particular important to determine 1-R with sufficient accuracy. For low temperature measurements, Eldridge[20] has developed a bolometric technique to do this directly.

EXPERIMENTAL EXAMPLES

Charge transfer bands

The characteristic features in the evolution of charge transfer bands from molecular spectra are illustrated by the absorption spectra in Fig.8. The lower are solution spectra of TTF^+ monomers and dimers. It should be noted that the intramolecular transitions (C and D) shifts up in energy on dimerisation[21]. The new dimer transition, B, corresponds to the creation of one doubly charged and one neutral molecule and is thus a crude measure of U-V. The two top spectra are obtained by powder absorption techniques. TTF-$Cl_{1.0}$ contains quasi-isolated doubly charged dimers. The spectrum is indeed a broadened version of the solution dimer spectrum. Finally, TTF-$Br_{0.79}$ is a single stack organic conductor. The main features to note are (1) the appearance of a new intense band, A, at low frequency, and (2) a very low remaining intensity in the B-band. The A-band is obviously what corresponds to the high dc conductivity (Drude type absorption), while the B-band can still be interpreted as a correlation band involving the creation of doubly occupied sites. The low intensity may seem surprising,

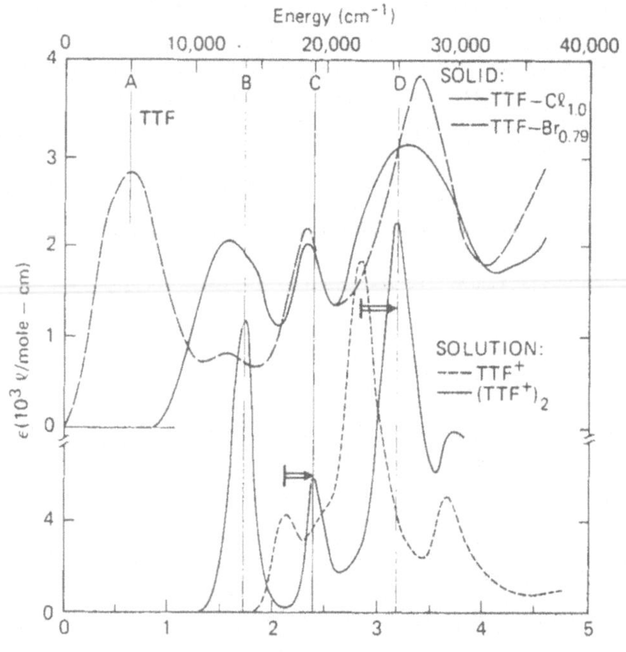

Fig. 8. Absorption spectra of TTF^+, $(TTF^+)_2$, TTF-Cl, and TTF-$Br_{.79}$. Reproduced after J.B. Torrance, B.A. Scott, B. Welber, F.B. Kaufman, and P.E. Seiden, Phys.Rev.B 19:730 (1979). Copyright, American Physical Society.

considering that $\rho \simeq 0.8$ (i.e. on average only every fifth site is neutral), but this is indeed predicted theoretically[10]. For lower band fillings the intensity should rapidly drop to negligible levels. An important structural property of TTF-$Br_{0.79}$ as well as of the TMTTF/$TMTSF_2X$ materials is that the polarization of the lowest molecular excitation in the monomer is approximately perpendicular to the stacking axis. In most TCNQ-salts, ET_2X-salts and many others this is not the case. Then the resulting mixing of the B and C bands leads to a redistribution of oscillator strength which is not completely understood, but experimentally a fair amount of strength is found at the B-position even for low band fillings (cfr. Fig. 1).

Plasma reflectance

In Fig. 9 we present an example of the polarized reflectance of a quasi-one-dimensional conductor (TMTSF-DMTCNQ). The reflectance spectra for the field perpendicular to the conducting stacks are remarkably flat with no important absorption bands, while the stacking axis spectrum shows a well-defined plasma edge around 7,000 cm^{-1} (0.87 eV) and an associated minimum of 0.6% (where $\varepsilon(\omega)$ passes the value 1) at 8,000 cm^{-1}. The main information contained in the position of the edge is the value of the plasmon frequency. With proper correction for the background polarizability, the bare plasma frequency may be obtained, and through that band structure parameters (transfer integrals) according to Eq. (18). The solid line in Fig. 9 represents a fit of the Drude model with one

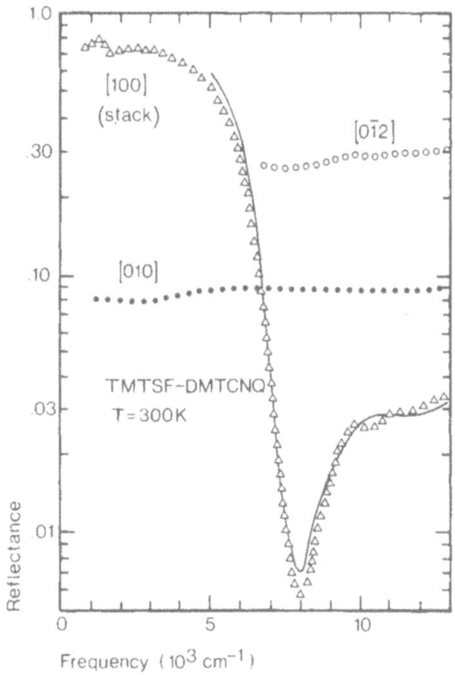

Fig. 9. Polarized reflectance of TMTSF-DMTCNQ at 300K. The solid line is a fit of the Drude model to the stacking axis spectrum with an extra oscillator added to account for the dispersion at 10,000 cm^{-1}.

extra oscillator to account for the B-band near 10,000 cm^{-1}. It is stressed that the fit is performed to the logarithm of reflectance. This is quite important for a reliable estimate of the background polarizability.

Although it is not necessary to understand the origin of the relaxation rate in this frequency range, it is noteworthy that the fit could be improved by allowing γ to slowly decrease with frequency. As the plasma edge in most of the organic conductors is situated near or even above frequencies corresponding to the one-electron bandwidth, this may be attributed to a decrease in the number of possible energy conserving phonon emission/absorption processes through the plasma range.

The Drude fitting procedure has been followed on numerous materials, and when carried out carefully, always leads to physically reasonable bandwidth estimates. The validity can be further supported by a number of examples. In Fig. 10(a+b) we show the temperature dependence of the plasma edge in [22] $TMTSF_2ClO_4$ and the pressure dependence of the edge in[23] TSF-TCNQ respectively. The blue shifts observed on lowering the temperature or increasing the pressure are reasonably consistent with expectations from typical data on thermal contraction[24], compressibility[25], and

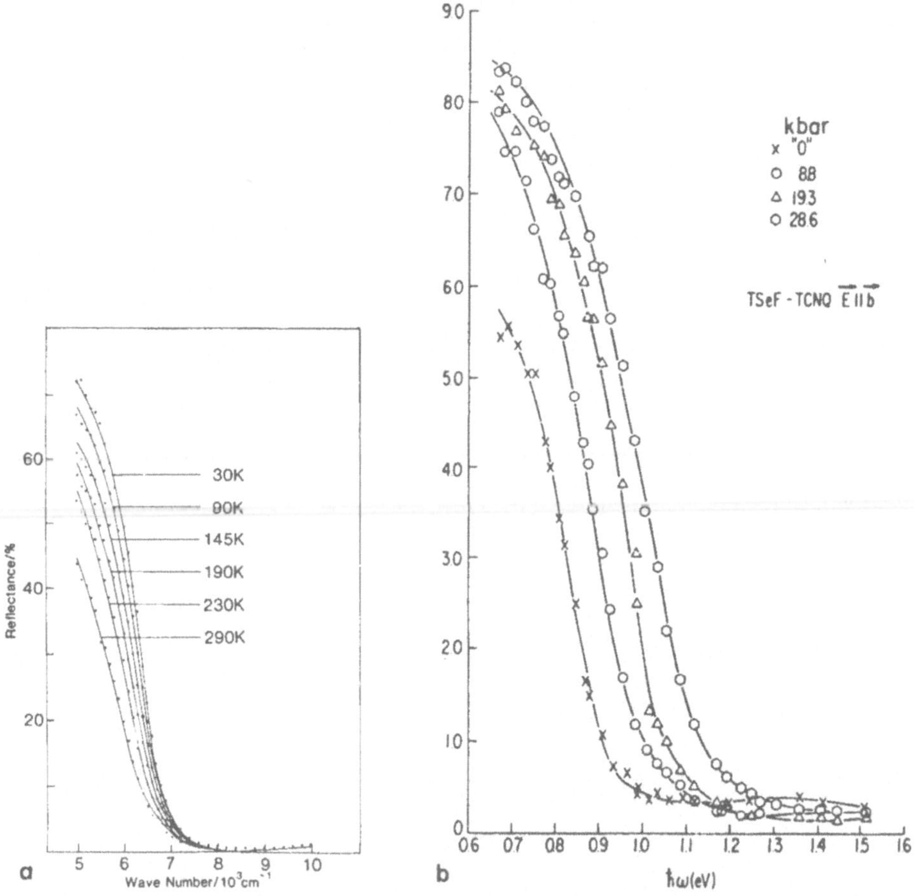

Fig. 10 (a) Temperature dependence of the plasma edge in $TMTSF_2ClO_4$. Reproduced after K. Kikuchi, Y. Ikemoto, K. Yakushi, H. Kuroda, and K. Kobayashi, Solid State Commun. 42:433 (1982). Copyright, Pergamon Journals, Ltd.
(b) Pressure dependence of the plasma edge in TSF-TCNQ. Reproduced after B. Welber, P.E. Seiden, and P.M. Grant, Phys.Rev.B 18:2692 (1978). Copyright, American Physical Society.

with molecular orbital calculations[26]. Both temperature and pressure dependence suggest that ω_p^2 (and hence t) increases about 5% for a decrease in the chain axis lattice constant of order 1%.

While it is reassuring that the Drude approach works well for these highly conducting materials, it is perhaps more important to check the consistency for materials which are moderately conducting or even semiconducting. We can for example compare the conductivity spectra of $TMTSF_2ClO_4$ and $TMTTF_2PF_6$ shown in Fig.11. $TMTSF_2ClO_4$ has a near Drude-like spectrum and is an excellent metal. $TMTTF_2PF_6$ has the oscillator strength centered at ~2,000 cm^{-1} (0.25 eV) and is indeed a semiconductor with $\sigma_{dc}(300K) \simeq 20\,\Omega^{-1}cm^{-1}$. Since there is no gap at the Fermi level in a one-electron band model for $TMTTF_2PF_6$ this behavior is usually attributed to strong electron-electron interactions. A Drude analysis of the plasma edge in these two materials[27] leads to estimates of t = 0.25 eV and 0.20 eV respectively. While these values may appear credible it is important to check their consistency against analysis on other materials. The isostructural TMTSF-DMTCNQ and TMTTF-DMTCNQ have overlap patterns and intermolecular distances close to those of the above mentioned compounds, but they are both metallic[28]. Assuming identical DMTCNQ bandwidths in the two materials analysis of the plasma edges[28,29] yields t(TMTSF) - t(TMTTF) = 0.055 eV in close agreement with the previous result. A more detailed discussion of the materials dependence further supports the validity of the Drude analysis approach[29].

An interesting direct demonstration that the plasmon frequency is not sensitive to non-Drude low frequency behavior comes from studies on the $(ET)_2X$ [BEDT-TTF_2X] materials. In Fig. 12 we show $\varepsilon(\omega)$ for $(ET)_2I_3$,

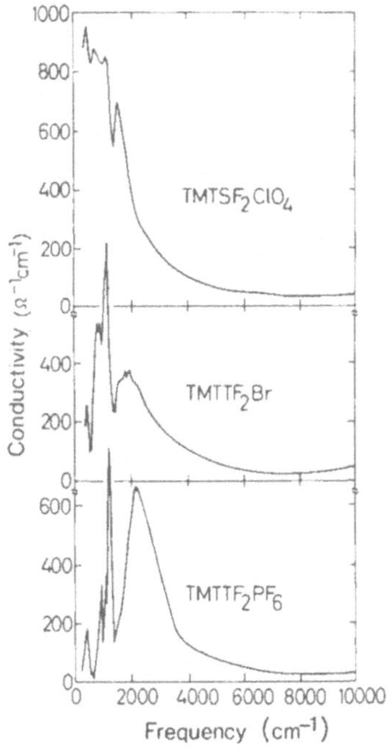

Fig. 11. Frequency dependent conductivities of 3 single-stack $TMTCF_2X$ conductors. $\overline{E}\parallel$ stacks, T = 300K.

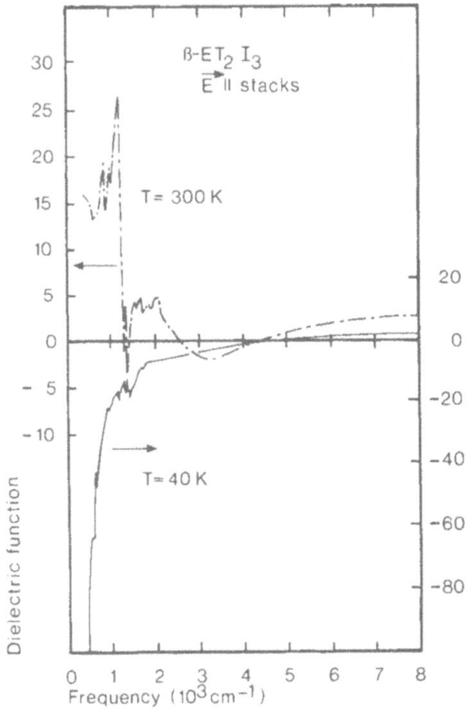

Fig. 12. $\varepsilon(\omega)$ for β-$(ET)_2I_3$ parallel to the stacks for T = 300K and 40K. Notice the different scales for the two curves[3].

$\bar{E}||$stacks[3]. Although a transition from strong non-Drude to near Drude behavior in the infrared is observed with decreasing temperature, the zero-crossing of $\varepsilon(\omega)$ at 4,000-5,000 cm^{-1}, which identifies the plasmon frequency, only shifts slightly, just as expected from the increase in t resulting from thermal contraction ($\Delta t/t \sim 10\%$).

Two-dimensional plasma behavior has been found in a few groups of materials, namely $TMTSF_2X$ and $(ET)_\alpha X$. As one example we show in Fig. 13 the polarized reflectance of $TMTSF_2PF_6$ as function of temperature[30]. The stacking axis spectrum is similar to what has been discussed above. In contrast the 300K b'-axis spectrum (corresponding to the direction of fairly strong interchain contacts in the group of materials) shows only a monotonic decrease in reflectance with no plasma minimum. Such behavior is characteristic for a metal where $\gamma > \omega_p$ (overdamped plasmons) and is typically found perpendicular to the stacks in organic conductors (although usually at much lower frequencies that in the present case). However, the

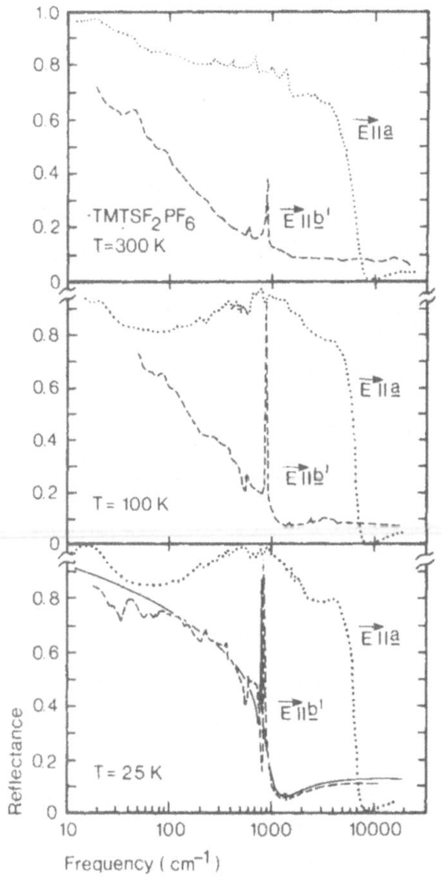

Fig. 13. Polarized reflectance of $TMTSF_2PF_6$ at 300K, 100K, and 25K. The stacking axis is **a**. The solid line is a Drude fit to the low temperature spectrum perpendicular to the chains[30].

remarkable feature in the $TMTSF_2X$ materials is that the frequency dispersion develops into a reasonably well-defined plasma edge at low temperature. Then a satisfactory Drude fitting procedure can be performed and the results given in Fig. 3 applied (ρ = 0.5). For $TMTSF_2PF_6$ one finds[27] $t_{b'}$ = 0.022 eV, about one twelfth of t_{stack}. The β-$(ET)_2X$ group[31,32] shows even smaller anisotropy (1:2), but also in that group the interchain plasmons are overdamped at 300K.

Interactions

We would like to illuminate the difficulties in sorting out the individual roles of electron-electron and electron-phonon interactions by a short discussion of data on a highly correlated [33], incommensurate[34] (ρ = 0.56 at 300K) conductor, DBTTF-$TCNQCl_2$. One would expect this material to be analogous to the excellent metal, TTF-TCNQ, but apparently due to rather small transfer integrals the system is correlation dominated with semiconducting behavior in the entire temperature range. This assessment is supported by (1) strong $4k_F$-scattering - 1D above 180K, 3D below, and (2) a high, Bonner-Fisher like, magnetic susceptibility down to ~40K, where a spin-Peierls transition accompanied by a $2k_F$-distortion quenches it.

The stacking axis conductivity spectrum[35] at 300K and 100K is shown in Fig. 14. The main features consist of a broad band around 2,000-2,500 cm^{-1} (~0.3 eV) and sharper structure below 1,500 cm^{-1}. The latter is resolved into several sharp bands at low temperatures, and can unambiguously be assigned to activated A_g-modes (see above). The semiconducting gap deduced from transport measurements in consistent with the onset of the broad band, which is therefore believed to be due to some sort of gap in the electronic spectrum.

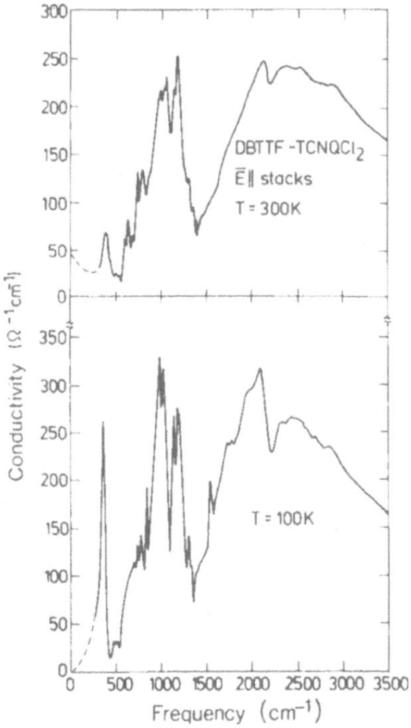

Fig. 14. Infrared conductivity of the semiconductor DBTTF-$TCNQCl_2$ at T = 300K and 100K[35].

There are two principally different interpretations of the observed behavior, both assuming strong Coulomb correlations. The first is due to Mazumdar and coworkers[4-5], and adopt a near-neighbor interaction, $V \simeq U/3$. They interpret the broad band as a correlation band, basically measuring either V or U-2V or both. For the near quarter-filled bands U and V can also account for the $4k_F$-CDWs in terms of a Wigner lattice type of model. The purely Coulomb induced broken symmetry is in turn assumed to activate the A_g-modes, i.e. the appearance of these is a secondary effect, and if the electron-phonon coupling could be switched off, the band at 0.3 eV would persist, but the A_g-modes would not be seen. The model immediately results in estimates of $V \simeq 0.3$ eV and $U \simeq 0.9$ eV.

The other approach[35] assumes $U \gg 4t$ and neglects V. In that case (ideally $U \to \infty$) the kinematic degrees of freedom are like those of spinless fermions in the usual tight-binding band[36]. Since the spin degeneracy is absent, the band is filled to twice the original wave-vector (k_F), hence the Peierls instability now occurs at[36] $4k_F$. In this scenario the gap is due to a multiphonon stabilized $4k_F$-CDW. Since a theory for the infrared properties of such a state exists, a fit can be made[35]. The conclusion is that the relative intensity of the A_g-modes as well as the gap-position can be reproduced reasonably well. The size of the gap, ~0.2 eV, corresponds to a mean field transition temepratura of order 600K, consistent with the persistence of the spectral structure to room temperature.

It is possible that the first model would give a somewhat similar spectrum. However, the point that the coupling to the A_g-modes (and other phonons) in the latter model greatly adds to the stability of the CDW can hardly be neglected. A better model should probably include both interactions simultaneously.

Based on the discussion on plasma behavior and the effects of electron-electron repulsion, we may also attempt to estimate the strength of the latter by comparing the actual oscillator strength to ω_p^2. A Drude analysis in the present material gives[9] $\omega_p^2 = 5.6 \times 10^7$ cm^{-2} and $4t = 0.41$ eV, while $(2/\pi\varepsilon_0) \int_{intra}\sigma(\omega)d\omega \simeq 2.6 \times 10^7$ cm^{-2}. The difference should then be interpreted as an effect of short range electron-electron interactions. Using Fig. 4 ($U=3V$, $\rho=0.5$), we get $U=1.7$ eV and $V=0.6$ eV. We note that these numbers do not quite agree with the Wigner lattice model which, however, neglects the effect of finite bandwidth.

Such sum rule estimates as compared with ω_p^2 may be useful for assessing the size of U and V. A survey of a number of materials[9] suggests that correlation effects are always important in organic conductors.

Instabilities: TTF-TCNQ

TTF-TCNQ[37] is metallic with a $\sim T^{-2}$ dependence of the dc conductivity down to below 100K. At 60K $\sigma(T)$ is sharply peaked, and from 53K and down a cascade of phase transitions destroys the metallic character. Structural studies have revealed the existence of 1D-$4k_F$ scattering from 300K and down to the phase transitions. Below 150K appreciable $2k_F$-scattering develops. The character is 1D down to 53K, where it becomes 3D. Several physical properties indicate that the 53K transition takes place on the TCNQ-stacks, and that the TTF-stacks follow at a lower temperature. This has for example been demonstrated by following absorption intensities of certain A_g-modes characteristic for each of the two stack-types[38].

Here we will discuss the infrared properties in general and in particular try to assign the $2k_F$- and $4k_F$-instabilities to individual stacks. Several authors have studied TTf-TCNQ. In Fig 15(a) is shown the temperature dependence of $\sigma(\omega)$ as obtained from polarized reflectance[39]. In Fig. 15(b) we reproduce low temperature results based on the bolometric technique[20] mentioned previously.

At low temperature (25K and 12K), $\sigma(\omega)$ displays a double peak structure with a narrow low-frequency band near 40 cm^{-1} ($\simeq$5 meV), and an intense band centered at 300 cm^{-1} ($\simeq$37 meV). The low-frequency band contains about 5% of the total oscillator strength. In view of the observed $2k_F$-superstructure and a single particle gap of order 300 cm^{-1} estimated from $\sigma_{dc}(T)$, the intense band may be ascribed to the $2k_F$-Peierls gap, while the 40 cm^{-1} band is assigned to the Fröhlich (CDW) pinned mode.

The only significant difference between the two sets of data is the degree of sharpness of the 300 cm^{-1} band. The oscillator strength and details like the indentation near 330 cm^{-1} (due to an A_g-mode on TCNQ) are comparable. However, the sharpness is not expected to vary in this way between 12K and 25K, and must either be due to difficulties with one or the other of the experimental techniques, or to differences in sample quality.

5% of the overall oscillator strength in the pinned mode suggests a CDW effective mass (per carrier) of 20 m^{x}_{opt}. From Eq. (24) we then have $\omega_{2k_F} \simeq 70\ cm^{-1}/\sqrt{\lambda}$. This value appears fairly reasonable, since presumably $\lambda \approx 1$, and ω_{2k_F} is really a weighted average of all involved phonons (most weight on lowest lying modes). The pinning frequency, ω_F, corresponds in a classical model of CDW depinning[40] to a threshold field of order 40 kV/cm. Such a high threshold field is believed to arise primarily from the Coulomb interaction between the two sets of oppositely charged density waves.

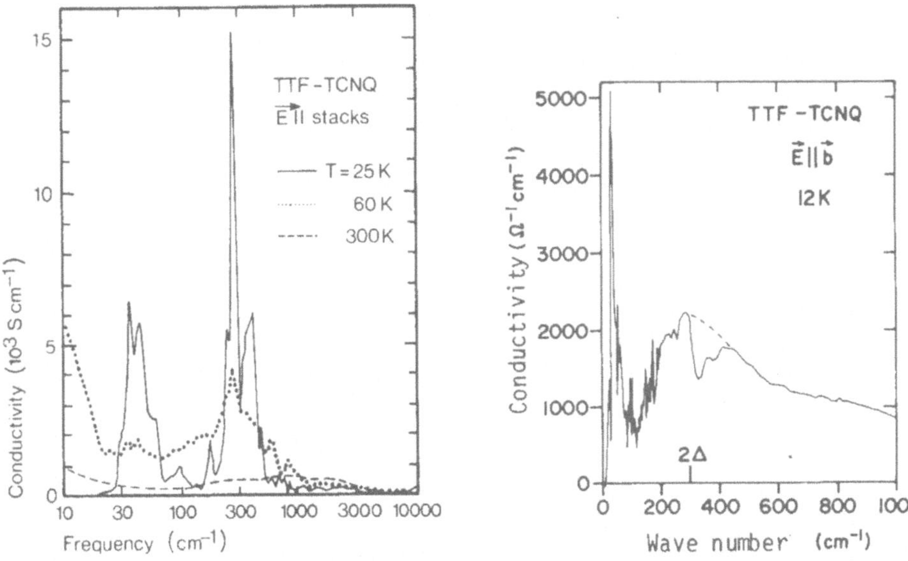

Fig. 15. (a) Stacking axis conductivity of TTF-TCNQ as obtained from direct reflectance studies[39]. T = 25K, 60K, and 300K. Notice the logarithmic frequency scale.
(b) Low temperature stacking axis conductivity of TTF-TCNQ. as obtained by measuring 1-R by a bolometric technique. Reproduced from J.E. Eldridge, Phys.Rev.B 31:5465 (1985). Copyright, American Physical Society.

Now turning to the $\sigma(\omega)$-spectra in the conducting phase (Fig. 15a), it is clear that the 60K data shows a broadening of the 300 cm^{-1} structure and apparently the oscillator strength of the low frequency band has moved to zero frequency. Since $\sigma_{dc}(60K) \simeq 10^4\ \Omega^{-1}\ cm^{-1}$, there is a sharp drop in $\sigma(\omega)$ in the millimetre and submillimetre range. Physically, this is consistent with a depinning of the CDWs at the phase transition. They appear to contribute to σ_{dc} and in the infrared a pseudogap induced absorption, broadened by fluctuations, may be followed to temperatures above 100K. Going to 300K the features are much broadened and the maximum in $\sigma(\omega)$ has moved to about 800 cm^{-1}. The latter change can not be understood in terms of the $2k_F$-instability.

Instead we may focus on the role of the $4k_F$-instability which gives rise to the only detectable superstructure at 300K. It may be associated with only one, or with both types of stack. In analogy with the case of DBTTF-TCNQCl_2, we may expect the infrared properties of the $4k_F$-CDW to consist of a maximum in $\sigma(\omega)$ corresponding to a pseudogap plus a number of sharper features near the A_g vibrational modes. TCNQ has such a mode near 2,200 cm^{-1} (C≡N stretch), which couples strongly to the electrons. Fig. 16(a) shows a high resolution study of the reflectance near this mode. At 300K and 200K the feature has the strength and shape of an ordinary infrared active mode superimposed on the metallic response of the conduction electrons. From 100K and down to 30K the oscillator strength increases and the shape is inverted (the original structure is presumably

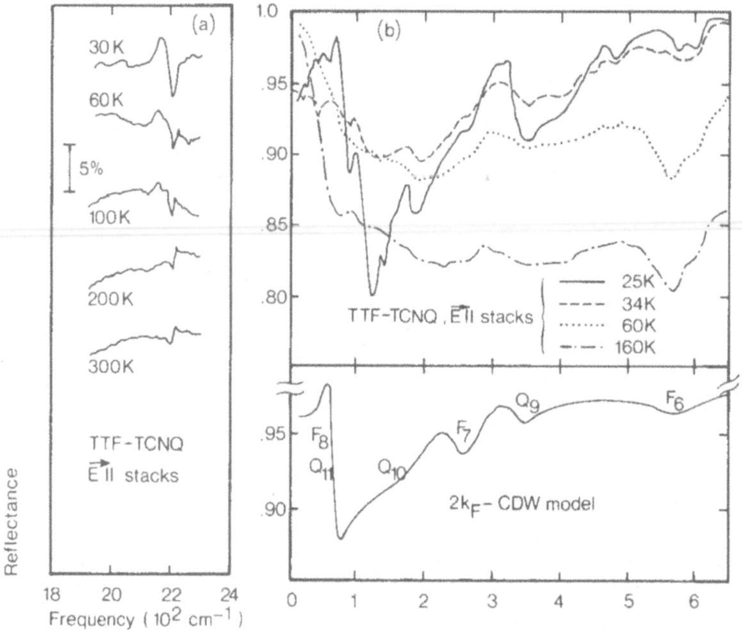

Fig. 16. Stacking axis reflectance of TTF-TCNQ.
(a) Temperature dependence in the vicinity of the C≡N stretch frequency[29].
(b) Temperature dependence in the far infrared[39]. At bottom a $2k_F$-CDW model with A_g-mode features identified. Q: TCNQ. F: TTF.

hidden behind the new band). The low temperature spectrum is exactly what is expected from a Peierls distorted semiconductor. The point is that the temperature dependence follows that of the $2k_F$-scattering. Thus the $4k_F$-instability must (at least largely) take place on the TTF-stacks. This conclusion agrees with other studies[41].

It is also corroborated by a careful inspection of the far infrared reflectance, shown in Fig. 16(b). Being raw data these curves should be reliable with respect to position and relative strength of fine structure. Also shown is a model calculation for a $2k_F$-CDW system with $2\Delta = 300$ cm^{-1}, using the known emv coupling constants of TTF and TCNQ[42]. The model mainly serves to identify the individual A_g-modes in the far infrared. It is again evident that TCNQ-modes (Q_{10}, Q_9) show a temperature dependence as that of the 2,200 cm^{-1} mode, while the TTF-modes (F_7, F_6) are different. F_6 corresponds to a strong, near temperature independent dip in the reflectance from 160K to 60K. Below 60K it moves slightly up in frequency and becomes weaker with a different shape. This behavior may tentatively be assigned to a cross-over from a $4k_F$- to a $2k_F$-CDW state on the TTF-stacks.

The above discussion should serve to illustrate how the A_g-mode features may be used to monitor instabilities on individual chains.

With this example our tour through the subject of infrared spectroscopy on low-dimensional conductors is over. Many other examples could have been given, but hopefully the reader has already got the impression of a powerful tool with a lot of potential applications.

ACKNOWLEDGMENTS

The author is grateful to Prof. K. Bechgaard for years of fruitful collaboration, and to Prof. D.B. Tanner for many discussions on infrared spectroscopy. The author is supported as a Niels Bohr Fellow of the Royal Danish Academy of Sciences and Letters.

REFERENCES

1. See, for example, F. Wooten, "Optical Properties of Solids", Academic Press, New York (1972).
2. W.P. Dumke, Phys.Rev. 124:1813 (1961).
3. C.S. Jacobsen, J.M. Williams, and H.H. Wang, Solid State Commun. 54:937 (1985) and erratum 57:no.8:i (1986).
4. A.N. Bloch and S. Mazumdar, J.Physique. 44: Colloque 3:1273 (1983); S. Mazumdar and A.N. Bloch, Phys.Rev.Lett. 50:207 (1983).
5. S. Mazumdar and S.N. Dixit, Phys.Rev.B 34:3683 (1986)
6. S. Mazumdar and Z.G. Soos, Phys.Rev.B 23:2810 (1981).
7. M.J. Rice, Solid State Commun. 31:93 (1979).
8. D. Baeriswyl, J. Carmelo, and A. Luther, Phys.Rev.B. 33:7247 (1986).
9. C.S. Jacobsen, J.Phys.C:Solid State Phys. 19:0000 (1986).
10. P.F. Maldague, Phys.Rev.B 16:2437 (1977).
11. E.H. Lieb and F.Y. Wu, Phys.Rev.Lett. 20:1445 (1968).
12. T. Holstein, Phys.Rev. 96:535 (1954); Ann.Phys.(N.Y.) 29:410 (1964).
13. P.B. Allen, Phys.Rev.B 3:305 (1971).
14. J.J. Hopfield, Comments Solid State Phys. 3:38 (1970).
15. See, for example, A. Painelli and A. Girlando, J.Chem.Phys. 84:5655 (1986).
16. P.A. Lee, T.M. Rice, and P.W. Anderson, Solid State Commun. 14:703 (1974).
17. M.J. Rice, C.B. Duke, and N.O. Lipari, Solid State Commun. 17:1089 (1975).

18. M.J. Rice, Solid State Commun. 25:1083 (1978).
19. R.H. Young and E.I.P. Walker, Phys.Rev.B 15:631 (1977).
20. J.E. Eldridge and G.S. Bates, Mol.Cryst.Liq.Cryst. 119:183 (1985).
21. J.B. Torrance, B.A. Scott, B. Welber, F.B. Kaufman, and P.E. Seiden, Phys.Rev.B 19:730 (1979).
22. K. Kikuchi, Y. Ikemoto, K. Yakushi, H. Kuroda, and K. Kobayashi, Solid State Commun. 42:433 (1982).
23. B. Welber, P.E. Seiden, and P.M. Grant, Phys.Rev.B 18:2692 (1978).
24. A.J. Schultz, G.D. Stucky, R.H. Blessing, and P. Coppens, J.Am.Chem.Soc. 98:3194 (1974).
25. D. Debray, R. Millet, D. Jerome, S. Barisic, J.M. Fabre, and L. Giral, J.Physique-Lettr. 38:L277 (1977).
26. F. Herman, Physica Scripta 16:303 (1977).
27. C.S. Jacobsen, D.B. Tanner, and K. Bechgaard, Phys.Rev.B 28:7019 (1983).
28. C.S. Jacobsen, K. Mortensen, J.R. Andersen, and K. Bechgaard, Phys.Rev.B 18:905 (1978).
29. C.S. Jacobsen, Mat.Fys.Medd.Dan.Vid.Selsk. 41:251 (1985).
30. C.S. Jacobsen, D.B. Tanner, and K. Bechgaard, Phys.Rev.Lett. 46:1142 (1981).
31. C.S. Jacobsen, D.B. Tanner, J.M. Williams, and H.H. Wang, Synthetic Metals, in press.
32. H. Tajima, H. Kanbara, K. Yakushi, H. Kuroda, and G. Saito, Synthetic Metals, in press.
33. C.S. Jacobsen, H.J. Pedersen, K. Mortensen, and K. Bechgaard, J.Phys.C: Solid State Phys. 13:3411 (1980).
34. K. Mortensen, C.S. Jacobsen, A. Lindegaard-Andersen, and K. Bechgaard, J.Physique. 44: Colloque 3:1349 (1983).
35. C.S. Jacobsen and K. Bechgaard, Mol.Cryst.Liq.Cryst. 120:71 (1985).
36. J. Bernasconi, M.J. Rice, W.R. Schneider, and S. Strässler, Phys.Rev.B 12:1090 (1975).
37. For a review, see for example, D. Jerome, and H.J. Schultz, Advances in Phys. 31:299 (1982).
38. R. Bozio and C. Pecile, Solid State Commun. 37:193 (1981).
39. D.B. Tanner, K.D. Cummings, and C.S. Jacobsen, Phys.Rev.Lett. 47:597 (1981); D.B. Tanner and C.S. Jacobsen, Mol.Cryst.Liq.Cryst. 85:137 (1982).
40. G. Grüner, A. Zawadowski, and P.M. Chaikin, Phys.Rev.Lett. 46:511 (1981).
41. See, for example, L. Forro, S. Bouffard, J.P. Pouget, J.Physique-lettr. 45:L543 (1984).
42. See, for example, R. Bozio and C. Pecile in: "The Physics and Chemistry of Low-Dimensional Solids", L. Alcacer, ed., Reidel, Dordrecht (1980).

QUASI ONE-DIMENSIONAL CONDUCTORS: THE FAR INFRARED PROBLEM

Thomas Timusk

Department of Physics
McMaster University
Hamilton Ont. Canada L8S 4M1

INTRODUCTION

In the metallic state the quasi one-dimensional conductors are characterized by very high values of dc conductivity equaling in some cases pure copper at room temperature. Thus one might expect their optical properties to follow the Drude free electron theory where the conductivity remains equal to the dc value up to a frequency of the order of the relaxation time. Instead the quasi one-dimensional conductors show a region of depressed optical conductivity in the far infrared. The dc conductivity[1] at low temperature of a material such as $(TMTSF)_2ClO_4$ is of the order of $5 \cdot 10^5$ $(\Omega \cdot cm)^{-1}$ whereas the optical conductivity[2] at 50 cm^{-1} is only a few hundred $(\Omega \cdot cm)^{-1}$. The change from high to low conductivity occurs in the experimentally difficult microwave region and very few experiments have been done at these limiting frequencies. The phenomenon of suppressed far infrared conductivity is almost universal in the quasi one-dimensional materials[2-7]. It was first observed in TTF-TCNQ but subsequently also reported in $(TMTSF)_2X$ family of compounds in their conducting state. This discrepancy has been generally explained in terms of a narrow collective mode with a very large effective oscillator mass centered at zero frequency. Such a model implies that the electrical current at low temperature is carried by the collective mode.

Another view that is widely accepted is the simple single particle picture of a free electron gas with a very long mean free path. The low far infrared conductivity must then be attributed to some experimental artifact such as a damaged surface layer of broken bonds that is not representative of the bulk properties of the material. Weger and Kahve[8] have proposed a model with broken strands that is able to explain most of the observed data with a suitable selection of fitting parameters.

In the group of materials with the octahedral anions a SDW transition takes place at 12 K and one would therefore expect a region of zero optical conductivity below about 30 cm^{-1} frequency. Observations show however that while a gaplike depression in the density of states can be observed to appear sharply at the transition temperature the conductivity remains at an elevated non-zero value well below this temperature. BCS-like calculations predict that there should be a gap in the spectrum of excitations below an energy of 2Δ at which point the conductivity should rise in a steplike fashion as shown by Psaltakis and Fenton[9].

Another puzzle is the appearance of strong phonon-like lines in materials in the SDW state superimposed on the continuous background. These lines are suggestive of the phase phonons that are activated by dimerization according to the mechanism proposed by Rice[10] but in $(TMTSF)_2SbF_6$ they can be observed[11] to be closely associated with the onset of the the spin density wave state, a state that is not usually associated with the lattice distortion required by the naive application of the Rice mechanism. The phonon like spectrum can also be observed[2,12] in the highly conducting state in $(TMTSF)_2ClO_4$ where the width of the phonon lines is strongly temperature dependent and one of the lines is strongly affected by a very modest magnetic field at low temperature.

DRUDE THEORY

Although reflectance or absorptance are the quantities that are most easily measured for a good conductor they are difficult to interpret physically. The optical conductivity or the imaginary part of the dielectric constant are direct measures of the spectrum of dissipative processes and can be derived from reflectance by a Kramers Kronig transformation. In simple systems where the lifetime of the electrons is governed by a single relaxation time τ the frequency dependent conductivity $\sigma(\omega)$ is given by the Drude formula:

$$4\pi\sigma(\omega) = \frac{\omega_p^2\tau}{(1+\omega^2\tau^2)} \tag{1}$$

Here $\omega_p^2 = 4\pi Ne^2/m$ where ω_p is the plasma frequency and N the number of electrons per unit volume and m the electron mass.

The Drude conductivity is constant at low frequency and equal in magnitude to the dc conductivity falling off as $1/\omega^2$ at frequencies larger than $1/\tau$. Given any two of $\sigma(0)$, τ or ω_p^2 the third one can be calculated. In the case of the (TMTSF) family of materials, for light polarized along the chain axis a very well defined plasma edge has been seen by Jacobsen et al. in the 1 μ spectral region[4] and a plasma frequency has been derived from this. For example for $(TMTSF)_2ClO_4$ ω_p = 10170 cm^{-1} and with a typical dc conductivity of 500 000 $(\Omega cm)^{-1}$ at low temperature one gets $1/\tau$ = 3.5 cm^{-1} (1 $(\Omega cm)^{-1}$= 4.78 cm^{-1}). This rather low relaxation rate is in reasonable agreement with other estimates of the scattering time at low temperatures and low frequencies[13]. It should be distinguished from high frequency relaxation rate of 680 cm^{-1} determined from a fit in the plasma frequency region[4]. At high frequency inelastic processes such as Holstein phonon emission contribute to the relaxation. These processes are frozen out at low temperature.

In principle $1/\tau$ can be obtained independently from a fit of eq. 1 to the far infrared spectrum. This is only possible in simple systems where $1/\tau$ is frequency independent and there are no interfering band-to-band transitions. The spectra of the quasi one-dimensional conductors show no such simple character: it has not been possible to isolate a 'relaxation region' where the conductivity varies as $1/\omega^2$ as demanded by eq. 1.

EXPERIMENTAL PROBLEMS

The principal difficulty in determining accurate optical constants of the organic conductors of both the TTF-TCNQ as well as the $(TMTSF)_2X$ family is the small size of the samples available. The long needle shaped crystals are at most several hundred microns in crossection -- just equal to the wavelength of infrared radiation in the energy range that encompasses their most interesting excitations. To avoid diffraction effects a reflectance sample should be larger than many wavelengths of the radiation used.

To overcome the problem a mosaic of small crystals is often assembled to form a larger sample perhaps several millimeters in size. This procedure increases sample size and permits the use of optical beams of larger

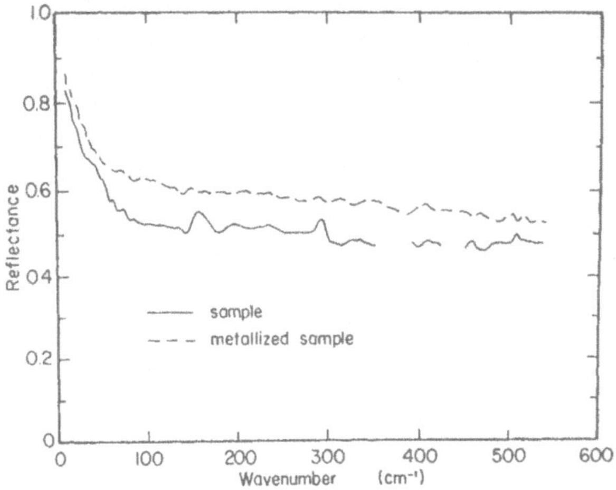

Fig 1. The reflectance of a mosaic of organic crystals. The lower curve is for the crystal mosaic as assembled, in the upper curve the crystals have been coated with a layer of gold that has a nominal reflectance of 99.5%. The rough surface structure of the mosaic is responsible for the large losses.

crossection and better control of polarization and collimation but the effects of diffraction from the gaps between the individual crystals remains. To correct for these effects the crystal is coated with an evaporated layer of some good metal that can then be used as a reference that, one hopes, includes most of the effects of diffraction. Figure 1 illustrates how large these effects can be. The coated sample shown shows an apparent loss of signal of some 40% in comparison to a flat optical mirror of the same size. An additional 10% absorption can be observed without the coating. It is this 10% that makes up the measured absorption signal and contains all the interesting physics.

A bolometric method, pioneered by Eldridge et al.[14], is very sensitive and can be used on very small single crystals with superior signal to noise ratio. However this method too is affected by diffraction but in principle at least diffraction effects can be estimated more accurately in the case of a single rectangular crystal than an irregular mosaic.

THE ZERO FREQUENCY MODE

The organic charge transfer salts behave like metals at high temperatures and generally undergo a transition to a state with a gap at low temperature. The gapped state is either an insulating CDW or SDW state or a superconducting one. We will first discuss the conducting high temperature state. It is this state that possesses the anomalous low frequency mode.

Figure 2 shows as an example the reflectance and the conductivity obtained by Kramers-Kronig transformation of $(TMTSF)_2ClO_4$ at 2 K[2]. At this temperature we are well above the superconducting transition and the sample has been cooled slowly to avoid the SDW state that occurs in quenched samples. Measured curves with the electric field parallel and perpendicular to the highly conducting a direction are shown. There are two effects that distinguish these curves from those of an ordinary highly conducting metal.

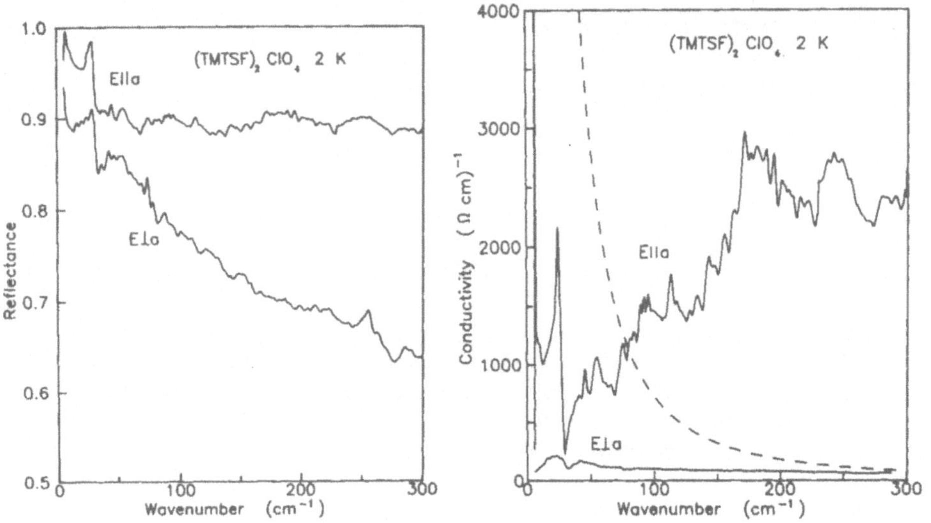

Figure 2. The reflectance (left) and the corresponding optical conductivity (right) of $(TMTSF)_2ClO_4$ in the far infrared. The dashed conductivity is calculated from the dc conductivity and the plasma frequency.

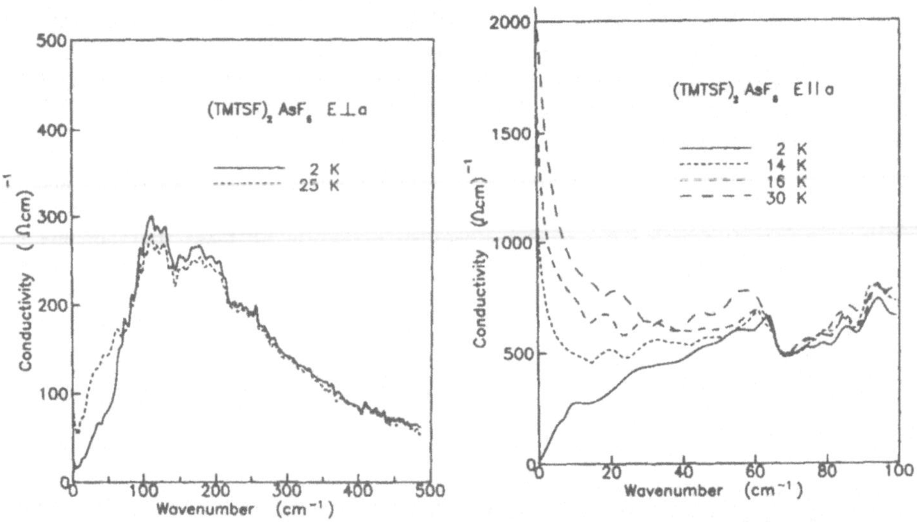

Figure 3. The optical conductivity of $(TMTSF)_2AsF_6$ for light polarized in two perpendicular directions. The optical conductivity is higher in the E||a, the highly conducting, direction. The effect of the SDW transition can be seen in the changes that take place at low temperature below 60 cm^{-1}.

In the highly conducting direction there is considerable absorption of the $(TMTSF)_2ClO_4$ in the 50 - 100 cm^{-1} region in contrast to an ordinary metal that follows Drude theory. In the optical conductivity this fact is reflected in the anomalously low conductivity in the 10 - 100 cm^{-1} region. The conductivity in the chain direction expected from an application of Drude theory is shown as a dashed line. It is clear that there is substantial discrepancy between the expected Drude or single particle result and the far infrared observations.

Another example[15] is shown in fig. 3 for $(TMTSF)_2AsF_6$. Again the reflectance and the conductivity are shown. This material undergoes a transition to the SDW state at 12 K[16-20] and curves are shown both above and below the transition temperature. In the sample used the transition is rather broad but at 30 K the sample is quite metallic and again would be expected to follow single particle Drude theory. As in the case of the ClO_4 compound there is instead a strong absorption in the conducting state which, in the conductivity plot, appears as an anomalously low conductivity in the 10 - 20 cm^{-1} region at 30 K. Extrapolating the observed far infrared conductivity to zero frequency the curve approaches about 2000 $(\Omega cm)^{-1}$ whereas the measured dc conductivity at the same temperature is of the order of 20·000 $(\Omega cm)^{-1}$, a factor 10 higher. At temperatures where the conductivity is lower there is less discrepancy between the extrapolated far infrared conductivity and the measured dc conductivity. For example at 14 K the far infrared extrapolation gives a value of 1000 $(\Omega cm)^{-1}$ about the same as the observed dc value.[20]

The $(TMTSF)_2AsF_6$ low temperature far infrared conductivity shown as a family of curves in figure 3 is very much what one expects for a free electron plasma with a relaxation time of the order of 10 cm^{-1} that is freezing out as the band gap of the SDW is developing. The discrepancy occurs at higher temperatures where the conductivity is much higher. The plasma absorption is much weaker than one would expect from the known density and conductivity.

The collective mode at zero frequency explains this lack of free electron response in the conducting state[6,7,21]. The parameters of the zero frequency mode can be found by assuming that the real part of the far infrared response arises from an oscillator at $\omega =$ 0 with a plasma frequency Ω_p:

$$\epsilon_1(\omega) = \epsilon_H - (\Omega_p/\omega)^2$$

where ϵ_H is the high frequency contribution to $\epsilon_1(\omega)$. Kramers Kronig analysis of $(TMTSF)_2ClO_4$ for example[22] shows that ϵ_H is 500 and ϵ_1 first crosses zero at $\omega = 18$ cm^{-1} which gives $\Omega_p = 400$ cm^{-1}. The effective mass of the oscillator can be calculated from the plasma frequency and we find M = 646. The width $1/\tau_c$ of the mode can now be estimated from a modified Drude conductivity of the mode:

$$4\pi\sigma_{dc} = \Omega_p \tau_c$$

The width found this way is very narrow, $1/\tau \leq 0.5$ cm^{-1} in most cases. This puts the high frequency edge of the mode into the range of frequencies that is difficult for conventional far infrared spectroscopy with small samples.

Attempts to observe the narrow mode in the conducting state predicted by this model have been made with microwave spectroscopy[16]. Javadi et al. used the cavity perturbation technique[23] to measure the microwave conductivity up to 35 GHz frequency of $(TMTSF)_2PF_6$. They conclude that the microwave conductivity is relatively frequency independent up to 35 GHz between the transition temperature of 11.5 K and 30 K. This would seem to be in apparent contradiction with the oscillator analysis that predicts a full width at half maximum of 0.32 cm^{-1} (10 GHz) for $(TMTSF)_2PF_6$ at 25 K[7].

However there is some room for a reconciliation of the two sets of experiments. The microwave experiments of Javadi et al. at 20 K do show a decrease in conductivity of the order of a factor of two between 4.5 GHz and 35 GHz. This difference translates to a width of about 1 cm^{-1} for the conductivity peak, a factor of three larger than what the far-infrared oscillator analysis gives for this quantity at 25·K.

A single particle relaxation time can also be obtained from the magneto resistance[13]. Depending on the residual resistance a value between 2.5 and 6 cm^{-1} is found for $(TMTSF)_2ClO_4$ in the relaxed state. These numbers are in good agreement with estimates from Drude theory. The oscillator analysis for this material predicts a very narrow collective peak of 0.005 cm^{-1} in clear contradiction of the single particle picture. Because of the very high dc conductivity of this material it is very difficult to perform meaningful microwave cavity perturbation experiments on it although they may be more conclusive here than in the case of the relatively low conductivity tetrahedral anion materials.

The question whether current in the 12 - 25 K temperature range is carried by single particles or by a collective mode must remain open at this stage. The far infrared measurements point to a collective mode with a width of 0.3 cm^{-1} for the SDW materials with the 12 K ordering temperature and a much narrower mode for $(TMTSF)_2ClO_4$. On the other hand there is plenty of evidence on the side of single particle transport particularly from magneto transport data. The case for the single particle point of view has been summarized recently by Green and Chaikin[24].

ENERGY GAPS

One of the most important applications of far infrared spectroscopy in the study of superconductors has been the establishment of the magnitude of the energy gaps. An ordinary superconductor is perfectly reflecting for all frequencies below the gap. As a consequence the conductivity is zero in the gap region as well. One would expect that in the case of the organic conductors, too, far infrared spectroscopy could be used to reveal gaps corresponding to the various electronic excitations that destroy the Fermi surface such as spin density waves, charge density waves or superconductivity.

While the superconducting energy gap is expected to be too small and measured easily in the $(TMTSF)_2X$ family the SDW gap should occur at a convenient $2\cdot\Delta$ = 30 cm^{-1} in the materials with the 12 K transition temperature if we assume the mean field value of $2\cdot\Delta = 3.5\cdot kT_c$. Psaltakis and Fenton[9] have shown that the optical conductivity of a system with a pinned SDW is zero up to the frequency $2\cdot\Delta$ where the conductivity rises abruptly to about 157 % of the normal value and then gradually decays to the 100 % value.

A number of reports of the observation of the CDW gap in the $(TMTSF)_2X$ materials exist in the literature[11,14] but it is evident from the published spectra that the gap is not well developed and the spectrum does not have the simple shape predicted. A necessary condition for identification of gap structure is that its advent be correlated with the transition temperature of 12 K.

Figure 3 shows the chain axis conductivity of $(TMTSF)_2AsF_6$ above and below the transition temperature[15]. There is a reduction of conductivity for frequencies less than 60 cm^{-1} but the conductivity does not go to zero as predicted nor is there an abrupt step that would allow one to obtain an accurate value of the energy gap. A similar spectrum for $(TMTSF)_2PF_6$ in the highly conducting direction is shown in fig 4.[25] In this case the overall reflectance was monitored carefully as a function of temperature. There was an abrupt change at 12 K confirming the hypothesis that the changes that take place between 15 K and 1.2 K are due to the SDW transition. Again the gap is poorly defined and the far infrared conductivity remains high down to very low frequencies whereas it is known that in this material the dc conductivity is very low in the SDW state, evidence that the gap that is formed at the transition spans the whole fermi surface.

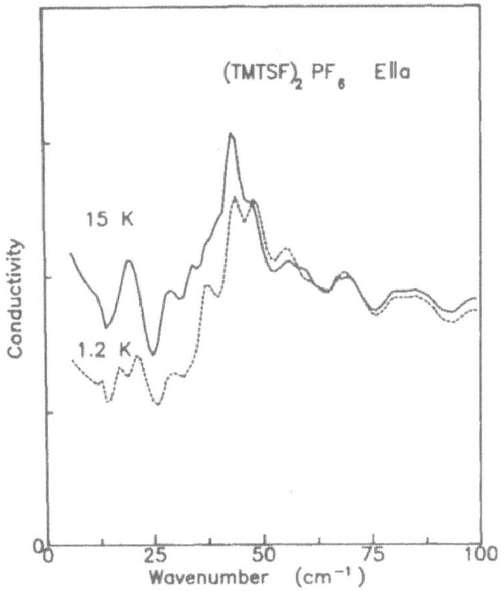

Fig. 4. The optical conductivity of $(TMTSF)_2PF_6$ above and below the SDW transition temperature. We associate the drop in conductivity that occurs in the 0-50 cm^{-1} region as the temperature is lowered through 12 K with the SDW gap.

Figures 3 and 4 suggest that the density of states is developing a gap with magnitude of the order of 40 - 60 cm^{-1} and that it is possibly quite anisotropic which would explain the absence of a sharp cutoff at $2 \cdot \Delta$ and that perhaps as a result of impurities and other damage there are gapless regions particularly for fields polarized in the highly conducting region where the radiation does not penetrate as deeply into the sample. The peaks that can be observed at 19.5 cm^{-1} and at 45 cm^{-1} in $(TMTSF)_2PF_6$ are probably due to phonons as they seem to change very little at the transition at 12 K.

INDUCED PHONON STRUCTURE

The electron phonon interaction in the organic quasi one-dimensional materials has received relatively little theoretical attention. There are nevertheless several very interesting new effects that seem to be related to phonons and closely associated with the one dimensional nature of these materials.

The best known example of unusual electron phonon mechanisms is the very strong optical response of totally symmetric phonons, transitions that are normally forbidden, but appear with enormous oscillator strength, of almost electronic magnitude. This process occurs in the charge density wave state in general and a detailed physical picture has been supplied by Michael Rice[10]. The Rice mechanism depends on the dimerization by the CDW. In simple terms the antisymmetric mode of the pair of molecules in the dimer pumps electronic charge back and forth along the chain direction giving rise to a very large fluctuating dipole moment at the frequency of the symmetric mode but polarized in the chain direction. The predictions of the model have been verified in detail in several CDW systems[26,27].

The Rice mechanism needs a CDW to function but strong totally symmetric vibrations of the TMTSF molecules can also be induced in a SDW transition as shown by Ng et al.[11] for $(TMTSF)_2SbF_6$. The optical conductivity of $(TMTSF)_2SbF_6$ is shown in fig. 5. It is a SDW system in many ways similar to $(TMTSF)_2PF_6$ and yet it is clear that a very strong line spectrum has been created by an electronic transition. The intensity and the shift of the lines follow the order parameter of the transition

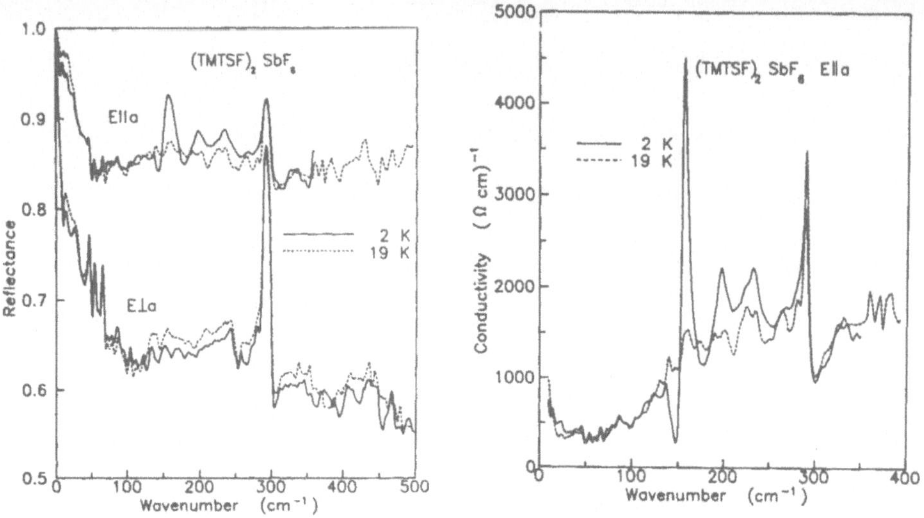

Fig. 5. The reflectance and the optical conductivity of $(TMTSF)_2SbF_6$ above and below the SDW ordering temperature. Strong lines from symmetric phonons appear in the SDW state.

very closely. It was pointed out by Psaltakis and Fenton[9] that for a commensurate SDW there is an accompanying CDW harmonic capable of inducing structure due to totally symmetric vibrations of the TMTSF molecule. Recent calculations[28] also show that the shape and position of the phonon lines are consistent with a SDW gap of the order of 50 cm^{-1}. Originally Ng. et al.[11] identified the rising absorption at 180 cm^{-1} as the SDW gap. In view of the unambiguous results for the other two anions where the gaps are in the more reasonable 50 cm^{-1}range (closer to the mean field value of $3.5 \cdot kT_c$) this identification seems now less compelling. The induced phonons in this material are polarized strictly in the chain direction (the line at 300 cm^{-1} in contrast is an anion line, independent of temperature, and visible in both polarizations). This further supports the CDW as the driving mechanism for the induced phonons. It appears that in $(TMTSF)_2SbF_6$ this CDW is mobile and capable of activating vibrations. Observable gap structure in the sliding, incommensurate case is not expected[29]. In the other two SDW compounds the wave is presumably pinned, incapable of inducing phonons but able to activate single particle electronic transitions across a gap[28].

The phonon spectrum associated with $(TMTSF)_2ClO_4$ is quite different. This material, in the annealed state, has shown no evidence of density wave transitions and yet there are sharp phonon peaks at 7 cm^{-1} and at 29 cm^{-1} that grow as the temperature is lowered[2,12] (Fig.6). Recent work by Eldridge et al. on powders[30] has identified the 29 cm^{-1} peak as due to a phonon that can also be seen in polarization perpendicular to the chain axis. The strength of this transverse mode is independent of temperature. The mode at the same frequency in the chain direction does not change in oscillator strength as the temperature is lowered (fig. 6). They also find that a group of modes in the transverse direction in $(TMTSF)_2PF_6$ also grow in amplitude as the temperature is lowered with integrated intensities following the magnitude of the dc conductivity rather than being associated with a density wave transition.

The close connection between the dc conductivity and the integrated intensity of certain phonon modes may provide a possible explanation of the magnetic field dependance of the 29 cm^{-1} mode first reported by Ng et al. who found that a field of 0.7 T would substantially destroy the mode[2].

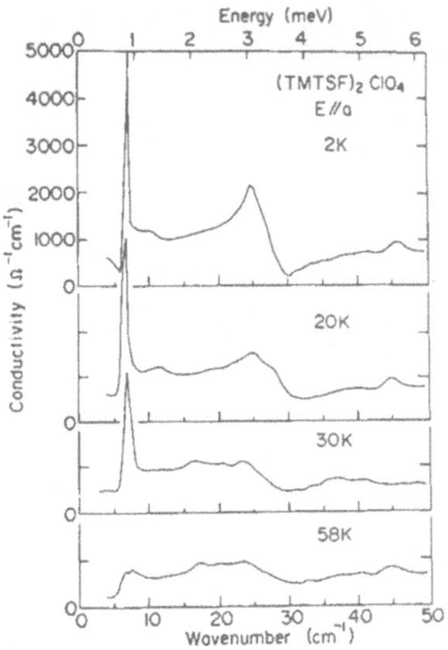

Fig. 6. Phononlike lines in the spectrum of $(TMTSF)_2ClO_4$. These lines are highly temperature dependent in width but do not seem to be associated with a particular phase transition.

SUMMARY

This review has emphasized some of the problems of understanding of far infrared spectra of the quasi one-dimensional conductors. The first two of these may possibly be related to sample surface quality and faulty experimental technique: with better methods perhaps the anomalous low far infrared response will disappear and the narrow mode at the origin and the extra conductivity in the SDW gap will all vanish along with it.

On the other hand we may also find that the phenomena seen here are real and point to new physics. The observation of the sharp phonon lines at least is on firm footing -- such measurements are easy in spectroscopy in general -- but a detailed theoretical understanding of their activation is missing.

With the discovery of high temperature organic superconductors it should be possible to study the superconducting energy gap by far-infrared absorption. The gap of $(ET)_2AuI_2$ with a transition temperature of 4-5 K[31], if it is BCS like, should occur in the 10 - 15 cm^{-1} region. Applications of the sensitive bolometric technique developed by Eldridge should help in solving some of the puzzles of low temperature properties of the organic conductors in the far infrared.

REFERENCES

1. D.U. Gubser, W.W.Fuller, T.O. Poehler, J. Stokes, D.O. Cowan, M. Lee, and A.N. Bloch, J. Mol. Cryst. Liq. Cryst. 79, 581, (1981)
2. H.K. Ng, T. Timusk and K. Bechgaard, J. Physique C3-44, 867, (1983)
3 D.B. Tanner, K.D. Cummings, and C.S. Jacobsen, Phys. Rev. Lett. 47, 597, (1981)
4. C.S. Jacobsen, D.B. Tanner and K. Bechgaard, J. Physique C3-44, 859, (1983).
5. P. Brüesch, S. Strässler, and H.R. Zeller, Phys. Rev. B12, 219, (1975)
6. M.J. Cohen, B. Coleman, and A.J. Heeger, Phys. Rev. B13, 5111, (1976)
7. C.S. Jacobsen, D.B. Tanner, and K. Bechgaard, Phys. Rev. B28, 7019, (1983)
8. M. Weger, and M. Kahve, J.Phys. C 12, 2567, (1979)
9. G.C. Psaltakis and E. Fenton, J. Physique C3-1125, 867, (1983)
10. M.J. Rice, Phys. Rev. Lett. 37, 36, (1976).
11. H.K. Ng, T. Timusk, and K. Bechgaard, Phys. Rev. B30, 5842, (1984)

12. W.A. Challener, P.L. Richards and R.L. Greene, J. Physique, C3-44, 873, (1983).
13. J.R. Cooper, L. Forro, and B. Korin-Hamzić, J. Mol. Cryst. Liq. Cryst. 119, 121, (1985)
14. J.E. Eldridge, G.S. Bates, J. Mol. Cryst. Liq. Cryst. 119, 183, (1985)
15. H.K. Ng, T. Timusk, D.Jérome, and K. Bechgaard, Phys. Rev. B32, 8041, (1985)
16. K. Bechgaard, C.S. Jacobsen, K. Mortensen, H.J. Pedersen, and N. Thorup, Solid State Commun. 33, 1119, (1980)
17. R. Brusetti, M. Ribault, D. Jérome, and K. Bechgaard, J. Phys. (Paris) 43, 801, (1982)
18. K. Mortensen, Y. Tomkiewicz, and K. Bechgaard, Phys. Rev. B25, 3319, (1982)
19. J.B. Torrance, H.J. Petersen, and K. Bechgaard, Phys. Rev. Lett. 49, 881, (1982)
20. J.B. Torrance, J. Phys. (Paris) Colloq. 44, C3-799, (1983)
21. P.A. Lee, T.M. Rice and P.W. Anderson, Solid State Commun. 14, 703, (1974)
22. H.K. Ng, T. Timusk, and K. Bechgaard, J. Mol. Cryst. Liq. Cryst. 119, 191, (1985)
23. H.H.S. Javadi, S.Sridar, G. Grüner, L.Chiang, and F. Wudl, Phys. Rev. Lett. 55, 1216, (1985)
24. R.L. Green, and P.M. Chaikin, Physica B & C, 126, 431, (1984)
25. D. Bonn, private communication.
26. M.J. Rice, L. Pietronero and P. Bruesh, Solid State Commun. 21, 757, (1977)
27. M.J. Rice, V.M. Yartsev, and C.S. Jacobsen, Phys. Rev. B21, 3437, (1980)
28. E. Fenton, these proceedings.
29. J.J. Hopfield, Phys. Rev. 139, 1214, (1964)
30. J.E. Eldridge, C.C. Homes, F.E. Bates, and G.S. Bates, Phys. Rev. B32, 5156, (1985)
31. K.D. Carlson, G.W. Crabtree, L. Nuñez, H.H. Wang, M.A. Beno, U. Geiser, M.A. Firestone, K.S. Webb, and J.M. Williams, Solid State Commun. 57, 89, (1986)

INFRARED CONDUCTIVITY DUE TO SPIN DENSITY WAVES

E.W. Fenton and G.C. Aers

Division of Physics
National Research Council of Canada
Ottawa, Canada K1A 0R6

Infrared conductivity due to a charge-density-wave (CDW) or spin-density-wave (SDW) condensation of electrons in a metal may include two contributions: $\sigma_{s.p.}$ from single-particle excitations of the electrons across a gap in the electron energy band due to the density wave; and $\sigma_{c.m.}$ from collective-mode motion (vibration) of the density wave. For the SDW states of the $(TMTSF)_2X$ superconductors, there may be some cases where the SDW is strongly pinned and other cases where it is very weakly pinned. When the SDW is strongly pinned, the conductivity will be similar to that of a semiconductor, arising almost entirely from $\sigma_{s.p.}$. In this case there is a conductivity edge and peak at the SDW gap 2Δ, with depletion of the conductivity at frequencies less than 2Δ in comparison to conductivity of the normal state of the metal. By very weak pinning we mean that $\sigma_{c.m.}$ contributes significantly to σ and in this case our theory shows that: (a) the single-particle conductivity $\sigma_{s.p.}$ commencing at 2Δ is almost entirely cancelled by $\sigma_{c.m.}$ and no conductivity edge should occur at 2Δ; (b) due to interaction with the lattice of a $2\underline{Q}$ CDW harmonic of the $1\underline{Q}$ SDW, phonon resonance peaks appear in $\sigma_{c.m.}$ for modes of intra-molecular vibrations with wave-vector $2\underline{Q}$; and (c) a peak near or overlapping zero frequency occurs, shifted from zero (as for the CDW) by weak commensurability, impurities, or by the stiffness of the density wave with regard to transverse distortions.

The present theory of infrared conductivity due to density waves began with the Lee, (Maurice) Rice, and Anderson (LRA) response theory.[1] The LRA theory for the CDW including an associated lattice distortion was extended by Michael Rice to include effects of more than one lattice distortional mode or phonon at wave-vector $\underline{Q}$.[2] With this addition, the LRA theory is in most respects adequate for the CDW. However for the SDW case, much of the physics is associated with interaction of the $2\underline{Q}$ CDW harmonic with the lattice, which was not included in the LRA theory. Response theory for the SDW with harmonics was developed by Fenton, Aers, and Psaltakis.[3,4]

The LRA response theory is a <u>complete</u> treatment of the Umklapp vectors associated with the density-wave condensate itself, and it is a <u>conserving</u> theory for the frequency-dependent conductivity. In general, treatment of Umklapp processes is not trivial. As one example, the theory of superconductivity co-existing with antiferromagnetism resulted in rather complicated equations, with complicated interpretations, when the magnetic Umklapp processes were treated incompletely. Treating them completely resulted in simple equations identical in form to the BCS theory.[5] The conserving character of the LRA theory is not trivial either. Response theories which are not conserving result in electron current being created or destroyed by the theory itself, rather than by physics incorporated into the original Hamiltonian. Even the BCS theory had trouble with this point, since singular response occurred for a time-independent longitudinal vector potential (gauge transformation) as well as for a transverse vector potential (magnetic field). More recently, following the famous or infamous observation of a large conductivity peak near 60 K in TTF-TCNQ, Feynman-diagram theories for the conductivity found that phonon softening near a Peierls transition would result in increased, decreased, or negligibly-changed conductivity, depending on which theory is chosen. The differences arose because the theories were not conserving.[6] Errors in the conductivity due to a non-conserving theory are particularly gross and misleading for electron condensates and/or near phase transitions.

Resorting to the jargon of theoreticians, in the LRA theory[1] the three-vertex describing scattering of an electron by the photon is obtained consistently from the self-energy diagram describing propagation of the electron in the metal with the photon absent. This procedure guarantees that the response theory is conserving. The LRA response diagrams for the CDW case are the two diagrams in Figure 1, where the solid line describes propagation of an electron in time and space, the dashed line describes a photon, and the wavy line represents a phonon. These diagrams are obtained from the first two of the Hartree-Fock diagrams in Figure 2 for the CDW electron state in the absence of the photon (by joining the two ends to a photon line and breaking an interior electron line with a second photon line). The dot-dashed line in the third and fifth diagrams of Figure 2 represents the coulomb electron-electron interaction. The third, fourth, and fifth diagrams of Figure 2 do not have corresponding response diagrams in the LRA response theory and so the LRA theory is not complete within the Hartree-Fock scheme. It is conserving however since all the response diagrams are used corresponding to a particular description of the electron in the density-wave system. The SDW with no harmonics is more simple than the CDW case because for the SDW, the second and third diagrams of Figure 2 do not contribute due to spin indices, eliminating associated response diagrams.

Figure 1: LRA response diagrams.

The 2Q harmonic of the SDW occurs because electron states k and k + 2Q are correlated with each other by the SDW correlation in second order, i.e. the k to k + Q times the k + Q to k + 2Q SDW correlations result in the square of a spin-space spinor which is diagonal and therefore represents a CDW correlation at wave-vector 2Q. Q is determined by Fermi surface nesting so that two of the three states k, k + Q, and k + 2Q are on the Fermi surface but the third is off the surface by an energy comparable to E_F. In this case, much like the Cooper pair amplitude in superconductivity theory, one of the two SDW correlations has magnitude $\sim 2\Delta/E_F$. The other SDW correlation has magnitude unity and so the 2Q CDW harmonic of the SDW has amplitude $\sim 2\Delta/E_F$ relative to the fundamental.[3]

When harmonics are included in the SDW case, the 2Q and other even-order CDW harmonics cause a self-energy diagram to occur which is analogous to the second diagram of Figure 2. A collective-mode contribution to the conductivity involving a 2Q phonon then occurs which is analogous to the second diagram in Figure 1 for the CDW case. For the SDW case, response diagrams corresponding to the fourth and fifth "exchange" diagrams of Figure 2 are represented by LRA[1] and by Fenton et al[3,4] as a single diagram similar to the second diagram of Figure 1. In this response diagram the phonon at frequency ω and wave-vector Q or 2Q is replaced by a total electron-electron exchange interaction averaged over frequency and wave-vector. This summed interaction is taken to good approximation as an interaction independent of wave-vector and frequency, since $2\Delta \ll E_F$ and the characteristic plasma frequency for the coulomb part causing the SDW is comparable to E_F. Including the third diagram of Figure 2 is equivalent to a change of the constant interaction in the exchange response diagram. The response theory of Fenton et al[2,3] including harmonics of the SDW is conserving, includes Umklapp processes completely, and like the LRA theory[1] is exact near the limit $\frac{2\Delta}{E_F} \to 0$.

The single-particle term in the conductivity, $\sigma_{s.p.}$, is shown schematically in Figure 3(a). Current relaxation occurs via the Umklapp vectors of the density wave which are responsible for the electron energy gap. However unlike a semiconductor where the source of the Umklapp vectors is fixed to the laboratory table, the incommensurate density wave may move with respect to the host lattice and the table. Density-wave recoil movement is expressed by the second diagram of Figure 1 in this

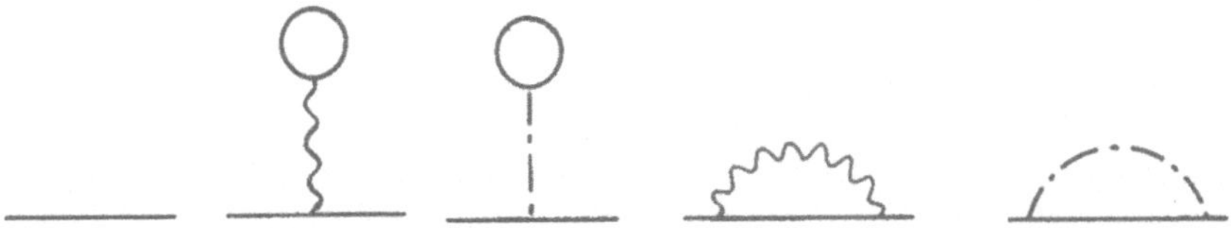

Figure 2: Hartree-Fock diagrams for the electron propagator in a density-wave system.

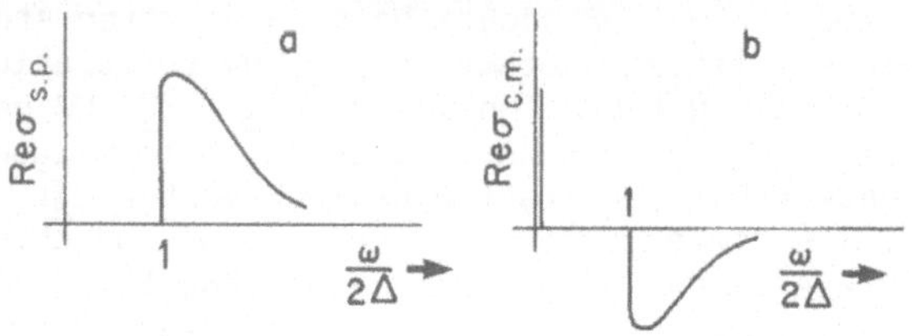

Figure 3: Schematic diagrams representing single-particle and collective-mode parts of conductivity versus frequency.

case. $\sigma_{c.m.}$ is shown schematically in Figure 3(b), where the weights of the broad peak starting at 2Δ and of the delta-function at the origin are both m_{band}/m^* compared to weight unity for the peak in Figure 3(a). For the CDW the effective mass m^* for collective movement of the electrons in the density wave is strongly enhanced, $m^* >> m_{band}$, because a large lattice distortion must follow this movement. In the total conductivity for the CDW case, a small delta-function conductivity peak of weight m_{band}/m^* occurs at zero frequency, and a large and broad conductivity peak of weight $(1-m_{band}/m^*)$ occurs commencing at 2Δ. With $m^* >> m_{band}$, the negative peak in $\sigma_{c.m.}$ starting at 2Δ in Figure 3(b) is only a small partial cancellation of the peak in $\sigma_{s.p.}$ starting at 2Δ in Figure 3(a). Because the total conductivity integrated over frequency is a fixed number (the conductivity sum rule), in σ the combined weight of the 2Δ peak and the peak at the origin is unity. If the density wave is strongly pinned, e.g. by thermal quenching of $(TMTSF)_2ClO_4$, then $\sigma \approx \sigma_{s.p.}$.

In the SDW case, the LRA theory with no $2\underline{Q}$ CDW harmonic leads to $m^* = m_{band}$. In this case the two peaks starting at 2Δ in Figure 3 exactly cancel each other and only a delta-function peak in σ with weight unity occurs at the origin.[1] The SDW with no harmonics does not distort the lattice. In this case, relaxation of the total current cannot occur since there is no interaction with the lattice. In theoretical language, the current operator commutes with the Hamiltonian and so the current is constant[7]: no conductivity occurs at non-zero frequency. When the $2\underline{Q}$ CDW harmonic of the SDW is included, m^*_{SDW} is not exactly m_{band} because of interaction with the lattice. However the electron-phonon coupling constant λ^{ph} for the CDW case becomes multiplied by the square of the relative amplitude of the $2\underline{Q}$ harmonic in the SDW case,[3,4] $\lambda^{ph} \rightarrow (\frac{2\Delta}{E_F})^2 \lambda^{ph} << \lambda^{ph}$, and m^*_{SDW} is not very different from m_{band}.

To this point, we have considered the density wave as interacting with the lattice only via a single phonon at wave-vector $\underline{Q}$ for the CDW case and wave-vector $2\underline{Q}$ for the SDW case. In organic materials with a large number of atoms in the crystal unit cell, the number n of phonons at wave-vector $\underline{Q}$ or $2\underline{Q}$ is large. One of these modes represents vibrations of organic molecules with respect to each other, an inter-molecular mode. If the molecules were rigid, the original LRA theory would result. Michael Rice showed how n-1 symmetric intra-molecular modes, where the molecule expands and contracts, are an important element in the second diagram of Figure 2 and in the condensation energy of the density wave. (In first approximation, frequencies of intra-molecular phonon modes are independent of wave-vector.) Much like the inter-molecular phonon mode, these intra-molecular vibrational modes are made infrared active by the density wave. The density wave mixes the inter-molecular and intra-molecular modes so that the completely soft zero-frequency mode that results (in the absence of pinning of the density wave) is a combination of n lattice distortions, in such a manner that the energy of the incommensurate density wave is independent of its position. The remaining n-1 vibrational modes of the density-wave system are internal vibrations of the non-moving condensate. In actual practice, a correspondence occurs of intra-molecular phonon peaks in the conductivity with nearby "bare" phonon frequencies. Mixing of the n phonon modes with each other and with the electrons depends on a function $f(\omega)$ derived by LRA[1] shown in Figure 4, which shows that effects of the density wave occur from zero frequency to many times 2Δ. The dotted line is Imf and the solid line Ref. (We have included a small imaginary part, $\omega \rightarrow \omega + 0.02\ \Delta i$, to represent removal of divergences by various lifetime effects.)

Physically, the mechanism by which phonon vibrations at wave-vector $\underline{Q}$ (CDW) or $2\underline{Q}$ (SDW) are made optically active in the infrared by the density wave can be understood in a one-dimensional model (although the theory is not restricted to quasi-one-dimensional materials). In this case, $Q = 2k_F$ and local change of $Q(r)$ in a phase oscillation means $2k_F \rightarrow 2k_F(r)$ and therefore $n_{el.} \rightarrow n_{el.}(r)$. Phase oscillation of the short-wavelength density wave produces a long-wavelength oscillation of

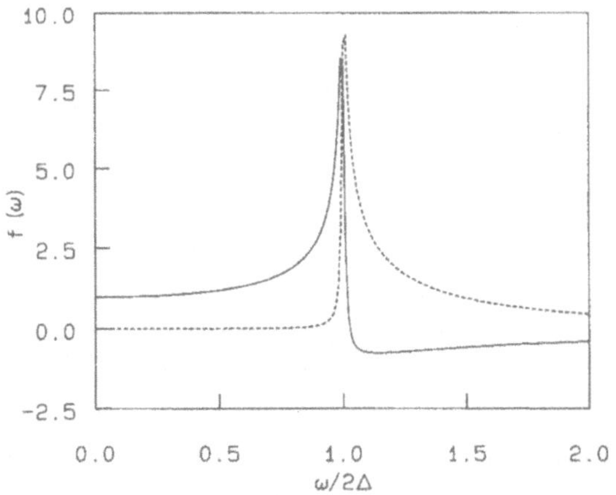

Figure 4: The LRA function $f(\omega)$ versus frequency.

the electron density. This motion is not followed by the compensating background charge and so a dipole moment occurs. Coupling to inter-molecular vibrations occurs because movement of the short-wavelength electron density peaks causes movement of electric-field maxima which causes movement of molecules. Coupling to symmetric intra-molecular vibrations occurs because change of the valence electron density at the $\underline{Q}$-wave maxima causes contraction or expansion of molecules at those maxima.

The mathematics for theory of the infrared conductivity due to density waves appears in detail in our earlier publications and in LRA.[1-4] The final result for the two contributions to the conductivity, $\sigma = \sigma_{s.p.} + \sigma_{c.m.}$, for an incommensurate CDW or SDW with no pinning of the density wave, is:

$$\sigma_{s.p.} = \frac{n_{el.} e^2}{i\omega m_{band}} \left[f\left(\frac{\omega}{2\Delta}\right) - f(0)\right]$$

$$\sigma_{c.m.} = \frac{-n_{el.} e^2}{i\omega m_{band}} \frac{\lambda\left(\frac{\omega}{2\Delta}\right)^2 f^2\left(\frac{\omega}{2\Delta}\right)}{\left[\sum_n \frac{\lambda_n}{\lambda} \frac{\omega_n^2}{\omega^2-\omega_n^2+i\omega\gamma_n}\right]^{-1} + 1 + \lambda\left(\frac{\omega}{2\Delta}\right)^2 f\left(\frac{\omega}{2\Delta}\right)}$$

where $f(x) = \left\{\pi i + \ln\left[\frac{1-(1-x^{-2})^{\frac{1}{2}}}{1+(1-x^{-2})^{\frac{1}{2}}}\right]\right\}/2x^2(1-x^{-2})^{\frac{1}{2}}$. For the CDW in the LRA theory, all the λ_n are dimensionless electron-phonon coupling constants λ_n^{ph} as defined in refs. 1-4. For the SDW[3,4] all but one of the λ_n are coupling constants $\lambda_n^{ph} \times \left(\frac{<2Q>}{<1Q>}\right)^2 \approx \lambda_n^{ph}\left(\frac{2\Delta}{E_F}\right)^2 << \lambda_n^{ph}$. $\lambda_{n=0}$ for the SDW case represents the electron-electron interaction in an exchange self-energy diagram, composed of the coulomb minus the phonon-mediated interactions, averaged over frequency and wave-vector and taken as a constant. The ω_n are frequencies of the $\underline{1Q}$ (CDW) or $\underline{2Q}$ (SDW) bare phonons (where bare means in the absence of the density wave) except for $\omega_{n=0}$ in the SDW case where the plasmon frequency $\omega_0 \approx E_F$ is taken as infinite. In both CDW and SDW cases, the total coupling constant for the density wave is $\lambda = \sum_n \lambda_n$. In mean-field theory, λ satisfies $\Delta = 2E_F e^{-1/\lambda}$ with $2\Delta = 3.5\ k_B T_c$. These relations will usually serve as an order-of-magnitude estimate for λ in actual systems. The γ_n are linewidths for the phonons in the absence of the density wave, except for $\gamma_0 = 0$ in the SDW case.

In Figure 5, we present for illustration model theoretical curves for incommensurate and unpinned CDW and SDW systems, using eqs. 1 and 2. The main differences between the CDW (LRA theory) and SDW cases are: (a) electron-phonon coupling is multiplied for the SDW case by the harmonic ratio factor squared, $(<2Q>/<1Q>)^2 << 1$; (b) the effective mass for electron movement in the density wave satisfies $m^* >> m_{band}$ for the CDW but $m^* \approx m_{band}$ for the SDW; and (c) one of the electron-electron interaction bosons for the SDW has a characteristic frequency which is effectively infinite, which affects shifts of phonon conductivity peaks from bare phonon frequencies. Three bare phonon frequencies ω_1, ω_2, and ω_3, for both CDW and SDW cases, are shown by the vertical dotted lines in the two upper figures. The bare frequency of a fourth phonon is taken as twice the gap frequency for the CDW case, $\frac{\omega_4}{2\Delta} = 2$. (Note the changes of both vertical and horizontal scales from figure to figure.) For the SDW

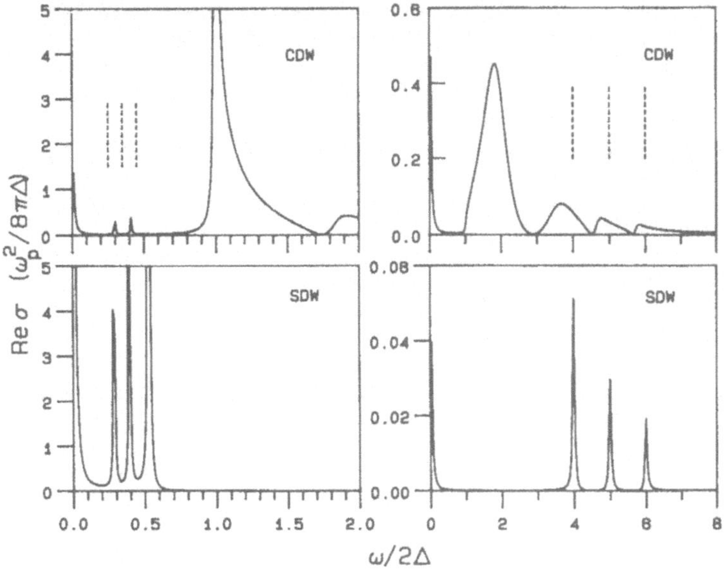

Figure 5: Theoretical curves of conductivity versus frequency using equations 1 and 2.

case, the fourth boson frequency ω_0 is taken as infinite. Pinning frequency of the density wave is taken as zero. λ_n^{ph} are all taken as 0.05. In $\lambda_n = \lambda_n^{ph} \times \left(\frac{<2Q>}{<1Q>}\right)^2$ for the SDW case, we have taken $\frac{<2Q>}{<1Q>} = 0.2$, the largeness of which will be discussed later herein. The coupling constant λ_0 for the coulomb-minus-phonon-mediated exchange interaction in the SDW case is taken as 0.192 so that $\lambda = \sum_n \lambda_n = 0.2$ for both SDW and CDW cases. With this condition, for the same band-structure parameters the mean-field energy gap 2Δ and transition temperature T_c are the same for SDW and CDW. The bare phonon lifetime frequency $\gamma_{n\neq 0}$ is taken as 0.02Δ. We have as well taken $\frac{\omega}{2\Delta} \rightarrow \frac{\omega}{2\Delta} + 0.01i$ in $f\left(\frac{\omega}{2\Delta}\right)$, which represents smoothing of the rise of the single-particle peak at 2Δ due to lifetime effects and an imaginary part in the gap frequency. In Figure 5, no pinning of the density wave occurs due to either intrinsic or impurity effects.

The first noticeable feature in Figure 5 is that a broad single-particle peak commencing at $\frac{\omega}{2\Delta} = 1$ occurs for both CDW figures, but is not visible at all for the SDW figures. As discussed regarding the schematic diagrams in Figure 3, this absence for the SDW is due to nearly-complete cancellation of broad peaks in $\sigma_{s.p.}$ and $\sigma_{c.m.}$, with the resulting broad peak having weight $(1 - m_{band}/m^*)$, which is extremely small for the SDW, even with the 2Q CDW harmonic included. The second noticeable feature is that, even though the effective coupling to phonons in the SDW case is reduced by a factor $(\langle 2Q\rangle/\langle 1Q\rangle)^2 \approx (0.2)^2$ from the CDW case, the phonon peaks in the SDW conductivity are stronger than those in the CDW conductivity. This perhaps-unexpected behaviour is because of the conductivity sum rule combined with absence of the broad single-particle absorption in the case of the unpinned SDW. For the CDW, 1Q phonon frequencies within the 2Δ gap are shifted downward from bare phonon frequencies, whereas for the SDW, 2Q phonons in the gap are shifted upwards. However for the SDW, phonon conductivity peaks above the gap are shifted downward from the bare 2Q phonon frequencies. For the CDW, bare 1Q phonon frequencies above the gap result not in conductivity peaks, but in Fano-resonance nulls of the conductivity due to interference of discrete levels with a continuum arising from the single-particle absorption.

Qualitatively, pinning of the incommensurate density wave will replace nearly delta-function peaks at the origin in Figure 5 with some event at higher frequency which has some broadness.[1,8] The "pinning frequency" for this event depends on the amplitude of the charge-density component of the density wave. This means that the pinning frequency for the SDW will be much lower than for a CDW with the same amplitude, since $\langle 2Q\rangle/\langle 1Q\rangle \ll 1$. The semi-phenomonological theories for pinning by impurities[8] appear questionable because reviewing all of the infrared experiments, generally speaking the entire gap of width 2Δ appears to fill due to impurity effects, rather than a well-defined peak appearing in the gap at some pinning frequency as predicted. The cause of this difficulty in the theory may be that in the semi-phenomonological theories,[8] the LRA function $f(\omega)$ of Figure 4, describing mixing of collective modes with electrons, is replaced by $f(0)$ for all ω.

Intrinsic pinning of the density wave in the response to incident light, where the polarization is transverse to the wave-vector, is associated with the elastic constant for transverse distortion of the density wave. In the phenomonological theory,[8] the square of the pinning frequency ω_p is taken as proportional to the elastic constant divided by m^*, and

$$\omega_p^2 \propto (\delta\rho)^2/m^*$$

The charge density wave amplitude $\delta\rho$ for the CDW is $N(0)\Delta/\lambda$, and $m^* \gg m_{band}$. $N(0)$ is the Fermi-level density of electron states. For the SDW, $\delta\rho \approx \frac{\langle 2Q\rangle}{\langle 1Q\rangle}\frac{N(0)\Delta}{\lambda}$, and $m^* \approx m_{band}$. Experimentally ω_p is comparable to 10 or 20 k_B in KCP and TTF-TCNQ.[8] Using parameters determined experimentally, we estimate $\delta\rho_{(TMTSF)_2PF_6} \approx \delta\rho_{KCP}/200$ and $m^*_{(TMTSF)_2PF_6} \approx m^*_{KCP}/1000$. In this case we expect ω_p to be comparable to 1 to 3 k_B for $(TMTSF)_2PF_6$. The microwave conductivity measurements of Javadi et al on the SDW state below $T_c = 12$ K in $(TMTSF)_2PF_6$ show a well-defined lowest-frequency peak with $\omega_p = 0.5\ k_B$ at $T = 3$ K and $\omega_p = 2k_B$ for $T = 6$ K.[9] The oscillator strength is much larger at 6 K than at 3 K. In view of the conductivity sum rule, which makes intuitive reasoning on oscillator strengths a largely-futile exercise, we would not necessarily expect oscillator strength in this mode to increase as the density wave amplitude increases with decreasing temperature, since collective-mode peaks at higher frequencies may grow rapidly. However it is expected that the intrinsic pinning frequency will increase as $\delta\rho$ increases with decreasing temperature, exactly opposite to what is observed.[9]

A detailed discussion of the relevance of the theory to the infrared optical experiments on SDW states in $(TMTSF)_2X$ appears in the article by Timusk.[10] It appears that the very-weak-pinning case mentioned at the start of our discussion above applies to $X = SbF_6$, and the strong-pinning case applies to $X = ClO_4$ (quenched), PF_6, and AsF_6, for the specimens on which optical infrared measurements were done. In the PF_6 salt, it is not known whether pinning is of comparable magnitude for the specimens in the microwave experiments as for specimens in the infrared experiments. Commensurability pinning should be negligible at least for $X = ClO_4$ and PF_6, because Delrieu et al[11] and Takahashi et al[12] have recently determined from NMR experiments that $\underline{Q}_{PF_6} = (0.5,\ 0.2 \pm 0.05)$ and $\underline{Q}_{ClO_4} = (0.5,\ 0.13 \pm 0.05)$. The SDW wave-vector is definitely incommensurate in the b-direction.[11,12] For an infinitely-stiff SDW, incommensurability in one direction means incommensurability in all directions, since proceeding sufficient length along a line in the incommensurability direction, exactly the same configuration is seen for any position of the SDW in the transverse directions. For the SDW, the b-axis stiffness length $\xi_0 = \hbar v_{bF}/\pi k_B T_c$ is much greater than the lattice constant and in this case any commensurability pinning energy should be small or negligible, in correspondence with zero pinning energy for the infinite-stiffness limit. The fact that a single wave-vector was sufficient to describe the NMR experiments tells us that distortion of the SDW due to commensurability energy is negligible.

Finally, for $X = PF_6$, ClO_4, ... the harmonic amplitude ratio for the SDW should be $\langle 2Q\rangle/\langle 1Q\rangle \approx 2\Delta/0.1\ E_F$ rather than $2\Delta/E_F$. The reason for this is that the a-component of $2\underline{Q}$ is a reciprocal lattice vector of the host lattice, and so $2\underline{Q}$ is parallel to the b-axis in $\underline{k}$-space. In this case, with electron states $\underline{k}$ and $\underline{k} + \underline{Q}$ both on the Fermi surface because of $\underline{Q}$-vector nesting, state $\underline{k} + 2\underline{Q}$ is off the Fermi surface by an energy comparable not to E_F, but to the $\underline{k}_{bF}$ component of E_F which is about $0.1\ E_F$. With a mean-field relation $2\Delta = 3.5\ k_B T_c$ and $T_c \approx 12$ K, and with $E_F \approx 0.25$ eV, this means that $\langle 2Q\rangle/\langle 1Q\rangle \approx 0.2$. Far-infrared measurements on $(TMTSF)_2SbF_6$ by Ng et al[13] show that as temperature is decreased through the SDW transition temperature, three new peaks in the conductivity appear at 154, 197, and 231 cm^{-1}. A theoretical fit to experiment with $\langle 2Q\rangle/\langle 1Q\rangle = 0.02$ in ref. 4 found a theoretical downward shift of the 154 cm^{-1} conductivity peak which was only a very small fraction of 1 cm^{-1}, in comparison to the experimental downward shift of roughly 15 cm^{-1}. Assuming $\underline{Q}$ for the SDW in $(TMTSF)_2SbF_6$ to be similar to $\underline{Q}$ for $X = PF_6$ and ClO_4, changing to $\langle 2Q\rangle/\langle 1Q\rangle = 0.2$ results in a theoretical downward shift which is roughly one third of the experimental shift. (The frequency shift is approximately proportional to $(\langle 2Q\rangle/\langle 1Q\rangle)^2$). With regard to the many uncertainties of experimental parameters, the theory being exact only near the limit $\frac{2\Delta}{E_F} \to 0$, and with the electron density of states taken as flat, order of magnitude agreement of theoretical and experimental frequency shifts is satisfactory.

The theory for the infrared conductivity due to a SDW is a mean-field theory, so far carried out only at zero temperature. There is no disagreement, even by seasoned advocates of the g-ology diagram and related one-dimensional theories, that the SDW state in $(TMTSF)_2X$ salts is almost certainly in the mean-field regime. A SDW state that is definitely in the mean-field regime is the prototype SDW state in chromium (body-centred cubic).[14] This is an example of a very-weakly-pinned SDW, with only an extremely weak conductivity edge and peak at the gap frequency. No sharp phonon peaks occur in conductivity versus frequency for this SDW because there are no intra-molecular or equivalent phonon modes in chromium.

REFERENCES

1. P.A. Lee, T.M. Rice, and P.W. Anderson, Solid State Comm. 14, 703 (1974).
2. M.J. Rice, Phys. Rev. Lett. 37, 36 (1976).
3. E.W. Fenton and G.C. Psaltakis, Solid State Comm. 47, 767 (1983).
4. E.W. Fenton and G.C. Aers, Mol. Cryst. Liq. Cryst. 119, 201 (1985).
5. E.W. Fenton, Solid State Comm. 54, 633 (1985).
6. E.W. Fenton, Solid State Comm. 18, 651 (1976).
7. J.J. Hopfield, Phys. Rev. 139, A419 (1965).
8. Reviewed by A.J. Berlinsky, Rep. Prog. Phys. 42, 1243 (1979).
9. H.H.S. Javadi, S. Sridhar, G. Grüner, L. Chiang, and F. Wudl, Phys. Rev. Lett. 55, 1216 (1985).
10. T. Timusk, Proceedings of the NATO Advanced Study Institute on Low-Dimensional Conductors and Superconductors, Mt. Orford, Quebec, 1986 (Plenum).
11. J.M. Delrieu, M. Roger, Z. Toffano, A. Moradpour, and K. Bechgaard, J. Physique 47, 839 (1986).
12. T. Takahashi, Y. Maniwa, H. Kawamura, and G. Saito, J. Phys. Soc. Japan 55, 1364 (1986).
13. H.K. Ng, T. Timusk, and K. Bechgaard, Mol. Cryst. Liq. Cryst. 119, 191 (1985).
14. E.W. Fenton and C.R. Leavens, J. Phys. F 10, 1853 (1980) and references therein.

RESONANCE RAMAN SPECTROSCOPY OF ORGANIC CONDUCTORS AND SEMICONDUCTORS

André D. Bandrauk

Département de chimie, Faculté des sciences
Université de Sherbrooke
Sherbrooke, Québec J1K 2R1

INTRODUCTION

As emphasized in our study of the phase transition of the quasi-metal TMA I TCNQ (trimethylammonium iodide TCNQ)[1], Resonance Raman Spectroscopy is an efficient method of probing properties of <u>excited</u> electronic states via the intensity dependence of Raman active modes, whereas <u>ground</u> state properties are manifested by frequency and linewidth variations of these modes. In addition, the experimental technique lends itself readily to study single crystals. We have previously done single crystal Raman spectroscopy of neutral D-A systems (eg perylene-TCNQ)[2], and a detailed study of the phase transition in single crystals of the semiconductor KTCNQ[3]. The latter study was instrumental in verifying theoretical models such as the dimerization model of metal-semiconductor transitions[4-8]. Further single crystal work involved a Raman spectroscopy investigation of trimerized systems[9-11] for which again simple models[12] predict similar electron-vibration coupling (vibronic) effects as occurs in dimerization.

All the systems investigated above are one-dimensional systems consisting of stacks or chains which exhibit well known instabilities to phase transitions, usually referred to as the <u>Peierls transition</u>, and are a function of the band filling of the system. Thus for 1/2 filled bands, one expects <u>dimerization</u> (2 k_F) as occurs in KTCNQ, polyacetylene, etc.; for 1/3 filled bands one expects <u>trimerization</u>(2 k_F); 1/4 filled bands give rise to <u>dimerized</u> (4 k_F) and <u>tetramerized</u> (2 k_F) structures[13-15] (as in MEM$(TCNQ)_2$)). We have designated here k_F as the Fermi energy electron wave vector = $\pi\nu/a$, with ν the partial filling, and a is the lattice spacing in the uniform, undistorted and therefore metallic phase. Recently TCNQ compounds with five fold commensurability, i.e., $\nu = 1/5$, have been discovered [16]. In these systems one would expect <u>pentamerization</u> to occur as a result of the Peierls instability of type 2 k_F, although this has not yet been verified spectroscopically.

In view of the high instability of one dimensional partially filled electronic bands to Peierls distortions, we would like to examine in the following what information can be extracted from Raman spectroscopy of typical systems cited above. The Raman information is obviously complementary to much of the infrared spectroscopy measurements usually performed on these systems, either via frequency dependent conductivity measurements[9,14] or direct infrared absorption[17]. Together, these measurements give information on electron-vibration coupling and serve to refine the models alluded to in the present introduction.

DIMER MODEL

A characteristic of organic semiconductors is the transfer of electrons from one molecule to the next along a linear chain of overlapping π molecular orbitals. It is this transfer which couples intramolecular (monomer) vibrations thus producing shifts in the frequencies of the monomer modes. The occupation by an electron of a monomer molecular orbital produces two effects: a) lengthening (shortening) of the bonds if the electron occupies an antibonding (bonding) orbital of the monomer; b) a change in the vibrational frequency. Most of the models cited in the introduction have so far neglected the last aspect, i.e., the frequency change[4-8]. One of the major uses of Raman spectroscopy has been the use of frequency shifts of Raman active modes of the organic ions with respect to the neutral molecules (eg of TCNQ[18] and TTF[19]) to measure the magnitude of the charge transfer. Clearly frequency charges are related to force constant changes. In fact a detailed study of force constants in single molecules from the view point of charge density distributions[20] shows that orbital occupations affect force constants significantly. This implies that a complete description of the effect of charge transfer on electron-vibration couplings must involve at least both linear and quadratic couplings between electron occupations and vibrational coordinates of the monomer molecules[21]. We shall therefore develop here the full dimer model used previously by us[10] to correlate infrared (IR) and Raman spectra in $M_2(TCNQ)_3$ and $M(TCNQ)_2$ systems, and thus examine the consequences of the linear and the quadratic electron-vibration (e-v) coupling approximation on the IR and Raman spectra of a dimer.

Let us consider a pair of identical neutral molecules, referred to as sites with intersite transfer integral $-t$. We assume harmonic vibrations Q_i on each site with linear e-v coupling $\gamma^o = \partial E/\partial Q_i$ and quadratic coupling $\delta = \partial^2 E/\partial Q_i^2$, where E is the energy of a site orbital (π molecular orbital). Then for two identical sites we write the total Hamiltonian as

$$H = -t\,[C_1^+C_2 + C_2^+C_1] + [\gamma^o Q_1 + \delta Q_1^2]C_1^+C_1 + [\gamma^o Q_2 + \delta Q_2^2]C_2^+C_2 + \frac{\omega_o^2}{2}[\dot{Q}_1^2 + Q_1^2 + \dot{Q}_2^2 + Q_2^2] \quad , \tag{1}$$

where the C's are the usual site electronic orbital operators obeying the fermion anticommutation rule $[C,C^+] = 1$, and ω_o is the frequency of isolated neutral molecule vibrations.

Defining symmetric, Q_g, and antisymmetric, Q_u, dimer vibrational modes by $Q_{g,u} = (1/2)^{1/2}[Q_1 \pm Q_2]$, we obtain a new electronic Hamiltonian,

$$H_{el} = -t\,[C_1^+C_2 + C_2^+C_1] + [C_1^+C_1 - C_2^+C_2]\,[\frac{\gamma^o}{2^{1/2}}Q_u + \delta Q_uQ_g] \quad , \tag{2}$$

and a new vibrational Hamiltonian,

$$H_Q = \frac{\omega_o^2}{2}[\dot{Q}_g^2 + Q_g^2 + \dot{Q}_u^2 + Q_u^2] + n_e[\frac{\gamma^o}{2^{1/2}}Q_g + \frac{\delta}{2}(Q_g^2 + Q_u^2)] \quad . \tag{3}$$

The symmetric dimer mode Q_g couples in first order to the total charge $n_e = C_1^+C_1 + C_2^+C_2$. Since this last quantity is a constant, this mode undergoes a linear displacement. The quadratic coupling changes the frequencies of the Q_g and Q_u modes linearly with the total charge n_e. Since δ is purely electronic, it will generally be negative[20]. This linear dependence in frequency as a function of degree of oxidation (loss of electrons) or reduction (acquisition of electrons) has been verified experimentally[18-19]. Neglecting the e-v couplings implicit in equation (2), both Q_g and Q_u modes remain degenerate albeit they are shifted with respect to the neutral monomer frequencies.

The e-v coupling which appears explicitly in equation (2) couples the charge fluctuation $\Delta n = n_1 - n_2$ linearly to the antisymmetric mode Q_u and quadratically to symmetric and antisymmetric modes via the force constant charge δ. This term has not been included in previous dimer[4-8], trimer[12] or tetramer[15] models of e-v couplings[1]. The dimer model represented by equation (1) has been extended to include on site electron repulsion (U) and nearest-neighbor electron repulsions (V), i.e., the extended Hubbard model, by Kral[22], in the site orbital representation (equation 2). This corresponds to a valence bond picture of electronic excitations[23].

We shall pass now from a site orbital representation to a dimer orbital representation via the symmetric and antisymmetric linear combinations of site orbitals, $C_{g,u} = (1/2)^{1/2}\,[C_1 \pm C_2]$, in order to emphasize the inherent symmetries of the dimer model. This is akin to the molecular orbital representation of the dimer, so that in the infinite chain limit one gets Bloch orbitals[23]. The electronic Hamiltonian now becomes in this dimer orbital representation

$$H_{el} = t\,[C_u^+C_u - C_g^+C_g] + [C_g^+C_u + C_u^+C_g]\,(\gamma^o/2^{1/2} + \delta\,\langle Q_g\rangle)Q_u \quad . \tag{4}$$

In equation (4) we have now replaced Q_g by its average $\langle Q_g\rangle$ which from equation (3) corresponds to the displacement $\langle Q_g\rangle = n_e\gamma^o/(2^{1/2}\omega_o^2)$. In this dimer orbital representation which we first used to analyze the Raman spectra of single crystals of KTCNQ[3], one sees that the Q_u mode couples the ground and first excited states which are of opposite symmetry.

The Hamiltonian (4) can be readily diagonalized to give the following adiabatic electronic states

$$\varepsilon_\pm(Q_u) = \pm\,(t^2 + \frac{\gamma^2}{2}\,Q_u^2)^{1/2} \quad , \tag{5}$$

which for the small ratio $\gamma/t < 1$, becomes

$$\varepsilon_\pm(Q_u) = \pm\,(t + \frac{\gamma^2}{4t}\,Q_u^2 + ..) \tag{6}$$

γ is now a <u>new</u> effective linear e-v coupling defined as

$$\gamma = \gamma^o + 2^{1/2}\,\delta\,\langle Q_g\rangle \tag{7}$$

where γ^o is the original "bear" linear e-v coupling obtained by neglecting quadratic couplings (see equation (1)). Since the Q_g mode is displaced by $\langle Q_g \rangle = \gamma^o n_e / 2^{1/2} \omega_o^2$, then we obtain $\gamma \simeq \gamma^o (1 + \delta n_e / \omega_o^2)$. The linear e-v coupling is thus modified to first order linearly as a function of the total charge and the force constant charge. Finally, the frequency of the Q_u modes, and hence the monomer Q_1 and Q_2 modes is shifted in the ground electronic state further by the quantity $- \gamma^2/2t$, i.e., the total expected frequency change with respect to the neutral monomer frequencies is given to first order by,

$$\Delta\omega = (\delta - \gamma^2/2t) \; / \; 2 \; \omega_o \tag{8}$$

where ω_o is the unperturbed neutral monomer frequency. We emphasize that the linear e-v coupling γ now depends on the force constant change δ.

We now return to the symmetry properties of the modes Q_1 and Q_2. The bare e-v coupling $\gamma^o = \partial E / \partial Q_i$, being an electronic force on the nuclei, must by symmetry involve only symmetric vibrations Q_i ($\partial E/\partial Q_i = \langle \partial H_{el} / \partial Q_i \rangle$) since H_{el} is invariant. Thus Q_i can only be a_g (totally symmetric) modes of the monomer. It is instructive at this point to emphasize that quadratic e-v couplings can involve symmetric and antisymmetric monomer modes Q_i, i.e., $\delta = \partial^2 E / \partial Q_i^2$ by symmetry can include antisymmetric Q_i's. In fact one can use frequency shifts of antisymmetric modes in the IR to monitor oxidation states of molecules[24] as is done in Raman spectroscopy for symmetric modes[18-19]. Thus, symmetric monomer modes Q_i (a_g) will couple linearly (γ^o) and quadratically (δ), whereas antisymmetric monomer modes (a_u) will only couple quadratically.

Using the dimer orbital representation explicit in equation (4), we can now write down the IR and Raman amplitudes following our previous work on multiphoton processes in molecules when small linear displacements are operative in the excited state[21]. These are given by the relations

$$T(IR) = \frac{\langle g | \vec{\mu} \cdot \vec{\varepsilon} | u \rangle \; \langle u | \partial H_{el} / \partial Q_u | g \rangle}{(E_g + \hbar\omega - E_u)} \; , \tag{9}$$

$$T(R) = \frac{\langle g | \vec{\mu} \cdot \varepsilon_1 | u \rangle \; \langle u | \partial H_{el} / \partial Q_g | u \rangle \; \langle u | \vec{\mu} \cdot \vec{\varepsilon}_2 | g \rangle}{(E_g + \hbar\omega_1 - E_u) \; (E_g + \hbar\omega_2 - E_u)} \; . \tag{10}$$

In expressions (9) and (10), $\vec{\mu}$ is the electronic dipole (transition moment) which has u (antisymmetric) symmetry, E_g and E_u are the energies of the electronic states with symmetry g and u; ω, ω_1, ω_2 are the IR, incident and scattered ($\omega_1 - \omega_2 = \omega$ (vibration)) Raman frequencies for the electric field polarizations ε, ε_1 and ε_2. Expression (10) also applies to electronic Raman amplitudes provided the final state $|g\rangle$ is now an excited electronic state $|g'\rangle$. The observed experimental intensities are proportional to the squares of the transition amplitudes T.

The amplitudes (9) and (10) show clearly the symmetries of the dimer modes which are expected to be observed. Thus, from (9), one sees that it is the antisymmetric combination Q_u of monomer a_g modes Q_1 and Q_2 which will be IR active as a result of "virtual" coupling of an excited state of opposite symmetry to the ground state. Furthermore, since the IR inversion symmetry center is in the middle of the stacking axis, then $\vec{\varepsilon}$ must be polarized along this axis, i.e., the Q_u mode is activated only by IR radiation polarized along the stacking axis. This is a clear signature of the effect of charge transfer between monomers on the activation in the IR of usually a_g, and therefore infrared inactive modes. Furthermore, as seen in equation (8), the frequency of the Q_u modes differs from the neutral monomer frequencies by the quantity $(\delta - \gamma^2/2t)/2\omega_o$, or from the ionic monomer frequencies by the quantity $- \gamma^2/4\omega_o t$. This last parameter depends on the square of a modified linear e-v coupling (equation (7)) and inversely on the magnitude of the transfer integral t.

Referring now to the Raman amplitude (10), we see that it is the symmetric combination Q_g of the monomer modes Q_i which is strongly activated by excitation into some excited electronic state $|u\rangle$. We note that if one could excite directly into the electronic state $|u\rangle$ generated by the ground state orbitals C_1 and C_2, i.e., $C_u = C_1 - C_2$, equation (4), one would again have access to the linear e-v coupling from the intensity dependence of the Raman amplitude (10), i.e., the matrix element[3] $\langle u|\partial H_{el}/\partial Q_g|u\rangle \simeq 2^{-1/2}(\partial\varepsilon/\partial Q_i)$. Neglecting the second order e-v coupling, equation (2), we notice that the Q_g modes will have the same frequency as the isolated ionic monomer a_g frequencies. Thus one should compare Raman frequencies of the ions in solution to the Raman frequencies in the solid to see whether the second order e-v coupling is negligible.

In summary, totally symmetric (a_g) modes of monomer molecules are activated in the IR by charge transfer processes. As a result, the observed Q_u frequencies are displaced from the ionic dimer symmetric modes Q_g by the quantity $- \gamma^2/2t$, equation (8). These Q_g symmetric frequencies will appear in the Raman spectrum of the solid close to the isolated (solution) ionic a_g modes provided crystal field effects are similar, and in the case of the dimer, provided the second-order e-v couplings arising from frequency changes upon variation in electron occupation can be neglected. The present treatment can be readily extended to trimers[12] and tetramers[15]. In the former, there are two different Q_g modes and one Q_u mode. In the latter there are now two different Q_g and two different Q_u modes, which implies splittings of the Q_i monomer modes in the solid state spectra. Finally, the effect of quadratic interactions is to modify the bare linear e-v coupling γ^o by the quantity $\delta n_e/\omega_o^2 = \omega^2/\omega_o^2 - 1 \simeq 2x$ where x would be the relative change (in %) of the isolated ionic molecule frequencies ω from the isolated neutral frequencies ω_o. Since $x \sim 5$-10%,[18-19] then the linear e-v coupling γ can differ up to 20% from its bare value γ^o.

We now present quantitative evidence of these vibronic effects in table 1 for the compound $NH_4Ni[S_2C_2(CN)_2]_2 \cdot H_2O$ where the nickel bisdithiolene anions

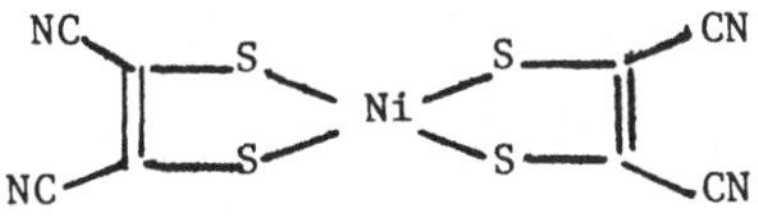

dimerize in a chain[25], the compound $TEDA_2TCNQ_3$ (TEDA = triethylenediamine) in which the TEDA cations form a hydrogenbonded chain parallel to the $(TCNQ_3)^{-2}$ chains[10], and finally KTCNQ whose phase transition is a well documented dimerization[3,17].

In all cases, there is a noticeable frequency shift of the infrared active Q_u modes which are out of phase linear combinations of the symmetric a_g modes shown in the assignment column. This frequency shift of the IR activated modes with respect to the Raman ionic modes is always of the order of 5%. This is the vibronic effect $- \gamma^2/(4t\omega_o)$ of equation (8), which is negative as observed in table 1. The quadratic term or force constant effect on the frequency is the term $\delta/2\omega_o$, which is also negative for anions with respect to neutral molecules. Thus comparing

Table 1. Observed Raman (a_g) modes and corresponding infrared modes. Frequencies in cm^{-1}.

Raman (Q_g)	IR (Q_u)		Assignment
	dimerization[25]: $\{Ni[S_2C_2(CN)_2]_2\}^{-1}$		
1467	1420		C=C
532	517		Ni-S
	trimerization[10]: $TEDA_2TCNQ_3$		
		$TCNQ^o$	
1601	1567	1602	C=C+C-H
1416	1343	1454	C=C
	dimerization[3,17]: KTCNQ		
1605	1580	1602	C=C+C-H
1390	1355	1454	C=C

neutral, $TCNQ^o$, frequencies to the Raman ionic frequencies, in plane modes such as C=C undergo shifts of about 65 cm^{-1}, which again is of the order of 5%. In fact in this case, the quadratic term, δ, has an effect twice that of the linear e-v coupling term γ.

These couplings are important for the stabilization of CDW's in the extended chain. Expanding the site fermion operators $C_j = N^{-1/2} \sum_k C(k) \exp(i\vec{k}\cdot\vec{R}_j)$ and the molecular vibrational displacements $Q_m = \sum_q (\hbar/2N\omega)^{1/2} [b(q) \exp(i\vec{q}\cdot\vec{R}_m) + b^+(q) \exp(-i\vec{q}\cdot\vec{R}_m)]$, the N-site Hamiltonian (1) becomes in the Bloch function representation,

$$H = \sum_k \varepsilon_k^o C_k^+ C_k + \sum_q \gamma(q)\, C^+(k+q)\, C(k)\, [b(q) + b^+(-q)] + \sum_{qq'} \delta(q,q')\, a^+(k+q+q')\, a(k)\, [b(q) + b^+(-q)]\, [b^+(q') + b^+(-q')] , \tag{11}$$

where

$$\gamma(q) = (\hbar/2N\omega)^{1/2} \sum_m (\partial E/\partial Q_m) \exp(-i\vec{q}\cdot\vec{R}_m)$$
$$\delta(q,q') = (\hbar/2N\omega) \sum_{mm'} (\partial^2 E/\partial Q_m \partial Q_{m'}) \exp[-i(\vec{q}\cdot\vec{R}_m + \vec{q}'\cdot\vec{R}_{m'})] . \tag{12}$$

When $q = 2k_F$, $\gamma(q)$ stabilizes $2k_F$ CDW's via linear e-v couplings[26]. One would also expect the quadratic e-v couplings to stabilize $4k_F$ CDW's via the term $\delta(2k_F, 2k_F)$ in view of the larger magnitude of δ with respect to γ.

An example of possible interference between site CDW's and bond alternation CDW's (due to modulation of the transfer integral t) is discussed by Kivelson[27].

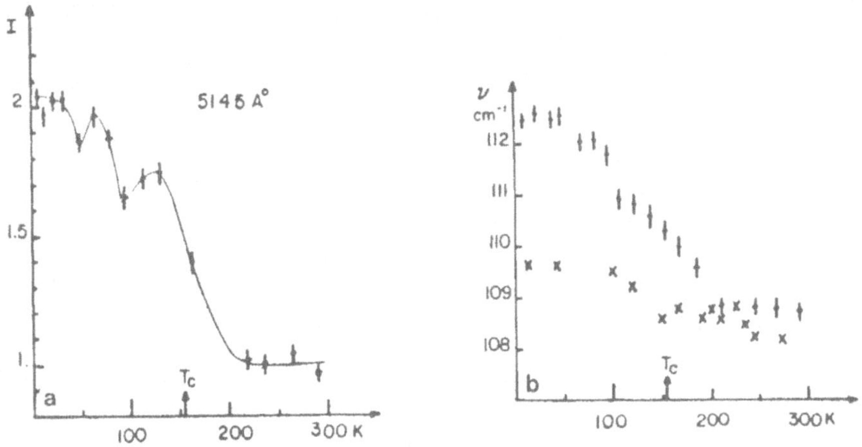

Fig. 1. Resonance Raman Spectroscopy of I_3 symmetric stretch in TMA I TCNQ as a function of temperature.
a) Integrated intensity.
b) Frequency variations for ordered (o) and disordered (x) crystals.

PHASE TRANSITIONS

Phase transitions such as induced by Peierls instabilities have been monitored in the IR by Pecile et al. for alkali TCNQ salts[17], TTF salts[28] and TMTTF salts[29] on powders dispersed in KBr or mulls, or on thin films. Intensity variations of IR active a_g modes as a function of temperature around the phase transition temperature T_c is indicative of vibronic coupling of these modes to the phases of CDW waves which generate the Peierls distortions[4]. In the following we shall give examples of Raman intensity variations of a_g modes in single crystal compounds which undergo metal-semiconductor transitions.

In figure 1 we present the integrated intensity and frequency variations of the symmetric vibration mode of the anion I_3^- in TMA I TCNQ. In both figures one detects three discontinuities corresponding to the three phase transitions known to occur in these compounds[1]. In particular, the main phase transition at 155 K which correlates with a large change of conductivity[30], manifests itself as a change in the vibrational frequency. As discussed in the previous section, this is usually related to a change in oxidation state (i.e. electron occupation) due to quadratic vibration coupling to the electrons in the system. These results therefore demonstrate that in this system, an electronic rearrangement in the I_3^- chain occurs at the phase transition. One possibility is perhaps a small charge transfer between the I_3^- and $TCNQ^{-2/3}$ chains. Thus in high T (metallic) phase, a lower I_3^- frequency would imply partial removal of electron from $TCNQ^{-2/3}$ chains, thus helping electron conductivities. It is normally suggested from X-ray results for this system that the iodides are disordered in the high T phase, thus producing a random potential on the conduction electrons, which does not inhibit delocalization. At low

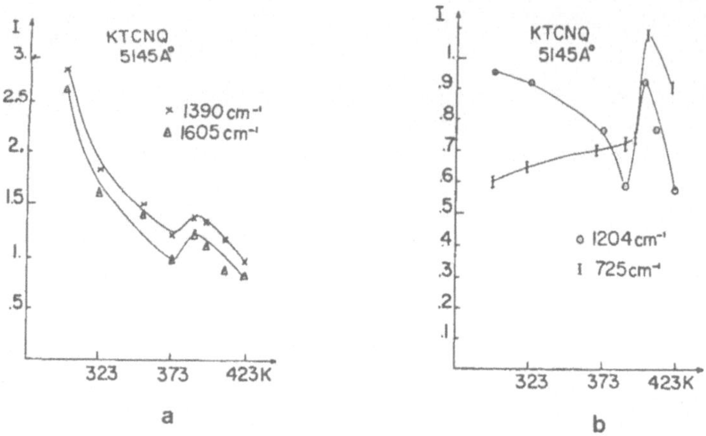

Fig. 2. Resonance Raman Spectrum of single crystal KTCNQ. Intensity variations of a) C=C modes; b) C-H and C-C modes.

T, ordering of the iodides introduces a 3-fold commensurate potential which opens a Fermi gap[30]. The Raman data in figure 1a and b clearly show that the I_3^- electrons are active in the phase transition. Oscillations in the intensities at phase transitions in organic conductors, also appears in other systems as discussed below.

Figure 2 shows the intensity variations of four $TCNQ^-$ modes in the 1:1 alkali salt KTCNQ as a function of temperature[31]. Oscillations in intensities of Resonance Raman spectra occur as a function of laser frequency due to the scanning of various vibrational resonances in the excited electronic states[21,32]. In the case of figures 2a and b, the laser frequency being fixed, the intensity variations must occur as a result of subtle changes in the excited electronic bands and electron-vibration couplings. The excitation frequency in figure 2 was taken to be 514.5 nm so that it would fall between the two localized (L) excitation bands of $TCNQ^-$ at 450 nm and 850 nm[33]. Furthermore the 1390 and 1605 cm^{-1} a_g modes are known to couple linearly (Franck-Condon type transitions[21]) to the red electronic band at 840 nm vhereas the a_g modes at 725 and 1204 cm^{-1} are known to also couple to the blue electronic band (450 nm). This is established by studying the intensity dependence as a function of laser excitation frequency of each mode[34], which behaviour can be predicted from the Raman amplitude, equation (10). One observes that the two pair of modes have the same intensity behaviour within each pair, but differ between pairs, as expected since they are sensitive to different electronic bands. The interesting intensity behaviour however has not been analyzed quantitatively for the simple reason that the excited electronic states and their appropriate vibrational frequencies are not well known. Thus resonance Raman spectroscopy remains at the moment a qualitative tool to examine mechanisms of phase transitions, because intensities depend on excited states. Nevertheless, frequency and linewidth variations reflect ground state properties and as discussed in the previous section are useful tools for examining symmetric modes[3].

Ideally one would like to probe properties which depend on the transfer integral t, equation 1. Clearly, the first excited electronic state, which is normally termed a charge transfer electronic state (CT) is the state to probe as it is a linear combination of the highest occupied ground state orbitals (equation (2)). These are the lowest lying electronic excitations which usually occur in the IR and are known to be extremely sensitive to pressure, contrary to the L bands[33]. Thus excitation with an IR laser in these lowlying bands would be sensitive to translational modes and their emv constants. We have previously suggested that this is the optimum Raman frequency region[5]. Recently Pick et al. reported the resonant enhancement in the red of a translational mode of the TMTSF cation in the PF_6 salt, which would imply resonance enhancement of the translation mode from an electronic excitation at 800 nm[35]. Excitation into the first transfer band of this compound and others would be even more interesting as this would allow one to measure the real coupling of this mode to the actual ground state electron or charge transfer.

Acknowledgments - We thank the Groupe de Recherche sur les Semiconducteurs et Diélectriques of the Faculté des sciences, Université de Sherbrooke, for financial support for the research reported in this article.

REFERENCES

1. A.D. Bandrauk, K.D. Truong, C. Carlone, S. Jandl, Phase transitions in TMA I TCNQ by Resonance Raman Spectroscopy, J. de Phys. 44 (C3): 1473 (1983).
2. A.D. Bandrauk, K.D. Truong, C. Carlone, A new donnor-acceptor system, $(Perylene)_3TCNQ$, Can. J. Chem. 60: 588 (1982); Chem. Phys. 66: 293 (1982).
3. A.D. Bandrauk, K.D. Turong, S. Jandl, Resonance Raman Spectroscopy of the Phase transition in an organic semiconductor, KTCNQ, Can. J. Chem. 60: 1881 (1982).
4. M.J. Rice, N.O. Lipari, S. Strässler, Dimerized organic Linear-Chain Conductors and the Unambiguous Experimental Determination of Electron-Molecular-Vibration Coupling Constants, Phys. Rev. Lett. 39: 1359 (1977).
5. M.J. Rice, Towards the Experimental Determination of the Fundamental Microscopic Parameters of Organic Ion-Radical Compounds, Solid State Commun. 31: 93 (1979).
6. M.J. Rice, V.M. Yartsev, C.S. Jacobsen, Investigation of the Nature of the Unpaired Electron States in the Organic Semiconductor N-methyl-N-ethylmorpholinium-TCNQ, Phys. Rev. B21: 3437 (1980).
7. K. Kral, Charge-Transfer Enhanced Infrared Absorption in Simple Alkaline TCNQ Salts, Czech. J. Phys. B27: 200 (1977).
8. Y. Toyozawa, Charge Transfer Instability with Structural Change I- Two Site-Two Electron Systems, J. Phys. Soc. Jap. 50: 1861 (1981).
9. K.D. Cummings, D.B. Tanner, J.S. Miller, Optical Properties of CS_2TCNQ_3, Phys. Rev. B24: 4143 (1981).
10. A.D. Bandrauk, K.D. Truong, C. Carlone, S. Jandl, K. Ishii, Hydrogen Bonding and Properties of Organic Conductors. I. Infrared and Raman Spectra of Some $M_2(TCNQ)_3$ and $M(TCNQ)_2$ Systems, J. Phys. Chem. 89: 434 (1985).

11. A.D. Bandrauk, K. Ishii, K.D. Truong, M. Aubin, A.W. Hanson, Hydrogen Bonding and Properties of Organic Conductors. II. Electronic Properties and Structure of $M_2(TCNQ)_3$ and $M(TCNQ)_2$ Systems, J. Phys. Chem. 89: 1478 (1985).
12. V.M. Yartsev, Electron-Molecular Vibration Coupling in Trimerized Organic-Ion-Radical Semiconductors, Phys. Stat. Sol. (b) 112: 279 (1982).
13. S. Huizinga, J. Kommandeur, H.T. Jonkman, Magnetic ($2k_F$) and Electronic ($4k_F$) Peierls Transitions from a Hubbard Hamiltonian Extended with Intersite-Dependent Transfer, Phys. Rev. B25: 1717 (1982).
14. V.M. Yartsev, C.S. Jacobsen, Infrared Properties of the Organic Semiconductor $MEM(TCNQ)_2$ in its High-Temperature Phase, Phys. Rev. B24: 6167 (1981).
15. V.M. Yartsev, Charge Transfer and Electron-Molecular Vibration Coupling in Tetramerized Quasi-1D Semiconductors, Phys. Stat. Sol. (b) 126: 501 (1984).
16. G.J. Ashwell, S.C. Wallwork, P.J. Pizkallah, Crystal Structure of $(DPPP)(TCNQ)_5(H_2O)$, Mol. Cryst. Liq. Cryst. 91: 359 (1983).
17. R. Bozio, C. Pecile, Phase Transitions of (1:1) Alkaline Salts of TCNQ-Vibronic Intensity Enhancement in the Infrared Spectra, J. Chem. Phys. 67: 3864 (1977).
18. S. Matsuzaki, R. Kuwata, K. Toyoda, Raman Spectra of Conducting TCNQ Salts; Estimation of the Degree of Charge Transfer from Vibrational Frequencies, Solid State Commun. 33: 403 (1980).
19. R.P. Van Duyne, T.W. Cape, M.R. Suchanski, A.R. Siedle, Determination of the Extent of Charge Transfer in TTF and TCNQ by Resonance Raman Spectroscopy, J. Phys. Chem. 90: 739 (1986).
20. R.F.W. Bader, A.D. Bandrauk, Relaxation of the Molecular Charge Distributions and the Vibrational Force Constant, J. Chem. Phys. 49: 1666 (1968).
21. A.D. Bandrauk, Scattering Theory of One and Two Photon Molecular Processes, Molec. Phys. 28: 1259 (1974).
22. K. Kral, Intermediate Infrared Absorption of Dianion $(2TCNQ)^{-2}$, Czech. J. Phys. B33: 791 (1983).
23. J.C. Slater, "Quantum Theory of Molecules and Solids" - vol. 1, McGraw Hill, N.Y. 1963.
24. J.S. Chappell, A.N. Bloch, W.A. Bryden, M. Maxfield, D.O. Cowan, Degree of Charge Transfer in Organic Conductors by Infrared Absorption Spectroscopy, J. Am. Chem. Soc. 103: 2442 (1981).
25. L.C. Isett, D.M. Rosso, G.L. Battger, Properties of $NH_4Ni[S_2C_2(CN)_2]_2 \cdot H_2O$, Phys. Rev. B22: 4739 (1980).
26. M.J. Rice, C.B. Duke, N.O. Lipari, Intramolecular Vibrational Stabilization of the CDW in Organic Metals, Solid State Commun. 17: 1089 (1975).
27. S. Kivelson, Solitons with Adjustable Charge in a Peierls Insulator, Phys. Rev. B28: 2653 (1983).
28. R. Bozio, C. Pecile, Experimental Study of the Role of Intramolecular Vibrations in the Peierls Transition of Organic Conductors, J. Phys. C13: 6205 (1980).
29. R. Bozio, M. Moneghetti, C. Pecile, Infrared Study of emv Interactions and Phase Transitions in the Organic Conductors $(TMTTF)_2X$, J. Chem. Phys. 76: 5785 (1982).
30. C. Coulon, S. Flandrois, P. Delhaes, C. Hauw, P. Dupuis, Effect of Interchain Coulomb Interactions on the Metal-Insulator Transition in Quasi-One-Dimensional Systems, Phys. Rev. B23: 2850 (1981).
31. K.D. Truong, "Etude par Spectroscopie Raman Resonnante des Semiconducteurs Organiques", Thèse de doctorat, Université de Sherbrooke (1982).

32. D.L. Jeanmaire, R.P. VanDuyne, Tunable Dye Laser Excitation of the Lowest $^2B_{1u}$ Excited State of $TCNQ^-$, J. Am. Chem. Soc. 98: 4034 (1976).
33. M. Tkacz, C.W. Jurgensen, H.G. Drickamer, The Effect of Pressure on Electronic Excitations in TCNQ and its Complexes, J. Chem. Phys. 84: 649 (1986).
34. W.T. Wozniak, G. Depasquali, M.V. Klein, Pre-Resonance Raman Intensity Studies of TCNQ and $TCNQ^-$, Chem. Phys. Lett. 40: 93 (1976).
35. M. Krauzman, H. Poulet, R.M. Pick, Resonant Raman Scattering in $[(TMTSF)_2PF_6]$ Single Crystal, Phys. Rev. B33: 99 (1986).

DISORDER IN ORGANIC CONDUCTORS

Libero Zuppiroli

CEA-IRDI-DMECN-DTech
Laboratoire des Solides Irradiés
Ecole Polytechnique
91128 Palaiseau Cedex

INTRODUCTION

The research devoted to low-dimensional conductors has become so active in the last ten years that rather all the main fields of solid state electronics have been extended to low dimensions. In particular, the role of disorder on the electronic properties of low-dimensional conductors and semi-conductors has been investigated in great detail. Since the year 1975 a large effort has been made by several groups of chemists to produce organic alloys, and the structural and electron properties of the quasi-one-dimensional conductors disordered by chemical doping have been reviewed several times since then (Jacobsen, Mortensen, Anderson and Bechgaard, 1978; Schultz and Craven, 1979; Cooper, Miljak and Korin, 1981; Tomic, Jerome, Mailly, Ribault and Bechgaard, 1983). Irradiation by ionizing particles was used for the first time in organic conductors around 1976 (Chiang, Cohen, Newman and Heeger, 1977; Zuppiroli, Ardonceau, Weger, Bechgaard and Weyl, 1978), and is now accepted as an excellent method for introducing disorder in a controlled way. Doping and irradiation are complementary because, as it will be shown further on, the chemical method produces a large number of weak random potentials, while irradiation creates strong perturbing potentials in concentrations as low as needed.

Organic conductors exhibit a large variety of electronic behaviours and transport properties. Depending on the compound, on the temperature, on the pressure, they can be found as metals, semi-metals, semi-conductors, superconductors and very often, they reach these peculiar electronic ground states through phase transitions some of which are strongly related to the quasi-one-dimensional character of the electron gas. In order to classify all these situations in connection with disorder, and to debate on the "good" questions, it may be of some use, within the present lectures, to remind briefly what are the consequences of disorder on the electronic properties of solids in general.

In a metallic phase characterized by a large number of carriers of the order of one per unit cell, foreign potentials in concentrations low enough to be considered as being independent from each other, mainly act as new scattering centres limiting the electronic mean free path. The Drude's (or Boltzmann's) behaviour is then characterized by the Mathiessen's rule: the total resistivity is the simple sum of a temperature dependent phonon resistivity and a temperature independent defect resistivity proportional to the concentration (Ashcroft and Mermin, 1976, chapter 16). When the random character of the potential chart increases markedly in the metal, the scattering events cannot be considered as independent anymore and the resistivity is dominated by interference effects between scattering events. This concept was used for the first time by Ziman (1961) in a calculation of the resistivity of liquid metals applied somewhat later to amorphous alloys (Sinha, 1970; Boucher, 1972). More recently, interferences between the wave functions of carriers diffusing between two points of the random medium through different paths, was recognised by Bergmann (1984) and others as a reason for quantum localization. Electronic transport in such a localized electronic system reveals some "granularity" in the spatial electron distribution: a charge carrier cannot cover a long distance (with respect to the localization length) without the help of phonons adjusting the electronic energy through inelastic collisions (Thouless, 1977). Thus at zero temperature when electrons can achieve elastic collisions only, the conductivity of a highly disordered metal of localized electrons is zero. In normal metals, the additional disorder associated with phonons causes the mobility to decrease with temperature; the new feature which appears in the high resistivity localized regime is the fact that phonons can increase rather than decrease the electron mobility, thanks to hopping mechanisms which will be examined more accurately further on (Overhof, 1976; Appendix II).

Contrary to a metal, a semi-conductor contains only few charge carriers which result from thermally activated processes dominating the transport. A single foreign potential is a well in which an electron (or a hole) can be localized. In the electronic excitation spectrum, this state appears as a descrete level in the gap (Ashcroft and Mermin, 1976, chapter 28). Disordering a semi-conductor produces nearly always some increase of the number of carriers called doping. But the introduction of too many localized states in the gap can lead finally to the localization of the intrinsic electron states themselves, at least partly, in the band edges. Electron localization in amorphous silicium, and other disordered semi-conductors can already be considered as an old subject: a large part of the good ideas in this field were introduced by Mott and his collaborators (Mott and Davis, 1979). Their approach of disordered systems concerns electrons with mean free paths of the order of one lattice distance. The shape of the wave function envelope determines the character of the state (extended or not): in the case of localized states wave functions and eigenvalues are exponentially decaying with distance, the same is the case of an electronic quantum state in a potential well.

A few questions rise from these scant recalls, when one tries to apply them to organic conductors: what is the role played by electron localization in these systems and to what extent are the quasi-one-dimensional processes different from the three-dimensional ones; what is it about the Drude's behaviour and the Mathiessen's rule; is there any evidence of the existence of states in the gap of an organic semi-conducting phase ?

Most of the time the modern solid state physicist cannot be content with single particles models only, like band theories. The presence of disorder creates more reasons for dissatisfaction and in many cases he (she) cannot interpret the results with models including the simple interplay of extended and localized states only: as a matter of fact disorder slows down electron propagation with the two following consequences: i/ the coupling to the lattice and to its distorsions becomes more efficient; ii/ electrons "feel much more each other" (enhanced Coulomb interactions). Recent theories have explained the deviations to the Boltzmann behaviour in dirty metals by an interference effect between impurity and phonon scattering (Girvin and Jonson, 1980; Belitz and Schirmacher, 1983): a new contribution called defect induced tunneling appears in the conductivity when the Boltzmann hypothesis of independent scattering events fails. In the Alt'shuler and Aronov (1979) interaction model, the interference between impurity scattering and Coulomb scattering is responsible for the anomalous behaviour of the resistivity: the electron impurity scattering process is influenced by the presence of interacting electrons and in turn Coulomb interactions are enhanced when these particles are substantially slowed down by frequent collisions with impurities.

When one approaches these problems of interactions from the side of the disordered semi-conductor rather than the side of the disordered metal, one discovers very soon the polaron and the bipolaron. The former accounts for electron-phonon interaction when it is well localized; it represents an electron self-trapped by a lattice distorsion (in many cases, the presence of polarons can be triggered by disorder (Emin, 1983; Cohen, Economou and Soukoulis, 1983) and the polaron-defect interaction is expected to be large. The latter, is the simplest theoretical object which accounts for electron-phonon interaction and electron-electron interaction at the same time: it is created when the gain in lattice energy due to the bonding of two polarons balances and even exceeds the localized Coulomb repulsion (Cohen, Economou and Soukoulis, 1984). Because of their relative simplicity bipolarons are often called upon for the interpretation of transport properties in narrow band solids including conducting polymers. (Chance, Brédas, Silbey, 1984). In such a case, conductivity is due to the hopping of a doubly charged carrier without spin.

Thus, coming back to organic conductors, it is clear that the introduction of a substantial disorder should revive the old but still important debate about the role of Coulomb interactions (large U and small U models). Of course, the present paper cannot claim to give actual solutions to these difficult problems but disorder, when introduced in a controlled way, creates a large variety of new situations which permit the clarification of a few points.

One has to consider, finally the group of properties which are more directly connected with the quasi-one-dimensional character of organic conductors (charge density waves, phase transitions, fluctuations and collectives modes). Many questions are to be solved, when disorder is added to the pure material. The charge density waves which are responsible of peculiar properties including non linear transport are surely affected by the presence of pinning centres: what does a charge density waves mosaïc look like ?

These are the problems that will be discussed in the present short review on the basis of experiments on alloys and on irradiated compounds. The alloys which have been studied until now are those for which good crystals were available. Actually they are all composed of more or less isostructural and isoelectronic molecules. Thus it is not surprising to find that an irradiation defect has much more effect on the transport and magnetic properties of a linear chain conductor than a foreign molecule of an alloy. It is striking to notice that in the 80 pages of the review paper of Schultz and Graven (1979) about $(TSF)_x\ (TTF)_{1-x}$ TCNQ, despite the variety of experiments and the richness of results, references to disorder and localization appear to be absolutely useless for understanding the properties of the alloy. Nowadays, extensive studies about Bechgaard salts $(TMTSF)_2X$ with non symmetric cations (ClO_4, ReO_4, BrO_4 ...) have revealed the important role of an order disorder transformation in the anionic system for the achievement of a metal to insulator phase transition (Jacobsen, Pedersen, Mortensen, Rindorf, Thorup, Torrance and Bechgaard, 1982). In the recent work of Tomic et al. (1983) concerning the alloy

Table 1. References of a few papers permitting a comparison between chemical disorder and irradiation disorder.

Compound	Chemical impurities	Irradiation
TTF-TCNQ	$(TSF)_x\ (TTF)_{1-x}\ TCNQ$ (Schultz and Craven, 1979; Engler, Scott, Etemad, Penney and Patel, 1977)	(Chiang et al., 1977; Bouffard et al., 1981; Gunning and Heeger, 1979; Sanquer, Bouffard and Forro, 1986)
HMTSF-TCNQ	$(HMTSF)_x\ (HMTTF)_{1-x}\ TCNQ$ (Cooper, Miljak and Korin, 1981)	(Cooper et al., 1981) (Zuppiroli et al., 1978)
TMTSF-DMTCNQ	$(TMTSF)\ (DMTCNQ)_x(MTCNQ)_{1-x}$ (Jacobsen et al., 1978)	(Forro, Janossy, Zuppiroli and Bechgaard, 1982; Forro, Zuppiroli, Pouget and Bechgaard, 1983)
N-propyl-quinolinium $(TCNQ)_2$	$(NPQn)_{1-x}(NEQn)_x(TCNQ)_2$ (Ero-Gecs, Forro, Vancso, Holczer, Mihaly, Janossy, 1979; Janossy, Mihaly, Forro, Cooper, Miljak, Korin, 1982)	(Ero-Gecs et al., 1979; Janossy et al., 1982)
TTT_2I_3	$(TTT)Br_xI_{1.5-x}$ (Mihaly, Zuppiroli, Janossy and Gruner, 1980)	(Mihaly et al., 1980)
$NbSe_3$	$Ta_xNb_{1-x}Se_3$ and $Ti_xNb_{1-x}Se_3$ (Glover, Clark, Azevedo, 1982)	(Fuller, Gruner, Chaikin and Ong, 1981)
$(TMTSF)_2$ ClO_4	$(TMTSF)_{1-x}(TMTTF)_xClO_4$ (Coulon, 1983) $(TMTSF)_2ClO_4)_{1-x}(ReO_4)_x$ (Tomic et al., 1983)	(Sanquer and Bouffard, 1985)

$(TMTSF)_2\ (ClO_4)_{1-x}\ (ReO_4)_x$ disorder appears only as an element of the competition between the antiferromagnetic semi-conducting phase and the superconducting one. But there is no reference to disorder or to localization in the interpretation of the transport properties in both phases. These are the reasons why, in most of the cases, the irradiation experiments give much more information about disorder effects than the results obtained with chemical doping.

A common feeling is that irradiation breaks everything in the crystal like a hammer used by a blind man. This is perfectly wrong: irradiation by ionizing particules (photons X, Γ or electrons) is a local chemistry of excited molecules (Zuppiroli, 1982). It produces usually molecular point defects in well defined configurations and in concentrations, the absolute values of which are usually known within a factor of two; the relative damages produced by a given particle in different targets of the same compound can, of course, be known very accurately because they are simply proportionnal to the dose; furthermore concentration of defects as low as required can be obtained very easily. Part of the most recent results regarding radiation damage are summarized in Appendix I: most of the defects consist in extra-bonds linking chains and molecules to each other. The references of a few papers permitting an interesting comparison between disorder introduced by alloying and by irradiation have been collected in table 1.

FROM QUINOLINIUM TO BECHGAARD SALTS

The room temperature conductivity of quinolinium $(TCNQ)_2$ or acridinium $(TCNQ)_2$ is of the order of 100 $(\Omega.cm)^{-1}$. Within a temperature range of 100 degrees their resistivity increases with increasing temperature like in a metal but the ratio $\rho(350\ K)/\rho(250\ K)$ is only 1.2 to 1.5. That's the way an organic metal looked like in the early seventies and many people, at that time, thought that these limitations were due to disorder. At lower temperatures the conductivity is more semi-conductor like, but the Arrhenius plots represented in figure 1 are curved. Usually, people involved in the study of the transport properties of disordered solids, always try to obtain straight lines from curved Arrhenius plots by changing the temperature scale from 1/T to $(1/T)^m$ with m = 1/4, 1/3 or 1/2. This was done in quinolinium $(TCNQ)_2$ by Bloch, Weisman and Varma (1972) who found that 1/2 was the good exponent while by fitting independently the same curve Brening, Döhler and Heyszenau (1972) found that 1/3 was better. An agreement with exponent 1/4 could have been possible too. But most people prefered 1/2 which is the exponent of variable range hopping in one dimension (see Appendix II). As a matter of fact, it was well known since 1961 that in a strictly one-dimensional system all electronic states are localised by any, arbitrarily small, random disorder and then low temperature conduction can only occur through hopping mechanisms (Borland, 1961; Mott and Twose, 1961). This feeling, that disorder was playing the most important role, rooted deeply in communist countries on the basis of theories (Berezinsky, 1974; Bulaevskii, Gusseinov, Evemenko, Topnikov and Schegolev, 1975; Gogolin, Mel'nikov and Rashba, 1976) and of new experiments, confirming the granularity of quinolinium $(TCNQ)_2$ and related compounds, from the point of view of electron propagation: the observation of dielectric constants of the order of several thousands was attributed soon to distributions of barriers or to localization (Gogolin, Zolotukhin, Mel'nikov, Rashba and Schegolev, 1975; Holczer, Gruner, Mihaly and Janossy, 1979; Gruner, 1981) while the existence of localized spins showing themselves in the $T^{-\alpha}$ tails of the magnetic susceptibility versus temperature curves, were attributed to disorder too (Schegolev, 1972; Cooper, Miljak and Korin, 1981; Tippie and Clark, 1981). Figure 2 shows such a tail illustrating this magnetic effect that will be discussed further on. Since the localized character of the electronic system was establihed so well, the metallic-like conductivity at higher temperatures had to be justified: this was done somewhat later by calling upon the delocalizing effect of phonons (Gogolin et al., 1976). Thus, for a few years, a coherent picture of transport and magnetism in organic conductors, based on disorder and localization was in fashion.

The later discovery in TTF-TCNQ of a more extended metallic state with a resistivity ratio $\rho(300\ K)/\rho(60\ K)$ of the order of 10, shifted the general interest towards new molecular conducting crystals which were probably less disordered than the former. The Orsay group revealed the existence of the 2k and 4k distortions due to charge density waves (CDW) and the tri-dimensional ordering of the 2k at low temperatures completing the Peierls transition. Then organic conductors were considered definitely as, charge density waves metals or Peierls insulators without any reference to disorder and the properties of $Qn(TCNQ)_2$ were completely revisited in terms of a Peierls semi-conductor: the curvature in the Arrhenius plot was attributed to a temperature dependent mobility of the carriers in a pure semiconductor (Epstein, Conwell, Sandman and Miller, 1977).

In the late seventies and the early eighties, Jerome and Bechgaard have not rested until they succeeded to extend the metallic state to lower and lower temperatures and to discover finally organic superconductivity in $(TMTSF)_2\ PF_6$. A typical Bechgaard salt like $(TMTSF)_2ClO_4$ does not exhibit the slightest low temperature upturn in the resistivity versus temperature curve attribuable to disorder but a ratio $\rho(300\ K)/\rho(4.2\ K)$ of the order of 100 and when defects are introduced at concentrations ranging from 10^{-5} to 10^{-3} mole fraction, the low temperature resistivity varies linearly with the concentration like in usual metals (see figure 3)

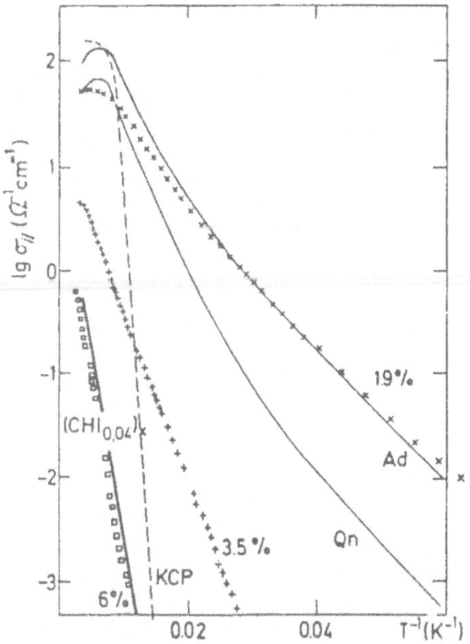

Figure 1 - Logarithm of absolute conductivity versus reciprocal temperature curves of various linear chain conductors (KCP, dotted line); $Qn(TCNQ)_2$, $Ad(TCNQ)_2$, iodine doped polyacetylene, full lines) compared to the curves of TMTSF-DMTCNQ containing various amouts of radiation induced defects (crosses and open squares). Several authors have tried to understand the properties of these linear chain conductors in terms of disorder (see text).

(Sanquer and Bouffard, 1987). These experimental facts seem to be in contradiction with the results of 1D localization theories that will be reviewed in the following section while the reasons of this contrast will be discussed further on.

ONE DIMENSIONAL LOCALIZATION THEORIES

It is easy to suspect that lowering the dimensionality below 3 has some important effect on the conditions of localization. This property is even reflected in the energy spectrum of a single electron in a potential well, as it is calculated through a straightforward integration of the Schrodinger's equation (Landau and Lifshits, 1956). The result is that the existence of a bounded state in a 3D well is bound to a condition on the width a and the depth U of the well ($Ua^2 > \pi^2 h^2/8m$), while the discrete spectrum exists for any well in one-dimension and even in two-dimensions.

The problem of an electron travelling through a collection of random wells forming a random potential with general form V(x) is less straightforward but has the following solution in 1D: the hamiltonian $-(d^2/dx^2) + V(x)$ has a surely pure point spectrum with exponentially decaying eigenfunctions when V(x) is any random potential describing any arbitrarily chosen small degree of disorder (Mott and Twose, 1961; Bush, 1972; Bentosela, Carmona, Duclos, Simon, Souillard and Weder, 1983). Usual experiments, even when performed at very low temperatures, fail to measure hamiltonian spectra or wave functions. Thus this kind of theoretical result has to be handled with some care by experimentalists measuring, for instance, transport properties; because, when one applies an electrical field F, the hamiltonian $-(d^2/dx^2) + V(x) + Fx$ has a purely

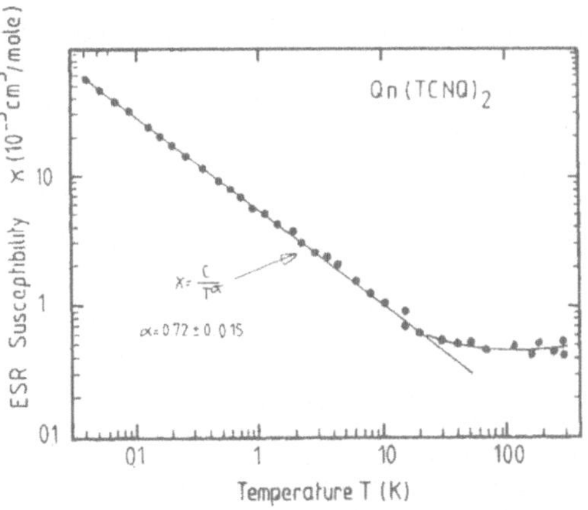

Figure 2 - Spin susceptibility as a function of temperature for a sample of $Qn(TCNQ)_2$. The power law at low T is evidence of the presence of localited spins (Experiments from Tippie and Clark, 1981).

absolutely continuous spectrum for any arbitrarily large disorder V(x) (Bentosela et al., 1983; Delyon, Simon and Souillard, 1984). Qualitative arguments can help to understand this apparent dilemma. Localization means that electron tunneling through the random potential has a limited range, called localization length L, but applying an electrical field is tilting the base line of the potential and then tunneling at long distances becomes possible; of course it can take a long time !

It is thus true that all states are localized in 1D but the "strength" of localization, reflected in the reciprocal length L^{-1} is essential for understanding real transport. When phonons are superimposed on this one-electron picture, it will be seen further on, that, from the point of view of transport, localized electrons are sometimes indistinguishable from a Drude's gas.

Gogolin (1982, 1983) had also the opportunity to distinguish between different regimes of exponential localization in his study of a 1D chain with identical impurities distributed at random, based on the Berezinsky (1974) method.

The interference effects between Coulomb and impurity scattering in 1D (i.e. the influence of Coulomb interactions on 1D localization) is a problem which has raised a recent theoretical interest (Chui and Bray, 1977, 1979; Apel and Rice, 1982, 1983; Saso, Suzumura and Fukuyama, 1985; Giamarchi and Schulz, 1987). Their results are not surprising:

- Cooper pariring, i.e. attractive interactions between electrons, when strong enough, can overcome the localizing effects of a random potential (pairs are not scattered by impurities);
- conversely disorder acts on retardation and prevents pairing when driven by too weak attractive interactions;
- on the contrary, repulsive interactions are always a help to localization because, if it is hard for a single electron to tunnel through a random potential, especially in 1D, this becomes even harder when the supplementary condition of electrons avoiding each other is imposed by the repulsive interactions.

These intuitive arguments can be summarized in an usual (g_1, g_2) phase diagram. Depending on the details of the calculations, the quantitative results are slightly different from author to author. Figure 4 showns, for example, the results of a simple extended Hubbard model calculation, with on site interaction U and nearest neighbour interaction V, taken out from a work of Suzumura (1985). In the japanese work (Suzumura and Fukuyama, 1984), the use of a phase hamiltonian, with phase variables representing charge and spin density fluctuations, encourages a representation of the disordered systems in terms of impurity pinning of either charge or spin density waves.

DELOCALIZING EFFECTS OF INTERCHAIN COUPLING AND PHONONS

At zero temperature all single electron states are localized in 1D and the introduction of Coulomb interactions eith realistic coupling constants localizes the states even more. How can one understand that Bechgaard salts are so good metals even at low temperatures and even in the presence of irradiation defects introduced in molecular concentrations of the order of 10^{-3} (Chaikin, Choi, Haen, Engler and Greene, 1982) ?

Part of the trouble is attribuable to the fact that, at least at low temperatures, Bechgaard salts are not one-dimensional at all. A transfer integral in the transverse direction of about t_b = 30 meV is not something negligeable and one may even wonder if it is justified to treat it as a simple perturbation of a 1D ground state. At higher dimensions than one, it is easier for an electron to escape localization and the potential due to disorder has to be random enough to force it to be still confined in space. In the Anderson model of localization (Mott and Davies, 1979) the "localizing strength" of the random potential is represented by the width δ of the random potential distribution measured in units of the transfer integral t; the localization condition in 3D is written $\delta/t \geq \zeta$ where ζ is a constant related to the topology of the system. The only study of Anderson localization in a strongly anisotropic system with two transfer integrals t and t' is due to Shante and Cohen (1976) and Shante (1978). They found localization to occur in the anisotropic system, when disorder δ exceeds the value $\zeta\ (tt')^{\frac{1}{2}}$. This threshold is, of course, lower than the 3D isotropic velue ζt, but larger than the 1D value ($\delta = 0$). Such an effect can surely explain the curve of figure 3 recorded in $(TMTSF)_2ClO_4$

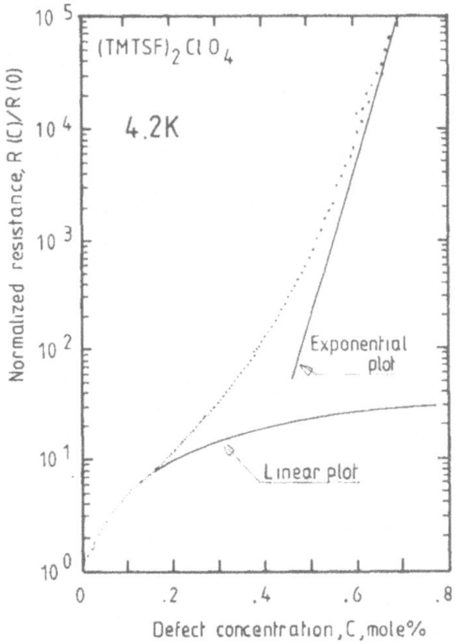

Figure 3 - The Bechgaard salt $(TMTSF)_2ClO_4$ has been irradiated at low temperatures in the metallic (relaxed) state. Until a defect concentration of 2.10^{-3} the resistance increase is linear with the concentration, (Boltzmann's behaviour) while it is exponential after 6×10^{-3}. The threshold to localization is due to the transverse coupling (Sanquer and Bouffard, 1985).

irradiated at low temperature (4.2 K). The x-axis just scales with the disorder δ the y-axis represents the resulting resistance increase at low temperature: a clear cross-over between a Drude's linear behaviour and an exponential behaviour is observed at concentrations of the order of 10^{-3} (molecular fraction of irradiation defects).

The effect of transverse coupling esplains the absence of localization in moderately disordered Bechgaard salts at low temperatures but the same argument cannot be applied to more one-dimensional systems at higher temperatures. How to understand for example the extended metallic state observed in TTF-TCNQ despite the fact that the mean free path is of the order of one to ten molecular distances only ? The delocalizing effect of phonons has been considered by several authors as the reason for the Drude's behaviour in systems such as metallic TTF-TCNQ and related compounds (Gogolin, Mel'nikov and Rashba, 1976; Kaveh, Weger and Gutfreund, 1979). All their arguments are based on the Rashba's calculation of the conductivity of a 1D electron gas coupled to 3D phonons. Disorder is introduced in the model in form of a concentration c of identical impurities with a short range foreign potential in form of a δ distribution. To the reflexion coefficient γ of the single impurity on the chain and to the concentration c, is associated an elastic mean free path $l = a/\gamma c$ (a is the lattice spacing on the chain). The localization length L is supposed to be large with respect to the distance between impurities a/c. Under these conditions, the result is that, despite the fact that all states are localized, one gets a sort of Drude's law for the conductivity: it is proportional to the elastic mean free path l and to a correcting factor l/l' which is the ratio of the elastic to the inelastic mean free path. We are just in a case where phonons help conduction in a very extreme way. The 1D calculation of Gogolin et al. has something similar, at least from the point of view of this result, with the 3D calculations of Belitz and Schirmacher (1983) mentioned in the introduction.

SPIN LOCALIZATION AT LOW TEMPERATURES

One important reason why there are no evidences of localization in the transport properties of organic metals and alloys is simply related to the trivial fact that, at low temperatures, the metallic state is unstable with respect to the opening of a gap at the Fermi level (CDW, SDW or super-

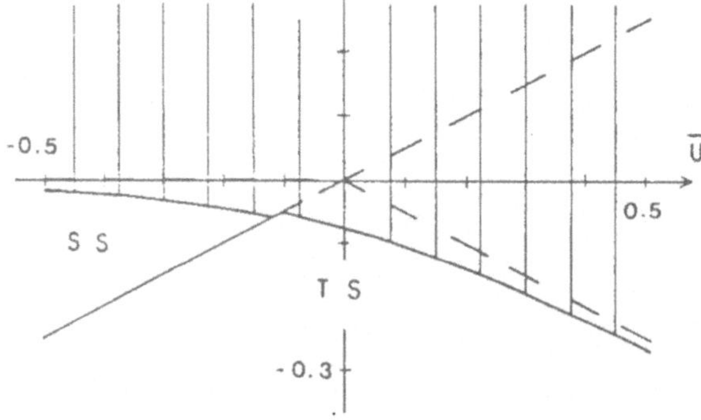

Figure 4 - Phase diagram for the one-dimensional half-filled extended Hubbard model in the presence of impurities where $\bar{U} = U/4\pi t$, $\bar{V} = V/4\pi t$, the shaded area corresponds to the localized region and the dashed line denotes the boundary for the clean system. (From Suzumura, 1985).

conductivity gap). For this reason the temperature range of exploration of the quasi-1D-metals is limited by a low bound, to temperatures where phonons are still too active and/or interchain coupling too efficient. Fortunately, when a gap is opened at the Fermi level in a non magnetic material, the magnetic susceptibility associated to the electron gas vanishes. It is thus possible to observe, at low temperatures, the possible magnetic contributions due to localized states. Indeed, this has been observed in most of the organic conductors and a lot of experimental work on these low temperatures magnetic proporties has been done in Zagreb (Cooper, Miljak and Korin, 1981; Korin-Hamzic, Miljak and Cooper, 1982) and in other groups (Tippie and Clark, 1981; Sanquer, 1986).

The presence of localized spins is reflected either in a sort of Curie tail ($T^{-\alpha}$ laws) in the susceptibility versus temperature curve (figure 2) or in the presence of a lorentzian line in the electron spin resonance (ESR) spectrum at low temperatures. These spins are, without doubt, directly related to disorder: in alloys as well as in irradiated samples their concentration is proportional to the defect concentration (Cooper et al., 1981, Korin et al., 1982; Zuppiroli, Delhaes and Amiel, 1982). They are not rigidity attached to a molecule (such as a dandling bond or a free radical) for the following reasons:

- in alloys such as $(HMTSF)_x(HMTTF)_{1-x}TCNQ$, for example (Cooper et al., 1981), they appear in low number (1 :100) with respect to the number of defects;
- more generally, no sign of hyperfine structure has ever been recorded on the ESR line even at very low temperatures (in $Qn(TCNQ)_2$, Tippie and Clark (1981) have recorded the ESR line until 0.01 K (figure 2)),
- the study of the spin dynamics, especially in $Qn(TCNQ)_2$ (Devreux, Nechtschein, Gruner, 1980; Glover, Clark and Azevedo, 1982) shows that indeed these spins are travelling very fast.

Thus, the low temperature localized spins owe their existence to the granalarity of the material and are not attached to a very precise location; a similar effect has been recorded in phosphorus doped silicon shere $T^{-\alpha}$ laws have been observed too (Bhatt and Lee, 1982).

The most simple granular material in which localized spins do appear in this way is LiF containing small metallic platlets of lithium (Taupin, 1967). Due to a size effect, the electronic levels are significantly separeted by an energy $\delta \geq k_B T$. The condition of charge neutrality of each particle within the insulator implies that, on average, half of the platlets contain an odd number of spins and half of them an even number. Because of the granularity reflected in the distribution of energies δ, even particles behave, as a whole, like an entity of spin 1/2 (spin flip involves an energy of the order of δ). This very simple idea has been applied to organic conductors: even magnetic segments bounded by defects can enclose an unpaired spin; these spins 1/2 interact with each other to give a $T^{-\alpha}$ law at low temperatures (Korin et al., 1982).

It is important to notice that the simgle electron character of this class of models is, in many cases, only a false pretence: the only way to maintain the separation between even and odd particles is to impose separately the electric neutrality of each of them and the only reason to do so, when tunneling between particles is possible, is to consider non negligeable Coulomb interactions preventing electron-hole pair creation (Coulomb gap). Thus except in the case of LiF where tunneling is impossible, because the particles are too far apart, this kind of magnetism requires two independent conditions: disorder as a source of granularity (or frustruation) and Coulomb interactions as a way of maintaining the electrons apart (when tunneling is permitted) and preventing spin pairing.

Both these ingredients are present in all the models which account for $T^{-\alpha}$ laws in organic conductors. Bondeson and Soos (1980) emphazised the role of Coulomb interactions by assuming (in a big U limit) that, basically, the spins are ordered within antiferromagnetic Heisenberg chains and that disorder can be viewed as the cause of random exchange between spins: in this picture, the unpaired spins of the $T^{-\alpha}$ laws are defects in the antiferromagnetic reference order. These models are called REHAC (i.e. random exchange Heisenberg antiferromagnetic chains). A moderate U model which seems more realistic has been developped by Gorkov, Dorokhov and Prigara (1978): they found a concentration of spins c_s, contributing to the Curie-like law, of the order of (U/t) (a/L), where U/t represents the strength of the Coulomb forces with respect to the transfer integral and L is the localization length which appears as the size of the magnetic segment.

FROM THE WEAK DISORDER IN ALLOYS TO THE STRONG GRANULARITY IN IRRADIATED SYSTEMS

The best way of comparing the strengths of the defects in different disordered systems is to look at the lattice parameters changes (or volume changes) of crystals when a known amount of these defects are introduced.

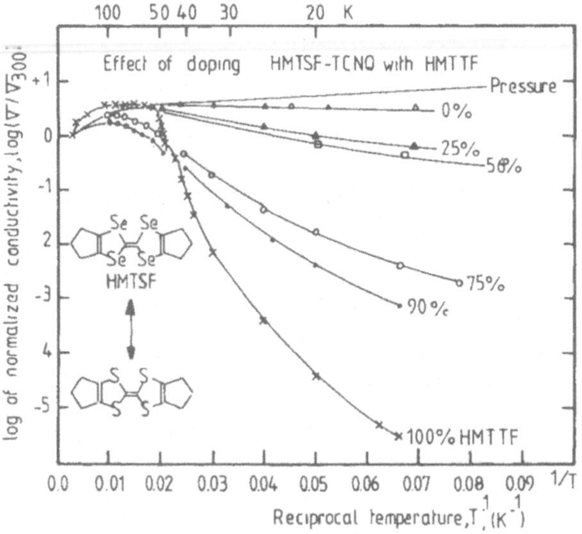

Figure 5 - Conductivity versus temperature curves for the alloy $(HMTSF)_{1-x}(HMTTF)_xTCNQ$ (unpublished results from K. BECHGAARD).

In the alloy $(TSF)_x(TTF)_{1-x}TCNQ$, Engler et al., (1977) have measured the lattice parameters very carefully and have found, per foreign molecule:

$$\frac{1}{a}\frac{da}{dx} = 3 \times 10^{-3} \qquad \frac{1}{b}\frac{db}{dx} = 3 \times 10^{-2} \qquad \frac{1}{c}\frac{dc}{dx} = 5 \times 10^{-4}$$

More recently, Trouilloud, Bouffard, Ardonceau and Zuppiroli (1982) measured the dilatation of TTF-TCNQ disordered by irradiation, and Zuppiroli, Housseau, Forro, Guillot and Pelissier (1986) measured the changes of lattice parameters in the same irradiated samples. The consistent results of these two independent experiments are the following:

$$\frac{1}{a}\frac{da}{dx} = 0.36 \qquad \frac{1}{b}\frac{db}{dx} = 0.5 \qquad \frac{1}{c}\frac{dc}{dx} = 0.1$$

These results on TTF-TCNQ give an idea on the respective strength of the forces deriving from the potential around type of defect. It is worth noticing that irradiation defects have important interstack effects while weak chemical impurities show only significant intrastack effects

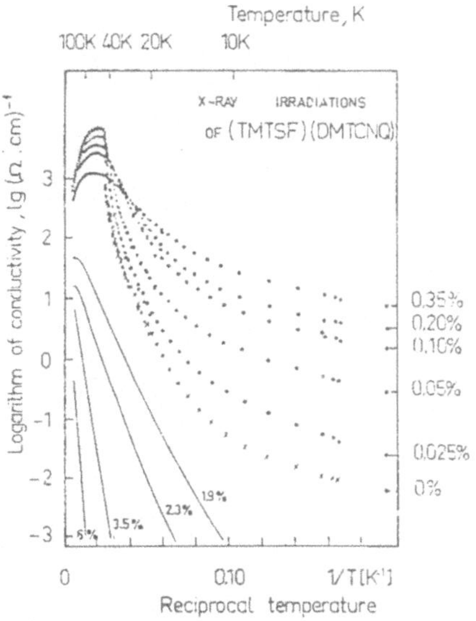

Figure 6 - Conductivity versus temperature curves of TMTSF-DMTCNQ submitted to x-ray irradiations. On each curve the defect concentration is expressed in mole percent. At high doses (continous curves) the conductivity becomes single activated over the whole temperature range.

(db/dx). Chemical alloying is simply replacing part of the molecules by isostructural and isoelectronic ones, while irradiating is primarily cross-linking the crystal in the three main directions, thanks to new chemical bonds between molecular units, as shown in Appendix I.

It is not surprising to realize that organic alloys do not exhibit transport anomalies due to disorder. Figure 5 shows once more, that there is no trace of localization in the transport properties of $(HMTSF)_x(HMTTF)_{1-x}$ TCNQ in the whole concentration range $0 \leq x \leq 1$. One can only observe, in the set of curves of figure 5, a smooth interpolation of the properties of the two pure compounds (x = 0; x = 1) by the curves of the alloys. On the contrary, as it has already been mentioned in the previous section about magnetic properties, localized spins do appear in this alloy but in a very small concentration (1 spin per 100 defects in the alloys instead of 1 spin per defect in irradiated samples).

The last part of this section will concern irradiation results only. By imposing the presence of these strong defects in concentrations from 10^{-3} to 10^{-1} mole fraction, it is possible to change drastically the transport properties. The signs of a new type of granularity emerge even at high temperatures and will be interpreted carefully in the following section.

There is no room in this short chapter for an extensive presentation of irradiation results. A more extended experimental review will be published elsewhere (Zuppiroli, 1987). The present summary will only stress the more general points, and will focus on relevant effects for the purpose of localization studies: the interesting defect molecular concentrations are of the order of 1 %. Figure 6 shows the conductivity versus temperature curves of TMTSF-DMTCNQ, over a large range of defect concentrations from 2.5×10^{-4} to 6 %. It is clear on these curves that there are interesting irradiation effects at low doses too, but here the interest will be primarily concentrated on the higher dose curves. Let us summarize the most important and general results.

i/ Except at low concentrations (c << 1 %) and low temperatures (figure 4) the resistivity of a sample irradiated in the metallic phase varies exponentially with the concentration of defects over several orders of magnitude (up to 6 orders); a typical rate of increase is of one order of magnitude per percent of defects. The transverse resistivity varies exponentially too (Mihaly, Bouffard, Zuppiroli and Bechgaard, 1980; Zuppiroli, Bouffard, Bechgaard, Hilti and Mayer, 1980).

ii/ Nothing drastic occurs in the reflectance spectra of irradiated samples in the vicinity of the Drude edge: the optical relaxation times, optical conductivities and plasma frequencies change by a few ten percent only when the dc conductivity demonstrates a several orders of magnitude variation (Zuppiroli, Jacobsen and Bechgaard, 1985).

iii/ The higher the temperature, the higher the damage rate $d\rho/dc$ for a given concentration c; this is a violation of the Mathiessen's rule (see the introduction) which predicts a rate independent of temperature (Chiang et al., 1977).

iv/ As soon as defect concentrations of a few percent are reached, the conductivity versus temperature curves are not metallic anymore but single-activated over the whole temperature range, with activation energies depending linearly on the defect concentration $\rho \sim \exp(\epsilon c/k_B T)$ (Zuppiroli et al., 1980; Mihaly et al., 1980; Holczer, Gruner, Mihaly and Janossy, 1979).

v/ Curie-like ($T^{-\alpha}$) laws do appear in the low temperature magnetic susceptibility. The ratio between the number of spins and the number of defects is approximately 1 to 1, when it is 1 to 100 in alloys (Korin-Hamzic et al., 1982; Zuppiroli et al., 1982).

vi/ The ESR linewidth (spin relaxation rate) is not increasing with the defect concentration as expected in usual metals but it is strongly decreasing with c (Forro, Bouffard and Zuppiroli, 1982; Sanquer, Bouffard and Forro, 1985).
vii/The giant dielectric constant measured in $Qn(TCNQ)_2$ and mentioned previously decreases with concentration like c^{-2} (Janossy, Holczer, Hsieh, Jackson and Zettl, 1982).

All these experimental facts were, for the most partm demontrated in several organic conductors. They reveal a very clear granularity of the material from the point of view of the transport and magnetic properties. Indeed the simplest way to account for this whole set of properties is an interrupted strands model with transverse hopping which is a model of 1D metallic particles separated by insulating barriers (Zuppiroli, 1984; Rice and Bernasconi, 1973) that will be presented in next section.

TRANSPORT AND MAGNETIC PROPERTIES OF A GRANULAR SYSTEM OF ONE-DIMENSIONAL METALLIC PARTICLES

Energy Separation and Charging Energy

According to the classical work of Abeles, Sheng, Coutts and Arie (1975) concerning 3D particles, the transfer of electrons between two grains of granular material separated by an insulating barrier is controlled by two different effects. The first one is what they call phonon assisted tunneling between the grains (hopping). This is the effect related to the energy separation as shown in figure 7. In this process the hopping frequency between the two grains can be written $1/\tau = (1/\tau).\ \exp -\delta/k_BT$ where δ is of the order of the average energy separation. The energy spacing δ is of the order of the ration of the bandwidth in the bulk, 4t, to the number of conducting electrons n in the cluster. A typical value for 100 A particles in 3D is δ = 0.1 meV. The exponential term in the previous expression of $1/\gamma$ is due to the phonon creation or annihilation probability (see Appendix II).

The second process controlling the electron transfer is related to the charging energy E_c required to transfer an electron between the two

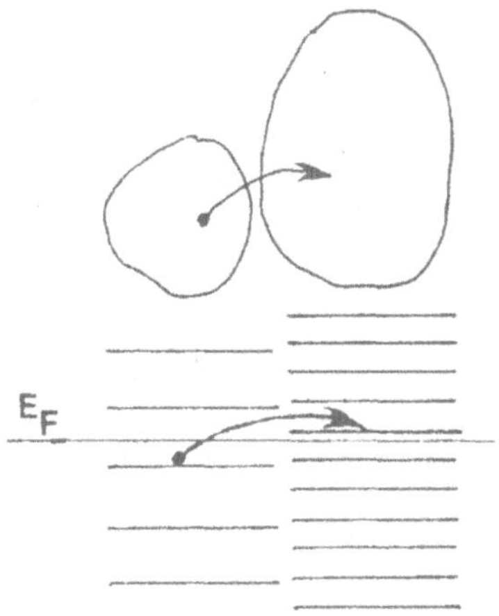

Figure 7 - Phonon assisted tunneling in a granular material.

neighbouring grains. E_c is of the order of $e^2/C_g = 2e^2/Kd$ where e is the electronic charge, C_g the effective grain capacitance taking into account the screening by surrounding grains, K is the effective dielectric constant and d the grain size. In typical 3D granular films d = 100 A and K = 10 so that E_c = 30 meV is very large compared to δ = 0.1 meV and dominates the tunneling rate $\gamma^{-1} = \gamma_0^{-1} \exp - E_c/2K_BT$.

The Conductivity of Irradiated Organic Conductors

In the case of segments produced by irradiation of quasi-1D-conductors, the energy separation δ reflecting the size effect is much larger (for a given particle size) than in 3D. For a particle of 100 A, for example, δ is ~ 10 meV in 1D instead of ~ .1 meV in 3D.

The charging energy $E_c(n)$ of a segment of size n is the Coulomb energy of an extra charge sitting in this segment. If U is the on-site Coulomb interaction, the charging energy of the segment is $E_c(n) \sim U/n$, the denominator expressing the fact that the larger the segment, the higher its capacitance C_g and the lower the total Coulomb energy. The charging energy $E_c(n)$ has to be compared to the energy separation $\delta(n) \sim 4t/n$ where 4t is the bandwidth. It is thus difficult to distinguish between the two proposed mechanisms: phonons assisted tunneling or excitation through the Coulomb gap $E_c(n)$, because finally the comparison between E_c and δ is equivalent to the comparison between U and t. Many experiments and theories seem to converge on the idea that U is at least of the same order than t (see the present proceedings, for example); but in any case the tunneling rate between segments can always be written

$$\gamma^{-1} = \gamma_0^{-1} \exp - \frac{\Delta E}{k_BT} \tag{1}$$

where ΔE is proportional to the larger of the two energies U and t.

The local problem of the electron jumping between two grains having been solved, the next step in the computation of transverse and longitudinal conductivities is to account for the electron percolation through the whole sample. This has been done in the paper of Zuppiroli et al., (1980) and in Appendix II; the final result is the following:

$$\rho_\perp (\bar{n}, T) = \rho_\perp (\infty, T) \exp (\varepsilon/\bar{n}\, k_B\, T) \tag{2}$$

$$\rho_{||} (\bar{n}, T) = \rho_{||} (\alpha/\bar{n})\, \rho_\perp (\infty, T) \exp (\varepsilon/\bar{n}\, k_BT) \tag{3}$$

where $\rho_\perp (\infty, T)$ and $\rho_{||} (\infty, T)$ are the resistivities of the pure sample with infinite chains, $\bar{n}$ is the average segment length expressed in number of molecules ($\bar{n} = 1/c$, where c is the defect molecular concentration), α is a constant of the order of one and ε an energy of the order of the bandwith or of the on-site Coulomb interaction U. The applicability of the above model to irradiated organic conductors has been demonstrated in the paper of Mihaly et al., (1980) and since that time in several other papers (Zuppiroli, 1987).

The Conduction Electron Spin Resonance

The ESR signal of small metallic particles was expected to reflect the quantum size effects. The theoretical aspects of the conduction electron spin resonance were revisited a few years ago by Buttet, Car and Myles (1982) who have reviewed most of the experiments on the subject too. What is qualitatively expected from the size effect on the ESR line ?

It is worth reminding first that in the bulk of a normal homogenous metal, the ESR linewidth ΔH is proportional to the spin relaxation rate that is to say the frequency of collisions which reverse the spin. Eliott (1954) demonstrated thirty years ago that the spin relaxation rate is in turn proportional to the total relaxation rate $1/\gamma$, the proportionality coefficient Δg^2 measuring the strength of the spin-orbit coupling $\Delta H \sim \Delta g^2/\gamma$. The consequence of the Eliott relation is as follows: increasing the disorder in a metal increases the total collision rate $1/\gamma$ and in turn the linewidth ΔH (Beuneu and Monod, 1976). Irradiated organic conductors behave exactly in the opposite way: the linewidth decreases with increasing disorder (Forro et al., 1982; Sanquer et al., 1985). In 3D metallic particles too the linewidth was demonstrated to decrease markedly with decreasing size of the clusters and was interpreted as the quenching of the Eliott mechanism by the level structure (Frolich, 1937; Taupin, 1967; Kawabata, 1970).

In 1D granular materials, the phenomenological theory of spin relaxation is based on the fact that in an isolated infinite chain no relaxation is possible. Any collision reversing the spin has to be associated with an escape from the chain. This has led Weger to replace the total collision rate in the Eliott formula by the transverse escape rate $1/\gamma_\perp$ (Weger, 1978) $\Delta H_{1D} \sim (\Delta g)^2\ \gamma_\perp$. When the assembly of metallic chains is transformed by irradiation in an assembly of segments, the transverse escape rate is decreased just as shown in formula 1 and the linewidth in the granular system follows $\gamma_\perp^{-1}$:

$$\Delta H\ (\bar{n},\ T) \sim \Delta g^2/\gamma_\perp \sim \Delta H\ (\infty,\ T)\ \exp -\ (\varepsilon/\bar{n}\ k_B\ T) \qquad (4)$$

The validity of the above formula is demonstrated by ESR experiments performed on several organic conductors (Forro et al., 1982, Sanquer et al., 1985).

Partial Conclusions

In the previous section entitled "Form the weak disorder in alloys to the strong granularity in irradiated systems" it has been shown that organic conductors containing irradiation defects at the level of a few mole percent are looking all the same from the point of view of their transport or magnetic properties. These large similarities between systems shich can be very different have suggested a very simple and general explanation in terms of granular materials of segments bounded by defects. Formulas 2-5 can account for the conductivities and the ESR linewidth very well. The presence of $T^{-\alpha}$ laws due to localized spins was explained in terms of magnetic segments bounded by defects in the section entitled "Spin localization at low temperatures". The divergence of the dielectric constant as n^2 can also be explained in terms of an interrupted strong model (Janossy et al., 1982; Rice and Bernasconi, 1973).

In the search for localization effects in the transport properties of quasi-1D-conductors one is disturbed by interchain coupling and phonons hiding the electronic granularity. Alloys do not contain defects strong enough to overcome these delocalizing effects. In irradiated samples the granularity is evident, but rigid, like in usual granular materials composed of mixtures of metallic and insulating particles, and can be interpreted without the help of the weak localization theories in dirty metals mentioned in the introduction.

THE STRUCTURAL EVIDENCES OF A CHARGE DENSITY WAVES 3D PINNING

Organic (and inorganic) quasi-1D-conductors are not simple metals or semi-conductors: their ground state is related to the existence of spin density waves, superconductivity or charge density waves always characterized by the opening of some king of gap at the Fermi level. The last part of this review will be devoted to the study of disorder in the charge density waves, and states in the gaps.

There are thousands of evidences of the CDW pinning from the studies of the CDW dynamics in the presence of a dc and/or an ac electric field, but from the point of view of the pinning process, these experiments are hardly ever direct; more direct evidences based on X-rays diffuse scattering and electron diffraction experiments are reported here.

From the theoretical point of view, most of the interpretations in the field of pinning or depinning are dominated by the phase equation and phase hamiltonian derived from the Fukuyama, Lee, Rice and Anderson approach (see for example the paper of Coppersmith in the present proceedings). This approach accounts for the gross features in the CDW dynamics but actually it is an over simplified description of the problem for the following reasons:

i/ The CDW is assumed to be an elastic (linear) medium: all the non linearities are included in the pinning and those coming from the spatial amplitude variations of the wave are neglected.
ii/ The usual calculations and numerical simulations are strictly limited to 1D: no transverse coupling between the CDW's and a strictly short range (δ distributions) 1D pinning.
iii/ The damping of the CDW is introduced in the equation in a completely phenomenological way and even the physical qualitative ideas of the origin of this overdamping are vague and controversial.
iv/ The possibilities of conversion between single carriers and the CDW condensate are not considered.

Such an oversimplified version which does not add very much to the classical overdamped oscillation model (see Gruner in the present proceedings) reveals to be appropriate for the study of "charge density waves in vacuum" but more sophisticated or alternative models (see Bjelis and Friedel in the present proceedings) will probalby permit to have a closer look at new aspects of CDW dynamics in real materials.

The aim of the present section is to show, at least in the case of strong pinning centers produced by irradiation that:

i/ pinning acts primarily on the transverse coherence lengths of the CDW and is thus a 3D process,
ii/ at least in 2D-CDW systems like TaS_2 or $TaSe_2$ phase defects, domains and boundaries can be identified,
iii/ at least in organic conductors and TaS_2, strong pinning produces a conversion between the condensed and free carriers through the presence of states in the Peierls gap.

The X-ray diffuse scattering patterns of TMTSF-DMTCNQ and TTF-TFCNQ or the electron diffraction patterns of TaS_2 and TaS_3 were followed and recorded as functions of the defect concentrations (Zuppiroli, Mutka and Bouffard, 1982; Forro, Zuppiroli, Poujet and Bechgaard, 1983; Forro, Bouffard and Poujet, 1984; Mutka, Bouffard and Zuppiroli, 1985). Despite the large differences between all these CDW systems the structural experiments give qualitatively the same results in all of them, simplified in the scheme of figure 8.

i/ Strong pinning centers, present in concentrations of the order of. 10^{-3} (at. fraction in inorganic and mole fraction in organic systems) prevent the low temperature ordering of the charge density waves and disturb their commensurability when they would like to lock on the lattice.
ii/ Defects disturb primarily the TRANSVERSE ordering of the charge density waves as shown in figure 8 (in the case of TMTSF-DMTCNQ, Forro et al., 1983, have measured both the transverse and longitudinal coherence lengths).
iii/ In the case of TaS_2 and $TaSe_2$ (Mutka and Housseau, 1983; Mutka et al., 1985; Bird, Eaglesham, Withers, Mc Kernan, Steeds and Wills, 1985) dark field and high resolution micrographs have shown that the number of microdomains and phase defects of the CDW is closely related to the number of lattice defects.

In 1D, the only possible singularity in the CDW due to non-linear effects is the commensurability soliton which does not seem to be really helpful in the description of the CDW dynamics; on the contrary, in 3D, the non-linear character of the CDW can be the source of many kind of singularities (discommensuration, phase dislocations etc...). The description of the CDW motion by the 3D motion of its singularities is probably a difficult but fruitful approach. It opens the possibility to take into account the interaction of the wave with single carriers emitted and absorbed at these singularities. As a matter of fact, the motion of a charge density wave through a forest of strong pinning centers defining rigidly the phase in several points of the sample, requires the conversion of a part of the condensate in free carriers and vice-versa in order to overcome the obstacles (Hall, Hundley and Zettl, 1986); some kind of conversion has also to occur at the contacts and can be described by phase slippages or cortices (Gorkhov 1983, Ong, Verma and Maki, 1984).

The introduction of strong pinning centers increases very much the number of free carriers within the CDW "insulating" phase. This fact was demonstrated several years ago in TMTSF-DMTCNQ, TaS_2, TTF-TCNQ and even in the spin density wave phse of $(TMTSF)_2$ PF_6 (Zuppiroli, Mutka and Bouffard, 1982; Forro, Janossy, Zuppiroli and Bechgaard, 1982; Forro, 1982; Mutka, Zuppiroli, Molinié and Bourgoin, 1981). Indeed, the hall coefficient is strongly decreasing with the concentration of strong pinning centers created by irradiation. This large increase of the number of carriers reflected in the several orders of magnitude increase of the low temperature conductivity visible in figure 6 (low defect concentrations), is due either to the defects themselves which create states in the gap or to the CDW discommensurations network induced by the strong pinning centers; indeed, charge density waves dislocation kinks are also doping centers creating free carriers through states in the gap.

STATES IN THE GAP : SOLITONS POLARONS AND BIPOLARONS

The experimental exploration of the gap of quasi-iD electronic systems is an open question, the interest of qhich is not restricted to CDW systems but to every kind of gap. In order to achieve an accurate exploration, it is easier to choose quasi-1D systems with gaps significantly larger than the usual hundreds of degrees of organic CDW gaps. Experiments have been performed with this purpose in N-Methyl derivatives of pyridinium with TCNQ (Zuppiroli, Przybylski and Pukacki, 1984).

N-Methyl derivatives of Pyridinium with TCNQ are 1: 2 charge transfer salts of tetracyanoquinodimethane. Despite the fact that the TCNQ chains should be a priori quarter filled bands, these systems are insulators with a gap of approximately 0.6 eV due to a large on site Coulomb interaction U. Conductivity and thermopower data of pure and irra-

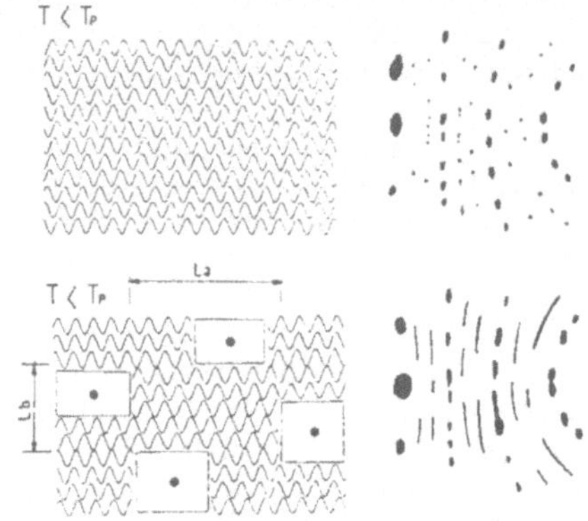

Figure 8 - Irradiation defects have been introduced in the ordered CDW phase of an organic conductor. The changes in the diffuse X ray scattering patterns, reflect the mosaicity of the wave (Forro et al., 1983, 1984).

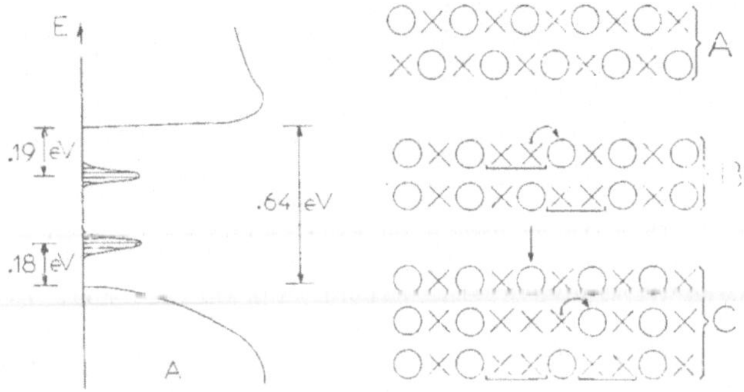

Figure 9 - Single particle excitation gap deduced from conductivity and thermopower experiments in N Me 4 Py $(TCNQ)_2$. The electron hole-symmetry is a general feature of transport in N-Methyl derivatives of pyridinium with TCNQ. It is true in nominally pure samples as well as in irradiated ones. The second part of the figure schematizes one of the possible excitation (soliton) accounting for such a gap: these are Wigner lattices in a quarter filled chain (Hubbard, 1976); the crosses are electrons and the circles holes; in A, the two configurations of the degenerate ground state are shown; B represents the soliton and its propagation. In C a supplementary charge has been injected at the position of the arrow; its propagation reveals the equivalence of this charge to two previous defects.

diated compounds can be successfully fitted to a simple semi-conductor model with deep lying localized impurity states which influence conduction at low temperatures by their thermal ionization. These fits, performed on three different derivatives, lead to a picture of the gap illustrated in figure 9. The numbers giving the position of the level are defined with an accuracy of 0.01 eV. Those on the figure refer to NMe 4MePy $(TCNQ)_2$. The surprising fact, is that the new levels introduced by irradiation in the 3 different compounds studied, are found to coincide, within the experimental accuracy, with the donor and acceptor levels present in the nominally pure samples. What is even more surprising is the fact that irradiation introduces donor and acceptor levels at the same time and that the electron hole symmetry of the gap is conserved. Thus the dominating defect in N-Methyl derivatives of pyridinium with TCNQ is not an ordinary state in the gap whose position depends very much on the precise irradiation defect configuration, but more likely some kind of soliton of charge 1/2 predicted by Hubbard (1978) and Rice and Mele (1981). Figure 9 gives a picture of this excitation in term of Wigner lattices in a quarter filled chain.

At the end of this section it is worth mentioning the short story of such intrinsic localized elementary excitations influencing the transport in quasi-1D semi-conductors.

i/ The soliton-antisoliton pair was first associated by Su, Schrieffer and Heeger (1980) to the degenerated ground state of trans-polyacetylene. This defect corresponds to a midgap state in the energy excitation spectrum. When charged, the soliton is a spinless excitation.
ii/ The search for charged but spinless excitations in other doped polymers than trans-polyacetylene with no degenerated ground state, leads to the introduction of the bipolaron instead of the soliton (Chance, Bredas and Silbey 1984). To this elementary excitation mentioned in the introduction, correspond two symmetric states in the gap of polyparaphenylene, for example.
iii/ Finally the soliton and more recently the bipolaron were "injected" in the gap of organic conductors, including $Qn(TCNQ)_2$ and even the Bechgaard salts (Conwell and Howard, 1985).

CONCLUSION

The main topics which are discussed in the present review about disorder in organic conductors are electron localization, spin localization, assemlies of one-dimensional metallic particles, charge density waves pinning and states in the gaps. All these questions are 1D applications of more general 2D or 3D problems evoked in the introduction. Thanks to many experimental results including irradiation experiments, and to a few pertinent theories, most of these problems are already understood, at least qualitatively. Perhaps the most urgent open questions in the field of disordered low-dimensional systems are still concerning charge density waves pinning and states in the gap.

APPENDIX I : RADIATION DAMAGE IN ORGANIC CONDUCTORS

The starting point of the defect production by ionizing particles, such as x-rays, Γ-rays or electrons, is a local ionization or excitation of an atom of a molecule. Because of the high polarizability of the molecules composing organic conductors, this kind of excitation can travel very easily and quickly within the molecule. It can also travel quickly in the crystal, from one molecular species to another through a particular form of charge transfer, or along the chain, because of the high electronic overlaps in this direction. Most of the excited states will recombine either radiatively, or non radiatively through multiphonon processes. A small part of them only will produce more or less stable free radicals, which will in turn recombine: there are no signs, in fact, of the presence of charged stable free radicals in irradiated organic conductors, but a small fraction of unstable radicals will remain in the structure long enough to participate in a local chemistry leading to the creation of a neutral stable defect. Some very important properties of irradiation defects in organic conductors are related to the radiolytic nature of the production process which results from the existence of an excited state and the competition between all the escape and recombination processes on the one hand and the limited possibility of local chemistry on the other hand:

i/ The concentration of defects c is proportional to the energy E absorbed by the crystal through electronic excitations. This energy can be calculated very easily using atomic considerations only (Mihaly and Zuppiroli, 1982); most of the equivalences between absorbed doses and defect concentrations can be found in the following papers: Zuppiroli, 1982; Zuppiroli, Jacobsen and Bechgaard, 1985; Zuppiroli, Bouffard and Jacob, 1985. For example, 10.8 MGy (mega Grey) of energy absorbed in $(TMTSF)_2ClO_4$ per unit mass (1 Gy = 1 Joule/kg) produces 1 mole % of spins as observed in the low temperature Curie like tail (Sanquer and Bouffard, 1985).

ii/ There are only limited possibilities for a local chemistry of excited molecules to take place ans this leads to a relatively low number of possible defect centers.

Two recent experiements have shown that most of the defects consist in extra-bonds linking chains and molecules to each other. It is worth summarizing the results of these experiments here. The first one has consisted in following the diffraction pattern of several organic crystals, including TTF-TCNQ, as a function of irradiation (Zuppiroli et al., 1986). The behaviour of Bragg spots is quite surprising: their intensity decreases ses to zero but during this process the spots remain sharp and are not displaced very much. Surprisingly, irradiation affects short range order only, and a reasonably long range order persists when the damage is sufficient to affect virtually all the molecules. Some kind of cross-linking is probably the main condition for such a process to occur in a damaged organic crystal because it attaches the molecules to each other and even at high doses, when the molecular crystal is "carbonized", it still retains several features of the initial crystalline structure.

The role of cross-linking was confirmed in the cas of $Qn(TCNQ)_2$ by a mas-spectrometry experiment of Mermilliod and Sellier (1983). They found that one of the most important defects in the structure, but presumably not the only one, results from the addition of a quinolinium radical to the cross-linking between two adjacent chains.

APPENDIX II : HOPPING BETWEEN LOCALIZED STATES ; HOPPING BETWEEN SEGMENTS

The inelastic process called hopping and dominating electron propagation is a phonon assisted tunneling between localized electronic states, (see figure 10a). Phonon assistance is required to provide the small energy difference between localized electrons on neighbouring sites. The hopping rate Γ_{ij} between two given sites i and j has been calculated by Miller and Abrahams (1960) (see also Ambegaokar, Halpering and Langer, 1971): it contains a spatial tunneling factor exp $(-2\alpha R_{ij})$ where α is the reciprocal localization length, and an activated factor exp $(-E_{ij}/K_BT)$ which gives the probability of a phonon to be created or absorbed with the necessary energy

$$\Gamma_{ij} = \gamma_0 \exp(-2\alpha R_{ij}) \exp(-E_{ij}/k_BT); \quad E_{ij} = 1/2\,(|E_i|+|E_j|+|E_i-E_j|) \quad (1)$$

where E_i and E_j are defined with respect to the Fermi level (origin of the energies.

Miller and Abrahams, and Ambegaokar, Halperin and Langer have demonstrated also the equivalence of the hopping problem, for a random distributition of impurities, with a resistance network with admittances (see figure 10b). The percolation aspect of this problem was emphazised by Pollak (1972) and Pike and Seager (1974). Once the hopping conductance is known between every given pair of sites, one has to find the conductance of the earier paths which cross all the sample and dominated the conductivity. It is useful to distinguish between two important cases, thus following the excellent review of Overhof (1976).

i/ Fixed range hopping in compensated, p-type germanium: when only hops between neighbouring sites can occur over a large temperature range, the critical paths do not change with temperature. In this case the conductivity versus temperature curve is finally activated with an activation energy which is a complex average of the values E_{ij} (Overhof, 1976).
ii/ Variable range hopping in amorphous silicon: when the critical path changes with the temperature, that is to say when further jumps than to the nearest neighbour are permitted, the percolation arguments give the same result as a very simple Mott's argument that it is worth repeating here: according to Mott, the decisive contribution to the conductivity results from hops with a minimal exponent in $\Gamma = \gamma_0 \exp(-2\alpha R - E/k_BT)$. The minimisation has also to account (in 3D) for the realtion $(4\pi/3)\, R^3 E N_F \sim 1$ representing the condition that, on average, one pair of sites with an energy difference E is present in the volume $(4\pi/3)R^3$ (N_F is the density of states at the Fermi energy). Combining these two relations one obtains

the condition : $\frac{d}{dR}\left|2\alpha R + \frac{3}{4\pi R^3 N_F k_B T}\right| = 0,$ which leads in 3D

to a conductivity : $\sigma = \sigma_0 \exp(-T_0/T)^{\frac{1}{4}}$

and in 1D : $\sigma \quad \sigma_0 \exp(-T_0/T)^{\frac{1}{2}}$ (Bloch et al., 1972).

$G_{ij} = (e^2/k_BT)\,\Gamma_{ij}$

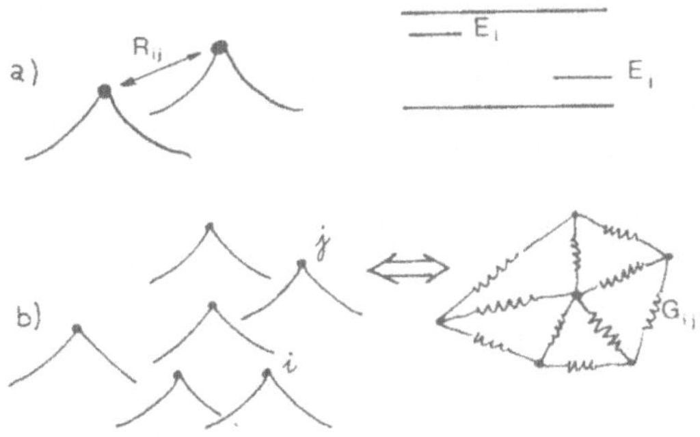

Figure 10a - Hopping between two energetically mismatched localized states.
Figure 10b - The resistive network equivalence of hopping between n sites.

A similar problem has been solved for the calculation of the conductivity of an assembly of 1D metallic particles produced by irradiation of an organic conductor (Zuppiroli et al., 1980). When strong random potentials are created by irradiation, the chains are made of segments of variable length; thus the energies of electrons in each 1D box. Consequently the transverse jumping becomes a phonon assisted hopping with a rate

$$\gamma^{-1} = \gamma_0^{-1} \exp - (\Delta E/k_B T)$$

The macroscopic transverse conductivity is determined by the percolation of electrons through this particular network, that is to say by the critical paths which cross all the samples and which carry most of the current. As only hops to nearest-neighbouring boxes can occur, this situation is very similar to the fixed range hopping and leads to formula (2) in the text. The longitudinal resistivity is deduced from the transverse by simply expressing that each defect interrupts the conducting chain and the electrons are forced to jump to a neighbouring one. The chain direction resistivity increases by the fraction of transverse resistivity mixed in the longitudinal path (relation 3).

REFERENCES

Abeles, B., Sheng, P., Coutts, M.D., and Arie, Y., 1975, Adv. Phys., 24:407.

Alt'Shuler, B.L., and Aronov, A.G., 1979, Zh. Eksp. Teor. Fiz., 77:2028 (Sov. JETP, 50: 968).

Apel, W., and Rice, T.M., 1982, Phys. Rev. B, 19:7063; 1983, J. Phys. C: Sol. State Phys., 16:L271.

Ambegaokar, V., Halperin, B.I., and Langer, J.S., 1971, Phys. Rev. B, 4:2612.

Ashcroft, N.W., and Mermin, N.D., 1976, in: "Solid State Physics", Holt-Saunders, London.

Belitz, D., and Schirmacher, W., 1983, J. Phys. C: Solid State Phys., 16:913.

Beuneu, F., and Monod, P., 1976, Phys. Rev. B, 13:3424.

Bentosela, F., Carmona, R., Duclos, P., Simon, B., Souillard, B., and Weder, R., 1983, Commun. Math. Phys., 88:387.

Berezinsky, V.L., 1974, Sov. Phys. JETP, 38:620.

Bergmann, G., 1984, Physics Reports, 107:1.

Bhatt, R.N., and Lee, P.A., Phys. Rev. Lett., 48:344.

Bird, D.M., Eaglesham, D.J., Withers, R.L., Mc Kernan, S., Steeds, J.W., and Wills, H.H., 1985, in: "Charge Density Waves in Solids" edited by Gy. Hutiray and J. Solyom, Lecture Notes in Physics, Springer-Verlag, Berlin, page 23.

Bloch, A.N., Weisman, R.B., and Varma, C.M., 1972, Phys. Rev. Lett., 28:753.

Bondeson, S.R., and Soos, Z.G., 1980, Phys. Rev. B, 22:1973.

Borland, R.E., 1961, Proc. Phys. Roy. Soc. London, 78:926.

Boucher, B., 1972, J. Non Cryst. Solids, 7:277.

Bouffard, S., Chipaux, R., Jerome, D., and Bechgaard, K., 1981, Solid State Commun., 37:405

Brening, W., Dohler, G.H., and Heyszenau, H., Phys. Letters, 39A:175.

Bulaevskii, L.N., Gusseinov, A.A., Evemenko, O.N., Topnikov, V.N., and Schegolev, I.F., 1975, Sov. Phys. Sol. State, 17:498.

Buttet, J., Car, R., and Myles, C.W., 1982, Phys. Rev. B, 26:2414.

Bush, R.L., 1972, Phys. Rev. B, 6:1182.

Chaikin, P.N., Choi, M.Y., Haen, P., Engler, E.M., and Green, R.L., 1982, Mol. Cryst. Liq. Cryst., 79:79.

Chance, R.R., Brédas, J.L., and Silbey, R., 1984, Phys. Rev. B, 29:4491.

Chiang, C.K., Cohen, M.J., Newman, P.R., and Heeger, A.J., 1977, Phys. Rev. B, 16:5163.
Chui, S.T., and Bray, J.W., 1977, Phys. Rev. B, 16:1329; 1979, Phys. Rev. B, 19:4020.
Cohen, M.H., Economou, E.N., and Soukoulis, C.M., 1983, Phys. Rev. Lett., 51:1302.
Cohen, M.H., Economou, E.N., and Soukoulis, C.M., 1984, Phys. Rev. B, 29:4496.
Conwell, E.M., and Howard, I.A., Phys. Rev. B, 1985, 31:7835.
Cooper, J.R., Miljak, M., and Korin, B., 1981, Chemica Scripta, 17:79.
Coulon, C., 1983, J. Physique C3, 44:88.
Delyon, F., Simon, B., and Souillard, B., 1984, Phys. Rev. Lett., 52:2187.
Devreux, F., Nechstein, M., and Gruner, G., 1980, Phys. Rev. Lett., 45:53.
Eliott, R.J., 1954, Phys. Rev., 96:266.
Emin, D., 1983, Comments Solids State Phys., 11:35, and 11:59.
Engler, E.M., Scott, B.A., Etemad, S., Penny, T., and Patel, V.V., 1977, J. Am. Chem. Soc., 99:5909.
Epstein, A.J., Conwell, E.M., Sandman, J., and Miller, J.S., 1977, Solid State Commun., 23:355.
Ero-Gecs, M., Forro, L., Vancso, G., Holczer, K., Mihaly, G., Janossy, A., 1979, Solid State Commun., 32:845.
Forro, L., 1982, Mol. Cryst. Liq. Cryst., 85:315.
Forro, L., Bouffard, S., and Poujet, J.P., 1984, J. Physique Lett., 45:L543.
Forro, L., Bouffard, S., and Zuppiroli, L., 1982, J. Physique, C3:927.
Forro, L., Janossy, A., Zuppiroli, L., and Bechgaard, K., 1982, J. de Physique, 43:977.
Forro, L., Zuppiroli, L., Pouget, J.P., and bechgaard, K., 1983, Phys. Rev. B, 27:7600.
Frohlich, H., 1937, Physica, 6:406.
Giamarchi, C., and Schulz, H.J., 1987, to be published.
Girvin, S.M., and Jonson, M., 1980, Phys. Rev. B, 22:358.
Gogolin, A.A., 1982, Phys. Reports, 86:3.
Gogolin, A.A., 1983, Solid State Commun., 45:361.
Gogolin, A.A., Mel'nikov, V.I., and Rashba, E.I., 1976, Sov. Phys. JETP, 42:168.
Gogolin, A.A., Zobotukhin, S.P., Mel'nikov, E.I., Rashba, E.I., and Schegolev, I.F., 1975, Sov. Phys, JETP Lett., 22:278.
Gorkov, L.P., 1983, JETP Lett., 38:87.
Gorkov, L.P., Dorokhov, O.N., and Prigara, F.V., 1978, Solid State Commun., 25:981.
Gunning, J., and Heeger, A.J., 1979, Phys. Stat. Sol. (b), 95:433.
Glover, K., Clark, W.G., and Azevedo, L.J., 1982, Mol. Cryst. Liq. Cryst., 85:1977.
Gruner, G., 1981, Chemica Scripta, 17:207.
Hall, R.P., Hundley, M.F., and Zettl, A., 1986, Phys, Rev. Lett., 56:2399.
Holczer, K., Gruner, G., Mihaly, G., and Janossy, A., 1979, Solid State Commun., 31:145.
Holczer, K., Gruner, G., Mihaly, G., and Janossy, A., 1979, Solid State Commun., 32:1233.
Hubbard, J., 1978, Phys. Rev. B, 17:494.
Jacobsen, C.S., Mortensen, K., Anderson, J.R., and Bechgaard, K., 1978, Phys. Rev. B, 18:905.
Jacobsen, C.S., Pedersen, H.J., Mortessen, K., Rindorf, G., Thorup, N., Torrance, J.B., and Bechgaard, K., 1982, J. Phys. C: Solid State Phys., 15:2551.
Janossy, A., Holczer, K., Hseih, P.L., Jackson, C.M., and Zettl, A., Solid State Commun., 43:507.
Janossy, A., Mihaly, G., Forro, L., Cooper, J.R., Miljak, M., Korin, B., 1982, Mol. Cryst. Liq. Cryst., 85:233.

Kaveh, M., Weger, M., and Gutfreund, H., 1979, Solid State Commun., 31:83.
Kawabata, A., 1970, J. Phys. Chem. Solids, 28:41.
Korin-Hamzic, B., Miljak, M., and Cooper, J.R., 1982, Mol. Cryst. Liq. Cryst., 85:1770.
Landau, L., and Lifshits, L., in: "Quantum mechanics", 1956, Moscow, page 135, 189, and 190.
Mermilliod, N., and Sellier, N., 1983, J. de Physique Coll., 44:C3-1353.
Mihaly, G., Bouffard, S., Zuppiroli, L., and Bechgaard, K., 1980, J. de Physique, 41:1495.
Miller, A., and Abrahams, E., 1960, Phys. Rev., 120:745.
Mott, N.F., and Twose, W.D., 1961, Adv. Phys., 10:107.
Mott, N.F., and Davis, E.A., 1979, in: "Electronic Processes in Non-Crystalline Materials", Second edition, Oxford.
Mutka, H., Bouffard, S., and Zuppiroli, L., 1985, in: "Charge Density Waves in Solids" edited by Gy. Hutiray and J. Solyom, Lecture Notes in Physics, Springer-Verlag, Berlin, page 55.
Mutka, H., and Housseau, N., 1983, Phil. Mag. A, 47:797.
Mutka, H., Zuppiroli, L., Molinié, P., and Bourgoin, J.C., 1981, Phys. Rev. B, 23:5030.
Ong, N.P., Verma, G., and Maki, K., 1984, Phys. Rev. Lett., 52:663.
Overhof, H., 1976, Festkorperprobleme, XVI.
Pike, G.E., and Seager, C., 1974, Phys. Rev. B, 10:1421.
Pollak, M., 1972, J. Non-Cryst. Solids, 11:1.
Rice, M.J., and Bernasconi, J., 1973, J. Phys. F: Metal Phys., 3:55.
Rice, M.J., and Mele, E.J., 1981, Mol. Cryst. Liq. Cryst., 77:223.
Saso, T., Suzumura, Y., and Fukuyama, H., 1985, Progress of Theoretical Physics, Supplement 84.
Sanquer, M., 1986, Rapport CEA-R-5335.
Sanquer, M., and Bouffard, S., 1985, Mol. Cryst. Liq. Cryst., 119:147.
Sanquer, M., and Bouffard, S., 1987, to be published in: "J. Phys. C: Solid State Physics".
Sanquer, M., Bouffard, S., and Forro, L., 1986, J. de Physique, 47:1035.
Shante, V.K.S., and Cohen, M.H., 1976, Bull. Am. Phys. Soc., 21:41.
Schultz, T.D., and Craven, R.A., 1979, in: "Highly Conducting One-Dimensional Solids", J.T. Devreese, R.P. Evrard, and V.E. Van Doren, eds., Plenum Press, New-York.
Sinha, A.K., 1970, Phys. Rev. B, 1:4541.
Su, W.P., Schrieffer, J.R., and Heeger, A.J., 1980, Phys. Rev. B, 22:2099.
Suzumura, Y., 1985, J. of the Phys. Soc. Japan, 54:2077.
Suzumura, Y., and Fukuyama, H., 1984, J. of the Phys. Soc. Japan, 53:3918.
Taupin, C, 1967, J. Phys. Chem. Solids, 28:41.
Thouless, D.J., 1977, Phys. Rev. Lett., 39:1167.
Tippie, L.C., and Clark, W.J., 1981, Phys. Rev. B, 23:5846.
Tomic, S., Jerome, D., Mailly, D., Ribault, M., and Bechgaard, K., 1983, Journal de Physique Coll., 44:C3-1075.
Trouilloud, P., Bouffard, S., Ardonceau, J., Zuppiroli, L., 1982, Phil. Mag. B, 45:277.
Weger, M., 1978, J. Physique Coll., C6:1456.
Ziman, J.M., 1961, Phil. Mag., 6:1013.
Zuppiroli, L., 1982, Radiation Effects, 62:53.
Zuppiroli, L., 1984, Berichte der Bunsen-Gesellshaft fur Physikalische Chemie, 88:304.
Zuppiroli, L., 1987, "Perfect Crystals and Real Crystals" in: Semi-Conductors and Semi-Metals Series, Quasi-One-Dimensional Organics (to be published).
Zuppiroli, L., Ardonceau, J., Weger, M., Bechgaard, K., and Weyl, C., 1978, J. de Physique (Lettres), 39:L-170.
Zuppiroli, L., Bouffard, S., Bechgaard, K., Hilti, B., and Mayer, C.W., 1980, Phys. Rev. B, 22:6035.
Zuppiroli, L., Bouffard, S., Jacob, J.J., 1985, Int. J. of Appl. Radiat. Isotopes, 36:843.

Zuppiroli, L., Delhaes, P., and Amiell, J., 1982, J. de Physique, 43:1233.
Zuppiroli, L., Housseau, N., Forro, L., Guillot, J.P., and Pelissier, J., 1986, Ultramicroscopy, 19:325.
Zuppiroli, L., Jacobsen, C.S., and Bechgaard, K., 1985, J. de Physique, 46:799.
Zuppiroli, L., Mutka, H., and Bouffard, S., 1982, Mol. Cryst. Liq. Cryst., 85:1.
Zuppiroli, L., Przybylski, M., and Pukacki, W., 1984, J. de Physique, 45:1925.

EFFECTS OF NON-MAGNETIC DISORDER IN ORGANIC SUPERCONDUCTORS

Silvia Tomić

Institute of Physics of the University
POB 304, 41001 Zagreb, Yugoslavia

Denis Jérome

Laboratoire de Physique des Solides
Bat 510, 91405 Orsay, France

Klaus Bechgaard

H.C.Oersted Institute
Universitetsparken 5, DK-2100, Copenhagen, Denmark

INTRODUCTION

The phenomenon of organic superconductivity was discovered for the first time in the organic cation-anion radical salts $(TMTSF)_2X$ in 1980, with characteristic critical temperatures of the order of one Kelvin ($T_{SC} \simeq 1K$).[1] Substantially higher critical temperature ($T_{SC} \simeq 8K$) was recently achieved in the organic compound belonging to the (BEDT-TTF) series: β-$(BEDT\text{-}TTF)_2I_3$.[2] Extensive amount of the research has been devoted to understand the nature and the origin of organic superconductivity. This paper will review a part of it performed on the compounds from the $(TMTSF)_2X$ family ($X^- = ClO_4^-$, ReO_4^- and FSO_3^-) and the solid solution $(TMTSF)_2(ClO_4)_{1-x}(ReO_4)_x$, $0<x<1$.

The organic conductors $(TMTSF)_2X$ are single chain systems in which a half-filled conduction band is created by a charge delocalization on the organic chain.[3] The spatial anisotropy of the overlapping molecular orbitals leads to an open Fermi surface ($t_a:t_b:t_c \simeq 10:1:1/30$, t_i is a transfer integral)[4] and is reflected by the strong anisotropy of the electronic properties.[4] The relevance of the one-dimensional (1d) physics is given by dimensionality crossover, which in the simplest estimate ($T_x^0 \simeq t_b/\pi$), is situated at about 100K,[5] but is substantially reduced if electronic correlations are taken into account: experimental evidence gives $T_x \simeq 10K$.[6] The ground states can be either insulating - due to the formation of a spin density wave (SDW) phase or to an anion ordering (AO), metallic or superconducting (SC). The phase transitions associated with the AO appear in the 1d region between 10K and 300K, while the antiferromagnetic (AF) and the SC ones happen below, but still in the proximity of the crossover temperature, at about $\lesssim$10K and $\simeq$1K, respectively.

Below the AO phase transitions,[7] which happen in the compounds with non-centrosymmetric anions, new superstructures with different wave vectors are established. Basically two of them appear to be the most significant, because they are associated with two extreme ground states. The first type (a, 2b, nc), n=1 or 2, is established in the compounds with the SC ground state ($(TMTSF)_2ClO_4$ at 1 bar and $(TMTSF)_2ReO_4$ at 12.5 kbar), while the other (2a, 2b, 2c) leads to the insulating non-magnetic state ($(TMTSF)_2ReO_4$ and $(TMTSF)_2FSO_3$ at 1 bar).

The purpose of this paper is to present and discuss experiments performed with the aim to follow changes in a low temperature (LT) behaviour of the electron gas with the increased amount of a disorder in, at the beginning, a pure system.

EXPERIMENTAL RESULTS

$(TMTSF)_2ClO_4$

Slowly cooled $(TMTSF)_2ClO_4$ undergoes the AO phase transition at T_{AO}=24K with wave vector $\vec{q}_2=(0,1/2,0)$ leading to the superstructure (a, 2b, c). Structural features of this transition[7] together with no dramatic change of the electronic properties[8] (Fig.1a.) suggest that its origin lies in the anion-anion interaction. However, the main consequence is the creation of a uniform Coulomb potential along conduction chains due to the equal orientation of the anions in respect to the organic molecules (Fig.2). In this so called Relaxed (R) LT state, the material becomes SC[9] at T_{SC}=1.2K with a characteristic transition width (ΔT_{SC}) more narrow than 0.1K.

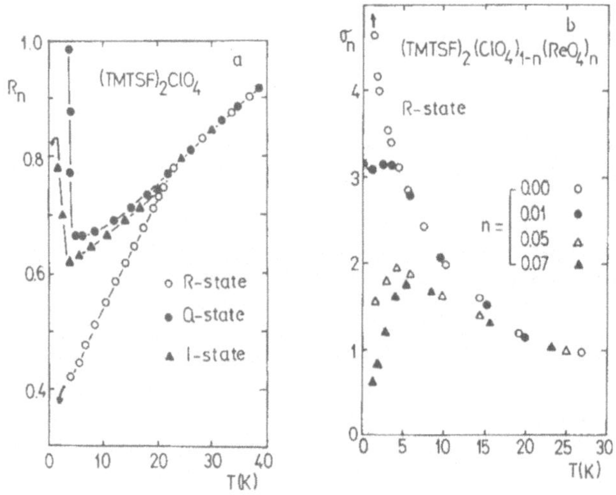

Fig.1. (a) Normalized resistance of $(TMTSF)_2ClO_4$ and (b) normalized conductivity of $(TMTSF)_2(ClO_4)_{1-n}(ReO_4)_n$.

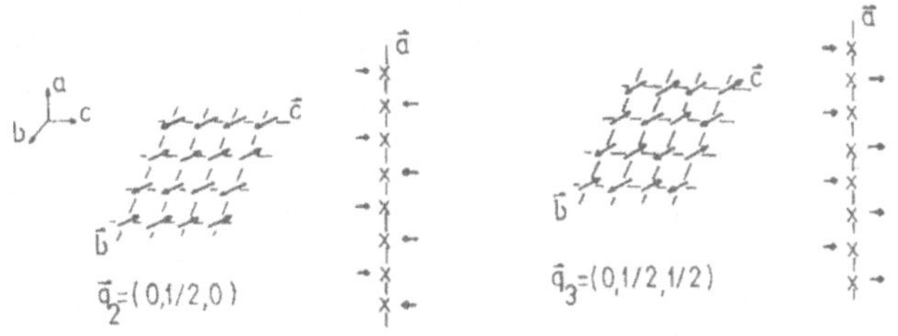

Fig.2. Schematic configuration of the anion orientation for $\vec{q}_2$ and $\vec{q}_3$ AO.

The simplest way to suppress this order is to quench the system in few seconds from 40K to 4.2K. Due to the finite kinetics, the anions are consequently frozen-in in disoreder at LT. In so obtained Quenched (Q) state, SC is suppressed and the SDW semiconducting state is established below 5K[8,10] (Fig. 1a).

Different cooling rates between extreme ones lead to a large scale of Intermediate (I) LT states[8] (Fig.1a.). Characterized by an incomplete AO,[11] they show mixed behaviour of a semiconducting one below about 4K, due to 3d AF correlations, preceeding a transition to SC ground state, which happens at lower temperatures and is broader than in the R-state ($0.85K<T_{SC}<1.2K$).[12] ESR[8,13] and X-ray[7] studies of the anion kinetics and annealing process clearly show that the disordered domains coincide with themagnetic regions. Moreover, a study of the cooling rate dependence of both the diamagnetic shielding and Meissner signals reveals that volume, i.e. bulk, SC diminishes in size and is finally eliminated above the critical cooling speed.[14] However, beyond this percolation threshold small SC regions are still observable.

<u>Solid solution $(TMTSF)_2(ClO_4)_{1-n}(ReO_4)_n$</u>. Already low levels (n>0.05) of ReO_4^- suppress simultaneously the long range (LR) AO with wave vector $\vec{q}_2$[15] and the SC ground state, and gives rise to 3d AF correlations, probably in analogy with quenching in the pure system[16,17] (Fig.1b.). In the case of the alloy with the critical concentration of 0.05 for which no Meissner effect has been detected, recent resistivity measurements seem to indicate the existence of a SC transition below 60mK[18] (Table 1.).

Even though AF correlations can be suppressed by moderate pressures (p) and metallic behaviour reappears (n=0.07, p≃3kbar), the SC ground state is not observed down to 50 mK.[13,16,17] Alloying appears to be more important than quenching because in pure $(TMTSF)_2ClO_4$ even though 3d AF correlations, with similar amplitude as in an alloy, appear in I-state, SC can still be observed.[16]

Table 1. $\vec{q}_2$ AO (ref. 15) and SC transition of $(TMTSF)_2(ClO_4)_n$ as a function of the ReO_4^- concentration. ξ_a is the AO correlation length.

n	0.00	0.01	0.03	0.05	0.07
T_{AO}(K)	24±0.5	22±1	23±1	22.5±1	23±1
ξ_a (Å)	≃500	>200	>200	≃55	≃40
T_{SC}(K)	1.2	0.9	0.78	?	—
ΔT_{SC}(K)	0.1	0.24	0.37		

$(TMTSF)_2ReO_4$

$(TMTSF)_2ReO_4$ exhibits a metal to insulator transition at 182K due to staggered AO with wave vector $\vec{q}_1=(1/2,1/2,1/2)$ which gives rise to the superstructure (2a,2b,2c). A large gap of 2000K opens at Fermi surface, the transition is of the first order (the hysteresis is about 0.8K.) and is insensitive on the cooling rate.[7,13] A strong anion-electron coupling in this compound seems to be in the origin of this phase transition.[5]

The phase diagram under pressure is displayed in Fig.3. In the low pressure region below 9.5 kbar, X-ray measurements[19] still show the existence of the LR $\vec{q}_1$ AO and the only effect of pressure is to diminish the transition temperature (T_{AO}).

The overall metallic behaviour is restored above 12.5 kbar, where a new AO with wave vector $\vec{q}_3=(0,1/2,1/2)$, corresponding to the superstructure (a,2b,2c), is established.[19] Simultaneously, the SC ground state is stabilized at T_{SC}=1.7K and with $\Delta T_{SC}\lesssim$0.1K.[13,20] It's worth of nothing that a periodicity of the uniform $\vec{q}_3$ AO (Fig.2.) corresponds to $4k_F$ (i.e. to spacing a in direct space), in contrast to $\vec{q}_2$ AO one which equals to 0 (i.e. a/2 in direct space). That means that the $4k_F$ component of the periodic anion potential is enhanced and consequently $4k_F$ dimerization gap and electron-electron Umklapp coupling (g_3) should be enhanced, as well. One would then expect that the SDW phase is favoured, rather than the SC one.[5] As this is not the case, it follows that the Umklapp coupling should not be relevant in $(TMTSF)_2X$ compounds, at least at high pressures.

In the intermediate pressure region, between 9.5 kbar and 12.5 kbar, the phase boundaries of $\vec{q}_1$ and $\vec{q}_3$ AO coexist, indicating that the pressure transforms one order ($\vec{q}_1$) into another ($\vec{q}_3$).[19] The main consequence of that is the strong dependence of the LT behaviour on the cooling rate.[13,21]

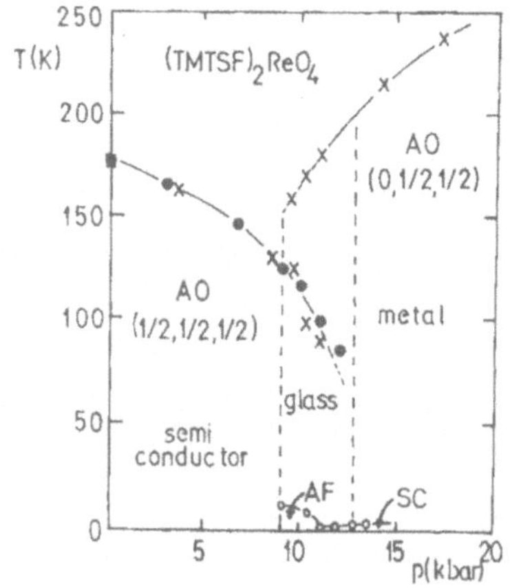

Fig.3. Phase diagram of $(TMTSF)_2ReO_4$. Circles and crosses for data from transport and X-ray (ref.19) measurements, respectively.

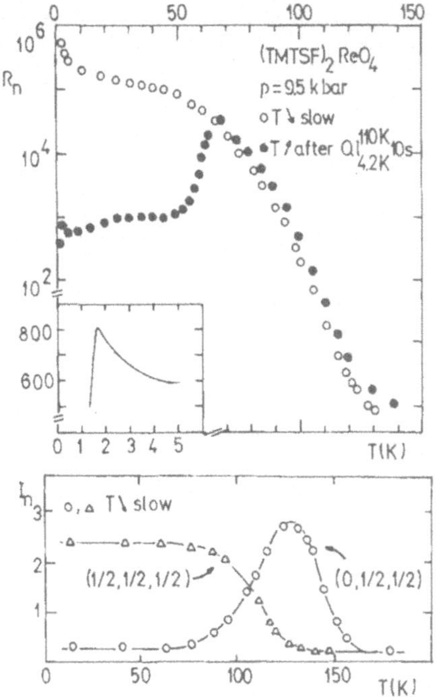

Fig.4. Normalized resistance and X-ray intensity (ref.19) of $(TMTSF)_2ReO_4$ at 9.5 kbar.

Between 9.5 kbar and 11 kbar (Fig.4.), $\vec{q}_1$ AO still prevails over $\vec{q}_3$ one leading to a very high resistance LT stable state. A LT upturn (below≃10K) in the resistivity is probably due to 3d AF correlations. However, after a quench in 10 secs from ≳110K, a temperature where $\vec{q}_3$ regions are more developed than $\vec{q}_1$ AO ones, the resistivity at 4.2K can be 3 orders of magnitude lower than in the stable state (obtained by slow cooling). Moreover, after a small upturn below 4K, the system enters in a probably partial SC state (the lowest attainable temperature was 1.2K). In warming it displays metallic behaviour up to about 50K, where it undergoes a transition to the high resistance stable state. This temperature (T_f≃50K) could be identified as a freezing temperature of classical movements of the anions at this particular pressure. (T_f is expected to increase with a pressure.)[19,21] Between 11 kbar and 12.5 kbar $\vec{q}_3$ AO starts to overcome $\vec{q}_1$ AO. The residual degree of the latter defines the magnitude of the anomaly in the resistivity and, also, influences the SC transition: T_{SC} is lower and ΔT_{SC} is broader.[19,21]

<u>Solid solution $(TMTSF)_2(ReO_4)_{1-m}(ClO_4)_m$</u>. Comparing data to that of the pure system $(TMTSF)_2ReO_4$ at 12 kbar, as far as SC is concerned, the result which follows is essentially the same as on the perchlorate side of the phase diagram: T_{SC} decreases, ΔT_{SC} broadens, and bulk SC is appearently replaced by a partial one before complete suppression (Table 2).[13,17]

Table 2. SC transition of $(TMTSF)_2(ReO_4)_{1-m}(ClO_4)_m$ at 12 kbar as a function of the ClO_4^- concentration.

m	0.00	0.01	0.03	0.07	0.35
T_{SC}(K)	1.7	1.5	1.2	1.15	–
ΔT_{SC}(K)	0.07	0.15	0.25	0.5	

$(TMTSF)_2FSO_3$

$(TMTSF)_2FSO_3$ is unique in the sense that the tetrahedral anion FSO_3^- carries a permanent dipole moment.[7] Another essential feature is that each of the two equivalent anion positions (as in the case of ClO_4^- and ReO_4^-) has four soubgroups depending on the position of the fluorine atom. Consequently, any order of tetrahedrons without the order of dipoles is not a perfect order of the anions.[22] A filamentary SC ground state is established at pressures above 7 kbar. The absence of the bulk effect is ascribed to a residual dipolar disorder.[23]

DISCUSSION

Presented data obtained from transport and X-ray measurements together with NMR[6] studies, clearly show that the applied pressure provides conditions necessary for SC coupling and transforms staggered AO into a uniform one compatible with metallic behaviour and the SC ground state. The anion order plays a determinant role in the competition between AF and SC phase: a disorder favours the former, while a uniform order is compatible with the latter. In the compounds with dipolar anions or in alloys, the SC ground state is easily suppressed, probably due to a strong random potential along the conduction chains which is produced by the anions. The extreme sensitivity of SC to the presence of non-magnetic impurities can be discussed in the framework of two theoretical models depending whether the SC ground state is singlet or triplet.[13]

In Larkin and Melnikov (LM) model,[24] valid for weakly coupled SC chains ($g_1<0$, $\Delta>t_\perp$, g_1 is backward scattering, Δ is pseudogap in the charge degrees of freedom, $t_\perp$ is transverse transfer integral), T_{SC} is defined by two parameters. First one is the interchain hopping amplitude of electron pair, and the other is the form of 1d correlation function. Then, already a small amount of impurities will induce a dephasing of electron pairs' wave functions between different chains and this will diminish T_{SC}. Moreover, above some critical impurity concentration (n_c) SC will vanish due to a decrease of the coherence length of pairs along the 1d chains. The experimental results are relatively in a good quantitative agreement with the predictions of this model only by taking into account the renormalization of the crossover temperature ($T_x \simeq \tilde{t}_\perp \simeq 10K$) due to cooperative 1d effects. Then it gives for a pure compound $T_{SC}<10K$ and for $n_c>0.03$. It's worth of noting that Josephson coupling such as used in LM model ($\simeq t_\perp^2/\Delta^2$) is valid only for negative g_1 and it is commonly excepted that $(TMTSF)_2X$ compounds are situated in the upper plane of (g_1,g_2) phase diagram ($\Delta=0$ there).[5,6] However, it was recently shown by C.Bourbonnais[6] that even for $g_1>0$ and for $T>T_x$ exists a coupling of Josephson type proportional to $\tilde{g}^2\tilde{t}_\perp^2/T^2$. According to his model, in which the exchange of 1d AF correlations is at the origin of the attractive interaction, one would also expect that singlet pairing prevails over triplet one, and that non-magnetic impurities have strong destructive influence on the SC ground state.

In the case of triplet pairing Anderson theorem does not hold even in 3d case, and therefore non-magnetic impurities act as pair breakers. Quantitatively, existing models give too strong effect and more work taking into account low dimensionality of these systems seems to be necessary. It's worth noting that the proximity of AF and SC in the phase diagram, as well as the non-activated magnetic susceptibility (i.e. $g_1>0$) are compatible (in the g-ology picture) with triplet pairing.[5,6]

In conclusion, we have tried to argue that, independently of the nature of the SC ground state, an analysis taking into account 1d electronic correlations is necessary to understand the unusual properties of the SC ground state in $(TMTSF)_2X$ compounds and their alloys. Once a prerequisite condition for SC pairing is realized,[5,6] it is the existence of the long range uniform anion potential that is a necessary condition for the establishment of the SC ground state.

ACKNOWLEDGEMENTS — This work has been done during the author's (S.T.) stay at Laboratoire de Physique des Solides, Orsay, in collaboration with L.Brossard, R.C.Lacoe, D.Mailly, M.Ribault and G.Rindorf. X-ray measurements have been performed by R.Moret, J.P.Pouget and S.Ravy. Numerous fruitful discussions with them, as well as with S.Barišić, C.Bourbonnais, F.Creuzet and V.J.Emery are greatly acknowledged.

REFERENCES

1. D.Jérome and H.J.Schulz, Adv.Phys. 31,299 (1982).
2. F.Creuzet, G.Creuzet, D.Jérome, D.Schweitzer and H.J.Keller, J.Physique 46,L-1079 (1985) and refs.there-in.
3. K.Bechgaard, Mol.Cryst.Liq.Cryst. 79,1 (1982).
4. P.M.Grant, Phys.Rev. B26,6888 (1982).
5. V.J.Emery, these Proceedings.
6. C.Bourbonnais, these Proceedings.
7. J.P.Pouget, these Proceedings.
8. S.Tomić, D.Jérome, P.Monod and K.Bechgaard, J.Physique Lett. 43,L-839 (1982).
9. K.Bechgaard, K.Carneiro, M.Olsen, F.B.Rasmussen and C.S.Jacobsen, Phys.Rev.Lett. 46,852 (1981).
10. T.Takahashi, D.Jérome and K.Bechgaard, J.Physique Lett. 43,L-565 (1982).
11. R.Moret, J.P.Pouget, R.Comes and K.Bechgaard, J.Physique 46,1521 (1985).
12. M.Ribault, J.Physique Coll. 44,C3-827 (1983).
13. S.Tomić, Thesis, Université de Paris-Sud, Orsay (1986), unpublished.
14. H.Schwenk, K.Andres and F.Wudl, Phys.Rev. B27,5846 (1983).
15. S.Ravy, R.Moret, J.P.Pouget and R.Comes, to be published in Physica B (1986).
16. S.Tomić, D.Jérome, D.Mailly, M.Ribault and K.Bechgaard, J.Physique Coll. 44,C3-1075 (1983).
17. S.Tomić, L.Brossard, R.C.Lacoe, D.Jérome, D.Mailly, M.Ribault, K.Bechgaard and G.Rindorf, to be published in Physica B (1986).
18. M.Petravić, J.R.Cooper and S.Tomić, unpublished.
19. R.Moret, S.Ravy, J.P.Pouget, R.Comes and K.Bechgaard, Phys.Rev.Lett. 57,1915 (1986).
20. S.Tomić, D.Jérome and K.Bechgaard, Mol.Cryst.Liq.Cryst. 119,241 (1985).
21. S.Tomić, D.Jérome and K.Bechgaard, J.Phys. C17,L-11 (1984).
22. S.Tomić, D.Jérome and K.Bechgaard, J.Phys. C17,L-655 (1984).
23. S.Tomić, D.Jérome and K.Bechgaard, Mol.Cryst.Liq.Cryst. 119,59 (1985) and refs. there-in.
24. A.I.Larkin and V.I.Melnikov, Sov.Phys. JETP 44,1159 (1976).

LOCALIZATION IN ONE-DIMENSIONAL INTERACTING ELECTRON SYSTEMS

Y. Suzumura

Department of Physics
Tohoku University
Sendai 980, Japan

INTRODUCTION

Recently it has been reported that low dimensional conductors are very sensitive to nonmagnetic impurities. Since this phenomenon may be related to the localization in low dimension,[1] we examine, for the first step, one-dimensional (1-D) electron systems in the presence of both impurities and mutual interactions. A competition between randomness and order arises because the impurity in the absence of interactions always leads to the localization[2] and the interaction in the clean system results in the divergence of various kinds of response function.[3] However both of these quantities cannot be treated by use of the usual perturbational method due to 1-D properties. There are two kinds of methods for studying the boundary between the delocalized state and the localized state in the plane of the coupling constant of the mutual interaction. One of them is the approach from the delocalized regime: the effect of impurities is examined by taking the clean system as an unperturbed state.[4-6] Another of them is the approach from the localized regime: the localization is regarded as the impurity pinning of the phase variable of charge density wave (CDW).[7,8]

The paper examines the boundary from the localized regime. After studying the clean system by applying the bosonization[9] to the phase hamiltonian,[10] the localization is studied in the limit of weak impurities.[8,11,12]

PHASE REPRESENTATION

The large 1-D fluctuation comes from the freedom of phase variables of order parameters. Therefore we derive microscopically phase variable and their relation to order parameters by use of the method of bosonization.[9] Four kinds of phase variables are defined as follows:[10]

$$\theta_{\pm} \text{ or } \phi_{\pm} = i \sum_{\substack{k\neq 0,p \\ s=\pm}} A_k(x)h(s)\{C^{+}_{1,p+k,s}C_{1,p,s} \pm C^{+}_{2,p+k,s}C_{2,p,s}\} , \qquad (1)$$

where $A_k(x) = \exp[-\alpha|k|/2-ikx]\pi/kL$. In eq.(1), s = +,- correspond to ↑, ↓ and $h(s) = 1$ for $\theta_\pm$ and $h(s) = s$ for $\phi_\pm$. The quantity L is the length of the system and $C^{+}_{1,p\uparrow}$ ($C_{2,p\uparrow}$) is the creation operator of the conduction electron with the positive (negative) momentum p and up spin. The quantity α is the cutoff parameter of the order of the lattice constant. The quantity $\theta_+(\theta_-)$ denotes the phase variable for the charge density (superconducting) fluctuation and θ_+ is complementary to θ_- due to the relation given by $[\theta_+(x),\theta_-(x')] = i\pi \mathrm{sgn}(x-x')$. When the fluctuation for θ_+ is large (small), the fluctuation for θ_- is small (large). The quantity ϕ_+ representing the spin density fluctuation is complementary to ϕ_-. By use of eq.(1), order parameters for CDW, spin density wave (SDW), singlet superconductivity (SS) and triplet superconductivity (TS) are expressed respectively as[9,10]

$$O_{\substack{CDW\\SDW}}(x) = (2\pi\alpha)^{-1} \exp[i(2k_F x + \theta_+ + \phi_\pm)] ,$$

$$O_{\substack{SS\\TS}}(x) = (2\pi\alpha)^{-1} \exp[i(\theta_- + \phi_\pm)] . \tag{2}$$

The degree of the fluctuation for these phase variables is determined by properties of mutual interactions. All fluctuations become equal in the absence of interactions. When the interaction is attractive, the fluctuation for θ_- becomes small and the corresponding fluctuation for θ_+ is enhanced due to the complementary relation between θ_+ and θ_- and then superconductivity is favorable compared with CDW or SDW.

STATE DIAGRAM FOR CLEAN SYSTEMS

For the simplicity, we consider the Tomonaga-Luttinger (TL) model, which is expressed in terms of phase variables as[9,10]

$$H_{TL} = \int dx\ [A_{\rho+}(\frac{d\theta_+}{dx})^2 + A_{\rho-}(\frac{d\theta_-}{dz})^2 + A_{\sigma+}(\frac{d\theta_+}{dx})^2 + A_{\sigma-}(\frac{d\theta_-}{dx})^2] , \tag{3}$$

where $A_{\rho\pm} = (v_F \pm g_2/\pi)/4\pi$, $A_{\sigma+} = A_{\sigma-} = v_F/4\pi$ and g_2 is the coupling constant for the forward scattering. For example, we consider the most favorable state when $g_2 < 0$. By noting that $A_{\rho+} < A_{\rho-}$ and that the complementary relation between θ_+ and θ_-, we find that the superconducting fluctuation is smaller than the charge density fluctuation and then we obtain the superconducting state in the case of $g_2 < 0$. Since the quantities $A_{\rho-}/A_{\rho+}$ and $A_{\sigma-}/A_{\sigma+}$ determine the properties of fluctuations of eq.(1), we define generally η_ρ and η_σ as[10],

$$\eta_\rho = (A_{\rho-}/A_{\rho+})^{1/2} \quad \text{and} \quad \eta_\sigma = (A_{\sigma-}/A_{\sigma+})^{1/2} , \tag{4}$$

which indicate the degree of the quantum fluctuation for the charge density and that for the spin density respectively. The classical case is given by $\eta_\rho = 0$ and $\eta_\sigma = 0$. Since the noninteracting case corresponds to $\eta_\rho = \eta_\sigma = 1$, the boundaries given by $\eta_\rho = 1$ and $\eta_\sigma = 1$ separate several kinds of states: CDW for $\eta_\rho < 1$ and $\eta_\sigma \le 1$; SDW for $\eta_\rho < 1$ and $\eta_\sigma \ge 1$; SS for $\eta_\rho > 1$ and $\eta_\sigma \le 1$; TS for $\eta_\rho > 1$ and $\eta_\sigma \ge 1$.

IMPURITY PINNING

Although the impurity scattering consists of the forward scattering and the backward scattering, we study only the effect of backward scattering which takes an essential role for the localization and the impurity pinning. The scattering couples with CDW and the correspoding hamiltonian is given by

$$H_{imp} = \frac{V_0}{\pi\alpha} \sum_j \int dx \cos\phi_+ \cos(\theta_+ + 2k_F x)\,\delta(x-R_j) \quad , \tag{5}$$

where R_j is the location of impurities. First we investigate the pinned state for the classical CDW and the effect of the quantum fluctuation by considering the system with only θ_+ and θ_-. Next, we show the impurity pinning for the TL model and the half-filled extended Hubbard model.

Impurity Pinning for Charge Density Fluctuation

By discarding the freedom of the spin density fluctuation in eqs.(3) and (5), we consider the hamiltonian:

$$H' = \int dx \, [A_+ (\frac{d\theta_+}{dx})^2 + A_- (\frac{d\theta_-}{dx})^2 + V \sum_j \cos(\theta_+ - \zeta(x))\,\delta(x-R_j)] \quad , \tag{6}$$

where $\zeta(x) = -2k_F x$ and V is proportional to V_0. Equation (6) is similar to the hamiltonian of the Peierls-Fröhlich system[7] except that the quantum fluctuation due to the second term must be taken account of. The energy for eq.(6) in the presence of the impurity pinning is calculated by writing $\theta_+ = \theta_s + \hat{\theta}$ where $\hat{\theta}$ denotes the quantum fluctuation around the classical value, θ_s. The quantity $\cos(\theta_+ - \zeta(x))$ in the third term is rewritten as $\cos\hat{\theta}\cos(\theta_s - \zeta(x))$. The energy gain from the classical part is obtained by the spatial variation of θ_s and is estimated as $-V(n_i/L_0)^{1/2}$ where L_0 is the pinning length introduced by Fukuyama and Lee.[7] The classical energy is reduced by the factor, $\cos\hat{\theta}$, due to the quantum fluctuation which is treated by use of the self-consistent harmonic approximation.[8] By minimizing the total energy which consists of the increase of the elastic energy ($\propto L_0^{-2}$) and the energy gain from the impurity pinning, one obtains that $n_i L_0 \propto [(n_i\alpha)^{\eta/2} V/n_i A_+]^{-2/(3-\eta)}$ where $\eta = (A_-/A_+)^{1/2}$. When the quantum fluctuation i.e., η increases, L_0 increases and becomes infinite at $\eta=3$. The pinned state is obtained for $0 \leq \eta < 3$ while the clean system belongs to superconductivity (CDW) for $\eta > 1$ ($\eta < 1$). Therefore it turns out that CDW is always pinned and superconductivity is pinned only for the weak attractive interaction.

Localization for the TL Model

In a similar way to eq.(6), we write $\theta_+ = \theta_s + \hat{\theta}$ and $\phi_+ = \phi_s + \hat{\phi}$ in order to study two kinds of quantum fluctuations for the total system given by eqs.(3) and (5). There are two types of pinning given by the distortion of θ_s (θ_s-glass) and the distortion of ϕ_s (ϕ_s-glass) which are determined by the magnitudes of $A_{\rho+}$ and $A_{\sigma+}$ for the TL model. It is obtained numerically that the θ_s-glass is realized for $A_{\rho+}/A_{\sigma+} \lesssim 2.5$ and the mixed state of the θ_s-glass and ϕ_s-glass is realized for $A_{\rho+}/A_{\sigma+} \gtrsim 2.5$.[12] The pinning length is calculated as[8]

$$n_i L_0 \propto [\frac{V_0}{\alpha n_i v_F}]^{-\frac{2}{3-\eta_\rho-\eta_\sigma}} [n_i\alpha]^{-\frac{\eta_\rho+\eta_\sigma}{3-\eta_\rho-\eta_\sigma}} \quad , \tag{7}$$

where η_ρ and η_σ are defined by eq.(4). When the quantum fluctuation increases, L_0 increases and becomes infinite at $\eta_\rho + \eta_\sigma = 3$ which is consistent with the localization-delocalization boundary obtained in the case of a spinless fermion from the delocalized regime.[4,5] It should be noted that eq.(7) becomes equal to the localization length obtained from the density distribution function when the mutual interaction is absent.[12]

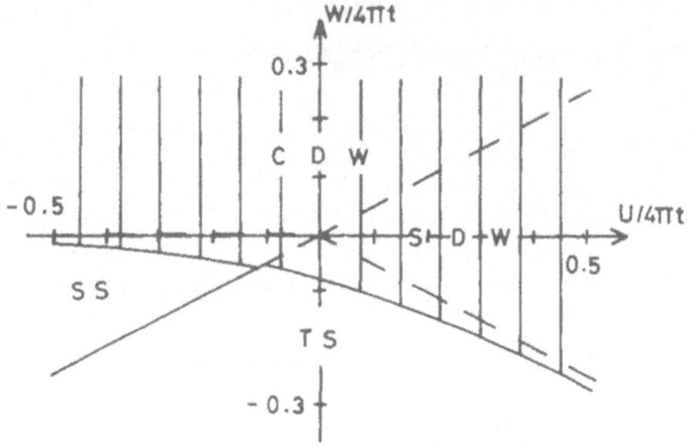

Fig.1 Localization for the 1-D half-filled extended Hubbard model[11]

Localization for the Half-Filled Extended Hubbard Model

As an example which includes the backward scattering and the Umklapp process, we study the half-filled extended Hubbard model given by $H_{EH} = -t\Sigma_{m,\sigma}(C^+_{m+1,\sigma}C_{m,\sigma}+h.c) + H_{int}$ where

$$H_{int} = U/2 \sum_{m,\sigma} n_{m,\sigma}n_{m,-\sigma} + W \sum_{m,\sigma,\sigma'} n_{m,\sigma}n_{m+1,\sigma'} \quad . \tag{8}$$

In eq.(8), $n_{m,\sigma} = C^+_{m,\sigma}C_{m,\sigma}$ and $C^+_{m\uparrow}$ is the creation operator of the electron with up spin at the lattice site m. In Fig.1, the localized region for eqs.(5) and (8) is shown by the shaded area on the U-W plane.[11] We obtain that CDW ($W > 0$ and $W > U/2$) and SDW ($-U/2 < W < U/2$) are always localized and that SS ($U/2 < W < 0$) and TS ($W < U/2$ and $W < -U/2$) are localized only for the weak interaction.

Discussion - The localization with the large value of the impurity potential is an interesting problem and has been recently studied both numerically[12,13] and analytically.[5,12,14]

REFERENCES

1. E. Abrahams, P.W. Anderson, D. C. Licciardello and T. V. Ramakrishnan, Phys. Rev. Lett. 42, 673 (1979).
2. V. L. Berezinskii, Zh. Eksp. Teor. Fiz. 65, 1251 (1973).
3. J. Sólyom, Adv. Phys. 28, 201 (1979)
4. S. T. Chui and J. W. Bray, Phys. Rev. B16, 1329 (1977).
5. W. Apel, J. Phys. C (Solid State Phys.) 15, 1973 (1982).
6. W. Apel and T. M. Rice, Phys. Rev. B26, 7063 (1982); J. Phys. C (Solid State Phys.) 16, L271 (1983).
7. H. Fukuyama and Lee, Phys. Rev. B17, 535 (1978).
8. Y. Suzumura and H. Fukuyama, J. Phys. Soc. Jpn. 52, 2870 (1983); 53, 3918 (1984).
9. A. Luther and V. J. Emery, Phys. Rev. Lett. 33, 589 (1974).
10. Y. Suzumura, Prog. Theor. Phys. 61, 1 (1979).
11. Y. Suzumura, J. Phys. Soc. Jpn. 54, 2077 (1985).
12. T. Saso, Y. Suzumura and H. Fukuyama, Prog. Theor. Phys. 84, 269 (1985).
13. T. Saso and Y. Suzumura, J. Phys. Soc. Jpn. 55, 25(1986); 55,No.9(1986).
14. T. Giamarchi and H. J. Schulz, preprint.

THE DYNAMICS OF CHARGE DENSITY WAVES

George Grüner

Department of Physics and Solid State Science Center
University of California
Los Angeles, CA 90024

1. INTRODUCTION

One of the interesting and novel concepts which are clearly related to lower dimensional metals is the appearance of various broken symmetry ground states such as superconductivity, charge density waves (CDW), and spin density waves (SDW). These have been readily found in various solids with linear chain structure and the phase transitions, ground states and the interplay between the various phases have been explored both by experiment and theory. The various ground states, all characterized by a complex order parameter, also couple to an applied electric field leading to collective transport phenomena due to the response of the collective modes. The electromagnetic properties of the superconducting phases are well understood, and little is known about the dynamics of the spin density wave ground state. A novel collective transport phenomenon, associated with the dynamics of charge density waves (CDW) is, however, well established by a rather broad range of experimental techniques.

The theoretical aspects of the field date back to more than thirty years. It has been suggested by Peierls[1] that a one dimensional metal is unstable towards a lattice distortion with a period $\lambda = \pi/k_F$ with k_F the Fermi wave vector. The ground state is characterized by a gap at the Fermi level and by a periodic charge density (the charge density wave)

$$\Delta\rho = \rho_0 \cos(2k_F r + \phi)$$

with ρ_0 the amplitude and ϕ the phase. As stressed in a brilliant paper by Fröhlich,[2] the transitional motion of the CDW leads to an electric current, which in the absence of damping and of interactions which break the translational invariance would lead to superconductivity. About twenty years later, these concepts became relevant to experiments when the first linear chain compounds (both organic and inorganic) were discovered and their properties were explored. Experiments on TTF-TCNQ and on the compound KCP suggested[3] a collective mode response at finite frequencies, but well below the single particle gaps. This led to the concept of collective mode pinning. Although nonlinear conduction has been found in the compound $K_{0.3}MoO_3$ (called blue bronze) earlier, the experiments on the electric field dependent conduction[4] in $NbSe_3$ were first clearly related to sliding charge density wave motion. The conductivity was also found to be strongly frequency dependent, and coherent current oscillations were observed in the nonlinear conductivity region.[5] Other materials, with TaS_3, $(TaSe_4)_2I$ and $K_{0.3}MoO_3$ as the most well known examples, have soon been discovered with dynamical properties similar to that found in $NbSe_3$. The behavior of other transport coefficients, such as Hall effect, thermopower, and also of the elastic properties have also been explored, and recent NMR experiments give direct evidence for CDW motion. Various models have also been proposed to account for the broad variety of experimental observations made in the presence of dc and/or ac electric fields. In spite of the significant progress in the area, many unresolved fundamental questions remain concerning the importance of quantum effects, the role of normal electrons and the internal degrees of freedom.

First, I will give a short overview of the field, introducing the basic notions and materials and summarizing the basic observations in the field. The next two lectures will deal with two different aspects of the dynamics. In certain situations, the response is highly coherent with the possibility of various interference phenomena. Lecture 2 will summarize experiments on interference effects together with their interpretation. In other cases, disorder is important, leading to a "glassy" dynamics with features similar to those found in spin glasses, random field systems, etc. These will be discussed later.

2. BASIC NOTIONS, MATERIALS, AND OBSERVATIONS

Consider a one-dimensional metal. In the absence of an interaction with the lattice, the ground state is as shown in Fig. 1(a), where the electron states are filled up to the Fermi level, and the underlying lattice is that of a periodic array of atoms with lattice constant a. As first pointed out by Peierls,[1] this state is not stable for a coupled electron-phonon system. In the presence of an interaction (of any strength) it is energetically more favorable to distort the lattice periodically with period λ related to the Fermi wave vector k_F

$$\lambda = \frac{\pi}{k_F} \quad . \tag{1}$$

A lattice distortion with this period opens up a gap at the Fermi level, as shown in Fig. 1(b) where the situation appropriate for a half filled band is drawn. As states only up to $\pm k_F$ are occupied, the opening of the gap leads to the lowering of the electronic energy. The lattice distortion leads to an increase of the elastic energy, but in 1D the total energy is lower than that of the undistorted metal (this is the consequence of the divergent Linhardt Function at $q = 2k_F$ in 1D). Consequently a distorted state is stable at $T = 0$. The state has a gap in the single particle excitation spectrum. The gap opening also leads to the modification of the electron density, much in the same way as in the nearly free electron theory of metals. The density $\rho = |\Psi|^2$ will be a periodic function of the of the position X with the period given by eq. (1), i.e., determined by the band filling. Thus for an arbitrary band filling the period of this modulated charge density [and the accompanying periodic lattice distortion, see Fig. 1(b)] will be incommensurate with the underlying lattice. At finite temperatures normal electrons, excited across the single particle gap Δ screen the electron-phonon interaction. This in turn leads to a reduction of the gap and eventually to a (second order) phase transition. Such semiconductor to metal transition is called the Peierls transition.

Calculations which lead to the Peierls transition are performed by using the 1D electron-phonon Hamiltonian

$$H = \sum_k \varepsilon_k C_k^+ C_k + \sum_q \omega_q^o b_q^+ b_q + \sum_{k,q} g\, C_{k+q}^+ C_k (b_q + b_q^+) \tag{2}$$

where C_k, C_k^+ and b_q, b_q^+ are the electron and phonon creation and annihilation operators with moments K and q, ε_k and ω_q^o are the electron and phonon dispersions and g is the electron-phonon coupling constant. In eq. (2) spin is omitted.

Defining a complex order parameter

$$\Delta e^{i\phi} = g(b_{2k_F} + b_{-2k_F}) \; , \tag{3}$$

where Δ and ϕ are real, the displacement field of the ions is

$$(b_{2k_F} + b_{-2k_F})\, e^{2ik_F x} + \text{H.C.} = \frac{2\Delta}{g} \cos(2k_F x + \phi) \quad . \tag{4}$$

One can diagonalize the electronic part of the Hamiltonian or setting up a self consistent equation for Δ by replacing b_{2k_F} with $\langle b_{2k_F} \rangle$. Such mean field approximation leads to a BCS gap equation

$$\Delta = 2D \exp\left(- \frac{1}{\lambda}\right) \tag{5}$$

where $\lambda = g^2 (\pi V_F)^{-1}$ is the dimensionless electron-phonon coupling constant and D is the electronic bandwidth. The temperature dependence of Δ also has the characteristic BCS form, and the transition temperature is given by $T_P = \Delta/1.76$. The spatially dependent electron density can also be evaluated, and

$$\rho(x) = \rho_0 + \frac{\rho_0 \Delta}{\lambda V_F k_F} \cos(2k_F + \phi) \tag{6}$$

where ρ_0 is the electron density in the absence of electron-phonon interactions. The second term on the right-hand side describes the CDW with period π/k_F and amplitude $\rho_0 \Delta/\lambda V_F k_F$.

The fact that the CDW ground state has always a lower energy than the undistorted metallic state is the consequence of the logarithmically divergent 1D Linhardt Function. It is not surprising, therefore, that CDW transitions have primarily been observed in materials which have a highly anisotropic crystal and electronic structure. Organic linear chain compounds, such as TTF-TCNQ are one class of materials which show Peierls transitions and a periodic lattice distortion (detected by X- rays, electrons, or by neutrons). The other group of materials, such as the transition metal trichalcogenides, MX_3, where M = Nb or Ta and X = S or Se, tetrachalcogens, such as $(TaSe_4)_2I$ or bronzes, like Ko_3MoO_3 all have a chain structure such as shown for $NbSe_3$ in Fig. 2. These materials undergo a Peierls transition at temperature somewhat below room temperature [note that the cut-off energy is the bandwidth and not the phonon frequency in eq. (4) and consequently the gap is much larger than the superconducting gap] as evidenced, for example, by the dc electrical resistivity $\rho(T)$. This is shown in Fig. 3 where the dc electrical conductivity measured on various materials is displayed. The arrows represent the temperatures T_p where a phase transition to the Peierls-state occurs. In the above cases the gap opens up over the entire Fermi surface, leading to a metal-insulator transition. $NbSe_3$ is an exception, here two transitions occur, one at T_1 = 145K, the other and T_2 = 59K, both removing only part of the Fermi surface.

In all of the materials mentioned above, the period lattice distortion has been observed by structural studies. In all of the cases, the CDW is incommensurate with the underlying lattice, and the measured scattering intensity as the function of temperature indicates a BCS-like order

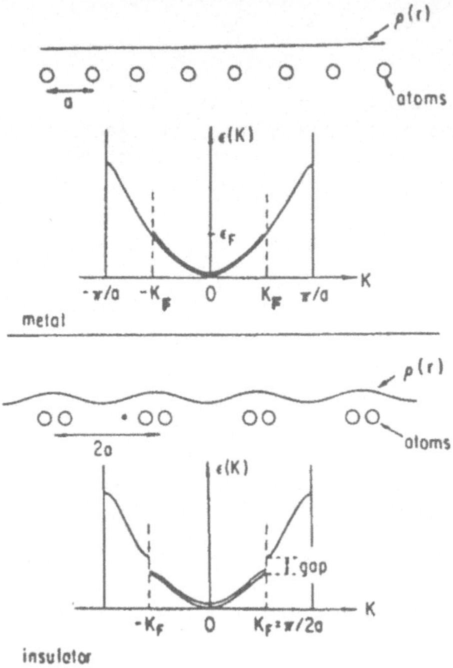

Fig. 1. Peierls distortion in a half filled band. In this case the period $\lambda = 2a$, is commensurate with the lattice.

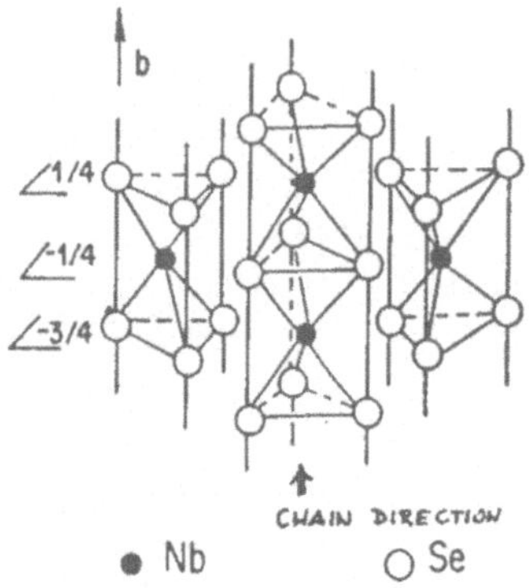

Fig. 2. Crystal structure of $NbSe_3$. The CDW develops along the chain direction b.

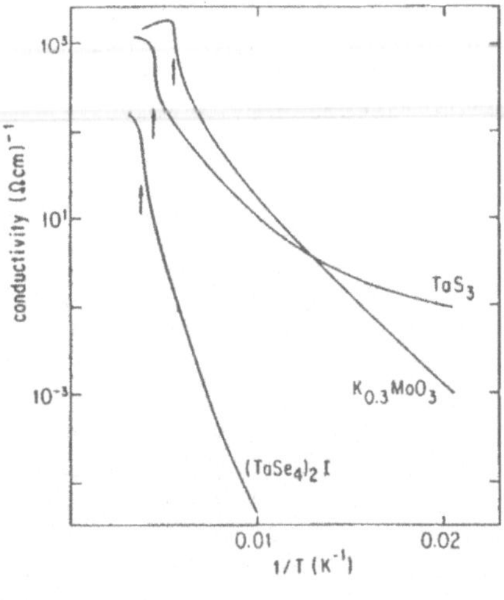

Fig. 3. Temperature dependence of the dc conductivity in several compounds which undergo a Peierls transition. The transition temperatures are indicated by the arrows.

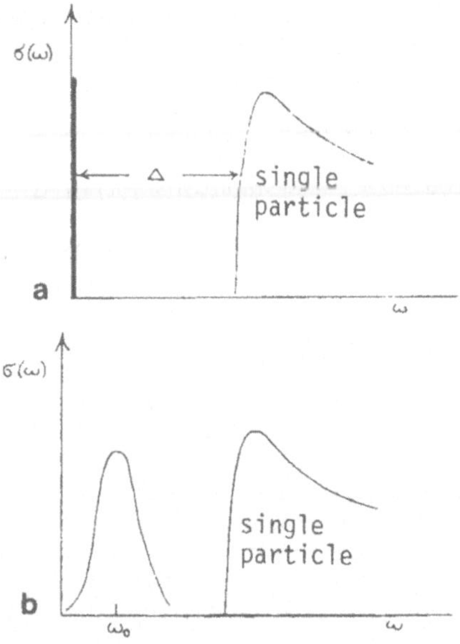

Fig. 4. Frequency dependent conductivity in the absence of pinning (a) and with pinning and damping (b).

parameter. The dynamics of the mode is described in terms of the time dependent order parameter, and both amplitude and phase fluctuations occur. These can be described by assuming that $\Delta(x,t)$ is given by[6]

$$\Delta(x,t) = (\Delta + \delta)\, e^{i\phi} \tag{7}$$

where Δ is the equilibrium order parameter, δ and ϕ are the fluctuations from the equilibrium value. To lowest order in δ and ϕ the amplitude mode corresponds to $\Delta_{2k_F} + \Delta_{-2k_F} = 2\Delta + 2\delta$, the phase mode corresponds to $\Delta_{2k_F} - \Delta_{-2k_F} = 2i\Delta\phi$. The dispersion relation of these modes has been evaluated by Lee, Rice, and Anderson[6] using the Hamiltonian eq. (2). The electron-phonon interaction transforms the phonons near ω_{2k_F} into an optical and an acoustic branch with frequencies:

$$\text{optical:} \quad \Omega_+^2 = \lambda\omega_{2k_F}^2 + \frac{1}{3}\frac{m}{m^*}(V_F q)^2 \tag{8a}$$

$$\text{acoustic:} \quad \Omega_-^2 = \frac{m}{m^*}(V_F a)^2 \tag{8b}$$

where the effective mass is given by

$$\frac{m^*}{m} = 1 + \frac{4\Delta^2}{\lambda\omega_{2k_F}^2} \quad . \tag{9}$$

In the absence of Coulomb effects, the optical mode has a gap, while the acoustic mode is gapless. The large effective mass is the consequence of the fact that in the total kinetic energy of the mode, the kinetic energy of both the electrons and of the oscillating ions has to be included.

Treating the phase as a classical field, the Lagrangian density associated with $\phi(x,t)$ is given by

$$L = n\left[\frac{m^*}{m}\left[\frac{1}{2k_F}\frac{d\phi}{dt}\right]^2 - \frac{\kappa}{2}\left[\frac{d\phi}{dx}\right]^2\right] , \tag{10}$$

where the first term is the kinetic energy of a line of mass m*n per unit length. The potential energy is the second term, with a phenomenological elastic constant κ. The dispersion relation corresponding to wave-like excitations of the form $\exp(-i\omega t - kx)$ is given by

$$\omega^2 = \left[\frac{\kappa}{m^*}\right]\left[2k_F a\right]^2 \tag{11}$$

and comparison with eq. (9) leads to

$$\kappa = m\left[\frac{V_F}{2k_F}\right]^2 \quad . \tag{12}$$

The excitations, which correspond to local modulation of the phase ϕ with a period q, are called phasons.

In the presence of an applied electric field the equation of motion

$$\frac{d^2\phi}{dt^2} + V_F\frac{m}{m^*}\frac{d^2\phi}{dx^2} = \frac{2k_F eE(\omega)}{m^*} \tag{13}$$

and the frequency dependent conductivity

$$\sigma(\omega) = \frac{j(\omega)}{E(\omega)} = \frac{m}{m^*}\frac{i\omega_p^2}{4\pi(\omega+i\delta)} \tag{14}$$

leading to

$$\mathrm{Re}\ \sigma(\omega) = \frac{ne^2}{m^*}\delta(\omega) \quad . \tag{15}$$

This feature, together with the gap Δ in the single particle excitation spectrum is reminiscent of superconductivity.

As in a superconductor, the phase $\phi(x,t)$ of the condensate plays an important role in the dynamics of the collective mode. A rigid displacement of the CDW leads to an electrical current. With $f = -eV_d = -e(dx/dt)$, $\phi = 2k_F x$ one obtains

$$j = -\frac{1}{\pi}\frac{d\phi}{dt} \quad . \tag{16}$$

A compression of the wave leads to the change of the electronic density, and therefore

$$n = -\frac{1}{\pi}\frac{d\phi}{dx} \quad . \tag{17}$$

The equation of continuity

$$\frac{dj}{dx} + \frac{dn}{dt} = 0$$

follows from the above equations. Eqs. (16) and (17) are different from those which relate $\phi(x,t)$ to the current and chemical potential in a superconductor, for which $j = -(d\phi/dx)$ and $\mu = d\phi/dt$. The above equations can also be derived by noting that the Fermi surface is tied to a moving CDW or to a compressed CDW and evaluating the total current or electron density using a band picture.

Experiments performed at low dc electric fields do not show evidence for the collective mode. This is because impurities, lattice imperfections, etc., interact with the CDW and they pin the phase to the underlying lattice. Representing this by an average pinning frequency ω_p and also including a phenomenological damping constant Γ, the phenomenological equation of q = 0 motion for the pinned mode in that of a harmonic oscillator

$$m^*\ddot{\chi} + \Gamma\dot{\chi} + \kappa x = eE\, e^{i\omega t} \tag{18}$$

where $\kappa = \omega_p^2 m^*$. This leads to a frequency dependent conductivity

$$\sigma(\omega) = \frac{ne^2}{i\omega m^*}\frac{\omega^2}{\omega_p^2 - \omega^2 - i\omega\Gamma} \quad , \tag{19}$$

with the low frequency dielectric constant $\varepsilon(\omega \to 0)$

$$\varepsilon(\omega \to 0) = \frac{4\pi ne^2}{m^*\omega_p^2} \quad . \tag{20}$$

The expected ω dependent response, for $\omega_0 \ll \Delta$ is shown in Fig. 4.

Indeed $\sigma(\omega)$ measured in materials at frequencies well below the single paticle gaps indicated strong absorptions in the millimeter wave spectral range. They can be associated with the response of the pinned collective mode. (Fig. 5.) Assuming a phenomenological damping constant $1/\tau$ and pinning frequency ω, the (classical) equation of motion for a uniform phase is given as

$$\sigma(\omega) = \frac{ne^2}{i\omega m^*}\frac{\omega^2}{\omega_0{}^2 - \omega^2 - i\omega/\tau} \quad . \tag{21}$$

Fits to the above equations lead to effective mass values in broad agreement with eq. (9), in particular, because λ and ω_{2k_F} do not change drastically from material to material, m* is approximately proportional to

Δ^2. The pinning frequencies are of the order of 10^{10} Hz, orders of magnitude smaller than the single particle gaps.

With small pinning energies as inferred from the frequency dependent conductivity studies, it is expected that moderate dc electric fields can provide an energy which is larger than the pinning energy. In this case the field would depin the mode inducing sliding CDW conduction. The electric field dependence of $\sigma = j/E$ observed in TaS_3 is shown in Fig. 6. There is a sharp threshold electric field below which σ is due to the normal electrons excited across the single particle gap. Above E_T the strongly nonlinear conduction is due to moving charge density waves; this is supported by several observations. First, the CDW is not destroyed by the applied electric fields as evidenced by structural studies. Second, recent NMR experiments[8] clearly demonstrate a motional narrowing due to the moving charge density waves, both in $NbSe_3$ and in $K_{0.3}MoO_3$.

Not only does the dc conductivity change when $E > E_T$, but there are accompanying field dependent thermoelectric and elastic phenomena. The thermoelectric power[9], measured in TaS_3 as the function of electric field is shown in Fig. 7, together with the measured differential conductivity $I'(0)/I'(e)$ where $I' = dI/dE$. The analysis of $s(E)$ is complicated, but the conclusion is that when the collective mode moves, very little heat is transported -- in accordance with the notion, that the CDW is a ground state condensate, and should carry no heat under ideal circumstances. Sliding CDW is also supported by Hall effect studies (although the issue is somewhat controversial). The elastic propeties are also field dependent, as shown in Fig. 8: the lattice gets "softer", and the internal friction increases in the nonlinear conductivity region. The phenomenon is not completely understood,[10] and experiments at various frequencies are required to clarify this point.

As both $\sigma(E)$ and $\sigma(\omega)$ are due to the dynamics of the collective mode, one expects a close relation between the L and E dependent response. In terms of the restoring force κ, the dielectric constant is given by

$$\varepsilon(\omega \to 0) = 1 + \frac{4\pi n e^2}{\kappa} \quad . \tag{22}$$

Assuming that a displacement of the collective mode by one wavelength leads to depinning, the condition $eE_T\lambda = (1/2)\, k\lambda^2$ leads to the relation

$$\varepsilon(\omega \to 0) E_T = 4\pi e \quad . \tag{23}$$

This is indeed observed (at least qualitatively), and ε and E_T values observed in several materials are displayed in Fig. 9. (Both ε and E_T are temperature dependent, but εE_T is nearly temperature independent except close to the transition temperatures.)

Although various interactions between the collective mode and the underlying lattice can lead to a restoring force κ, it is usually assumed that interactions with the impurities are most important.

Fukuyama and Lee[11] considered the problem of impurity pinning in detail. The Hamiltonian in one dimension

$$H = \frac{v_F}{4k_F} \int d\vec{r}\, (\nabla\phi)^2 - V_0 \rho_0 \sum_i \cos[2k_F X_i + \phi(X_i)] \tag{24}$$

where the first term represents the elastic energy of the deformable CDW, and the second term describes the interaction between the CDW and the impurities distributed at random at positions i. In less than four dimensions, eq. (24) leads to a finite phase-phase correlation length and

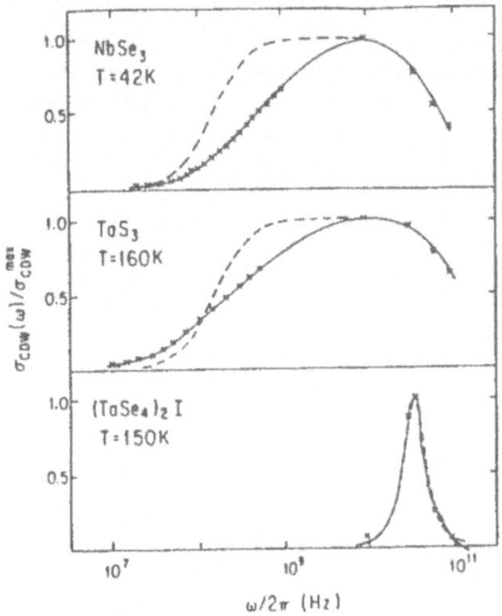

Fig. 5. Frequency dependent conductivities of various compounds in the CDW ground state. The full lines are guides to the eye.

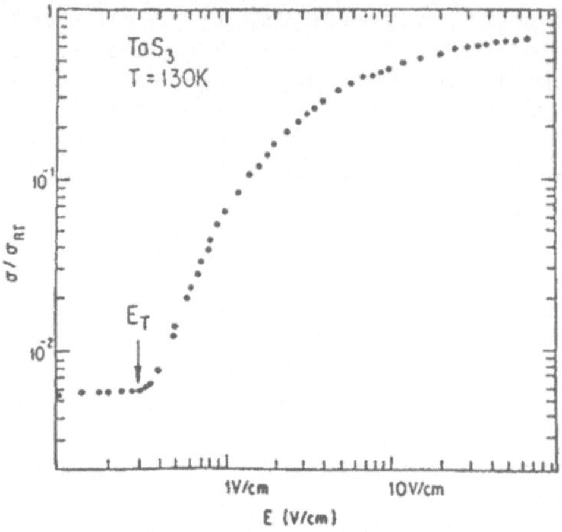

Fig. 6. Electric field dependent conductivity in TaS_3.

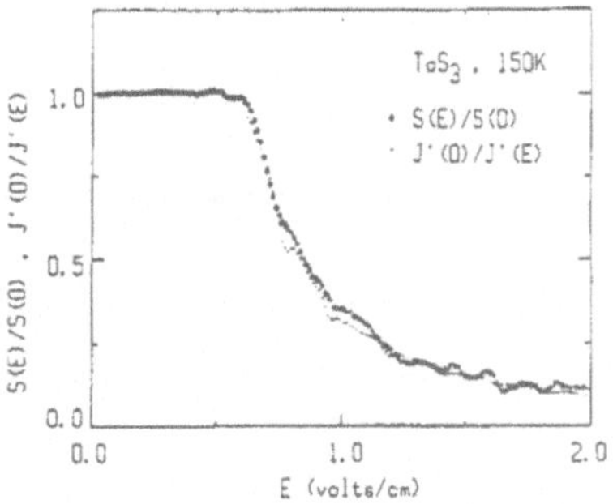

Fig. 7. Electric field dependent thermopower, S(E) and differential conductivity J'(E) = dJ/dE in TaS_3. Ref. 9.

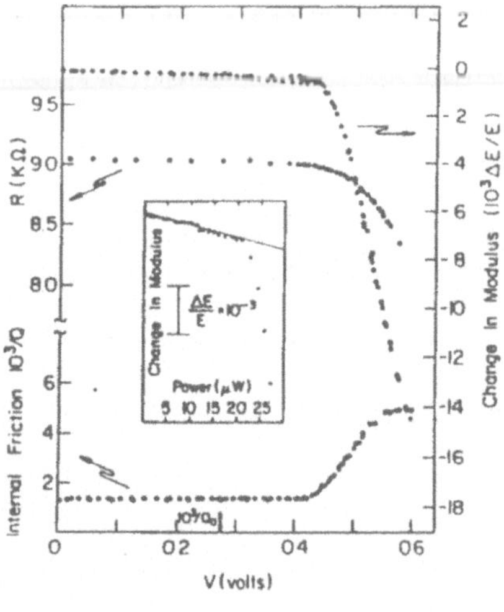

Fig. 8. Electric field dependent resistance R, elastic modulus E and internal friction 1/Q in TaS_3. Ref. 10.

also to a finite pinning energy in the thermodynamic limit.

The first term in eq. (24) favors a uniform phase, the second favors local distortions around impurities. A dimensionless parameter

$$e = \frac{V_0 \rho_0}{V_F n_i}$$

tells us which of these is more important. For $\varepsilon \gg 1$ (strong pinning) the phase of the CDW is fully adjusted to every impurity site to obtain a maximum potential energy gain, the cost in elastic energy for doing this is negligible. For $\varepsilon < 1$ (weak pinning) the phase cannot be adjusted fully at every impurity site but only over a length scale L_0 longer than $1/n_i$, the average distance between the impurities. These situations are schematically shown in Fig. 10.

For strong impurity pinning, where the elastic energy of the CDW is neglected, the pinning energy is trivially given by

$$E_{PIN} = \Delta E_{POT} = - V_0 \rho_0 n_i \quad , \tag{25}$$

where in 1D n_i is the number of impurities in a unit length, or in 3D in a unit volume. As the phase is fully adjusted at every impurity site, the phase-phase correlation length

$$L_0 \sim \frac{1}{n_i} \; . \tag{26}$$

The case of weak impurity pinning is more interesting. In 3D, assuming a volume L_0 over which the phase is constant but is adjusted to the impurity fluctuations, one obtains a potential energy gain

$$\Delta E_{POT} = - V_0 \rho_0 \left(\frac{n}{L_0^3}\right)^{1/2} \; . \tag{27}$$

The elastic energy, from eq. (24)

$$E_{EL} = \frac{3\pi}{4} \frac{V_F}{L_0^2} \; . \tag{28}$$

Minimizing the total energy with respect to L_0 leads to

$$L_0 = \left[\frac{\pi V_F}{V_0 \rho_0}\right]^2 \frac{1}{n_i} \tag{29}$$

and

$$E_{PIN} = - \frac{1}{4} \frac{V_0^4 \rho_0^4}{\pi^3 V_F^3} n_i^2 \; , \tag{30}$$

i.e., the pinning energy is proportional to the square of the impurity concentration. As discussed before, an applied electric field E can depin the mode, if the energy gain for displacement of the CDW by one period, $eE\lambda$, exceeds E_{PIN}. This crude argument leads to a threshold field

$$E_T = \frac{1}{\pi e} V_0 \rho_0 n_i \qquad \text{strong pinning} \tag{31}$$

$$E_T = \frac{1}{4\pi e} \frac{V_0^4 \rho_0^4}{V_F} n_i^2 \qquad \text{weak pinning} \tag{32}$$

i.e., for strong pinning E_T is proportional to the impurity concentration for weak pinning $E_T \sim c^2$.

While the fact that the charge density waves are pinned by impurities in these inorganic linear chain compounds is well confirmed by

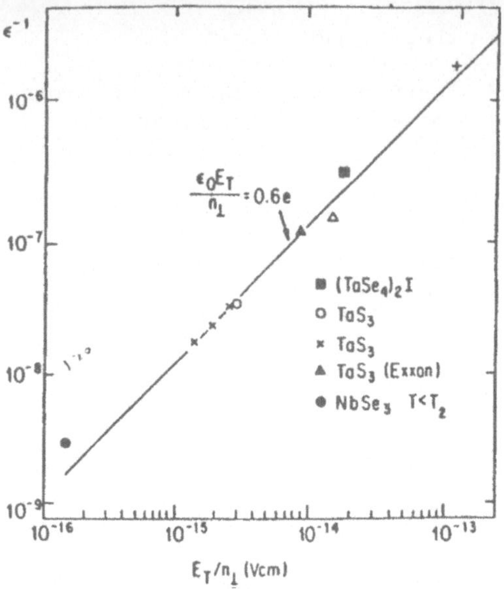

Fig. 9. The relation between the low frequency dielectric constant ε and threshold field E_T in various materials.

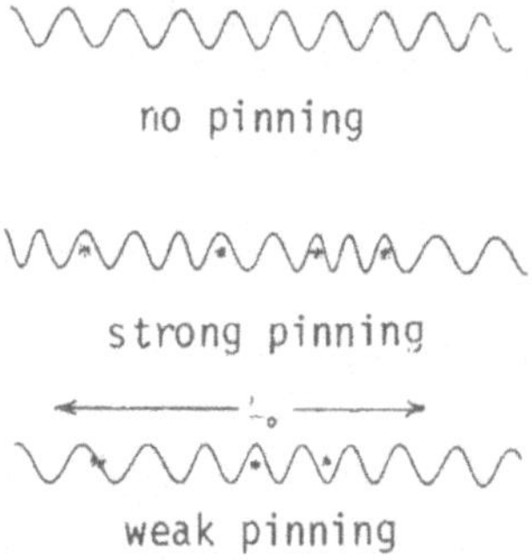

Fig. 10. Strong and weak impurity pinning. For strong pinning the local phase is adjusted to lead to maximum energy gain (here with an attractive interaction the amplitude of the CDW is maximum at the impurity site) at every impurity site. For weak pinning the average phase is adjusted over the length L_o much larger than the distance between impurities.

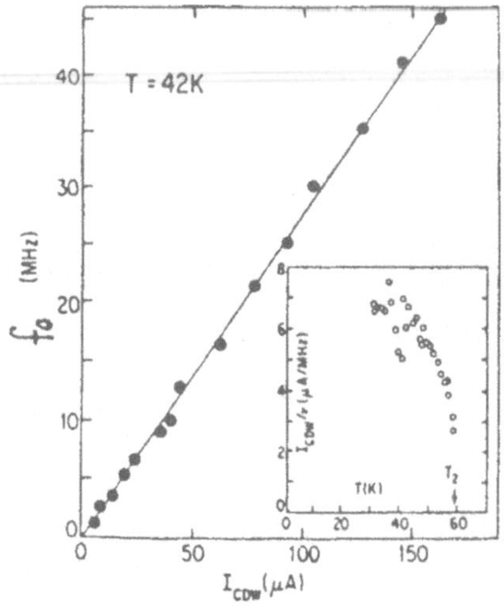

Fig. 11. Relation between the current carried by the CDW and fundamental frequency. The insert shows the ratio I/f_o as the function of temperature.

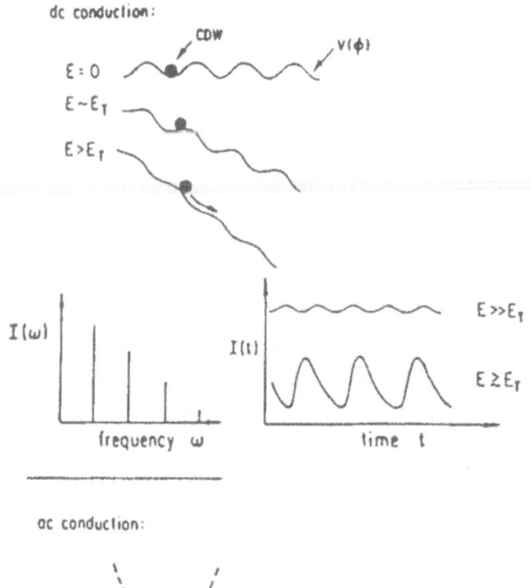

Fig. 12. Classical particle model of CDW transport.

experiments on alloys or on materials where lattice defects are induced by irradiation, the detailed concentration dependence is not clear at present. For materials where impurities are created by irradiation, $E_T \sim c$, showing that such defects act as strong pinning centers. Early experiments[12] on alloys of $NbSe_3$ suggested a c^2 concentration dependence, while subsequent experiments[13] give linear relation between E_T and the concentration, measured by the residual resistivity ratio: this then would suggest strong impurity pinning in all cases studied so far. Also, for small concentration of impurities there is no correlation between E_T and c, suggesting that other sources of pinning, such as contacts, grain boundaries, may also be important.

3. COHERENT EFFECTS, INTERFERENCE EXPERIMENTS

A spectacular phenomenon, current oscillations in the presence of dc applied fields, accompanies the nonlinear conductivity.[14] These can be studied both by looking at the time dependent response directly or by performing a Fourier analysis of the signal using a spectrum analyzer. The frequency of the oscillations is proportional to the excess current[15] $j_{CDW} = j(E) - j(\text{ohmic})$ and

$$\frac{j_{CDW}}{f_0} = Ae\,\frac{n(T)}{n(0)} \quad , \tag{33}$$

where j_{CDW} is the current density per chain, n(T) is the temperature dependent condensate density, and A is a numerical factor clsoe to two. This relation, observed in $NbSe_3$, is shown in Fig. 11 with the radio j_{CDW}/f_0 is the function of temperature in the insert. Simple arguments, assuming the translational motion of the condensate, can account for this observation. As every period λ of the CDW contains exactly 2e electrons (per chain), assuming that the frequency f_0 of the current oscillations corresponds to a displacement of the CDW by λ (leading to the same energy configuration) leads to eq. (33) with A = 2. Alternatively one can look at the Fermi surface displaced by $\delta k = mV_d/\hbar$ in the presence of a drifting CDW. The energy difference between the two sides of the Fermi surface is given by

$$\Delta E = \frac{\partial\varepsilon}{\partial V} V_d = \frac{\partial\varepsilon}{\partial K}\frac{\partial K}{\partial V} V_d = 2V_F m V_d \tag{34}$$

as $V_F = h\pi/m\lambda$, $\delta E = 2k_F h V_d$. Associating a frequency $\hbar\omega_0 = \delta E$ with transitions across the Fermi surface and noting that $j_{CDW} = neV_d$, eq. (34) is recovered, with A = 2.

A highly oversimplified model,[16] which treats the phase as a classical variable, and assumes that the internal degrees of freedom of the condensate can be neglected, accounts for the qualitative features of the ω and E dependent response. Assuming that the pinning can be represented by an average potential, which, as the consequence of the periodic CDW is assumed to have a periodic form

$$V(\phi) = \frac{\omega_0^2}{2k_F}(1 - \cos 2k_F x) \quad . \tag{36}$$

The equation of motion

$$\frac{d^2x}{dt^2} + \frac{1}{\tau}\frac{dx}{dt} + \frac{\omega_0^2}{2k_F}\sin 2k_F x = \frac{eE}{m^*} \quad , \tag{37}$$

which in terms of the phase $\phi = 2k_F x$ is given by

$$\frac{d^2\phi}{dt^2} + \Gamma'\frac{d\phi}{dt} + \sin\phi = \frac{E}{E_T} \quad , \tag{38}$$

with $\Gamma' = (\omega_0\tau)^{-1}$, $E_T = (\lambda/2\pi)(m^*\omega_0^2/e)$ and time is measured in units of

ω_0^{-1} . Here ω_0 and τ are phenomenological parameters. Equation (37) represents the damped motion of a particle in a tilted sinusoidal potential as shown in Fig. 12. The low field ac response is the same as eq. (21), and a fit of $\sigma(\omega)$ to a simple harmonic oscillator response is shown in Fig. 13. Nonlinear conduction occurs for $E > E_T$ when the particle moves down in the staircase potential. The onset of nonlinear conductivity is sharp, and for an overdamped motion [as evidenced by the $\sigma(\omega)$ studies – see Fig. 13],

$$\sigma(E) = \left\{ \begin{array}{ll} 0 & E < E_T \\ \frac{ne^2}{m^*\Gamma}\,[1 - (E_T/E)^2]^{1/2} & E > E_T \end{array} \right\} . \quad (39)$$

Thus, the model leads to a threshold field E_T and to saturation at high electric fields. The observed behavior, however, is more gradual than that given by eq. (39), and while the model predicts a divergent differential conductivity at E_T, this has not been observed.[5] The model, however, naturally leads to current oscillations with $f_0 \sim I$ as observed.

Many experiments have also been performed to study the current oscillations in detail. These involve the joint applications of dc and ac fields

$$E = E_{dc} + E_{ac}\cos\omega t \quad . \quad (40)$$

To study these it is useful to recall that eq. (38) is identical to that of a physical pendulum in a gravitational field, or a rigid particle moving in a "washboard" potential. Equation (38) is also well known in the Josephson junction literature, where it describes the dynamics of a Josephson junction in the resistively shunted junction (RSJ) model. In the later case θ represents the superconductor phase difference across the junction, $\Gamma = (\omega_J RC)^{-1}$ with ω_J the Josephson plasma frequency, and R and C respectively the resistance and capacitance of the junction. Also, if eq. (38) is used to describe the Josephson junctions, E refers to the current through the junction, and E_T to the critical current of the junction. With this analogy, the current oscillations described by Eq. (33) are analogous to the ac Josephson effect, with the generated frequency linear in time averaged junction voltage.

Several features, such as the modification of the ac response by the application of dc fields, or ac field induced dc conduction can be described by eq. (38) at least qualitatively (or, also with electronic circuit analogs, such as the relaxation oscillator[17]). Equally interesting are experiments where interference effects, arising from coupling between the externally applied ac fields and the intrinsic oscillation with $\omega_{ext}/2\pi$ is close to f_0 are studied. Effects analogous to Shapiro steps in Josephson jucntions – i.e., steps in the dc I-V curves due to the application of ac fields – have been observed.[15,18] An example of this is shown in Fig. 14, where I-V recordings are displayed for ac voltages of various amplitude, is a $NbSe_3$ specimen.

In the absence of ac electric fields the current-voltage characteristics is smooth, with a well defined onset for nonlinear conduction. Increasing V_{ac}, the threshold field is reduced (and also becomes sharper), and well defined steps in the nonlinear conductivity region. The steps are first broad, but become progressively narrower with increasing V_{ac}. The definition of the step height δV is shown in the figure.

The step height δV is a strong function of V_{ac}. The step defined as n = 1 corresponds to an interference between the 100 MHz r.f. field and

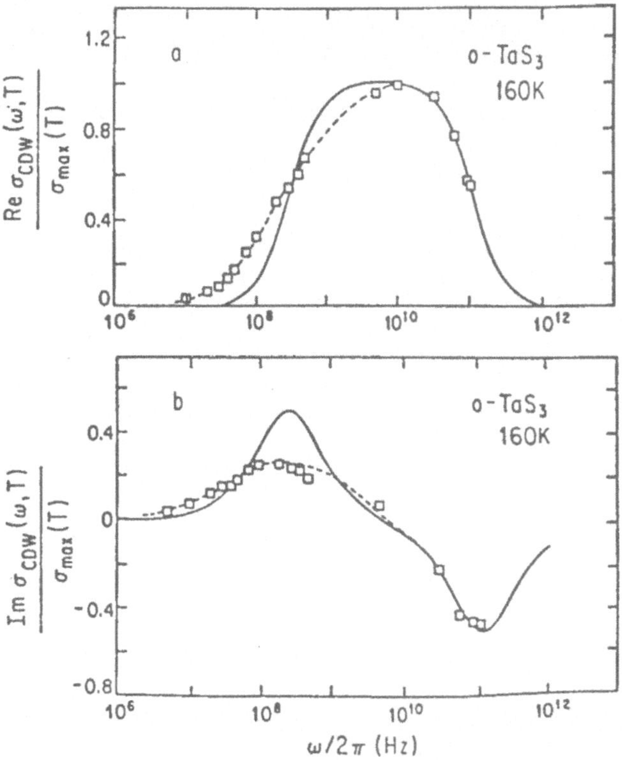

Fig. 13. Frequency dependent conductivity of TaS_3. The full line is a fit to Eq.(19). D.Reagor, S.Sridhar, and G. Gruner, Phys. Rev.B 34, 4 (1986).

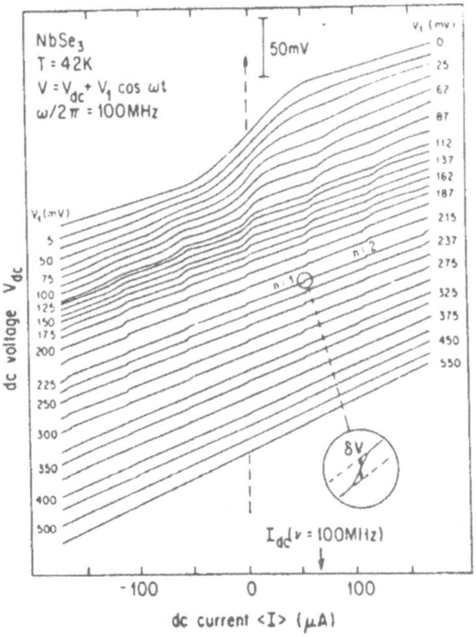

Fig. 14. ac field induced steps in the d.c. I-V curves in $NbSe_3$. Ref. 18.

the fundamental of the intrinsic oscillation. Harmonic and subharmonic steps with smaller amplitude can be observed.

Equation (38) leads, in a high frequency ($\omega_{ext} > \omega_0^2 \tau$) limit to $\delta V = \alpha 2V_T(V_{ac} = 0)J_n|V_{eff}|$:

$$V_{eff} = \frac{V_{ac}\omega_0^2 \tau}{\omega V_T(\omega=0)} , \qquad (41)$$

where α represents the volume fraction of the specimen which responds coherently to the external fields, V_T the threshold voltage, and in J_n the Bessel function of order n. Computer simulations in the frequency region $\omega_{ext} \lesssim \omega_0^2 \tau$ also lead to oscillations of δV as function of V_{eff}, closely resembling Bessel functions behavior. The Bessel function behavior (also observed in Josephson jucntions) is clearly recovered in the experiment with $\omega_0^2 \tau$ values close to those observed in $\sigma(\omega)$ studies. The characteristic values of α are between 0.1 and 1, suggesting highly coherent response. Experiments performed as the function of frequency are also in agreement with the model. Interference effects show up also in the ac response.[18]

The sensitivity of detection of the interference phenomena can be enhanced by displaying the differential resistance dV/dI. Peaks in the derivative correspond to steps in the I-V curve, with width of the peak in voltage units corresponding to the height of the step.

Figure 15 shows the differential resistance measured with and without the externally applied r.f. voltage $\omega_{ext}/2\pi$ = 25 MHz. There appears a series of peaks at applied dc current such that

$$pf_0 = q\frac{\omega_{ext}}{2\pi} \qquad (42)$$

with p and q integers.[19] Some of the peaks are identified on the figure. This identification was made by plotting I_{CDW} vs. ω_{ext} and checking that the slopes were p/q times that for the fundamental. Qualitatively, the width of the peaks decreases with increasing p and q as expected for mode locking between two frequencies both with slowly decaying harmonic content. The peaks conseqauently correspond to regions in which any harmonic of the internal frequency locks to any harmonic of the external field. Such mode locking has been extensively studied for the equation of motion (38) and both numerical and analytical solutions are available. In general, a Devil's staircase behavior in which interference occurs for any p and q integer is recovered by such calculations for a weakly damped system. For strong damping, interference is weak and no subharmonics are expected for an overdamped response. Devil's staircase behavior is clearly implied by the experimental result, but due to noise effects higher order interference peaks cannot be resolved.

Several explanations have been advanced to account for the observed subharmonic peaks. It has been assumed[20] that the response in the nonlinear region is underdamped – a highly unlikely situation. A nonsinusoidal potential also leads to a rich subharmonic structure in qualitative agreement with experiments,[21] and internal degrees of freedom[22] also account for this behavior. The calculations also give clear predictions concerning the amplitude and form of the interference peaks. These could be further tested by experiments.

The appearance of current oscillations with a high quality factor Q suggest a highly coherent response, with the whole specimen responding as one coherent domain. This situation is nearly appropriate for rather small $NbSe_3$ specimens where the small theshold fields indicate larger coherent Lee-Rice domains. Even here, the detailed examination of the ω

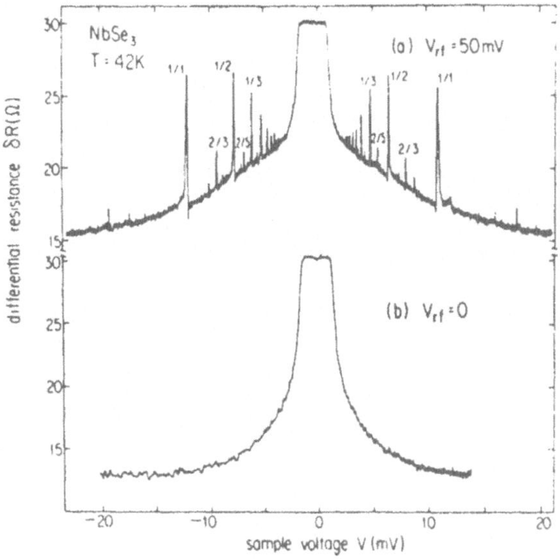

Fig. 15. Subharmonic steps observed in interference experiments. Ref. 19.

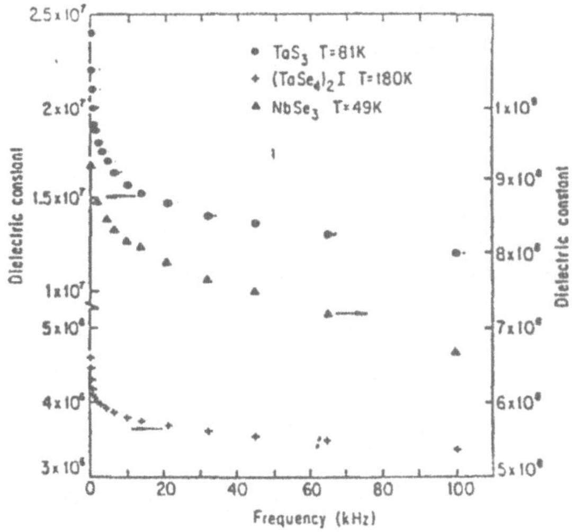

Fig. 16. Frequency dependent dielectric constant in several materials in the CDW state.

dependent response suggests[23] that disorder is important, and the dynamics of the local deformations of the collective mode play a significant role in $\sigma(\omega)$ and $\sigma(E)$. Disorder effects are even more significant in other materials, especially at low temperatures, when the normal electrons cannot screen the impurity potentials.

4. DISORDER, THE DYNAMICS OF INTERNAL DEFORMATIONS

The dynamics of internal modes are important both in the pinned and in the current carrying state. Here I will focus mainly on the properties of the pinned mode, reviewing experiments performed on $(TaSe_4)_2I$ and on $K_{0.3}MoO_3$ at low temperatures.

The following topics will be discussed:

(a) ac response in the small amplitude limit; frequency dependent conductivity;

(b) nonlinear ac response and third harmonic generation;

(c) time dependent response in the long time limit; and,

(d) approaching the threshold from below: polarization divergence and critical slowing down.

4a. *Small Amplitude ac response*

As discussed earlier, a single harmonic oscillator fit provides an adequate description of $\sigma(\omega)$ (see the full line of Fig. 13) in several materials, at temperatures not far below the Peierls transition temperature. A much better fit is obtained by assuming a distribution of relaxation times. Such a fit is shown by the dotted line in Fig. 13.

The low frequency tail of $\sigma(\omega)$ can be well described by the empirical formula

$$\sigma(\omega) = A\left(\frac{i\omega}{\omega_0}\right)^{\alpha}, \qquad \text{with} \qquad \alpha \approx 0.8 \,. \tag{43}$$

Similar behavior has been found in other materials like $NbSe_3$, $K_{0.03}MoO_3$, and $(TaSe_4)_2I$. The dielectric constant, $\mathrm{Re}\,\varepsilon(\omega) = 4\pi\,\mathrm{Im}\,\sigma(\omega)/\omega$ diverges as $\omega \to 0$. This behavior is shown for several compounds in Fig. 16. Recent experiments[24] suggest that at low frequencies the ω dependence is better represented by a so-called Cole-Davidson formula:

$$\varepsilon(\omega) = \frac{\varepsilon_0}{(1 + (i\omega\tau_0)^{\alpha})^{\beta}}, \tag{44}$$

with α and β less than one. Both eqs. (43) and (44) are typical of random conductors and, consequently, it has been argued that disorder plays an important role in the dynamic of the collective mode.

4b. *Nonlinear Response and Third Harmonic Generation*

The ac conductivity, as displayed in Fig. 13, represents the response of the system to small amplitude excitations. The response, however, strongly depends on the ac amplitude.[25] Also, higher order harmonics at $\Omega = 3\omega$, $\Omega = 5\omega$, etc., appear in the response to a sinusoidal excitation, $E = E_0 \cos \omega t$, even for drive amplitudes far below the threshold field of the nonlinear conduction.[26]

For a single harmonic oscillator in the nonlinear potential, $V(x) = (1/2)kx^2 - (1/4)bx^4$, the equation of motion is

$$1/\tau\ \dot{x} + \omega_0^2\, x + (b/m)\, x^3 = eE_0/m^* \cos \omega t \,. \tag{45}$$

The linear response is given by the conductivity, eq. (21). The first order correction due to the nonlinearity leads to an amplitude-dependent first harmonic and nonzero higher harmonics, given by

$$\varepsilon^{(1)} = \varepsilon_0^{(1)} + b\zeta(\omega)\, E_0^2 \;, \qquad \varepsilon^{(3)} = b\eta(\omega)\, E_0^2 \;, \tag{46}$$

where $\zeta(\omega)$ and $\eta(\omega)$ are simple, analytical functions of the frequency. Equation (45) leads to specific predictions concerning the field and frequency dependence of the nonlinear response. At low frequencies, $\omega < \omega_0^2 \tau$, both the first and third harmonic nonlinear response are independent of frequency and both are proportional to the square of the drive amplitude. The leading term in the frequency dependence of the linear response Re $\varepsilon_0^{(1)} \sim (1-a\omega^2)$, the leading frequency dependence in the third harmonic is nonsingular, Re $\varepsilon^{(3)} \sim (1-b\omega^2)$.

The excitation amplitude dependence of $\varepsilon^{(1)}$ measured in $K_{0.3}MoO_3$ is displayed in Fig. 17 for various applied V_0 voltages, normalized to the threshold voltage V_T for the appearance of nonlinear dc conduction. Although $\varepsilon^{(1)}$ appears to saturate as $V_0/V_T \to 0$, the response is strongly nonlinear and frequency dependent, with $\varepsilon^{(1)}$ slowly increasing with decreasing frequencies. Similar enhancement is observed in the out-of-phase component. The thrid harmonic response $\varepsilon^{(3)}$ is displayed in Fig. 18 in the same parameter range. Again, $\varepsilon^{(3)}$ is both V_0 and ω dependent with increasing response with increasing V_0 and decreasing ω.

The amplitude dependent fundamental response has been noted earlier.[25] The response at any given excitation amplitude V_0 can be analyzed by assuming that linear response theory is appropriate, and increasing V_0 leads to changes in the parameters which characterize the response. Such an approach was adopted by Cava et al.,[27] who argued that the mean relaxation time τ_0 which appears in a Cole-Davidson expression, eq. (44), increases with increasing ac drive. The interpretation implies that linear response theory is relevant, and that the response to an excitation $V = V_0 \cos\omega t$ is given by a sinusoidal function with no higher harmonics, in contrast to our findings concerning the third harmonic generation: for linear response, the third harmonic is identically zero.

The behavior, as displayed in Figs. 17 and 18, has not been accounted for yet, but calculations based on random field Ising model, lead to qualitatively similar behavior, i.e., both linear and nonlinear response, described in certain frequency ranges as having a fractional power frequency dependence.

I also note that, just as the frequency dependent small amplitude response is different from that of a simple harmonic oscillator, the third harmonic response is also different from that given by eq. (46). Instead of smooth frequency dependence, predicted by the simple model based on eq. (45), the third harmonic response diverges as $\omega \to 0$.

4c. *Time Dependent Response*

While a single degree of freedom system is expected to give an exponential approach to equilibrium in response to a pulse excitation, a broad variety of experiments indicate much more complicated time dependences. We have observed a stretched exponential dependence[28]

$$P(t) = P(0)\left[1 - \exp\left(-\frac{t}{\tau}\right)^{\alpha'}\right] \;, \tag{47}$$

with $\alpha' < 1$ in response to a sudden electric field dependent pulse; but

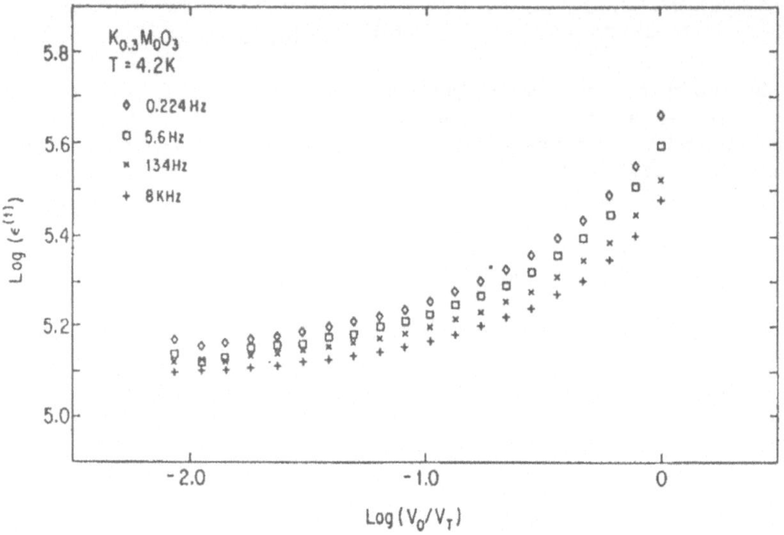

Fig. 17. Nonlinear fundamental response in $K_{0.3}MoO_3$. V_T is the threshold field for the onset of nonlinear conduction.

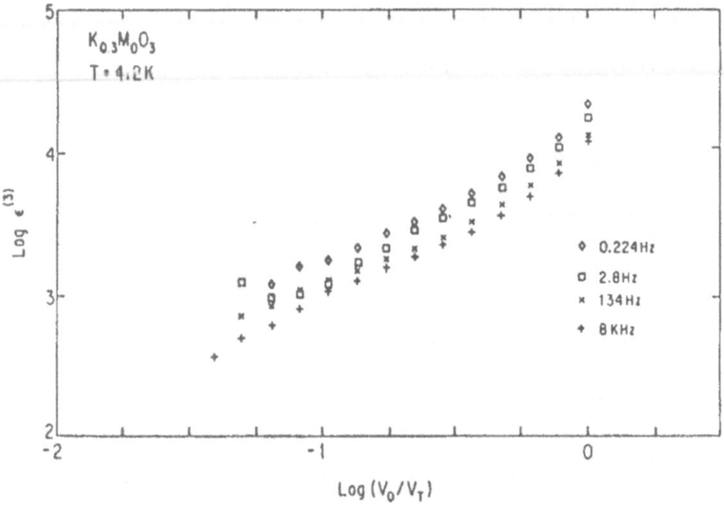

Fig. 18. Third harmonic response in $K_{0.3}MoO_3$. Ref. 26.

other forms, such as logarithmic, also work in limited time regimes. The above expression is again characteristic of that of random systems,[29] and has also been derived by assuming hierarchical relaxation. While at high temperatures the polarization decays to zero, in the liquid He^4 temperature range a remanent polarization is observed.[28] This is demonstrated in Fig. 19, where the polarization measured after the field was increased step by step is displayed. The hysteresis loop is reminiscent to that observed in random magnetic systems (and also in ferromagnets), and is clearly related to the local deformations of the collective mode, which remain frozen after the electric field is turned off.

4d. *Polarization Effects and "Critical Phenomena"*

The classical particle model leads to various divergencies when E_T is approached from below or from above. In the pinned state, the differential dielectric constant (i.e., the out-of-phase component of the ac response for small amplitude ac excitation in the presence of a dc field E_{dc})

$$\frac{d\varepsilon(\omega\to o)}{dE_{ac}} \sim (E_T - E_{dc})^{-1/2} \tag{48}$$

while above threshold the differential conductivity

$$\frac{d\tau(\omega\to o)}{dE} \sim (E_{dc}-E_T)^{-1/2} \quad . \tag{49}$$

The polarization $P = ne\lambda$ at threshold, i.e., does not diverge as $E \to E_T$ from below. The experiments performed in $K_{0.3}MoO_3$ at low temperatures are in clear contradiction with this single particle picture. The dielectric constant ε does not show the divergence predicted by eq. (47).[30] A closer look at the behavior below E_T indicates that the critical parameter may be the static polarization, which is both hysteretic and divergent as $E \to E_T$.

In Fig. 20 a quantity characteristic of the saturation polarization, $P\alpha'$, is plotted for several electric fields. The data were evaluated from relaxation measurements similar to those discussed before. The exponent α' is not expected to change dramatically; therefore most of the variation is due to the field dependence of the polarization. As the voltage approaches V_T, the polarization seems to diverge, and close to E_T the experimental results can be described with the expression:

$$P_0 \approx (E_T - E)^{-\beta} \, , \qquad \beta \approx 3 \quad . \tag{50}$$

Whether β can be regarded as a critical exponent remains to be seen. It is obvious, however, that depinning is preceded by a polarization divergence at low temperatures.

The concept, that depinning can be described as a dynamical critical phenomenon has been advanced by Fisher,[31] who also discusssed a mean field treatment of the problem. Many of the observations discussed in this section have also been recovered by numerical simulations,[32] with the zero frequency cusp of $\varepsilon(\omega)$ as one example.

I have not discussed the interplay of coherence and incoherence in the current carrying state, such as the relation between broad band and narrow band noise, and other related phenomena. One should mention, however, that the current oscillations disappear in the thermodynamic limit,[33] in agreement with the expectation, and also with model calculations.

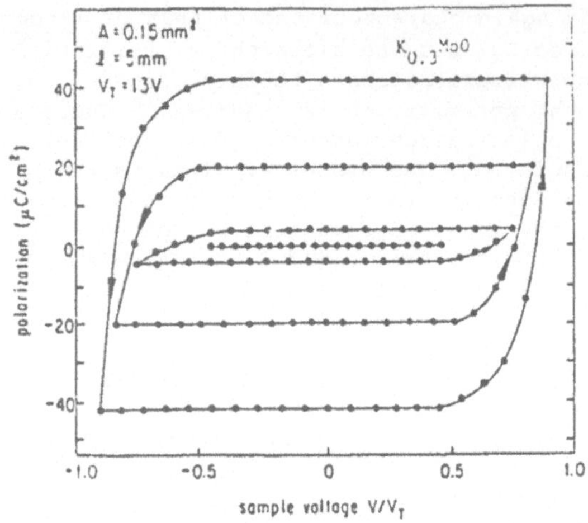

Fig. 19. Dielectric polarization loops observed in $K_{0.3}MoO_3$ at low temperatures. V_T refers to the threshold voltage for the onset of nonlinear conduction. Ref. 28.

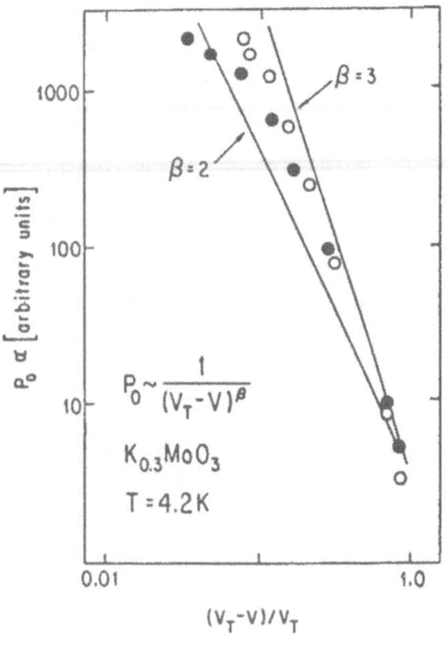

Fig. 20. Pα' (for the definition of P and α' see Eq. (47))versus applied electric field in $K_{0.3}MoO_3$.

5. CONCLUSIONS

While the existence of Fröhlich conduction is by now well established by a broad range of experiments, several fundamental questions remain concerning the dynamics of the collective CDW mode. It appears that the field is considerably broader than envisioned a few years ago, and depending on the nature of the pinning sources (whether extended defects, randomly positioned impurities, etc., are dominant) on the temperature, or the dimensions of the specimens, etc., rather different dynamical behavior can be observed. This also possibly accounts for some of the experimental controversies concerning the origin of current oscillations, the importance of the internal degrees of freedom, etc.

Most of the coherent effects discussed earlier have been observed mainly in $NbSe_3$, while incoherent phenomena are most prominent in $K_{0.3}MoO_3$. The most likely reason for this is that the phase-phase coherence length is comparable with the sample dimensions in $NbSe_3$, while it is much smaller in $K_{0.03}MoO_3$. This difference should also show up in current-voltage characteristics and in the frequency dependent conductivity. Extended defects can also lead to the break of phase coherence. This is responsible for switching and hysteresis phenomena.

In this review I did not attempt to make a detailed comparison with current theories. These have been summarized by John Bardeen and Sue Coppersmith at this meeting. Many of the observations can be accounted for both by a classical description of CDW dynamics and by concepts where genuine quantum phenomena (such as tunneling) are important and further experiments appear to be important to clarify this point.

Also, it has been demonstrated that the organic compound TTF-TCNQ also displays nonlinear transport. Intercalated graphite at high magnetic fields and the compound HgCdTe was also shown to have nonlinear transport properties with a sharp threshold field.

With important experimental and theoretical questions to be resolved, and with emerging new materials, it is expected that the field will broaden considerably, evolving into one of the major areas of condensed matter science.

6. ACKNOWLEDGMENTS

Many of the experiments summarized here were performed by my students and postdoctoral associates. The contributions by Stuart Brown, Laszlo Mihaly, David Reagor, and Srinivas Sridhar were most important. This research was supported by the NSF grants DMR 81-31294 and DMR 84-06896.

REFERENCES

1. R. E. Peierls, "Quantum Theory of Solids," Oxford University Press (1955).
2. M. Fröhlich, Proc. Roy. Soc. A223, 296 (1954).
3. See, for example, "Highly Conducting One Dimensional Solids," I. T. Decrease, et al., eds., Plenum Press, New York (1979).
4. P. Monceau, N. P. Ong, A. M. Portis, A. Meerschant, and J. Rouxel, Phys. Rev. Lett. 37, 6902 (1976).

5. For a review, see G. Grüner and A. Zettl, Physics Reports 119, 117 (1985); P. Monceau, in "Electronic Properties of Inorganic Quasi-One-Dimensional Materials," P. Monceau, ed., Riedel (1985).
6. P. A. Lee, T. M. Rice, and P. W. Anderson, Solid State Comm. 14, 703 (1974).
7. S. Sridhar, D. Reagor, and G. Grüner, Phys. Rev. Lett. 55, 1196 (1985); D. Reagor and G. Grüner, Phys. Rev. Lett. 56, 6 (1986).
8. P. Sergasan, et al., Phys. Rev. Lett. 56, 1854 (1986); J. H. Ross, Z. Wang, and C. P. Slichter, Phys. Rev. Lett. 56, 663 (1986).
9. J. Stokes, A. N. Bloch, A. Jánossy, and G. Grüner, Phys. Rev. Lett. 52, 372 (1984).
10. J. W. Brill and W. Roark, Phys. Rev. Lett. 53, 846 (1984); G. Mozurkewich, P. M. Chaikin, W. G. Clark, and G. Grüner, Solid State Commun. 56, 421 (1985).
11. H. Fukuyama and P. A. Lee, Phys. Rev. B 17, 535 (1977).
12. N. P. Ong, et al., Phys. Rev. Lett. 42, 811 (1979).
13. P. Monceau, Physica B + c.
M. Underweiser, B. Alavi, M. Maki, and G. Grüner (unpublished).
14. R. M. Fleming and C. C. Grimes, Phys. Rev. Lett. 42, 1423 (1979).
15. P. Monceau, J. Richard, and M. Renard, Phys. Rev. Lett. 45, 43 (1981).
16. G. Grüner, A. Zawadowski, and P. M. Chaikin, Phys. Rev. Lett. 46,
17. M. Weger, G. Grüner, and W. G. Clark, Solid State Comm. 35, 243 (1980).
18. A. Zettl and G. Grüner, Phys. Rev. B 29, 755 (1984).
19. S. E. Brown, G. Mozurkewich, and G. Grüner, Phys. Rev. Lett. 54, 2272 (1984).
20. P. Bak and M. Azbel (to be published).
21. R. E. Thorne, J. R. Tucker, John Bardeen, S. E. Brown, and G. Grüner, Phys. Rev. B 33, 7342 (1986).
22. P. F. Tua and J. Ruvalds, Solid State Comm. 54, 471 (1985).
23. Wei-yu Wu, G. Mozurkewich, L. Mihály, and G. Grüner, Phys. Rev. Lett. 52, 2382 (1984).
24. R. J. Cava, et al., Phys. Rev. B 30, 3228 (1984).
25. R. J. Cava, P. Littlewood, R. M. Fleming, R. G. Dunn, and E. A. Rietmann, Phys. Rev. B 33, 2439 (1986); J. P. Stokes, M. O. Robbins, and S. Bhattachaya, Phys. Rev. B 32, 6939 (1986).
26. T. Cheng, L. Mihály, and G. Grüner (to be published).
27. R. J. Cava, et al., Phys. Rev. B 30, 3228 (1985).
28. L. Mihály and G. X. Tessema, Phys. Rev. B 33, 5858 (1986); L. Mihály and G. Grüner (to be published).
29. R. Kohlrausch, Ann. Phys. 12, 393 (1947); R. G. Palmer, D. L. Stein, E. Abrahams, and P. W. Anderson, Phys. Rev. Lett. 53, 958 (1984).
30. Wei-yu Wu, L. Mihály, and G. Grüner, Solid State Comm. 55, 663 (1985).
31. D. S. Fisher, Phys. Rev. B 31, 1396 (1985).
32. P. Littlewood, Phys. Rev. (to be published).
33. G. Mozurkewich and G. Grüner, Phys. Rev. Lett. 55, 2206 (1983).

FRÖHLICH CONDUCTIVITY IN TRANSITION METAL TRI- AND TETRACHALCOGENIDES

Pierre Monceau

Centre de Recherches sur les Très Basses Températures, CNRS
BP 166 X, 38042 Grenoble-Cédex, France

In the last fifteen years, intensive experimental and theoretical work has been undertaken in the understanding of the physical properties of systems with a restricted dimensionality. In a great number of such compounds the interaction between ions and electrons, the so-called electron-phonon interaction, can cause a modulated collective deformation of the electronic charge density to give a lower energy at low temperature, as firstly pointed out by Peierls.[1] This lecture gives a general survey of the dynamical properties of such a collective state.

PEIERLS TRANSITION AND FRÖHLICH CONDUCTIVITY

As is well known from band theory, every Brillouin'zone is generally a locus of discontinuity for the electronic energy. If, in a one-dimensional electronic system with a Fermi vector of k_F, a periodic lattice distortion (PLD) of wave-vector $2k_F$ occurs, the band structure has to be re-calculated because this new periodicity. A new Brillouin zone appears at $|k_F|$ and so, each occupied electronic energy for $|k| < k_F$ decreases, giving the new ground state of the system called a charge density wave state (CDW) of wave vector $q = 2k_F$. The occupied electronic states are the Bloch wave functions of the superlattice periodicity :

$$\psi_k = e^{ikr} \sum_n V_{k,n} \, e^{inqr}$$

and consequently the electronic density has components in e^{inqr}, especially on wave vectors $\pm q$:

$$\rho_{el} = \rho_\circ + 2\rho_q \cos(qr+\phi) + \ldots \qquad (1)$$

where $\rho_\circ$ is the uniform electronic density, $2\rho_q$ the charge modulation amplitude. The phase, ϕ, specifies the place of the CDW with regard to the lattice ions. The electronic charge density is locally partially neutralized by the displacement of each ion in a new equilibrium position, the displacement of the nth ion, initially at $nr_\circ$, being

$$u_n = u_\circ \sin(nqr_\circ+\phi) \qquad (2)$$

Since a gap, Δ, is opened at the Fermi level, the CDW state has a lower energy that the metallic one. Other electronic transitions may appear in low dimensional systems such as spin density wave or superconductivity but we will restrict our discussion only to CDW systems. Such a CDW formation has also been observed in two-dimensional compounds, namely transition metal dichalcogenides. The Fermi surface of these compounds shows a nearly flat cylindrical shape, in such a way that many states located on either side of the Fermi surface cylinder are connected by the same vector $q = 2k_F$ (the nesting condition). The low-temperature ground state remains metallic. For a strictly one-dimensional conductor the Fermi surface consists of two parallel planes separated by q. All the states are connected by the same q. The energy gap involves the whole Fermi surface and the low temperature ground state is insulating.

Below the temperature, called the Peierls temperature transition, at which the lattice distortion occurs (in the strictly one-dimensional case, it is known that, because of fluctuations, no long range order can be established and there is no phase transition down to T = 0. A finite transition temperature implies some coupling between chains and therefore the compounds have to be pseudo one-dimensional) the condensed state can be described by an order parameter which can be chosen as $\rho_q e^{i\phi}$ (see equation 1) or as the lattice distortion which is proportional to ρ_q. The modulation of the ion positions can be detected by X-rays, neutrons or electron diffraction measurements. Superlattice spots appear near the main Bragg spots corresponding to the unmodulated structure. Measurements of the components of these superlattice spots along the reciprocal axis give the CDW wavelength. In real space pictures of the CDW have been obtained using a high resolution technique in electronic diffraction studies. This method is very powerful to study defects in the CDW lattice induced, for instance, by electron irradiation. Very recently the use of a tunneling microscope on a cleaved surface of the two-dimensional 1T-TaS_2 CDW compound has revealed the real CDW structure formed by hexagonal arrays of mounds with the CDW wavelength spacing.[2]

The opening of a gap below the Peierls transition temperature gives to physics involved with the CDWs some analogy with semiconductors. But the essential feature is that the CDW wavelength, $\lambda_{CDW} = 2\pi/2k_F$, is controlled by Fermi surface dimensions and generally is unrelated to lattice periodicities, i.e. the CDW is incommensurable with the lattice. Consequently the crystal no longer has a translation group which means that, contrarily to semiconductors, the phase, ϕ, of the lattice distortion is not fixed relatively to the lattice and is able to slide along q. This phenomena is easy to understand if we think of the CDW as resulting from an electronic interaction via the lattice phonons ; this interaction is the same, in every galilean frame provided that its velocity is small compared to the sound velocity (in which case the interaction would be strongly modified). So that the CDW condensation may arise in any set of galilean axis with uniform velocity, v, giving in the laboratory frame an electronic current density,

$$J = -n_o ev \qquad (3)$$

where n_o is of the order of the electron number condensed in the band below the CDW gap. This model of a sliding CDW was proposed by Fröhlich[3] in 1954 as a mechanism which could lead to a superconducting state. This Fröhlich mode is a direct consequence of the translation invariance. In fact, in real systems, as shown by Lee, Rice and Anderson,[4] this translation invariance is broken because the phase, ϕ, is pinned to the lattice. The pinning can be provided by impurities, commensurability between the CDW wavelength and the lattice. Oscillations of the CDW pinned mode is expected to lead to large low frequency ac conductivity and to a large dielectric constant. An applied dc electric field, however, can supply the CDW with an energy higher than the pinning one and above a threshold electric field, the CDW can slide and carry a current but damping prevents superconductivity.

MATERIALS

This new type of conductivity associated to the collective CDW mode, called the Fröhlich conductivity, has been recently observed [5] in three families of inorganic compounds, namely transition metal trichalcogenides as $NbSe_3$, TaS_3 with monoclinic or orthorhombic structures, NbS_3, molybdenum oxydes called blue bronzes $K_{0.30}MoO_3$ and $Rb_{0.30}MoO_3$ and halogened transition metal tetrachalcogenides as $(NbSe_4)_2I$, $(NbSe_4)_{10}I_3$, $(TaSe_4)_2I$. The structure of these compounds can be described in chains of trigonal prisms stacked on the top of each other with a cross-section close to an isosceles triangle in the case of $NbSe_3$, of layers of infinite chains of MoO_6 octahedra with K ions separating the layers in the case of $K_{0.3}MoO_3$ and of parallel $(TaSe_4)$ chains with iodine atoms lying between them in the case of $(TaSe_4)_2I$. Except $NbSe_3$ which remains metallic at low temperature all the other compounds exhibit a semiconducting behaviour below the Peierls transition temperature. According to the compounds, this transition temperature is spread between 330 K for NbS_3 down to 59 K for $NbSe_3$. The wavelengths of the CDW distortion appear to be incommensurate, very often near of four lattice distances, along the chain direction. A temperature dependence of the CDW wavelength has only been detected in orthorhombic TaS_3 and blue bronze with an apparent commensurability at low temperature. The amplitude of the gap is found much larger than the value expected from a mean field theory ($2\Delta/kT_c = 3.5$) (see Table 1).

PROPERTIES OF THE CDW CURRENT-CARRYING STATE

Since the first observation in 1976[6] of non-linear transport properties when a d.c or a microwave field was applied to $NbSe_3$, the properties of this new current-carrying state have been largely studied and they can be summarized as follows :

- The d.c electrical conductivity increases above a threshold field E_T.
- The conductivity is strongly frequency dependent in the range of 100 MHz - a few GHz.
- Above the threshold field, noise is generated in the crystal which can be analysed as the combination of a periodic time dependent voltage and a broad noise following a 1/f variation.
- Interference effects occur between the ac voltage generated in the crystal in the non-linear state and an external rf field.
- Hysteresis and memory effects are observed, principally at low temperature.

Fig. 1 shows a typical variation of the electrical conductivity (normalized to the ohmic value) as a function of the reduced electric field. The V(I) characteristics are drawn in the insert ; the deviation from the Ohmic law is observed above a critical current I_T which leads to a threshold field defined as $E_T = RI_T/\ell$ where R is the resistance of the sample, ℓ : the distance between voltage electrodes. E_T varies typically from a few mV/cm in $NbSe_3$ to a few tenths of a volt for the other compounds. It is experimentally found that E_T strongly increases when T is lowered. The temperature variation of E_T is drawn in Fig. 2 for all the compounds exhibiting non-linear properties. Several phenomenological laws have been tried to fit the $\sigma(E)$ variation. One particularly has been largely used which is reminiscent of some kind of Zener-tunneling process such as :

$$\sigma(E) = \sigma_a + \sigma_b\left(1-\frac{E_T}{E}\right)\exp\left(-\frac{E_o}{E}\right) \tag{4}$$

Table 1. Data concerning Peierls or superconducting transitions in transition metal tri- and tetrachalcogenides. Listed are : the resistivity at room temperature, the Peierls or the superconducting temperature, the amplitude of the Peierls gap and its ratio with the Peierls temperature, the nature of the ground state at low temperatures and the components of the superlattice structure on the reciprocal axes.

		$\rho(\Omega \times cm)$ at room temperature	Peierls temperature (K)	2Δ (K)	$\frac{2\Delta}{kT_c}$	Surstructure a^*	b^*	c^*	Superconducting temperature	Ground state at low temperature
NbS_3	type I	80								Insulating
	type II	8×10^{-2}	330	4400	13.3	{0.5 0.5	0.298 0.352	0 0		Semiconducting
$TaSe_3$		6×10^{-4}							2.0	Superconducting
$NbSe_3$		2.5×10^{-4}	{145 59	700	11.9	0 0.5	0.24117 0.26038	0 0.5	3.5 under 5.5 kbar	Metallic
TaS_3	orthorhombic	3.2×10^{-4}	215	1600	7.44	{? 0.5	0.1 0.125 (T < 130 K)	0.255 0.250		Semiconducting
	monoclinic	3×10^{-4}	{240 160	1900	11.9	0 0.5	0.253 0.247	0 0.5		Semiconducting
$(Fe_{1+x}Nb_{1-x})Nb_2Se_{10}$		10^{-3}	~140	360	2.55	{0 0.5	0.27 0.33	0 /		Semiconducting
$(TaSe_4)_2I$		1.5×10^{-3}	263	3000	11.4	{0.05 1	0.05 0	0.084 0.943		Semiconducting
$(NbSe_4)_{10}I_3$		1.5×10^{-2}	285	3900	13.7	0	0	0.487		Semiconducting

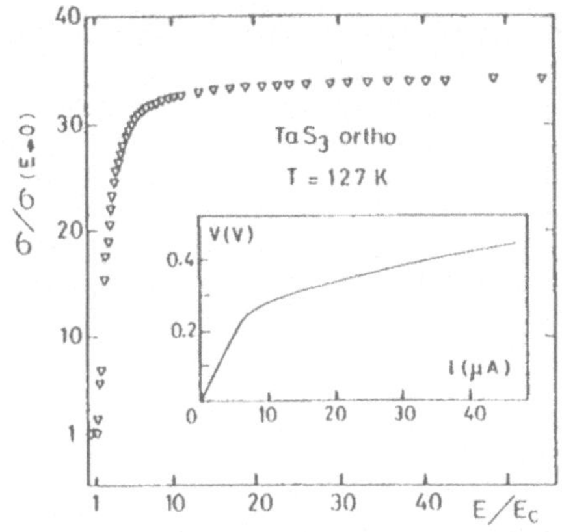

Fig. 1. Variation of the non linear electrical conductivity (normalized to the Ohmic value) as a function of the reduced electric field for an orthorhombic TaS_3 sample. The insert shows the V(I) characteristics.

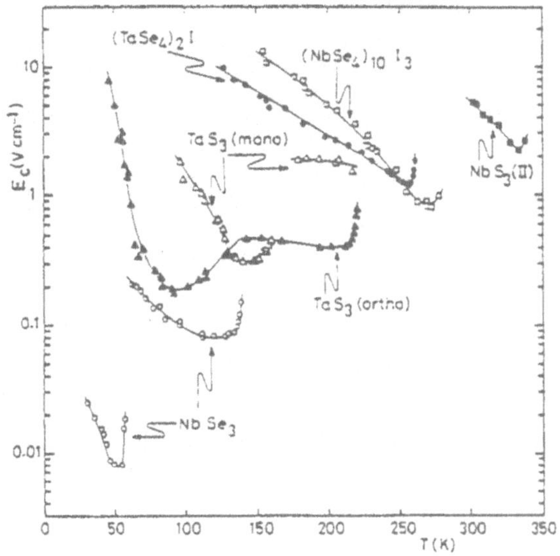

Fig. 2. Variation of the threshold field E_c (logarithmic scale) as a function of temperature for NbS_3, $NbSe_3$, both forms of TaS_3, $(TaSe_4)_2I$ and $(NbSe_4)_{10}I_3$.

with σ_a : the ohmic conductivity, $E_o = kE_T$ with k between 2 and 5. The so small activation field E_o precludes a single electron process because the gap which can be derived from such a Zener expression is several orders of magnitude lower than the thermal energy kT. When $E \to \infty$ the conductivity saturates to the value $\sigma_a+\sigma_b$ which is of the order of the metallic conductivity extrapolated from above the critical temperature. E_T is also seen to largely increase when the crystals are doped with impurities.

The low field ac conductivity shows a strong increase in the frequency range of 100 MHz-1 GHz and a saturation at a value close to the dc one for infinite fields. This behaviour is described in terms of a harmonic oscillator response due to the oscillations of the pinned CDW mode. For $NbSe_3$ and TaS_3 the response is overdamped. Recent measurements in the range of 10-100 GHz have revealed the inertial term and have allowed to estimate the CDW effective mass, the pinning frequency and the damping constant. However at low temperatures the description with a single oscillator fails and a distribution of pinning frequency has to be taken into account.

When E passes beyond E_T, a time-dependent voltage is generated in the crystal which can be studied with a spectrum analyser. Besides a broad band noise the frequency dependence of which varies as 1/f, the Fourier transformed voltage spectra, as shown in Fig. 3 for $NbSe_3$, reveal a fundamental frequency and many harmonics. The fundamental frequency appears at E_T and decreases with the current applied to the sample.

Steps can be observed in the d.c V(I) characteristics if a rf current is superposed to a d.c current exceeding the critical one. Such frequency synchronization is expected in non-linear phenomena such as Josephson junctions (Shapiro steps), for rf frequencies in the immediate vicinity of harmonics or subharmonics eigenfrequencies of the phenomena. It is to be noticed that during synchronization, the sample recovers nearly the Ohmic differential conductivity implying (see below) that the whole sample is oscillating nearly coherently.

Because of the strong interaction of the CDW with impurities, it is unlikely that the CDW can be described by a unique ground state and many metastable states have to be taken into account. Deformation of the CDW phase can be induced by a current and by temperature. The time scale for the metastable states to relax to the ground state can be very broad depending of the materials and of the temperature ; it is found that this decay time increases strongly at low temperature.

CURRENT MODELS FOR NON LINEAR TRANSPORT PHENOMENA IN CDW systems

Bardeen[7] was the first to interpret the non-linear conductivity in the CDW systems described above as the Fröhlich conduction induced by the

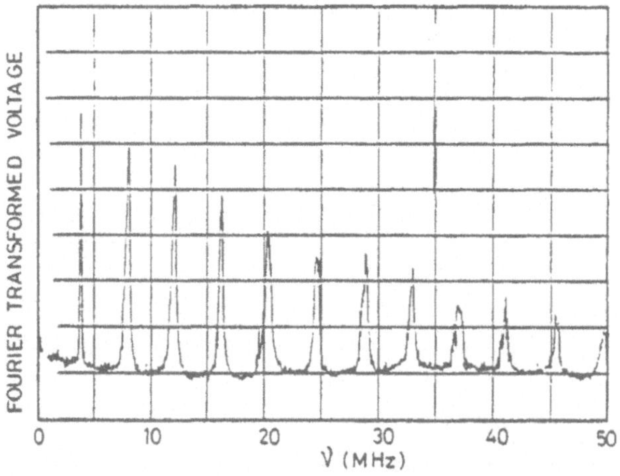

Fig. 3. Fourier transformed voltage spectra as a function of frequency for a $NbSe_3$ sample (T = 42 K) in the non-linear state.

CDW motion. The major part of the theoretical work has been carried in the incommensurate case considering the extra conductivity as due to a collective motion of the CDW phase.

CDW domains

However the existence of a threshold field, E_T, leads to a problem. As for the vortex pinning in type II superconductors a true threshold can only be accounted if we introduce some elasticity of the phase : a random distribution of impurities or dislocations, is the more probable cause of such a pinning force. But by its interaction with a completely rigid lattice the force summation will be random leading to a $V^{1/2}$ resultant (V being the volume $\sim$ number of impurities), and to E_T going to zero in the thermodynamic limit. To the contrary, some elasticity allows for a deformation of the phase, and a finite second order effect due to the individual pinning forces.

A very popular model by Lee and Rice[8] shows that the phase coherence between distant points tends to zero with the distance if an arbitrarily small elasticity is introduced in the random pinning problem :

$$<[\rho(\vec{r}+\vec{L})-\rho(\vec{r})-\vec{q}\vec{L}]^2>$$

extrapolating to zero when L increases.

This is due to the accumulation of small phase disturbances over a great number of disturbing centers. They defined a "domain size" such that the phase deviation to ideality is of the order of π : the domain gives an internal random summation, but in actual problems L is only a few microns and any measurement is concerned with a many domain problem. *If it can be assumed that the domains* act independently, the result will be the sum over a great number of domains and the total pinning force be proportional to the total length : E_T is independent of the length. One of the consequences of this model is an E_T dependence with the square of the impurity concentration.

It have to be said that even now, the experimental status of the thermodynamic limit of E_T is not clear, but as many results give finite values for relatively macroscopic samples, we shall develop some theories concerning the domain motion.

The classical domain motion

A domain is associated with an equivalent mass, some dissipative mechanism (thermalization of the phase motion by the phonon bath), a net charge, and a resultant pinning force which is of course periodic in ϕ, since if ϕ is increased by 2π each impurity sees the same charge distribution in its vicinity. This leads to an equation of motion :

$$M\ddot{\phi} + \Gamma\dot{\phi} + F_o \sin\phi = QE \qquad (5)$$

where E is the applied field, Q the charge.

This simple equation leads to at least qualitative explanations of many CDW phenomena :

- A threshold field defined by : $E_T = \frac{F_o}{Q}$.
- If $E = E_o \cos\omega t$ and $E_o \ll E_T$ a linearization of the $\sin\phi$ term gives a linear response theory, leading to a good agreement with the complex conductivity measured in low fields : $\sigma(\omega)$ and $\varepsilon(\omega)$ (overdamped oscillator).

For a dc field E higher than E_T, the "sinϕ" force term gives rise to a velocity modulation at a fundamental frequency, ν, and its harmonics which can be considered as the origin of the ac voltage generated in these systems. It has to be noted that the λ_{CDW} assumed periodicity for the force (where λ_{CDW} is the CDW wavelength), means that the fundamental frequency is linked to the mean CDW velocity by :

$$v_{CDW} = \lambda_{CDW}\nu \tag{6}$$

Therefore according to Eq. 4 the extra current carried by the CDW into motion is given by :

$$J_{CDW} = n_\circ e\dot{v}_{CDW} = n_\circ e\lambda_{CDW}\nu \tag{7}$$

A consequence of the classical equation of motion (Eq. 5) is that for E slightly higher than E_T, the extra d.c current varies as :

$$J_{CDW} \sim (E-E_T)^{1/2} \tag{8}$$

However, experimental results show a nearly 3/2 power law as shown in Fig. 4 where is drawn the variation of the fundamental frequency for an orthorhombic TaS_3 sample as a function of $(E-E_T)$ at different temperatures. Some attempts have been made to explain the regime near E_T such as the calculation of Fisher establishing some analogy between the vicinity of E_T and the critical behaviour of a second order phase transition, leading to the 3/2 exponent.

According to Eq. 7 the slope of J_{CDW}/ν is a measurement of the number of electrons condensed below the CDW gap. The extra-current J_{CDW} is measured from the non-linear V(I) characteristics. When ν is plotted as a function of J_{CDW}, all the curves drawn in Fig. 4 collapse in a unique straight line (except for temperatures near the Peierls transition temperature). Fig. 5 shows the linear relationship between J_{CDW} and ν for an orthorhombic sample which is still valid with a CDW current density of 30 000 A/cm^2. The number of electrons deduced from the ν/J_{CDW} slope is for any CDW compound of the order of the electron concentration in the bands affected by the CDW condensation as it can be calculated from band struc) tures or from chemical bonds. This result is thought to be the proof of

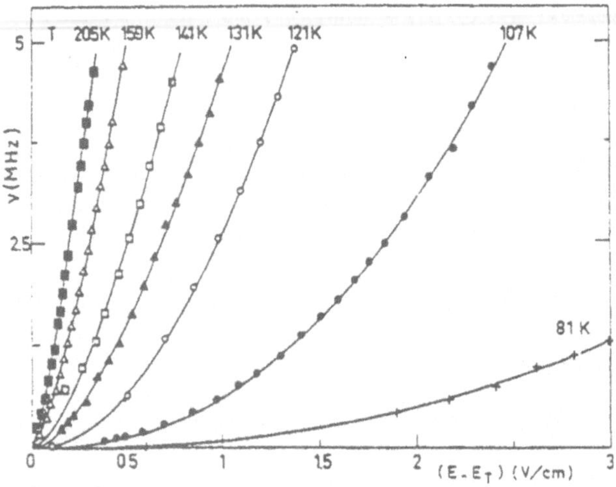

Fig. 4. Variation of the fundamental noise frequency, ν, measured in the Fourier-transformed voltage as a function of $E-E_T$ at different temperatures for an orthorhombic TaS_3 sample. The Peierls transition temperature is T_P = 215 K. The lines are fits to the expression $(E-E_T)^\gamma$ with $\gamma \sim 1.5$.

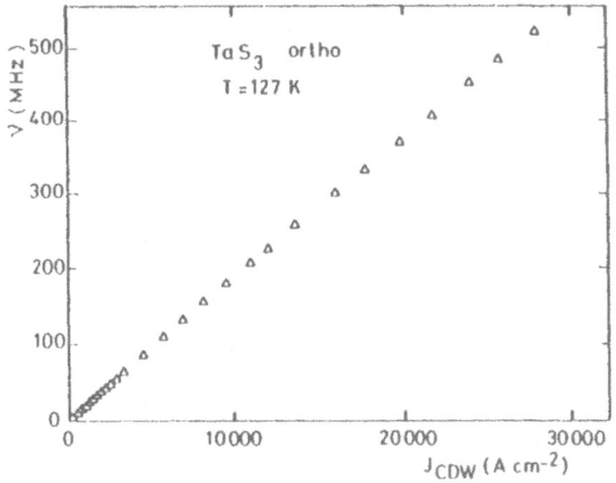

Fig. 5. Variation of the current J_{CDW} carried by the CDW as a function of the fundamental frequency measured in the Fourier-transformed voltage for an orthorhombic TaS_3 sample at T = 127 K. The slope $J_{CDW}/\nu = ne\lambda_{CDW}$ leads to the number of electrons condensed below the CDW gap.

the Fröhlich conductivity : when the field overcomes the threshold one, the electrons, which were trapped below the CDW gap, coherently participate to the electrical conductivity.

If, in Eq. 5, E is the superposition of a dc field $E > E_T$ and a small ac field $E_o \cos\omega t$, the non-linear $\sin\phi$ term gives a frequency linking between the sliding CDW wave and the applied ω, if the eigenfrequency is near of ω. During this synchronization the wave velocity is independent of the continuous field, and therefore the differential conductivity equals the linear Ohmic value.

The quantum tunnel explanation

This model has been put forward by Bardeen.[7] He also starts with a domain model and with the same hamiltonian that this one leading to Eq. 5 but in a quantified version : the presence of a kinetic plus a periodic potential energy allows for an analogy with band theory, the system being described by a wave function $\psi(\phi)$.

The periodic potential opens a "pinning gap" in the kinetic energy spectrum, and the wave functions are Bloch functions

$$\psi_k(\phi) = e^{ik\phi}\left[\sum_n e^{i\pi n\phi} \times a_n\right] \quad \text{with many electrons}$$

Contrarily to metals, a single k state is occupied for one domain. As for usual conductivity an applied electric field leads to :

$$\hbar \frac{dk}{dt} = QE$$

But when k hits the gap, a Bragg reflection sends it on the other side. The result is an alternative phase motion with no d.c current, unless a Zener breakdown occurs : the phase tunnels through the pinning gap, δ, on a length, ℓ, such that :

$$QE\ell = \delta$$

Of course this tunneling is meaningless if $\ell > L$, the domain size, or if $E < \delta/QL = E_T$.

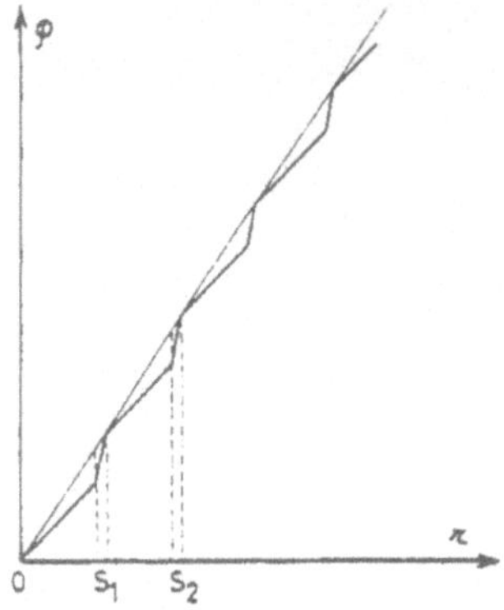

Fig. 6. Schematic variation of the phase $\phi(r)$ in the nearly commensurate case showing a discommensuration lattice (S_1, S_2, ...) as a function of the distance r. The straightline corresponds to $\phi = 2k_F r$ and the heavy line to the commensurate phase.

This field is interpreted as the threshold field. Some improvements have been done by taking into account the dissipative terms. Predictions have been made on the frequency mixing in the non-linear state, which differs from the classical model, but up to now, no completely convincing proof has been experimentally obtained.

Discommensuration lattice

Although the non-linear conductivity is generally accepted as the evidence of the Fröhlich conductivity, an alternative explanation, however, has been suggested : the motion of a discommensuration lattice.

If the wave vector q is near enough to some rational fraction of the reciprocal lattice vector, the CDW wavelength is very near of an integer number of atomic spacings. To minimize the interaction energy between the periodicity of the lattice and that of the wave, during "long" paths, ϕ follows exactly the commensurate slope, these paths being separated by compression places S_1, S_2, ... : the discommensuration lattice (Fig. 6). By analogy with the polymer $(CH)_x$ where stacking faults in the alternate C-C bindings, are either magnetic or charged, the discommensurations are thought as bearing a charge. A collective motion of the discommensuration lattice will carry a current. The cristallographic evidence of such a super-superlattice asks for high order Bragg reflections of the superlattice spots : these one are so weak than this idea can be considered as hopeless.

Injection problems

In CDW systems electrons are either in the condensed state, or may be present as excitations. If a current lead brings electrons in some place, they are unable to add directly some oscillations to the phase (Remember that one wavelength contains one electron for each spin per metallic chain). They have to be injected as excitations. The conversion of these electronic excitations into wavelengthes, can only occur at places where $\Delta = 0$.

By analogy with type II superconductors Maki has described vortices around which the phase rotates of 2π. If we draw (Fig. 7) the planes corresponding to $\phi = 2n\pi$ (n algebraic integer), these vortices looks like edge dislocations. Around the dislocation core the phase gradient is gigantic : the core is a normal area with $\Delta = 0$. If we suppose for example

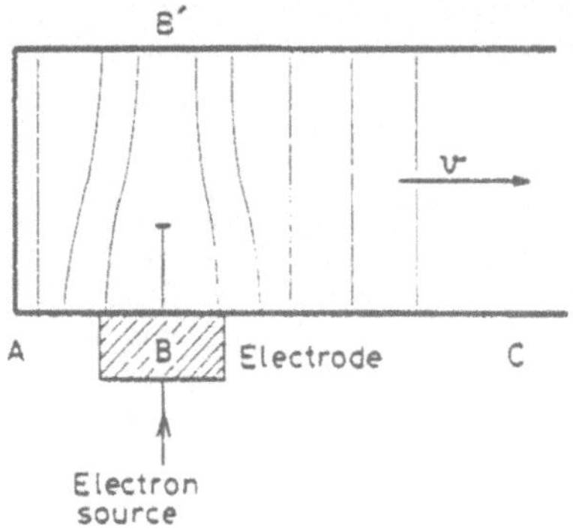

Fig. 7. Schematic topological defect in the CDW lattice to accomodate the phase gradient between parts in the sample in which the CDW moves and is at rest.

that A is fixed by the end of the sample, if a vortex is created at B and climbs to B' every line at the right of B has been translated by one wavelength. A continuous flow of vortices may assure a transition between a moving part BC and a static one AB. The electrons injected at the electrode, travel as excitations to the vortex core where they can condense easily ($\Delta = 0$), and when two electrons per chain have been condensed the core is translated to the next chain ...

If Maki has given an equivalent description of the Abrikosov-Gor'kov vortices in superconductors, Gor'kov has treated the equivalent of the superconducting weak links. Neglecting the transverse variations for a very thin film of CDW sample he has shown that phase slippage centers are necessary to accomodate the electrons arriving (or leaving) from an electrode.

These pulsatory phenomena are probably at the origin of at least a part of the observed periodic noise.

CONCLUSIONS

Description of one-dimensional systems has been for a long time a field only interesting for theoreticians. In the 70's inorganic as well as organic chemists have started to succeed to grow some compounds relevant for this kind of physics. Non-linear properties associated with the Peierls transition has also been considered as a curiosity as long as $NbSe_3$ was the only compound to exhibit such properties. Now, because the ability of chemists to grow many new compounds, it can be said that this behaviour is in fact the general property of pseudo one-dimensional systems as described above.

The sliding of the CDW nearly explains every feature at least qualitatively. However some problems remain to be solved. One among them is the explanation for the increase of the threshold field at low temperature : the only energy scale is the gap so it is expecting that the low temperature regime is reached a few K below the Peierls transition temperature. As shown in Fig. 5 the apparent viscosity of the CDW is seen to strongly increase when the temperature is lowered. The mechanism of the CDW damping and its temperature dependence has to be clarified.

A definitive evidence of the sliding CDW would be a direct observation of the velocity. Recent NMR measurements[9] showing a motional narrowing of the NMR lineshape when the CDW is put into motion gives the microscopic proof of the motion. Further experiments with the help of the tunneling microscope would be very convincing.

Acknowledgements - I wish to thank M. Renard, J. Richard, M.C. Saint-Lager, H. Salva, Z.Z. Wang for their cooperation in this work. This research has been possible only thanks to the possibility for chemists to synthetize new families of inorganic low-dimensional conductors. I acknowledge P. Gressier, L. Guemas, F.A. Levy, A. Meerschaut and J. Rouxel for providing us with the samples.

REFERENCES

1. R.F. Peierls, "Quantum Theory of Solids", Oxford University Press (1955) p. 108.
2. R.V. Coleman, B. Drake, P.K. Hansma and G. Slough, Phys. Rev. Lett. 55: 394 (1985).
3. H. Fröhlich, Proc. Royal Soc. A 223:296 (1954).
4. P.A. Lee, T.M. Rice and P.W. Anderson, Solid State Commun. 14:703 (1974) (1974).
5. For a review see Proceedings of the International Conference on Charge Density Waves in Solids, held in Budapest in August 84, "Lecture Notes in Physics", edited by Gy Hutiray and J. Solyom, Springer-Verlag Publishing Company, Berlin (1985), Vol. 217.
   "Electronic Properties of Inorganic Quasi One-Dimensional Compounds" Parts I and II, edited by P. Monceau, D. Reidel Publishing Company, Dordrecht, Holland (1985).
   "Crystal Chemistry and Properties of Materials with Quasi One-Dimensional Structures", edited by J. Rouxel, Reidel Publishing Company, Dordrecht, Holland (1986).
   G. Grüner and A. Zettl, Phys. Rev. 119:117 (1985) ; G. Grüner, Rev. Mod. Phys. (1986).
6. P. Monceau, N.P. Ong, A. Portis, A. Meerschaut and J. Rouxel, Phys. Rev. Lett. 37:602 (1976).
7. J. Bardeen, Phys. Rev. Lett. 42:1498 (1979) and 45:1978 (1980).
8. P.A. Lee and T.M. Rice, Phys. Rev. B 19:3970 (1979).
9. J.H. Ross Jr., Z. Wang, and C.P. Slichter, Phys. Rev. Lett. 56:663 (1986).
   P. Segransan, A. Janossy, C. Berthier, J. Marcus and P. Butaud, Phys. Rev. Lett. 56:1854 (1986).

SIZE EFFECTS IN CHARGE DENSITY WAVE DEPINNING

Pierre Monceau

Centre de Recherches sur les Très Basses Températures, CNRS
BP 166 X, 38042 Grenoble-Cédex, France

This paper is devoted to a review of some size effects in charge density wave (CDW) transport.

Below the Peierls transition temperature the electronic density is defined as :

$$\rho = \rho_o + 2\rho_q \cos(qr + \phi) + \ldots$$

where ρ_o is the uniform electronic density, $2\rho_q$ the charge modulation. The phase ϕ specifies the place of the CDW with regard to the lattice ions. The CDW order parameter can be defined as $\rho_q e^{i\phi}$. Non-linear transport properties occur when an electric field is applied with a sufficient strength to overcome the pinning of the CDW phase provided by impurities.[1] Properties of this non-linear state have been reviewed in the precedent lecture. One of the more puzzling effect, not yet understood, is the long range order which manifests itself in the oscillatory behaviour of the voltage across the sample and conjointly the disorder resulting from randomly distributed impurities.

LENGTHS INVOLVED IN THE CDW DEPINNING

Let define lengths involved in the CDW dynamics.

CDW amplitude coherence length

It is the length over which the order parameter can vanish. This length is equivalent to the Pippard coherence length for a superconductor and is equal for an isotropic material to :

$$\xi \sim \frac{\hbar v_F}{\pi\Delta}$$

where Δ is the CDW gap ; typically with $\Delta = 10^3$ K, $v_F = 5\times10^7$ cm/s, $\xi \sim$ 30 Å. When the anisotropy is taken into account, the transverse coherence lengths are reduced by the anisotropy ratio.

CDW wavelength

For almost all the compounds exhibiting CDW transport, the bands are nearly a quarter-filled ; therefore the CDW wavelength is ~ 4 times the distance between transition metal atoms along the chain direction and $\lambda_{CDW} \sim 14$ Å.

Discommensuration lattice

It has been noted that the CDW wavelength is in the vicinity of commensurability of order M = 4. So the ground state, instead of being homogeneous, may consist of commensurate regions separated by defects called discommensurations. The distance between discommensurations is a measure of the incommensurability. This distance is given by $\ell = (\pi/k_F - Md)^{-1}$ where d is the distance between atoms along the chain axis. For $NbSe_3$ this distance is 180 Å for the upper CDW transition and 174 Å for the lower one.

Fukuyama-Lee-Rice domain size

Fukuyama, Lee and Rice[2,3] have shown that the interaction between the CDW and impurities involves two competitive energies :

. An elastic energy, $1/2\ \kappa\int[(\partial\phi/\partial z)]^2 dz$, which indicates the ability of the CDW to be stretched or compressed by pinning centers.

. An impurity energy given by the Hamiltonian :

$$-V\rho_q \sum_i \cos\left[Qr_i + \phi(r_i) - \phi_i\right]$$

where the summation is over impurities at sites r_i and ϕ_i is the phase that the impurity tends to accomodate. The strength of the pinning has to be defined : if $V\rho_q$ is large compared to Δ (strong pinning case) the CDW phase is fixed at each pinning center and the characteristic length is the average distance between impurities. In the case of weak pinning, each individual impurity is unable to fix the phase but the phase coherence between distant points tends to zero with the distance. The phase correlation function can be written as :

$$\langle\phi(0)\ \phi(x)\rangle \sim e^{-x/L_\circ}$$

Fukuyama, Lee and Rice define a "domain size" such that the phase deviation to ideality is of the order of π. This distance is obtained by minimizing the energy of a domain by unit volume :

$$\frac{\kappa}{L^2} - \frac{V\rho_q (nL^3)^{1/2}}{L^3}$$

where the first term is the elastic energy paid by distributing the phase and the second term the impurity pinning energy gained by adjusting the phase. This term results from a random walk summation over the nL^3 impurities included in the domain (n : impurity concentration). The domain length is found to be :

$$L_\circ = \frac{\kappa^2}{n(V\rho_q)^2}$$

It is generally accepted that a typical value for $L_\circ$ is $\sim$ 1-10 μm. So any measurement is concerned by a problem with many problems. Therefore size effects on the threshold value are expected only if the sample size is comparable to a unique or a very small number of domains.

Velocity-velocity correlation distance

Whereas L_o represents the static characteristic length scale which is the smallest distance over which the CDW phase can vary, Fisher[4] has shown the existence of another length, ξ, which diverges at threshold. This correlation length measures the decay rate of the local CDW velocity-velocity correlation function. When the electric field is reduced from infinity, the CDW slows down and when $E \to E_T$ larger and larger regions of the CDW will spend most of their time relatively stationary at the values they will have when E is decreased below threshold. The linear dimension of these semicoherent regions is a measure of the correlation length which diverges as :

$$\xi = L_o\left(\frac{E - E_T}{E_T}\right)^{\nu}$$

with $\nu = 1/2$ in the mean field approximation. However when E is large, let say $E > 2E_T$, ξ and L_o become comparable.

Finally all these lengths have to be compare with the size of the sample : its length and its cross-section.

LENGTH DEPENDENCE OF THE THRESHOLD FIELD

Measurements have been reported which show that the threshold electric field value, defined as $E_T = V_T/\ell$, strongly increases when the distance, ℓ, between electrodes is reduced below 100 μm.[5-8] It has been first reported by Gill[5] that for a given geometry the threshold is larger when the 4 contact probes have a transposed configuration than in a normal configuration (in a transposed configuration the inner electrodes are used for current). He suggested that in the transposed configuration the CDW coherence has to be broken which requires a breaking force represented by the increase in the threshold field. Furthermore the role of contacts has been stressed by Ong et al[9] who show that the release of the CDW phase accumulation below electrodes may take place through the formation of dislocations or phase slip centers or vortex in the CDW lattice. The breaking of the CDW coherence by lateral injection has been recently studied.[10] The (b,c) plane of a $NbSe_3$ crystal (typical dimension, 3 mm × 50 μm × 3 μm) was gently pressed on a sapphire on which gold strips have been evaporated. Strips for current contacts are 300 μm large. The voltage can be measured through several electrodes of 8 μm width separated by 25, 50, 100, 200, 500 μm (Fig. 1a). The anisotropy of $NbSe_3$ perpendicular to the (b,c) plane was first estimated by measuring the V/I ratio, V the voltage between leads separated of 25 μm, as a function of the distance at which the current is injected. At small distances, V/I is strongly enlarged. A model calculation and a rheostatic simulation yields a transverse anisotropy value, $\sigma_\perp$, of $\sim$ 200. Consequently the effective thickness is larger by $(\sigma_\perp)^{1/2}$ and the 8 μ width electrodes only play a small perturbation on the current distribution : they are "non-injecting" or "non-shunting".

Then the depinning current, I_T, is measured by recording the differential resistance dV/dI as a function of the applied current I (I is the current injected at the ends of the sample) between several lengths. The Ohmic resistance is shown to scale the length between the middle of electrodes. Fig. 1 shows that I_T (and therefore E_T) is independent of the length at least down to 25 μm which proves that the voltage probes do not perturb the CDW motion and that the pinning is homogeneous along the length of the sample. But now if the current (called i) injection is made through the voltage leads (a transposed configuration) the depinning occurs for $i_T = I_T + i_o$, i_T increasing when ℓ is reduced. The electrode now is "injecting" or "shunting". The threshold voltage $V_T = Ri_T$ as a function

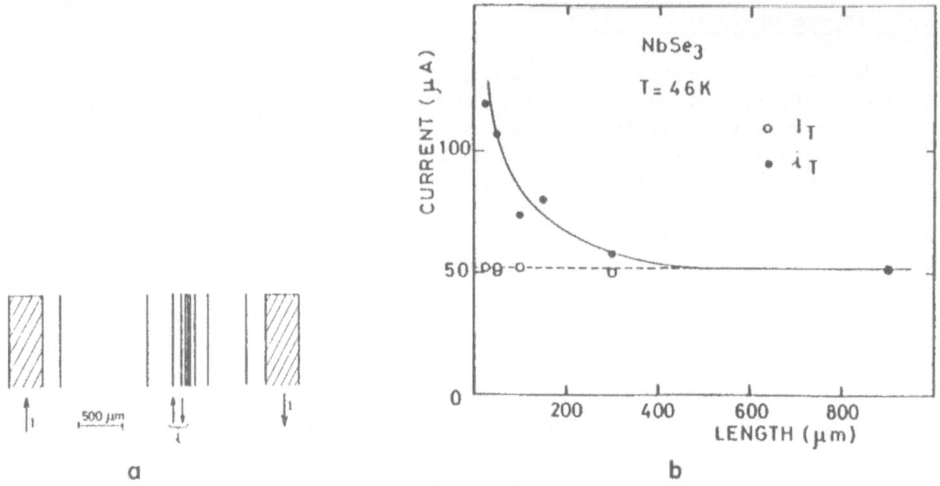

Fig. 1. a) Schematic description of electrode distribution.
b) Variation of the depinning current as a function of the length between electrodes. The current I is applied at the ends of the $NbSe_3$ sample ; the current i is applied through the voltage electrodes. The width of the electrodes is 8 μm.

of the length shows a linear variation with a finite intercept, $V_\circ$, for $\ell \to 0$. Therefore we can write, $V_T = RI_T + V_\circ = E_c\ell + V_\circ$ where $E_c\ell$ is the threshold for the sample with infinite length. $V_\circ$ is temperature dependent and typically 0.3 mV at 46 K. $V_\circ$ is interpreted as the field for breaking the longitudinal CDW coherence. In CDW systems electrons are either in the condensed state, or may be present as excitations. If a current lead brings electrons in some place, they are unable to add directly some oscillations to the phase (one CDW wavelength contains one electron for each spin per metallic chain). They have to be injected as excitations. The conversion of these electronic excitations into wavelengths can only occur at places where the gap $\Delta = 0$. The voltage to create a transverse sheet with $\Delta = 0$ has been shown to be too large.[11] Therefore the conversion of normal electrons into the condensate may take place through the phase-slip centers or vortices as discussed by Ong and Maki.[12]

VOLUME DEPENDENCE OF VOLTAGE OSCILLATION AMPLITUDE

The amplitude of the voltage oscillations has been studied as a function of the volume of the sample. The initiation of this study was the theoretical result found by Sneddon[13] indicating the absence of voltage oscillations in the thermodynamic limit and the experimental results showing an ac voltage in a $(TaSe_4)_2I$ sample with a cross-section of 12 μ^2 whereas no ac voltage was detected in another $(TaSe_4)_2I$ with a cross-section approximately two orders of magnitude larger.[14] Moreover, if the ac voltage has its origin at the electrodes, its amplitude is expected to be independent of the length of the sample. Two contradictory experimental results have been obtained on $NbSe_3$ at the same temperature by Mozurkewich and Grüner and Ong et al.[9] The former authors have found that the oscillation amplitude, ΔV_1, measured with a Frouier spectrum analyser increases as the root of the length and decreases as the root of the cross-section, A, of the sample (See Fig. 2). From ΔV_1, they define an oscillating current density, $\Delta J_1 = \Delta V_1/RA$ where R, proportional to ℓ, is the non-linear dc resistance for the applied field. Consequently, $\Delta J_1 \propto \Omega^{-1/2}$ and the current oscillation is vanishing in the infinite volume limit. In order to

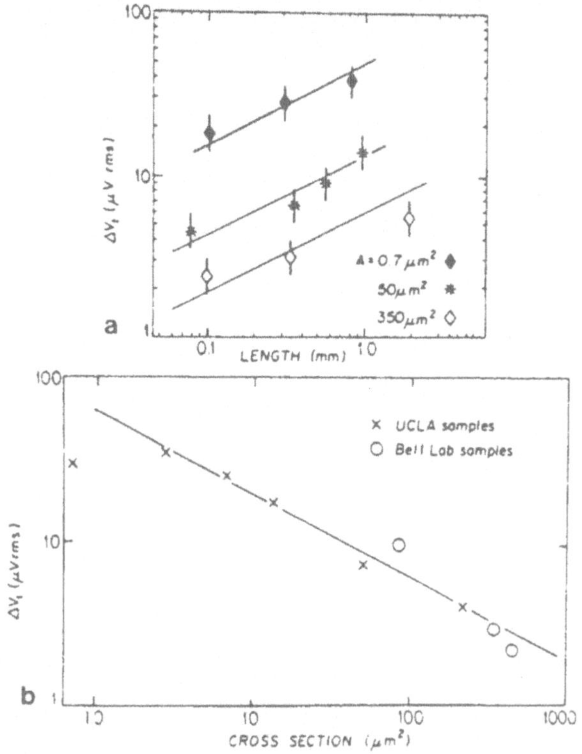

Fig. 2. Oscillation amplitude ΔV_1 in $NbSe_3$ at 42.5 K, measured between 10 and 20 MHz, for samples of different dimensions. (a) ΔV_1 vs length for samples of three cross-sections A. The lines have slope 1/2. (b) ΔV_1 vs A for samples of fixed length ℓ = 0.3 mm. The line has slope -1/2 (from ref. 15).

interpret these amplitude measurements, the instrumental bandpass of the spectrum analyser has to be considered : Mozurkewich and Grüner have set the analyser band-width to 100 kHz, a value larger than the width of the frequency peak response, whereas Ong et al. have used a resolution of 1 kHz and integrated the area under the fundamental peak of the spectrum. The latter authors have firstly found[9] that the integrated ac voltage is independent of the length of the sample with a variation in the length by a factor 60. Further work[16] including extensive computer data acquisition and averaging has shown that the voltage oscillation amplitude ΔV_1 varies as $[1-\exp(-L/\xi)]$ with ξ approximately 200 μm and that for ℓ larger than 0.9 mm the voltage oscillation is independent of the sample length. For a given length ΔV_1 is also shown to vary as $v_o[1-\exp(-\omega/\omega_o)]$. The variation of the amplitude v_o as a function of ℓ is shown in Fig. 3.

In ref. 9 and 16 the role of pinning of the CDW at the electrodes is strongly stressed. The effect of this pinning propagates into the bulk up to a distance ξ which means that the current carried by the CDW in the bulk is influenced by the carrier conversion occuring at the ends. In the vortex model of Ong and Maki[12] this distance is the screening distance of vortex creation. At the electrode the CDW velocity undergoes a discontinuity. This perturbation in velocity is screened in the interior of the condensate by the motion of mobile pre-existing dislocations in the bulk.

Moreover the local nature of the CDW oscillations has been thought to be proved by the study depinning in a thermal gradient. While it has been reported[9,17] that the fundamental noise frequency splits only into two

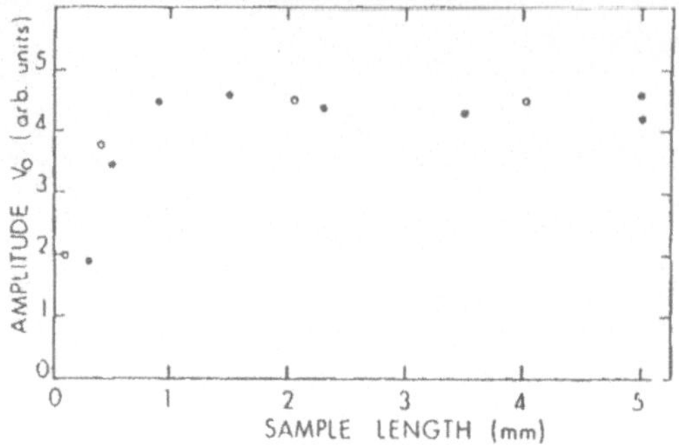

Fig. 3. Variation of v_o vs. the sample length ℓ for samples 1 (solid) and 2 (open circles) where v_o is the value of ΔV_1 in the large-ω limit. Each data point represents the average of 2.000 measurements of ΔV_1 (20 values of $\omega \times 100$ samplings per ω value) (from ref. 16).

frequencies when ΔT is applied across the sample, however many reports show that the splitting occur via two, three or more frequencies indicating that the sample breaks into separate macroscopic domains[10,18,19]. Another way to suppress the vortex creation at the electrodes is the injection of current at a temperature above the Peierls transition temperature. With a reentrant temperature configuration, the electrodes being warmed above T_P and the middle of the sample kept cold, noise frequency has been still detected.[10,18] So it looks like wortices are un-escapable to accomodate the conversion of normal carriers $\rightleftarrows$ CDW condensate as it was shown from results in Fig. 1 but the role of these vortices in the noise generation does not seem essential.

STEP MODE-LOCKING AS A FUNCTION OF LENGTH

Interference effects have been shown to occur between the internal frequency of the voltage oscillations and an external rf signal of frequency ω_{ac}.[20] Peaks appear in the differential dV/dI[20,21] or steps[22] in the dc V(I) characteristics at frequencies $n\omega = p\omega_{ac}$. These interferences are similar to Shapiro steps observed in Josephson superconducting junctions. Studies of these steps as a function of the frequency ω_{ac} and of the amplitude of the rf signal have been performed.[22] When the amplitude is large enough the locking is complete and a plateau equal to the normal differential resistance is measured in the dV/dI(I) characteristics. Then the sample acts as a single domain. Recently it has been shown that the complete mode locking occurs only over sufficiently short distances.[23] The degree of locking is measured as the ratio between the height h of the interference peak and the maximum height between the Ohmic differential conductivity and the low field dynamic impedance. When the measurement exceeds a frequency-dependent correlation length, complete locking is lost. Fig. 4 shows the variation of the locking as a function of the distance for a fixed frequency ω_{ac} = 5 MHz; complete locking disappears when d > 400 µm. This correlation length has also been shown to decrease sharply when the frequency increases. Incomplete locking indicates that the time averaged velocity of the CDW vary into the crystal but discontinuity in this averaged velocity requires the existence of phase slip. Also Hall et al[23] suggest that the correlation length they measure is the distance between phase-slip centers in the bulk of the sample the origin of which might arise from inhomogeneities in impurity concentration or in the cross-section of the sample.

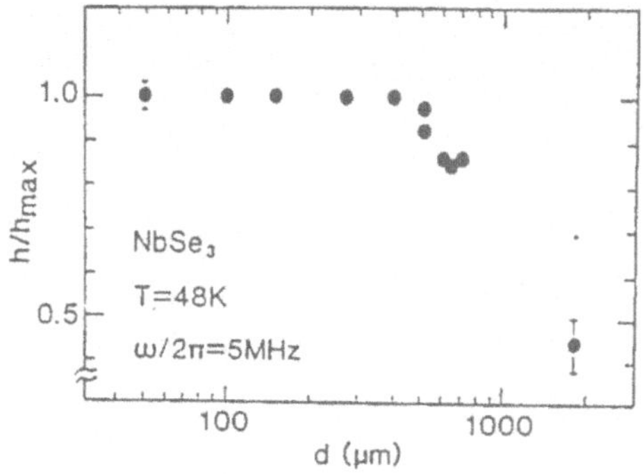

Fig. 4. Variation of the degree of locking as a function of the probe separation for a fixed rf frequency of 5 MHz (from ref. 23).

CONCLUSIONS

The threshold field has been shown to be length independent at least down to 25 μm indicating that the longitudinal size of CDW domains with a nearly constant phase is much less, probably in the micron range. In dynamical properties correlation lengths of 200-500 μm are obtained. These correlation lengths may result from the "rigidity" of the CDW current but the experimental values are much larger than those deduced from existing theories. Coupling between domains has to be analysed in detail for a better understanding of the coherent CDW motion.

REFERENCES

1. For a review see Proceedings of the International Conference on Charge Density Waves in Solids, held in Budapest in August 84, "Lecture Notes in Physics", edited by Gy Hutiray and J. Solyom, Springer-Verlag Publishing Company, Berlin (1985), Vol. 217.
"Electronic Properties of Inorganic Quasi One-Dimensional Compounds" Parts I and II, edited by P. Monceau, D. Reidel Publishing Company, Dordrecht, Holland (1985).
"Crystal Chemistry and Properties of Materials with Quasi One-Dimensional Structures", edited by J. Rouxel, Reidel Publishing Company, Dordrecht, Holland (1986).
G. Grüner and A. Zettl, Phys. Rev. 119:117 (1985) ; G. Grüner, Rev. Mod. Phys. (1986).
2. H. Fukuyama and P.A. Lee, Phys. Rev. B 17:535 (1978).
3. P.A. Lee and T.M. Rice, Phys. Rev. B 19:3970 (1979).
4. D.S. Fisher, Phys. Rev. B 31:1396 (1985).
5. J.C. Gill, Solid State Commun. 44:1041 (1982).
6. A. Zettl and G. Grüner, Phys. Rev. B 29:755 (1984).
7. G. Mihaly, Gy Hutiray and L. Mihaly, Phys. Rev. B 28:4896 (1983).
8. M. Prester, Phys. Rev. B 32:2621 (1985).
9. N.P. Ong, G. Verma and K. Maki, Phys. Rev. Lett. 52:663 (1984).
10. P. Monceau, M. Renard, J. Richard and M.C. Saint-Lager, Physica B, in press.
11. M.C. Saint-Lager, Thesis 3° Cycle, University of Grenoble (1983), unpublished.

12. N.P. Ong and K. Maki, Phys. Rev. B 32:6582 (1985).
13. L. Sneddon, Phys. Rev. B 29:719 (1984).
14. M. Maki, M. Kaiser, A. Zettl and G. Grüner, Solid State Commun. 46:497 (1983).
15. G. Mozurkewich and G. Grüner, Phys. Rev. Lett. 51:2206 (1983).
16. T.W. Jing and N.P. Ong, Phys. Rev. B 33:5841 (1986).
17. N.P. Ong, G. Verma, and X.J. Zhang, Charge density waves in solids, "Lecture Notes in Physics" Vol. 217 p. 296, edited by Gy Hutiray and J. Solyom.
18. S.E. Brown, A. Janossy and G. Grüner, Phys. Rev. B 31:6869 (1985).
19. J.W. Lyding, J.S. Hubacek, G. Gammie and R.E. Thorne, preprint.
20. P. Monceau, J. Richard and M. Renard, Phys. Rev. Lett. 45:43 (1980).
21. J. Richard, P. Monceau and M. Renard, Phys. Rev. B 25:948 (1982).
22. A. Zettl and G. Grüner, Phys. Rev. B 29:755 (1984).
23. R.P. Hall, M.F. Hundley and A. Zettl, Physica B, in press.

CDW TRANSPORT IN TTF-TCNQ : IMPURITY, COULOMBIC AND COMMENSURABILITY PINNING

J.R. Cooper, F. Creuzet, L. Forro,
D. Jérome, R.C. Lacoe and H.J. Schulz

Laboratoire de Physique des Solides
Université Paris-Sud
91405 Orsay

We review the salient results of a recently performed investigation of TTF-TCNQ showing the existence of non-linear conductivity related to CDW transport below the Peierls phase transition.

After the discovery of non linear electrical transport in $NbSe_3$ [1], the search for and study of charge density wave (CDW) conduction shifted from organic to inorganic systems.

In the inorganic materials [2], non linear effects are generally interpreted as resulting from CDW depinning by moderate electric fields.

A CDW can be pinned by impurities [3,4], lattice potential (commensurability pinning) or by Coulomb interactions between oppositely charged chains in two-chain systems. However, if the energy associated with an external electric field (E) coupled to the CDW is sufficient to overcome the pinning energy, the CDW will carry a current which will contribute to the conductivity in a non-linear way.

The two-chain organic conductor TTF-TCNQ [5] is a unique system in several respects :

(i) Peierls transitions have been observed by X-ray diffuse scattering and neutron diffractions [6]. At ambient pressure, there are three successive phase transitions : the upper transition at T_H = 54 K which involves distortion of the TCNQ stack , while the lower two transitions at T_M = 49K and T_L = 38K involve the TTF stack.

The wave vector of the periodic lattice modulation associated with the Peierls distortion is incommensurate with the underlying lattice at ambient pressure, $q_b = \rho \frac{b^*}{2}$ (where ρ is the charge transfer). The charge transfer increases from .59 at 1 bar to 2/3 at $P_c \approx 19$ kbar, corresponding to a third-order commensurate q_b.

(ii) Studying TTF-TCNQ, one can select either impurity, commensurability, or Coulombic pinning as the potential pinning mechanism. At 1 bar TTF-TCNQ represents a single-chain incommensurate CDW state in the narrow temperature range $T_H > T > T_M$. Below T_M, Coulombic coupling between CDW's developing on TCNQ (electron-like) and TTF (hole-like) chains must be considered. In the intermediate pressure range $4 < P < 15$ kbar, the single-chain incommensurate CDW state on the TCNQ stacks extends fairly low in temperature as compared to that at 1 bar. Then, impurity pinning should prevail. Near 19 kbar, one expects commensurability pinning to be important.

(iii) As known from transport and EPR studies, TTF-TCNQ is a system in which the amount of impurities (thus, the concentration of pinning centers) can be adjusted by proper irradiation [8]. All above-mentioned features have been studied at Orsay on TTF-TCNQ [9,10,11] by a short pulse technique together with a bridge circuit used to subtract the linear component of the conductivity. At all pressures we find deviations from Ohm's law with a threshold field E_T. An example of such non linear conduction is displayed in figure 1 for P = 4 kbar. The normalized σ/σ_o versus log E follows roughly a linear law at all temperatures.

The T-dependence of E_T, normalized to its minimum value versus T/T_P is shown on figure 2 along with the data for $NbSe_3$ below the CDW transition at 59 K [12]. At 4 and 8.9 kbar, the T-dependence of E_T is about similar to that of $NbSe_3$. Only a slow increase of E_T is observed upon cooling below $T/T_P \approx 0.9$. As shown in fig. 2, the T-dependence of E_T at low temperature and 1 bar is much faster for TTF-TCNQ than for $NbSe_3$. It was suggested that this reflected the growth of the CDW on the TTF stack below 49 K at 1 bar [9]. Due to the Coulomb coupling between TCNQ (e-like) and TTF (h-like) CDW's it is not possible with the E-field used to excite the mode where the two chains move in opposite directions, which carries the maximum CDW current. Instead, the E-field depins both CDW's from impurities but the non-linear current becomes the algebraic sum of the currents due to the motions of both TCNQ and TTF condensates in the same direction. This is only possible when $\rho_{cQ} - \rho_{cF} \neq 0$, i.e. the net charge of the two-chain system is non zero, where $\rho_{cQ(F)}$ is the fraction of condensed carriers in the respective chains [13].

Assuming that the pinning is mainly due to the TCNQ chain [9]:

$$E_T = E_T^o \rho_{cQ}/\rho_{cQ} - \rho_{cF} \qquad (1a)$$

where E_T^o is the threshold field for an isolated chain and [9]:

$$J_{ex} = J_{ex}^o/(1 - \rho_{cF}/\rho_{cQ})$$

where J_{ex}^o is the single chain excess current.

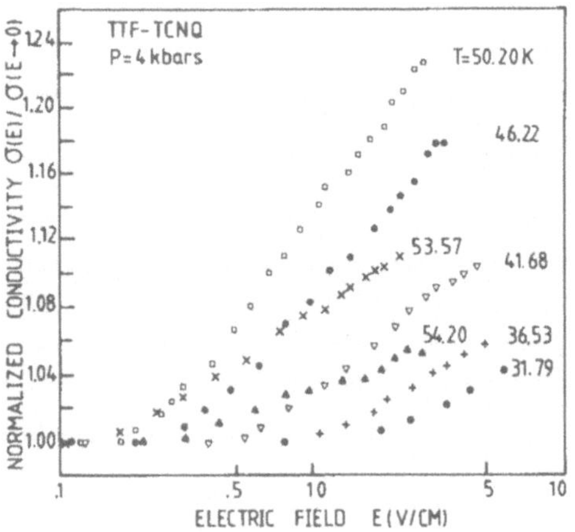

Fig. 1. Normalized conductivity σ/σ_o versus log E for TTF-TCNQ at 4 kbar and various temperatures below T_P.

J_{ex} is derived by the relation :

$$J_{ex} = J_{tot} - J_o \tag{2}$$

where J_{tot} and J_o are respectively the total and normal currents.

At ambient pressure, the T-domain in which TTF-TCNQ behaves as a single-chain CDW material is restricted to the narrow range 49 K < T < 54 K. The strong T-dependence below 49 K is due to the additional contribution provided by the dependence of the denominator $\rho_{cQ} - \rho_{cF}$ in eq. (1a). Between 4 and 12 kbar, the single-chain behaviour of TTF-TCNQ ($\rho_{cF} = 0$) extends down to about 35 K (i.e. $T/T_p = 0.63$) and the T-dependence of E_T becomes somewhat closer to that observed in $NbSe_3$.

In order to investigate the effect of commensurability on the field-dependent σ, measurements were made in the pressure range 16-24 kbar [10] where both chains undergo a Peierls transition at the same temperature T_p. Figure 3 displays the pressure dependence of E_T. As TTF-TCNQ is driven through third order commensurability at 19 kbar a non monotonous behaviour of E_T versus P is noticed. At 18.8 kbar only a lower limit estimate of E_T can be given (E_T > 20 V/cm) since i) deviations from linearity are found to be small ($\sim$ 2.3% at 18.8 kbar and 30 V/cm) and ii) heating effects are still possible at such large values of the pulsed electric field. The plot of E_T in fig. 3 establishes that $E_T(18.8\text{kbar}) > 20E_T(13.7\text{kbar})$. The measured values of E_T at commensurability are far from those calculated from commensurability pinning ($\approx 10^4$ V/cm). [3] However, the high pressure data clearly suggest that commensurability enhances significantly the pinning mechanism. Also shown in figure 3 is the pressure dependence of the excess current J_{ex} at E = 30 V/cm for T/T_p = 0.9 and 0.7. In some cases this

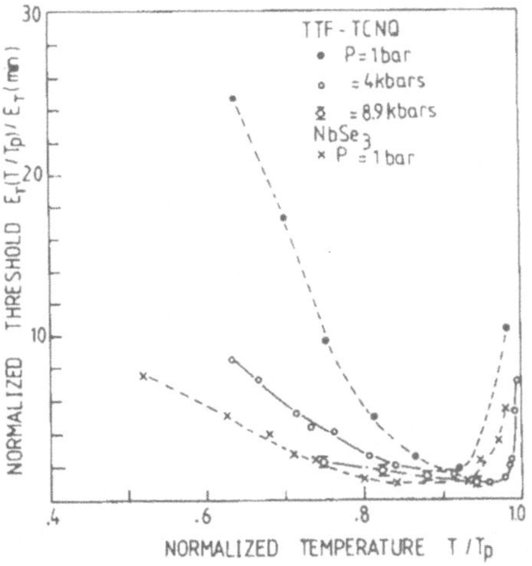

Fig. 2. Normalized threshold field E_T/E_T^{min} versus normalized temperature T/T_p for TTF-TCNQ at 1 bar, 4-8.9 kbar and for $NbSe_3$ at ambient pressure.

required extrapolation of the low-field σ/σ_0 versus log E curves. In the low-P range, J_{ex} is not much pressure dependent. The small value of J_{ex} measured under ambient pressure at $T/T_p = 0.7$ (T = 38K) reflects the compensation effect (eq. 1b) due to the onset of the CDW on the TTF-chain. However, as driven through commensurability there is a dip in J_{ex} by a factor about 4 and 10 at $T/T_p = 0.9$ and 0.7 respectively in the P-domain 16-22 kbar. For systems near commensurability, the ground state is not necessarily the incommensurate CDW state but rather commensurate CDW segments separated by solitons [14,15]. These solitons whose density depends on the deviation from commensurability can be charged and carry a current. It follows that for a truly commensurate case, the soliton density is zero. However, soliton transport is not likely to be effective in the present situation since the pinning between solitons of opposite charge on near neighbour chains is ex-ected to be strong on account of the Coulomb interaction [16]. Therfore, the overall charge of the soliton pair is zero and cannot contribute to the conduction. It is clear that a cancellation between ρcF and ρcQ could explain both an increase

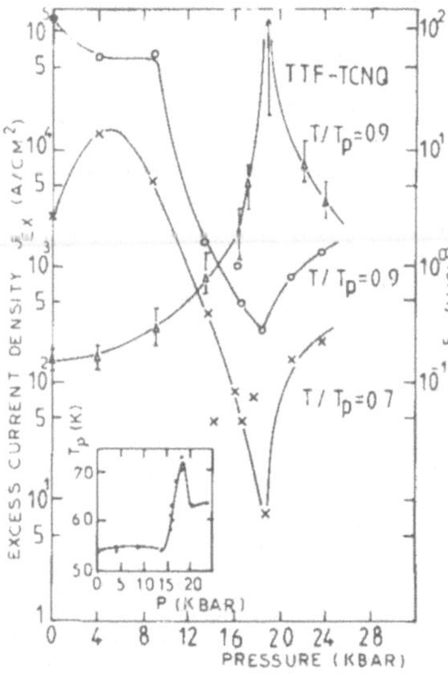

Fig. 3. Log J_{ex} versus P for TTF-TCNQ at E = 30 V/cm for two values of the reduced temperature $T/T_p = 0.9$ and 0.7 (left scale). Log $E_T(T/T_p = 0.9)$ versus P (right scale). At commensurability only a lower limit estimate of E_T is obtained. The insert shows the pressure dependence of the upper Peierls transition observed in the present work (also in good agreement with data of reference (7)).

of E_T and a dip of J_{ex} in the commensurability domain (according to eqs. 1). This effect can be taken into account (in first order) by plotting the pressure dependence of $J_{ex} \cdot E_T$. Data in figure 3 show that the product $J_{ex} \cdot E_T$ (at $T/T_P = 0.9$) exhibits a maximum at commensurability. This is a confirmation that the peak of E_T at 19 kbar is not merely a result of $\rho_{cF} - \rho_{cQ}$ going to zero in the commensurability region (see eq. 1a) but probably E_T^{Ω} in eq. 1a passes through a maximum as well.
Non linear conductivity studies have also been performed on TTF-TCNQ samples in which a known concentration of defects has been introduced by high energy electron irradiation [11]. Figure 4 shows that irradiation strongly affects the main features of the threshold field. The transitions of TTF-TCNQ at 54 and 38 K are also smeared out by the presence of defects. As shown in the inset of figure 4, the minimum threshold field follows a linear dependence versus defect concentration. These features suggest that these defects act as strong pinning centers on the motion of the CDW's [17].

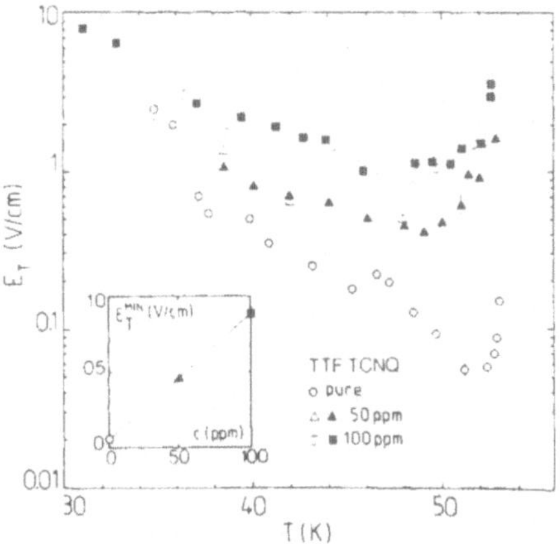

Fig. 4. Log E_T versus temperature (P = 1 bar) for pristine TTF-TCNQ and samples containing 50 and 100 ppm defect concentration. The insert shows the concentration dependence of the minimum threshold field.

Summarizing, the renewed interest for TTF-TCNQ has shown that properties of sliding CDW's so far well characterized in quasi 1-D inorganic compounds only can be observed in the Peierls state of an organic compound. Furthermore, the original structure of the two-chain compound TTF-TCNQ makes it possible a study of the CDW pinning by commensurability effect with the underlying lattice.

We thank S. Bouffard for his contribution towards the calibration of impurity concentration, K. Bechgaard and I. Johannsen for the supply of samples.

REFERENCES

1. P. Monceau, N.P. Ong, A.M. Portis, A. Meerschaut and J. Rouxel Phys. Rev. Lett. 37, 161 (1976).
2. "Charge Density Waves in Solids", Lecture Notes in Physics 217, edited by G. Hutiray and J. Solyom (Springer Verlag, 1985).
3. P.A. Lee, T.M. Rice and P.W. Anderson, Solid State Comm. 14, 703 (1974).
4. H. Fukuyama and P.A. Lee, Phys. Rev. B17, 535 (1978).
5. For a review of properties of TTF-TCNQ, see for example D. Jérome and H.J. Schulz, Adv. in Physics, 31, 299 (1982).
6. For a review of X-ray and neutron diffraction experiments, see R. Comès and G. Shirane, "Highly Conducting One-Dimensional Solids" edited by J.T. Devreese (Plenum, New-York, 1978).
7. R.H. Friend, M. Miljak and D. Jérome, Phys. Rev. Lett. 40, 1048 (1978).
8. L. Zuppiroli, Radiation Effects, 62, 53 (1982).
9. R.C. Lacoe, H.J. Schulz, D. Jérome, K. Bechgaard and I. Johannsen, Phys. Rev. Lett. 55, 2351 (1985).
10. R.C. Lacoe, J.R. Cooper, D. Jérome, F. Creuzet, K. Bechgaard and I. Johannsen, Phys. Rev. Lett., to be published.
11. L. Forro, R.C. Lacoe, S. Bouffard, D. Jérome, Phys. Rev. B to be published.
12. P. Monceau in Electronic Properties of Inorganic Quasi-One-Dimensional Materials II, 139, D. Reidel Publishing Company (1985).
13. P.A. Lee, T.M. Rice, Phys. Rev. B19, 3970 (1978), T.M. Rice, P.A. Lee and M.C. Cross, Phys. Rev. B20, 1345 (1979).
14. W.L. McMillan, Phys. Rev. B12, 1185 (1975) and Phys. Rev. B14, 1496 (1976).
15. W.P. Su, J.R. Schrieffer and A.J. Heeger, Phys. Rev. Lett. 42, 1698 (1979).
16. We thank S. Barisic for pointing out this remark.
17. H. Fukuyama and P.A. Lee, Phys. Rev. B17, 535 (1978).

COULOMB FORCES IN QUASI ONE-DIMENSIONAL

CDW SYSTEMS

Slaven Barišić

Department of Physics, Faculty of Sciences
POB 162, Zagreb, Croatia, Yugoslavia

I INTRODUCTION

The aim of this text is to review briefly some theoretical ideas developed in the physics of the (quasi) 1d metals. The restriction to "metals" means here that we shall consider the system of extended (although tightly-bound) electrons which interact weakly with phonons and also one with another. At the outset such a system is described in terms of large energy scales such as the electron band width 4t, the phonon Debye frequency ω_D, and the plasma frequency ω_0. Owing to weak interactions the smaller and smaller energies appear in the course of the analysis. In this sense the present approach can be viewed as a "first principle" determination of the parameters usually used in the phenomenological description of the low frequency (or temperature) behaviour of the metallic 1d systems.

In two recent review papers[1] a similar approach was used to discuss the equilibrium properties of the conducting trichalcogenides and of the TTF-TCNQ salts. In contrast to those papers the present text insists more on the unity of the theoretical description of 1d metals than on the particularities of various materials. In this sense the present paper is close to another recent brief survey[2], but extends it in several respects, and in particular in those which concern the effects of long range Coulomb forces on the static and low frequency properties of the CDW systems.

II HAMILTONIAN

The microscopic discussion of 1d metals starts usually from the SSH hamiltonian[3]

$$H_{SSH} = \Sigma t c^{+}_{i+1} c_i + \Sigma q_o t(u_{i+1}-u_i)\, c^{+}_{i+1}\, c_i + \frac{K}{2}\,\Sigma(u_{i+1}-u_i)^2 + \frac{M}{2}\,\Sigma \dot{u}_i^{\,2} . \quad (1)$$

This however is only a part of the tight-binding hamiltonian[4] known from the early examinations of the transition metals and their compounds. The full hamiltonian is[4]

$$H = H_{SSH} + H_g + H_c \quad . \quad (2)$$

Here H_c represents the long-range Coulomb part of the TB hamiltonian,

$$H_c = \frac{e^2}{2} \sum_{i \neq j} \frac{\delta\rho_i \; \delta\rho_j}{|\vec{R}_i - \vec{R}_j|} \qquad (3)$$

important in phenomena involving the redistributions of the charge $\delta\rho_i$ over long distances. H_g is the short range part of the Coulomb hamiltonian. In its simplest version[5] it is parametrized in terms of two matrix elements g_1 and g_2 associated with the backward and forward scatterings which involve respectively the $2k_F$ and $q \approx 0$ momentum transfers. It should be noted in this respect that the separation between H_g and H_c is at present arbitrary. The two "independent" parameters g_1 and g_2 are usually associated through the Fourier transform with the two interactions U,V of the extended Hubbard model. However the Coulomb interactions between further neighbors can be transferred from H_c to H_g contributing strongly to g_2. It will turn out later that a natural separation between H_g and H_c occurs at the Thomas-Fermi screening distance k_{TF}^{-1} $(\omega_o = v_F k_{TF})$ beyond which the interactions are screened off in most (but not in all) processes.

III SYSTEM OF COUPLED ELECTRONS AND PHONONS (H_{SSH})

3.1. Cooper pairing

H_{SSH} of Eq.(1) describes the electrons with the band width $4t$ $(\varepsilon_F = v_F k_F)$ which interact with phonons characterized by the Debye frequency $\omega_D = (K/M)^{1/2}$ through the coupling constant $2\pi\lambda \approx t/M\omega_D^2 q_o^{-2}$. The Slater coefficient q_o of Eq.(1) is of the order of $1Å^{-1}$ so that the quantity in the denominator of λ ressembles the Lindemann melting temperature,[4] i.e. λ is smaller than unity in the narrow band materials.

The interaction λ can lead to Cooper pairing[6]. As is well known the two electrons in a Cooper pair are correlated over the distance $\xi = v_F/T$. The time required by the second electron to reach the position where the first one has triggered the piling up of the positive charge, is $\xi/v_F \simeq T^{-1}$. The time needed by the positive charge to pile up is $2\pi\omega_{ph}^{-1}$, so that it is seen by the incoming electron only if[4] $2\pi T < \omega_{ph}$. Then the process can be continued ad infinituum. The lowest order process is shown in Fig.1, where the double exchange of $2k_F$ phonons results in the (forward) contribution to the original (backward) interaction λ, which is of the order of $\lambda^2 \log(\omega_D/T)$, (where $\omega_{ph} \approx \omega_D > 2\pi T$). The two become comparable at

$$T_c \approx \omega_D \, e^{-1/\lambda} \qquad (4)$$

On the order hand the $q \approx 0$ acoustic phonons do not contribute to the Cooper pairing[5] because $\omega_{ph}(q \approx 0) \quad 0 < 2\pi T$ at any T. In other words the SSh hamiltonian is characterized by the effective interactions

$$\gamma_1 = -\lambda \text{ (cut-off } \omega_D) \; ; \; \gamma_2 = 0 \qquad (5)$$

Fig. 1. Cooper pariring of a k_F electron (full line) with a $-k_F$ electron (dashed line) through $2k_F$ phonon fluctuations (wavy lines) coupled to electrons by the electron phonon couplings (heavy dots).

3.2. Electron - hole pairing

According to Eq.(1), the deformation U_i modulates locally the band width and the electron flow from the region where the states are expelled from below the Fermi level ε_F into the region where they are introduced below it, in the attempt to keep the chemical potential ε_F constant (cf Fig.2). The resulting linear response CDW $\delta\rho^{(1)} \sim u_{2k_F}$ to the $2k_F$ deformation field u_{2k_F} is shown in Fig.3a. Assuming that the deformation is slow enough for the electrons to follow it, the corresponding harmonic deformation energy $q_o t\ u\ \delta\rho^{(1)} \sim |u|^2$ can be defined within the Born-Openheimer picture[4]. It is shown schematically in Fig.3b. The second order perturbation energy is always negative, and when added to the bare term $M\omega_D^2|u|^2$ it decreases the effective phonon frequency. Alternatively, through Fig.3c, we can understand[5] the e-h pairing as giving rise to the correction to the bare interaction λ. Again it amounts to the frequency renormalization since $\lambda \sim \omega_D^{-2}$. In 1d the bubble in Figs.3 is logarithmic at $q \approx 2k_F$ and Fig.3c gives a contribution of the order of $\lambda^2 \log(\varepsilon_F/T)$. The latter is comparable to the bare interaction λ at

$$T_p \approx \varepsilon_F\, e^{1/\lambda} \quad , \qquad (6)$$

when the renormalized phonon frequency reaches zero.

3.3. Mixing of pairings

The interpretation of the e-h pairing in terms of the vertex correction, Fig.3c, has the advantage that in contrast to the deformation energy, Fig.3b, it does not depend on the validity of the adiabatic approximation. This approximation breaks if

$$2\pi T_p < \omega_D \quad , \qquad (7)$$

in particular because the Cooper pairings are present then at T_p (cf §3.2.). In fact both the e-e and e-h corrections to λ are logarithmic in temperature i.e. comparable with logarithmic accuracy, and must[5] be treated on equal footing. Apparently the analysis in terms of vertex corrections is particularly suitable for this purpose. It not only allows to compare the (geometric) series in e-e and e-h channels term by term but also to include on equal

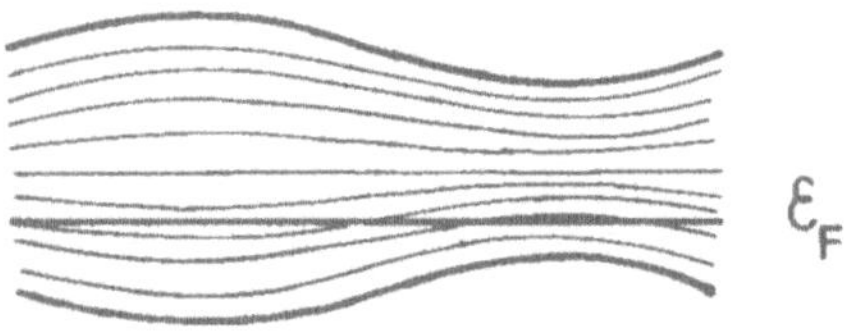

Fig.2 Deformed band, with local variation of the number of states below the Fermi level: CDW

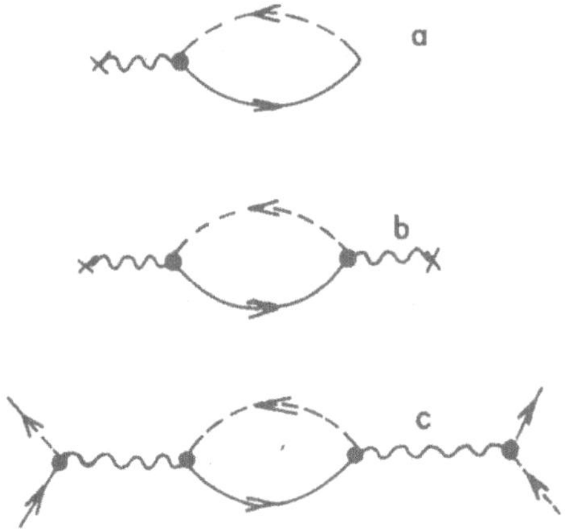

Fig.3 Electron hole pairing in (a) $2k_F$ CDW response (b) $2k_F$ deformation energy, (c) interaction renormalization. Wavy line with a cross in (a) and (c) denotes the deformation field U_{2k_F}. Other notations as in Fig.1.

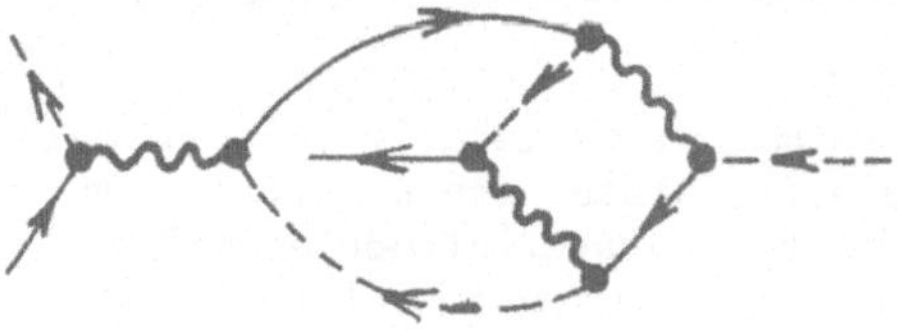

Fig.4. Example of mixing of e-e and e-h pairings. Notations as in Fig.1.

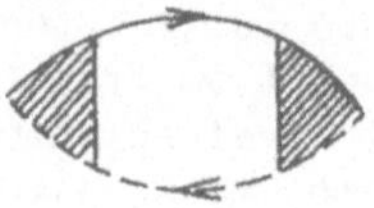

Fig.5. Correlation function χ in terms of the triangular vertices Δ_3 (shaded triangles).

footing the interference, parquet[5] terms between the two channels, such as the (lowest order) term shown in Fig.4. It can be said that Cooper pairs carry the interaction in e-h chanel and vice versa. The resulting parquet series determines the dressed vertex provided that $\lambda<1$.
Once the dressed interaction is known it is straightforward to determine various correlation functions. An important technical point[5,7] must be made here: It is possible to symmetrize the parquet theory so that an appropriate triangular vertex[7] Δ_3 appears on both sides of the correlation function as shown in Fig.5.

$$\chi \approx \Delta_3^2 \tag{8}$$

Such triangular vertex depends only on one variable and differs therefore from the definition of triangular vertices usual in the many body theories. The distinctive property of Δ_3 is that in terms of its unique variable it has simple scaling properties, taken for granted in the multiplicative renormalization group scheme[8], but proven in the parquet approach. The scaling shows that depending on the values of the coupling constants γ_1, γ_2 various types of fluctuations dominate, as illustrated in Fig.6. According to Eq.(5) the SSH model lies on the negative γ_1 axis of this diagram. The corresponding behaviors of the uniform magnetic susceptibility χ_H are also displayed in Fig.(6) illustrating that the negative γ_1, goes with the behaviour of χ_H which is exponentially activated[9] at low T.

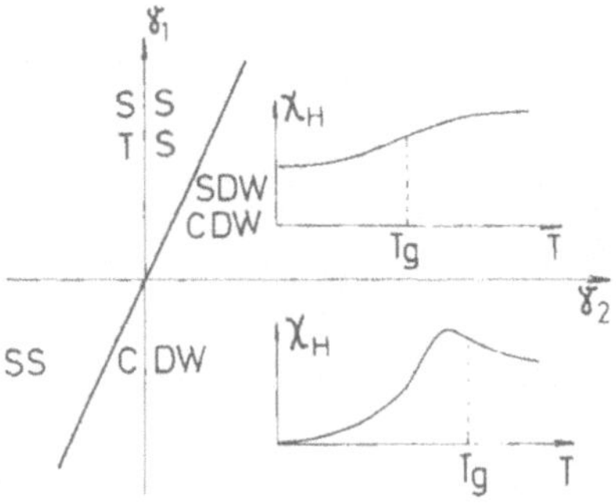

Fig.6. Dominant fluctuations for given backward γ_1 and forward γ_2 interactions. Inserts show the behaviours of the uniform magnetic susceptibility for $\gamma_1 \gtrless 0$.

In the limit $2\pi T_p < \omega_D$ the SSH model is thus associated with strong CDW fluctuations, indicating the tendency to develop a corresponding long-range order in quasi 1d systems. These CDW fluctuations are inextricably linked with the lattice fluctuations, i.e. they are reflected in phonon correlators[5]. From the lattice point of view the situation is highly nonadiabatic in the sense that the Migdal theorem is broken and the Cooper pairing plays an important role. Consistently with this the electron redistribution frequency ω_c is small ($\omega_c \simeq T < \omega_D$). Not surprising then the central peak is predicted at low $T(T < \omega_D)$ in the $2k_F$ scattering cross section.

3.4. Kohn-Peierls limit

This limit corresponds to the situation opposite to that considered in the preceding section, Eq. (7), i.e. now

$$2\pi T_p > \omega_D \qquad (9)$$

As argued in §3.1. the Cooper pairing, or more generally any virtual exchange of phonons is absent in the regime (9). On the other hand the e-h pairing discussed in §3.2 is present. The corresponding electron redistribution frequency is easily shown[10] to be $\omega_c \approx T > T_p$ i.e. larger than ω_D due to the requirement $2\pi T_p > \omega_D$. In fact rather than to ω_D, ω_c should be compared[10] to $\omega_{LO} = \lambda^{1/2}\omega_D$ which is even smaller. ω_{LO} is the natural frequency scale in presence of the e-h pairing. Indeed when the adiabatic condition $\omega_c > \omega_{LO}$ holds the definition of the harmonic deformation energy $E^{(2)}$ in Fig. 3b becomes meaningful. Together with the bare lattice energy $E^{(2)}$ gives the total deformation energy $E^{(2)}_{TOT} = M\,\tilde{\omega}^2\,|u_{2k_F}|^2$ which defines the renormalized phonon frequency $\tilde{\omega}$ as

$$\tilde{\omega}^2 = \omega_D^2\left[1 - \lambda \log\frac{\varepsilon_F}{T}\right] \qquad (10)$$

Such $\tilde{\omega}^2$ can be rewritten as $\tilde{\omega}^2 = \lambda\omega_D^2 \log T_p/T$ exhibiting $\omega_{LO} = \lambda^{1/2}\omega_D < \omega_c$. Eq.(10) describes the logarithmic behavior of the high temperature Kohn anomaly: the $2k_F$ phonon frequency softens towards T_p, the spectrum being dispersive in the whole harmonic regime[10].

When parquet correlations are absent at T_p they do not show up below this temperature due to the development of the large (pseudo) gap in this temperature range. This leads to treating the electrons at low temperatures as a free gas with the (pseudo) gap

$$\Delta_p \approx T_p \qquad (11)$$

i.e. within the usual Peierls scheme[11] (see also §4.2.).

Although most of the quasi 1d metals of interest satisfy the condition (9) and some of them develop 1d regimes over a considerable range of temperatures the logarithmic behavior of the elastic $2k_F$ cross section $S92k_F)^{-2}$ is not frequently observed in the harmonic regime. The actual temperature dependences are more pronounced[12]. This suggests that the Coulomb interactions may play an important role, the problem to which we turn now.

IV SHORT RANGE COULOMB INTERACTIONS

This section is devoted to the properties of the hamiltonian $H_{SSH} + H_g$. For the time being let H_g be described by g_1, g_2 postponing further the precise definition of those parameters.

In presence of the Coulomb interactions it is still possible to have the phonon field essentially classic. The three following subsections are devoted to this case. The essential quantum effects of the phonon field are considered in the last subsection.

4.1. Harmonic theory

Due to g_1, g_2 the electron which rolls down the valley of the deformation potential in Fig.2 encounters the other electrons and the bubbles of Figs.3 are filled up with interactions. As a result the free electron bubbles are replaced by the CDW correlation functions of Fig.(5) evaluated with g_1, g_2 i.e. $\log(\varepsilon_F/T)$ is replaced by χ_{CDW} of Eq.(8), everywhere and in particular in Eq.(10) for $\tilde{\omega}$. χ_{CDW} differs however from $\log(\varepsilon_F/T)$ only at low temperatures [13] when the Coulomb correlations are developed, i.e. below the temperature

$$T_g = \varepsilon_F \, e^{-1/|g|} \tag{12}$$

at which the parquet corrections to g's are of the same order of magnitude as g's themselves. Above Tg the electrons act as nearly free.[13]

The renormalized phonon frequency $\tilde{\omega}$ may reach zero above or below T_g. If this occurs above T_g the corresponding temperature is equal to T_p, because the electrons act as free [13]. $T_p > T_g$ means $\lambda > g$. If $\lambda < g$ $\tilde{\omega}$ reaches zero at T_L below T_g, when the Coulomb correlations are developed. The classic treatment of phonons implicit here is justified provided that $2\pi T_L(T_p) > \omega_D$. If on the contrary $2\pi T_L(T_p)$ turns out to be smaller than ω_D the alternative approach is required, to be discussed in §4.4. as was announced at the beginning of this section.

Assuming thus $g > \lambda$ and $2\pi T_L > \omega_D$ two different situations $g_1 \gtrless 0$ have to be considered, according to the diagram of Fig.6 used now with $\gamma_{1,2} = g_{1,2}$. All CDW metals known to us exhibit a weakly temperature dependent uniform magnetic susceptibility, which is consistent (either with the Pauli behavior ($\lambda > g$) or) with the $\gamma_1 = g_1 > 0$ behavior of Fig.6. In this latter case $\chi_{CDW} \approx (\varepsilon_F/T)^{2g_2-g_1}$ instead of $\log(\varepsilon_F/T)$ in Eq.(10) for $\tilde{\omega}$. The knowledge of χ_{CDW} allows us to verify first the validity of the adiabatic approximation for electrons, implicit in (the modified) Eq.(10). Indeed the redistribution frequency of electrons ω_c which characterizes the frequency dependence of χ_{CDW}, turns out to be of the order of $T \gtrsim T_L$. Analagously to the Peierls case [10] (§3.4., $\lambda > g$) the requirement that the phonon field is classic $2\pi T_L > \omega_D$ makes the electron response adiabatic, $\omega_c > \omega_D$. Furthermore the modofied Eq.(10) determines T_L through $\tilde{\omega} = 0$, and by a simple generalization the length scale $\xi = v_F/T_L$, in addition to the already mentioned time scales. The values of T_c, ω_c and ξ are given in Table I.

The $2k_F$ scattering cross section $S(2k_F) \sim \tilde{\omega}^{-2}$ becomes thus strongly temperature dependent in the correlated regime, although the theory is harmonic in lattice coordinates i.e. the phonons which soften are dispersive. The well developed power laws have indeed been observed[12] at high temperatures in some organic metals.

4.2. Anharmonic coupling

The decrease of the phonon frequency $\tilde{\omega}$ is accompanied by the increase of the average of the squared $2k_F$ displacements since $M\,\tilde{\omega}^2 \langle |U|^2 \rangle = T$. Therefore the anharmonic terms have to be taken into account. The leading interaction is mediated by electrons in the way shown in Fig.7b and the higher order terms can be constructed along the similar lines.

The activation of these terms goes together with the development of the (pseudo) gap[10,14,15] in the electron spectrum. This (pseudo gap prevents the development of parquet (Cooper) correlations via phonons at low temperatures as already explained in §3.4. for the $\lambda > g$ case. For $\lambda < g$ it stops in addition the further development of the Coulomb correlations. This means that the renormalization of the electron-phonon coupling $\lambda^{1/2}$ (denoted by a dot in diagrams) by the shaded vertex Δ_3^{CDW} stops at approximately the value reached at T_L. This value is defined by $\chi_{CDW} = 1$, or, according to Eq.(8) $\lambda^{1/2}\,\Delta_3^{CDW} = 1$. Everything happens as if we are dealing

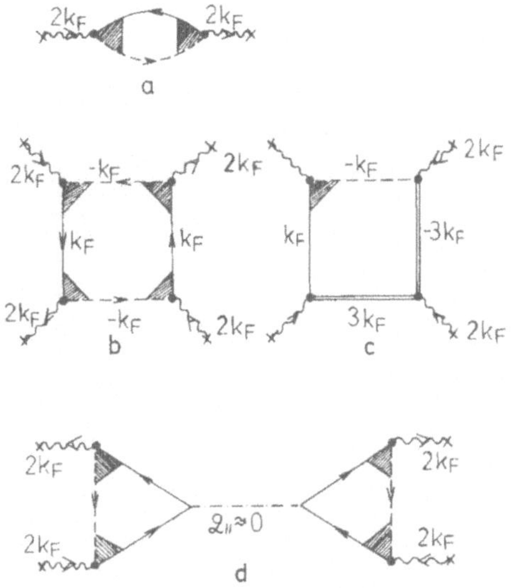

Fig.7. Expansion of the deformation energy in terms of $2k_F$ displacements. (a) harmonic (b) direct anharmonic (c) Umklapp (d) long range Coulomb contribution.

Table I

limit / quantity	$2\pi T_{p,L} > \omega_D$: $\lambda > g$	$2\pi T_{p,L} > \omega_D$: $\lambda < g$
mean field transition temperature	$T_p \simeq \varepsilon_F e^{-1/\lambda}$	$T_L \simeq \left(\frac{\lambda}{2g_2-g_1}\right)^{\frac{1}{2g_2-g_1}}$
characteristic distance	$\xi = \frac{v_F}{T_p}$	$\xi = \frac{v_F}{T_L}$
redistribution frequency of electrons	$\omega_c \approx T_p$	$\omega_c \approx T_L$
characteristic frequency of vibrations	$\omega_{LO} \approx \lambda^{\frac{1}{2}}\omega_D$	$\omega_{LO} \approx \omega_D$
law of corresponding states	$\Delta_p \approx T_p$	$\Delta_L \approx T_L$
relation between gaps and displacements	$n_F\Delta_p^2 \approx \lambda K\|u_{2k_F}^2\|$	$n_F\Delta_L^2 \approx K\|u_{2k_F}\|^2$

with the Peierls theory in which the bare electron-phonon coupling $\lambda^{1/2}$ is replaced by the Coulomb-enhanced value $\lambda^{1/2}\Delta_3^{CDW} \approx 1$. Figs.7a,b are suggestive in this sense. The Peierls laws of the coresponding states (e.g.Eq.(11)) are independent of the value of the electron-phonon coupling constant and we immediately have $\Delta_L \approx T_L$. Furthermore the quantities of the $\lambda < g$ theory differ from the corresponding values of the Peierls ($\lambda < g$) theory by the replacement of λ by unity. This is shown in Table I which also includes the quantities involved in the LRA extension[16] of the Peierls theory, namely the phason velocity v_{phs} or the corresponding frequency $\omega_{LO} = v_{phs}\xi^{-1}$.

At low enough temperatures where the phasons form the well defined excitations they can (and should be) treated as bosons. However the corresponding breakdown of the classic approximation for the lattice motion[15] is unessential because it is not accompanied by the activation of (additional) Cooper pairings or by any other kind of breakdown of the adiabatic approximation. The reason is that due to the law of corresponding states $\Delta_{p,L} \approx T_{p,L}$ the condition $2\pi T_{p,L} > \omega_D$ transforms at $T \approx 0$ in $\Delta_{p,L} > \omega_D$.

At intermediate temperatures $T \lesssim T_{p,L}$ the phasons are expected to merge into a central peak associated with the classical anharmonic motion of the lattice. This peak should persist up to the temperatures $T \gtrsim T_{p,L}$, where it should transform continuosly into the dispertive harmonic regime of §4.1.

4.3. Umklapp coupling

This term, shown in Fig.7c for the fourfold commensurability $2k_F M d_{\parallel} = 2\pi$, $M = 4$, is smaller than the direct quartic term of Fig.7b for two reasons. First it involves the $3k_F$ electron states which lie far from the Fermi level. The T^{-2} singularity of the direct quartic term is thus removed i.e. the reduction factor is $\delta = (T^2_{p,L}/\varepsilon_F^2)\log(\varepsilon_F/T_{p\ L})$ (one logarothmic integration remains). An additional effect occurs in the $g > \lambda$ limit[1,2] Only one of four electron-phonon couplings $\lambda^{1/2}$ in Fig.7c is enhanced by the Coulomb renormalization Δ_3^{CDW}, i.e. the additional reduction factor with respect to the direct term is $(\Delta_3^{CDW})^3$. For $g << 1$ this factor is not important and the main effect of Coulomb interactions is to increase T_L with respect to T_p (at a given λ), resulting in an overall increase of Umklapp coupling. However δ remains small i.e. it leads to the sine-Gordon commensurability transition only for very small departures of $\kappa = 2k_F - 2\pi/Md_{\parallel}$ from the commensurability, $\xi^2\kappa_c^2 \approx \delta$. $d_{\parallel}\kappa_c$ is therefore of the order of $T^2_{p,L}/\varepsilon_F^2$. (The long-range interactions will increase somewhat κ_c, as will be discussed in §5.4.). This extremely small value of κ_c probably explains the absence of commensurability effects in blue bronzes[12].

4.4. Quantum limit

As argued in §4.1. the essential quantum effects appear if $2\pi T_{p,L}$ turns out to be smaller than ω_D. Below ω_D the phonon-mediated Cooper pairing are activated in addition to the Coulomb induced terms since there is no large (pseudo) gap to counter both of them. The Coulomb and the phonon mediated interactions act additively, setting aside (with logarithmic accuracy) the difference in the corresponding cut offs, ω_o or ε_F and ω_D respectively. The diagram of Fig.6 can now be used with $\gamma_1 = g_1 - \lambda$, $\gamma_2 = g_2$ in both $\lambda \lessgtr g$ regimes. If neccessary, the difference between cut-offs can be introduced in a way similar to that used in the BCS theory of superconductivity[17].

The nature of the CDW/lattice state in the $\lambda < g$ limit does not differ in an essential way from its nature in the $\lambda > g$ limit which was described at the end of §3.3: The system is highly nonadiabatic and characterized therefore by the central peak in the inelastic cross section at $2k_F$. The main difference between $g \lessgtr \lambda$ is associated with the transition $\gamma_1 \lessgtr 0$ in Fig.6 which is relevant when $g > \lambda$ but concerns the magnetic degrees of freedom only.

V LONG RANGE COULOMB FORCE

In this section we wish to discuss the modifications of the results presented above which are due to the long range Coulomb forces. These modifications, quantitative in character in the metallic regime, become essential at low temperatures when the screening is absent.

5.1. Coulomb matrix element and metallic screening

H_c of Eq.(3) can be rewritten in terms of sinusoidal CDW's characterized by the wavevector $\vec{q}$. $q_{\parallel}$ determines then the CDW wavelength along the chain and $q_{\perp}$ the dephasing of CDW's between the chains. When the CDW wavelength $2\pi q_{\parallel}^{-1}$ along the chain is small with respect to the interchain distance $d_{\perp}$ the potential created by the CDW on the neighboring chain is (exponentially) weak[18] due to the compensation of the contributions from the oppositely charged parts of the CDW. The Coulomb matrix element is then weakly dependent on $q_{\perp}$ (and $q_{\parallel}$). On the contrary when $2\pi q_{\parallel}^{-1} > d_{\perp}$ the Coulomb matrix element develops the familiar $1/|\vec{q}|^2$ singularity.

Assuming that $k_F \sim d_{\parallel}^{-1}$ this singularity falls in the range of forward scattering whereas the Coulomb matrix element at $2k_F$ can well be described by the constant g_1. Since the $1/|\vec{q}|^2$ singularity is the dominant singularity in the problem it requires first of all the summation of the RPA series. Unlike at $q_{\parallel} \approx 2k_F$ the RPA bubble is non-logarithmic in the forward $q_{\parallel} < k_F$ range. It corresponds to the (inverse) dielectric constant[19]

$$\varepsilon_{RPA}^{-1} = \frac{\omega^2 - v_F^2 q_{\parallel}^2}{\omega^2 - \omega_p^2} \quad , \quad \omega_p^2 = v_F^2 q_{\parallel}^2 + \omega_o^2 \frac{q_{\parallel}^2}{q_{\parallel}^2 + q_{\perp}^2} \quad , \tag{13}$$

where ω_o is the plasma frequency $\omega_o \simeq v_F k_{TF}$ and k_{TF} is the Thomas-Fermi screening wave-vector $k_{TF}^2 d_{\perp}^2 \approx n_F e^2/d_{\parallel}$.

The screened Coulomb interaction ($\sim \varepsilon_{RPA}^{-1}$) describes in particular the process of scattering of two electrons on a given chain by the fluctuation of polarization on the neighboring chain. To this extent it ressembles the Little mechanism of superconductivity[20]. This leads us to the next stage of the theory, the consideration of the logarithmic singularities, i.e. of e-e and e-h pairings by the screened Coulomb interaction.

5.2. Parquet theory

It follows from the above considerations that the backward $q_{\parallel} \sim 2k_F$ scattering is well represented by a coupling constant g_1, because weakly dependent on $\vec{q}$. On the other hand g_2 is replaced by the screened Coulomb interaction, an essential function of $\vec{q}$ and ω. However in absence of backward interchain scattering ($q_{\perp}$ dependence of g_1) the $q_{\perp}$ dependence of forward scattering integrates out in the parquet theory. The resulting on-chain interaction does not exceed the distances larger than k_{TF}^{-1} (or times larger than ω_o^{-1}), the Little-like mechanism being neither attractive nor repulsive at average. The screening distance k_{TF}^{-1} introduces thus a natural separation between H_g and H_c of Eq.(2): The first neighbor interaction $e^2/d_{\parallel}$ of the extended Hubbard model is complemented in g_2 by the further terms along the same chain, up to k_{TF}^{-1}. The Fourier transform of the 1d Coulomb interaction is logarithmic and hence[21]

$$g_2 \approx n_F \left[U + \frac{e^2}{d_{\parallel}} \log \frac{\varepsilon_F}{\omega_o} \right] \quad . \tag{14}$$

To the same degree of accuracy $g_1 \approx n_F U$.

Although the resulting theory maps therefore on the 1d parquet theory the importance of 3d effects can be appreciated in the following way. The described procedure is meaningful provided that $g_2 < 1$ in spite of the assumption $\varepsilon_F > \omega_o$. In systems of dense chains this holds provided that $n_F e^2/d_{\parallel} < 1$ (and $n_F U < 1$). For distant ($d_{\perp}^{-1} >> d_{\parallel}$) chains $\omega_o << \varepsilon_F$ due to $\omega_o \sim 1/d_{\perp}$ and the requirement $g_2 < 1$ breaks down ultimatelly, modifyng the nature of the theory. This limit was recently considered both numerically[22] and analytically[23], with the results that differ considerably from the conventional parquet behaviors.

A different problem is related to the $2\pi T_L > \omega_D$ limit discussed in §§4.1. and 4.2. Namely it is not clear that g_2 of Eq.(14) can be used at low temperatures, §(4.2.) because the (pseudo) gap Δ_L inhibits then the screening to some extent, on replacing the RPA dielectric constant (13) used in the definition (14) of g_2 by[23]

$$\varepsilon_{\Delta}^{-1} = \frac{\omega^2 - v_F^2 q_{\parallel}^2 - \Delta_L^2}{\omega^2 - \omega_p^2} \quad , \quad \omega_p^2 = \Delta_L^2 + \left[v_F^2 + \frac{\omega_o^2}{|\vec{q}|^2} \right] q_{\parallel}^2 \quad . \tag{15}$$

Indeed, for $\vec{q}$ and ω small $\varepsilon_\Delta \approx 1 + \omega_o^2/\Delta_L^2$ is of the semiconductor type but coincides otherwise with ε_{RPA} of Eq.(13). However if $\Delta_L << \omega_o$ the metallic behavior of ε_Δ dominates in parquet diagrams because the electrons also have the gap Δ_L in their spectrum[24] and do not see it in ε_Δ: In all parquet diagrams ε_Δ is averaged over $q_{\shortparallel}$ and ω; g_2 is always a virtual interaction. Consequently all the results of section IV remain valid with g_1 and g_2 evaluated here. In fact this result was derived by bosonization[23].

In conclusion, to this level of approximation the effect of H_c is only quantitative since the metallic screening renders the theory short ranged both in the adiabatic and in the nonadiabatic limit. In the adiabatic limit the short range theory with $g > \lambda$ does not differ qualitatively from the Peierls theory since only the scales are renormalized, according to Table I. In fact the qualitative effects of H_c appear only in the adiabatic limit but beyond the main parquet approximation, as will be discussed below.

5.3. Long CDW's

The long range Coulomb interaction appears explicitly in Fig.7d. This term, unimportant in the short range theory when the dashed line is associated with $g_2 < 1$ may become important when the small momentum Coulomb interaction is screened by ε_Δ of Eq.(15). In contrast to the parquet diagrams (involved in the definition of Δ_3) which integrate the forward interaction over all moments and frequencies, Fig.7d keeps the frequency and the momentum fixed. Those can be chosen so small ($v_F q$, $\omega < \Delta_{p,L}$) that ε_Δ corresponds to the semiconductor screening, i.e. the dashed line (of Fig.7d.) to the singular long-range interaction.

The further understanding of Fig.7d can be achieved on noting that each triangle with two external wavy lines proportional to U_{2k_F} represents a second order CDW response $\delta\rho^{(2)} \sim u^2$ to the deformation field[12,25]. At low enough temperatures $\delta\rho^{2}$ adds to $\delta\rho^{(1)}$ of Fig.3b. At $T \approx 0$

$$\delta\rho^{(2)} \approx \frac{1}{2\pi}\left[\frac{\partial\phi}{\partial x} - \overline{\frac{\partial\phi}{\partial x}}\right] \qquad (16)$$

where $\partial\phi/\partial x$ is the derivative along the chain of the $2k_F$ CDW phase and $\overline{\partial\phi}/\partial x$ its average along the chain. This latter term accounts for the charge conservation on the chain[25]. Fig.7d describes then the Coulomb interaction between phase gradients.

This allows us to write down the closed equation for the motion of the phase at $\omega, v_F q < \Delta_{p,L}$

$$\frac{1}{\omega_{LO}^2}\ddot{\phi} = \xi^2\phi'' + \delta\sin 4\phi + \frac{\omega_o^2}{T_L^2}\frac{\partial}{\partial x}\int\frac{d\vec{r}'\,\delta\rho^{(2)}(\vec{r}')}{\left[(x-x')^2 + \varepsilon_\Delta(x_\perp - x_\perp')^2\right]^{1/2}} \qquad (17)$$

In a way (at low T) this equation summarizes the adiabatic part ($\Delta_{p,L} > \omega_n$) of the theory expounded here. Indeed all the coefficients involved in Eq. (17) are given in Table I, as determined from the microscopic approach. Eq.(17) can be generalized to include the amplitude fluctuations important at intermediate temperatures. This is easy in the clean limit (note that $\tau_\phi^{-1}\dot{\phi}$ is omitted in Eq.(17)) considered here, but much harder[26] in the dirty limit.

If the Umklapp δ term is omitted the theory of Eq.(17) becomes harmonic giving rise to the phason spectrum[2,27]

$$\omega_{phs}^2 = \omega_{LO}^2\left(\xi^2 q_{\shortparallel}^2 + \frac{q_{\shortparallel}^2}{q_{\shortparallel}^2 + q_\perp^2/\varepsilon_\Delta}\right) \qquad (18)$$

This corresponds to the dielectric constant for the longitudinal electric fields

$$\varepsilon_{phs} = \frac{q_{\parallel}^2\omega_o^2 + q_{\perp}^2\Delta_{p,L}^2}{\vec{q}^{\,2}\Delta_{p,L}^2} \quad \frac{\omega^2 - \omega_{phs}^2}{\omega^2 - v_{phs}^2 q_{\parallel}^2} \tag{19}$$

valid at $\tau_{\phi}^{-1} < \omega < \omega_{LO}$ and again at $\omega = 0$. Eq.(19) completes the evaluation of the dielectric constant which started with plasma frequencies ω_o in Eq.(13) went through $\Delta_{p,L} < \omega_o$ in Eq.(15) and finished with $\omega_{LO} \lesssim \Delta_{p,L}$ in Eq.(19). The small values of the "Coulomb frequency" $\omega_{LO} \lesssim \Delta_{p,L}$ (Table I) show up due to the effect of the semiconducting screening ε_{Δ}, which makes therefore the whole approach internally consistent.

ω_{phs} of Eq.(18) has two $\vec{q} = 0$ limith. For $q_{\perp} > q_{\parallel}$ $\omega_{phs} = \omega_T = 0$ and for $q_{\parallel} > q_{\perp}$ $\omega_{phs} = \omega_{LO}$. Although ω_T appears in the nominator of Eq.(10) it is also the zero of the denominator of the component along the chain

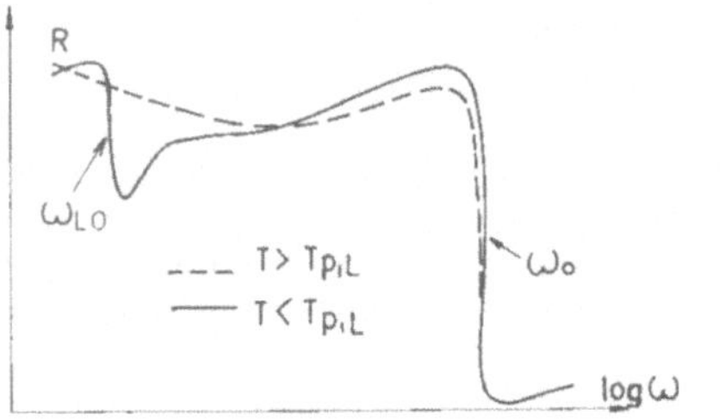

Fig.8 Characteristic behavior of the reflectivity in the $2\pi T_{p,L} > \omega_D$ limit with primary (ω_o) and secondary (ω_{LO}) plasma edges.

$\varepsilon_{\parallel}$ of the uniaxial dielectric tensor. It characterizes therefore (as a "pinning" frequency) the Fröhlich conduction. On the other hand ω_{LO} is the unique zero of $\varepsilon_{\parallel}$. Consequently it appears as a secondary plasma edge in the low T reflectivity at incidence normal to the chain axis. This edge is observed[28,29,30] in many materials showing at first that $\omega_{LO} > \tau_{\phi}^{-1}$. Second, the value of ω_{LO} observed in $\lambda < g$ materials tests our prediction that the small $\lambda^{1/2}$ in the LRA relation $\omega_{LO} = \lambda^{1/2}\omega_D$ must be replaced by unity due to the Coulomb enhancement Δ_3^{CDW}. This discussion is illustrated in Fig.8.

The static screening in Eq.(19) is characterized by two Thomas-Fermi lengths $\lambda_{TF}^{\parallel} = \varepsilon_{\Delta} \cdot \lambda_{TF}^{\perp} = \xi$. The large electron charge e^2 (i.e. the large $e^2/d_{\parallel}$) appears in $\lambda_{TF}^{\perp}$ as it does in the transverse dispersion of the phasons Eq.(18), but is removed from $\lambda_{TF}^{\parallel}$ and ω_{LO} due to ε_{Δ}. $\lambda_{TF}^{\perp}$ is thus of the order of the metallic k_{TF}^{-1} (Eq.(13)), but the screening length along the chain is increased in the CDW state. The smallness of $\lambda_{TF}^{\perp}$ can have the far reaching consequences in the definition of the Lee-Rice domains, as will be discussed elsewhere.

5.4. Charged solitons

Eq.(17) with the Umklapp δ term included leads first to the investigation of the ground state ($\omega = 0$) properties in the nonlinear regime. So far this has been carried out only approximately[25]. The treatment is based on the observation that according to Eq.(16) the single harmonics $\phi \sim x$ do not induce the long range forces. It is therefore possible to expand around the commensurate solution ϕ = cte or around the pure $2k_F$ state $\phi = \kappa x$ (for κ see §4.3.). The former is appropriate if $n_F e^2/d_{\parallel}$ is small and the latter if it is large.

The limit of $n_F e^2/d_{\parallel}$ small is physically somewhat more transparent. One starts from the dilute lattice of sine-Gordon solitons (nearly commensurate state). The soliton width ℓ and the intersoliton distance d are treated as variational parameters. The corresponding Coulomb energy is

$$U_c = n_S \left[U_S(\ell) + \frac{1}{2} V_S(d) \right] \qquad (20)$$

In turns out that V_S is negative. The effective attraction between charged solitons is due to the compensating charge $\overline{\partial\phi}/\partial x$ of Eq.(16), of the opposite sign. This is apparent in Fig.9. As a result the system jumps discontinu-

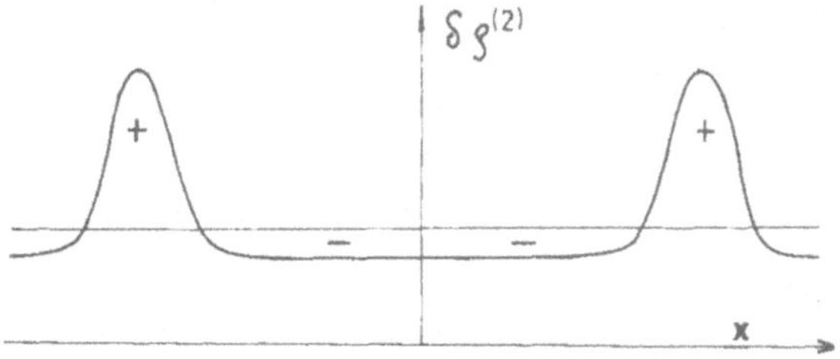

Fig.9 Charge distribution for the soliton lattice with overall neutrality

osly from the commensurate state into the state with finite soliton density (1st order transition). The equilibrium distance between solitons results from the balance of the short range sine-Gordon repulsion between solitons and their long range attraction. For reasonable values of $n_F e^2/d_{\parallel}$ $d \approx \ell$, i.e. the solitons are ill defined, making the limit of large $n_F e^2/d_{\parallel}$ relevant. The 1st order transition from the commensurate state occurs practically into the pure $2k_F$ harmonic, increasing somewhat the range of stability of the commensurate state with respect to the sine-Gordon value κ_c given in §4.3.

The long range forces are not activated if the charged solitons and antisolitons stand close one to another. This can occur in the chain systems with weakly splitted electron bands or with the overlapping electron and hole bands. In the former case the electrons transferred by the band splitting to the neighboring chain can form a lattice which is compensated by the lattice of holes on their chain of origin. Such configuration is stabilized by the interchain coupling bilinear in deformations[31]. The bilinear coupling may originate from the Coulomb backward interchain terms

($q_\perp$ dependence) of g_1) neglected in 5.2. Whatever its origin the bilinear coupling gives rise to an effective[32] sine-Gordon term, circumventing thus the problem of a too small Umklapp.

Umklapp is the largest in the three-fold commensurate systems such as TTF-TCNQ which also posesses the overlapping electron and hole bands. In this case the soliton lattice may appear on one chain either due to the metallic screening on the other chain or due to the formation of the compensating lattice on this latter chain if it undergoes the structural transition too. These possibilities are currently under investigation.

ACKNOWLEDGMENT: Some of the new results mentioned here have been obtained in collaboration with I.Batistić, A.Bjeliš and E.Tutiš.

REFERENCES

1. S.Barišić, Electronic properties of inorganic quasi 1d materials (ed. P.Monceau), 1, 1985 Reidel.
S.Barišić, A.Bjeliš, Theoretical aspects of band structures and electronic properties of pseudo 1d solids (ed. H.Kaminura), 49, 1985 Reidel.
2. S.Barišić, Mol.Cryst.Liq.Cryst. 119, 413 (1985)
3. W.P.Su, J.R.Schrieffer, A.J.Heeger, Phys.Rev. 322, 2099 (1980)
4. S.Barišić, J.Labbé, J.Friedel, Phys.Rev.Lett. 25, 919 (1970)
S.Barišić, Phys.Rev. B5, 932,941, (1972)
5. Yu.A.Bychkov, L.P.Gorkov, J.E.Dzyaloshinskii, Zh.Eksp.Theor.Fiz. 50, 738 (1966) (Sov.Phys.JETP 23, 489 (1966))
6. J.Bardeen, L.N.Cooper, J.R.Schrieffer, Phys.Rev. 106, 162 (1957); 108, 1175 (1957).
7. I.E.Dzyaloshinskii, A.Y.Larkin, Zh.Eksp.Teor.Fiz. 61, 791 (1971) (Sov.Phys.JETP 34, 422 (1972)); L.P.Gorkov, I.E.Dzyaloshinskii, Zh.Eksp.Teor.Fiz. 67, 397 (1974), (Sov.Phys.JETP 40, 198 (1975)
8. N.Menyhard, J.Solyom, J.Low Temp.Phys. 12, 529 (1973)
J.Solyom, J.Low Temp.Phys. 12, 547(1973) and Adv. in Phys. 28, 201 (1971)
9. A.Luther, V.J.Emery, Phys.Rev.Lett. 33, 589 (1974)
V.J.Emery, Highly Conducting 1d solids (eds. J.T.Devreese, R.P.Evrard, V.E. van Doren), 247, (1979) Plenum N.Y.
10. A.Bjeliš, K.Šaub, S.Barišić, Il Nuovo Cimento 23B, 102 (1974)
11. R.E.Peierls, Quantum Theory of Solids, 108 (1955) Oxford Univ.Press, London
12. J.P.Pouget, this issue
13. S.Barišić, J.Physique, Coll.C2, 39, 262 (1978)
14. P.A.Lee, T.M.Rice, P.W.Anderson, Phys.Rev.Lett., 31, 462 (1973)
15. S.A.Brazovskii, I.E.Dzyaloshinskii, Zh.Eksp.Teor.Fiz. 71, 2338 (1976) (Sov.Phys. JETP 44, 1233 (1976))
16. P.A.Lee, T.M.Rice, P.W.Anderson, Sol.St.Comm. 14, 703 (1974)
17. J.W.Garland, Phys.Rev.Lett. 11, 114 (1963)
18. K.Šaub, S.Barišić, J.Friedel, Phys.Lett. 56A, 302 (1976)
19. I.E.Dzyaloshinskii, A.Y.Larkin,Zh.Eksp.Teor.Fiz. 65, 411 (1973)
20. W.A.Little, Phys.Rev. A134, 1416 (1964)
21. S.Barišić, J.Physique 44, 119 (1983)
22. S.Mazumdar, D.K.Campbell, Phys.Rev.Lett. 55, 2607 (1985)
23. S.Barišić, E.Tutiš, J.Phys.C. 19, 6303 (1986)
24. V.Ya Krivnov, A.A.Ovchinnikov, Mol.Cryst.Liq.Cryst. 119, 435 (1985)
25. S.Barišić, I.Batistić, J.Physique Lett., 46, 819 (1985)
26. A.Bjeliš, this issue
27. Z.Nakano, S.Takada, J.Phys.Soc.Japan 56, 977 (1985)
28. P.Brüesch, Lecture Notes in Physics 34, 194 (1974)
29. C.S.Jacobsen, this issue
30. T.Timusk, this issue
31. Z.Z.Wang, H.Salva, P.Monceau, M.Renard, C.Rouceau, R.Ayroles, F.Levy, L.Guemas, A.Meerschant, J.Physique Lett.44, 311 (1983)
32. A.Bjeliš, S.Barišić, J.Phys.C 19, 5607 (1986)

DYNAMICS OF QUASI-ONE-DIMENSIONAL CHARGE DENSITY WAVES

Aleksa Bjeliš

Institute of Physics of the University
POB 304, 41001 Zagreb, Croatia, Yugoslavia

1. INTRODUCTION

The collective conduction in the charge density wave (CDW) systems was proposed about thirty years ago by Fröhlich[1] as a possible mechanism of superconductivity. Already then it was realized that this phenomenon is restricted to highly anisotropic materials which have an almost flat Fermi surface and undergo the Peierls[2] instability. In the idealized situation considered in Sec.2, the corresponding Goldstone mode is associated with the bodily and frictionless translational motion of the whole Fröhlich condensate including the CDW modulation.

The direct experimental evidences in trichalcogenides and tetrachalcogenides,[3,4] blue bronzes[3,5,6] and the organic compound TTF-TCNQ[7] indicate that the dynamics of CDWs in real systems is considerably more complex. The modern studies which were initiated by the early measurements on TTF-TCNQ,[8] show that the collective motion is strongly affected by pinning and relaxation processes.[9-12] The phenomenological approach to the deformable and dissipative CDW[12] as well as the related microscopic aspects are considered in Sec.3.

In this approach the dynamics of CDW is reduced to the variations of its phase which is the essential variable for the description of the Fröhlich condensate. In Sec.4 we consider dynamical phase slippages in which variations of CDW amplitude are equally important.[13] These processes are shown to generate coherent current oscillations,[14] which are among the most interesting features of collective conduction.

2. PERFECT FRÖHLICH CONDUCTOR

Peierls semiconductor

The main property of the Peierls ground state is the macroscopic condensation of electron-hole pairs with a common wavevector. This is made possible by the nesting property of the Fermi surface shown in Fig.1. After the linearization in the longitudinal direction, the tight binding dispersion for k close to $\pm k_F$ is given by

$$\varepsilon(\mathbf{k}) \simeq E_F + v_F(\pm k - k_F) - 2t_\perp \cos(\mathbf{k}_\perp \mathbf{a}_\perp), \qquad (1)$$

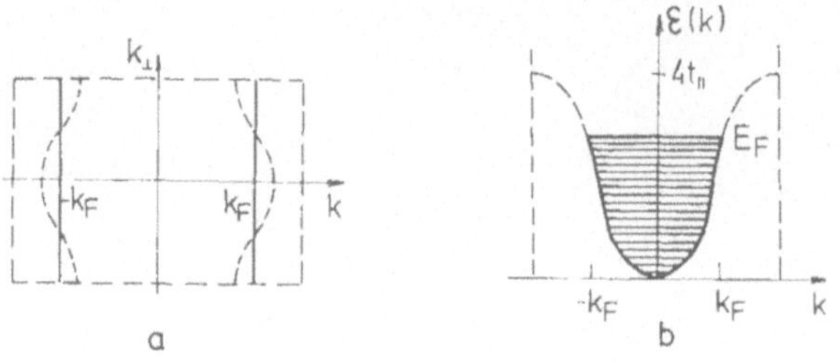

Fig. 1. The tight binding electron spectrum.
(a) Fermi surface with the nesting wave vector $(2k_F, \pi/a_\perp)$;
(b) The dispersion curve for $k_\perp = \pi/2a$.

where $v_F \simeq 2t_{\parallel}a \cos(k_F a)$ is the Fermi velocity and $t_\perp$ is the transverse overlap integral. The corresponding term in the hamiltonian is

$$H_{el} = iv_F \int dx \left[\Psi_1^+ \Psi_1' - \Psi_2^+ \Psi_2'\right] \quad , \tag{2}$$

where the prime denotes d/dx, and the electron field operator $\Psi(x)$ is conveniently written in terms of two components,

$$\Psi(x) = \Psi_1(x) \exp(ik_F x) + \Psi_2(x) \exp(-ik_F x) \quad , \tag{3}$$

representing the left and right sides of the Fermi surface from Fig.1. The spin indices and transverse coordinates are omitted for simplicity.

There are two interactions which may give a finite condensation energy to electron-hole pairs, namely the direct Coulomb interaction, and the indirect interaction via the $2k_F$-oscillations of crystal lattice.[15] The latter mechanism dominates in trichalcogenides and blue bronzes.[3-6] With the lattice deformation written as

$$w(x) = u(x) \exp(i2k_F x) + u^+(x) \exp(-i2k_F x) \quad , \tag{4}$$

the electron-phonon and phonon terms in the hamiltonian are respectively given by

$$H_{el-ph} = g \int dx (u\Psi_2^+ \Psi_1 + u^+ \Psi_1^+ \Psi_2) \tag{5}$$

and

$$H_{ph} = \int dx \left[\frac{1}{2} M\omega_0^2 \, u^+ u + \frac{1}{2M} \Pi^+ \Pi\right] \quad . \tag{6}$$

$g \simeq \partial t_{\parallel}/\partial x|_a$ is the electron-phonon coupling constant, while Π, ω_0 and M are respectively the momentum conjugate to u, the bare phonon frequency at $q \simeq 2k_F$ and the ionic mass.

The Peierls solution[2] of the hamiltonian (2,5,6) follows after supposing that the deformation (4) is static and sinusoidal, i.e. after replacing u(x) by $\langle u(x) \rangle \equiv u = cte$. The Heisenberg equations for the electron operators (3) then read

$$i\dot{\Psi}_1 = iv_F\Psi_1' + gu\Psi_2 \quad ,$$
$$i\dot{\Psi}_2 = -iv_F\Psi_2' + gu^*\Psi_1 \quad . \tag{7}$$

Their solutions are Bloch functions

$$\Psi_{1k}^{(\pm)}(x,t) = \frac{\sqrt{E(k) \pm \varepsilon(k)}}{\sqrt{2E(k)}} \exp\left[iE_\pm(k)t + i(|k|-k_F)x\right]$$
$$\Psi_{2k}^{(\pm)}(x,t) = \frac{\sqrt{E(k) \mp \varepsilon(k)}}{\sqrt{2E(k)}} \exp\left[iE_\pm(k)t + i(|k|-k_F)x\right] \tag{8}$$

with the spectrum

$$E_\pm(k) = \pm E(k) = \pm\left[\varepsilon^2(k) + |\Delta|^2\right]^{1/2} \quad . \tag{9}$$

The equilibrium value of the gap

$$\Delta \equiv g\,u \tag{10}$$

follows from the minimization of the total energy

$$\Delta F = \sum_{k_{occ}} E_-(k) - \varepsilon(k) + |\Delta|^2/\lambda E_F \tag{11}$$

and is equal to $\Delta_0 \equiv 4\,E_F \exp(-1/\lambda)$. Thus, the conducting state is unstable no matter how small is the coupling $\lambda \equiv g^2/M\omega_0^2 E_F$. This is the consequence of the anomalous density of states $n(E) \sim (1-|\Delta|^2/E^2)^{-1/2}$ for the dispersion (9), which is just another way to express the nesting nature of the band (1).

The temperature scale for the Peierls instability is defined through the BCS relation $T_P^{MF} = |\Delta|/1.76$. Note however that the real transition temperature, T_P, is due to the quasi one-dimensionality of the problem determined by the transverse interactions, and may considerably differ from T_P^{MF}.[15,16]

The finite deformation u induces also the sinusoidal modulation of the electron density, as follows from the summation over occupied states (8). The amplitude of this charge density wave is

$$\rho_0 = en\Delta/\lambda E_F \quad , \tag{12}$$

where $n \equiv 2ak_F/\pi$ is the number of electrons per unit cell.

The lattice deformation (4), the gap in the electron spectrum (9) and the CDW (12) are three manifestations of the Peierls instability. Any of the respective quantities, u, Δ and ρ may serve as the order parameter in further considerations.

Elastic properties of CDW

Let us now allow for lattice deformations which slightly differ from the sinusoidal form (4) and are not necessarily static. With $\Delta(x,t)$ chosen as the order parameter, the Landau expansion for the deformation energy reads

$$\Delta F_{def} = \int d^3x \{ a'\big((T/T_P - 1)|\Delta|^2 + \xi_0^2 |d\Delta/dx|^2 + \\ + \xi_\perp^2 |d\Delta/dx_\perp|^2\big) + b |\Delta^4| + \dots \} , \tag{13}$$

with the coefficients $a' \approx n_F \sim 1/v_F$, $b \approx 0.1\, n_F/T_P^2$ and $\xi_0 \approx v_F/T_P$.[17] Note that only contributions coming from the electrons are taken into account in b and $\xi_{\shortparallel}$. The corresponding lattice contributions are usually much smaller. E.g. the contribution from the phonon dispersion to the longitudinal correlation length, $v_{phon}/\sqrt{\lambda}\,\omega_0$, is much smaller than ξ_0 in eq.(13). Few mechanisms may however dominate in the transverse correlation $\xi_\perp$. These are, beside the electron transverse hopping which gives $\xi_\perp \simeq t_\perp a_\perp/T_P$, the Coulomb interaction between neighboring CDWs and the transverse dispersion of the bare phonon.[15]

The variations of Δ in time lead also to the finite kinetic energy due to the motion of lattice

$$\Delta F_{kin} \approx \frac{n_F}{\omega_0^2 \lambda} \int |\dot{\Delta}|^2 \, dx \quad . \tag{14}$$

From (13) and (14) one may determine the harmonic collective excitations of the Peierls ground state. After introducing

$$\Delta(x,t) = \big[\Delta\,(T) + \delta(x,t)\big]\exp\big[i\phi(x,t)\big] \quad , \tag{15}$$

with $\Delta^2(T) \equiv \dfrac{a'(1-T/T_P)}{2b}$, and keeping only gradient terms, we come to[18]

$$\Delta F = \int dx\, n_F \{ \frac{\Delta^2(T)}{T_P^2} [\frac{T_P^2}{\omega_0^2 \lambda} (\dot{\phi})^2 + v_F^2 (\phi')^2] + \\ + \frac{1}{\omega_0^2 \lambda} (\dot{\delta})^2 + (1 - T/T_P)\, \delta^2 \} \quad . \tag{16}$$

It is immediately seen that there are two eigenmodes which represent the oscillations of the phase ϕ and the amplitude δ, and respectively have the dispersions

$$\omega_{ph}^2(q) = \frac{\lambda \omega_0^2}{T_P^2} (v_F q)^2 \equiv \frac{m_{el}}{m^*} v_F^2\, q^2 \equiv v_{ph}^2\, q^2 \tag{17}$$

and

$$\omega_{amp}^2 = (1 - T/T_P) \cdot \lambda\, \omega_0^2 \quad . \tag{18}$$

It follows from eq.(17) that the characteristic velocity for the propagation of the acoustic phason mode is, due to the lattice kinetic energy (14), much smaller than the Fermi velocity. The ratio $(v_{ph}/v_F)^2$ defines the effective CDW mass m*, which is at T=0 given by[9]

$$\frac{m^*}{m_{el}} = 1 + \frac{4\Delta_0^2}{\lambda \omega_0^2} \quad . \tag{19}$$

The electronic contribution which is added in eq.(19) is obviously negligible with respect to the lattice term (17), i.e. $m^*/m_{el} \approx 10^2 - 10^3$.

The acoustic nature of the long wavelength phase oscillations is the consequence of the independence of the free energy (13) on the phase ϕ. This is the essential property of the Peierls-Fröhlich ground state. As will be seen later, phase dependent contributions to (13) which come from the lattice discreteness, impurities etc., modify this result[9] and lead to a finite gap in the long-wavelength part of the dispersion $\omega_{ph}(q)$.

The phase oscillations go together with finite local values of ϕ', i.e. to local variations of the Fermi wave number k_F, as is seen from eqs. (4) and (15). That means that the local filling of the band, i.e. the mean number of electrons per new $(2k_F)^{-1}$ unit cell varies as

$$\rho(x) = en\left[1 + \phi'/2k_F\right] \quad . \tag{20}$$

Thus, spatial phase variations imply the redistributions of the total band charge. In that sense the Fröhlich condensate at T=0 comprises all the band electrons.

The mode (18) which describes the small oscillations of the amplitude has the finite frequency at q=0. This frequency increases by lowering the temperature, and is equal to $\sqrt{\lambda}\,\omega_0$ at T=0.[9] The optic nature of the ampliton comes from the short-range redistribution of the charge from the CDW on the $(2k_F)^{-1}$ - scale. Note in this respect that long range charge polarizations due to phasons may induce the strong Coulomb effects.[19] This leads to the observation of plasmon-like excitations at finite frequencies in the optical measurements, but does not invalidate the mechanism of Fröhlich conduction.

Collective charge transport

The spatial variations of the band charge density attributed to long-wavelength phasons, oscillate also in time with the frequency $\omega_{ph}(q)$. In other words, phasons induce also the local electrical currents. In order to find out what is the current which corresponds to a finite temporal variations $\dot{\phi}$, let us consider another simple limit defined by $\dot{\phi}$ = const., ϕ' = 0. As will be seen later, this situation is realized by putting the CDW in a finite external dc electric field. The CDW then translates with a finite velocity

$$v_{CDW} = \dot{\phi}/2k_F \quad , \tag{21}$$

as follows from eqs (4) and (15).

With $\dot{\phi}\neq 0$ we encounter the problem of time dependent potential in the Heisenberg equations (7). It can be straightforwardly solved after introducing new electron operators[18]

$$\Psi_{1,2}(x,t) = \tilde{\Psi}_{1,2}(x,t)\ \exp\ (\pm\dot{\phi}t/2) \quad . \tag{22}$$

This is just an example of the chiral transformation.[20] The solutions of the Heisenberg equation have again well-defined wave numbers and are given by

$$\tilde{\Psi}^{(\pm)}_{1k}(x,t) = \frac{\sqrt{\tilde{E}(k) \pm \left[\varepsilon(k) - \dot{\phi}/2\right]}}{\sqrt{2\,\tilde{E}(k)}}\ \exp\left[i\tilde{E}_{\pm}(k)t + i(|k| - k_F)x\right]$$

$$\tilde{\Psi}^{(\pm)}_{2k}(x,t) = \frac{\sqrt{\tilde{E}(k) \mp \left[\varepsilon(k) - \dot{\phi}/2\right]}}{\sqrt{2\,\tilde{E}(k)}}\ \exp\left[i\tilde{E}_{\pm}(k)t + i(|k| - k_F)x\right] \ , \tag{23}$$

with

$$\tilde{E}_{\pm}(k) = \pm\tilde{E}(k) = \pm\left\{\left[\varepsilon(k) - \dot{\phi}/2\right]^2 + |\Delta|^2\right\}^{1/2} \quad . \tag{24}$$

Having the time-dependent original hamiltonian (2,5,6) we have arrived to the states (23) which are non-stationary, as is seen explicitly from their form. The question which now arises is how to generalize the notion of band spectrum, given in the static situation by eq.(9). The most natural way is to start from the electron propagators defined by $G_{ij}(x,t) = -i < \mathbf{T}[\Psi_i(x,t)\Psi_j^+(0,0)]>$, where $i,j = 1,2$ and $\mathbf{T}$ is the usual operator of time ordering.[21] The quasiparticle excitation spectrum is then given by poles of their Fourier transforms in the complex ω-plane. These poles are given by[17]

$$E_{1,2}^{lab}(k) = \tilde{E}(k) \pm \dot{\phi}/2 \quad , \tag{25}$$

and are shown in Fig.2. They are to be associated to the separate propagation of the components $\Psi_{1k}(x,t)$ and $\Psi_{2k}(x,t)$, as is seen from eq.(23).

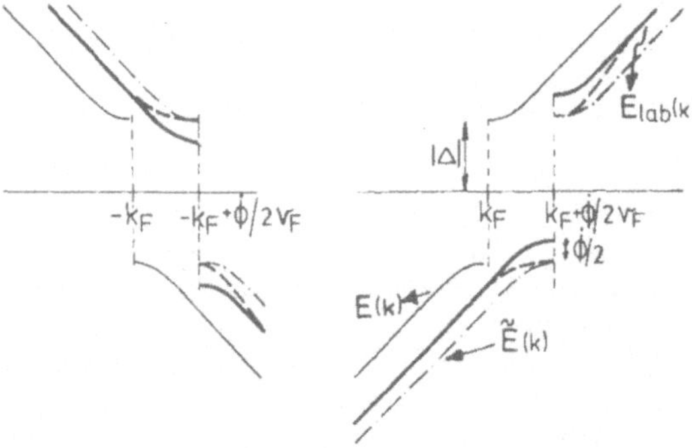

Fig. 2. The left ($k \approx -k_F$) and right ($k \approx k_F$) side of the excitation spectrum for finite $\dot{\phi}$. $E(k)$, $\tilde{E}(k)$ and $E_{lab}(k)$ are given by eqs (9), (24) and (25) respectively. The dashed line corresponds to the expression (26).

Another quantity which is also used in literature[12] as the characterisation of the spectrum, is the average of two excitations in eq.(25), multiplied by the weighting factors from eq.(23),

$$\tilde{E}(k) \pm \frac{\varepsilon(k) - \dot{\phi}/2}{\tilde{E}(k)} \cdot \dot{\phi}/2 \quad , \tag{26}$$

(see also Fig.(2)). This expression is equal to <id/dt>, i.e. to the mean value of the hamiltonian in the given non-stationary state (23).

It is clear from eq.(24) and Fig.(2) that, due to the finite value of $\dot{\phi}$, the occupied states are shifted in the k-space. This results in the net current

$$j_c = e \sum_{k_{occ}} \frac{\partial E_{lab}(k)}{\partial k} = en\dot{\phi}/2k_F = env_{CDW} \quad . \tag{27}$$

The same result follows from eq.(26). Thus the velocity per band electron in the laboratory frame (i.e. the frame of the crystal lattice) is at T=0 equal to v_{CDW}. On the other hand in the CDW frame, which moves with velocity v_{CDW}, there is no Fröhlich current, and the band spectrum is again (9). Note however that the inclusion of the lattice kinetic energy associated to finite $\dot{\phi}$ leads to the corrections in the expressions (23) and (25) which are of the second order in $\dot{\phi}/v_{phon}$. In particular, one has an increase of the effective gap in eq.(25) (and also of the corresponding CDW amplitude) in the laboratory frame.[1] However, these contributions are much weaker than those linear in $\dot{\phi}$, since usually $\dot{\phi} << \omega_0 << \Delta_0$ in experiments.[3-6] Furthermore, the relative shifts of gap edges with respect to the Fermi level are also small, so that the spectrum in the laboratory frame is far from being semimetallic, i.e. there are no low energy quasiparticle excitations in the Fröhlich condensate. This is the essential difference between the Fröhlich conductor and the perfect metal at T=0.

The equation of continuity for the Fröhlich condensate follows directly from the results (20) and (27), and the equality $\partial^2\phi/\partial x\partial t = \partial^2\phi/\partial t\partial x$. This result is valid for perfect Fröhlich conductor, when there is no excitations across the gap due to finite temperature and residual scatterings on crystal imperfections.

3. COLLECTIVE CONDUCTION IN REAL MATERIALS

The central problem of the CDW dynamics is the formulation of the equation of motion for the Fröhlich condensate in the finite external field. We start from the phenomenological approach which was developed in Refs.(10,12), and is reminescent to the two-fluid model for the superconductors.

In order to determine the fraction of electrons which participate in the collective transport at finite temperatures, it is most convenient to look for the drift velocity of normal carriers in the CDW frame in which the band behaves just like in ordinary semiconductor.[12] The mean velocity in this frame is

$$\langle v\rangle_{CDW} = -\frac{k_F}{v_F}(v_N - v_{CDW}) \sum_k \left[\frac{\partial E(k)}{\partial k}\right]^2 \frac{\partial f}{\partial E} \equiv (v_N - v_{CDW})\,\rho_N \quad , \tag{28}$$

where v_{CDW} is already introduced in eq.(21), v_N is the drift velocity of normal carriers measured in the laboratory frame, and ρ_N is the fraction of carriers participating in the normal (i.e. semiconducting) transport. It is then natural to represent the mean velocity in the laboratory frame by

$$\langle v\rangle_{lab} = v_{CDW} + \langle v\rangle_{CDW} \equiv \rho_N v_N + \rho_C\, v_{CDW} \quad , \tag{29}$$

where

$$\rho_C = 1 - \rho_N = 1 + \frac{k_F}{v_F} \sum_k \left(\frac{\partial E}{\partial k}\right)^2 \frac{\partial f}{\partial E} \tag{30}$$

is the fraction of electrons which effectively belong to the Fröhlich condensate. The expression (27) for the collective current is then replaced by

$$j_C = en\,\rho_C\, v_{CDW} \quad . \tag{31}$$

At T=0 two expressions coincide, while by increasing temperature j_C gradually decreases, vanishing as $\Delta(T)/T$ as $T \to T_P$.[12] This latter law will be however modified after taking into account relaxation processes (see eq. (39)).

Relaxation processes

With the two mentioned types of carriers, one may distinguish on the phenomenological level among the three types of scattering processes, and introduce correspondingly the following relaxation times:[12] τ_N and τ_C for the relaxation of normal and collective carriers respectively on the laboratory frame (thermal phonons, impurities, etc), and τ for the relaxation processes with a momentum transfer between normal carriers and the condensate. The underlying assumptions are that there is no strong mixing between these mechanisms, and that all relaxation times are long enough ($\tau_i > \Delta, T$), so that the quasiparticle approach is justified. With this picture in mind, one may establish the equations for momentum rates for each type of carriers.

Since the relaxation of normal carriers on the condensate is proportional to the relative velocity $v_N - v_{CDW}$, the equation for normal carriers reads

$$m_{el}\rho_N\dot{v}_N = e\rho_N E - \tau_N^{-1} m_{el}\rho_N v_N - \tau^{-1} m_{el}\rho_N(v_N - \dot{\phi}/2k_F) \quad , \tag{32}$$

where E is the external electric field, and the relation (21) is taken into account.

Let us now formulate the equation for the CDW phase ϕ. The electric field and the relaxation terms enter into the equation analogously as into eq.(32). The additional terms however appear due to the presence of impurities, lattice discreteness (especially if $2k_F$ is close to some commensurate value), etc. The effective force coming from these sources, $F(\phi,x)$ causes local deformations of CDW, in particular of its phase. Therefore, the equation must be also completed by the elastic term introduced already in the expression (13). Altogether one has

$$\begin{aligned} m^*\rho_C\ddot{\phi}/2k_F = e\rho_C E - \tau_C^{-1} m^*\rho_C\, \dot{\phi}/2k_F + \tau^{-1} m_{el}\rho_N(v_N - \dot{\phi}/2k_F) + \\ + m^*\rho_C v_{ph}^2\, \phi''/2k_F + F(\phi,x) \quad . \end{aligned} \tag{33}$$

In the more complete equation one should also allow for the dependence of ϕ in the transverse direction and include the transverse elastic term from the eq.(13).

As long as the variations of CDW amplitude are not important, the eq. (33) represents the basis for a number of theoretical approaches to the various aspects of CDW dynamics. Before mentioning some of them, let us continue with the discussion of eqs (32) and (33) by considering two opposite limiting situations.[12] In the first one the CDW is pinned, so that one may put $\dot{\phi}=0$ in eq.(32). The second limit corresponds to the motion of CDW which is so fast that the spatial dependent terms in eq.(33) may be neglected.

Eliminating v_N from eq.(33) by imposing the condition of stationary flow, $\dot{v}_N=0$, in eq.(32), one comes in both limits to the equation of motion for ϕ of the same type,

$$m^*\rho_C\, \ddot{\phi}/2k_F \approx e\rho_{eff}E - \tilde{\tau}^{-1}\, m^*\rho_C\, \dot{\phi}/2k_F \tag{34}$$

with

$$\rho_{eff} = \rho_C + \frac{\rho_N}{1 + \tau/\tau_N} \quad , \tag{35}$$

in both cases. Thus, the presence of normal carriers leads to the increase of effective density of the condensate which couples to the external field.[12] Furthermore, the effective relaxation time in eq.(34) is different in two limits.[12] For the pinned CDW it is given by

$$\tilde{\tau}^{-1} = \tau_C^{-1} \quad , \tag{36}$$

while for the fast CDW

$$\tilde{\tau}^{-1} = \tau_C^{-1} + \frac{m_{el}}{m^*}\frac{\rho_N}{\rho_C} \cdot \frac{1}{\tau + \tau_N} \quad . \tag{37}$$

The depinning of CDW leads to the decrease of the effective relaxation time, i.e. the damping term in eq.(34) becomes larger.

Pinning and depinning

Let us now consider two important points which follow from eq.(33) and concern the spatial variations of ϕ.

In the absence of electric field, eq.(33) is just the more complete equation for phase oscillations from Sec.2. The term F leads in general to the already mentioned gap in the $q \to 0$ limit of the dispersion $\omega_{ph}(q)$.[9] This result follows on expanding $F(\phi,x)$ with respect to the stable (or metastable) configuration $\phi(x)$ which minimizes the total deformation energy. The damping term gives a finite lifetime to these phasons. Note that the force F and the relaxation time $\tilde{\tau}$ may originate from different physical mechanisms.

The second question concerns the threshold field below which the CDW is pinned. It was estimated[12] by calculating the energy gain from the external field, necessary to overcome the energy stored in the deformed stable configuration $\phi(x)$. This static solution of eq.(33) defines the Lee-Rice domain as the characteristic length scale for the spatial variation of phase. The problem of the CDW dynamics closely above the threshold field is still very difficult, and is usually treated by simplifying eq.(33) in various ways (see review[3] and proceedings[32,33]).

Microscopic aspects

The phenomenological approach which resulted in eq.(33) opens some important questions. The first one concerns the microscopic origin of relaxation times introduced in eqs (32) and (33). Apart from the well understood mechanisms of the relaxation of normal carriers on the laboratory frame, little is known about the more complex processes which involve the Fröhlich condensate. There are indications that the transverse correlations are essential for such relaxations. E.g. the contribution to τ_C coming from the scattering of phasons on thermal phasons was shown[24] to be dependent on the transverse phason dispersion $\omega_\perp$, i.e. $1/\tau \simeq \lambda^2 m_{el} T^2/m^*\omega_\perp^2$. Similarly, the analysis of the scattering of normal carriers on phasons[12] showed that at low temperatures τ is influenced by the transverse hopping $t_\perp$.

The microscopic justification of the time-dependent Landau expansion (13,33) is closely related to the above questions. The CDW motion was treated microscopically by using the diagrammatic expansion in Δ developed earlier by Eliashberg and Gor'kov[25] for superconductors with impurities. This method is applicable to CDW systems in which the relaxation of normal electrons is dominating, so that $\tau_C^{-1} m^* \ll \tau_N^{-1} m_{el}$. It is known that the expansion of Refs. (25) does not reduce to the usual Ginzburg-Landau dynamic equation in the case of "pure" systems for which $\tau_N^{-1} < \Delta, T$. The same conclusion holds for the CDW systems. On the other hand, the expansion of Ref.(25) leads to the well defined equation of motion for the "dirty" gapless systems[13] which were excluded from the phenomenological approach (33) due to $\tau_N^{-1} > \Delta, T$.

The limit of weak scatterings, when all reciprocal relaxation times are very small is particularly interesting. The system then may have an underdamped behavior in which the pinning centers (or potentials) are also responsible for the CDW relaxations. The relaxation length may become extremely large, of the order of Lee-Rice domain. In this limit one might expect the macroscopic quantum effects considered extensively in Refs.(26).

4. COHERENT OSCILLATIONS AND THE ROLE OF CDW AMPLITUDE

Coherent oscillations

Well above the threshold field the Fröhlich transport shows an unusual phenomenon observed in numerous CDW systems, namely the dc current becomes noisy.[27] This apparent narrow band noise consists in fact of highly coherent and nonsinusoidal oscillations, as evidenced by the spectral analysis.[3-6] An early experimental result[28] of this kind is shown in Fig.3.

The early attempts to explain these oscillations are again based on the eq.(33).[3] It was shown in the previous Section that after neglecting the terms depending on ϕ and ϕ' the Fröhlich current has only a dc contribution. In the simplest approximation these terms were replaced by an effective force $F(\phi)$ which is periodic in ϕ and does not depend on x. By choosing $F(\phi) = \sin\phi$ one arrives to the so called damped oscillator model[29] in which the CDW reduces to the single degree of freedom, represented in Fig.(4a) by an endless inhomogeneous spring. Due to this inhomogeneity its screw rotation which corresponds to the Fröhlich current is not monotonous, but is periodically (and nonsinusoidally) modulated with the fundamental frequency $f_0 = k_F j_C / e\pi\rho_C$.

The force $F(\phi)$ is well defined in the case of single Lee-Rice domain in which the phase is almost homogeneous. However, this force vanishes after averaging over many domains in the sample with randomly distributed pinning centers. The macroscopic system may be represented more realistically as an ensemble of loosely linked domains, shown in Fig.(4b). This ensemble however

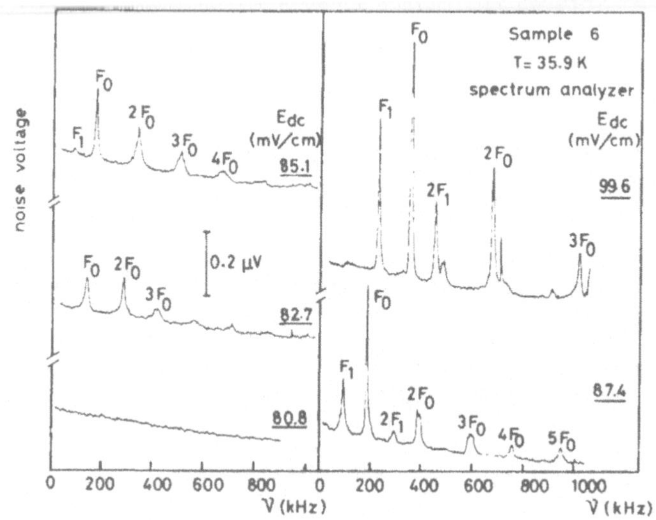

Fig. 3. The spectral function of the "periodic noise" of $NbSe_3$ for several values of the external dc field (after Ref.28).

cannot oscillate coherently due to the destructive coupling of randomly phased oscillations from neighboring domains.[30] The internal oscillations are then induced only by applying an external ac field.[31]

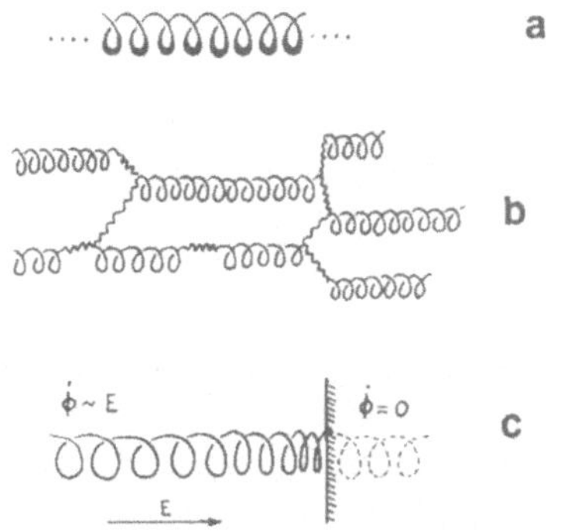

Fig. 4. Schematic presentation of the classical damped oscillator model for the CDW motion (a), the ensemble of Lee-Rice domains (b), and the stoppage of the CDW at the strong obstacle.

The flow in Figs (4a,b) is assumed to be total, so that the CDW does not remain static in any part of the sample. Let us now introduce strong obstacles through which the CDW cannot move even when the depinning in volume is achieved. As is shown in Fig.(4c), after imposing $\dot{\phi}=0$ at the obstacle and having the Fröhlich current $\dot{\phi} = 2k_F e\rho_{eff}\tilde{\tau}E/\rho_C m^*$ (see eq.(34)) far enough from it, the number of loops in the spring (i.e. the charge density of the condensate) in the vicinity of the obstacle increases (or decreases for the opposite direction of E) in time. Such deformations induce strains which may be relaxed only by successive destruction (or creation) of loops, i.e. by the conversion of the collective Fröhlich flow into the normal (ohmic) current. This is possible through the process of phase slippage[13,32] in which the CDW amplitude passes through zero.[33] Thus, the local stoppage of the Fröhlich flow on the obstacles is the feature in which the spatial and temporal variations of the CDW amplitude have to be taken into account.

Gor'kov's model

As it was seen in Sec.2, after assuming that the amplitude $|\Delta|$ is constant, the variations of phase can be directly related to the local redistributions of charge in the Fröhlich condensate. The characterization of the collective carriers however becomes more complex if the amplitude also varies, so that the reasoning leading to the eq.(33), especially to terms describing the coupling to the external electric field and the relaxation of CDW, cannot be simply generalized.

The simultaneous variations of $|\Delta|$ and ϕ were taken into account in Refs (13,32) in which the CDW dynamics was considered in the limit of strong single particle scattering on impurities. Close to the critical temperature the application of the perturbative method[25] to the CDW order parameter leads to the equation

$$\dot{\Delta} + \frac{4}{9}\pi^2 \tau_s (T^2 - T^2_{P_{imp}})\,\Delta + \frac{10}{27}\tau_s |\Delta|^2\,\Delta + \frac{8}{3}\tau_s v_F\, i\, e\, E\, \Delta - \tag{38}$$

$$- 6\,\tau_s\,(\xi_0^2\, d^2\Delta/dx^2 + \xi_\perp^2\, d^2\Delta/dx_\perp^2) = 0 \quad ,$$

and to the accopmanying expression for the total current

$$j = \frac{\sigma}{e}\left\{ e\,E - \frac{8}{3} e\,E\,\tau_s^2|\Delta|^2 + \frac{8}{9}\, i\,\tau_s \frac{\overline{v}^2}{\overline{v^2}}(\dot{\Delta}\Delta^* - \Delta\dot{\Delta}^*)\right\} \quad . \tag{39}$$

Here $T_{P_{imp}}$ is the critical temperature which is assumed to be much smaller than T_P, i.e. τ_s is close to the critical value $\tau_{crit} \simeq 4\gamma/3\pi T_P$ (γ - Euler constant) for which the CDW ordering is completely washed out.

The comparison of eqs (33) and (38) for $|\Delta|$ = const shows that m^* and τ_c from eq.(33) are replaced here by m_{el} and the single particle relaxation time τ_s (which corresponds to τ_N from Sec.3). This again indicates that the two equations describe different microscopic situations, as already stated. This is also clear from the expression for the current (39) in which the two first terms represent the metallic current with the conductivity σ, somewhat reduced due to the formation of the gap in the density of states (second term).[13] The third term is the collective current ($\overline{v^2}$ and $\overline{v}^2$ are the band averages of the velocity in the transverse direction). For $|\Delta|$ = const it is proportional to $|\Delta|^2$ (see also Refs 34,35), and not to $|\Delta|$ as was obtained in the limit $|\Delta|\tau >> 1$ (eq.(31)).[35]

Diffusion of phase slippages

The eq.(38) was analyzed numerically[14] after introducing simplified boundary conditions of Fig.(4c) and assuming that the solutions $\Delta(x,t)$ do not depend on the transverse coordinate $x_\perp$. This analysis shows (Fig.5) that the generation of phase slippages proceeds as a slow diffusion of a weak minimum in $|\Delta(x,t)|$ followed by its abrupt decrease in a very short time interval, and its fast disappearance after the moment of the phase slippage. In the external dc field this process repeats periodically in time with the frequency $\omega_f = 2\pi/t_{ps} = (16\gamma/9\pi^2)\xi_0 eE/\hbar$. The position of the phase slip center is given by (units in eqs (40) and (41) are defined below Fig.5)

$$x_{ps} \simeq 1.15\, E^{-0.284} \quad . \tag{40}$$

For the experimental values of the electric field x_{ps} is rather far from the boundary (about 10^2-10^3 ξ_0) so that the above results are to a great extent independent of the details of the boundary.

The highly nonlinear and dissipative production of phase slippages is possible only if the energy provided from the external field is sufficiently large. In other words, in finite samples phase slippages occur only above

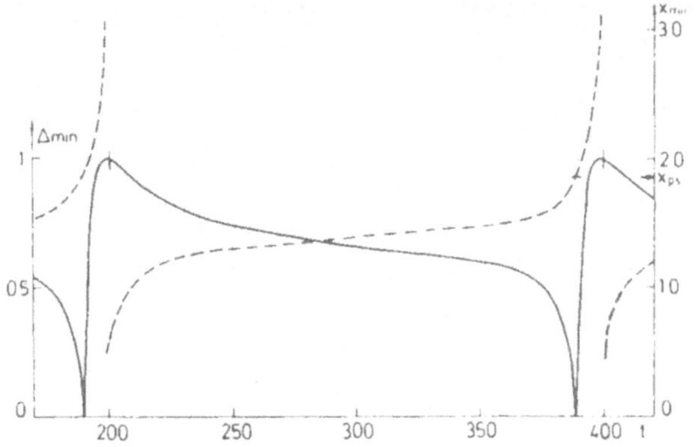

Fig. 5. The time dependence of the minimum of CDW amplitude (full line) and of its position (dashed line) for E = 0.175 in the unit of $\pi^2(T^2_{Pimp} - T^2)/6\ \bar{v}e$. The units for t, x and $|\Delta|^2$ are $9/[4\pi^2\tau_s(T^2_{Pimp} - T^2)]$, ξ_0 and $\pi^2(T^2_{Pimp} - T^2)$ respectively.

some threshold field, while below this field the CDW remains fixed and deformed. The value of this threshold field is given by[14]

$$E_T(L) \simeq 2.55\ L^{-1.23} \quad , \tag{41}$$

where L is the length of the sample. The length dependent E_T was indeed observed in specimens of $NbSe_3$[37-39] and TaS_3[40] shorter than ∿ 100 μm. Some of the measurements are in a good agreement with the eq.(41), as is shown in Fig. 6. For longer specimens $E_T(L)$ of eq.(41) is overcome by contributions from the bulk pinning.

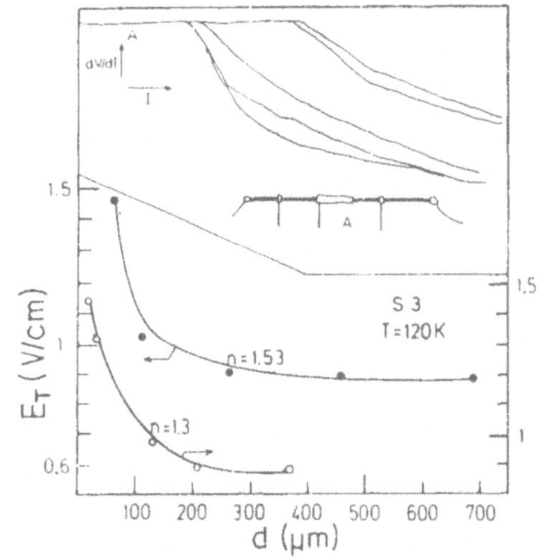

Fig. 6. The threshold field vs the distance between contacts for two samples of $NbSe_3$ (after Ref. 39).

The "finiteness" of the sample in the above experiments is in fact defined by the distance between two electrical contacts. Since the current is usually shunted through such contacts, the collective transport in the part of the sample below them is not possible since locally $E < E_T$, and the stoppage from Fig.4c is realized. Although the experimental results are rather controversial,[3,4,22,23] there are many indications[41,42] that the contacts play an essential role in the generation of coherent current oscillations. We note that phase slippages should be expected close to any strong obstacle, like e.g. at the segment boundaries in samples with switching current--voltage characteristics[43-45], etc.

Resonance effects

The fast variations of the order parameter in the region of phase slippages give pulses in the current (39) and in the corresponding voltage shown in Fig.7. For experimental values of the electric field (10^{-4} to 10^{-3} in the units of Fig.7) these pulses are very sharp, so that many higher harmonics together with the fundamental frequency ω_f appear in the spectral distribution.

The rich multiharmonic content of the current pulses lead to significant interference effects in the external field with a finite ac component, $E + E_1 \cos(\omega_0 t)$.[46] The resonances appear not only at the harmonics of the external frequency, $\omega_f = n\omega_0$, but also at its subharmonics, i.e. for $\omega_f = n\omega_0/m$,[47,3] where n and m are integers. These resonances are manifested as the steps in the V(E) curve. Such steps are observed in e.g. $NbSe_3$[48,43,47] and TaS_3,[49] and are usually called Shapiro steps, in analogy with the similar phenomenon in Josephson junctions.[50]

The numerical study shows also an interesting feature in the vicinity of resonances.[51] The interval between two successive phase slippages, t_{ps}, becomes then modulated with the modulation frequency $\Omega = |\omega_f - n\omega_0/m|$ (Fig. 8). Although the mean value $\langle t_{ps}\rangle$ over the long period $2\pi/\Omega$ remains equal to $2\pi/\omega_f$, t_{ps} is in the long intervals almost locked to $2\pi m/n\omega_0$. The compensation is realized in fast beats, as shown in Fig.8. By approaching a given resonances, these beats become sharper and sharper. Such highly nonsinusoidal modulation is expected to give additional peaks in the low frequency range, at the frequency Ω and its higher harmonics.

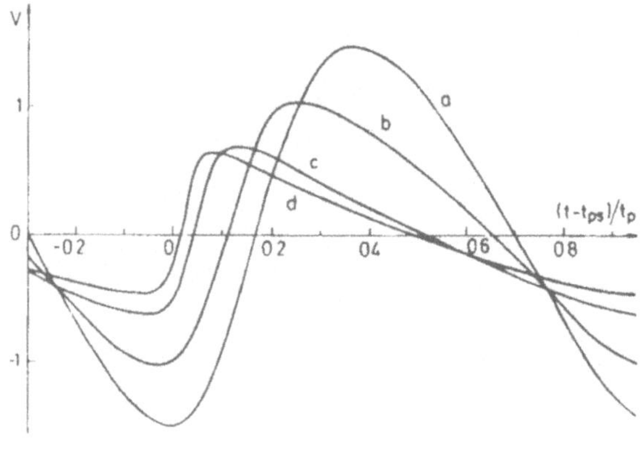

Fig. 7. The voltage pulses for the electric fields E = 1.0 (a), 0.5 (b), 0.2 (c) and 0.13 (d) in the units defined in the caption of Fig.5. The phase slippage occurs at $t = t_{ps}$.

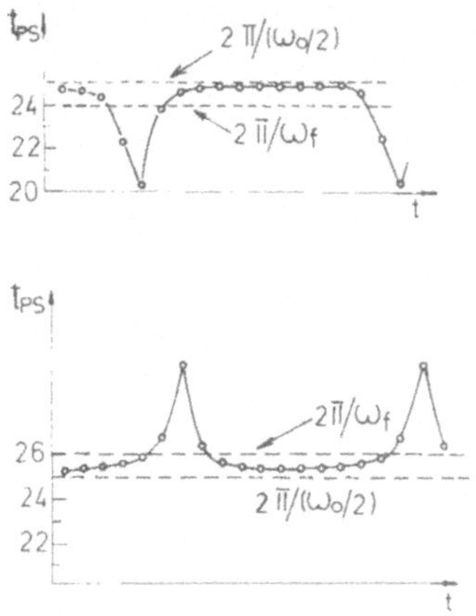

Fig. 8. Beats in the modulation of intervals between two successive phase slippages before (a) and after (b) the resonance $\omega_f = \omega_0/2$, for $\omega_0 = 0.5$ and $E_1 = 0.3$. Units are as in Figs. 5 and 7.

ACKNOWLEDGEMENTS

The author is grateful to S.Barišić, I.Batistić and D.Jelčić for enlightening discussions, and to Lu Yu and H.J.Schulz for useful comments. He acknowledges the hospitality of the International Centre for Theoretical Physics in Trieste where this manuscript was partially prepared.

REFERENCES

1. H.Fröhlich, Proc.Roy.Soc. A223,296 (1954).
2. R.E.Peierls, Ann.Phys., Lpz., 4,121 (1930); Quantum Theory of Solids, p.108, Clarendon Press, Oxford (1955).
3. G.Grüner and A.Zettl, Physics Reports 119,117 (1985); G.Grüner, to be published.
4. P.Monceau, in Electronic Properties of Inorganic Quasi-One-Dimensional Materials, II, p.139, ed.P.Monceau, D.Riedel, Publ.Comp., Dordrecht (1985).
5. R.M.Fleming, Synthetic Metals 13,241 (1986).
6. C.Schlenker, J.Dumas, in Crystal Chemistry and Properties of Materials with Quasi-One-Dimensional Structures, p.135, ed.J.Rouxel, D.Riedel Publ.Comp., Dordrecht (1986).
7. R.C.Lacoe, H.J.Schulz, D.Jérome and L.Johannsen, Phys.Rev.Lett. 55,2351 (1985).
8. P.M.Chaikin, J.F.Kwak, T.E.Jones, A.F.Garito and A.J.Heeger, Phys.Rev. Lett. 31,601 (1973); D.Jérome and H.Schulz, Adv.in Phys. 31,299 (1982).
9. P.A.Lee, T.M.Rice and P.W.Anderson, Solid State Commun. 14,703 (1974).
10. H.Fukuyama and P.A.Lee, Phys.Rev. B17,535 (1978); P.A.Lee and H.Fukuyama, ibid. B17,542 (1978).
11. K.B.Efetov and A.I.Larkin, Zh.Eksp.Teor.Fiz. 72,2350 (1977) (Sov.Phys. JETP 45,1236 (1977)).
12. P.A.Lee and T.M.Rice, Phys.Rev. B19,3970 (1979); T.M.Rice, P.A.Lee and M.C.Cross, Phys.Rev. B20,1345 (1979).
13. L.P.Gor'kov, Zh.Eksp.Teor.Fiz. 86,1818 (1984) (Sov.Physics JETP 59,1057 (1985)).
14. I.Batistić, A.Bjeliš and L.P.Gor'kov, J.Physique 45,1049 (1984).
15. S.Barišić, in Electronic Properties of Inorganic Quasi-One-Dimensional Materials, I, p.1, ed. P.Monceau, D.Riedel Publ.Comp., Dordrecht (1985).

16. S.Barišić and A.Bjeliš, in Theoretical Aspects of Band Structures and Electronic Properties of Pseudo-One-Dimensional Solids, p.49, ed. H.Kamimura, D.Riedel Publ. Comp., Dordrecht (1985).
17. D.Allender, J.W.Bray and J.Bardeen, Phys.Rev. B9,119 (1974).
18. S.A.Brazovskii and I.E.Dzyaloshinskii, Zh.Eksp. Teor.Fiz. 71,2338 (1976) (Sov.Phys. JETP 44,1233 (1976)).
19. S.Barišić, these Proceedings.
20. Z.B.Su and B.Sakita, Phys.Rev.Lett. 56,780 (1986).
21. A.A.Abrikosov, L.P.Gor'kov and I.E.Dzyaloshinskii, Methods of Quantum Field Theory in Statistical Physics, Dower Inc, New York (1963).
22. Proc.Int.Conf. on Charge Density Waves in Solids, Budapest, Sept. 1984, ed.G.Hutiray and J.Solyom, Lecture Notes in Physics, Springer-Verlag, vol.217, Berlin, Heidelberg, New York, Tokyo.
23. Proc.Int.Symp. on Non-Linear Transport and Related Phenomena in Organic Quasi-One-Dimensional Conductors, Sapporo, Oct. 1983, Hokkaido University.
24. S.Takada, M.Wong and T.Holstein, Ref.22, p.227.
25. L.P.Gor'kov and G.M.Eliashberg, Zh.Eksp.Teor.Fiz. 54,612 (1968) (Sov. Physics JETP 27,328 (1968)).
26. J.Bardeen, Phys.Rev.Lett. 55,1010 (1985), and these Proceedings.
27. R.M.Fleming and G.G.Grimes, Phys.Rev.Lett. 42,1423 (1979).
28. J.Richard, P.Monceau and M.Renard, Phys.Rev. B25,948 (1982).
29. G.Grüner, A.Zawadowski and P.M.Chaikin, Phys.Rev.Lett. 46,511 (1981).
30. L.Sneddon, M.C.Cross and D.S.Ficher, Phys.Rev.Lett. 49,292 (1982). L.Sneddon, Phys.Rev. B29,719 (1984).
31. S.N.Coppersmith and P.B.Littlewood, Phys.Rev.B31,4049 (1985); S.N.Coppersmith, Phys.Rev. B34,2073 (1986).
32. L.P.Gor'kov, Pisma Zh.Eksp.Teor.Fiz. 38,76 (1983) (Sov.Phys.JETP Letters 38,87 (1983)).
33. The conversion to the ohmic current may also proceed through the formation of transverse phase vortices and their propagation along the obstacle. This mechanism was proposed by N.P.Ong, G.Verma and K.Maki, Phys.Rev.Lett. 52,663 (1984). See also K.Maki, Phys.Rev. B33,2852 (1986).
34. M.J.Rice, A.R.Bishop, J.A.Krumhansl and S.E.Trullinger, Phys.Rev.Lett. 36,432 (1976).
35. H.J.Schulz, Solid State Commun. 34,455 (1980).
36. The common expression which covers both limits has the form $j \sim \sim |\Delta|^2\tau\dot{\phi}/(1 + |\Delta|^2\tau^2)^{1/2}$. See L.P.Gor'kov and E.N.Dolgov, Zh.Eksp.Teor. Fiz. 77,396 (1979)(Sov.Phys.JETP 50,203 (1979)).
37. M.C.Saint-Lager, Thèse, Grenoble 1983; P.Monceau, M.Renard, J.Richard and M.C.Saint-Lager, Physica B (1986), to be published.
38. J.C.Gill, Solid State Commun. 44,1041 (1982); Ref.22, p.377, Ref.23, p.139.
39. M.Prester, Phys.Rev. B32,2621 (1985).
40. G.Mihaly, Gy.Hutiray and L.Mihaly, Phys.Rev. B28,4896 (1983).
41. N.P.Ong and G.Verma, Phys.Rev. B27,4495 (1983); Ref.23, p.115.
42. X.J.Zhang, N.P.Ong and J.C.Eckert, Phys.Rev.Lett. 56,1206 (1986); N.P.Ong, G.Verma and X.J.Zhang, Ref.22, p.296.
43. A.Zettl and G.Grüner, Phys.Rev. B26,2298 (1982); Phys.Rev. B29,755 (1984).
44. J.W.Lyding, J.S.Hubacek, G.Gammie and R.E.Thorne, Phys.Rev. B33,4341 (1986).
45. R.P.Hall, M.F.Hundley and A.Zettl, Phys.Rev.Lett. 56,2399 (1986).
46. A.Bjeliš and D.Jelčić, J.Physique Lettres 46,L283 (1985).
47. S.E.Brown, G.Mozurkewich and G.Grüner, Phys.Rev.Lett. 52,2277 (1984).
48. P.Monceau, J.Richard and M.Renard, Phys.Rev.Lett. 45,43 (1980); Phys.Rev. B25,931 (1982).
49. S.E.Brown and L.Mihaly, Phys.Rev.Lett. 55,742 (1985).
50. S.Shapiro, Phys.Rev.Lett. 11,80 (1963); S.Shapiro, A.R.Janus and S.Holly, Rev.Mod.Phys. 36,223 (1964).
51. D.Jelčić, I.Batistić and A.Bjeliš, to be published.

COLLECTIVE EFFECTS IN CHARGE DENSITY WAVES

S. N. Coppersmith

Physics Department, Jadwin Hall
Princeton University
Princeton, NJ 08544 USA

INTRODUCTION

In this talk I will discuss the classical deformable model of charge density wave (CDW) transport, which has been investigated extensively since its introduction by Fukuyama and Lee (1978) and Lee and Rice (1979) (here referred to as FLR). In a one hour talk it is totally impossible to review the field adequately, so these notes will necessarily present a small (and biased) selection of the work that has been done.

First I will discuss how the model provides a framework that can be used to interpret consistently a wide range of experimental data using phenomenological parameters that can be estimated microscopically and that vary slowly as the experimental conditions are changed. The model can be used to understand experiments both above and below the threshold field over a wide range of frequencies. After introducing the model, I will discuss the existence of a threshold field, the dc current-voltage characteristic near threshold, the history dependence of the transient response to current and voltage pulses, and the response to the application of combined *ac* and *dc* fields ("Shapiro steps").

THE FLR MODEL

Fukuyama, Lee, and Rice model the CDW as an elastic medium in the presence of random impurities that is coupled to an externally applied electric field. It is assumed that the charge density $\rho(\vec{r}) = \rho_0 \cos(\vec{Q} \cdot \vec{r} + \phi)$ is weakly perturbed by the impurities, so that ρ_0 is basically constant and ϕ is slowly varying on the scale of the CDW wavevector $\vec{Q}$. The energy of the CDW has three contributions, one from the elasticity tending to make deformations energetically unfavorable, one from the impurity potential, which encourages deformations, and finally the contribution caused by the coupling to the externally applied electric field. Thus, the potential energy U is written:

$$U = \int d^3x \{ \tfrac{K}{2} |\Delta\phi|^2 + V(x) \cos(\vec{Q} \cdot \vec{r} + \phi(\vec{r})) - F\phi(\vec{r}) \} . \qquad (1.1)$$

Here, K describes the elasticity, V the impurity potential, and F the coupling to the externally applied field. In order to discuss the CDW current-carrying state, one must specify the dynamics of the phase variable. Sneddon, Cross, and Fisher (1982) (referred to here as SCF) introduced a very natural equation of motion in which the dynamic characteristics are contained in a damping constant and inertia term. They carefully took into account the fact that the correct coordinate frame to use is that which is moving along with the CDW. The equation of motion for the distortion $u(\vec{r}, t)$ of the CDW at point $\vec{r}$ at time t is:

$$m\ddot{u} + \gamma\dot{u} = -K\,\nabla^2 u + \frac{\partial V\,(\vec{r}+u\,(\vec{r})\hat{z})}{\partial z}\cos\vec{Q}\cdot\vec{r} + F\,. \tag{1.2}$$

Here, γ is the damping constant, and m is the inertia. Experimental measurements of the linear *ac* conductivity in zero field indicate that in the frequency range of up to 100 MHz the inertial term is totally negligible (see, e.g., Gill (1981a)). Therefore, m will be set to zero in the following discussion.

Equation (1.2) is, if we could solve it, all that is necessary to interpret a wide variety of experiments that probe the CDW's response to applied fields. However, it is a nonlinear differential equation coupling a very large number of degrees of freedom, and though some progress has been made, our understanding of the model is still incomplete. The main sources of insight have been perturbation expansions valid for very high driving fields, numerical simulations of discretized (mostly one-dimensional) models, and analytic and numerical work on modifications of the FLR model that have proven more amenable to analysis. Using the results from all of these approaches, a picture of the model's behavior has emerged that appears to be internally consistent and which has many features that appear also in the experimental data. Below, a few properties of the model are summarized.

EXISTENCE OF A THRESHOLD FIELD

Fukuyama and Lee demonstrated that this model has a finite threshold field in less than four dimensions (and it has been argued by Fisher (1985) that Griffiths-like singularities lead to a finite threshold field in any finite number of dimensions). By this one means that a stationary metastable state of the CDW exists even in the presence of a finite field. Once the field exceeds threshold, the CDW can no longer remain stationary and it starts to slide. The argument that a threshold field exists is qualitative, relying on the fact that if the CDW can lower its energy by distorting, then it finds a metastable state that is correlated with the impurities. When the field is applied, if there is no continuous path in configuration space to another low energy state, then the CDW remains pinned. We assume that no such path exists, so that the order of magnitude of the threshold field is determined by the energy gained by deforming.

Therefore, the question is whether it is energetically favorable for the CDW to distort in the absence of an applied field. Clearly, if the impurity potential is strong compared to the CDW elasticity, then deformations are energetically favorable, so the only nontrivial case is if the impurity potential is weak. Experimentally, the energy imparted by the applied field is always much smaller than the other characteristic energies in the problem (such as the Fermi energy, the Peierls temperature and the Debye temperature), so it reasonable that the weak pinning limit applies.

One thus considers a uniform state and asks whether a deformation with characteristic size ξ is energetically favorable (in zero field). The elastic energy cost of distortion per unit volume is K/ξ^2. Because the deformations are correlated with the impurity potential, the $V(\vec{r})$ term leads to a negative contribution to the energy proportional to the square root of the number of impurities in the region, so the energy gain per unit volume is $V_0 n_i^{1/2} \xi^{-d/2}$, where V_0 is the scale of the impurity potential and n_i is the impurity density. One then minimizes the total energy of the distorted state and finds that it is lower than that of the uniform state in all dimensions less than four. The length scale $\xi \sim (V_0^2 n_i/K^2)^{1/(d-4)}$ is a measure of the decay of phase correlations; I will refer to it as the FLR length.

By setting the energy gain equal to the energy imparted by an applied field, one finds that the threshold field depends on the impurity concentration n_i as $F_T \sim n_i^{(2/(4-d))}$ In principle, this dependence can be tested experimentally, and George Grüner has presented data at this summer school that is consistent with $F_T \propto n_i$. This result is consistent with both the strong pinning limit of FLR and the weak pinning limit with effectively two-dimensional fluctuations.

THE THRESHOLD FIELD: A CRITICAL POINT

The analytic and numerical results of many workers (see, e.g. Sokoloff, 1981, 1985; Fisher 1985; Littlewood (1986)) point to the conclusion that the threshold field F_T divides two qualitatively different phases. Most trivially, in the unpinned phase the CDW has a nonzero time-averaged velocity, while in the pinned phase the velocity is zero. In addition, in the pinned phase, exponentially many metastable states exist (basically, each region of linear dimension ξ has a finite number (> 1) of stable configurations). On the other hand, in the moving phase, the evidence indicates that only one steady-state solution to the equations of motion exists.

Intuitively, one can understand this result by noting that the elasticity imparts a negative feedback on the system. For instance, imagine that the pinning potential is much stronger in one section of the system. Then one would expect the CDW to move slower in that region. However, as the process proceeds the slow region falls farther and farther behind the rest of the CDW and eventually the elasticity "pulls" the slow region hard enough so that it moves with the same velocity as the rest of the CDW. Thus, one finds that for periodic boundary conditions, the system eventually settles into an apparently unique sliding state.

Further evidence that the threshold field is a critical point is provided by perturbation theory in powers of the impurity potential strength V_0 (SCF, 1982; Sneddon, 1984; Fisher,

1985). One finds that the lowest order correction to, for example, the current-voltage characteristic involves an integral of the form

$$V_0^2 \int d^d\, q\, \frac{1}{K^2 q^4 + v^2}.$$

This integral has an infrared divergence as the velocity v tends to zero in less than four dimensions. This type of behavior is reminiscent of that found in statistical mechanical systems near a critical point.

Further examination of this analogy has revealed both similarities and differences with the critical phenomena seen at second order phase transitions, though here too our knowledge is incomplete, since the threshold behavior of any three-dimensional model with short-range interactions has yet to be determined. Exponents are known for the zero-dimensional case, models with infinite-range interactions (mean field theory; see Fisher, 1983, 1985; Sneddon, 1984; Littlewood and Varma, 1986) and for a one-dimensional model with nearest-neighbor interactions in an incommensurate potential (Coppersmith, 1984; Coppersmith and Fisher, 1986). One finds nontrivial critical exponents for the current-voltage characteristic, the polarizability, and the a.c. conductivity, and the exponents appear to obey scaling relations. However, unlike in usual critical phenomena, one finds two diverging time scales (described by two different critical exponents), so any renormalization group description of the transition will have to be more complicated than that used at simple critical points.

Experimentally, most "well-behaved" samples exhibit a current-voltage characteristic near threshold that appears to be well-described by a power law:

$$I \propto (V - V_c)^{\varsigma}, \quad \text{with} \quad \varsigma > 1.$$

This result is not compatible with the zero dimensional model, (Grüner et al., 1981) and it is consistent with the theoretical result that ς tends to increase as the number of dimensions is raised, up to the mean field result $\varsigma = 3/2$. However, broad band noise, which is enhanced above threshold, is not a feature of the model. This noise may be associated with defects in the CDW or with contact effects.

The description of the critical behavior near threshold remains incomplete and offers many difficult and exciting theoretical challenges. The problem is an example of a bifurcation of a dynamical system with infinitely many nontrivial degrees of freedom, so studying it may yield insight into a large class of infinite-dimensional nonlinear equations.

METASTABILITY AND HISTORY DEPENDENCE

Gill (1981b) apparently was the first to notice that the transient response of a CDW to current pulses depends on the sample history. This observation is quite easy to reconcile with the FLR picture of CDW's. The logic is that the response depends on the initial conditions (e.g. the metastable state the CDW is in at the moment the pulse is applied), which in turn depend on the history of the sample. Figure 1 shows some numerical results of P. Littlewood (1986) on a one-dimensional discretized version of the FLR model. The CDW is started from a "virgin", nearly undistorted state. A voltage pulse is applied, and the CDW exhibits a slow transient response. When the voltage is removed, the CDW relaxes to a vastly different metastable state, with much more polarization built in. The response to the second pulse is much faster because the CDW starts out in a metastable state that is much closer to the eventual moving configuration.

Thus, the FLR model provides a simple and intuitive explanation for the history dependence of the CDW transient response. However, more detailed and quantitative characterizations of this behavior would sharpen the comparison to experiment.

MODE-LOCKING AND COMPETING PERIODICITIES

When the CDW moves with velocity v, an internal "washboard" frequency $\omega = \vec{Q} \cdot \vec{v}$ is generated, where Q is the CDW wavevector. This is true whether or not the CDW is effectively infinite in extent. If, in addition to the dc voltage that causes the CDW to move, one applies an ac voltage, then one sees the effects of two competing periodicities. In particular, there is a tendency for the frequencies to lock in, so that they are rational multiples of each other.

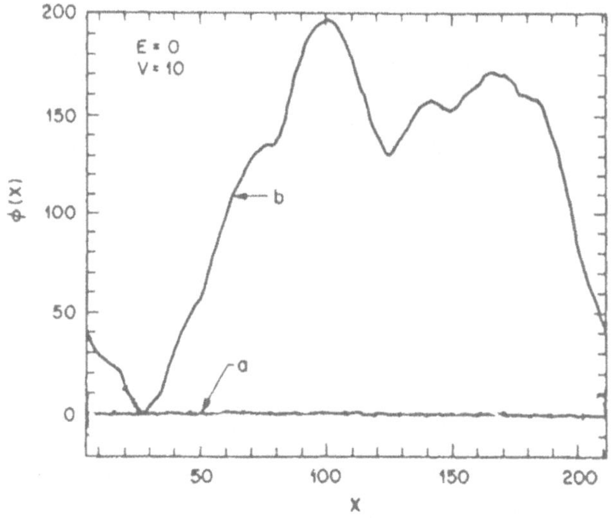

Figure 1: Phase of a 1-dimensional model of CDW's as a function of position. Plot (a) is a "virgin" state obtained by relaxation of a uniform configuration. Plot (b) is the metastable state reached after application of a voltage pulse just above threshold.

This type of effect was observed in the early 1960's in Josephson junctions (Shapiro, 1963), and many experiments on CDW's have revealed qualitatively similar phenomena (see e.g., Brown, Mozurkewich and Grüner, 1984). However, close examination of the experiments reveals features that are not simply explained in terms of a single driven nonlinear oscillator. For instance, figure 2 shows the differential resistance as a function of dc voltage for two different frequencies (Brown and Gruner, 1986). The curves are clearly qualitatively different; the differential resistance curve taken at 25 MHz appears to be a series of rather smooth peaks, while the 6.4 MHz curve exhibits a much more jagged structure, with "locked" steps, or regions where dV/dI takes on the value associated with the uncondensed electrons only.

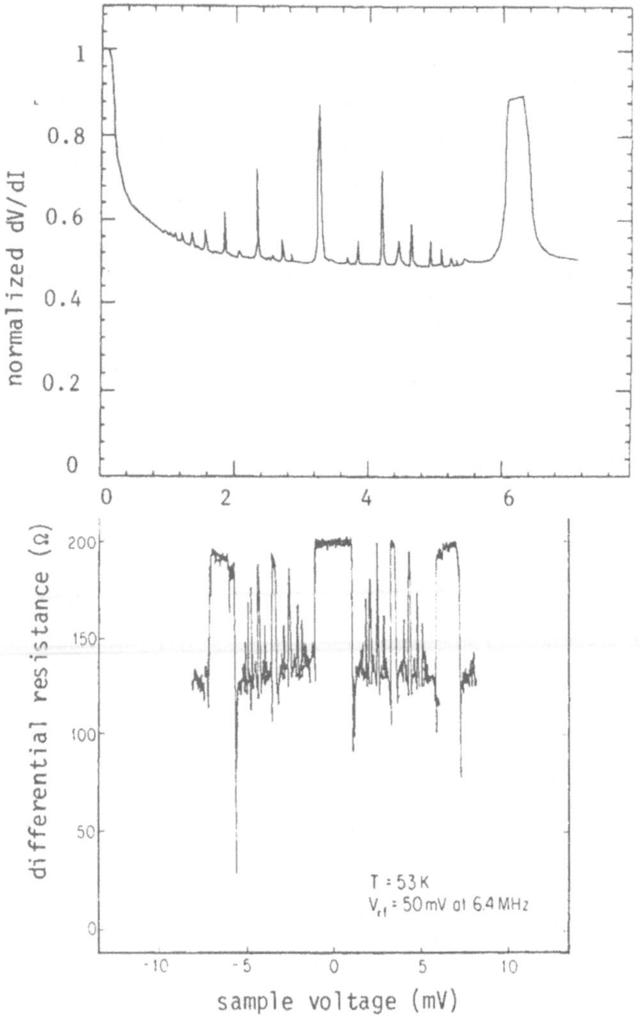

Figure 2: Plots of differential resistance versus dc voltage for fixed ac amplitude and frequency.
a) ac frequency = 25 MHz, ac amplitude = 100 mV
b) ac frequency = 6.4 MHz, ac amplitude = 50 mV

Peter Littlewood and I have examined the implications of the SCF equation of motion to this type of experiment. Our main result is that the FLR model actually predicts that figures 2a and 2b should be different. We find that whether the CDW locks with the driving frequency or not depends crucially on the amount of time the CDW spends below threshold. If the CDW spends enough time below threshold to relax substantially towards a metastable state, then locking is observed, but if the frequency is high then sufficient relaxation does not occur and only peaks in dV/dI are observed. We have also found that at very high frequency and ac amplitude the high order subharmonics are suppressed relative to the low order ones.

That the threshold field plays a special role in mode locking was noted experimentally by Brown et al. (1986). More recently, Bhattacharya et al. (1986) studied the systematics of mode locking as the amplitude of the ac field is varied. They found that for very small amplitude, when the applied field is always greater than threshold, then mode-locking does not occur. For intermediate amplitude ac fields, mode-locking is observed, but when the ac amplitude is made very high, then once again smooth peaks in dV/dI are observed rather than true locking. This "reentrant" behavior is seen to be consistent with the predictions of the FLR model, and is hard to reconcile with simple single-particle equations of motion.

A SIDE NOTE

There are two separate issues that need to be addressed when evaluating the classical deformable model. The first is what the microscopic acceleration and damping mechanisms are, and whether they are truly constant over the experimental range, and the second is whether the phase variations play the vital role in determining the dynamic response. I espouse the viewpoint that the second question is the "interesting" one, if only because the SCF equation of motion is certainly vastly oversimplified, so that the microscopic damping constant, for instance, will depend on frequency, temperature, and applied field. Whether the model is useful depends on whether these variations are slow and smooth. For instance, the differential resistance near threshold can be fitted using a one-particle model, but only if the damping or effective charge exhibits a singularity near the threshold field. This situation will be considered unacceptable. On the other hand, the high-field asymptotic behavior of the current-voltage characteristic could be changed substantially by small, smooth changes in the damping constant. I consider this an "acceptable" variation, since my main concern is with qualitative properties of the model that can be correlated with qualitative features of experiment. An extremely large quantity of data on CDW's can be fit using an "effective" one particle equation of motion with parameters and an external noise term that depend on voltage, frequency, etc. The work described here could be viewed as a means to calculate the effective parameters of the simple theory starting from a hydrodynamic model. Our results indicate that the parameters and the noise have nontrivial dependences on the experimental conditions.

CONCLUSIONS

The main conclusion I wish to make is that the FLR model does have some relevance to CDW experiments as well as theoretical interest in its own right. I hope I have given a flavor of the type of results that have been obtained as well as the large number of unsolved problems that remain.

REFERENCES

Bhattacharya S., Stokes J., and Higgins M., 1986. To be published.

Brown S., and G. Grüner 1986, unpublished.

Brown S., G. Grüner, and Mihaly L., *Sol. St. Comm* **57**, 165 (1986)

Brown S., Mozurkewich G., and G. Grüner, 1984. *Phys. Rev. Lett.*, **52**, 2277.

Coppersmith S.N., and Fisher D.S., 1984. *Phys. Rev.* **B 30**, 410.

Coppersmith S.N. and Fisher D.S., 1986. To be published.

Fisher D.S., 1983. *Phys. Rev. Lett.*, **50**, 1486.

Fisher D.S., 1985. *Phys. Rev.* **B 31**, 1396.

Fukuyama H., and Lee P.A., 1978. *Phys. Rev.* **B 17**, 535.

Gill J.C., 1981a. *Sol. St. Comm.* **37**, 459.

Gill J.C., 1981b. *Sol. St. Comm.* **39**, 1203.

Grüner G., Zawadowski A., and Chaikin P., 1981. *Phys. Rev. Lett* **46**, 511.

Lee P.A., and Rice T.M., 1979. *Phys. Rev.* **B 19**, 3970.

Littlewood P.B., and Varma C.M., 1986. To be published.

Littlewood P.B., 1986. *Phys. Rev.* **B 33**, 6694.

Shapiro S., 1963. *Phys. Rev. Lett.*, **11**, 80.

Sneddon L., Cross M.C., and Fisher D.S., 1982. *Phys. Rev. Lett.*, **49**, 292.

Sneddon L., 1984. *Phys. Rev.* **B 29**, 719,725.

Sneddon L., 1984 *Phys. Rev.* **B 30**, 2974.

Sokoloff J., 1981. *Phys. Rev.* **B 23**, 1992

Sokoloff J., 1985. *Phys. Rev.* **B 31**, 2270.

MOVING CHARGE-DENSITY WAVES: THE LOW FREQUENCY RESPONSE

R. M. Fleming

AT&T Bell Laboratories
Murray Hill, N. J. 07974
U.S.A.

INTRODUCTION

Since the original observation of charge-density waves (CDW's) in layered compounds in 1975,[1] a number of interesting CDW phase-transitions and physical phenomena related to CDW's have been observed. These include commensurate-incommensurate transitions,[2,3] the observation of a "striped" CDW phase[3] and unusual first-order transitions to a metallic state.[4] In all of these materials the phase of the CDW is determined solely by the impurity potential and the temperature. No extra conductivity from field-induced motion of the CDW is observed. In 1976 however, Monceau and co-workers observed that the conductivity of $NbSe_3$ is field dependent at temperatures below each of the two CDW phase transitions in that material.[5] It is now believed that the enhancement of the conductivity is a result of a charge-density wave that moves in the applied electric field, a phenomenon now seen in about ten compounds.[6–9] The non-linear electrical response resulting from CDW motion can be observed in several different ways. For example, the dc response shows increased conductivity at fields above an impurity dependent threshold field, E_T. The low-frequency ac conductivity shows a broad increase in the conductivity that can be crudely approximated as the overdamped response of a pinned mode, however, more detailed measurements have shown that the response is better described with a distribution of pinning energies. The ac conductivity remains enhanced for many decades of frequency but eventually rolls off for frequencies on the order of 100GHz. A moving CDW also exhibits several surprising features not expected from simple models. At least in small samples, the average motion of the CDW is not uniform. Even for a dc bias, the electrical response may contain both a periodic and a broad-band component.[10] There is general agreement that the periodic response to a dc bias reflects the periodicity of the CDW, however there is controversy over the origin of the response.[11,12] Does the periodic response come from the bulk or is it generated at the contacts?

A second unexpected feature of CDW motion is metastability. Experimentally, one finds that the ohmic and the non-linear response are a function of the electrical and the thermal history of the sample.[13–19] Such hysteresis would not be expected if the CDW moved without deformations, i.e. if the CDW had only a single degree of freedom. After moving, the pinned CDW never relaxes completely to the ground state. Instead, it finds a metastable state that depends on the previous history of the sample. As shown in Sec. 3, a metastable state is structurally characterized by local deformations in the phase of the CDW. A metastable state results in a macroscopic polarization of the sample and the decay of this polarization may be infinitely long. Depending on the temperature, the dipole moment associated with the polarization may be screened by the normal carriers.

I will also address the low-frequency dynamic response of the charge-density wave and discuss the relationships between CDW dynamics and metastable states. The term "low-frequency" means frequencies well below the mean pinning frequency of the CDW, frequencies on the order of the mean dielectric relaxation frequency. Dielectric relaxation frequencies in materials with moving CDW's are temperature dependent, but are typically less than 10MHz. I will also describe the influence of impurities on metastable states and low-frequency CDW dynamics. I will primarily discuss our results on the "blue-bronze", $K_{0.30}MoO_3$, however, the conclusions are applicable to all materials with moving CDW's.

MODELS OF CHARGE-DENSITY WAVE MOTION

Both classical[20,21] and tunneling models[22] have been proposed to describe the motion of a CDW in an electric field. The models differ, of course, on a microscopic level in describing the coupling of an electric field to the CDW condensate, however for the most part, the experiments discussed in this paper do not directly address the microscopic issues. The two models in their later forms also differ on a macroscopic level. The classical models generally describe the motion of a deformable object and the many degrees of freedom are incorporated into the problem by allowing the phase of the CDW to distort. Although most present classical models only consider phase fluctuations of the CDW, amplitude fluctuations could also be treated as well. The tunneling model in its original form has only a single degree of freedom and would have difficulty describing the metastable, long-time phenomena. However, more recent applications of the tunneling theory invoke coupled domains of CDW (each with a single degree of freedom) and this description presumably could show metastability .[23] In this paper I will only compare the data with classical models of CDW transport, however, in at least one other paper,[24] our data have been also fit with models based on the tunneling theory.

The simplest classical description of CDW motion is a uniform pinning model[25] where the random impurity pinning potential is replaced by a sinusoidal potential of constant amplitude, V_o and wave vector **Q**. The uniform pinning model describes the motion of a rigid CDW that moves with no distortions. This model has only one degree of freedom, the position of the CDW, **u**, and is sometimes referred to as a "single particle model". The equation of motion for a CDW moving in a uniform, periodic potential may be written as

$$m^{*}\ddot{\mathbf{u}} + \lambda_o\dot{\mathbf{u}} + V_o\sin(\mathbf{Q}\cdot\mathbf{u}) = \rho_c E. \qquad (1)$$

Here m^{*} is the collective mass density of the CDW, λ_o is the effective CDW viscosity, V_o is the pinning force per unit volume, ρ_c is the collective CDW charge density and E is the electric field. In the limit of small displacements where $\sin(\mathbf{Q}\cdot\mathbf{u}) \approx |Q|u$, one can derive an expression for $\sigma(\omega)$ by taking the Fourier transform of Eq. 1 and by writing the CDW polarization as $P(\omega) = \rho_c\mathbf{u}(\omega) \approx \epsilon(\omega)E/4\pi$. Combining the two equations and using the relation $\sigma(\omega) = i\omega\epsilon(\omega)/4\pi$ leads to

$$\sigma(\omega) = \left[\frac{\rho_c^2}{m^*}\right] \frac{i\omega}{[\omega_o^2 - \omega^2 + i\omega/\tau_o]} \tag{2}$$

where the pinning frequency is given by $\omega_o^2 = |Q|V_o/m^*$ and the inertial damping frequency by $1/\tau_o = \lambda_o/m^*$. In the overdamped limit where there is no inertia ($m^* \approx 0$), it useful to rewrite Eq. 2 as the dielectric constant

$$\epsilon(\omega) = \left[\frac{4\pi\rho_c^2}{|Q|V_o}\right] \frac{1}{1 + i\omega\tau_R} \tag{3}$$

where $\tau_R = \lambda_o/|Q|V_o$. For comparison with experiment, the pinning potential V_o can be approximated as $V_o \approx \rho_c E_T$. Equation 3 is the form a Debye relaxation originally used to describe the dielectric relaxation of polar molecules. The relaxation frequency $1/\tau_R$ marks the point where Re ϵ relaxes from its zero-frequency value $\approx 4\pi\rho_c/|Q|E_T$ to zero. There is a corresponding peak in Im ϵ at $1/\tau_R$. For the case of the conductivity described by Eq. 2 with small but finite effective mass, the conductivity first rises at the relaxation frequency $1/\tau_R$ and saturates with a maximum value of $\sigma_{max} = \rho_c^2/\lambda_o$. The conductivity remains saturated until the inertia forces a roll-off of the conductivity at a frequency of $1/\tau_o$. One expects a peak in the imaginary part of σ at the relaxation frequency and a negative peak in Im σ at the roll-off frequency. Note that the relaxation frequency can be expressed in terms of the roll-off frequency and pinning frequency, $1/\tau_R = \omega_o^2\tau_o$. In the literature $1/\tau_R$ is sometimes referred to as the "cross-over frequency".

Another attraction of the uniform pinning model has been its ability to produce an ac response to a dc bias. The effect of an electric field is to "tilt" the periodic pinning potential and allow the CDW (thought of here as a "particle") to "roll down" the corrugated potential. One can intuitively see that such motion will have a periodic component, the so-called "washboard frequency" of the CDW. The washboard frequency, given by $\mathbf{Q}\cdot\mathbf{v}$ has been associated with the periodic response to a dc bias discussed in Sec. 1.[26] The periodic response is a measure of the CDW velocity, but the "washboard" mechanism has difficulty explaining the periodic response of a real material where the pinning is random. Even if one ignores other difficulties with the uniform pinning model, this description would only be valid in the limit that the entire sample is one domain with motion described by a single coordinate. In the thermodynamic limit of large samples, one would not expect a bulk mechanism to produce a periodic response to a dc bias.

Early low frequency measurements indicated that Eq. 3 was a reasonable approximation of the ac conductivity.[27,28] More recent measurements have shown that the uniform pinning model has several serious deficiencies that make it inappropriate for any detailed description of CDW motion.[29–31] The most dramatic weakness is the inability for a model with a single degree of freedom to account for the metastable and history-dependent phenomena described in Sec. 3. Also, the model fails to predict the correct shape of the dc current-voltage (I-V) response. Experimentally, I-V curves have upward curvature ($d^2I/d^2V > 0$) but Eq. 1 leads to an I-V with a singularity at E_T and downward curvature. In addition, the uniform pinning model only describes the gross features of the frequency dependent conductivity. As will be disussed in Sec. 4, Eq. 3 fails to adequately describe the low-frequency ac response. The problem relates to the inclusion of only one degree of freedom, and thus only one relaxation time in Eq. 3 and one pinning energy in Eq. 1. As we will see in the following discussion, a better description of CDW motion describes a CDW that moves with deformations. Time-dependent CDW fluctuations lead to screening currents, a subject discussed more detail in Scc. 5.

We therefore see that the uniform pinning model contains at least three assumptions which are an over-simplification of the actual CDW response. The first assumption is uniform, periodic pinning rather than random pinning. The second assumption is the inclusion of a single pinning potential, V_o, rather than a distribution of pinning energies. The third assumption is the use of the applied electric field in Eq. 1 rather than a field modified by screening. Despite these problems, the uniform pinning model is a useful pedagogical tool for understanding the basic issues of CDW transport, particularly since more appropriate theories such as those discussed below only supply analytic expressions for the CDW conductivity in limiting cases.

A more realistic treatment of impurity pinning has been described by Fukuyama and Lee[20] and by Lee and Rice[21] (FLR). FLR consider pinning by random impurities and allow for phase fluctuations of the charge-density wave while ignoring amplitude fluctuations. This approach is different from the uniform pinning model since it describes the motion of a deformable object. The FLR model inherently allows for many degrees of freedom and thus it can be used to describe a hysteretic response. FLR write a Hamiltonian as

$$H = \int \left[\frac{1}{2}\kappa|\nabla\phi|^2 + \sum_i V_i(\mathbf{r}-\mathbf{R}_i)\rho(r) - \rho_c E\phi(\mathbf{r})/Q \right] d^3r. \quad (4)$$

Here κ is the CDW elasticity, V_i is the pinning of an impurity at R_i and $\rho(r) = \rho_o\cos[\mathbf{Q}\cdot\mathbf{r}+\phi(\mathbf{r})]$ is the CDW modulation. Because of the non-linear nature of the response, Eq. 4 does not easily lead to a general analytic solution for $\sigma(\omega)$. Several workers[30–32] have investigated the dynamics using the overdamped equation of motion

$$\lambda_o d\phi/dt = -\partial H/\partial\phi \quad (5)$$

in the limit of large electric fields >> E_T. A mean-field theory valid for many degrees of freedom has also been proposed.[35] In addition, numerical models based on the response of a CDW moving in a one-dimensional FLR Hamiltonian using a variety of pinning potentials have also been proposed to describe the CDW response at intermediate fields.[36–41] Some of thc numerical results will be presented in the next section.

METASTABLE STATES

Metastability in the charge-density wave response was first recognized by Gill in 1981 who measured the sample resistance following a pulse large enough to depin the charge-density wave.[13] Gill noticed that whenever the sign of a pulse was reversed, the sample conductivity at the leading edge of the pulse was initially higher than the steady-state value. This manifestation of metastability is now known as the "pulse sign memory effect". The conductivity decay occurs on time-scales of tens of microseconds in $NbSe_3$[13,14] and time-scales of a several hundred microseconds in $K_{0.30}MoO_3$.[16] The time required to lose the memory of the current direction is a function of temperature and it can exceed many hours at low temperatures. Gill proposed that the structure of a material with a moving CDW was metastable, it did not decay to the ground state following a pulse but remained in a metastable state that was structurally different for the two current directions.

Experimentally, metastability also takes other forms. For example, the pulse sign memory effect results in large "capacitive" hysteresis in low-frequency I-V traces.[16] If the maximum value of the voltage exceeds threshold, the I-V curves are in the shape of distorted loops that are traversed in the same direction as loops seen in a capacitive circuit. Loops in the I-V response can be observed at extremely low frequencies, as low as a few Hz in $K_{0.30}MoO_3$. The hysteresis is the largest when the bias is bi-polar and crosses both positive and negative threshold. Much less hysteresis is seen for uni-polar I-V curves and no hysteresis is seen for I-V curves that do not cross threshold. As we will see later, these loops arise because of a depolarization current as the CDW switches from one polarization state to another.

One can also view hysteresis by observing the behavior of the material after a current or a temperature pulse. For example, if $K_{0.30}MoO_3$ is pulsed from a negative current less than the negative threshold to a positive current greater than the positive threshold, the conductivity (which is initially larger than the steady-state value) decays logarithmically in time.[16] Similarly, if one pulses the temperature of TaS_3 and observes the time dependence of the low-field conductivity, one finds that $\sigma \propto \log(\text{time})$.[17] The latter experiment illustrates that hysteresis is not only associated with the transient properties of the moving CDW state, it is also observed in the normal-state resistance. This feature has also been observed in thermal hysteresis experiments in TaS_3.[18] The normal state conductivity of TaS_3 is different for warming and cooling. At any temperature, if the field is pulsed to a high value, one obtains a conductivity mid-way between the warming and cooling value.

For other experimental configurations, time dependencies other that logarithmic can be observed, e.g. in depolarization experiments a depolarization current proportional to a stretched exponential $I \propto \exp\{-(\text{time})^n\}$ has been observed.[19] The appearance of logarithmic time dependencies or time dependencies related to a sum over exponentials suggests that metastability can be schematically represented by the potential well shown in Fig. 1. The potential well is characterized by a succession of local minima. The closer one moves to the ground state, the deeper the local minima become. Of course, Fig. 1 represents only one degree of freedom, so the real energy surface will have a separate branch of the type shown in Fig. 1 for each degree of freedom. One can see that this will lead to a large or even an infinite number of metastable states.

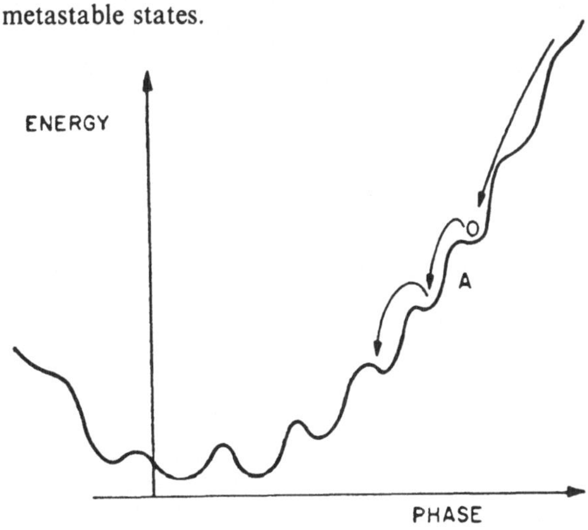

Fig. 1. Schematic view of a metastable energy surface showing the relaxation of one degree of freedom. As the ground state is approached, the barriers get progressively higher. (From ref. 36)

We have thus far discussed some of the experimental manifestations of metastable states. We have seen that metastable states result in transient changes in the moving state and long term changes in the properties of the pinned CDW. We now turn to a description of some experiments that give information on the structure of metastable states. The first experiment, which shows that a metastable CDW state is characterized by a frozen dipole moment, is known as thermally stimulated depolarization (TSD).[42] TSD experiments were originally performed on insulating polymers and are used to measure the thermal relaxation of induced dipole moments. (Materials that have field induced dipole moments that thermally relax are known as "electrets".) The basic TSD experiment consists of inducing a dipole moment by applying an electric field and then cooling the sample in a constant field. At low temperatures the field is removed, but the dipole moment remains because of a frozen-in polarization causing image charges to collect on the electrodes at each end of the sample. As the sample is warmed and the induced dipoles thermally relax, the image charges flow through the external circuit producing a small current (typically $\sim 10^{-11}$A in non-CDW materials). One can use TSD to measure of the induced moment (by measuring the integrated current) and the activation energy of the dipole relaxation (by measuring the temperature dependence of the TSD current).[43]

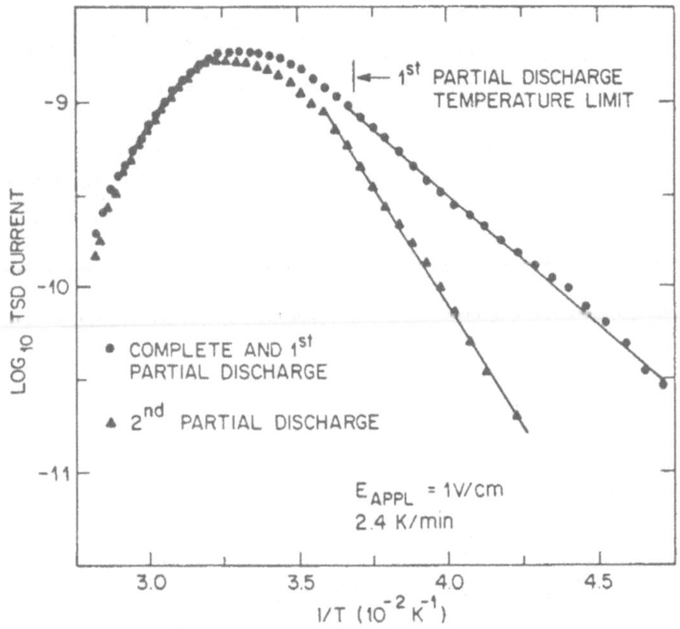

Fig. 2. The temperature dependence of the discharge current from a field-induced polarization of the CDW in $K_{0.30}MoO_3$ measured at a constant heating rate. At low temperatures, the current is thermally activated (330K). If the sample is only partially discharged and re-cooled with E=0, a larger activation energy is obtained (600K) indicating a distribution in the relaxation times of the CDW polarization. (From ref 42)

True TSD experiments can only be performed on insulating materials however all CDW conductors are semiconductors except for $NbSe_3$ which is semimetal. In semiconductors, the TSD experiment is complicated by the presence of the activated carriers, however at low temperatures where few carriers are present, one can still perform the measurement. The presence of conduction electrons at higher temperatures means that one cannot measure a magnitude of the induced moment since the image charges can relax by flowing through the sample as well as the external circuit. Nevertheless, one can use TSD to show that a metastable CDW state is characterized by a macroscopic polarization and one can also use the temperature and the time dependence of the TSD current to obtain information on the distribution of metastable states. One finds that in $K_{0.30}MoO_3$, TSD currents are large (10^{-9}A in a sample volume of $10^{-3}cm^3$) implying enormous polarizations when one considers that only part of the induced image charges relax through the external circuit. Furthermore, one can use TSD to measure the activation energy of the dipole relaxation process as shown in Fig. 2. At low temperatures, the relaxation of the induced dipole moment shows activated behavior, independent of the heating rate. If the discharge of the induced moment is interrupted before completion and the sample is cooled again with no additional field, a second activation energy higher than the first is obtained. This is direct evidence that the decay of metastable states in CDW materials is associated with a distribution of energies, a feature that will be further emphasized in the discussion of the low-frequency ac conductivity. One should not infer from Fig. 2, however, that the CDW may be thermally depinned. As we will show later, all relaxation processes in CDW materials may be related to the flow of screening currents. The activation energies obtained are therefore related to the Peierls gap in the normal electron band structure.

Strictly speaking, metastable states are properties of the pinned CDW. One expects that the equilibrium configuration of the moving CDW is unique and is only a function of the electric field. However, at low temperatures where the CDW relaxation time is long, the evolution to dynamic equilibrium can be long. This can result in non-equilibrium properties of the moving state that are a function of the starting phase configuration.[44,45] One example of hysteresis in the non-equilibrium, moving state is the "pulse-duration memory effect" seen both experimentally in $K_{0.30}MoO_3$[44] and in numerical simulations.[36] There is presently no explanation of the mechanism of the pulse duration memory effect, although it is probably related to the mode-locking seen in other CDW materials.[46] The effect is characterized by an anomalous oscillatory response that depends on the duration of the previous pulse.

Early attempts to observe the structure of a metastable state by x-ray scattering were unsuccessful.[47] These experiments on $NbSe_3$ were only able to probe the longitudinal coherence parallel to the directions of the non-linear charge transport because of the flexible morphology and the large mosaic spread of $NbSe_3$ crystals. The experiments showed that despite the pinning of the CDW by impurities, the CDW has long-range order (> 4000Å) in a direction parallel to the non-linear current flow. No field-induced change in the longitudinal line widths was detected. Later experiments on $K_{0.30}MoO_3$ confirmed that the charge-density wave retains longitudinal long range order in the presence of an electric field.[48] Unlike the transition metal trichalcogenides, $K_{0.30}MoO_3$ samples are of excellent quality with small mosaic spreads and x-ray scattering in the transverse direction of $K_{0.30}MoO_3$ showed that there is a metastable, field-induced loss of order.[48–50] For a sample that is cooled in a field, the transverse coherence can be as short as 700Å. Figure

3 shows the x-ray analogy of a TSD experiment where the transverse width as a function of temperature is measured in a field-cooled specimen. For temperatures less than about 100K, the relaxation time from metastable states exceeds hours and an excess transverse width is observed in the x-ray scans (which take about an hour). Above 100K the relaxation time is short and the zero-field width is recovered before an x-ray scan is completed. The field induced disorder and subsequent broadening is inhomogeneous, i.e. the disorder is characterized by local motion of the CDW rather than a microscopic loss of long range order. If one applies asymmetric contacts, the peak broadening will be asymmetric or even shifted in a transverse direction.[49,50]

Numerical simulations have provided insight in understanding the structural character of a metastable state. Using one-dimensional numerical models of the FLR Hamiltonian, Littlewood has shown that the application of a field below threshold produces local motion of the CDW.[36] Figure 4 shows the CDW phase plotted as a function of a transverse direction in the sample for several electric fields (labeled in the right portion of the figure). The horizontal axis represents a direction transverse to the CDW motion, the vertical axis represents a direction parallel to the CDW motion. The initial ground state is represented by a uniform phase difference of zero. Note that the phase of the CDW begins to move locally as

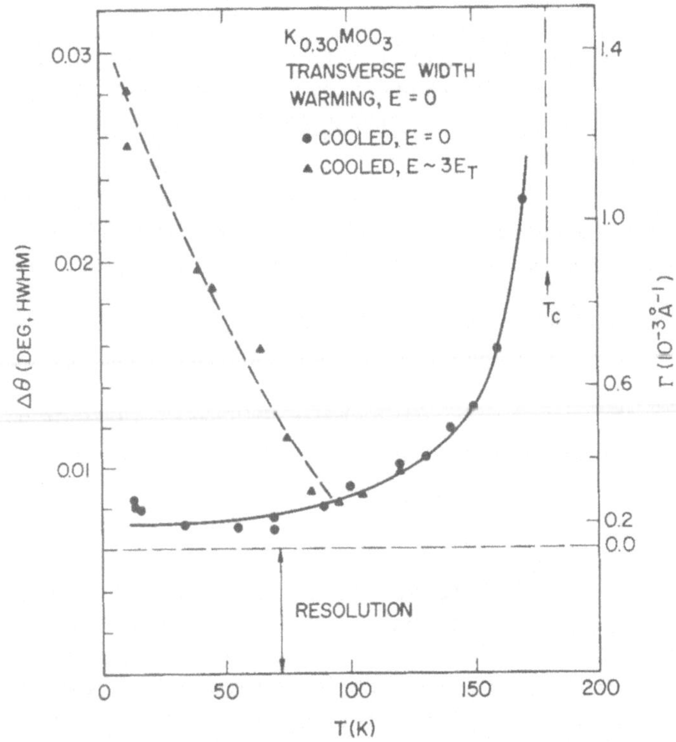

Fig. 3. The temperature dependence of the transverse width of an x-ray superlattice peak of the CDW in $K_{0.30}MoO_3$. Cooling in a field introduces a loss of order in a direction perpendicular to the non-linear current flow. Data taken on warming after field-cooling (triangles) show that the loss of order is metastable at temperatures below about 100K. (From ref 48)

soon as the field is applied. This is a result of differences in the local pinning strength between one part of the sample and another. If the field is removed at any time, the CDW phase does not relax the ground state, instead it remains pinned in a distorted state. A snapshot of the moving state is shown for an electric field E = 0.025 (solid line) and the pinned state reached if the field is turned off (dashed line). Note that the polarization that was established in the pinned states does not continue to grow while the CDW is moving. In other words, differences in pinning allow for the local motion of one part of the CDW at fields below threshold, but once the CDW is moving, all parts of the sample move with the same *average* velocity. (We will see later that fluctuations are important in the moving phase.) From Fig. 4 one can see that a metastable CDW state corresponds to a distortion of the phase of the pinned CDW. These phase distortions result in charge displacements and a induce a macroscopic polarization of the sample. In addition, the distortion shown in Fig. 4 causes an inhomgeneous loss of order in the transverse direction in agreement with x-ray scattering measurements. This picture of metastability implies that many if not all the unusual non-linear effects seen in materials with moving CDW's are a result of the *lack* of quasi-one-dimensional ordering and the transverse coupling present in these materials.

The previous discussion assumes only phase fluctuations of the CDW are important. Experiments have shown that in some samples amplitude fluctuations likely play a role. A class of materials where amplitude fluctuations are probably important are the so-called "switching samples".[51–54] Switching occurs in at least some of the samples in all the materials with moving CDW's.[8] In switching

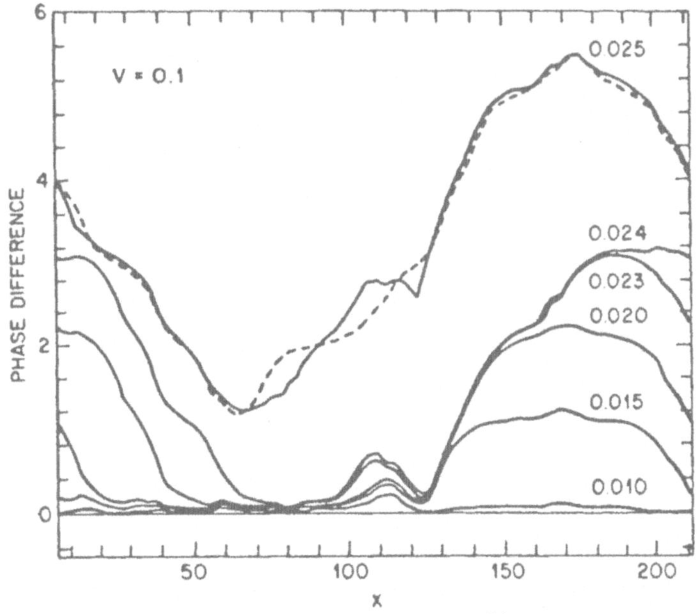

Fig. 4. The results of numerical calculations[36] showing the phase of the pinned CDW as a function of a dimension perpendicular to the non-linear current flow for varying electric fields. The top curve is a snapshot of a moving CDW showing the pinned configuration (dashed line) obtained by removing the field.

samples, the CDW depinning introduces a discontinuity in the I-V curve, a "switch" from low field conductivity to high-field conductivity. Several switches can sometimes be seen in the same sample and sometimes a normal, continuous threshold and a switch can be observed in the same sample. Switching has been identified as coming from a phase-slip center, a local area in the sample where the CDW amplitude is locally driven to zero.[55] Switching samples have double valued I-V curves in the region of the switch and considerable voltage or current fluctuations.[9] Amplitude fluctuations probably play an important role in switching samples, however, it is not presently clear whether or not they are important in other samples as well. There is evidence that phase-slip centers can be generated with at steep temperature gradient,[56] and Lee and Rice[21] have used a model of CDW vortices pinned to impurities as a model for producing phase slip. However, the discussions in this paper assume that only phase fluctuations are important in describing low-frequency CDW dynamics. The experimental observations will be limited to samples where switching is plays a minor role in the low-frequency CDW dynamics.

The identification of a metastable state with phase distortions of the pinned CDW allows one to understand the pulse sign memory effect discussed earlier. Figure 4 shows that the pinned CDW reached after the application of a field is characterized by a polarization resulting from local motion. If we start from the pinned configuration obtained after moving the CDW in one direction and reverse the field, we assume that the final phase configuration of the moving state will look like the dashed line reflected through the horizontal axis. If the reversed field is

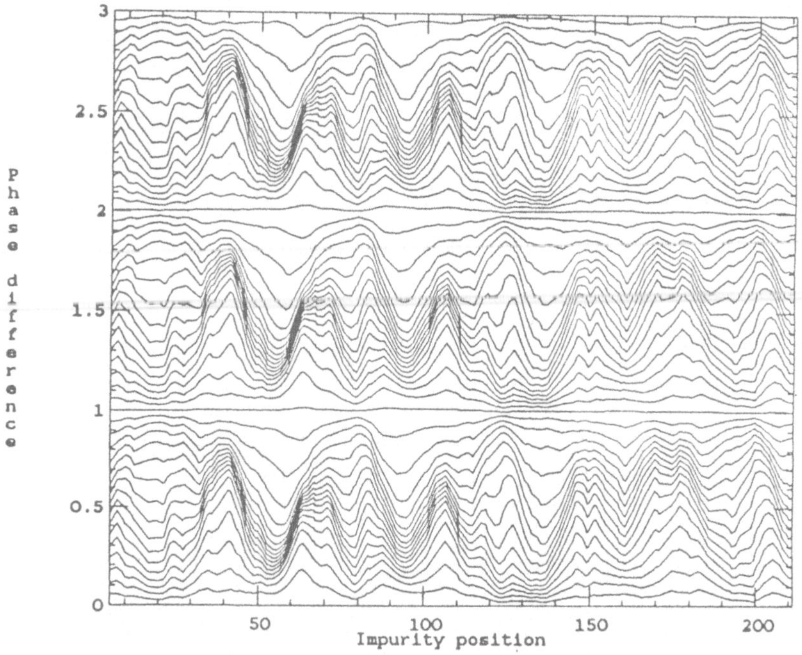

Fig. 5. The results of numerical calculations[36] showing the phase of the moving CDW as a function of a dimension perpendicular to the non-linear current flow. Each curve is a snapshot of the moving CDW taken at equal time intervals. The temporal periodicity comes from the finite size of the numerical calculation. Note that dc motion of the CDW is accompanied by large fluctuations.

abruptly turned on, the moving state will initially have a polarization that is in the opposite sense to the CDW motion. The current will therefore have an extra contribution, dP/dt, coming from the change in the polarization from one direction to the other. The time scale for the polarization reversal will be the dielectric relaxation time, τ_R, mentioned in Eq. 2 and discussed in more detail in Sec. 4.

The numerical models also give insight into the low frequency dynamics of the moving CDW. Figure 5 shows a sequence of snapshots of the CDW phase measured in units of 2π versus the transverse position. The snapshots are taken at regular times and show the spatial as well as the time evolution of the CDW phase in response to a dc voltage above threshold. The most striking feature of Fig. 5 is the temporal periodicity of the CDW motion, however this occurs because the numerical simulation is performed on a finite size sample. (Note, however, that the simulation shows no broad-band noise. After the phase is advanced by 2π, exactly the same phase configuration is obtained.) The principal point of Fig. 5 is to illustrate that the CDW moves forward by local motion, first in one area of the sample and then another. This means that at any point in the sample, the gradient of the phase (and hence the local CDW energy gap) is constantly changing. A local change in the CDW energy gap will result the flow of normal electrons into or out of the area. This leads to the important conclusion that dc motion of the CDW at moderate values of the field is accompanied by fluctuations and screening currents. We will see in the next section that electron screening is also important in understanding the temperature and the frequency dependence of the conductivity.

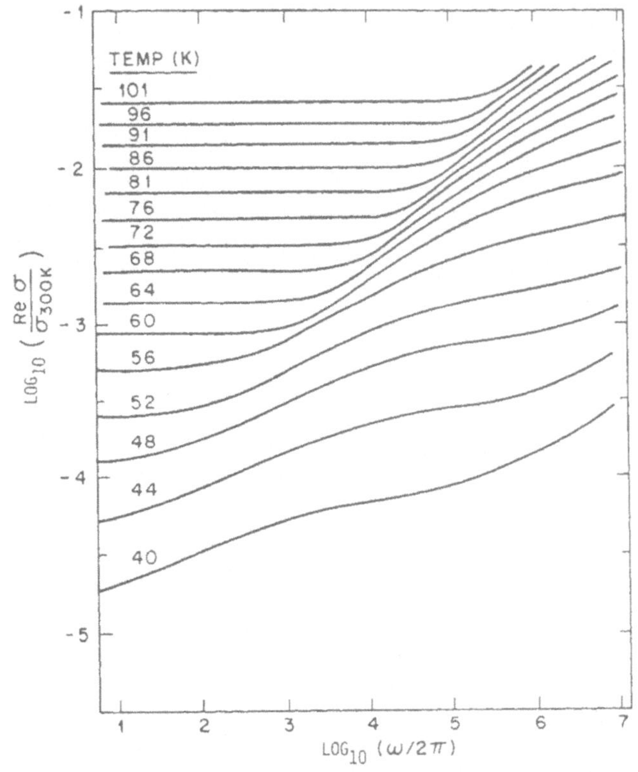

Fig. 6. The real part of the conductivity in $K_{0.30}MoO_3$ as a function of frequency for temperatures between 40 and 101K. (From ref. 29)

The low-frequency response predicted by the uniform pinning model is given by Eq. 3. From Eq. 3, one expects that Re $\epsilon \sim$ constant at low frequencies and Re $\epsilon \sim 0$ at frequencies above $1/\tau_R$ with a peak in Im ϵ peak at $1/\tau_R$. Re σ will rise and saturate at $1/\tau_R$ with a peak in Im σ at the same frequency. The real and the imaginary parts of the conductivity of $K_{0.30}MoO_3$, shown in Figs. 6 and 7 for several temperatures and a frequency of 5Hz - 13MHz,[29] are quite different from the prediction. There is no plateau of Re σ for frequencies above a relaxation frequency and there is no clear peak in Im σ. As discussed in Sec. 3, the problem occurs because the true dynamic response of the CDW is described by a Hamiltonian with many degrees of freedom.

A proper comparison of the ac conductivity with theory would compare the data with an expression for the conductivity derived from the FLR Hamiltonian (Eq. 4,5). As mentioned in Sec. 2, analytic expressions for $\sigma(\omega)$ based on Eq. 5 only exist in perturbation theory for certain limiting values. One must therefore turn to empirical descriptions of the ac response to parameterize the data. Consider the case where the CDW pinning is characterized by a distribution of pinning energies. Since $\tau_R = \lambda_o/|Q|V_o$ from Eq. 2,3, this is equivalent to saying that an induced polarization decays with a distribution of relaxation times, $g(\tau)$. One could write the time-dependent polarization as

$$P(t) = P(0)\int_0^\infty g(\tau)e^{-t/\tau}d\tau. \qquad (6)$$

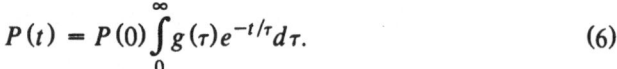

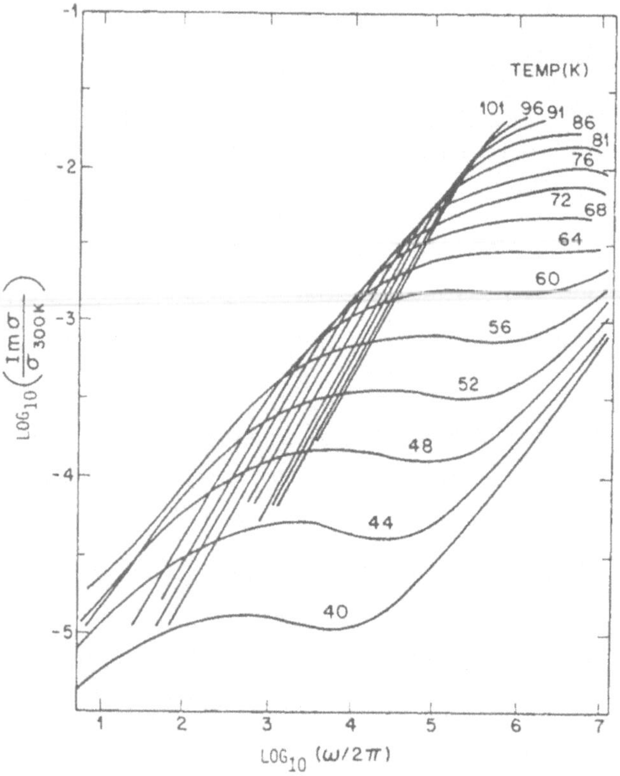

Fig. 7. The imaginary part of the conductivity in $K_{0.30}MoO_3$ as a function of frequency for temperatures between 40 and 101K. (From ref. 29)

The problem is to derive an expression for the dielectric function from Eq. 6 using an appropriate expression for $g(\tau)$. One could substitute various empirical expressions for $g(\tau)$ into Eq. 6, and calculated $\epsilon(\omega)$, however an approach that has been found to be more generally useful for the description of a variety of molecular relaxations[57] is to write an empirical generalization of the Debye relaxation, (Eq. 2) as

$$\epsilon(\omega) = \epsilon_\infty + \frac{\epsilon_o - \epsilon_\infty}{[1 + (i\omega\tau_{avg})^{1-\alpha}]^\beta} \tag{7}$$

where τ_{avg} is the mean relaxation time of the distribution $g(\tau)$. (Note that $\alpha = 0$ and $\beta = 1$ corresponds to a single Debye relaxation (Eq. 3)). In this approach, one calculates an analytical expression for the distribution of relaxation times, $g(\tau)$, that will produce the dielectric function given by Eq. 7. An examination of the analytic expression for $g(\tau)$ reveals that the parameters α and β characterize the width and the skewness of $g(\tau)$ respectively.[55] An important feature of this approach is that it allows for the description of different behavior for frequencies above and below the mean relaxation frequency, τ_{avg},

$$\epsilon \sim [1-\beta(i\omega)^{(1-\alpha)}], \qquad \omega << 1/\tau_{avg}, \tag{8}$$

$$\epsilon \sim [i\omega]^{\beta(1-\alpha)}, \qquad \omega >> 1/\tau_{avg}, \tag{9}$$

i.e. a $g(\tau)$ which is not logarithmically symmetric about τ_{avg} can be described.

Equation 7 is not intended to be a model of low-frequency dynamics, however the parameterization does provide useful information. The distribution of relaxation times, $g(\tau)$ can be identified as the density of pinned CDW states.[45] Fits to the data taken at different temperatures allow one to obtain qualitative information about the shape of the density of pinned CDW states.[29,45,58–60] In particular, we have shown that the spectral weight of $g(\tau)$ shifts to lower frequencies as the temperature is lowered or following the initial voltage application after cooling in zero field.[45]

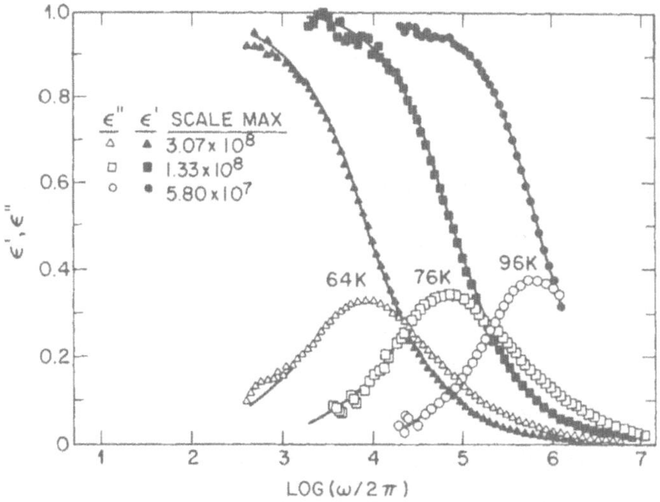

Fig. 8. The real and the imaginary parts of the dielectric function of $K_{0.30}MoO_3$ as a function of log(frequency) for three temperatures. (From ref. 29)

When analyzing the low frequency response of a CDW conductor where the response is determined by a distribution of relaxation times, the information is clearer if one plots the data as the dielectric function rather than the conductivity. This is because the decrease in Re ϵ and the peak in Im ϵ at $1/\tau_{avg}$ are always easily observable while the structure in $\sigma(\omega)$ may be masked as in the data shown in Figs. 6 and 7. To calculate $\epsilon(\omega)$ from $\sigma(\omega)$ in a conductor one must first subtract the dc conductivity from Re σ and use the relation $\sigma(\omega) = i\omega\epsilon(\omega)/4\pi$. Figure 8 shows Re ϵ and Im ϵ as a function of frequency for three different temperatures. The solid lines are fit to Eq. 7 and the agreement is excellent (within a few percent). The fall-off of Re ϵ and the peak of Im ϵ at $\omega = 1/\tau_{avg}$ are immediately apparent. Notice also that the static dielectric constant, which measures the polarizability of the sample, is enormous. The static dielectric constants of moving CDW materials are the largest known for any material. Also, as the temperature is decreased, the mean relaxation time τ_{avg} moves to lower temperatures indicating that the CDW response becomes sluggish and the static dielectric constant becomes larger. We have already seen from TSD experiments that the mean relaxation frequency goes to zero at low temperatures. Later we will show that both ϵ_o and τ_{avg} show thermally activated behavior resulting from the temperature dependence of screening currents.

The deviation of the low-frequency data from the predictions of Eq. 3 are more dramatically illustrated if one plots Re ϵ versus frequency on a linear scale as in Fig. 9. The expected form of Re ϵ from the uniform pinning model is a Lorentzian centered at zero frequency as shown by the dashed line in Fig. 9. The deviation of the data from the Lorentzian line-shape becomes more dramatic as the temperature is lowered. By 60K the data have developed a "cusp" characteristic of glassy systems. If there is an energy gap in the density of metastable states, one expects Re $\epsilon(\omega)$ to be a constant at low frequency. The presence of finite slope of $\epsilon(\omega)$ at zero frequency indicates that the broad distribution of relaxation times extends to

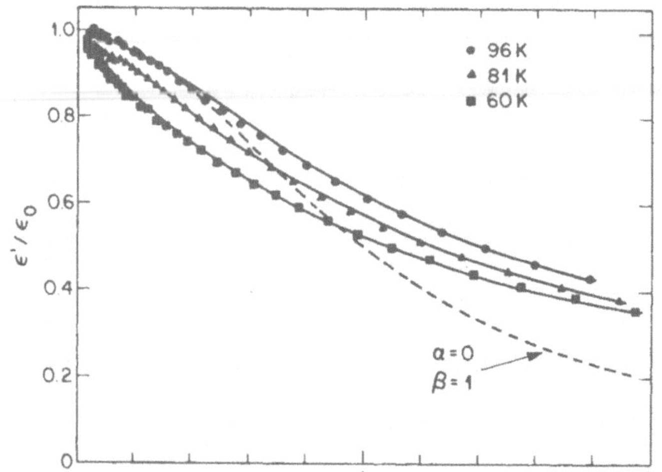

Fig. 9. A linear plot of the real part of the dielectric function of $K_{0.30}MoO_3$ for three temperatures. The dashed line is the Lorentzian lineshape predicted by the uniform pinning model. (From ref. 29)

infinite time.[36] This is in agreement with our previous observation that one can freeze a polarization into the material by cooling in a field.[42] The ac measurements show the existence of a polarization even at temperatures where there are significant screening electrons. The same general behavior of the low frequency cusp in the dielectric function is found in all moving CDW systems including $NbSe_3$[29]. We have found no evidence of a reported power-law dependence[61] (Re $\epsilon \sim \omega^{\alpha}$, $\alpha > 1$) except when the ac amplitude is intentionally made large.[62]

The temperature dependence of α and β contains information about the temperature dependence of the distribution of relaxation times. In $K_{0.30}MoO_3$, the parameter α increases while the quantity $\beta(1-\alpha)$ remains about constant as the temperature is lowered. For $K_{0.30}MoO_3$ in the temperature range 100 - 60K, α changes from about 0.07 to 0.25 while $\beta(1-\alpha) \approx 0.70$.[29] If we think of this in terms of the density of pinned metastable states, the effect of decreasing the temperature is to increase the relative number of low energy metastable states available to the CDW. A similar effect occurs when after one applies a field for the first time after cooling in zero field.[44] Another effect of lowering the temperature is to increase the mean dielectric relaxation time, τ_{avg} in all semiconducting CDW compounds. The relaxation time τ_{avg} shows activated behavior with an activation energy within a factor of two of the band gap in $K_{0.30}MoO_3$[29], and equal to the band gap in $(TaSe_4)I_2$.[60] In TaS_3 τ_{avg} shows activated behavior over a limited temperature range.[58] We suspect that in that compound, there may be a CDW phase transition at about 80K (e.g. a commensurate-incommensurate transition[15]) causing a loss of the low-frequency oscillator strength.[63] The static dielectric constant ϵ_o also increases as the temperature is lowered in $K_{0.30}MoO_3$, TaS_3 and $(TaSe_4)I_2$. Typically ϵ_o is activated over a limited temperature range and it saturates at low temperatures. The temperature dependence of τ_{avg} and ϵ_o suggest a connection between the dielectric relaxation frequency, the polarizability and the number of normal electrons. A similar conclusion is supported by measurements of the dc conductivity as a function of temperature.[63] In nearly all CDW materials it has been shown that the extra contribution to the conductivity from CDW motion scales with the normal state conductivity.[63,64] In the case the semiconducing compounds, this conclusion implies that the CDW viscosity, λ_o, diverges as the temperature is lowered. This is contrary to intuition because damping that is thermal in origin would be expected to decrease as the temperature is lowered. An explanation for the temperature dependence of the damping is via the screening currents of normal carriers as discussed in the next section.

Impurity pinning plays an important role in CDW transport. Impurities determine the threshold field for dc conduction as well as dynamic properties related to the pinning potential, V_o. The random impurity potential is responsible for the rich variety of metastable effects seen in these compounds. Theoretically, two types of pinning are thought to occur, strong and weak.[21] In the strong pinning case, the local CDW phase is a function of the pinning strength of a single impurity, while in the weak pinning case the local CDW phase is a compromise over the phase preferred by several impurity sites. There has been some controversy over the role of strong versus weak pinning sites. This is complicated by the fact that all samples do not show the same types of pinning. As mentioned above, amplitude fluctuations and strong pinning are probably important in "switching" samples. For other samples the results are mixed depending on the experiment.

$K_{0.30}MoO_3$ shows evidence of both strong and weak pinning depending on the dopant.[59,65] In $K_{0.30}MoO_3$ one can put the impurities at one of two different places in the structure, the alkali sublattice (Rb doping) or the transition metal sublattice (W doping). In $K_{0.30}MoO_3$ the conduction band has largely Mo 4d character,[66,67] the only role of the potassium atoms is to transfer charge to the conduction band. Consequently, we expect that Rb doping will result in weak pinning impurities while W doping will result in strong pinning impurities. In $K_{0.30}MoO_3$ we observe $E_T \sim n^2$ for Rb dopants and $E_T \sim n$ for W dopants in agreement with theory.[65] The strongly perturbing W impurities also have an enormous effect on the static dielectric constant and the dielectric relaxation frequency.[59] The substitution of 1% W on the Mo sublattice decreases ϵ_o by 3½ orders of magnitude while 50% Rb on the K sublattice only decreases ϵ_o by about one order of magnitude. Similar effects are observed for τ_{avg}, a five order of magnitude increase in $1/\tau_{avg}$ for 1% W and a three order of magnitude increase for 50% Rb. We believe this clearly establishes the existence of two types of pinning, strong and weak. Given the dramatic effects of experimentally induced strong pinning centers, we believe that the dominant pinning in "pure" $K_{0.30}MoO_3$ is weak pinning.

ELECTROSTATIC SCREENING

We have seen in the preceding sections that the pinned CDW possesses metastable states consisting of distortions of the CDW phase. We have also seen that the moving state is characterized by time-dependent fluctuations of the CDW phase. Both of these situations produce gradients of the phase of the CDW which, when viewed locally, are indistinguishable from a shift of the CDW wave vector. Thus one can view gradients of the CDW phase as changes in the CDW energy gap which will be compensated by flow of normal carriers into or out of the local area. Given the large changes in the density of electrons as a function of temperature in the semiconducting compounds, is therefore no surprise that the low-frequency charge-density wave dynamics are strongly temperature dependent. In this section we will explore the role of electrostatic screening in low-frequency CDW dynamics.

The first study of CDW screening was a time-dependent Landau theory by McMillan.[68]. McMillan assumed that the normal electrons relax to the frame of the moving CDW and he calculated a damping parameter which scales as ρ_c^2/σ. Later Sneddon independently developed a hydrodynamic model ($Q, \omega = 0$) based on the FLR Hamiltonian and dissipation via the normal carriers.[69] Sneddon's approach was different from McMillan's in that he assumed that the normal electrons relax to the frame of the lattice rather than the moving frame. Sneddon reached a similar conclusion to McMillan, the damping due to screening is temperature-dependent and scales as ρ_c^2/σ. Screening has also been phenomenologically addressed within the tunneling model of CDW transport by Tucker, et al.[24]

The low-frequency results presented in Sec. 4 describe CDW dynamics that become more and more sluggish as the temperature is lowered. Measurements of the dc conductivity indicate that the CDW viscosity is diverging at low temperatures at a rate proportional to $1/\sigma$.[63] High frequency measurements, on the other hand, present a different picture.[70–74] The high frequency damping ($1/\tau_o = \lambda_o/m^*$ from Eq. 2) is nearly temperature independent at frequencies above about 4GHz in all CDW materials. In $(TaSe_4)I_2$[73] and $K_{0.30}MoO_3$,[74] low temperature conductivity measurements show that the response is dominated by a high-frequency underdamped mode. This suggests that the damping mechanism is frequency dependent with the temperature dependence of the damping scaling as

$1/\sigma$ at low frequencies and a constant at high frequencies. For intermediate frequencies there is less information on the temperature dependence of $\sigma(\omega)$. The suggestion is that the overall CDW response is overdamped at high temperatures but at low temperatures the low-frequency contribution drops out leaving an underdamped peak in Re σ.

If one uses Sneddons model[69] to calculate $\sigma(\omega)$ in the presence of screening, the result is the same as Eq. 2 with the inertial damping frequency $1/\tau_o$ replaced by a frequency dependent damping $1/\tau(\omega)$ where $1/\tau(\omega)$ is given by

$$\frac{1}{\tau(\omega)} = \frac{1}{m^*}\left[\lambda_o + \frac{\lambda}{1+i\omega/\omega_s}\right] \tag{10}$$

where

$$\lambda = \rho_c^2/\sigma \tag{11}$$

$$\omega_s = 4\pi\sigma/\epsilon_s \tag{12}$$

Note the difference between Eq. 14 and the previous result for uniform pinning, $\lambda_o = \rho_c^2/\sigma_{\max}$. In the screened limit $\omega << \omega_s$, if we assume that the damping from electrostatic screening, λ, is much larger than other sources of damping, λ_o, one may write the overdamped response in the same form as Eq. 3 with $1/\tau_R \approx \lambda/m^* \approx \rho_c/\sigma|Q|E_T$. This expression for the dielectric relaxation frequency has a temperature dependence consistent with observations. In $K_{0.30}MoO_3$,[59] experimental values of τ_{avg} have been compared with values calculated from other measured parameters using a slightly different expression for τ_R. In TaS_3,[63] $1/\tau_{avg}$ scales exactly with measured values of σ. These results suggest that the basic assumptions concerning electrostatic screening are correct, screening by normal carriers is apparently the dominant loss mechanism for low-frequency CDW motion.

A detailed comparison of high-frequency[70–73] and infra-red[74] results with calculations is still in a preliminary stage.[75] Equation 10 predicts an underdamped response for $\omega >> \omega_s$ in qualitative agreement with low-temperature, high frequency measurements.[73,74] (One would interpret the underdamped response predicted from Eq. 10 when $\omega >> \omega_s$ as a charge-density wave plasmon.) It is not clear, however, if Eq. 10 is a complete expression for electrostatic screening as there are subtle issues in the problem. For example, Sneddon's calculations are valid in the q=0 limit of longitudinal excitations, but in the case of infra-red measurements where the light propagates in a direction which is perpendicular to the CDW displacements, one would a response which is predominantly transverse. Experimentally this can get even more complicated because in an anisotropic material with a distribution of pinning energies, the transverse and the longitudinal modes are likely to be coupled and it may be difficult to separate the two responses. To proceed further with the analysis, one needs a theory which includes the long-wavelength limit of both longitudinal and transverse excitations as well an explicit treatment of a distribution of pinning energies.

SUMMARY

In most quasi-one-dimensional materials, interest has centered around the low-dimensionality of the material and the unique properties of a one-dimensional electronic system. Materials with moving charge-density waves are an interesting intermediate state between low dimensional materials and isotropic materials. There must be enough anisotropy to produce a charge-density wave, but without transverse coupling, the rich physical phenomena associated with metastability would not be present. In fact, it is likely that the CDW would not move at all since even dc motion is accompanied by transverse fluctuations and screening currents.

We have seen that the "sliding" CDW moves in a way that is very different from that originally envisioned by Fröhlich. Motion of the CDW is not a sliding at all but a piecewise motion involving alternate motion of local areas of the sample. In most samples, one can assume that the piecewise motion introduces distortions of the CDW phase. In other samples, e.g. "switching samples", we have seen that local strong pinning and amplitude fluctuations are important. In either case the local motion of the CDW implies that the the energy gap, is locally changing resulting in normal electron screening currents. Motion of the CDW is therefore a cooperative phenomenon between the condensed electrons and free carriers. At low temperatures where there are no free carriers in the semiconducting compounds, there is no motion of the CDW except possibly at high fields where the CDW may move with no distortions.[76,77] Screening currents result in a temperature dependence of the damping that is different for high frequencies and low frequencies and a low-frequency viscosity which scales as ρ_c^2/σ.

The similarity of the non-linear response among the materials with moving CDW's is remarkable. All materials show an extremely broad distribution of pinning energies, a distribution that extends all the way to zero frequency. The primary differences between materials relate to differences that are extrinsic to the CDW such as the sample size, the temperature and the sample conductivity. There is often confusion around this fact. For example, in $K_{0.30}MoO_3$ the slow relaxation times and the long-lived metastable phenomena can be interpreted as a CDW with more disorder and more "incoherent effects". However, if one compares $K_{0.30}MoO_3$ with TaS_3, the distribution of pinning energies is narrower and the sample quality is much higher in the case of $K_{0.30}MoO_3$. Similarly, the beautiful current oscillations seen in $NbSe_3$ have lead to the interpretation that $NbSe_3$ is more "coherent" with a narrower distribution of pinning energies. However, the distinguishing characteristics of $NbSe_3$ are its semimetallic conductivity and its small size and relatively good crystal quality. This results in a short relaxation time, but $\epsilon(\omega)$ still shows a low-frequency cusp characteristic of a distribution of pinning energies that extends to zero frequency.[29]

ACKNOWLEDGEMENTS

The results presented in this paper are the result of numerous collaborations. I would like to particularly acknowledge the contributions of R. J. Cava, S. B. Coppersmith, R. G. Dunn, P. B. Littlewood, E. A. Rietman, and L. F. Schneemeyer. I would also like to thank P. B. Littlewood for numerous discussions on electrostatic screening.

REFERENCES

1. J. A. Wilson, F. J. DiSalvo and S. Mahajan, Adv. Phys. 24:117 (1975) and P. M. Williams, G. S. Parry, and C. B. Scruby, Philos. Mag. 29:695 (1975).
2. R. M. Fleming, F. J. DiSalvo, R. J. Cava and J. V. Waszscak, Phys. Rev. B 24:2850 (1981).
3. R. M. Fleming, D. E. Moncton, D. B. McWhan and F. J. DiSalvo, Phys. Rev. Lett. 45:576 (1980).
4. R. M. Fleming, L. W. ter Haar, and F. J. DiSalvo, Phys. Rev. B (to be published).

5. P. Monceau, N. P. Ong, A. M. Portis, A. Meerschaut, and J. Rouxel, Phys. Rev. Lett. 37:602 (1976), and N. P. Ong and P. Monceau, Phys. Rev. B 16:3443 (1977).
6. "Charge Density Waves in Solids", Gy. Hutiray and J. Solyóm, ed., Springer-Verlag, Berlin, (1985).
7. "Proceedings of the Yamada Conference XV on the Physics and Chemistry of Quasi One-Dimensional Conductors", Lake Kawaguchi, Japan, May 1986 (to be published in Physica B).
8. P. Monceau in "Electronic Properties of Quasi-One-Dimensional Materials", Vol II, P. Monceau, ed., Reidel, Dordrecht, The Netherlands, (1985).
9. C. Schlenker, and J. Dumas in "Crystal Chemistry and Properties of Materials with Quasi-One-Dimensional Structures", J. Rouxel , ed., Reidel, Dordrecht, The Netherlands, (1986).
10. R. M. Fleming and C. C. Grimes, Phys. Rev. Lett. 42:1423 (1979).
11. N. P. Ong, G. Verma and K. Maki, Phys. Rev. Lett. 52:663 (1984).
12. George Mozurkewich and George Grüner, Phys. Rev. Lett. 51:2206 (1983).
13. J. C. Gill, Solid State Commun. 39:1203 (1981).
14. R. M. Fleming, Solid State Commun. 43:167 (1982).
15. A. W. Higgs and J. C. Gill, Phys. Rev B 47:737 (1983).
16. R. M. Fleming and L. F. Schneemeyer, Phys. Rev. B 28:6996 (1983).
17. G. Mihály and L. Mihály, Phys. Rev. Lett. 52:109 (1984).
18. J. W. Brill and S. L. Herr, Solid State Commun. 49:265 (1984).
19. G. Kriza and G. Mihály, Phys. Rev. Lett. 56:2529 (1986).
20. H. Fukuyama and P. A. Lee, Phys. Rev. B 17:535 (1977).
21. P. A. Lee and T. M. Rice, Phys. Rev. B 19:3970 (1979).
22. John Bardeen, Phys. Rev. Lett. 42:1498 (1979), 45:1978 (1980), 55:1010 (1985) and John Bardeen and J. R. Tucker, in Ref. 6, p 155.
23. R. E. Thorne, W. G. Lyons, J. W. Lyding, J. R. Tucker and John Bardeen, Phys. Rev. B, to be published.
24. J. R. Tucker, W. G. Lyons, J. H. Miller, Jr., R. E. Thorne and J. W. Lyding, to be published.
25. G. Grüner, A. Zawadowski, and P. Chaikin, Phys. Rev. Lett. 46:511 (1981) and P. Monceau, J. Richard and M. Renard, J. Phys. C 15:931 (1982).
26. P. Monceau, J. Richard and M. Renard, Phys. Rev. Lett. 45:43 (1980).
27. G. Grüner, L. C. Tippie, J. Sanny, W. G. Clark, and N. P. Ong, Phys. Rev. Lett. 45:935 (1980).
28. A. Zettl and G. Grüner, Phys. Rev. B 25:2081 (1982).
29. R. J. Cava, R. M. Fleming, P. Littlewood, E. A. Rietman, L. F. Schneemeyer, and R. G. Dunn, Phys. Rev. B 30:3228 (1984).
30. N. P. Ong, D. D. Duggan, C. B. Kalem, and T. W. Jing, in Ref. 6, p 387.
31. S. Battacharya, J. P. Stokes, and Mark O. Robbins (unpublished).
32. L. Sneddon, M. C. Cross and D. S. Fisher, Phys. Rev. Lett. 49:292 (1982).
33. L. Sneddon, Phys. Rev B 29:719 (1984), 29:725 (1984).
34. S. N. Coppersmith and P. B. Littlewood, Phys. Rev. B 31:4049 (1985).
35. D. S. Fisher, Phys. Rev. Lett. 50:1486 (1983) and Phys. Rev. B 31:1396 (1985).
36. P. B. Littlewood, in Ref. 6, p 369, also in Proc. of the Conference on Spatio-Temporal Dynamics, Los Alamos, NM, Jan 1986, (to be published in Physica D), also Phys. Rev. B 33:6694 (1986).
37. S. N. Coppersmith and D. S. Fisher, Phys. Rev. B 28:2566 (1983).
38. S. N. Coppersmith, Phys. Rev. B 30:410 (1984).
39. S. N. Coppersmith and P. B. Littlewood, Phys. Rev. Lett. 57:Oct. (1986).
40. S. Abe, J. Phys. Soc. Japan 55:1987 (1986).

41. H. Matsukawa and H. Takayama, in Ref. 7.
42. R. J. Cava, R. M. Fleming, E. A. Rietman, R G. Dunn and L. F. Schneemeyer, Phys. Rev. Lett. 53:1677 (1984).
43. J. van Turnhout, "Thermally Stimulated Discharge of Polymer Electrets", Elsevier, Amsterdam, (1975).
44. R. M. Fleming and L. F. Schneemeyer, Phys. Rev. B 33,:2930 (1986).
45. R. J. Cava, P. B. Littlewood, R. M. Fleming, L. F. Schneemeyer and E. A. Rietman, Phys. Rev. B 34:1184 (1986).
46. A. Zettl and G. Grüner, Solid State Commun. 46:501 (1983); A. Zettl, M. Sherwin and R. P. Hall, in Ref. 6, p 333.
47. R. M. Fleming, D. E. Moncton, and D. B. McWhan, Phys. Rev. B 18:5560 (1978), and R. M. Fleming and D. E. Moncton, Phys. Rev. B 30:1877 (1984).
48. R. M. Fleming, R. G. Dunn and L. F. Schneemeyer, Phys. Rev. B 31:4099 (1985).
49. T. Tamegai, K. Tsutsumi, S. Kagoshima, Y. Kanai, M. Tani, H. Tomozawa, M. Sato, K. Tsuji, J. Harada, M. Sakata, and T. Nakajima, Solid State Commun. 51:585 (1984).
50. T. Tamegai, K. Tsutsumi, S. Kagoshima, Y. Kanai, H. Tomozawa, M. Tani, Y. Nogami and M. Sato, Solid State Commun. 56:13 (1985); K. Tsutsumi, T. Tamegai, S. Kagoshima and M. Sato, Ref 6. p 17.
51. A. Zettl and G. Grüner, Phys. Rev. B 26:2298 (1982).
52. M. P. Everson and R. V. Coleman, Phys. Rev. B 28:6659 (1983).
53. J. Dumas, C. Schlenker, J. Marcus and R. Buder, Phys. Rev. Lett. 50:757 (1983).
54. K. Tsutsumi, T. Tamegai, S. Kagoshima and M. Sato, J. Pys. Soc. Japan 54:3004 (1985).
55. R. P. Hall, M. F. Hundley and A. Zettl, Phys. Rev. Lett. 56:2399 (1986).
56. G. Verma and N. P. Ong, Phys. Rev. B 30:2928 (1984).
57. For a review of dielectric relaxation see for example, C. J. F. Böttcher and P. Bordewijk, "Theory of Electric Polarization", Vol. II, Elsevier, Amsterdam, (1978).
58. R. J. Cava, R. M. Fleming, R. G. Dunn and E. A. Rietman, Phys. Rev. B 31:8325 (1985).
59. R. J. Cava, L. F. Schneemeyer, R. M. Fleming, P. B. Littlewood and E. A. Rietman, Phys. Rev. B 32:4088 (1985).
60. R. J. Cava, P. Littlewood, R. M. Fleming, R. G. Dunn and E. A. Rietman, Phys. Rev. B 33:2439 (1986).
61. Wei-yu Wu, L. Mihály, George Mozurkewich and G. Grüner, Phys. Rev. Lett. 52:2382 (1984).
62. R. J. Cava, R. M. Fleming, R. G. Dunn, E. A. Rietman and L. F. Schneemeyer, Phys. Rev. B 30:7290 (1984).
63. R. M. Fleming, R. J. Cava, L. F. Schneemeyer, E. A. Rietman and R. G. Dunn, Phys. Rev. B 33:5450 (1986).
64. X. J. Zhang and N. P. Ong, Phys. Rev. Lett 55:2919 (1985).
65. L. F. Schneemeyer, R. M. Fleming and S. E. Spengler, Solid State Commun. 53:505 (1985).
66. M. -H. Whangbo, and L. F. Schneemeyer, Inorganic Chem. 25:2424 (1986).
67. G. K. Wertheim, L. F. Schneemeyer, and D. N. E. Buchanan, Phys. Rev. B 32:3568 (1985).
68. W. L. McMillan, Phys. Rev. B 12:1197 (1975).
69. Leigh Sneddon, Phys. Rev B 29:719 (1984).
70. S. Sridhar, D. Reagor and G. Grüner, Phys. Rev. Lett. 55:1196 (1985).

71. D. Reagor, S. Sridhar and G. Grüner, Phys. Rev. B 34:2212 (1986).
72. S. Sridhar, D. Reagor and G. Grüner, Phys. Rev. B 34:2223
73. D. Reagor, S. Sridhar, M. Maki and G. Grüner, Phys. Rev. B 32:8445 (1985).
74. H. K. Ng, G. A. Thomas and L. F. Schneemeyer, Phys. Rev. B 33:5358 (1986).
75. P. B. Littlewood, to be published.
76. A. Maeda, T. Furuyama, and S. Tanaka, Solid State Commun. 55:951 (1985).
77. L. Mihály and G. X. Tessema, Phys. Rev. B 33:5858 (1986).

NMR STUDIES OF CHARGE DENSITY WAVES IN LOW DIMENSIONAL CONDUCTORS

C. Berthier and P. Ségransan

Laboratoire de Spectrométrie Physique
Associé au CNRS, Université Scientifique
Technique et Medicale de Grenoble
BP. 87, 38402 Saint Martin d'Hères
Cedex, France

INTRODUCTION

Since its early beginnings the NMR technique has shown to be well suited to the study of the metallic state [1]; the first NMR study in superconducting aluminum is dated as early as 1959 [2]. Quite naturally the discovery of new synthetic metals, with very anisotropic electronic properties, brought a renewal in this field and NMR has rapidly become an important tool for the investigation of the dimensionality of the electron gas, and for the study of the various possible ground states of quasi 1D conductors: Charge Density Wave (CDW), Spin Density Wave (SDW) or superconducting state (SS), as well as their precursor fluctuations.

This paper will be strictly devoted to those systems which undergo a Peierls distortion, or a periodic lattice distortion (PLD) accompanied by a CDW with partial destruction of the Fermi surface. Nevertheless, all the discussion of the NMR lineshape in presence of an incommensurate modulation of the hyperfine coupling can be applied to the SDW state also. In the first part, we shall briefly recall some basic notions of NMR in solids and discuss the hyperfine coupling mechanisms in metals. The second part is devoted to the information one can obtain from NMR on the static propreties of the CDW (symmetry of the q star, incommensurate commensurate transition, temperature dependence of the order parameter, interaction of the CDW with charged defects). The third part will be concerned with the study of the dynamics of the CDW (critical fluctuations, amplitudons and phasons) and the last one will be devoted to the recent NMR studies of CDW motion under applied electric field (direct evidence of the Fröhlich mode) and their possible developments.

NMR IN SOLID STATE PHYSICS : BASIC NOTIONS

Each nuclear isotope is characterized by the value of its spin I, its gyromagnetic ratio γ_n (in MHz/Tesla) which defines its magnetic moment, its quadrupole moment Q (for values of I larger than 1/2) and its natural abundance (Table I)

NMR spectroscopy in solids [3,4] consists of looking for the perturbative effects induced by the nuclei environments of the Zeeman resonance of a specific isotope defined by the main Hamiltonian :

$$\mathcal{H}_Z = - \gamma_n \hbar H_o I_z = \hbar \omega_L I_z \tag{1}$$

where H_o is the external applied field, $I_z = \sum_i I_z^i$ (i denotes a summation over all lattice sites occupied by the considered isotope) and ω_L the Larmor angular frequency.

The magnitude of these perturbations usually rank in the following order

The quadrupolar Hamiltonian $\mathcal{H}_Q$ $(I>1/2)$

describes the electrostatic coupling to the electrons and other ions in the solids ; since nuclei have zero electric moment, only coupling to the electric field gradients (EFG) tensor generated by the charges external to the nuclei is relevant.

The magnetic hyperfine Hamiltonian $\mathcal{H}_M$

describes the magnetic coupling of the nuclei to the electrons in the solids ; in a metal, this coupling is dominated by the contribution of the conduction electrons.

The nuclear dipole-dipole Hamiltonian $\mathcal{H}_D$

describes the dipolar coupling between the magnetic moments of the nuclei. Notice that in organic compounds, for nuclei of spin 1/2 like ^1H and ^{13}C, for which $Q = 0$, one can have the situation of $\mathcal{H}_D >> \mathcal{H}_M$ which requires the use of High Resolution in Solids NMR techniques [5].

Let us now consider the effect of the magnetic hyperfine Hamiltonian. This coupling may be written :

$$\mathcal{H}_M = -\gamma_n \hbar \vec{H}_o \overline{\overline{K}} \vec{I} = - \gamma_n \hbar H_o K_{zz} I_z \qquad (2)$$

where $\overline{\overline{K}}$ is the magnetic hyperfine tensor, called the Knight tensor (or shift) in the case of metals. This hamiltonian simply leads to an effective value of $\gamma = \gamma_n [1+K(\theta,\varphi)]$ where θ and φ denote the orientation of H_o with respect to the principal axis of K. This Hamiltonian does not lift the degeneracy between the (m,m+1) transition resonance lines.

Let us now consider the effect of the quadrupolar Hamiltonian $\mathcal{H}_Q$. In the framework of the principal axis of the EFG tensor, $\mathcal{H}_Q$ can be written

$$\mathcal{H}_Q = \frac{e^2 q Q}{4I(2I-1)} \{3I_z^2 - I(I+1) + \eta (I_+^2 + I_-^2)\} \qquad (3)$$

where $eq = V_{zz}/2$

and the asymmetry parameter $\eta = \frac{V_{yy} - V_{xx}}{V_{zz}}$ $\quad |V_{zz}| > |V_{yy}| > |V_{xx}|$

Table I

Characteristics of some isotope nuclei frequently encountered in low dimensional conductors. Remember that the sensitivity of an experiment is proportional to the square of the Larmor frequency.

Isotope	^1H	^2D	^{13}C	^{14}N	^{39}K	^{77}Se	^{87}Rb	^{93}Nb	^{97}Mo	^{181}Ta
Gyromagnetic ratio $\gamma_n/2\pi$ (MHz/Tesla)	42.6	6.5	10.7	3.0	1.9	8.1	13.9	10.4	2.2	5.1
Nuclear spin I	1/2	1	1/2	1	3/2	1/2	3/2	9/2	5/2	7/2
% isotropic abundance	100	0.015	1.1	99.6	93	7.5	28	100	9.5	100
quadrupole moment Q 10^{-24} cm^2		0.0028		0.071	-0.07		0.14	-0.16	0.14	4.8

To first order in the perturbation, the frequency of the (m,m+1) transition is shifted by an amount :

$$\Delta\nu_{m\to m+1}^{(\theta,\varphi)} = \frac{(2m+1)\ \nu_Q}{4}\left[(3\cos^2\theta-1) + \eta\sin^2\theta\ \cos 2\varphi\right] \quad (4)$$

where θ, φ denotes the orientation of the external magnetic field H_o with respect to the principal axis of the EFG and $\nu_Q = \frac{3e^2qQ}{2I(2I-1)h}$ (a few MHz is a typical value of ν_Q).

The (-1/2,1/2) transition, unperturbed in first order, is shifted in second order of perturbation by the amount :

$$\Delta\nu_{(-\frac{1}{2}\to\frac{1}{2})}(\theta,\varphi) = -\frac{\nu_Q^2}{16\nu_L}\left(I(I+1)-\frac{3}{4}\right)\left[(1-\cos^2\theta)\ (9\cos^2\theta-1) + \eta f(\theta,\varphi)\right] \quad (5)$$

Fig.1 shows NMR spectra corresponding to an I =3/2 spin in presence of an EFG tensor (η=0). for the case of a single crystal, of a powder, and of a single crystal undergoing a spatial modulation of the EFG (due to a CDW for example): because of the angular dependence of the position of the resonance lines, the powder spectrum which is an average over all possible values of θ and φ, is considerably broadened. One sees that to study a small spatial modulation of ν_Q it is necessary to work in a single crystal.

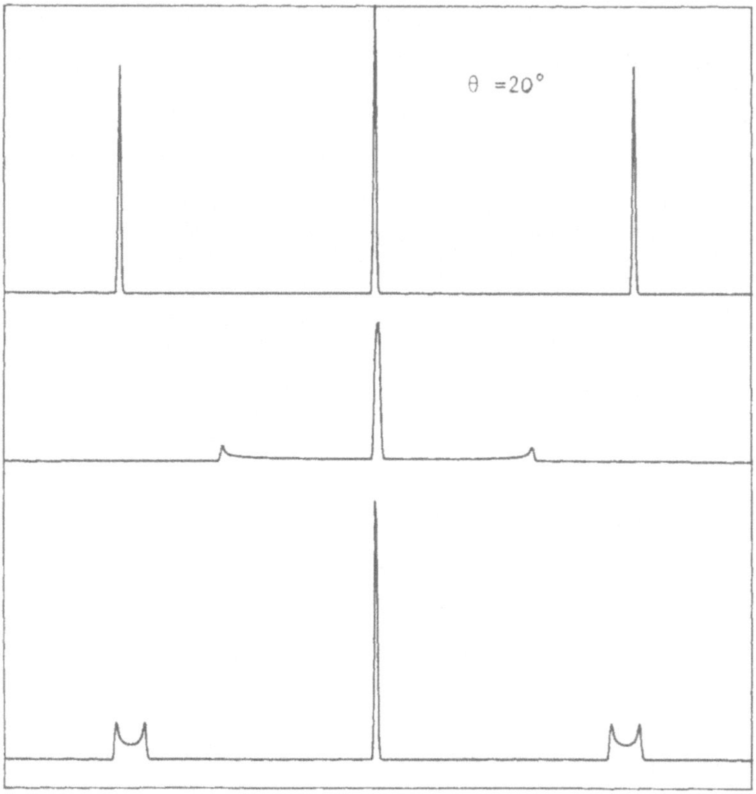

Fig.1 NMR spectra corresponding to an assembly of spin I =3/2 in presence of an EFG tensor (η =0). ν_L = 80.000 MHz, ν_Q = 4.5 MHz. a) single crystal. b) a powder. c) single crystal undergoing a small sinusoidal spatial modulation (5%) of ν_Q.

Conventional pulsed NMR spectroscopy deals with the following quantities

- The resonance frequencies of the various (m,m+1) transitions : their knowledge as a function of θ and φ allows the determination of the parameters of $\mathcal{H}_M$ and $\mathcal{H}_Q$.

- The line shape of a particular transition (m,m+1)

The measured quantity is $\langle I_x(t)\rangle = e^{i\omega_L t}G(t)$ the correlation function of the transverse nuclear magnetization, which is the Fourier transform of the lineshape, according to the fluctuation-dissipation theorem. This lineshape is said to be homogeneous when all the interactions contributing to the time decay of G(t) are time irreversible (dipole-dipole interaction, lifetime broadening). The spin-spin relaxation time T_2 is defined as the correlation time of G(t). The lineshape may also be inhomogeneously broadened, by for example, a distribution of the hyperfine field (Fig. 1)

In this case $\langle I_x(t)\rangle = e^{i\omega_L t}G^*(t)$ is the Fourier transform of the inohomogenous lineshape, nevertheless the function G(t) can still by measured by the spin-echo technique [3, 4].

-The spin lattice relaxation time T_1 describes the return to thermal equilibrium of the population of the Zeeman levels. When these levels are equidistant, T_1 is defined by

$$\frac{d\langle I_z(t)\rangle}{dt} = -\frac{1}{T_1}\langle I_z(t)\rangle$$

The origin of the spin-lattice relaxation lies in the fluctuations of the hyperfine coupling. Let us suppose that this latter can be simply described as :

$$\mathcal{H} = \sum_i \gamma_n \hbar \vec{h}^i(t)\, \vec{I}^i = \sum_i (\gamma_n \hbar h_z^i(t)\, I_z^i + \gamma_n \hbar \delta \vec{h}^i(t) \vec{I}^i) \qquad (6)$$

where $h^i(t)$ is a local random field (semi-classical approximation[3]). The spin-lattice relaxation rate (SLRR) can then be written as

$$\frac{1}{T_1} = \int_0^\infty \langle \delta h_+(R_i,t),\, \delta h_-(R_i,0)\rangle e^{i\omega_L t}\, dt \qquad (7)$$

If we Fourier transform h(R,t) in the q space, and within the RPA approximation which decouples fluctuations of different q vectors, we obtain, using the fluctuation-dissipation theorem :

$$\frac{1}{T_1} = \sum_q \int_0^\infty \langle \delta h_q^+(t),\, \delta h_q^-(o)\rangle\, e^{i\omega_L t} = \sum_q \frac{kT}{\omega_L}\, \chi''_q(\omega_L) \qquad (8)$$

where $\chi_q''(\omega)$ is the imaginary part of the susceptibility associated with the random variable $h_q(t)$. When this variable h_q is coupled to the order parameter of a phase transition, the spin-lattice relaxation rate allows to study critical fluctuations and the excitations of the condensed phase at low temperature, with the restriction that the SLRR always include a sommation over all the q vectors [6].

HYPERFINE INTERACTION IN METALS

a) Normal state

Knight Shift

The dominant contribution to the SLRR in a metal is due to the conduction electrons ; in the framework of the two s-d bands model, introduced by Jaccarino and coworkers[7] to describe Knight shifts and spin lattice relaxation in transition metals, the magnetic hyperfine interaction can be decomposed in 4 contributions : (neglecting electronic spin-orbit coupling).

- The contact term is the hyperfine field due to the conduction electrons having a non-vanishing wave function at the nuclei (s-like conduction electrons). It may be expressed as :

$$K_s = \frac{\Delta H}{H} = 8\pi/3 < |\psi_s(0)|^2 >_{E_F} \chi_s \Omega \qquad (9)$$

This isotropic contribution is proportionnal to the density of states at the Fermi level of the "s" band.

- The core polarisation term is the hyperfine field due to the polarisation of the internal "s" electrons by the electrons of the "d" or "p" band. Usually negative, it may be expressed as :

$$K_{cp} = \frac{\Delta H}{H} = \frac{8\pi}{3H_o}\sum_n \left(\left[\psi_{n\uparrow}(0)\right]^2 - |\psi_{n\downarrow}(0)|^2\right) \frac{H_{cp}\chi_d}{\mu_B} \qquad (10)$$

This isotropic term is proportionnal to the density of states at the Fermi level of the "d" ("p") band ($N_d(E_F)$) and is usually temperature dependent.

-The dipolar term is due to the dipolar magnetic interaction between the conduction electrons ("p" or "d") and the nuclei. It is anisotropic, but also proportionnal to $N_d(E_F)$ and vanishes if the symmetry is cubic.

- Finally nuclear spins interact not only with the spin of the electrons, but also with their orbital motion. To second order of perturbation, this gives rise to the so-called orbital or Van Vleck contribution to the Knight shift, which is always positive.

In the tight binding approximation, it can estimated as [9] :

$$K_{orb} \propto \frac{(N-n)n}{N\,\Delta E} (\mu_B)^2 <r^{-3}> \qquad (11)$$

where N is the total number of electrons in the "d" (or"p") bands,
n is the number of occupied states in the bands,
ΔE is the effective width of the bands
$\langle r^{-3}\rangle$ is averaged over the "d" or "p" wave function.

It is very important to notice that this term is not proportionnal to the density of states at the Fermi level.

Quadrupole interaction

The EFG tensors experienced by nuclei in metals are due both to the conduction electrons and the ions. First principles calculations are hardly possible even in simple metals, since they require a precise knowledge of the wave functions of the conduction electrons, which are usually ill-defined in band structure calculations.

Spin lattice relaxation

The SLRR in metals is usually dominated by the magnetic hyperfine interaction with the conduction electrons. Whatever is the mechanism, -contact, core polarisation, dipolar or orbital - T_1T is always proportional to the square of the density of states at the Fermi level. For those Knight shifts contributions which are proportionnal to $N(E_F)$, this leads to partial Korringa relationships [10,11]

$$(K_i)^2 T_1^i\, T = cte$$

where i stands for the contact, core polarisation and dipolar interaction. The relationship is not valid for the orbital-Van Vleck contribution.

For quasi-1D metals, the SLRR can be written as [12,13]

$$\frac{1}{T_1 T} = \sum_q \frac{\chi''_q(\omega_L)}{\omega_L} = \frac{1}{\omega_L}\{\chi''(q=0,\omega_L) + \chi''(q = 2k_F,\omega_L)\} \qquad (12)$$

The first term, which describes the diffusion behavior of the conduction electrons along the chains gives rise to a frequency dependence of T_1, the cutoff of which can be used to determine the transverse hopping rate between neighbouring conducting chains.

b) PLD/CDW state

Magnetic hyperfine interaction - Knight shift

Those contributions to the Knight shift (contact, core polarisation, dipolar) which are proportionnal to $N(E_F)$ consequently should vanish below the Peierls transition temperature T_P. In the picture of the pseudo-gap [13],they desappear as the inverse of the longitudinal correlation length ξ [14] of the CDW state.

In contrast, the orbital Knight shift, which arises from conduction electrons states within the whole band, is barely affected by the opening of a gap, since usually kT_P is much smaller than the bandwidth ΔE. Furthermore, it is a local quantity, as it depends on distance as $\langle r^{-3}\rangle$ and in presence of a CDW, where the electronic density n becomes $n+\delta n(r)$, we see that this term is linearly coupled to the order parameter of the Peierls transition.

Quadrupolar coupling

In the PLD/CDW state, the electronic density becomes modulated in the crystal

$$\rho(\vec{R}) = \rho_o[1+\alpha\cos(2\vec{k}_F\vec{R}+\phi)] = \rho(\vec{R}) + \delta\rho(\vec{R}) \qquad (13)$$

Consequently, the EFG tensor experienced by the nuclei becomes also spatially modulated, with the same period. Let us first neglect the tensorial character of quadrupolar coupling and consider only the variation of $V_{zz}(R) = V_{zz} + \delta V_{zz}(R)$. In principle, both the ionic distortion, and the modulation of conduction electron density contribute to the modulation of the EFG.

As the CDW wavelength is usually only of the order of a few interatomic distances, it is not obvious that the relationship between $\delta V_{zz}(R)$ and the order parameter $\delta\rho(R)$ is local.

Nevertheless, if we consider only the electronic contribution, we can express it as :

$$V_{zz}(\vec{R}) = \sum_{\vec{k}} \frac{\langle\psi_{\vec{k}}(\vec{r})|P_2(\cos\theta)|\psi_{\vec{k}}(\vec{r})\rangle}{|\vec{r}-\vec{R}|^3} \qquad (14)$$

where $P_2(\cos\theta)$ is the Legendre polynomia.

In the tight binding approximation, and neglecting the two center integrals, we then find $V_{zz}(R)$ proportionnal to $\delta\rho(R)$.

In the following we shall confine ourselves to this local approximation [15], even when the observed nuclei are not on the metallic chains, but are located in between like in $Rb_{0.3}MoO_3$. In such a case, the modulation of the conduction electron density is reflected in the NMR spectra not only through the Coulomb interaction, but also through the hybridization of the"s" and "p" wave functions of the alkali atoms with those of the conduction electrons ; such an hybridization is usually important even when the so-called "charge transfer" in intercalation compounds is considered as nearly complete [16,17].

Spin lattice relaxation in CDW systems

All the contribution to the nuclear spin-lattice relaxation due to conduction electrons (single particule excitations) are proportionnal to the square of the density of states at the Fermi level, and exponentially vanish below T_P if the Fermi surface is totally destroyed (quasi ID conductors). Thus the critical fluctuations of the CDW, and its excitations below T_P (phasons and amplitudons) are expected to become the dominant contribution to the SLRR, provided their coupling to the nuclei is strong enough. These effects are discussed in section III.

II. STATIC PROPERTIES OF THE CDW

The unusual character of NMR lineshapes in presence of an incommensurate modulation of the hyperfine coupling has been first recognized in CDW systems [18,19]. Nevertheless, the most extensive NMR studies of incommensurate phases have been performed in dielectrics compounds [20]. The first NMR studies of the CDWs dealt with quasi 2D transition metal dichalcogenides : $2H-NbSe_2$ [18,19,21,22,23] $2H-TaSe_2$ [21,24,25], $1T-TaS_2$ [26], VSe_2 [27,28] and $TiSe_2$ [29]. As to the quasi 1-D conductors, the earlier studies on $NbSe_3$ were performed on bunches of crystals oriented along the $\vec{b}$ direction [30,31]. Recently NMR lineshapes have been studied in single crystals of $(NbSe_4)_{10}I_3$ [32], $Rb_{0.3}MoO_3$ [33,34,35] and $NbSe_3$ [36].

Let us examine the theoretical NMR lineshape of a given (m,m+1) transition of a single crystal below the onset of the CDW. Within the local approximation, one has for a nucleus at site $\vec{R}$:

$$\nu_{m\to m+1}(\vec{R}) = \nu_o(\vec{R}) + F\left[\delta\rho(\vec{R})\right] = \nu_o(\vec{R}) + \alpha\delta\rho(\vec{R}) + \beta\left[\delta\rho(\vec{R})\right]^2 \qquad (15)$$

where we have expanded $\nu(\vec{R})$ in powers of real part of order parameter. The coupling constants α and β originate in the modulation of either the electric field gradient or the orbital Knight shift tensor. In CDW systems, one usually has $\alpha \gg \beta$, even in the case of (-1/2,1/2) transition subject to second order quadrupolar perturbation. *

The lineshape $g(\nu-\nu_o)$ is simply given by

$$g(\nu-\nu_o) = \iint_{\nu=cte} \frac{dS}{\left|\vec{\nabla}_{\vec{R}}\,\nu(\vec{R})\right|} \qquad (16)$$

Naturally, $g(\nu-\nu_o)$ has to be convoluted with some nuclear dipole-dipole homogeneous broadening. The major effect of incommensurateness is to give rise to an infinite number of inequivalent sites and to spread the intensity of the (m,m+1) transition over a continuous range of frequencies, which can be as large as a few hundred of kHz.

At the reverse, in presence of commensurability of order p, the supercell contains a finite number of inequivalent sites (atmost p for a one component q vector) giving rise to intense narrow resonance lines.

In the case of a single incommensurate q vector $-q=2k_F-$ and in the plane wave approximation, one has:

$$\delta\nu(\vec{R}) = \alpha\delta\rho(\vec{R}) + \beta\left[\delta\rho(\vec{R})\right]^2 = \nu_1 \cos(2\vec{k}_F\vec{R}+\phi) + \nu_2 \cos^2(2\vec{k}_F\vec{R}+\phi) \qquad (17)$$

If $\nu_2=0$, $g(\nu)$ takes the simple analytical form :

$$g(\nu) \propto (\nu_1^2 - \nu^2)^{-1/2} \qquad \nu_1 > \nu > -\nu_1$$
$$g(\nu) = 0 \qquad \nu > |\nu_1| \qquad (18)$$

*This is true as soon as the modulation of the quadrupolar coupling $\delta\nu_Q$ is small compared to the quadrupolar coupling ν_Q.

As a result of the local approximation, the lineshape is independent of the value of q, provided it is incommensurate. This is no longer true in the case where non-local interactions become dominant, like the electron nucleus dipolar magnetic interaction in some SDW systems [37,38].

Typical computer simulation of NMR lineshapes for various values of ν_1, ν_2, and q in the plane wave model are given in Fig. 2

On the other hand, in systems with multiple deformations, the lineshape is very sensitive to the symmetry of the q star; for example, the lineshape due to a triple incommensurate CDW[18,25] such as :

$$\delta\rho(\vec{R}) = \delta\rho\left[\cos(\vec{q}_1\vec{R}+\phi) + \cos(\vec{q}_2\vec{R}+\phi) + \cos(\vec{q}_3\vec{R}+\phi)\right]$$

$$|\vec{q}_1|=|\vec{q}_2|=|\vec{q}_3| \qquad \vec{q}_1 + \vec{q}_2 = \vec{q}_3 = 0 \tag{19}$$

is quite different from the lineshape resulting from the superposition of the spectra given by formula (18) due to three domains each having CDW with a differently oriented single q wavevector. (Fig. 2c)

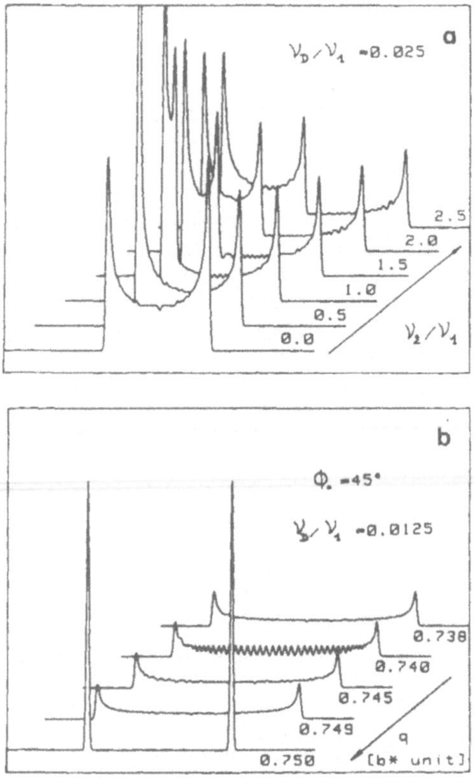

Fig. 2 Theoretical NMR lineshapes in the plane wave limit : $\nu(R) = \nu_1 \cos(q.R+\phi) + \nu_2 \{\cos(q.R+\phi)\}^2$. All frequencies distributions are convoluted with a Lorentzian of HWMH=ν_D a) as a function ν_1/ν_2 the ratio of the coupling constants linear and quadratic in the order parameter. b) in the linear case (ν_2=0) as a function of wave vector q.q=0.75 and q=0.74 corresponds to commensurate structures of order 4 and 50 respectively.

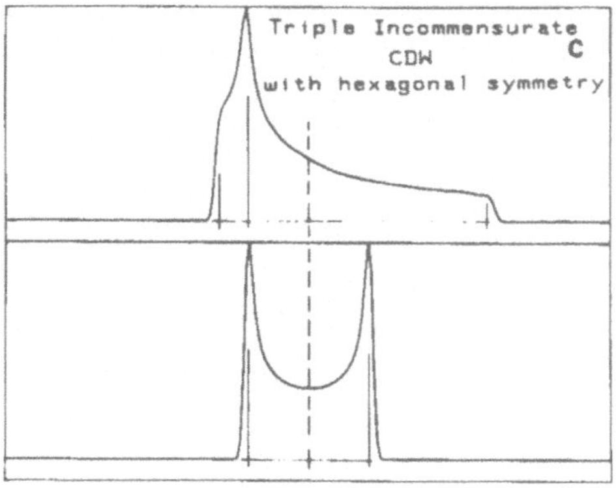

Fig. 2c Comparison between the theoretical lineshapes corresponding to a triple incommensurate CDW with hexagonal symmetry (top) and the superposition pattern of three equivalent domains with $2\pi/3$ rotated single q wavevector. (bottom)

Temperature dependence of the order parameter

We have seen in formula (17) that ν_1 is proportional to the modulus of the order parameter. More generally if $g(\nu-\nu_o,T)$ is the lineshape as a function of the temperature below T_P, one has in the linear approximation:

$$g(\nu-\nu_o, T) = g\left(\frac{\nu-\nu_o}{|\psi_{(T)}|}\right) \tag{20}$$

where $|\psi(T)|$ is the amplitude of the order parameter. One can thus extract the temperature dependence of the order parameter from that of the lineshape [20,22,33]. In Fig. 3a the temperature dependence of the (-1/2,1/2) transition of ^{87}Rb in $Rb_{0.3}MoO_3$ [33] is shown. The corresponding temperature dependence of the order parameter, shown in Fig. 3b, deviates from the mean field BCS theory ; a critical exponent $\eta = 0.32 + 0.03$ is found [39] in good agreement with the expected value $\eta = 0.35$ for a 3D phase transition with a two component order parameter.(Fig. 3c) Close to the transition, within a temperature range of about 5-10K, an anomalous behavior is observed, corresponding to a partial motional narrowing of the line, due to thermally induced depinning of the CDW. A quite similar behavior was also observed in the insulating compound Rb_2ZnCl_4 [40] close to the paraelectric incommensurate phase transition.

Incommensurate-commensurate transition

When the q vector of the CDW becomes close to a simple rational fraction of a lattice reciprocical vector

$$\vec{q} = \frac{m}{p}\vec{b^*}(1-\delta)$$

as shown by McMillan,[40] the CDW gains energy by the creation of commensurate domains ($q=q_c = mb^*/p$)separated by walls called discommensurations (DC). Within the DC, the phase varies rapidly by an amount of $2\pi/p$. The existence of discommensurations was first demonstrated by NMR in 2H - $TaSe_2$ [24]. Extensive studies of the incommensurate commensurate transition (ICT) by NMR have been performed in the dielectrics [20,41]. Let us briefly describe the case of 1D incommensurate modulation [20] for which

$$\delta\rho(R) = \delta\rho \cos(q_c R + \phi(R) + \phi_o) \tag{21}$$

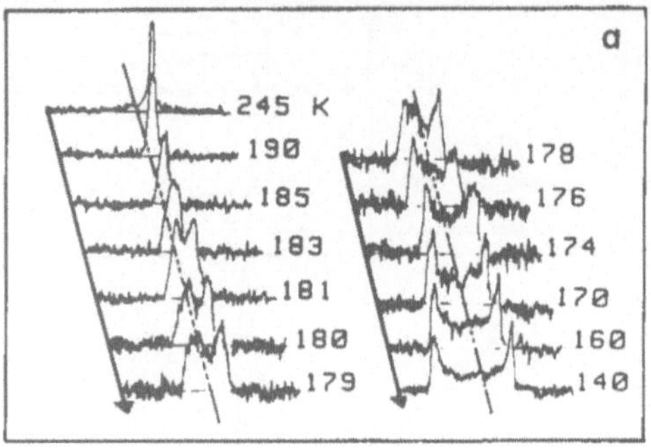

Fig. 3a Temperature dependence of the (-1/2,1/2)NMR line of ^{87}Rb in $Rb_{0.3}MoO_3$ [33,39]

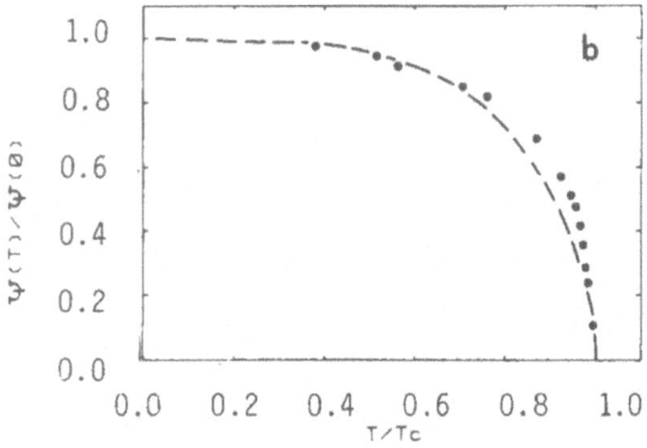

Fig. 3b Temperature dependence of the order parameter $\psi(T)/\psi(0)$ extracted from the ^{87}Rb NMR lineshape [33,39]. The dashed curve corresponds to the BCS mean field theory.

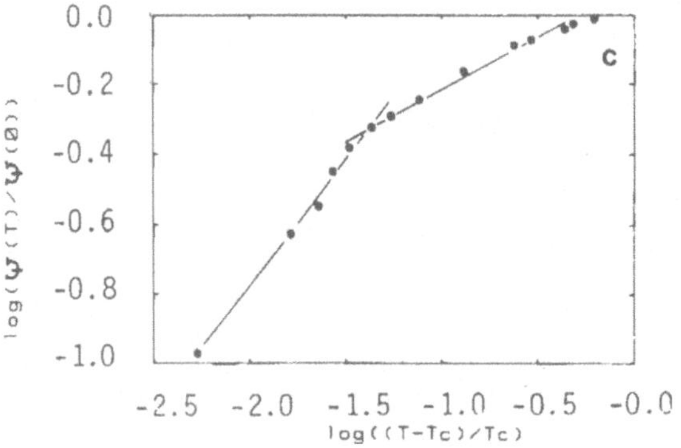

Fig. 3c Log-Log plot of the reduced order parameter as a function of $(T-T_C)/T_C$. A critical exponent $\eta=0.32 \pm 0.03$ is found [39]. The anomalous behavior close to the transition is due to a partial depinning of the CDW [40].

where $q_c = \frac{b^*}{p}$ is the commensurate wave vector. The incommensurate plane wave limit corresponds to

$$\phi(R) = (q-q_c)R$$

In presence of a lock-in potential, and in the continuum limit, $\phi(R)$ is shown [40,42] to obey the Sine-Gordon equation

$$\frac{d^2\phi}{dR^2} = \alpha^2 p \sin(p\,\phi) \tag{22}$$

A particular solution of this equation is given by the "monosoliton"

$$\phi(R) = \frac{4}{p} \operatorname{Arctg}(\exp(-\alpha\sqrt{p}\,R)) \tag{23}$$

where α^{-1} correspond to the width of the wall.
As long as the distance ℓ between solitons is large compared to the width α^{-1}, $\phi(R)$ of the DC phase can be expressed as:

$$\phi(R) = \sum_{n=-\infty}^{+\infty} \frac{4}{p} \operatorname{Arctg}(\exp(-\alpha\sqrt{p}\left[R - (n+\tfrac{1}{2})\ell\right])) \quad (\mathrm{mod}\ 2\pi) \tag{24}$$

When $\alpha\ell$ becomes of the order of 1, one has to solve the Sine-Gordon equation which can be rewritten in the form:

$$\frac{d\theta}{dx} = 2(\delta^2 + \cos^2(\frac{\theta}{2}))^{1/2} \qquad \theta = p\phi\ ,\ x = \sqrt{p}\,\alpha R \tag{25}$$

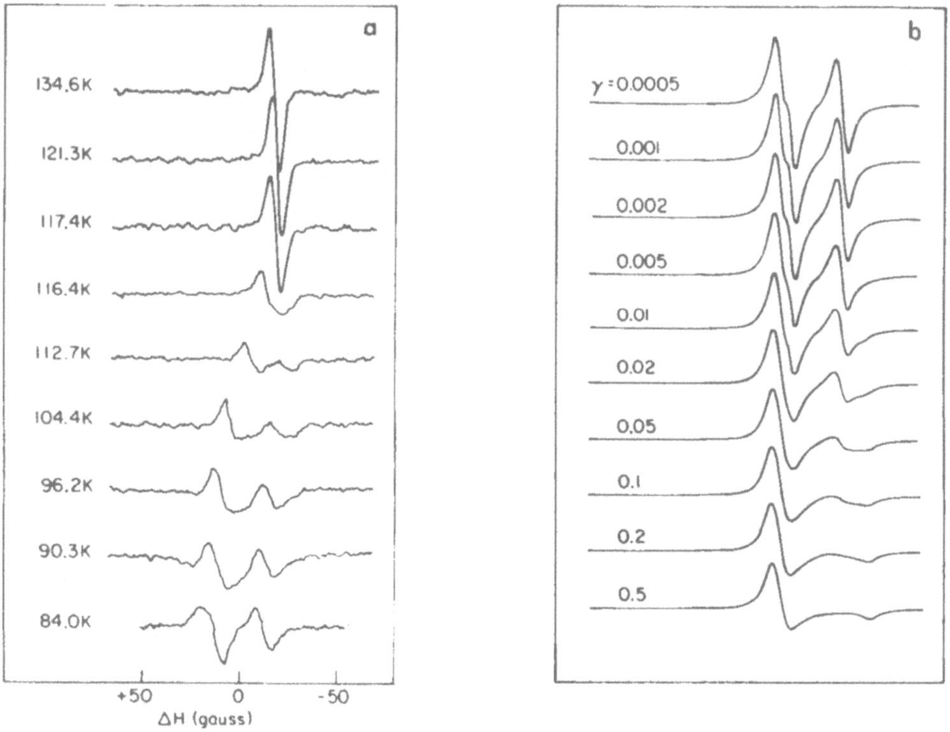

Fig. 4 a, left) NMR derivative spectra of ^{77}Se in $2H\text{-}TaSe_2$ as a function of temperature.(24)
b, right) Theoretical NMR spectra in the presence of CDW at various solitons density.(24)

where δ is an integration constant. ($\delta=0$ corresponds to the monosoliton limit, $\delta=\infty$ to the plane wave limit). A soliton density can be empirically defined [20] by comparing the slope $d\theta/dx$ at the center of a wall with the average slope of phase $2\pi/L$ where L is the period of $\theta(x)$ One thus obtains

$$n_s = \frac{\pi}{2} / K\left[(1+\delta^2)^{-1}\right] \tag{26}$$

where K is the elliptic function of 1st kind [43]

Clear evidence of DC and incommensurate-commensurate transition in CDW systems have been obtained in quasi-2D systems [24,25] only.(Fig. 4a,b). In quasi-1D conductors like $Rb_{0.3}MoO_3$, in which the wave number remains very close to 3/4 below 100K, an NMR lineshape quite close to the plane wave limit has been found [33,34,35]. Fig.4c shows the variation of the lineshape with $\alpha\ell$ for a value of q=0.749 in the soliton lattice

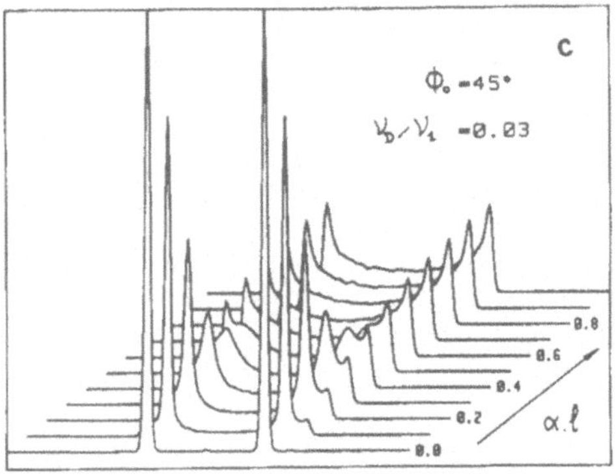

Fig. 4c Computer simulation of NMR lineshapes for a 1-dimensional soliton lattice (q=0.749) as a function of the ratio of the soliton width α^{-1} to the solitons spacing ℓ. (Equ. 24).

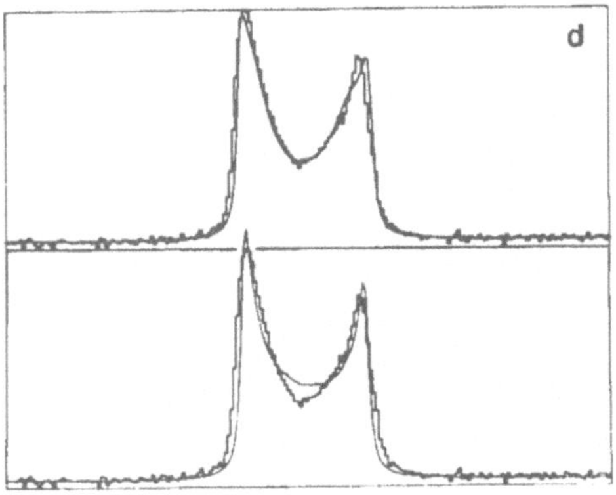

Fig. 4d Comparison of computer simulations to the experimental lineshape of ^{87}Rb in $Rb_{0.3}MoO_3$ (52). Bottom : plane wave limit, ν_1=4.9kHz, ν_2=0.7kHz. Top : multisoliton lattice with $\alpha\ell$=0.56

approximation. In Fig. 4d, the experimental lineshape of ^{87}Rb in $Rb_{0.3}MoO_3$ [39] is compared with simulations within the plane wave and the soliton lattice approximation. A term quadratic in $\delta\rho(R)$ is also included. It turns out that DCs, if they exist in this system at all, have a width of the order of their spacing.

Interaction between CDW and charged impurities

- Anomalous Friedel oscillations.

A strongly temperature dependent anomalous broadening of NMR lines beginning well above the normal incommensurate CDW transition temperature has been reported in some low dimensional conductors ($2H-NbSe_2$[18,19], $Rb_{0.3}MoO_3$[33]). This broadening, initially attributed to dynamic CDW fluctuations [18] has been shown to be inhomogeneous and static on the NMR time scale (10^{-3} s^{-1}) [22] and thus corresponds to a static distribution of quadrupolar couplings. Such an effect may be readily understood by considering the nature of the Friedel oscillations around an impurity in a low dimensional conductor. The electronic susceptibility of the electron gas in the RPA approximation is given by

$$\chi(\vec{q}, T) = \frac{\chi^o(\vec{q}, T)}{1-U(\vec{q})\,\chi^o(\vec{q}, T)} \tag{27}$$

where $U(\vec{q})$ represents both the electron-electron and the electron phonon coupling. The q component of the screening charge around an impurity defined by a bare potential $w^o(q)$ is given by :

$$\delta\rho(q) = \frac{\chi^o(q,T)w^o(q)}{1-(U(q)+4\pi/q^2)\chi^o(q,T)} \tag{28}$$

The component at $q=2k_F$ which corresponds to the Friedel oscillation in 3D, is seen to be strongly temperature dependent especially in the quasi 1D system. These anomalous Friedel oscillations are expected to be quite anisotropic. Fig. 5 shows the broadening of the (1/2,3/2) ^{93}Nb NMR line in $2H-NbSe_2$ corresponding to these oscillations. This effect deserves further investigation; it may contribute to the central peaks observed in X rays or neutrons scattering in CDW systems [44].

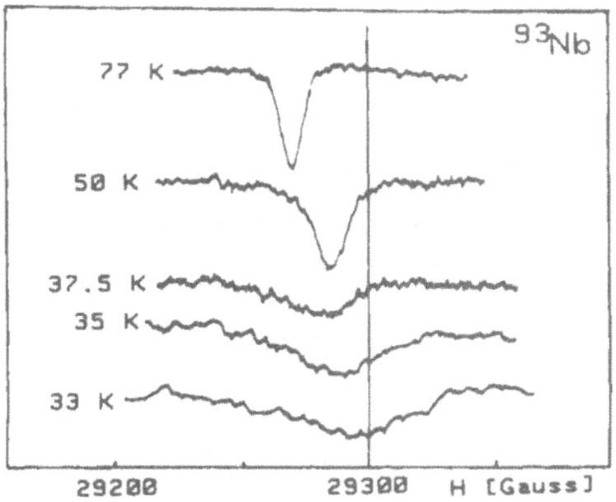

Fig. 5 Temperature dependence of the ^{93}Nb $(-\frac{1}{2}, \frac{1}{2})$ NMR lineshape in $NbSe_2$ above the Peierls transition temperature T=33K[22]. The broadening is due to anomalous Friedel oscillations around impurities.

Defect density waves

It has been suggested by Lederer et al[45] that mobile charged defects could partially order because of their interaction with the CDW, considered as rigid. Such an effect has indeed been observed in $Rb_{0.3}MoO_3$ just below the Peierls transition, through a time dependent broadening of the NMR line on a line scale of a few hours. This phenomenon is well known in incommensurate dielectrics[46].

III - DYNAMICS OF THE CDW

Up to now spin lattice relaxation studies in CDW systems have been mainly concerned with the modification of the density of states at the Fermi level as the gap opens[14,22,29,48]. Contributions to the SLRR by the critical fluctuations of the CDW around the Peierls transition were found to be unobservable in 2D systems[22] and barely observable in $NbSe_3$ [30]. However, the contributions from both critical fluctuations and excitations of the condensate below T_P (phasons and amplitudons) have been recently observed in $Rb_{0.3}MoO_3$ [39]. We shall illustrate by this study the power of NMR techniques to investigate CDW dynamics, and shall also widely refer to the extensive work on the incommensurate dielectrics [20,49]. Let us give a brief account of the theory. As seen in section II, the quadrupolar frequency at site $\vec{R}$ can be written as

$$\omega_Q(\vec{R}) = \omega_Q^o(\vec{R}) + \omega_1 \psi(\vec{R}) + \omega_2[\psi(\vec{R})]^2 \tag{29}$$

where $\psi(\vec{R})$ is the real part of the order parameter.
Let us write

$$\begin{aligned} \psi(\vec{R}, t) &= \overline{\psi(\vec{R})} + \delta\psi(\vec{R},t) \\ \omega_Q(\vec{R}, t) &= \overline{\omega_Q^o(\vec{R})} + \delta\omega_Q(\vec{R},t) \end{aligned} \tag{30}$$

One has :

$$\frac{1}{T_1(\vec{R})} \propto \int_0^\infty \langle\delta\omega_Q(\vec{R},t)\,\delta\omega_Q(\vec{R},0)\rangle\, e^{i\omega_L t}\, dt \tag{31}$$

Around T_P, the spin lattice relaxation rate is due both to the usual Korringa relaxation (wich rapidly vanishes below T_P), and to the critical fluctuations of the CDW. Thus

$$\frac{1}{T_1(\vec{R})T} = A \sum_q \frac{\chi^{CDW}_{2k_F+q}(\omega_L)}{\omega_L} + B(\chi^M_{exp} - \chi_{orb})^2 \tag{32}$$

where χ^M_{exp} is the measured magnetic susceptibility and χ^{CDW} the electrical susceptibility, and A and B are two constants. Below T_P, the phase and the amplitude fluctuations of ψ become decoupled

$$\begin{aligned} \delta\psi(R,t) = \sum_q \{(A_q e^{i(qR+\omega_A(q)t)} + cc)\cos(2k_F R+\phi_o) + \cdots \\ \cdots + (\varphi_q e^{i(qR+\omega_\varphi(q)t)} + cc)\sin(2k_F R + \phi_o)\} \end{aligned} \tag{33}$$

where A_q (φ_q) is the anihilation operator of an amplitudon (phason) of wave vector q. One has the dispersion relation

$$\omega_\alpha^2(\vec{q}) = (\omega_\alpha^o)^2 + K_{/\!/} q_{/\!/}^2 + K_\perp q_\perp^2 \quad \alpha = A, \varphi \tag{34}$$

where ω_α^o is the pinning frequency.

We must distinguish between two relaxation processes :
The direct process, which corresponds to the emission (absorption) of one excitation at the Larmor frequency. Since $\omega_L << \omega_\alpha^o$ this process is efficient only if the modes are overdamped. In this case

$$\frac{1}{T_1} \propto (\omega_1 + \omega_2 \overline{\psi})^2 \{X^2 J_A(\omega_L) + (1-X^2) J_\varphi(\omega_L)\} \tag{35}$$

where

$$X = \cos(2k_F R + \phi_o)$$

and

$$J_\alpha = \frac{kT}{\omega_L} \sum_q \chi''_q(\omega_L) = \frac{kT}{\omega_L} \sum_q \frac{\Gamma_q \omega_L}{(\omega_\alpha^2(q) - \omega_L^2)^2 + \Gamma_q^2 \omega_L^2} \tag{36}$$

Let us suppose that $\Gamma_q = \Gamma$ for $q_{/\!/} < \Lambda_{/\!/}$, $q_\perp << \Lambda_\perp$ and $\Gamma = 0$ elsewere . In the limit where $K_{/\!/}\Lambda_{/\!/}^2 >> K_\perp \Lambda_\perp^2 >> \Gamma\omega_L$ we obtain :

$$J_\alpha = M^{-1} K_\perp K_{/\!/}^{1/2} \Gamma/\omega_\alpha^o \tag{37}$$

At this stage, two considerations have to be made. First, the amplitudon gap ω_A^o is much larger than the phason gap, ω_φ^o , and as a result, $J_\varphi >> J_A$. Secondly, the relative weight of the contribution of the phasons and the amplitudons to the SLRR varies with X, i.e within the linewidth

$$g\left(\frac{\nu-\nu_o}{\nu_1}\right) = g(X) \alpha (1-X^2)^{-1/2}$$

The phasons are inefficient in relaxing nuclei contributing to the two singularities in g(X) at X=+1,-1 but fully affect those at the center of the line at X=O. This has been experimentally observed in incommensurate dielectrics [20,49] and in $Rb_{0.3}MoO_3$ [39]. Another process to consider is the Raman process, which corresponds to the scattering of excitations by the nuclei from wave vector q to q', with an energy change $_L$. The Raman process only involve terms bilinear in A_q or $_q$, and can be written [49] as

$$\frac{1}{T_1} = \omega_2^2 \left[X^2 J_{AA}(\omega_L) + (1-X^2) J_{\varphi\varphi}(\omega_L)\right] \tag{38}$$

with

$$J_{\alpha\alpha} = \frac{1}{M^2} \int_{\omega_\alpha^o}^{\omega_c} \frac{[\mathcal{D}(\omega)]^2}{\omega^2} \frac{e^{\frac{\hbar\omega}{kT}}}{\left(1-e^{\frac{\hbar\omega}{kT}}\right)^2} d\omega \tag{39}$$

If assuming a density of states for the phasons (amplitudons)

$$\mathcal{D}(\omega) = K^{-3/2} \omega(\omega^2 - \omega_o^2)^{1/2} \qquad \text{(isotropic dispersion)}$$

one finds :

$$J_{\alpha\alpha} \propto \frac{1}{M^2 K^{3/2}} (kT)^3 \int_{x_o}^{x_1} (x^2 - x_o^2) \frac{e^x}{(1-e^x)^2} dx \qquad x = \frac{\hbar\omega}{kT}$$

A low temperature $\hbar\omega >> kT$, we obtain

$$J_{\alpha\alpha} \propto M^{-2} K^{-3/2} \; kT \; e^{-\frac{\hbar\omega_{\alpha}^{o}}{kT}} \qquad \alpha = A,\varphi \tag{40}$$

Let us consider the experimental results in $Rb_{0.3}MoO_3$[39]. In Fig. 6 is shown the temperature dependence of $(T_1T)^{-1}$ for the ^{87}Rb (T_1^{-1} is averaged over the whole line). For T > 200 K, the SLRR is dominated by the Korringa process and scales with the squared experimental spin susceptibility $\chi^2(T)$ (equation 32). In the temperature range 170–200 K, a peak is observed, corresponding to the critical fluctuations around T_P=181 K. Between 170 and 100 K, the SLRR is nearly flat, and then exponentially decreases below 100K. Several explanations are possible for this decrease.

a) In presence of an incommensurate–commensurate transition, ω_{φ}^{o} increases by an order of magnitude, giving a sudden decrease of the SLRR, as observed in Rb_2ZnCl_4[20]. However, the lineshape at low temperature in $Rb_{0.3}MoO_3$ is in contradiction with the existence of commensurate domains.

b) The overdamping of the phason modes, that is the value of Γ could decrease at low temperature, in connection with the decreasing number of free carriers available for screening. From equ. 37, this should lead to a decrease of the SLRR.

c) As far as the Raman processes are concerned, a thermally activated behavior is also expected at temperatures lower than the excitations gap.

Further data are needed for a more definite conclusion ; it nevertheless illustrates how sensitive the SLRR is to CDW fluctuations for nuclei coupled to the CDW by quadrupole couplings.

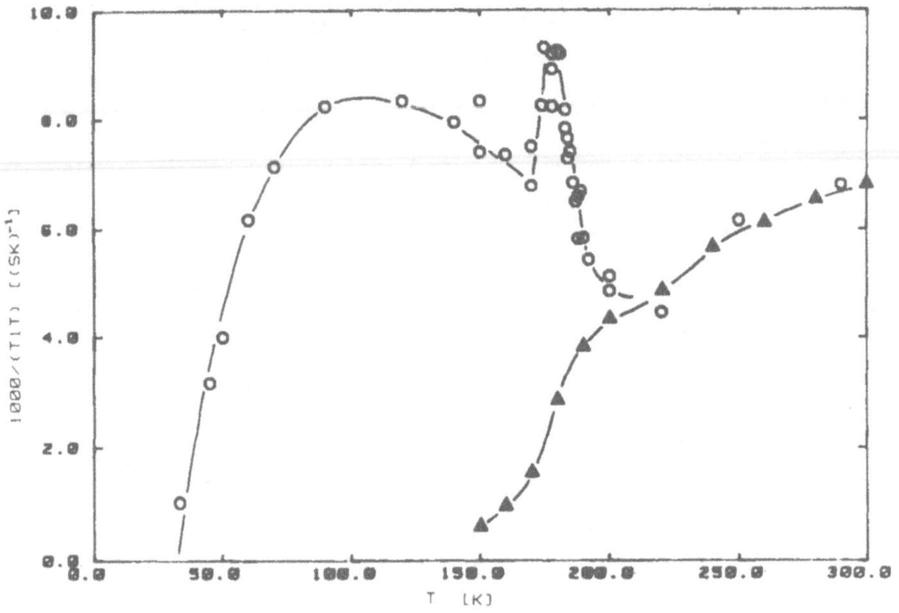

Fig. 6 ^{87}Rb spin lattice relaxation rate as a function of temperature in $Rb_{0.3}MoO_3$(39). The full triangles are the squared experimental spin susceptibility reduced at an appropriate scale.

IV - NMR UNDER BIAS CURRENT

Non-linear transport properties and large voltage fluctuations induced by a current referred to as "noise" have been discovered in a number of low-dimensional conductors exhibiting a Peierls transition [50,51]. Theories attempting to explain this phenomenon are based on Fröhlich's idea of a collective motion of the electrons which may be envisaged in its simplest form as a rigid sliding motion of the CDW coupled to an oscillating coherent motion of the lattice. Although these ideas are generally accepted, a direct proof of a motion of the electron condensate as a whole under an electric field has been lacking. Recently, three NMR experiments in $NbSe_3$[36] and $Rb_{0.3}MoO_3$ [52,53] have given evidence for CDW motion. In Ref. 52 electric conductivity noise and NMR of ^{87}Rb in a $Rb_{0.3}MoO_3$ crystal under current were measured at 77 K on the same sample, and unambigbuously showed that the appearance of the extra conductivity and noise above a threshold electric field is induced by a sliding motion of the CDW.

Let us briefly describe the theory of the NMR lineshape in presence of a sliding CDW[54,39].

Within the adiabatic approximation, the correlation function of the nuclear transerve magnetization G(t) can be written [3] as

$$G(t) = \frac{1}{N} \sum_j e^{i\int_0^t \omega(R_j,t')\, dt'} \tag{41}$$

with

$$\omega(R_i,t) = \omega^0(R_i) + \omega_1 \cos\left[2k_F R_i + \phi(t)\right] + \omega_2 \cos^2\left[2k_F R_i + \phi(t)\right] \tag{42}$$

In the static case, ($\phi(t) = \phi_0$) and assuming $\omega_2 = 0$, (41) becomes :

$$G(t) = \frac{1}{N} \sum_i \{J_0(\omega_1 t) + 2 \sum_n J_n(\omega_1 t) \cos n(qR_i+\phi)\} \tag{43}$$

where J_n is the Bessel function of nth order.
Using the fact that q is incommensurate

$$\frac{1}{N} \sum_i \cos n(qR_i+\phi) = \delta(nq-K) \qquad (K = mb^*) \tag{44}$$

(43) reduces to $G(t) = J_0(\omega_1 t)$

Let us now assume and $\phi(t) = \Omega t$, which corresponds to a sliding CDW with uniform velocity $v_{CDW} = \lambda\Omega/2\pi$. Using again the condition (44) a straightforward calculation leads to

$$G(t) = \sum_{n=-\infty}^{+\infty} \left[J_n\left(\frac{\omega_1}{\Omega}\right)\right]^2 e^{in\Omega t} \qquad \text{i.e.}$$

(45)

$$g(\omega) = \left\{\sum_{n=-\infty}^{\infty} \left[J_n\left(\frac{\omega_1}{\Omega}\right)\right]^2 \delta(\omega - n\Omega)\right\} * h(\omega)$$

where $h(\omega)$ is a Lorentzian lineshape corresponding to the secular spin-spin interaction. In Fig.7a the variation of the lineshape as a function of ω_1/Ω according to (45) is shown. Well defined sidebands appear at $\omega = \pm n\Omega$ for a uniform motion of the CDW leading to a straightforward determination of v_{CDW}. However, if there is a distribution in v, these sidebands will be smeared and hidden by the noise (Fig.7b).

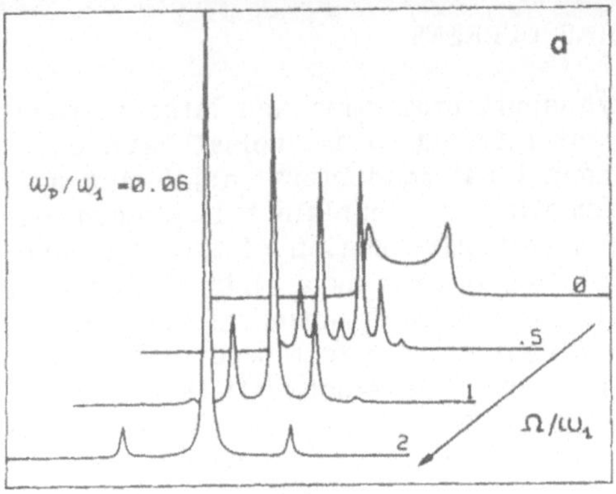

Fig. 7a Theoretical NMR lineshape in presence of a CDW sliding with uniform velocity $v = \lambda_{CDW}\,\Omega/2\pi$. The local quadrupole frequency is $\omega(R) = \omega_1 \cos(qR + \Omega t + \phi o)$

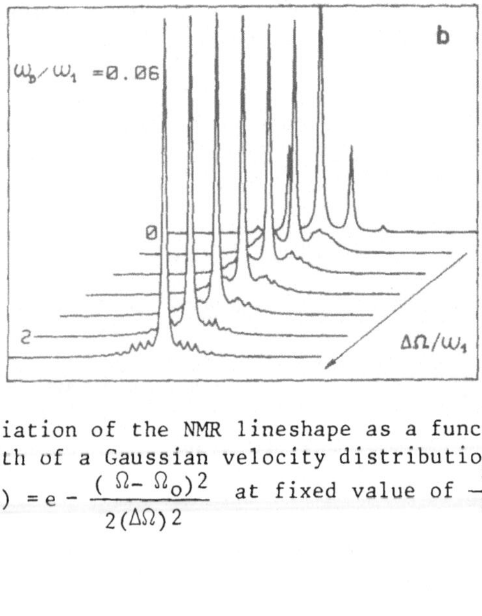

Fig. 7b Variation of the NMR lineshape as a function of the width of a Gaussian velocity distribution $p(\Omega) = e^{-\frac{(\Omega-\Omega_o)^2}{2(\Delta\Omega)^2}}$ at fixed value of $\frac{\Omega_o}{\Omega_1} = 1$

The main effect of the CDW motion on the lineshape is then the appearance of a central line corresponding to n = 0 in (45).

The position and the width of the central line n = 0 - contrary to the sidebands - is independent of v. Moreover, calculations show that for $\Omega/\omega_1 > 2$, 90% of the intensity is included in the central line. The effect of a finite value for ω_2 simply results in a shift by an amount of $\omega_2/2$ of the position of the central line. In Fig.7c three NMR spectra at different values of the bias current (I=0, 17 and 165 mA) are shown .

The NMR spectra contain a static component which decreases with increasing bias current, but does not disappear even for the largest current, implying that the CDW remains pinned in part of the sample. A fit of the NMR spectra to the expression

$$g(\omega) = F_s\,\mathcal{G}_s(\omega) + (1-F_s)\,\mathcal{L}_D(\omega)$$

where $\mathcal{G}_s$ is the lineshape for I=0, F_s is the fraction of sample in which

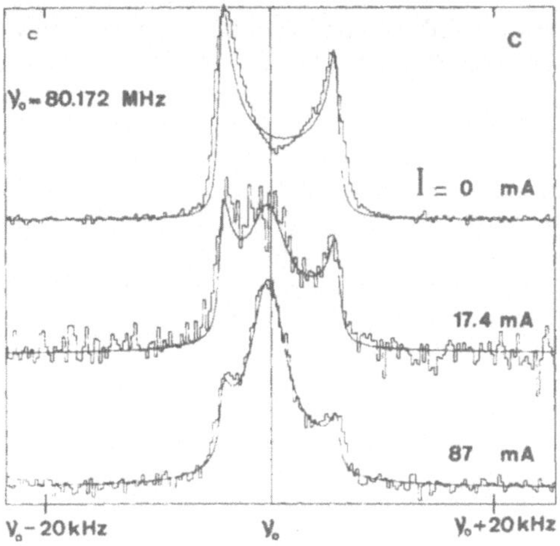

Fig. 7c Experimental NMR lineshapes of ^{87}Rb in $Rb_{0.3}MoO_3$ under bias current (I=0, 17,87 mA).

the CDW is static, and $\mathcal{L}_D(\omega)$ is a Lorentzian of constant width allowed the determination of the dynamic fraction $F_D = 1-F_S$ as a function of the bias current (Fig. 7d).

The main conclusion of this study, as well as those of Ref. 36,53, is a direct evidence of the sliding of the CDW. But it also demonstrates that the current density is highly inhomogeneous in the sample, and that the onset of the Fröhlich mode is far from being simultaneous in the whole sample. Careful analysis of the NMR data in $NbSe_3$[36], where a collection of single crystals was used, leads to a similar conclusion.

The knowledge of the dynamic fraction of the sample allows a better determination of the effective CDW current density, j_{CDW} than the ratio of the current over the full sample cross-section and therefore, a safer determinations of the relationship between j_{CDW} and the whashboard frequency [50] of the voltage oscillations associated to the non linear trans-

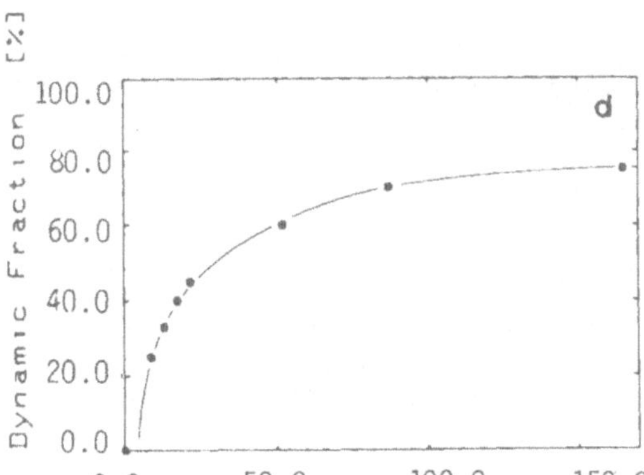

Fig. 7d Fraction of the sample in which the CDW is moving as a function of the intensity of the bias current.

port.Contrary to earlier studies the washboard frequency determined in this way was found in agreement with the theoretical prediction of the classical model [56] in $Rb_{0.3}MoO_3$.

The central line corresponding to the part of the crystal in which the CDW is moving shows a dynamical broadening indicating that the CDW motion is not entirely periodic, its phase fluctuates in time with a correlation time τ_c of the order of 200 μs. The failure of the earliest attempt [55] may have been partly associated to the wide distribution of the CDW velocity, but also to the finite correlation time τ_c. In this study a spin echo technique was used in which the amplitude and the shape of echo at time much larger than τ_c were detected, thus missing the contribution of nuclei due to the dynamic fraction of the sample.

Finally, NMR has also been used to checked the displacement of the CDW for applied electric field below the threshold field[36]. Surprisingly only small displacements were found at field close to the threshold. This was recently explained [57] in the framework of the classical deformable fluid model of Fukuyama Lee and Rice [58] as a consequence of the slow relaxationof the phase for low fields. In this NMR study a sequence of unipolar voltage pulses was applied; quite different results are predicted for a bipolar pulse sequence. This should be checked in a near future.

Acknowledegments

The authors are grateful to Professor C.P. Slichter for giving permission to reproduce some of his NMR data and indebted to P. Butaud and A. Jannossy for helpful discussions.

REFERENCES

1. J. Winter, Magnetic Resonance in Metals, Oxford Clarendon Press (1971)
2. L.C. Hebel and C.P. Slichter, Phys. Rev. 113,1504 (1959).
3. A. Abragam, Principles of Nuclear Magnetism, Clarendon Press, Oxford (1961)
4. C.P. Slichter, Principles of Magnetic Resonance, Springer Series in Solid State Science, Vol.1, (Springer Verlag, Berlin Heidelberg New York 1978)
5. M. Mehring, High Resolution NMR Spectroscopy in Solids, Series NMR : Basic Principles and Progress, Vol.11, Springer Verlag, (Berlin, Heidelberg, New York, 1976).
6. A. Rigamonti, Adv. in Physics, 33 115 (1984).
7. A.M. Clogston, V. Jaccarino and Y. Yafet, Phys. Rev. 134 A650 (1964).
8. V. Jaccarino, Proceedings of International School of Physics E. Fermi, Varenna 1966, ed. W. Marshall, New York, Academic Press Vol. 37, pp 335-85.
9. J. Korringa, Physica 16 (1950)
10. A. Narath, Phys. Rev. 162 320-32 (1967)
11. F. Devreux, Phys. Rev. B13 4651 (1976).
12. G. Soda, D. Jerome, M. Weger, J. Alizon, J. Gallice, H. Robert, J.M. Fabre and L. Giral, J. Phys. 38 931 (1977).
13. P.A. Lee, J.M. Rice and P.W. Anderson, Phys. Rev. Lett. 31 462 (1963)
14. H. Niedoba, H. Launois, D. Brinkman and H.U. Keller, J. Phys. Lett. 35 L251 (1974) ; E.F. Rybaczewski, L.S. Smith, A.F. Garito, A.J. Heeger and B.G. Silbernagel, Phys. Rev. B14 2746 (1976).
15. Non local corrections have been considered in the case of the incommensurate dielectrics ; see R. Blinc, J. Seliger and S. Zumer, J. Phys. C ; Solid State Phys. 18 2313 (1985)
16. B.R. Weinberger, Phys. Rev. B17 566 (1978).

17. P. Molinie, L. Trichet, J. Rouxel, C. Berthier, Y. Chabre, P. Ségransan, J. Phys. Chem. Solids 45 105 (1984).
18. J.A.R. Stiles and D.L.G. Williams, J. Phys. C, Solid State Phys. 9 3941 (1976).
19. C. Berthier, D. Jerome, P. Molinie and J. Rouxel, Solid State Comm. 19 131 (1976).
20. R. Blinc, Phys. Rep. 79 331 (1981) ; R. Blinc, P. Prelovsek, V. Rutar, J. Seliger and S. Zumer in "Incommensurate Phases in Dielectrics" eds. R. Blinc and A.P. Levanyuk, Elsevier Science Publishers, B.V. (1986), chap. 4.
21. F. Borsa, D.R. Torgeson and H.R. Shanks, Phys. Rev. B15 4576 (1977)
22. C. Berthier, D. Jerome and P. Molinie, J. Phys. C ; Solid State Phys. 11 797 (1978).
23. A.V. Skripov and A.P. Stepanov, Solid State Comm. 53 469, (1985)
24. B.H. Suits, S. Couturie and C.P. Slichter, Phys. Rev. Lett. 45 194 (1980) ; Phys. Rev. B23 5142 (1981).
25. L. Pfeiffer, R.E. Walstedt, R.F. Bell and T. Kovacks, Phys. Rev. Lett. 49 1162 (1982).
26. M. Naito, H. Nishihara, S. Tanaka, J. Phys. Soc. Japan 54 3946 (1985)
27. A.H. Thompson and B.G. Silbernagel, Phys. Rev. B19 3240 (1979).
28. A.V. Skripov, A.P. Stepanov, A.D. Shevchenko and Z.D. Kovalyuk, Phys. Stat. Solidi (b) 119 401 (1983).
29. R. Dupree, W.W. Warren and F.J. Disalvo, Phys. Rev.B16 1001 (1977)
30. F. Devreux, J. Physique 43 1489 (1982).
31. B.H. Suits and C.P. Slichter Phys. Rev.29 41 (1984)
32. P. Butaud, P. Ségransan, C. Berthier, and A. Meerschaut in Charge Density Waves in Solids, ed. by G. Hutiray and J. Solyom, Lectures Notes in Physics, Vol. 217 (Springer Verlag, Berlin 1985) p. 71.
33. P. Butaud, P. Ségransan, C. Berthier, J. Dumas and C. Schlenker, Phys. Rev. Lett. 55 253 (1985).
34. D.C. Douglass, L.F. Schneemeyer, S.E. Spengler B.A.P.S. 30, 465 (1985) and to be published.
35. K. Nomura, K. Kume and M. Sato, Solid State Comm. 57 611 (1986)
36. J.H. Ross, Z. Wang and C.P. Slichter, Phys. Rev. Lett. 56.
37. J.M. Delrieu, M. Roger, Z. Toffano, A. Moradpour and K. Bechgaard, J.Physique 47 839 (1986) ; Z. Toffano, Thesis, Orsay (1986).
38. T. Takahashi, Y. Maniwa, H. Kawamura and G. Saito, J. Phys. Soc. Japan, 55 1364 (1986).
39. P. Butaud, P. Ségransan, C. Berthier (to be published).
40. W.L. McMillan, Phys. Rev. B12 1187 (1975) ; B14 1496 (1976)
41. R. Blinc, F. Milia, B. Topic and S. Zumer, Phys. Rev. B29 4173 (1984).
42. A.D. Bruce and R. Cowley, J. Phys. Solid St. Phys. II 3609-30; A.D.Bruce, J. Phys.C Solid St. Phys. 13 4615 (1980)
43. M. Abramowitz and J.A. Stegun, "Handbook of Mathematical Functions", Dover, N.Y. (1970).
44. J.P. Pouget, This Institute
45. P. Lederer, G. Montambaux and J.P. Jamet, Mol. Cryst. Liqu. Cryst. 12 99 (1985).
46. G. Errandona et al., Ferroelectrics 53 247 (1984).
47. S. Wada, R.Aoki, and O. Fujita, J. Phys. F. Met. Phys. 14 1515 (1984)
48. T. Takahashi, D. Jerome, F. Masin, J.M. Fabre and L. Giral, J. Phys. C, Solid St. Phys. 17 3777 (1984).
49. S. Zumer and R. Blinc, J. Phys.C ; Solid State Phys. 14, 465 (1981).
50. e.g. See, Charge Density Waves in Solids, eds. Gy. Hutiray and J. Solyom, Lecture Note Series in Physics, 217, (Springer Verlag New York 1985).
51. R.C. Lacoe, H.J. Shulz, D. Jerome, K. Bechgaard and I. Johannsen, Phys. Rev. Lett. 55 2351 (1986).

52. P. Ségransan, A. Janossy, C. Berthier, J. Marcus and P. Butaud, Phys. Rev.Lett. 56 1854 (1986).
53. K. Nomura, K. Kume and M. Sato, J. Phys. C, Solid State Phys. 19 L289 (1986).
54. M. Kogoj, S. Zumer and R. Blinc, J. Phys. C, Solid State Phys., 17 2415 (1984).
55. D.C. Douglas, L.F. Schneemeyer, and S.E. Spencer, Phys. Rev. B32, 1813 (1985)
56. G. Gruner, A. Zawadowski and P.M. Chaikin, Phys. Rev. Lett. 46 511 (1981) : P. Monceau, P. Richard, Phys. Rev. B25, 981 (1982).
57. S.N. Coppersmith, Phys. Rev. Lett. 57 1191 (1986).
58. Fukuyama and P.A. Lee, Phys. Rev. B17 535 (1977) ; P.A. Lee and T.M. Rice, Phys. Rev. B19 3970 (1979).

CHARGE DENSITY WAVE PROPERTIES OF THE

MOLYBDENUM BLUE BRONZES $A_{0.30}MoO_3$

Claire Schlenker

Laboratoire d'Etudes des Propriétés Electroniques des Solides*
C.N.R.S. - B.P. 166
38042 Grenoble Cedex - France

I - INTRODUCTION

The transition metal oxide bronzes belong to the large family of transition metal oxides which often show unusual electronic properties such as metal-non metal transitions or charge density wave (CDW) instabilities.

The oxide bronzes are compounds with the general formula A_xTO_m where A is a monovalent metal, often an Alkali metal and T a transition metal. In the oxide TO_m, the d states are usually empty and the A metal donates its outer electron to the conduction band which becomes partially filled in the bronze. The physical properties thus depend strongly on the extension of the d wave function. While the 5d tungsten bronzes are metallic at all temperatures, the 3d vanadium bronzes such as $Na_xV_2O_5$-β, rather show a semiconducting behavior. The molybdenum bronzes are intermediate as far as the 4d electron localization and the width of the conduction band are concerned[1]. The so-called red bronzes $A_{0.33}MoO_3$ are semiconductors, the blue bronzes of formula $A_{0.30}MoO_3$ show a metal to semiconductor transition and the purple bronzes $A_{0.9}Mo_6O_{17}$ are metallic at all temperatures. In this case, one should note that the partial filling of the conduction band results both from the charge transfer from the alkali metal and from the lack of oxygen, as compared to the oxide MoO_3. From this point of view, the molybdenum oxides Mo_4O_{11} are similar to the purple bronzes. In the two compounds, the monoclinic η and the orthorhombic γ phases, only the lack of oxygen accounts for the metallic conductivity[2]. All these materials, bronzes and oxides, are well-defined compounds in which the alkali or oxygen concentration cannot be changed in a continuous way.

An important property common to these bronzes and oxides is their anisotropic, layer-type crystal structure. This leads to anisotropic electrical properties : the purple bronzes and Mo_4O_{11} oxides are quasi two-dimensional metals, while the blue bronzes, because of the existence of infinite chains of MoO_6 octahedra in the structure, are quasi one dimensional metals. In all cases, the Fermi surface is expected to be anisotropic and to provide the possibility of "nesting", giving rise to CDW instabilities[4].

In a quasi one-dimensional (1 D) metal, the Fermi surface may be described in a first approximation by two parallel planes, distant of $Q=2k_F$, k_F being the Fermi wave vector. In a quasi two-dimensional metal the Fermi surface is quasi-cylindrical with respect to an axis perpendicular to the two-dimensional plane and will often show nesting with a wave vector Q parallel to this plane. Such materials are unstable towards a lattice distortion of the wave vector Q, which leads to a new Brillouin zone and opens a gap at the Fermi surface. This decreases the electronic energy of the system. The competition between the increase of elastic energy associated to the lattice distortion and the corresponding decrease of electronic energy may favour the distorted state. This so-called Peierls transition can be described in the mean field theory. The transition (or Peierls) temperature is then expected to be simply related to the zero temperature gap[5]. In a quasi one-dimensional metal, the gap opening at the Fermi surface is normally complete and the Peierls transition is a metal to semiconductor transition. In a quasi two-dimensional metal, the Peierls transition is associated with partial gap openings only and is therefore a metal-metal transition.

In the Peierls distorted state, the lattice distortion is accompanied by a modulation of the electronic density, with a periodicity $2\pi/Q$. This is the charge density wave. This new periodicity, determined only by the degree of filling of the conduction band may be incommensurate with the inital lattice periodicity The CDW may then be either commensurate or incommensurate and the physical properties are expected to be sensitive to this property.

In spite of some characteristics common to all the compounds mentioned above, this course will concern only the quasi one-dimensional systems, the blue bronzes. Details concerning the quasi two-dimensional systems can be found in ref. 2 and a review in ref. 3.

The potassium blue bronze has been synthetized for the first time in 1964 by Wold et al.[6] by the electrocrystallisation technique. The crystal structure was refined by Graham and Wadsley[7] in 1966. Physical studies by Bouchard et al.[8] established that the blue bronze shows a semiconductor-to-metal transition in the vicinity of 180 K. Later, Perloff et al. noticed a large anisotropy of the electrical conductivity in the plane of the layers[9]. Detailed studies of the transport properties were then performed by Fogle and Perlstein[10]. They especially measured the low-temperature behavior of the conductivity σ and reported a non-ohmic behavior at $T < 20$ K, when the sample is insulating ($\sigma \approx 10^{-14}\ \Omega^{-1}\ cm^{-1}$). They proposed a model of an excitonic insulator for the semiconducting state.

It is only recently that the anisotropy of the conductivity was rediscovered and studied in greater details by Brusetti et al.[11]. Optical reflectivity measurements showed indeed that $K_{0.30}MoO_3$ is a quasi-one-dimensional (quasi-1 D) metal in the high-temperature phase [12]. Later, X-ray diffuse scattering studies by Pouget et al.[13] led to the conclusion that the metal-to-semiconductor transition is a Peierls transition towards an incommensurate charge density wave (CDW) state. At the same time, in a search for non-ohmic conductivity in the temperature range where the conductivity is not vanishingly small ($T > 50$ K), Dumas et al.[14] established that the blue bronze shows nonlinear transport, due to the sliding of the CDW.

The possible sliding of incommensurate CDW had been predicted long ago by Fröhlich and is known as the Fröhlich mechanism. Nonlinear conductivity attributed to this mechanism was first reported for niobium triselenide in 1976 by Monceau et al. [15]. Since then, a considerable amount of work has been devoted to this property and to the related phenomena first in $NbSe_3$ and TaS_3, then in the blue bronzes and in some transition metal tetrachalcogenides[16]. Reviews on inorganic quasi-one-dimensional compounds have lately been published [17,18].

Since the discover of non linear transport, the blue bronzes have been the object of extensive studies. Their physical properties have been surveyed in a recent article[19]. The present course will try to up-to-date part of these properties. Other properties related to the dynamics of the CDW are described in ref. 20. The nuclear magnetic resonance results and the optical properties are reviewed in ref. 21 and 22 respectively, and neutron inelastic scattering data in ref. 23.

II - CRYSTAL STRUCTURE

Three compounds which show the same crystal structure and similar properties have now been synthetized and studied : two alkali metal bronzes $K_{0.30}MoO_3$ and $Rb_{0.30}MoO_3$ and the thallium compound $Tl_{0.30}MoO_3$ [24,25].The crystal structure is monoclinic, space groupe C 2/m

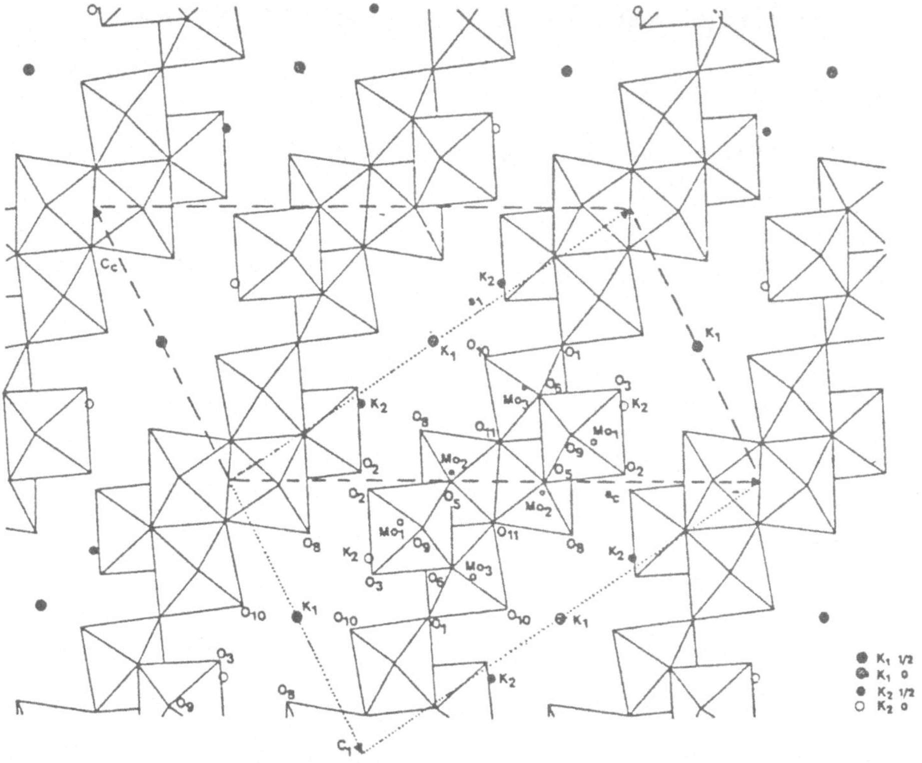

Fig. 1. Crystal Structure of the blue bronze - Projection on the (010) plane showing the octahedra slabs. The C-centered and I-centered cells are both indicated (Ref. 26).

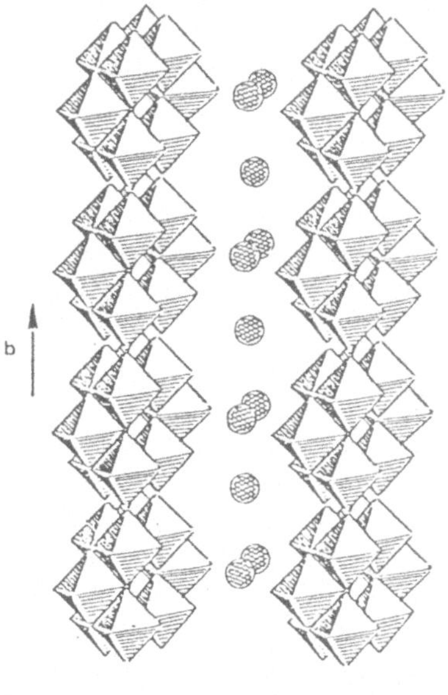

Fig. 2. Crystal structure showing the infinite slabs separated by the alkaline ions (●) and the infinite chains of MoO_6 (Mo(2) and Mo(3)) octahedra parallel to the b axis.

Table 1 : Lattice parameters (C-centered unit cell) (Ref. 7, 24, 25 and 26).

	a(Å)	b(Å)	c(Å)	β(degrees)	A^+ ionic rad. (Å)
$K_{0.30}MoO_3$	18.25	7.56	9.85	117.5	1.33
$Rb_{0.30}MoO_3$	18.94	7.56	10.04	118.8	1.47
$Tl_{0.30}MoO_3$	18.54	7.57	10.07	118.4	1.47

with 20 formulae per unit cell. Table I shows that the three compounds have very similar lattice parameters, in spite of the differences in size of the ion A^+. This shows that the lattice parameters are mainly determined by the MoO_3 skeleton.

The structure is built with clusters of ten distorted MoO_6 octahedra. These clusters are linked together via corner sharing along the b axis and the [102] direction (C-centered unit cell)and therefore form infinite slabs (Fig. 1). The alkali atoms lie between the slabs (Fig.2). The crystal structure is primarily a layered type structure. It contains three independent Mo sites and the 4d electron distribution over the 20 Mo sites [4 Mo(1), 8 Mo(2), 8 Mo(3)] of the unit cell are 10, 45 and 45 % for $K_{0.30}MoO_3$ and 14, 43 and 43 % for $Rb_{0.30}MoO_3$ [25]. The 4d electrons are therefore mainly located on the Mo(2) and Mo(3) sites. One should note that these sites are involved in infinite chains of MoO_6 octahedra, parallel to the b axis, as shown in figure 2. However, the blue bronze should rather be viewed as containing infinite chains of clusters of 10 MoO_6 octahedra.

III - THE PEIERLS INSTABILITY

Fig. 3 shows the electrical resistivity of $Rb_{0.30}MoO$ single crystals as a function of inverse temperature for several crystallographic orientations. The resistivity is found to be, in

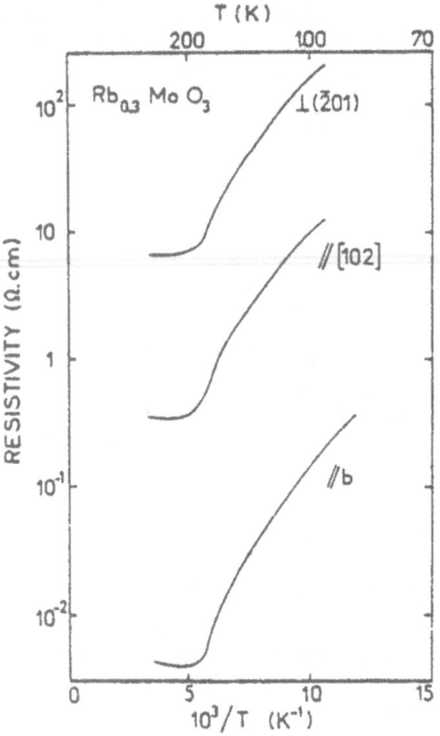

Fig. 3 . Electrical resitivity (logarithmic scale) of $Rb_{0.30}MoO_3$ versus inverse temperature along b, [1 0 2] and perpendicular to the octahedra layers $(\bar{2}\ 0\ 1)$ plane .

the plane of the layers, one order of magnitude larger along the [102] direction than along the b axis. It is still higher along the [201] axis perpendicular to the layers. The metal to semiconductor transition occurs in the three compounds in the vicinity of 180 K.

Optical reflectivity studies with polarized light established that above 180 K, $K_{0.30}MoO_3$ is a truly quasy-1D metal, since the plasma frequency was found to be 2,7 eV along b and smaller than 0.03 eV along [102][12]. In the semiconducting state, the gap was found to be 0.15 eV . One should note that the electrical resistivity does not show a simple activated behaviour, except in a narrow temperature interval (30 K < T < 70 K), where the activation energy is found ≈ 0.03 eV [11,27]. Below 30 K - 20 K, the crystals are highly resistive and a non ohmic behaviour is found for very low values of dc currents [10,28]. The ohmic value of the resistivity is therefore difficult to measure. Also, side effects such as the presence of a more conductive surface layer should be taken into account. However, the resistivity data show that the blue bronze is not an intrinsic semiconductor in the low temperature state. The conductivity is associated to the presence of defects or impurities levels in the gap. EPR studies establish the presence of Mo^5 ($4d^1$) localized states at low temperature. These states might result from a stoechiometry defect and could well be responsible for the transport properties below 100 K.

Other transport properties corroborate the quasi 1D character above 180 K and the extrinsic nature of the transport below. The thermopower shows a strong anisotropy in the plane of the layers above 180 K and no simple thermal dependance below : it does not obey a 1/T law between 38 K and 180 K, thus corroborating the presence of defects levels in the gap. At T = 38 K, it shows a peak probably due to the competition between two mechanisms of transport[11]. The Hall coefficient is positive above 180 K and changes sign at ≈180 K[8,28,29](Fig. 4).

The carriers concentration and the Hall mobility have been obtained in a simple model involving only one type of carriers, holes above 180 K and electrons below (Fig. 5). At room temperature, the carrier concentration (3.2×10^{21} cm^{-3}) is consistent with a complete charge transfer from the A metal to the conduction band. An activation energy, ~ 0,06 eV may be found in a small temperature interval (50 K < T < 100 K) for the carriers concentration[29]. It is not simply consistent with the resistivity data and probably not significative because of the narrowness of the temperature interval. It is more interesting to note that the Hall mobility

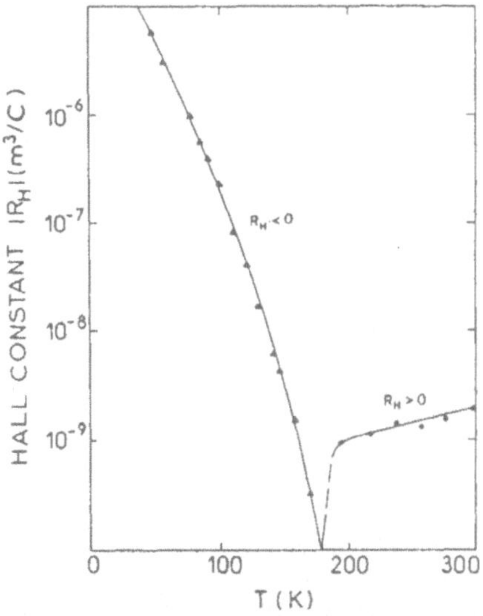

Fig. 4 . Hall constant as a function of temperature for $K_{0.30}MoO_3$.(Ref.28)

shows a maximum near 100 K, corresponding again to some competition between several transport mechanisms.

X-ray diffuse scattering have established that the 180 K transition in the blue bronze is a Peierls transition[13]. Well-defined satellite reflections have been found below 180 K, with a reduced wave vector $Q = (0,q_b,0.5)$. The in-chain component q_b is incommensurate and temperature dependent [30,31,32] (Fig 6a). Above 180 K, the structural phase transition is announced by anisotropic diffuse scattering which corresponds to the Kohn anomaly[23]. The anisotropy of the diffuse scattering, which provides at 300 K a correlation length along [102] of ~ 8Å, confirms that the blue bronze should be considered as built with chains of clusters of MoO_6 octahedra. The thermal variation of the satellite intensity follows approximately the BCS law predicted by the mean field theory (Fig 6b). However, more refined analysis of the X-ray data and NMR data show that this variation is well described by a critical exponent of 0.32, close to the value of 0.35 characteristic of an XY Heisenberg model[21]. One should also note that the ratio of the Peierls gap $2\,\Delta$ to the Peierls temperature kT_p is experimentally found to be ~ 8 and is not consistent with the value of ~ 3.5 predicted by the simple mean field theory. This indicates that fluctuations play an important role in the Peierls transition.

Detailed analysis of the X-ray data showed that the wave vector of the Peierls distortion should be taken as $q_0 = (0, 1\text{-}q_b, 0.5)$. This corresponds to a value for $2k_F \approx 3/4\ b^*$. If there is a complete charge transfer from the A metal to the conduction band, one expects 3 conduction electrons per $K_{0.3}Mo_{10}O_{30}$ primitive cell. The k_F value is then consistent with either a doubly degenerate conduction band or more likely two bands overlapping at the Fermi level.

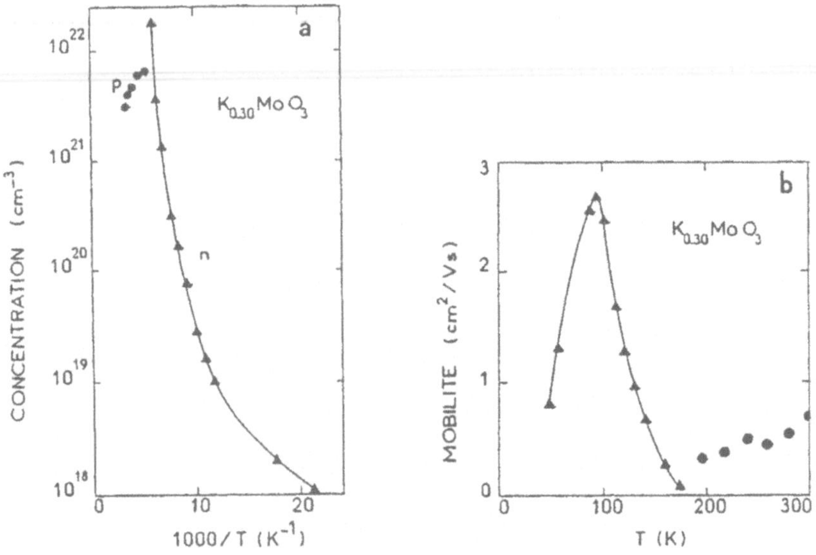

Fig. 5 --(a) Carrier concentration vs temperature obtained with a model involving one type of carriers , holes above 180 K, electrons below --(b) Hall mobility vs T (Ref 28)

The band structure was earlier speculated, by analogy with ReO_3, as being built on hybridized Mo and O states leading to filled σ and π bonding bands and empty antibonding bands[12]. A rough LCAO caculation performed for one chain with two Mo sites only , is consistent with this speculation[33].(Fig 7a).

Tight binding calculations for real chains and for a slab of $Mo_{10}O_{30}$ clusters have also been performed for the bands close to the Fermi level[24]. (Fig. 7b). They are consistent with the presence of two overlapping bands roughly three quarters filled. They also show that a third band lies above, close to the Fermi level. Fig. 7c, d , e show the Fermi surfaces (FS) deduced from this calculation for the two bands and the predicted nesting of the FS along b* : the upper FS of the first band is nested to the lower FS of the second band and reciprocally, with a wave vector 0.75 b*. This is in agreement with the existence in the blue bronze of one CDW only.

The temperature dependence of the q_0 wave vector has to be attributed to a change of the total number of electrons in the two conduction bands. It has been proposed that this change is due to the presence of a third band just above the Fermi level[30]. The experimental data would be consistent with a narrow band located ~ 650 K above the Fermi level , in agreement with the band structure calculation .

X-ray photoemission spectroscopy studies corroborate the band structure calculation[35].Moreover, angular resolved UV photoemission spectra show an angular dependence consistent within 15 % with the $2k_F$ value obtained from X-ray data[36]. This may be the first direct measurement of the Fermi wave vector in a low dimensional solid.

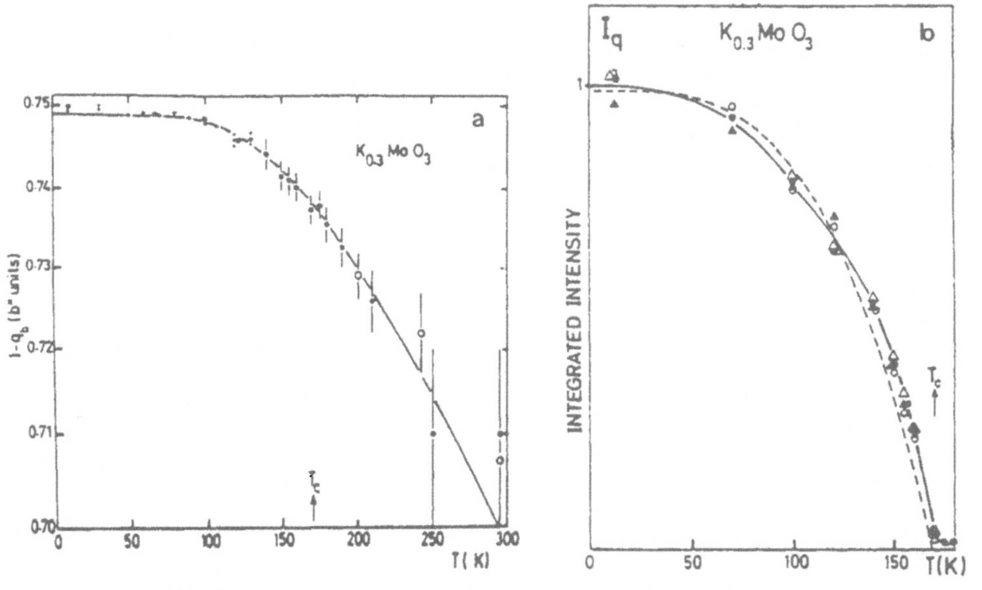

Fig. 6 . --(a) Temperature dependence of the incommensurate component $1\text{-}q_b$ ($\approx 2k_F$) of the satellite reflections (below T_c) and of the diffuse scattering (above T_c).
--(b) Normalized integrated intensity of the (17, $\pm$ 1 $\pm$ q_b, $\overline{8.5}$) and 17, $\pm 1 \pm q_b$, $\overline{8.5}$) satellite reflections, as a function of the temperature. The dotted line gives the temperature dependence of the square of the B.C.S. order parameter (Ref 30).

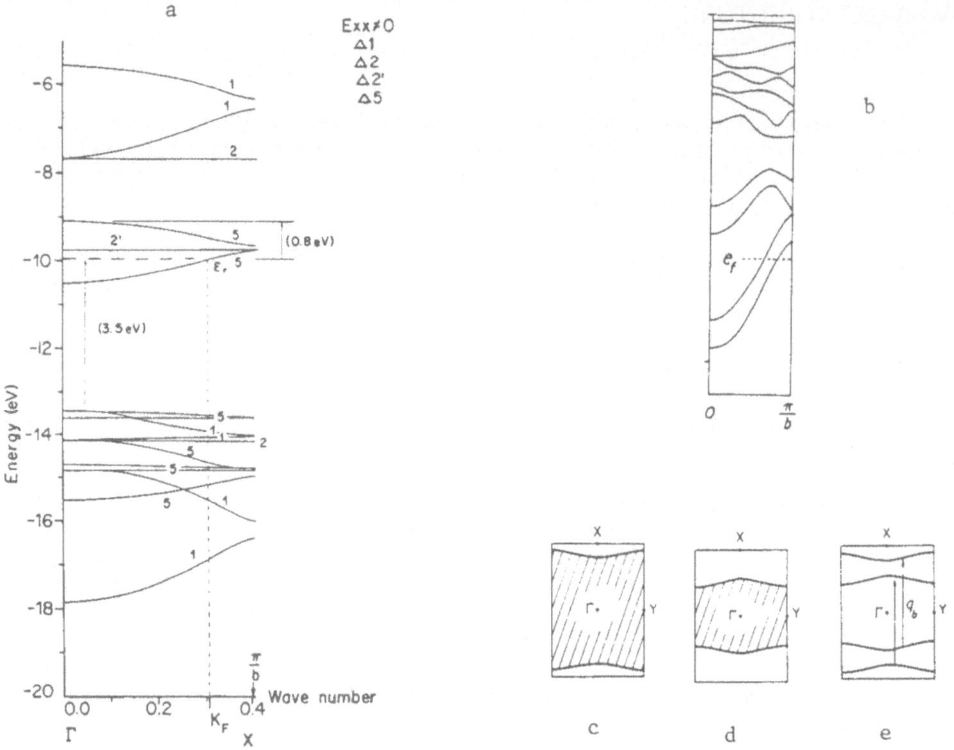

Fig. 7 -- (a) Band structure calculated with the LCAO method for one chain with 2 Mo sites (Ref 33) --(b) Tight binding calculation for a chain of clusters of MoO_6 octahedra showing two bands overlapping at the Fermi level (ref 34) --(c) (d) (e) Fermi surfaces associated with the two partially filled d-block bands of a real $Mo_{10}O_{30}$ slab. (c) Fermi surface of the first band, where the wave vectors of the shaded and unshaded regions lead to occupied and unoccupied band levels, respectively ; (d) Fermi surface of the second band ; (e) nesting of the Fermi surfaces of the first and second bands.

V - MAGNETIC AND THERMAL PROPERTIES

The magnetic susceptibility of $K_{0.30}MoO_3$ has been measured by several authors [8,11,37,38]. It is found anisotropic between the (2 01) plane of the layers and the perpendicular orientation (Fig. 8). The observed susceptibility is the result of several contributions : $\chi = \chi_{dia} + \chi_{Pauli} + \chi_{Van\ Vleck}$. The anisotropy is mainly due to the anisotropy of the Van Vleck paramagnetism, related to the anisotropy of the band structure. The decrease of χ with decreasing temperature corresponds to the opening of the Peierls gap and to the vanishing of the Pauli contribution. This result has been analysed by taking into account the fluctuations . A pseudogap of 200 K above 180 K is found. The obtained zero temperature Peierls gap is of the order of 1200 K, not too far from the value obtained from optical reflectivity studies.

Specific heat measurements have been performed both at low temperatures and in the vicinity of the Peierls transition. An anomaly, corresponding to an entropy change of ~ 150 mJ $mole^{-1}K^{-1}$, is found at T_p. This, compared with the estimated value of the density of states at the Fermi level and with the prediction of the mean field theory, also indicates that fluctuations play an important role at the transition[19,37]. Between 0.1 K and 1 K, an excess contribution linear in temperature has been attributed to some disorder[39]. This disorder may be due to randomly pinned CDW, in agreement with the hysteresis properties found by transport measurements and described in Section VII.

The blue bronze single crystals can be grown with a comparatively large size , typically 3x1x0.1 mm^3 . The CDW transport studies are therefore performed on large samples, while in the case of the transition metal trichalcogenides, they are done on very small crystals, whisker-like. The experiments may therefore be casier on the blue bronze, but on the other hand the problem of the homogeneity of the current density in large samples becomes critical. We will see that it is difficult to depin the CDW in the whole sample or to obtain a uniform CDW velocity. This does not imply that the quality of the crystals is questionable. In fact, as well the samll mosaic angle detected by neutron inelastic scattering [40] as preliminary x-ray topography experiments [41] show that the crystals are often of excellent structural quality.

We will discuss in this section two important aspects of CDW transport in the blue bronze : the threshold field, behaviour and thermal variation and the dynamics of the CDW, reflected in the current oscillations. The hysteresis and metastability properties will be reported in section VII.

1. Threshold behaviour

Fig. 9 shows de voltage current characteristics obtained at different temperatures on a $K_{0.30}MoO_3$ crystal. Some samples exhibit in a given temperature range a switching from the ohmic regime to the non ohmic one, while others always show a smooth threshold behaviour. In all cases, there is no switching above ~ 100 K and the threshold field is not well defined and is difficult to measure in this temperature range. Fig. 10 shows the thermal variation of the threshold field E_t . While all authors find an increase of E_t with T below 100 K, various results have been reported in the literature above 100 K. It seems now established that some samples show a maximum at 100 K (Fig. 10 a) while others do not (Fig. 10 b). At low temperatures (T ≤ 20 K), extremely large threshold fields (E_t ~ 100 V/cm) have been found. This corresponds to the temperature region where the samples are highly resistive and where they show induced dielectric polarization [20,44,45].

It has been proposed that the maximum of the threshold field near 100 K could be related to the coupling between CDW and mobile defects [46]. Other experiments, such as NMR[21,47], suggest that time-dependent properties just below 180 K have to be accounted for by diffusion of defects. These defects might be related to some stoechiometry deviation and to A metal

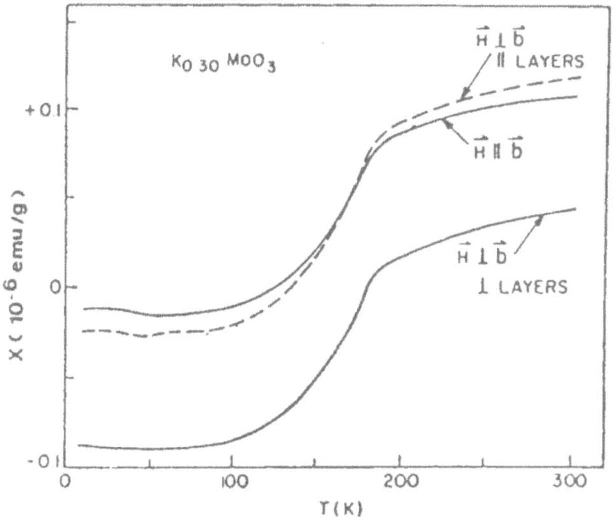

Fig. 8 Magnetic susceptibility of $K_{0.30}MoO_3$ as a function of temperature for different orientations of the magnetic field (Ref 36)

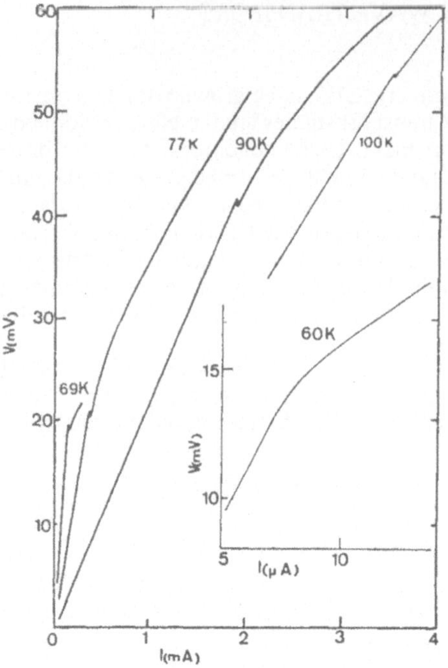

Fig. 9 Voltage - current characteristics at different temperatures for a $K_{0.30}MoO_3$ crystal exhibiting switching below 100 K (Ref 19)

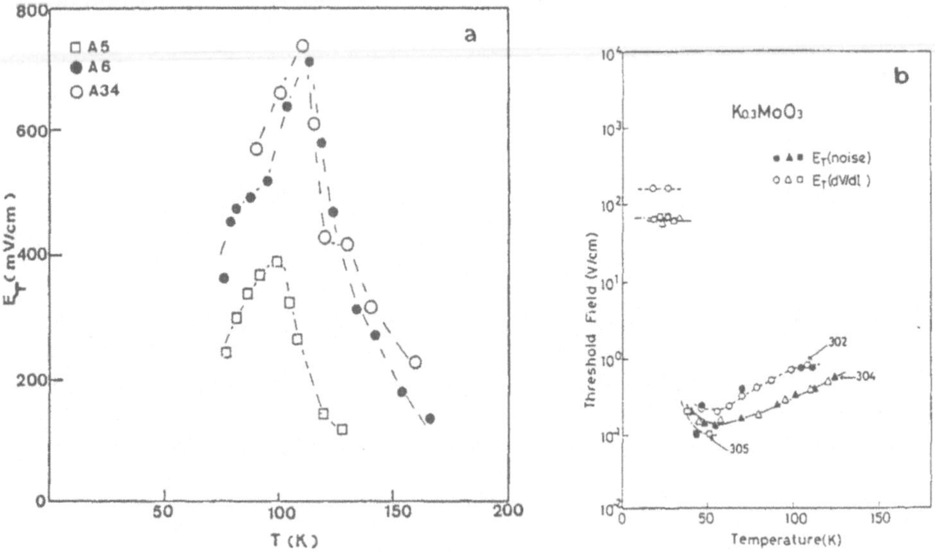

Fig. 10 -- (a) Threshold field vs temperature for several $Rb_{0.30}MoO_3$ samples (Ref 42) -- (b) Same for $K_{0.30}MoO_3$ samples (Ref 43)

vacancies, inducing electronic defects on the Mo sites. In a model of CDW transport involving discommensurations (DC) (see section IV) coupled to mobile point defects, the threshold field depends on the relative velocity of DC compared to the diffusion time of point defects. Such a mechanism could induce a maximum in the threshold field at a temperature where both processes have similar time scales and is analogous to the so-called "hardness peak" in metallurgy[46].

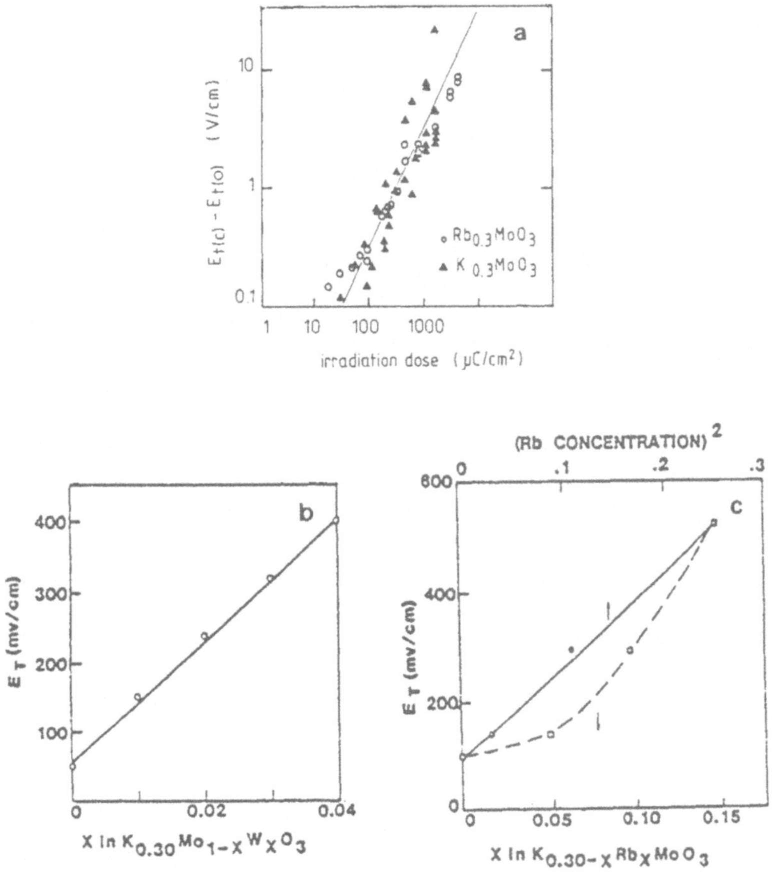

Fig. 11 -- (a) Increase of the threshold field as a function of the irradiation dose for $Rb_{0.30}MoO_3$ and $K_{0.30}MoO_3$ crystals (Ref 48) -- (b)Variation of the dc threshold field with tungsten concentration in $K_{0.30}Mo_{1-x}W_xO_3$-- (c) Same with rubidium concentration in $K_{0.30-x}Rb_xMoO_3$ (Ref 49)

The large threshold fields found below ~ 20 K are related to large dielectric polarizations in this temperature range. However, the detailed mechanism accounting for these results is not clear at the moment.

The threshold field has been studied as a function of the concentration of defects created either by electron irradiation or by doping. For electron irradiation E_t is found to increase linearly with the concentration of defects[48](Fig.11a). Fig.11 b and c show that doping with isoelectronic impurities has a strong effect if the impurities are located on the chains of MoO_6 clusters (case of W impurities substituted on Mo sites) and a weak effect if they are located on other sites (Rb impurities substituted on alkali sites)[49]. One expects, in the Lee and Rice model[50], a linear variation of E_t vs concentration for weak pinning centers and a quadratic variation for strong pinning centers. The experimental results show that electron irradiation defects and W impurities are strong pinning centers while Rb impurities are weak pinning ones.

2 - Current oscillations

It is well established that the depinning of CDW is usually accompanied by current oscillations with a frequency F approximately proportional to the excess CDW current density Jcdw. In the quasi-classical model of CDW transport the so-called washboard frequence F is expected to be given by $F = Vd / \lambda$, where Vd is the drift velouty of the CDW and λ the superlattice period. As $J_{CDW} = n_c (T)\, e\, Vd$, where $n_c (T)$ is the concentration of electrons condensed in the CDW at the temperature T, a simple relation between F and J_{CDW} is predicted : $S = J_{CDW} / F = n_c(T)\, e\, \lambda$ [16,17]. This relation is often well obeyed, within a factor of 2, in the trichalcogenides. In the case of the blue bronze, current oscillations are also found : Fig. 12 a shows the spectrum analysis of the voltage obtained from a $Rb_{0.30}MoO_3$ crystal above threshold. A fundamental frequency and several harmonics are observed[51]. Fig. 12 b shows the frequency F vs the average CDW current : a linear law is approximately obeyed, but not for small values of J_{CDW} and the above relation is not consistent with the expected value of $n_c(T)$. In fact, it is now clear that the slope S varies from sample to sample and depends for a given sample on the geometry and quality of the current contacts and on the temperature. This

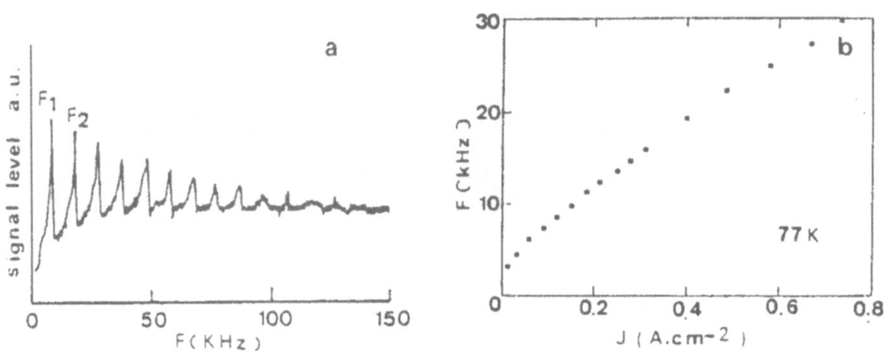

Fig. 12 -- (a) Spectrum analysis of the voltage from a $Rb_{0.30}MoO_3$ crystal above threshold at 77 K. (size of the sample : 4 x 1.4 x 0.34 mm^3)
-- (b) Washboard frequency as a function of the average CDW current in a $Rb_{0.30}MoO_3$ sample (Ref 51).

indicates that the CDW current is not homogeneous. A fraction only of the crystal volume, probably in shape of platelets or filaments, participates into the current oscillations at a given frequency and this fraction is temperature dependent. In the best cases, this fraction is found to be ~ 30 %. Furthermore, NMR data are consistent with a whole distribution of CDW velocities in the samples [47]. Let us point out that for transport studies on the blue bronze and especially for non linear transport, a great care should be taken in the current contacts : the end faces of the crystals should be carefully polished before the contacts, either evaporated indium or electroplated copper, are realized. Otherwise, the large anisotropy of the resistivity leads to complicated and inhomogeneous current paths.

Current oscillations with very low frequencies (in the Hz range) have also been found in the blue bronze [19]. These frequencies are probably due to the existence of CDW conducting filaments and may correspond to locally very slow CDW velocities. One should also note that friction between regions with different CDW velocities should be taken into account.

VII - HYSTERESIS AND METASTABILITY

Several of the transport properties involving hysteresis and time-dependent effects, strongly suggest that, especially below ~ 100 K, the CDW state of the blue bronze should be rather viewed as a CDW glass. These properties involve mainly the low field resistance R and the CDW excess current.

1 - Low field resistance

Fig. 13 shows the time dependance of R et various temperatures T between 77 K and 210 K after a cooling process between 300 K and T, with a cooling rate of 5 K/minute : this effect is maximum at a temperature between 100 and 140 K and still persists above the Peierls temperature[42]. Fig. 14 shows the thermal hysteresis curve obtained for the low field resistance on an electron irradiated crystal of $K_{0.30}MoO_3$. All samples show such an hysteresis, but the effect is larger on an irradiated crystal. If the thermal cycle includes a relaxation for several hours at a given temperature during the cooling process, a subsequent heating cycle is characterized by a step in the curve in the vicinity of the temperature T : this is a remarkable memory effect, which has to be attributed to rearrangement and trapping of some point defects. As the low field resistance in this temperature range is extrinsic and associated to levels in the Peierls gap, these properties are clearly due to metastable populations (and possibly positions) of these levels. The same kind of mechanism very likely accounts for the data of Fig. 13.

Hysteresis phenomena are also connected to the electrical "history" of the samples.
Fig 15a shows the diffential resistance dV/dI as a function of the dc current : 1 (R_1 value of R) corresponds to the "virgin" state obtained after cooling from 300 K. When the current has been swept above threshold, a different value of the low field resistance R_2 is obtained. For subsequent current loops, the value of R is approximately reproducible. These properties have been found by various authors and are obviously related to the thermal hysteresis. This has been studied in great details on TaS_3 in ref. 53. One can define an isothermal remenent (relative) resistance (IRR) by $\Delta R/R = (R_2-R_1)/R_1$. The IRR depends on the maximum dc current passed through the sample, as shown on fig. 15 b. If now the sample is cooled from 300 K to the measuring temperature with an applied dc current, the resistance R_{th} is different from R_1 obtained under zero current cooling. The thermoremanent (relative) resistance TRM defined as $R_{th}-R_1/R_1 = \Delta R/R$ depends on the dc cooling current (Fig. 15 b). These properties are analogous to those of the remanent magnetizations in spin glasses : they are characteristic of a system which can reach a great number of nearly equivalent metastable states. Numerical studies of the CDW state corroborate this point [54]. Since the low field resistance is related to point defects, they also establish that there is a coupling between CDW and point defects : the CDW state is strongly correlated to the configuration of these point defects.

2 - CDW current

Time dependent properties are also found on the excess CDW current. Fig. 16 indicates that this effect is larger in a sample containing a large concentration of defects, such as a

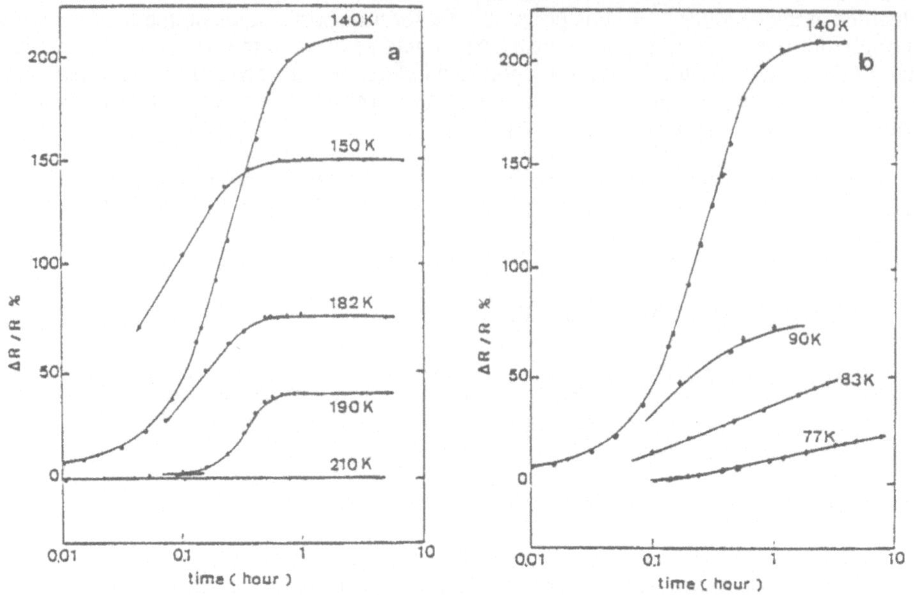

Fig. 13 Variation of the low field resistance vs time at various temperatures T after cooling from 300 K down to T with a cooling rate of 5 K/minute (Ref 42).

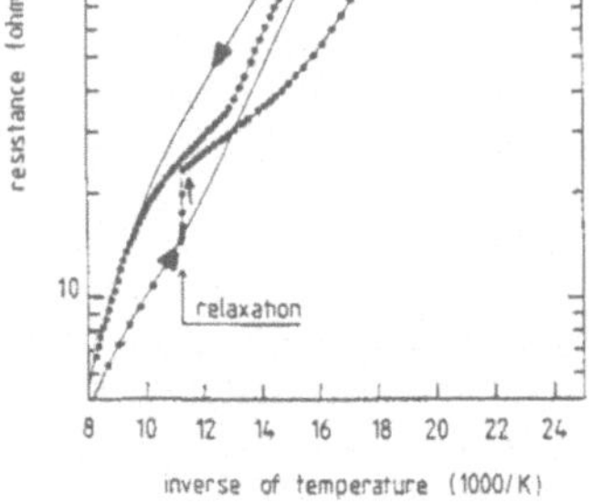

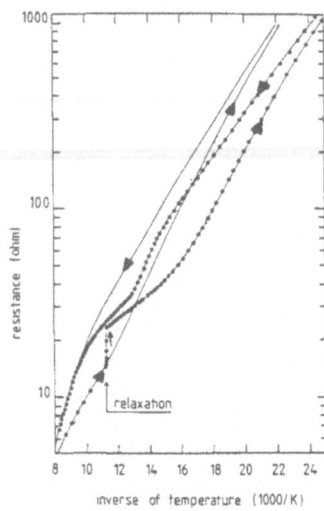

Fig. 14 Log R vs 1000 T for a $K_{0.30}MoO_3$ sample irradiated with 2.5 Mev electrons at a dose of 12 mC/cm^2 . The continuous line corresponds to a normal cooling and heating cycle. The points correspond to a cycle including a relaxation at 90 K for several hours (Ref 52).

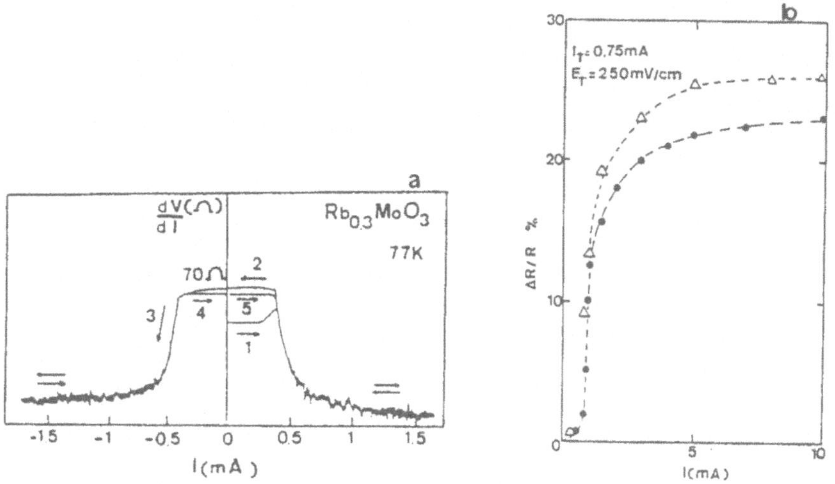

Fig. 15 --(a) Differential resistance dV/dI (at 33 Hz) as a function of the dc current for $Rb_{0.30}MoO_3$ at 77 K. 1 refers to the virgin state obtained after cooling from 300 K. --(b) Variation of the thermoremanent resistance (Δ) vs the current applied during cooling and of the isothermal remanent resistance (o) vs the maximum applied current (see text) (Ref 19).

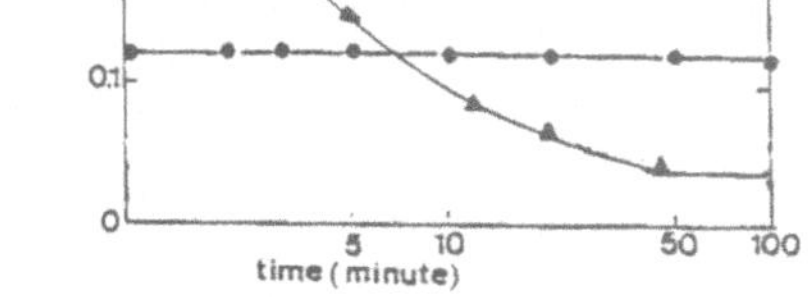

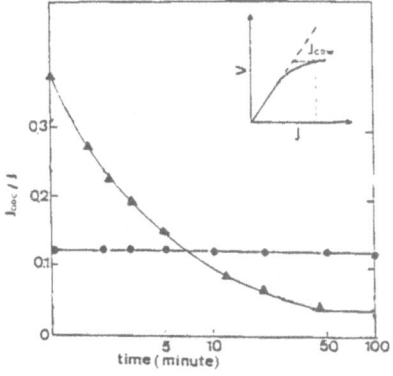

Fig. 16 Time dependance of the CDW average current density for a $Rb_{0.28}MoO_3$ (▲) and a $Rb_{0.30}MoO_3$ (●) crystal (Ref 42). The inset shows how the CDW current density is obtained on the V-I characteristic.

highly non-stoechiometric Rb blue bronze crystal, than in a stoechiometric one. These properties very likely show that the fraction of the volume of the crystal participating in CDW transport is decreasing with increasing time. In a picture involving CDW domains, it is consistent with the trapping of more and more domain walls by impurities or other point defects. One therefore expects a larger effect in a less "pure" sample.

Low frequency current oscillations also depend on the thermal and electrical "history" of the samples, as discussed in details in ref. 54. This is again consistent with various metastable CDW states leading to slightly different dynamic properties.

VIII - LOCAL PROPERTIES : MOSSBAUER AND EPR STUDIES

A detailed understanding of the CDW state and of the CDW transport requires the knowledge of local and microscopic properties ^{87}Rb NMR studies, especially under dc current, have been reported recently for the blue bronze : the narrowing of the NMR line under current above threshold, does establish that non linear transport is due to the depinning of the CDW[47]. Mossbauer studies have also been performed on ^{57}Fe doped $K_{030}MoO_3$ crystals[56]. The spectra which show three quadrupolar doublets corresponding to Fe substituted on K and Mo sites, are temperature dependent. Fig. 17 a shows typical data for the Mossbauer linewidth as a function of temperature. While, in a simple picture, the linewidth is expected to increase in the CDW incommensurate phase below T_p, the experiment shows that it is increasing steeply only below ~ 120 K. This indicates that the Fe ions act as strong pinning centers between 180 K and ~ 120 K and that, in this temperature range, the phase of the CDW at the impurity site is determined by the Fe itself. Below ~ 120 K, there is a change of behaviour, as if the Fe^{3+} would become progressively weak pinning centers. This is indeed consistent with the predictions of the Lee and Rice model for impurity pinning [50]. The pinning behaviour depends in this model on the relative values of the CDW order parameter and of the parameter ε measuring the ratio of the pinning energy to the elastic energy associated to the pinning : the strong pinning case corresponds to $\varepsilon > [(Tp - T)/Tp]^{1/2}$. The Mossbauer data therefore indicate that ε is of the order of 0.5 to 0.7 for Fe impurities in the blue bronze : they give direct microscopic information on the pinning strength of the impurities.

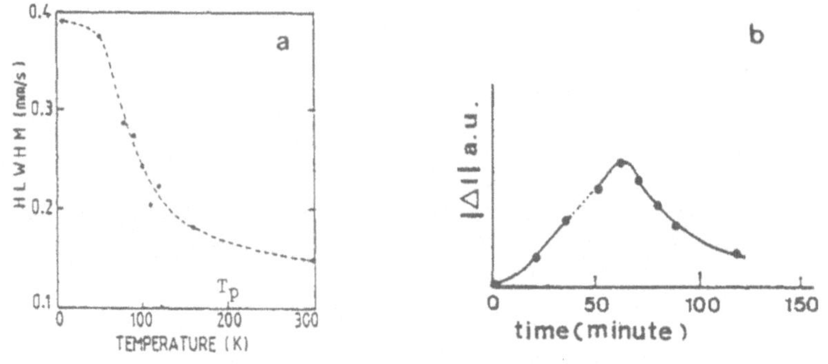

Fig. 17 -- (a) ^{57}Fe Mossbauer linewidth vs temperature for ^{57}Fe-doped $K_{0.30}MoO_3$ crystal (Ref 56) -- (b) Parameter related to the intensity of the extra EPR line as a function of time at T = 4.2 K, after quenching the crystal from 300 K (Ref 57).

Electron Paramagnetic Resonance (EPR) is another local technique which may give microscopic information on eventual paramagnetic defects and on their coupling with CDW. Blue bronze single crystals normally show at low temperatures an EPR spectrum which consists of several lines. One of these lines has to be attributed to Mo^{5+} ($4\,d^1$) paramagnetic centers, very likely due to a stoechiometry defect. When the crystals are now quenched from 300 K down to 4.2 K, the EPR spectra are time dependent.[57] Fig. 17 b shows typical data for the intensity of the extra EPR line as a function of time. The evolution of the EPR spectrum corresponds to the relaxation of the CDW state and should be compared to the dielectric relaxation also found in the blue bronze [58]. The EPR data show that the long time relaxations are associated to the successive creation and vanishing of magnetic centers, probably of Mo^{5+} type. This suggests that some CDW structural defects are correlated to localized, unpaired electrons.

IX - CHARGE DENSITY WAVE DEFECTS

Most of the current models describing CDW transport do not take into account explicitly the existence of CDW structural defects such as discommensurations or phase dislocations. The so-called classical model[17] describes the CDW as a single overdamped particle and is oversimplified. The microscopic model most commonly invoked was proposed by Fukuyama and Lee [59] and consists of an incommensurate CDW with a slowly varying phase interacting weakly with impurities located at random positions of the underlying lattice. A similar model has been recently used to describe the behavior of the CDW near threshold, considered as a dynamic critical phenomenon[60]. Another semimicroscopic model describes CDW transport in terms of a coherent macroscopic quantum tunneling phenomenon over large volumes[61].

However, it is well known that an incommensurate state can be described in terms of commensurate domains separated by walls or discommensurations (DC)[61]. The phase slips necessary to account for the deviation of the Fermi wave vector from the commensurate values are located in the DC which have to be electrically charged. These DC, as suggested by Lee and Rice[50], are delimited by phase dislocation loops. Similar CDW defects have been proposed independently by Ong and Maki in terms of phase vortices[63] and by Gorkov et al. [64] (phase slippage centers) to describe the conversion from electrons condensed in the CDW into normal carriers at the contacts and to account for the current oscillations above threshold. But, it is only recently that the analogy between the depinning of CDW involving the motion of CDW defects and the mechanical deformation of crystals has been worked out in details[65,46]. In this model, the CDW is elastically displaced below threshold, thus inducing large dielectric polarizations. At the threshold field, CDW defects are nucleated and start to move. The permanent plastic flow of these defects would lead to the excess CDW current. The voltage current characteristic would be analogous to the stress-strain curve in the plastic regime of metals. Several properties of non-linear transport in the blue bronze support this analogy : for example, the low frequencies (~ 1Hz) voltage pulses observed above threshold are similar to the serrations in stress-strain curves, known as the Portevin-Le Chatelier effect[66]. While the serrations are attributed to propagating bands of slipped crystal, the CDW low frequencies pulses suggest the existence of current inhomogeneities in shape of filaments[65].

One of the objections to the model of moving CDW defects lies in the large electrostatic forces between charged DC[67]. In the case of the blue bronze, this objection is serious only at low temperatures ($T < 40$ K), when the normal carriers concentration is too small to provide on efficient screening mechanism. It has therefore been proposed recently that another type of CDW defect could be present : walls parallel to the one-dimensional axis (b-axis in the case of the blue bronze) would carry no charge. Such walls would be analogous to Bloch walls in ferromagnets and would move perpendicular to b when an electric field is applied along b, thus inducing a CDW current[68]. In this context, one should note that X-ray studies show that the transverse width of the satellite depends on the thermal and electrical history of the sample: the application of an electric field along b during cooling causes a loss of transverse order at low temperatures[69]. It has also been found that the transverse component of the satellite wave vector is a function of the applied electric field and shows some hysteresis properties[70]. These results which establish that the electric field induces metastable deformation of the CDW support a model of longitudinal (// b) domain walls.

It is now clear that some more theoretical work is needed in relation with the CDW structural defects. One should also emphasize that there is presently a great lack of direct studies of these defects. An experimental effort in that direction, using tools such as electron microscopy would be now extremely welcome.

X - CONCLUSION

It is now well established that non linear transport in several types of materials including the tri and tetra chalcogenides and the blue bronzes is due to the depinning of the CDW. Metastability associated to the CDW state and to CDW transport is present in all these materials and is especially important in the blue bronze. Although the main features of CDW transport are well known, the microscopic mechanisms of the CDW depinning are still controversial. Lots of studies have now been performed, especially on the hysteresis properties and the related phenomena are well documented. A microscopic understanding now requires techniques other than transport, providing local information. NMR is being used successfully. Other resonance methods, such as EPR, should be developed further, when possible, in these materials. Finally, the phase slip centers or domain structures, so often invoked in the discussions of the data, should be studied directly by the available observation techniques.

Acknowledgements

Part of the results reported here have been obtained at Laboratoire d'Etudes des Propriétés Electroniques des Solides - C.N.R.S. Thanks are due to J. Marcus for the crystal growth, to J. Dumas, C. Escribe-Filippini, D. Feinberg, J.Y. Veuillen and R. Chevalier for various studies and helpful discussions. The author has also benefited from stimulating discussions with S. Aubry, J. Friedel and many other colleagues who cannot all be listed here.

REFERENCES

*Laboratoire associé à l'Université Scientifique Technologique et Medicale de Grenoble.

1. M.J. Sienko in "Non Stoichrometric Compounds, Advances in Chemistry Series" (Ed. R.F. Gould), American Chem. Soc., Washington DC, p. 224 (1963).
   P. Hagenmuller in Progress in Solid State Chemistry (Ed. H. Reiss) Pergamon, vol. 5, p. 71 (1971).
2. C. Schlenker, J. Dumas, C. Escribe-Filippini, H. Guyot, J. Marcus and G. Fourcaudot, Phil. Mag. B 52, 647 (1985).
   H. Guyot, Thèse de Doctorat d'Etat, Univesité Scientifique et Médicale de Grenoble (1986).
3. C. Schlenker, J. Dumas and C. Escribe-Filippini in "Low-dimensional properties of molybdenum bronzes and oxides" C. Schlenker ed. (D.Reidel Publ. Comp.) (to be published).
4. See for example J. Friedel (Lecture I), this volume and references therein.
5. See for example V.J. Emery, this volume and references therein.
6. A. Wold, W. Kunnmann, R.J. Arnott and A. Ferreti, Inorg. Chem. 3, 545 (1964)
7. J. Graham and A.D. Wadsley, Acta Cryst. 20, 93 (1966).
8. G.H. Bouchard Jr , J.H. Perlstein and M.J. Sienko, Inorg. Chem. 6 , 1682 (1967).
9. D.S. Perloff, M. Vlasse, and A. Wold, J. Phys. Chem. Solids 30, 1071 (1969).
10. W. Fogle and J.H. Perlstein, Phys. Rev. B 6, 1402 (1972).
11. R. Brusetti, B.K. Chakraverty, J. Deven yi, J. Dumas, J. Marcus and C. Schlenker in "Recent Development in Condensed Matter Physics (eds J.T. Deevreese, L.F. Lemmens. V.E. Van Doren and J. Van Royen), Plenum, New York, Vol. 2,p. 181 (1981).
12. G. Travaglini, P. Wachter, J. Marcus and C. Schlenker, Solid State Commun. 37,599(1981)
13. J.P. Pouget, S. Kagoshima, C. Schlenker, and J. Marcus, J. Physique Lett. 44, L113 (1983).
14. J. Dumas, C. Schlenker, J. Marcus and R. Buder, Phys. Rev. Lett. 50, 757 (1983).

15. P. Monceau, N.P. Ong, A.M. Portis, A. Meerschaut and J. Rouxel, Phys. Rev. Lett, 37 602 (1976).
16. For recent reviews see : Proceedings of the Yamada Conf. XV on Physics and chemistry of quasi -one-dimensional conductors (May 1986) Lake Kawaguchi, Japan (to be published in Physica B) - J.C. Bill - Contemp. Phys. 27, 37 (1986).
17. P. Monceau (ed.) Electronic properties of inorganic quasi-one-dimensional compounds (D. Reidel Publ. Comp.) (1985) and this volume. G. Gruner and A. Zettl Phys. Rep. C 119, 117 (1985). G. Gruner, this volume.
18. J. Rouxel (ed.) Crystal chemistry and properties of materials with quasi-one-dimensional structures (D. Reidel Publ. Comp.) (1986).
19. C. Schlenker and J. Dumas in 18 - p. 135.
20. R.M. Fleming, this volume and references therein.
21. C. Berthier, this volume. P. Butaud, P. Segransan, C. Berthier, J. Dumas and C. Schlenker, 55, 253 (1985).
22. S. Jandl, this volume and references therein.
23. J.P. Pouget, this volume and references therein.
24. M. Ganne, A. Boumaza, M. Dion and J. Dumas. Mat. Res. Bull, 20, 1297, (1985)
25. B.T. Collins, K.V. Ramanujachary and M. Greenblatt, Solid State Comm., 56, 1023, (1985).
26. M. Ghedira, J.Chenavas, M. Marezio and J. Marcus, J. Solid State Chem., 57, 300 (1985).
27. L.F. Schneemeyer, F.J. Di Salvo, S.E. Spengler and J.V. Waszczak, Phys. Rev. B 30, 4297 (1984).
28. E. Bervas Thesis, Université Scientifique et Médicale de Grenoble (1984).
29. L. Forro, J.R. Cooper, A. Janossy and K. Kamaras (to be published).
30. J.P. Pouget, C. Noguera, A.H. Houdden and R. Moret, J. Physique, 46, 1731 (1985).
31. M. Sato, H. Fujishita and S. Hoshino, J. Phys. C. Solid State Phys., 16 L 877 (1983).
32. R.M. Fleming, L.F. Schneemeyer and D.E. Moncton, Phys. Rev.B 31, 899 (1985).
33. G. Travaglini and P. Wachter in "Charge density waves in solids" Lecture notes in Physics 217, 115 (1985).
34. M.H. Whangbo and L.F. Schneemeyer, Inorg. Chem. 25, 2424 (1986).
35. G.K. Wertheim, L.F. Schneemeyer and D.N.E. Buchanon, Phys. Rev. B. 32, 3568, (1985).
36. J.Y. Veuillen, R. Cinti and E. Al Khoury Nemeh (to be published) - J. Dumas et al. Int. Conf. on Synthetic Metals, Kyoto (June 1986) (to be published).
37. D.C. Johnston, Phys. Rev. Lett. 52, 2049 (1984).
38. L.F. Schneemeyer, F.J. Di Salvo, R.M. Fleming and J.V. Waszczak, J. Solid State Chem. 54 358 (1984).
39. K.J. Dahlhauser, A.C. Anderson and George Mozurkewich, Phys.Rev.. B 34, 4432 (1986)
40. J.P. Pouget and C. Filippini, Private Communication.
41. J. Baruchel, J. Marcus and M. Schlenker, Private Communication
42. A. Arbaoui, Thesis, Université Scientifique et Médicale de Grenoble (1985).
43. A. Maeda, T. Furuyama and S. Tanaka, Solid State Commun. 55, 951 (1985).
44. G. Kriza and G. Mihaly, Phys. Rev. Lett. 56, 2529 (1986).
45. L. Mihaly and G.X. Tessema, Phys. Rev. B 33 5856 (1986).
46. J. Friedel, Lecture II this volume.
47. P. Segransan, A. Janossy, C. Berthier, J. Marcus and P. Butaud, Phys. Rev. Lett. 56 1854 (1986).
48. H. Mutka, S. Bouffard, J. Dumas and C. Schlenker, J. Physique Lettres 45, L 729 (1984)
49. L.F. Schneemeyer, S.E. Spengler, F.J. Di Salvo and J. Waszczak, Mol. Cryst. Liq. Cryst. 125, 41 (1985).
50. P.A. Lee and T.M. Rice, Phys. Rev. B 19,3970 (1977).
51. C. Schlenker and J. Dumas; P. Beauchène, J. Dumas, A. Janossy, J. Marcus and C. Schlenker in Ref. 16.
52. H. Mutka, F. Rullier-Albenque and S. Bouffard ,J. Physique (to be published)
53. D.M. Duggan and N.P. Ong, Phys. Rev. B 34 ,1375 (1986).
54. P.B. Littlewood, Phys. Rev. B 33, 6694 (1986)

55. J. Dumas, A. Arbaoui, H. Guyot, J. Marcus and C. Schlenker, Phys. Rev. B 30 2249 (1884).
56. J.Y. Veuillen, R. Chevalier, J. Marcus and C. Schlenker in Ref 16.
57. J. Dumas, R. Buder, J. Marcus, C. Schlenker and A. Janossy in ref. 16.
58. R.J. Cava, R.M. Fleming, E.A. Rietman, R.G. Dunn and L.F. Schneemeyer, Phys. Rev. B 53, 1677 (1984). G. Kriza and G. Mihaly, Phys. Rev. Lett. 56, 2529 (1986)
59. H. Fukuyama and P.A. Lee, Phys. Rev. B 17 , 535 (1977).
60. D.S. Fisher, Phys. Rev. B 31, 1396 (1985).
61. J. Bardeen In ref 16; Phys. Rev. Lett. 45, 1978 (1980)This volume and references therein.
62. W.L. McMillan, Phys. Rev. B 14 ,1496 (1976).
63. N.P. Ong and K. Maki, Phys. Rev. B 32, 6582 (1985).
64. I. Batistic, A. Bjelis and L.P. Gorkov, J. Physique 45 1049 (1984) and references therein.
65. J. Dumas and D. Feinberg, Europhys. Lett. 2, 555 (1986)
66. See for example L.P. Kubin and Y.J. Estin, J. Physique 47, 497 (1986).
67. S. Barisic, this volume.
68. S. Aubry, to be published.
69. R.M. Fleming, R.G. Dunn and L.F. Schneemeyer, Phys. Rev. B 31, 4099 (1986).
70. T. Tamegai et al. Solid State Comm. 51, 585 (1984); 56, 13 (1985).

INFRARED AND RAMAN MEASUREMENTS OF CDW IN KCP, $K_{0.3}MoO_3$ AND $Rb_{0.3}MoO_3$

S. Jandl

Université de Sherbrooke
Département de physique
Sherbrooke, Québec, Canada

New excitation are observed once a charge density wave (CDW) is generated in a system following a Peierls transition. In the commensurate case, the following features can be observed:

a) Phonon branches folding with possibly new Raman and infrared active phonons [1].

b) Raman active amplitude modes of the CDW that behave like soft modes below the point of commensurate second order phase transition.

c) Infrared active phase modes of the CDW whose frequency is different from zero for q = o. However, this frequency equals zero in the incommensurate phase unless trere is pinning of the CDW [2].

In infrared reflectivity measurements, the response of the pinned CDW including fluctuations into an unpinned state is discussed phenomenologically in terms of an oscillator model where the dielectric constant is [3]

$$\varepsilon = \varepsilon_\infty + \varepsilon_\infty \omega_p^2 \Big/ \left[\left(\frac{\omega^2}{\omega^2 + \gamma^2} \right) \omega_o^2 - \omega^2 - i\omega \left(\Gamma_o + \frac{\omega_o^2 \gamma}{\omega^2 + \gamma^2} \right) \right], \quad [1]$$

where ε_∞ is the high-frequency dielectric constant due to transitions across the Peierls gap, ω_p, the plasma frequency of the CDW, is related to the effective mass m^* of the collective excitations by

$$\omega_p^2 = 4\pi N e^2 / m^* \varepsilon_\infty,$$

N being the density of condensed electrons. γ is a characteristic frequency which defines the transition from diffusion controlled to oscillation controlled behavior. Γ_o is the scattering rate in the absence of pinning and ω_o measures the effectiveness of pinning and the phason frequency. Γ_o and γ are the only temperature dependent parameters. The contribution of the single particle conductivity and the phonons to the dielectric constant is respectively

$$\varepsilon_{sp} = \frac{4\pi i \sigma_{sp}}{\omega} \text{ and } \varepsilon_p = \varepsilon_\infty \prod_{J=1}^{M} \frac{\omega_{jL}^2 - \omega^2 - i\gamma_{jL}\,\omega_{jL}\,\omega}{\omega_{jT}^2 - \omega^2 - i\gamma_{jT}\,\omega_{jT}\,\omega}, \quad [2]$$

ε_∞ is the high frequency dielectric constant, ω_{jT}, γ_{jT} and ω_{jL}, γ_{jL} are the frequency and the damping of the jth oscillator transverse and longitudinal phonon respectively.

In the case of a Peierls transition the electron-phonon coupling hamiltonian takes the form [4] :

$$H_{e-p} = \frac{1}{\sqrt{N}} \sum_{k,q} \frac{g(q)}{\sqrt{2\Omega_q}} C^+_{k+q} C_k (a_q + a^+_{-q}), \qquad [3]$$

where C^+_k (C_k) and a^+_q (a_q) are creation (annihilation) operators for 1-D Bloch electrons and for a longitudinal phonon q of frequency Ω_q respectively. g is the strength of the electron-phonon interaction. The renormalized frequency becomes:

$$\omega_q^2 = \Omega_q^2 [1 - \lambda\chi(q,T)] \qquad [4]$$

with λ a dimensionless electron-phonon coupling parameter and $\chi(q,T)$ a response function. Two modes arise from linear combinations of $\pm 2k_F$ phonons: The amplitude and the phase modes.

In the case of KCP no phonon folding is observed at room temperature since the Peierls distorsion is already present [5]. At low temperature a phason is observed at 15 cm^{-1} [3]. In both $K_{0.3}MoO_3$ and $Rb_{0.3}MoO_3$, as shown by C. Schlenker in these proceedings, an incommensurate CDW occurs at 180 K and a quasi commensurate CDW at 100 K. The following features are detected

a) possibly a phonon folding [6]
b) a phason at 3cm^{-1} and 22cm^{-1} [7]
c) a peak attributed to a soliton [8,9].

In the following figures, infrared reflectivity of $K_{0.33}MoO_3$ and $Rb_{0.3}MoO_3$ at 300 K and 55 K are shown. There is an increase in the number of oscillators used to fit the experimental points (10 at 300 K and 15 at 55 K for $K_{0.3}MoO_3$, 11 at 300 K and 19 at 55 K for $Rb_{0.3}MoO_3$). The crystal structure of these two materials is face centered monoclinic ($C_{2/m}$) and there are 86 atoms per unit cell. One should expect some 60 A_u and 60 B_u infrared active modes excited with $\vec{E}$ respectively parallel and perpendicular to the chains.

The number of the observed phonons is clearly smaller than expected and the increased number at low temperature could not be unequivocally attributed to phonon branches folding as it is clearly demonstrated in the case of layer compounds [1]. Typically for these materials few optical active phonons are predicted and observed at room temperature, while many phonons become active following the folding at low temperature.

Figure 1 (a)

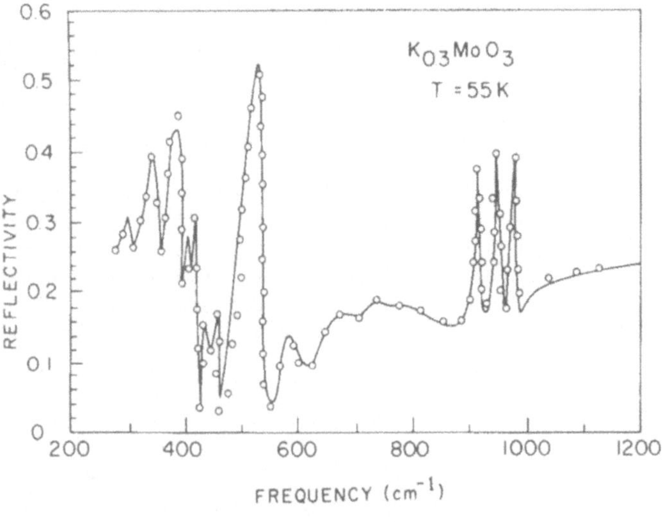

Figure 1 (b)

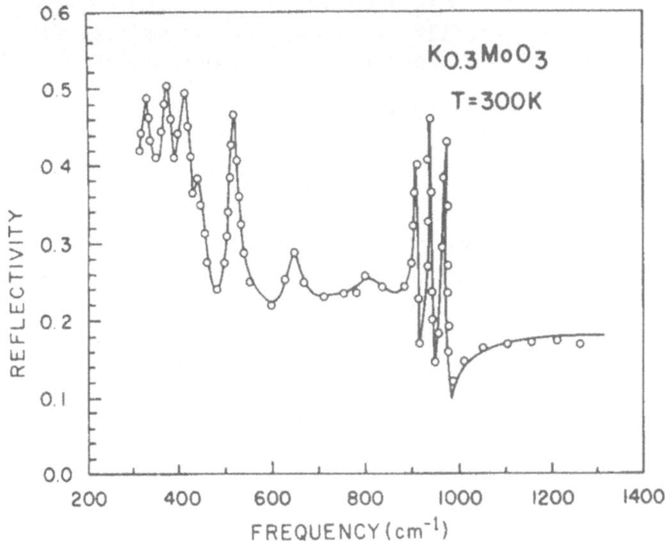

Figure 1 (c)

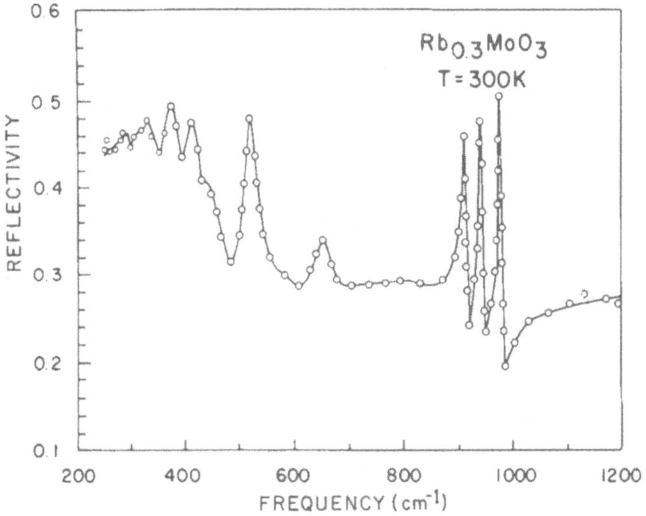

Figure 1 (d)

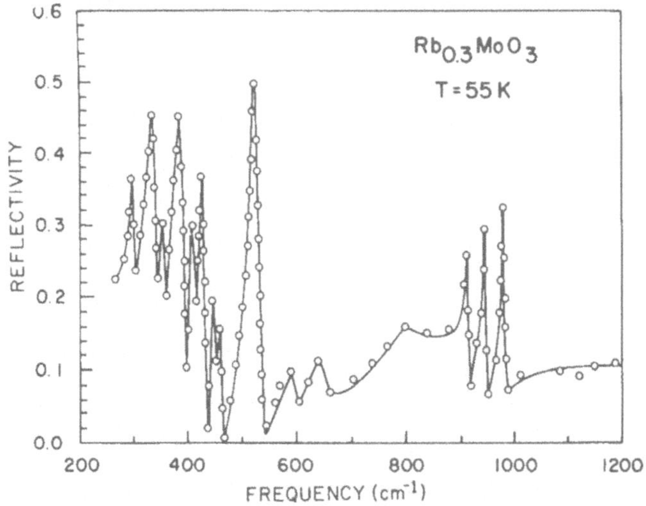

Figs. 1 (a), (b), (c), (d). Infrared reflectivity spectra of $K_{0.3}MoO_3$ and $Rb_{0.3}MoO_3$ at 300K and 55K respectively circles are experimental data and solid lines oscillator fits.

Table I. Oscillator parameters (in cm^{-1}) to fit reflectivity spectra of $K_{0.3}MoO_3$ and $Rb_{0.3}MoO_3$

$K_{0.3}MoO_3$ (T = 55 K)

ω_T	ω_L	γ_T	γ_L
979	983	3	9
945	948	4	5
913	916	6	6
769	757	84	104
645	649	21	28
597	601	15	15
519	535	8	10
461	465	12	5
446	447	4	10
425	435	10	7
409	414	8	11
381	397	16	9
355	357	6	7
333	343	11	15
296	297	5	10

$Rb_{0.3}MoO_3$ (T = 55 K)

ω_T	ω_L	γ_T	γ_L
978	980	5	6
945	946	4	4
911	912	5	6
787	779	128	156
728	729	17	17
700	700	11	11
693	693	13	12
655	626	117	103
630	628	17	18
566	541	57	138
527	542	15	25
462	463	5	6
435	434	6	4
425	425	5	7
401	400	13	8
384	365	31	48
353	353	2	2
315	309	30	15
308	268	28	40

$K_{0.3}MoO_3$ (T = 300 K)

ω_T	ω_L	γ_T	γ_L
972	979	6	8
939	943	4	7
909	913	7	7
644	645	30	44
517	518	11	41
439	445	22	39
422	427	31	18
410	395	26	74
374	385	16	56
330	344	24	67

$Rb_{0.3}MoO_3$ (T = 300 K)

ω_T	ω_L	γ_T	γ_L
976	981	4	8
941	944	4	8
911	915	6	9
655	661	38	51
596	601	61	58
518	523	17	51
441	474	42	95
413	431	32	32
374	394	30	35
350	359	76	36
300	301	15	14

In Table I using formula [2], the fitting parameters of the infrared reflectivity at 55K and 300K are given. The difference in phonon frequencies between the two materials is very slight, reflecting the fact that the Molybden-Oxygen bonding is dominant in the force constants.

In references [3] and [8] the far infrared reflectivity measurements of KCP and $K_{0.3}MoO_3$ were analyzed following formula [1] and calculating the different parameters included within. Table II summarizes the obtained values at 5K.

Table II. Parameters (in cm^{-1}) of the oscillator to fit the far-infrared reflectivity data of KCP and $K_{0.3}MoO_3$ at 5K.

KCP (ref.3)	$K_{0.3}MoO_3$ (ref.8)
$\omega_p = 58$	$\omega_p = 59.7$
$\omega_o = 15$	$\omega_o = 14.5$
$\Gamma_o = 8$	$\Gamma_o = 9.7$
$\gamma = 0$	$\gamma = 0$

In both fittings of KCP and $K_{0.3}MoO_3$ infrared measurements, the single particle contribution to d.c. conductivity is considered important at all temperatures. If the Fermi energy ε_F is known, the dimensionless electron-phonon coupling constant λ is determined by mean field theory [10] with

$$2\Delta = 16\varepsilon_F e^{-1/\lambda} \qquad [5]$$

where 2Δ is the Peierls gap energy and corresponds to 0.2eV for KCP (4) and 0.15 eV for $K_{0.3}MoO_3$ [8]. The calculated λ is then 0.2 and 0.3 respectively for the two materials. According to Lee et al [11], the unrenormalized frequency Ω_q of the LA phonon which couples to the electron density is related to the CDW amplitude mode frequency ω_a and λ, actually $\Omega_q = \lambda^{-\frac{1}{2}}\omega_a$. The CDW effective mass calculated with the mean-field theory relation is:

$$m^* = m_e \left(1 + \frac{4\Delta^2}{\lambda\Omega_q^2}\right) \qquad [6]$$

$m^* \sim 10^3\, m_e$ for KCP and 800 m_e for $K_{0.3}MoO_3$ which is in good agreement with the deduced value from the plasma frequency of Table II. Nevertheless, in the case of $K_{0.3}MoO_3$ this agreement is no more evident with the recent work of Ng et al [7] that infers an effective mass $m^* \sim 10^4\, m_e$, an order of magnitude larger than calculated previously. Ng et al attribute this large effective mass to increased interchain coupling in the CDW at low temperature.

In the infrared study of $K_{0.3}MoO_3$ [8] and $Rb_{0.3}MoO_3$ [9], the authors attribute the strong peak around 500 cm^{-1} to a soliton midgap state resulting from dislocations in the charge density wave lattice. Machida and Nakano [12] predicted the existence of exponentially localized midgap states in nearly commensurate CDW systems and referred to its observation in TaS_3 [13]. It should be pointed out that the nature of the peak attributed to the midgap state is controversial since it could simply be a phonon as recently mentioned in a recent study on TaS_3 [14].

The amplitudon Raman active mode was detected at about 44 cm^{-1} in the case of KCP [15] and 50 cm^{-1} in the case of $K_{0.3}MoO_3$ [16] and $Rb_{0.3}MoO_3$ [17]. The main characteristic of this mode is an anomalous temperature dependent damping constant behaviour with however a non softening frequency. It is not absolutely clear that the detected excitation is a CDW amplitude mode rather than a phonon vibration sensitive to the environmental conditions as pointed out in the case of KCP [18].

In conclusion, non organic CDW systems like KCP, $K_{0.3}MoO_3$ and $Rb_{0.3}MoO_3$ are interesting materials to study the eventual CDW excitations. Still more work is needed in order to assign without equivoke the nature of optically observed vibrations.

References

1. J.E. Smith Jr, J.C. Tsang and M.W. Shafer, Sol. St. Comm. 19, 285 (1976).

2. Light scattering near phase transition. Edited by H.Z. Cummins and A.P. Levanyuk in Modern problems in condensed matter Sciences vol. 5 (North-Holland).
3. P. Brüesch, S. Strässler and H.R. Zeller, Phys. Rev. B, 12, 219 (1975).
4. W. Gläser, Festkörperprobleme 14, 205 (1974).
5. R. Comès, M. Lambert and H. Launois, phys. Rev. B, 8, 57 (1973).
6. S. Jandl, M. Banville, J. Marcus, C. Schlenker to be published.
7. H.K. Ng, G.A. Thomas and L.F. Schneemeyer, phys. Rev. B, 33 8755 (1986).
8. G. Travaglini and P. Wachter, phys. Rev. B, 30, 1971 (1984).
9. N. Massa, to be published in phys. Rev. B.
10. H. Fröhlich, proc. R. Soc. Lond. A223, 296 (1954).
11. P.A. Lee, T.M. Rice, and P.W. Anderson, Sol. St. Commun. 14, 703 (1974)
12. K. Machida and M. Nakano, phys. Rev. lett. 55, 1927 (1985).
13. M.E. Itkis and F. Ya. Nad, JETP Lett. 39, 448 (1984).
14. S.L. Herr, G. Minton and J.W. Brill, phys. Rev. B, 33, 8851, (1986).
15. E.F. Steigmeier, R. London, G. Harbeke and H. Anderset, Sol. St. comm. 17, 1447 (1975).
16. G. Travaglini, I. Mörke and P. Wachter, Sol. St. comm. 45, 289 (1983).
17. S. Jandl, G. Bernier, J. Marcus and C. Schlenker (to be published)
18. M. Futamata, Y. Morioka and I. Nakagawa, Proceedings of the IXth int. conf. on Raman Spectroscopy p. 212-213 (low dimensional and amorphous materials).

SOLID FRICTION ON LATTICE AND SPIN MODULATIONS

J. Friedel

Université Paris-Sud
91405 Orsay (France)

INTRODUCTION

In many cases, lattice or spin modulations show a hysteretic motion. The possible origins of the solid friction responsible for this behaviour are two-folds :

- lattice friction on weakly commensurate modulations.

- friction due to lattice defects on uncommensurate modulations.

The modulations considered here can have various origins. They can be driven by an externally applied electric field or by an internal thermodynamic force due to their change of period with, say, temperature, pressure or magnetic field.

I will try and review in a very qualitative way this well trodden field, insisting on the more speculative aspects. For simplicity sake, I will restrict myself to quasistatic and classical phenomena at low temperatures. Kinetic effects, quantum and critical thermal fluctuations will not be considered. In this limit, most of the phenomena are fairly direct extensions of behaviours known and described for over 30 years in the context of dislocations in crystals. I will in particular point to a possible coupling of the two effects mentioned above, i.e. the influence of lattice defects on lattice friction

I - LATTICE FRICTION OF WEAKLY COMMENSURATE MODULATIONS

A straight extension of the Peierls friction for-say-epitaxial dislocations, this question has been mentioned many times for 2d and 3d modulations, but not much studied, especially from an experimental point of view. We analyse the increasing complexity with dimensionality of the mechanisms involved, and show why this regime is difficult to observe.

a - One dimension

We consider modulations with increasing deviations from a pure sinusoïdal behaviour (Figure 1). Early historical examples are periodic epitaxial dislocations in the Frenkel-Kontorova model (elastic chain of atoms on a rigid substrate, (Figure 2) and periodic Bloch walls in the helicoïdal arrangement of a chain of spins rotating in the xy plane perpendicular to the axis z of the chain and submitted to a crystal field along x' x (Figure 3). In the first case, the period λ is related to the difference between the equilibrium distance a' of the atoms along the chain and the period a of the substrate potential : $\lambda = a^2/|a'-a|$ for weak potentials ; in the second case, λ is the half pitch of the helix : $\lambda = \pi a/\theta$, if θ is the equilibrium angle between successive spins. The thickness ξ of each dislocation core or Bloch wall decreases with increasing deviations from a pure sinusoïdal behaviour, thus with increasing ratio of the amplitude A of the perturbing potential, due to the substrate or to the crystal field, to an elastic modulus K of the chain.

Each period λ of the modulation has an interaction w with the underlying lattice which depends on its phase z_o along the axis of the chain, and oscillates with the period a of the lattice. The amplitude δw of this oscillation obviously increases with decreasing thickness ξ of the dislocation or wall (Figure 1). A classical computation by Peierls shows that, in the limit $\xi \gg a$ where the form and amplitude of the modulation can be taken independent of its phase z_o, δw decreases exponentially as $-\frac{\pi\xi}{2a}$ where $\xi \cong a(K_a/A)^{1/2}$ for $\xi < \lambda$ and is vanishingly small for $\xi \gtrsim \lambda$, when the dislocations or walls have lost their individuality (cf Appendix).

If the period λ is commensurate with a, all periods can sit at a minimum of δw and add their blocking effects. The solid friction force on each period is related to the maximum slope of $\delta w(z_o)$.

If the period λ is uncommensurate, all periods sample all values of $\delta w(z_o)$, so that the total energy of the modulation is phase independent and the friction diappears.

The physical reason for uncommensuration is that the chain is too rigid to accept deviations due to lattice friction. Indeed, with increasing ratio of the external potential A over the chain rigidity K, thus decreasing ratio of ξ/a, the Peierls friction of commensurate modulations increases, as also the size of the domain of values of λ near a commensurate value where the incommensurate modulations are less stable than the commensurate one.

This classical argument neglects two effects :

1 - It assumes that the form of the modulation does not change with its phase z_o. This is obviously wrong for very thin walls ($\xi \cong a$), as the example of the helical spin chain shows (Figure 4). However, in such a case, the long range behaviour of the wall is vanishingly small, so that the interaction between walls only takes place for very small values of λ, a case to be treated separately. We shall then assume $\xi > a$ in what follows.

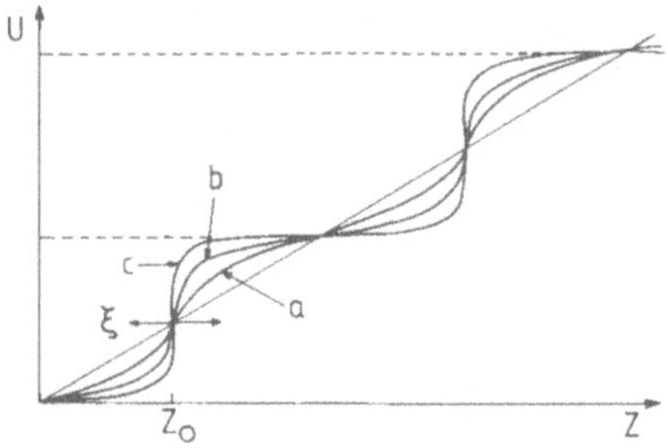

Fig. 1. Modulations u(z) with increasing deviations from pure sinusoïdal behaviour (u relative displacement or cosine of spin rotation)

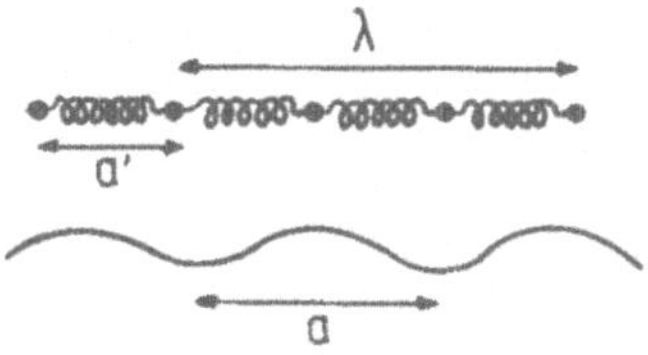

Fig. 2. Frenkel Kontorova model.

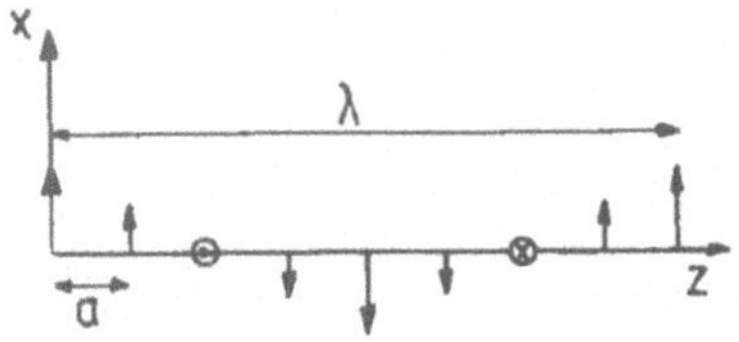

Fig. 3. Helical chain of spin in crystal field.

Fig. 4. Two configurations for the thin Bloch wall of a helical spin model.

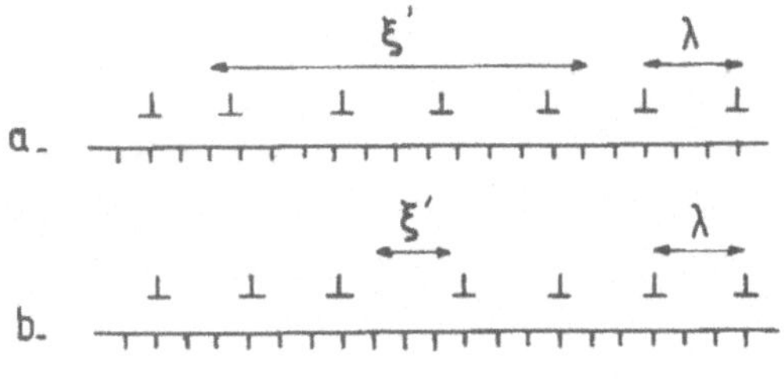

Fig. 5. Broad and narrow shifts s(a).

2 - Only periodic arrays of walls or dislocations have been considered. We can however imagine, in the commensurate case, that the modulation is shifted, on half the chain, by one or a few periods a with respect to the other half. This introduces a local dilatation or compression of the lattice of dislocations or walls. We will call shifts (+ na) and antishifts (-na) these local discommensurations. Such excited states will be metastable if far from each other, as the Peierls friction of the chain will always be larger than the long range elastic reaction to the core distortion of the shift or antishift.

These discommensurations are partial dislocations of the lattice of dislocations or walls. Their own Peierls friction will decrease exponentially as $-\frac{\pi\xi'}{2\lambda}$, where ξ' is their own width, which decreases with increasing strength of the commensuration. Figure 5 pictures elementary shifts +a in the two extreme limits $\xi' \gg \lambda$ and $\xi' \cong \lambda$. As usual for dislocations, such discommensurations have energies which vary in proportion to the rigidity of the modulation against a variation of λ, and to the square of the Burgers vector na.

Then for $\lambda \cong \xi$ one expects the modulation to be rigid, thus shifts and antishifts to have sizeable energies and large width ξ', and to be easily mobile. For $\lambda \gg \xi$, one expects the modulation to be easily compressed or extended, thus shifts and antishifts to have small energies, but to be sessile, i.e. difficult to move.

In all cases, the force necessary to create at OK a pair of shifts and antishifts in a perfect modulation is near to the friction force to move the modulation rigidly. However, at T $\neq$ OK, thermal excitation of these pairs are possible, under smaller forces. Pairs of elementary shift and antishift ($n = \pm 1$), of lowest energy, will dominate. If $\lambda \cong \xi$, one expects the shifts and antishifts created at finite temperature under an applied force to move rapidly until they annihilate each other : the chain remains nearly perfect all the time, but for temporary fluctuations. For $\lambda \gg \xi$, a larger concentration of shifts and antishifts would be obtained at thermodynamical equilibrium, because of their low energy of formation U_2. However, because of their large energy of motion U_m, equilibrium is rarely reached. At low concentrations, the concentration of shifts and antishifts should then be a function of previous history of the sample, whether heat treatment of applied force. In these conditions of solid friction, the elementary walls or dialocations would no longer be periodic, but present a frozen cahotic behaviour, with many strongly metastable states of energies near that of the stable unperturbed state. In both cases, and neglecting quantum fluctuations that could dominate the low temperature regime, the average motion of the chain would be thermally activated. At high temperatures and small applied forces, the energy of activation should be equal to U_2, the energy to create a pair of separated shifts and antishifts, or to U_m, the energy to move shifts and antishifts, whatever is the slower of these two processes in series.

Finally if a lattice modulation is coupled with a charge density wave, the creation of shifts and antishifts introduces inner polarisation effects which will only be screened by independent electrons if the modulated crystal is a conductor. If ze is the charge carried by each wave length λ of the modulation in its motion, each shift has therefore an effective charge $-\frac{na}{\lambda}$ ze, and each antishift a charge $\frac{na}{\lambda}$ ze. The corresponding forces due to an applied electric field E are thus

$\pm \frac{na}{\lambda}$ ze E.

b - Two dimensions

We consider a collection of modulations on parallel and equidistant chains in a plane, and increase progressively their interactions, which tends to block their relative phases into 2d modulations.

1 - For weak interactions, shifts and antishifts can still be produced independently under thermal agitation. They are still charged in the presence of a CDW. Two types of coupling will develop.

Intrachain couplings. The existence of an isolated shift in one chain creates a phase mismatch of the modulation of that chain with respect to neighbouring chain, on one side of the shift (Figure 6a). One can say that the shift s is bordering two stacking faults ff' which will pull it on one side to decrease their area. One can also say that f, f' and s are part of the same loop of partial dislocation with Burgers vector + $\underset{\sim}{a}$ along the chain. In the presence of a CDW the loop is charged on the shifts and neutral along the stacking faults.

Then for $\lambda \cong \xi$, the existence of f and f' will offer a further difficulty to the separation of shifts and antishifts ; they will occur at thermal equilibrium only as closed pairs, i.e. closed loops of partial dislocations. For $\lambda \gg \xi$, the existence of f and f', when weak, only adds a small bias to the creation and motion of shifts and antishifts, which will remain very disordered.

Interchain couplings. At large enough concentrations of shifts and antishifts, one can expect a tendency for them to align from chain to chain, so as to form 1d positive and negative discommensurations (Figure 6b). This can also be looked at as the creation of one large partial dislocation loop by condensing a number of small ones. For $\lambda \cong \xi$ this process is expected at high temperatures, where the concentration of (a+s) pairs is larger. For $\lambda \gg \xi$, it is expected when large concentrations of shifts and antishifts are moved by applied forces on neighbouring chains.

These partial dislocation loops of Burgers vector na are but simple examples of lines of more general forms. Such lines, if straight, can be in metastable equilibrium (Figure 6c). Finite segments of such dislocations can in principle meet at triple nodes to build a twodimendional (Frank) network of dislocations.

Another possible regroupment of 1d positive and negative discommensurations is pictured figure 6d : n half walls of shifts and n' half walls of antishifts are equivalent to a dislocation of Burgers vector (n+n')a. If this is equal to

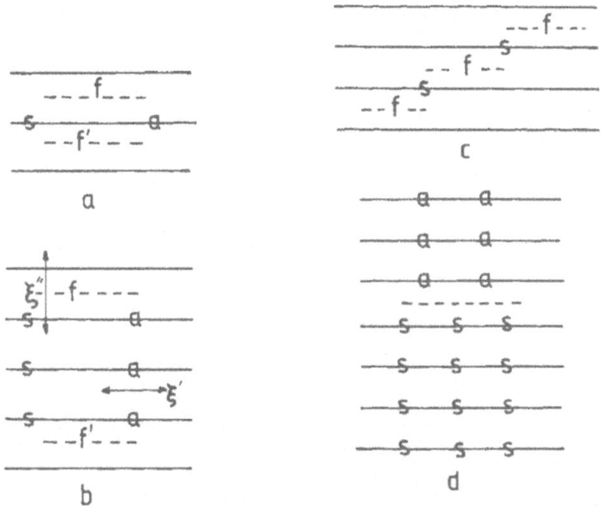

Fig. 6. Coupling of shifts s and antishifts a: a) intrachain; b) c) d) interchain.

Fig. 7. Dislocation in the modulation. Continuous lines: surfaces of constant phases. a) easy glide; b) easy climb.

λ, one has a perfect dislocation point of the modulation, with a core of large width of order ξ', parallel to its Burgers vector (Figure 7a). It is obviously a defect of fairly large energy. It can glide easily parallel to its Burgers vector $\underset{\sim}{\lambda}$ if its constitutive shifts are mobile, i.e. if $\lambda \cong \xi$. To make it climb perpendicular to $\underset{\sim}{\lambda}$ would however require the creation of new shifts and destruction of antishifts (or vice versa), with a lateral displacement of the stacking fault. Such a process would be more difficult as well as producing large electrical charges in the core of the dislocation in the case of a charge density wave.

It is fairly obvious that such a perfect dislocation point is most stable if its dilatation and compression parts compensate at long range, thus if $n' = n$ (if $n + n' = \lambda/a$ is even) or $n' = n \pm 1$ (if λ/a is odd). In the first case, such a dislocation is neutral in the presence of a CDW and does not interact directly with an external electric field ; for weak interactions, such a field would tend any way to destroy this perfect dislocation by pulling the walls of shifts and antishifts apart by developing a stading fault. In all cases, such perfect dislocation point can interact with the distortions of the modulations, i.e. with the partial dislocations na discussed above.

2 - For <u>large interactions</u>, the core of <u>perfect dislocation points</u> of the modulation is expected to change the nature of its splitting, from parallel to perpendicular to the chain direction (Figure 7). In the first case ($\xi' >> \xi'' \cong b$), these dislocations were 'glissile' but could not 'climb' ; in the second case ($\xi' < \xi'' >> b$), these dislocations might be less 'glissile' but could 'climb' normal to the chain direction, if the charges developed in such climb are compensated. The same distinction is classical for dislocation points observed in 2d lattice of hydrodynamic vortices. As in that case, one expects an increase in the interchain coherence length ξ'' to stabilize such dislocation points against dissociation into partial dislocations under suitable forces (here the electrical field for CDW).

Nevertheless one expects individual <u>partial dislocations</u> to play still a leading role in the motion of the modulation. Such partial dislocations, with elementary Burgers vector $\pm$ a, of lowest energy, will be the most likely to occur. The topology is that of figure 6. But the phase shift varies continuously over ξ'' across f,f' : these dislocations can also have a finite width ξ''normal to the chains, and the lattice friction to their glide normal to a might be weak, for large enough interchains interactions. These 1d discommensurations then take more or less curved shapes, with <u>kinks</u> and <u>antikinks</u> which are weak ; these are easy to form if ξ'' is large, but also easy to move along the discommensuration if ξ' is large.

We will not discuss here their creation or multiplication which have to rely on stress concentrations mentioned in section III below. We only discuss here their motion. Figure 8 pictures the double grid of stable (S) and

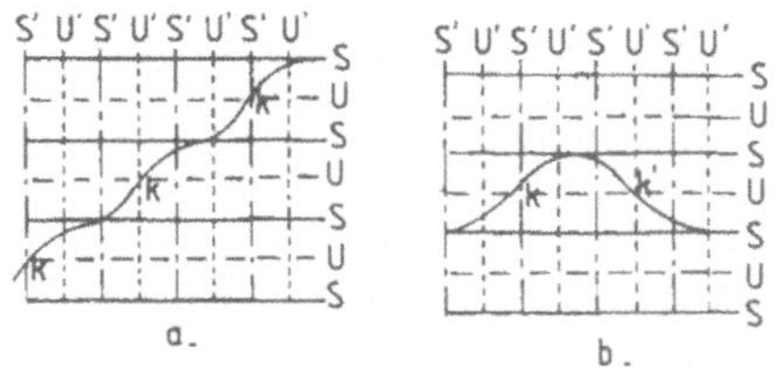

Fig. 8. Kink, on 1d discommensurations. SS', UU' respectively stable and unstable positions in two closepacked directions.

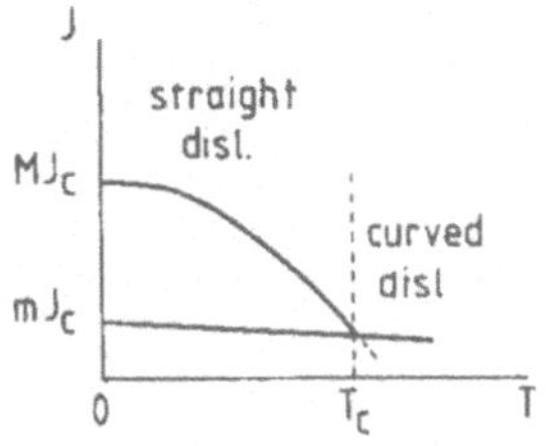

Fig. 9. Thermal variation of the critical microcurrent mI_c and of the critical macrocurrent mI_c.

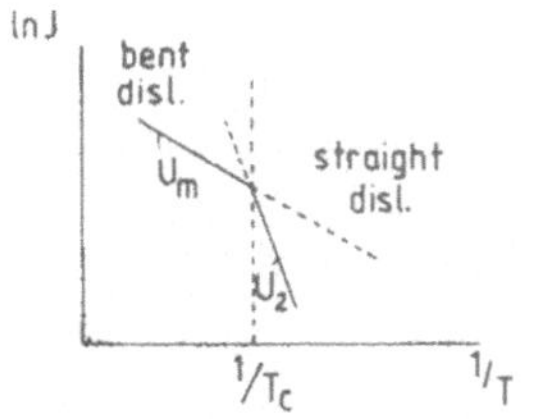

Fig. 10. Two thermally activated regimes.

and unstable (U) positions associated with the Peierls friction along the chains and across the chains. We assume that the interactions between chains are still weak enough for the Peierls friction across the chains to dominate on that along the chains (thus $\xi''/\xi' < 1$). The continuous kinks and antikinks then relate portions of the partial dislocations which are parallel to the chains ; they replace the abrupt shifts and antishifts discussed so far.

At OK, these discommensurations will move if the applied force overcomes the lattice friction in the direction of weakest lattice friction. If the modulation contains initially a (Frank) network of discommensurations at arbitrary angles to the chains, such as in figure 8a, the kinks will move above a critical force until they disappear at surfaces or pile up at defects. The discommensurations will then become mostly parallel to the chains (Figure 8 b) and will only be able to move under a larger critical force. These two critical forces will be called the microcurrent and macrocurrent limits in Figure 9.

At finite temperatures, the macrocurrent limit decreases because the modulation can slip by thermally activated creation and motion of pairs of kink and antikink, as pictured Figure 8b. Because these two processes are in series, it is the slower of the two which limits the flow of modulation. In the high temperatures limit, this flow occurs under small forces. The activation energy U_2 for creation of pairs of kink and antikink is by definition larger than that U_m for their motion, seeing that the lattice friction against U_2 is assumed larger than that against U_m. There will then be a critical temperature T_c below which it is the creation of kinks and antikink, which limits the flow : as soon as created, these will rapidly move along the discommensuration and disappear or get blocked. The straight discommensurations of the macrocurrent limit will therefore move but remain essentially straight, parallel to the chains (Figure 10). Above T_c, kinks and antikinks move and disappear less rapidly than they are created ; the discommensurations become curved ; in this second regime, the macrocurrent limit is replaced by the extrapolation of the microcurrent one (Figure 9).

It is finally clear that, with increasing interactions between chains and thus increasing ξ''/ξ' ratio, both T_c and the difference between macro and microcurrent limits decrease. If ξ''/ξ' become larger than unity, the roles of the two lattice frictions, Figure 8, are interchanged ; when straight, the discommensurations are then normal to the chains direction. Thus a pile up of straight discommensurations parallel to the chain (for $\xi''/\xi' < 1$) corresponds to a bending of the modulations (Figure 11 a) ; a pile up of straight discommensurations normal to the chains (for $\xi''/\xi' > 1$) corresponds to a pile up of the modulations (Figure 11 b). Such pile ups can indeed be produced by the activation of Frank Read sources. At high temperatures or if $\xi''/\xi' \cong 1$ ($T > T_c$), both types of distortions of the modulations are expected to be possible.

Remarks : 1) In case of a CDW, the force due to an applied electrical field E produces a work $\frac{na}{\lambda} \rho e \mathcal{A}$ if such a partial dislocation sweeps an area $\mathcal{A}$. $\rho e = \frac{ze}{b}$ is the charge per unit length of a modulation (ze charge of a shift, b interchain distance). The electrical field exerts therefore a configurational force per unit length normal on each point to the partial dislocation and of value $\frac{na}{\lambda} \rho e = \frac{na}{\lambda}\frac{ze}{b}$.

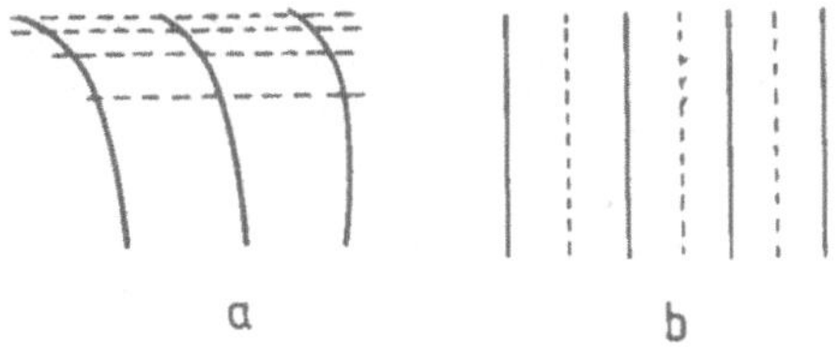

Fig. 11. Low temperature regime of straight discommensurations: a - parallel to the chains ($\xi' > \xi''$); b - normal to the chains ($\xi' < \xi''$). Continuous lines: periods of the modulation; punctuated lines: discommensurations (the chains are horizontal).

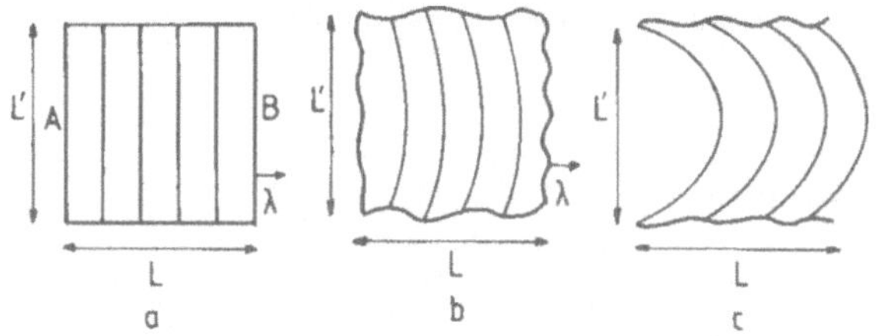

Fig. 12. Blocking of the phase by surfaces. Brute force mechanisms of deblocking.

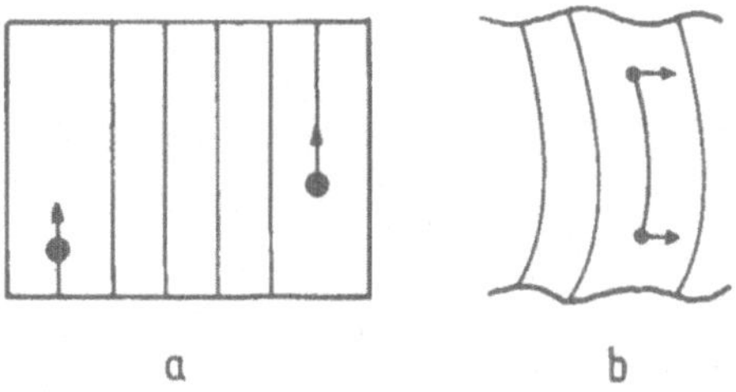

Fig. 13. Blocking of the phase by surfaces. Dislocation mechanisms of deblocking: a - climb (front and rear blocking); b - glide (lateral blocking).

2) We have seen that $\xi'' \cong b$ if the interaction between chains is weak, the normal condition for 1d organic conductors. For 2d compounds, it is known from the study of dislocations in crystals that ξ' and ξ'' are of order a and b, and the two Peierls frictions large for covalent bonding ; ξ' and ξ'' are both much larger than a and b and the two Peierls frictions small for metallic, ionic or Van der Waals bonding.

3) Complications associated with 2d geometry can lower still further the Peierls friction, as stressed in Part I.

The dislocations or walls constituting the modulation (see figures 1, 2, 3) can, is some cases, split into partial dislocations separated by ribbons of stacking faults. Because of their smaller Burgers vectors, these partial dislocations have smaller Peierls frictions; because of their proximity, their mutual interaction usually prevents them from sitting at the same time at minima of the Peierls potential, thus lowering still further the total Peierls friction.

The lattice geometry can sometimes accept several modulations with non parallel reciprocal vectors. This can produce a 2d network of dislocations. Splitting, if possible in this case, usually produces non parallel partial dislocations, when successive nodes of the network have different properties.

c - Three dimensions

We can again consider a family of parallel chains with increasing interactions between chains, which locks the relative phases of the modulations on the different chains.

1 - For weak interactions, pairs of shifts and antishifts on one chain can condense now into 2d discommensuration planes related by cylinders of stacking faults. Solid friction occurs for narrow modulations walls ($\lambda \gg \xi$).

2 - The 2d discommensuration surfaces and stacking faults can take more continuously curved shapes : the 2d discommensuration surfaces can take families of parallel steps which are the id equivalent of the kinks. With increasing interactions between chains, the core width ξ'' of these steps normal to λ increases and the lattice friction against their glide normal normal to λ decreases.

Again several families of modulations with non colinear wave vectors can possibly coexist, leading to structures that can be rather complex and have not been studied in detail.

Perfect dislocation lines of the basic modulations have cores which are preferentially split parallel to λ for weak interactions between chains, but might be split perpendicular to λ when the interaction between chains is so large that it is more physical to describe the structure as a piling of planes normal to λ. In the first case, dislocations are glissile but do not climb ; in the second case, then should climb more easily but do not glide. All intermediary situations can occur, with dislocations lines pointing in all directions and building a 3d (Frank) network.

In presence of a CDW, the force due to an applied electrical field is a direct and obvious extension of that described in 2d: normal to the discommensuration surface, proportional to the electric field and to the charge per unit area of the modulation.

In conclusion, the solid friction of lattice origin requires rather stringent conditions of weak commensurability. The motion of the modulation involves the development and motion of partial or full dislocations of the modulation. If in 1d or 2d some of these defects can be created by thermal activation, this is not so usually in 3d, where these defects must preexist or be created using stress concentrations on a defective part of the sample.

II - FRICTION OF UNCOMMENSURATE MODULATIONS DUE TO LATTICE DEFECTS

Because they are uncommensurate, the modulations are rigid enough for their compressions or dilatations along their wave vector to be treated as continuous : the shifts and antishifts loose their identity because their width ξ' becomes infinite. Contrary wise, in 2d or 3d, the phase of the modulation can vary abruptly ($\xi'' \cong b$) or smoothly ($\xi'' \gg b$) normal to its wave vector. we shall talk of weak or strong interactions for a chain model as before.

The defect considered are surfaces and interfaces, dislocations and point defects.

a - Surfaces and interfaces

They are known to block strongly the phase of the modulations, for lattice modulations by electrostatic effects, for spin modulations by local changes in spin orbit couplings. Under small applied forces, only a limited displacement is possible : for a lattice modulation, it corresponds to a charge polarisation, but no permanent electric current.

For large enough forces, a permanent motion becomes possible . 3 mechanisms can be imagined.

- The modulation can become unpinned at the surfaces. In the exceptionnal case of crystals with flat surfaces just parallel and perpendicular to $\underset{\sim}{\lambda}$ (figure 12 a), only the perpendicular surfaces block the phase. The compression of the modulation by the applied field produces a stress concentration factor inversely proportionnal to the number of periods thus to the length L of the crystal. In the more usual case of rough surface (Figure 12), the roughness of the end surfaces increases the stress concentration ; but the roughness of the lateral surfaces blocks the oscillations so as to make the wave vector locally rotate out of its best direction. The local stress concentration to free the phase on the lateral surfaces is then inversely proportional to the width L' of the crystal.

- For the weak interaction case, the bending of the modulations near the lateral surfaces should be easy. Under the applied force, one could then reach the critical condition where the modulations become parallel to the surfaces near the lateral surfaces (Figure 12 c).The lateral friction then disappears in this Orowan configuration. This mechanisms can produce a shift of the modulation on cylinders parallel to λ that have to be long (large L) but can be thin (small L').

- Besides these 'brute force' processes, one can consider the possibility of multiplication and motion of dislocations of the modulation. In the simple case of figure 12 a, one needs to create dislocations of the modulation near both ends A and B of the crystal, and to make them climb normal to λ so as to remove periods of the modulation at one end and add periods at the other (Figure 13 a). In the more realistic case 12 b, the roughness on the end surfaces can help and develop such dislocations ; the pinning on the lateral surfaces can be relieved if dislocations of the modulation are created near them that can glide parallel to λ (Figure 13 b). These processes however assume a preexisting (Frank) network of dislocations of the modulation, or roughnesses important enough to create easily new dislocations.

It must be pointed out that, as in the case of commensurate modulations described Figure 7, the climb of a dislocation of the modulation involves the local suppression or addition of part of a period of the modulation. In case of a CDW, a suppression of a modulation involves the emission in the conduction band of the free electrons locally condensed in the CDW, or the absorption of positive holes in equal number ; the addition of a modulation involves symmetrically the emission of positive holes in the valence band or the absorption of free electrons to be condensed in the CDW. This means that the climb of such a dislocation involves a special solid friction, independent of speed, and equal to

$$F = \frac{ze}{b} \Delta$$

where Δ is the energy involved in emitting a free electron or a hole. When occuring near an electrode, such a process can produce the current involved in the motion of the CDW. When occuring in the middle of a sample, e.g. near a grain boundary, the dislocation climb can produce free carriers that can be at least temporarily captured by impurity centers. There is an obvious parallelism of the role of source or sink of electric carriers in the climb of a dislocation of a CDW and the role of source or sink of points defects in the climb of a crystal dislocation.

b - Dislocations of the lattice

They are known to block the walls of the modulations by three mechanisms :

- their long range strains.
- their core structure.
- their topology.

It is well known that these effects are rather modest in the usual case where the width ξ of the walls of the modulation, and a fortiori their period λ is large compared with the average distance between random dislocations, a case met in heavily deformed crystals.

A case where dislocations are more effective is if they are polygonised into walls of low angle boundaries. They then act as grain boundaries if their density is sufficient.

Finally if, for instance by bending a crystal, one introduces over a macroscopic region an excess dislocations all of one sign, the curvature of the crystal will produce a distortion of the modulations which can be sometime relieved by creating dislocations of the modulation. But this is rather rare. Somewhat similar effects are expected in the presence of a thermal gradient, which usually acts most strongly on the value of λ.

c - Point defects

Point defects will act on the local amplitude of the modulation and usually try pin down their phase, to optimise their coupling.

As a result, dilute point defects at random should distort an otherwise stable modulation, and thus decrease their stability as measured by the temperature T_1 where it vanishes as a static distortion. They can however produce above T_1 local distortions, which can be looked at as fluctuating modulations locally stabilised by the coupling with a point defect. This last effect is especially large but anisotropic for weakly coupled chains.

Contrarywise point defects at random but in large concentrations cannot distort effectively the modulation over the short distances involved. They then can only change locally the amplitude of the modulation in a way which depends on its local phase.

Finally if mobile, point defects can, by changes of position over distances less than λ, optimise their coupling with a given modulation. This increases the stability of the modulation ; and, if the point defects are not too mobile, this blocks very effectively the phase of the modulation. To be effective, the point defects must be able to move by distances less than λ during the heat treatment preparing the sample, but not able to move by macroscopic distances during the time the external force is applied. These conditions are very easily fullfilled. Furthermore, contrary to the boundary blocking by surfaces, this is a blocking in the volume. It is valid for modulations with broad as well as thin walls ($\xi \cong \lambda$ or $\xi << \lambda$). The coupling strength can originate from size effects, electrostatic or magnetic couplings. It is usually strong enough to be stable in all conditions where the modulation considered is stable. Two characteristics met in the plasticity of alloys and explained in this way are also met in the present case : a peak of the critical field for the motion of CDW, at temperatures between those where the defects are not mobile enough to adapt to the CDW and those where they are too mobile to pin down the CDW ; a heterogeneous mode of motion where slipping bands develp and travel through the sample, giving repeated peaks in the critical field.

III - LATTICE FRICTION AND LATTICE DEFECTS IN WEAKLY COMMENSURATE MODULATIONS

They are not expected to be stricly additive.

It is well known that the Peierls friction of a dislocation can be lowered by its short range interaction with lattice defects : surfaces, other dislocations, point defects. The reason is that these perturbations can help to nucleate the kinks by the motion of which the Peierls friction can be overcome.

One can thus expect in some cases the solid friction due to lattice friction to be lowered by the presence of some lattice defects, by the same kind of mechanism. There is no evidence that I know of such effect.

It is also wellknown that stress concentrations due to surface roughness or coherency at interfaces can help and create new dislocations. They should also help to create the partial dislocations or perfect dislocations of the modulation.

IV - NEARLY COMMENSURATE MODULATIONS

This is a specific case of incommensurate modulations, when the wave length λ is near to a simple commensurate value λ_c. If the commensurate modulation is strongly pinned down by a large Peierls friction, the motion of the incommensurate modulation can be considered as due to that of a periodic family of discommensurations of the commensurate lattice.

In other words, the 'substrate' lattice is not the basic lattice of period a, but the (perfect) superlattice of period λ_c corresponding to the commensurate modulation. The discommensurations of the periodic lattice are the equivalent of the basic 'dislocations' or 'walls' of figure 2, 3 in 1d, and their straightforward extensions in 2d and 3d, as discussed in chapter II.

APPENDIX: PEIERLS FRICTION IN THE FRENKEL KONTOROVA MODEL

Let x_n be the abcissa of the nth atom and

$$x_n = (n + u_n)a .$$

The total energy of the chain is, if the outside potential is sinusoïdal,

$$U = A \sum_n (1 - \cos \frac{2\pi x_n}{a}) + \frac{1}{2} K a \sum_n (x_{n+1} - x_{n-1} - a')^2$$

$$= A \sum_n (1 - \cos 2\pi u_n) + \frac{1}{2} K a^3 \sum_n (\frac{a-a'}{a} + u_{n+1} - u_n)^2 .$$

The equilibrium of the nth atom is given by

$$2\pi A \sin 2\pi u_n + K a^3 \left[\frac{a-a'}{a} + 2 u_n - u_{n-1} - u_{n+1} \right] = 0$$

If u_n varies slowly enough with n, a continuous limit is valid. Then

$$K a^3 \frac{d^2 u}{dn^2} \cong 2\pi A \sin 2\pi u \qquad (A.1)$$

If furthermore the distance λ between successive dislocations is large enough, each dislocation (soliton) can be treated as independent, as if a' = a. Then the solution of (A1) is

$$\operatorname{tg} \frac{1}{2} \pi u = \exp \left[\pm \pi (n - \upsilon)/\ell_o \right]$$

where υa is the position of the center of the dislocation and its width is given by

$$\xi = 2\,\ell_o\, a = \sqrt{\frac{Ka}{A}} \qquad (A.2)$$

The conditions of validity are finally $a \ll \xi \ll \lambda$ or $1 < \frac{Ka}{A} < \frac{\lambda^2}{a^2}$.

The corresponding energy is, taking (A.1) into account and for a' = a,

$$W + \delta w = 2A \sum_n (1 - \cos 2\pi\, u_n).$$

Strictly speaking, it is not coherent to keep in this expression the summation over n while computing u(n) in the continuous limit. As stressed in the text, this is equivalent to assuming that the form of the modulation is independent of its phase, as in the continuous limit, an assumption valid in the limit of weak friction, thus for $\lambda \gg \xi \gg a$.

Then, using

$$\sum_{n=\infty}^{\infty} f(n) = \sum_{s=-\infty}^{\infty} \int_{\infty}^{\infty} f(x) \cos 2\pi\, x\, s\, dx,$$

One obtains

$$w + \delta w = 2\,A \sum_{s=-\infty}^{\infty} \frac{\cos 2\pi\, x\, s\, dx}{ch\ 2\pi \frac{x-\upsilon}{\ell_o} + 1}$$

A straighforward integration in the complex plane shows that the term in s = O gives the constant energy w and that the leading term in δ w is given by the term in $s = \pm 1$. The dominating poles of $\left[ch\ 2\pi \frac{z-\upsilon}{\ell_o} + 1\right]^{-1}$ at $z = \upsilon \pm i\,\frac{\ell_o}{2}$ give

$$\delta w \propto A\, e^{-\pi \ell_o} \sin 2\pi\, \upsilon \qquad (A.3)$$

Similar formulae obtain for epitaxial dislocations between two thick elastic media. The main difference is then that $\ell_o \propto Ka/A$.

Acknowledgements

The author acknowledges with thanks fruitful discussions with Drs. S. Barisic, C. Berthier, S. Brazowski, P. Lederer, P. Monceau, T. M. Rice, C. Schlenker and H.J. Schulz. The concepts of neutral discommensurations, parallel to their Burgers vector and the formula for the applied force were independently proposed by S. Aubry and D. Feinberg, respectively.

INDEX